ADVANCES IN THE THEORY OF QUANTUM SYSTEMS IN CHEMISTRY AND PHYSICS

Progress in Theoretical Chemistry and Physics

VOLUME 22

Honorary Editors:

Sir Harold W. Kroto *(Florida State University, Tallahassee, FL, U.S.A.)*
Pr Yves Chauvin *(Institut Français du Pétrole, Tours, France)*

Editors-in-Chief:

J. Maruani (formerly *Laboratoire de Chimie Physique, Paris, France*)
S. Wilson (formerly *Rutherford Appleton Laboratory, Oxfordshire, U.K.*)

Editorial Board:

V. Aquilanti *(Università di Perugia, Italy)*
E. Brändas *(University of Uppsala, Sweden)*
L. Cederbaum *(Physikalisch-Chemisches Institut, Heidelberg, Germany)*
G. Delgado-Barrio *(Instituto de Matemáticas y Física Fundamental, Madrid, Spain)*
E.K.U. Gross *(Freie Universität, Berlin, Germany)*
K. Hirao *(University of Tokyo, Japan)*
E. Kryachko *(Bogolyubov Institute for Theoretical Physics, Kiev, Ukraine)*
R. Lefebvre *(Université Pierre-et-Marie-Curie, Paris, France)*
R. Levine *(Hebrew University of Jerusalem, Israel)*
K. Lindenberg *(University of California at San Diego, CA, U.S.A.)*
R. McWeeny *(Università di Pisa, Italy)*
M.A.C. Nascimento *(Instituto de Química, Rio de Janeiro, Brazil)*
P. Piecuch *(Michigan State University, East Lansing, MI, U.S.A.)*
M. Quack *(ETH Zürich, Switzerland)*
S.D. Schwartz *(Yeshiva University, Bronx, NY, U.S.A.)*
A. Wang *(University of British Columbia, Vancouver, BC, Canada)*

Former Editors and Editorial Board Members:

I. Prigogine (†)
J. Rychlewski (†)
Y.G. Smeyers (†)
R. Daudel (†)
M. Mateev (†)
W.N. Lipscomb (†)
H. Ågren (*)
D. Avnir (*)
J. Cioslowski (*)
W.F. van Gunsteren (*)

H. Hubač (*)
M.P. Levy (*)
G.L. Malli (*)
P.G. Mezey (*)
N. Rahman (*)
S. Suhai (*)
O. Tapia (*)
P.R. Taylor (*)
R.G. Woolley (*)

† deceased, * end of term

For further volumes:
http://www.springer.com/series/6464

Advances in the Theory of Quantum Systems in Chemistry and Physics

Edited by

PHILIP E. HOGGAN
Université Blaise-Pascal, Clermont-Ferrand, France

ERKKI J. BRÄNDAS
Department of Quantum Chemistry, University of Uppsala, Sweden

JEAN MARUANI
Laboratoire de Chimie Physique, CNRS and UPMC, Paris, France

PIOTR PIECUCH
Michigan State University, East Lansing, Michigan, USA

and

GERARDO DELGADO-BARRIO
Instituto de Física Fundamental, CSIC, Madrid, Spain

 Springer

Editors

Philip E. Hoggan
Pascal Institute
Labex IMOBS3, BP 80026
F-63171 Aubière Cedex
France
pehoggan@yahoo.com

Jean Maruani
Laboratoire de Chimie Physique
CNRS & UPMC
11 Rue Pierre et Marie Curie
F-75005 Paris
France
marjema@wanadoo.fr

Gerardo Delgado-Barrio
Instituto de Física Fundamental
CSIC
Serrano 123
E-28006 Madrid
Spain
gerardo@imaff.cfmac.csic.es

Erkki J. Brändas
Department of Quantum Chemistry
University of Uppsala
S-751 20 Uppsala
Sweden
erkki@kvac.uu.se

Piotr Piecuch
Department of Chemistry
Michigan State University
East Lansing, Michigan 48824
USA
piecuch@chemistry.msu.edu

ISSN 1567-7354
ISBN 978-94-007-2075-6   e-ISBN 978-94-007-2076-3
DOI 10.1007/978-94-007-2076-3
Springer Dordrecht Heidelberg London New York

Library of Congress Control Number: 2011942474

Printed on acid-free paper

Springer is part of Springer Science+Business Media (www.springer.com)

PTCP Aim and Scope

Progress in Theoretical Chemistry and Physics

A series reporting advances in theoretical molecular and material sciences, including theoretical, mathematical and computational chemistry, physical chemistry and chemical physics and biophysics.

Aim and Scope

Science progresses by a symbiotic interaction between theory and experiment: theory is used to interpret experimental results and may suggest new experiments; experiment helps to test theoretical predictions and may lead to improved theories. Theoretical Chemistry (including Physical Chemistry and Chemical Physics) provides the conceptual and technical background and apparatus for the rationalisation of phenomena in the chemical sciences. It is, therefore, a wide ranging subject, reflecting the diversity of molecular and related species and processes arising in chemical systems. The book series *Progress in Theoretical Chemistry and Physics* aims to report advances in methods and applications in this extended domain. It will comprise monographs as well as collections of papers on particular themes, which may arise from proceedings of symposia or invited papers on specific topics as well as from authors' initiatives or translations.

The basic theories of physics – classical mechanics and electromagnetism, relativity theory, quantum mechanics, statistical mechanics, quantum electrodynamics – support the theoretical apparatus which is used in molecular sciences. Quantum mechanics plays a particular role in theoretical chemistry, providing the basis for the valence theories, which allow to interpret the structure of molecules, and for the spectroscopic models employed in the determination of structural information from spectral patterns. Indeed, Quantum Chemistry often appears synonymous with Theoretical Chemistry: it will, therefore, constitute a major part of this book series. However, the scope of the series will also include other areas of theoretical chemistry, such as mathematical chemistry (which involves the use of algebra and topology in the analysis of molecular structures and reactions); molecular mechanics, molecular dynamics and chemical thermodynamics, which play an important role in rationalizing the geometric and electronic structures of molecular assemblies and polymers, clusters and crystals; surface, interface, solvent and solid-state effects; excited-state dynamics, reactive collisions, and chemical reactions.

Recent decades have seen the emergence of a novel approach to scientific research, based on the exploitation of fast electronic digital computers. Computation provides a method of investigation which transcends the traditional division between theory and experiment. Computer-assisted simulation and design may afford a solution to complex problems which would otherwise be intractable to theoretical analysis, and may also provide a viable alternative to difficult or costly laboratory experiments. Though stemming from Theoretical Chemistry, Computational Chemistry is a field of research in its own right, which can help to test theoretical predictions and may also suggest improved theories.

The field of theoretical molecular sciences ranges from fundamental physical questions relevant to the molecular concept, through the statics and dynamics of isolated molecules, aggregates and materials, molecular properties and interactions, and to the role of molecules in the biological sciences. Therefore, it involves the physical basis for geometric and electronic structure, stales of aggregation, physical and chemical transformations, thermodynamic and kinetic properties, as well as unusual properties such as extreme flexibility or strong relativistic or quantum-field effects, extreme conditions such as intense radiation fields or interaction with the continuum, and the specificity of biochemical reactions.

Theoretical Chemistry has an applied branch – a part of molecular engineering, which involves the investigation of structure–property relationships aiming at the design, synthesis and application of molecules and materials endowed with specific functions, now in demand in such areas as molecular electronics, drug design and genetic engineering. Relevant properties include conductivity (normal, semi- and supra-), magnetism (ferro- and ferri-), optoelectronic effects (involving nonlinear response), photochromism and photoreactivity, radiation and thermal resistance, molecular recognition and information processing, biological and pharmaceutical activities, as well as properties favouring self-assembling mechanisms and combination properties needed in multifunctional systems.

Progress in Theoretical Chemistry and Physics is made at different rates in these various research fields. The aim of this book series is to provide timely and in-depth coverage of selected topics and broad-ranging yet detailed analysis of contemporary theories and their applications. The series will be of primary interest to those whose research is directly concerned with the development and application of theoretical approaches in the chemical sciences. It will provide up-to-date reports on theoretical methods for the chemist, thermodynamician or spectroscopist, the atomic, molecular or cluster physicist, and the biochemist or molecular biologist who wish to employ techniques developed in theoretical, mathematical and computational chemistry in their research programmes. It is also intended to provide the graduate student with a readily accessible documentation on various branches of theoretical chemistry, physical chemistry and chemical physics.

Obituary – W.N. Lipscomb (1919–2011)

On 14 April, 2011, Nobel Laureate William Nunn Lipscomb Jr. passed away at Mount Auburn Hospital in Cambridge, Massachusetts. He died from pneumonia and complications from a fall he suffered several weeks earlier. Lipscomb was Abbott and James Lawrence Professor of Chemistry at Harvard University, Emeritus since 1990.

Lipscomb was born on 9 December, 1919 in Cleveland, Ohio, but his family moved to Lexington, Kentucky, when he was one year old. His mother taught music and his father practiced medicine. They "stressed personal responsibility and self reliance"[1] and created a home in which independence was encouraged. A chemistry kit that was offered him when he was 11 years old kindled Lipscomb's interest in science. He "recalled creating 'evil smells' using hydrogen sulfide to drive his two sisters out of his room"[2]. But it was through a music scholarship (he was a classical clarinetist) that he entered the University of Kentucky, where he eventually earned a bachelor of science degree in chemistry in 1941.

As a graduate student at the California Institute of Technology, Lipscomb was a protégé of Nobel Laureate Linus C. Pauling, whose famous book *The Nature of the Chemical Bond* was to revolutionize our understanding of chemistry. Lipscomb records[1] that

Pauling's course in The Nature of the Chemical Bond was worth attending every year, because each lecture was new...

In 1946, Lipscomb gained a Ph.D. degree in chemistry from Caltech with a dissertation in four parts. The first two were entitled: Electron Diffraction Investigations of Vanadium Tetrachloride, Dimethylketene Dimer, Tetrachloroethylene, and Trichloroethylene, and: The Crystal Structure of Methylammonium Chloride. Parts

---

[1] Process of Discovery (1977): an Autobiographical Sketch, in: *Structures and Mechanisms: from Ashes to Enzymes*, G.R. Eaton, D.C. Wiley and O. Jardetzky, ACS Symposium Series, American Chemical Society, Washington, DC (2002).

[2] *The New York Times*, 15 April, 2011.

3 and 4 were classified work for W.W.II. His thesis ends with a set of propositions, the last of which display his sense of humor:

(a) Research and study at the Institute have been unnecessarily hampered by the present policy of not heating the buildings on weekends.
(b) Manure should not be used as a fertilizer on ground adjacent to the Campus Coffee Shop.

Before eventually arriving at Harvard, Lipscomb taught at the University of Minnesota from 1946 to 1959. By 1948, he

> had initiated a series of low temperature X-ray diffraction studies, first of small hydrogen bonded systems, residual entropy problems and small organic molecules [and] later ... [of] the boron hydrides B_5H_9, B_4H_{10}, B_5H_{11}, B_6H_{10}, B_9H_{15}, and many more related compounds in later years (50 structures of boron compounds by 1976).

Lipscomb authored two books, both published by W.A. Benjamin Inc. (New York). The first (1963) was entitled *Boron Hydrides*. The second (1969), co-authored with G. Eaton, was on *NMR Studies of Boron Hydrides and Related Compounds*. He published over 650 scientific papers between 1942 and 2009. His citation for the Nobel Prize in chemistry in 1976, "for his studies on the structure of boranes illuminating problems of chemical bonding", echoes that of his mentor Linus Pauling in 1954, "for his research into the nature of the chemical bond and its application to the elucidation of the structure of complex substances". It is for his work on the structure of boron hydrides that Lipscomb is most widely known.

The field of borane chemistry was established by Alfred Stock, who summarized his work in his Baker Lectures[3] at Cornell in 1932. As early as 1927, it had been recognized that there exist relatively simple compounds which defy classification within the Lewis-Langmuir-Sidgwick theory of chemical bonding[4]. A particularly outstanding anomaly is the simplest hydride of boron, which Stock's pioneering work[4] established to be the dimer B_2H_6:

> The electronic formulation of the structure of the boron hydrides encounters a number of difficulties. The ordinary concepts of valence will not suffice to explain their structure; this is shown by the fact that in the simplest hydride, diborane B_2H_6, which has $2 \times 3 + 6 = 12$ electrons, as many bonds must be explained as are required for C_2H_6 which has two more ($2 \times 4 + 6 = 14$) electrons available. Thus it is that any structural theory for these compounds requires new hypotheses.

Diborane is said to be electron deficient, since it has only 12 valence electrons and appears to require 14 to form a stable species.

After some years of uncertainty, the structure of diborane was definitively settled by the infrared studies of Price[5] (in 1940–41, Stitt had produced infrared and

---

[3] A. Stock, *Hydrides of Boron and Silicon*, Cornell University Press (1933).

[4] G.N. Lewis, J. Am. Chem. Soc. 38, 762 (1916); I. Langmuir, J. Am. Chem. Soc. 41, 868, 1543 (1918); N.V. Sidgwick, *The Electronic Theory of Valency*, Oxford University Press (1927); L. Pauling, *The Nature of the Chemical Bond*, Cornell University Press (1939).

[5] W.C. Price, J. Chem. Phys. 15, 614 (1947); ibid. 16, 894 (1948).

thermodynamic evidence for the bridge structure of diborane[6]) and the electron-diffraction study of Hedberg and Schomaker[7]. The bridging structure of the diborane bonding was confirmed by Shoolery[8] from the ^{11}B NMR spectrum.

The invariance of the single-determinant closed-shell molecular orbital wave function under a unitary transformation of the occupied orbitals was exploited by Longuet-Higgins[9] to show that for a minimal basis set the molecular orbitals involved in the B-H-B bridge could be localized to form two three-centre two-electron bonds. Lipscomb, W. H. Eberhardt and B. L. Crawford[10] demonstrated how this simple procedure could be extended to higher boron hydrides. Noticing the similarity of bonding in B_2H_6 and in the bridge regions of B_4H_{10}, B_5H_9, B_5H_{11}, and $B_{10}H_{14}$ led Lipscomb to write[11]:

> These ideas suggest that ... the hybridization about boron in many of these higher hydrides is not greatly different from the hybridization in diborane. In addition, the probable reason for the predominance of boron triangles is the concentration of bonding electron density more or less towards the center of the triangle, so that the bridge orbitals (π-orbitals in B_2H_6) of the three boron atoms ... overlap. It does seem very likely ... that the outer orbitals of an atom are not always directed toward the atom to which it is bonded. This property is to be expected for atoms which are just starting to fill new levels and therefore may be a general property of metals and intermetallic compounds.

In the early 1960s, Edmiston and Ruedenberg[12] placed the localization of molecular orbitals on a somewhat more objective foundation by transforming to that basis in which interorbital exchange is a minimum. Lipscomb and coworkers[13] found that when applied to diborane this approach indeed leads to localized three-centre bonds for the B-H-B bridge. Lipscomb recalls[1] how the localization of molecular orbitals

> ... produced a vivid connection between the highly delocalized symmetry molecular orbitals and the localized bonds in which chemists believe so strongly.

He also records[1]:

> One disappointment was that the National Science Foundation refused to support the work started by J. Gerratt and me on spin-coupled wave functions.

Gerratt and Lipscomb introduced spin-coupled wave functions in 1968[14]. The energy expression for spin-coupled wave functions

---

[6] F. Stitt, J. Chem. Phys. 8, 981 (1940); ibid. 9, 780 (1941).

[7] K. Hedberg and V. Schomaker, J. Am. Chem. Soc. 73, 1482 (1951).

[8] J. Shoolery, Discuss. Faraday Soc. 19, 215 (1955).

[9] H.C. Longuet-Higgins and R.P. Bell, J. Chem. Soc. 250 (1943); H.C. Longuet-Higgins, J. Chim. Phys. 46, 275 (1949); Rev. Chem. Soc. 11, 121 (1957).

[10] W.H. Eberhardt, B. Crawford and W.N. Lipscomb, J. Chem. Phys. 22, 989 (1954).

[11] W.N. Lipscomb, J. Chem. Phys. 22, 985 (1954).

[12] C. Edmiston and K. Ruedenberg, Rev. Mod. Phys. 35, 457 (1963); J. Chem. Phys. 43, 597 (1965).

[13] E. Switkes, R.M. Stevens, W.N. Lipscomb and M.D. Newton, J. Chem. Phys. 51, 2085 (1969).

[14] J. Gerratt and W.N. Lipscomb, Proc. Natl. Acad. Sci. U.S. 59, 332 (1968).

... does not assume any orthogonality whatsoever among the orbitals and, depending upon which kinds of restrictions are placed upon [them], ... may be made to reduce to the energy expression for any of the orbital-type wave functions commonly used. Thus, if one specifies the [orbitals] to be atomic orbitals, then [the energy expression] is the general valence-bond energy. Other commonly used approximations ... may be embraced by imposing ... orthogonality restrictions ...

The theory of spin-coupled wave functions was developed by Gerratt et al.[15] and applied to a wide range of molecular systems including diborane[16].

Lipscomb also studied the structure and function of large biomolecules. He wrote[1]:

My interest in biochemistry goes back to my perusal of medical books in my father's library and to the influence of Linus Pauling from 1942 on ....

He used X-ray diffraction methods to determine the three-dimensional structure of proteins and then analyzed their function. Among the proteins studied by Lipscomb and coworkers were carboxypeptidase A[17], a digestive enzyme, and aspartate carbamoyltransferase[18], an enzyme from E. coli.

Lipscomb was invited to a large number of scientific conferences. In 1986 he chaired the Honorary Committee of the Congress *Molecules in Physics, Chemistry, and Biology* organized by Jean Maruani and Imre Czismadia in Paris, and in 2002 the *Fourth International Congress of Theoretical Chemical Physics (ICTCP-IV)* organized by Jean Maruani and Roland Lefebvre in Marly-le-Roi. He enthusiastically supported the foundation of this bookseries: *Progress in Theoretical Chemistry and Physics*, for which he has been an Honorary Editor from the very beginning. Editor-in-Chief Jean Maruani remembers he could always get his cheerful and friendly voice on the phone when he needed him.

William Lipscomb will be remembered as a scientist, an educator (three of his students received the Nobel Prize), and an inspiration to all. He is survived by his wife, Jean Evans, and three children – including two from an earlier marriage, as well as by three grandchildren and four great-grandchildren.

Stephen Wilson
Editor-in-Chief of
Progress in Theoretical Chemistry and Physics

[15] J. Gerratt, Adv. At. Mol. Phys. 7, 141 (1971); J. Gerratt, D.L. Cooper, M. Raimondi and P.B. Karadakov, in: Handbook of Molecular Physics and Quantum Chemistry, vol. 2, ed. S. Wilson, P.F. Bernath and R. McWeeny, Wiley (2003).

[16] S. Wilson and J. Gerratt, Molec. Phys. 30, 765 (1975).

[17] W.N. Lipscomb, J.A. Hartsuck, G.N. Reeke, Jr., F.A. Quiocho, P.H. Bethge, M.L. Ludwig, T.A. Steitz, H. Muirhead, J.C. Coppola, Brookhaven Symp. Biol. 21, 24 (1968).

[18] R.B. Honzatko, J.L. Crawford, H.L. Monaco, J.E. Ladner, B.F.P. Edwards, D.R. Evans, S.G. Warren, D.C. Wiley, R.C. Ladner, W.N. Lipscomb, J. Mol. Biol. 160, 219 (1983).

Obituary – Matey Mateev (1940–2010)

On July 25, 2010, world-renowned Bulgarian scientist, professor and academician Matey Dragomirov Mateev died in a car accident. Late on Sunday afternoon, on the way back to Sofia from his country house, at the foot of Stara Planina Mountain, he lost control of his vehicle and crashed against a tree. His wife Rumiana, sitting on the passenger's seat, died instantly, while Mateev died on the way to the hospital.

Matey Mateev was one of the most prominent Bulgarian physicists, with significant achievements in the fields of theoretical, mathematical, and nuclear physics. He was also known for his ethical and moral values and service to his community. In Bulgarian circles he was called 'the noble man of science'. His relatives were former Sofia physicians, intellectuals and public figures. His father, Pr. Dragomir Mateev, was also a prominent scientist as well as the Director of the Institute of Physiology of the Bulgarian Academy of Sciences, and for many years the Rector of the Higher Institute for Physical Culture (presently National Sports Academy *Vassil Levski*).

Matey Mateev had a son living in Sofia and a daughter in Barcelona. His tragic death came a few days after the happiest moment in his life: on July 22, his daughter gave birth to a girl – and his wife was planning to travel from Sofia to Barcelona to see her granddaughter, on July 29.

As a tragic coincidence, the funeral service was held on July 29, in the church St. Sofia – an annex to the cathedral Alexander Nevski. Hundreds of people came to pay their respects to Matey Mateev and his wife: relatives, friends, colleagues, public figures in the arts and in the media, members of parliament. Bulgarian president Georgi Purvanov sent a letter of condolences to his family, reading: "I was very grieved to learn about the unexpected death of the outstanding Bulgarian scientist and public figure Matey Mateev. We lost one of our prominent physicists, an internationally recognized authority, a loved lecturer, and a reputable leader in our system of science and education".

Academician Matey Mateev had an beautiful career. Born on April 10, 1940 in Sofia, he graduated in 1963 from the Faculty of Physics at University St. Kliment Ohridski, majoring in nuclear physics. Right after his graduation, he began working as a physicist and later as an assistant professor at the same faculty. In 1967 he won a one-year scholarship to the newly-established International Centre for Theoretical Physics in Trieste, Italy. He came back to Bulgaria and, soon afterwards, left again to the Joint Institute of Nuclear Physics, where he worked at the Laboratory of Theoretical Physics from 1971 to 1980, where he defended his Ph.D. dissertation. He came back to Bulgaria to work as an associate professor and, starting 1984, a full professor at the Faculty of Physics of Sofia University. In 1996 he was appointed Head of the Department of Theoretical Physics. During his career he has also been Dean of the Faculty of Physics and Vice-Rector of Sofia University.

Matey Mateev was loved by his students in physics and, from 1980 onwards, he lead one of the most attended courses at the Faculty of Physics. A generation of physicists has matured under his supervision and leadership. He authored over 100 major scientific publications in hot topics in physics.

Matey Mateev was elected a member of the Bulgarian Academy of Sciences in the Physical Sciences in 2003. As President of the Union of Bulgarian Physicists for many years, he was a champion for the establishment of a National Foundation for Fundamental Research. He has also been a Chairman of the Expert Committee for Physics at the National Science Fund and a Vice-President of the Balkan Physics Union. Between 1997 and 2003 he was a Chairman of the Committee for Bulgaria's Cooperation with the Joint Institute of Nuclear Physics.

In 1999 Matey Mateev became a member of the European Center for Nuclear Research (CERN) in Switzerland. Bulgaria's active participation in CERN's experimental and theoretical research was one of his major services to science and to his country. He became a member of the Committee for Bulgaria's Cooperation with CERN and of the Board of CERN, where he represented Bulgaria throughout the period 1999–2000 and was the team leader of Bulgarian scientists invited to work at the Large Hadron Collider on its activation in CERN.

Matey Mateev has gone all the way up to the top of the scientific and administrative ladder. Until the democratic changes that occurred in Bulgaria in 1989, he was the Chairman of the Science Committee at the Council of Ministers. In 1990 he was appointed Deputy Minister (and in 1991 Minister) of Public Education. He remained in office for three successive terms. It was under his guidance and supervision that the Public Education Act was drawn up, as well as texts on education in Bulgaria's Constitution, which were adopted by the National Assembly in 1991.

Matey Mateev supported the organization of the *Sixth European Workshop on Quantum Systems in Chemistry and Physics* (QSCP-VI) in Sofia in 2001, by Alia Tadjer and Yavor Delchev, and the first award of the *Promising Scientist Price of CMOA*, which he attended at Boyana Palace (the Bulgarian President Residence), where the protocol of the ceremony was established.

Matey Mateev was Editor-in-Chief of the *Bulgarian Journal of Physics*, member of the Board of *Balkan Physics Letters*, and member of the Board of *Progress in Theoretical Chemistry and Physics*.

In spite of his wide reputation and prestige, Matey Mateev remained a warm-hearted and broad-minded person. We will always remember him, not only for his achievements in research, education, science policy and public service, but also for his friendly attitude towards colleagues, his overall dedication, and his readiness to help in any situation.

Rest in peace!

Rossen Pavlov
Senior Scientist at INRNE
Bulgarian Academy of Sciences

Preface

This volume collects 32 selected papers from the scientific contributions presented at the 15th International Workshop on *Quantum Systems in Chemistry and Physics* (QSCP-XV), which was organized by Philip E. Hoggan and held at Magdalene College, Cambridge, UK, from August 31st to September 5th, 2010. Participants at QSCP-XV discussed the state of the art, new trends, and the future of methods in molecular quantum mechanics, and their applications to a wide range of problems in chemistry, physics, and biology.

Magdalene College was originally founded in 1428 as a hostel to house Benedictine monks coming to Cambridge to study law. Nowadays it houses around 350 undergraduate students and 150 graduate students reading towards Masters or Ph.D degrees in a diverse range of subjects. The College comprises a similarly diverse set of architectures from its medieval street frontage through to the modern Cripps Court – where the scientific sessions took place, which blends modern design with traditional materials.

The QSCP-XV workshop followed traditions established at previous meetings:

QSCP-I, organized by Roy McWeeny in 1996 at San Miniato (Pisa, Italy);
QSCP-II, by Stephen Wilson in 1997 at Oxford (England);
QSCP-III, by Alfonso Hernandez-Laguna in 1998 at Granada (Spain);
QSCP-IV, by Jean Maruani in 1999 at Marly-le-Roi (Paris, France);
QSCP-V, by Erkki Brändas in 2000 at Uppsala (Sweden);
QSCP-VI, by Alia Tadjer in 2001 at Sofia (Bulgaria);
QSCP-VII, by Ivan Hubac in 2002 at Bratislava (Slovakia);
QSCP-VIII, by Aristides Mavridis in 2003 at Spetses (Athens, Greece);
QSCP-IX, by Jean-Pierre Julien in 2004 at Les Houches (France);
QSCP-X, by Souad Lahmar in 2005 at Carthage (Tunisia);
QSCP-XI, by Oleg Vasyutinskii in 2006 near St Petersburg (Russia);
QSCP-XII, by Stephen Wilson in 2007 near Windsor (England);
QSCP-XIII, by Piotr Piecuch in 2008 at East Lansing (Michigan, USA);
QSCP-XIV, by Gerardo Delgado-Barrio in 2009 at El Escorial (Spain).

Attendance of the Cambridge workshop was a record in the QSCP series: there were 138 scientists from 32 countries on all five continents.

The lectures presented at QSCP-XV were grouped into the following seven areas in the field of *Quantum Systems in Chemistry and Physics*:

1. Concepts and Methods in Quantum Chemistry and Physics;
2. Molecular Structure, Dynamics, and Spectroscopy;
3. Atoms and Molecules in Strong Electric and Magnetic Fields;
4. Condensed Matter; Complexes and Clusters; Surfaces and Interfaces;
5. Molecular and Nano Materials and Electronics;
6. Reactive Collisions and Chemical Reactions;
7. Computational Chemistry, Physics, and Biology.

There were sessions where plenary lectures were given, sessions accommodating parallel talks, and evening sessions with posters preceded by flash oral presentations. We are grateful to all plenary speakers and poster presenters for having made this QSCP-XV workshop a stimulating experience and success.

The breadth and depth of the scientific topics discussed during QSCP-XV are reflected in the contents of this volume of proceedings in *Progress in Theoretical Chemistry and Physics*, which includes five sections:

I. General: 1 paper;
II. Methodologies: 10 papers;
III. Structure: 8 papers;
IV. Dynamics and Quantum Monte-Carlo: 6 papers;
V. Reactivity and Functional Systems: 7 papers;

The details of the Cambridge meeting, including the complete scientific program, can be obtained on request from pehoggan@yahoo.com.

In addition to the scientific program, the workshop had its fair share of other cultural activities. One afternoon was devoted to a visit of Cambridge Colleges, where the participants had a chance to learn about the structure of the University of Cambridge. There was a dinner preceded by a tremendous organ concert in the College Chapel. The award ceremony of the CMOA Prize and Medal took place in Cripps Lecture Hall. The Prize was shared between three of the selected nominees: Angela Wilson (Denton, TX, USA), Julien Toulouse (Paris, France) and Robert Vianello (Zagreb, Croatia), while two other nominees (Ioannis Kerkines – Athens, Greece – and Jeremie Caillat – Paris, France) received a certificate of nomination. The CMOA Medal was then awarded to Pr Nimrod Moiseyev (Haifa, Israel). Following an established tradition of QSCP meetings, the venue and period of the next QSCP workshop was disclosed at the end of the banquet that followed: Kanazawa, Japan, shortly after the ISTCP-VII congress scheduled in Tokyo, Japan, in September, 2011.

We are pleased to acknowledge the support given to QSCP-XV by Trinity College (Gold sponsor), Q-Chem (Silver sponsor) and the RSC (Bronze sponsor) at Cambridge. The workshop was chaired by Pr Stephen Elliott (Cambridge, UK) and cochaired by Pr Philip E. Hoggan (Clermont, France). We are most grateful to all members of the Local Organizing Committee (LOC) for their work and dedication. We are especially grateful to Marie-Bernadette Lepetit (Caen, France) for her efficiency in handling the web site, and to Jeremy Rawson, Alex Thom, Aron Cohen, Daniel Cole, Neil Drummond, Alston Misquitta (Cambridge, UK), and Joëlle Hoggan, who made the stay and work of the participants pleasant and fruitful. Finally, we would like to thank the Honorary Chairs and members of the International Scientific Committee (ISC) for their invaluable expertise and advice.

We hope the readers will find as much interest in consulting these proceedings as the participants in attending the workshop.

The Editors

Contents

Part I
Fundamental Theory

Chapter 1
Time Asymmetry and the Evolution of Physical Laws

Erkki J. Brändas

Abstract In previous studies we have advocated a retarded-advanced sub-dynamics that goes beyond standard probabilistic formulations supplying a wide-range of interpretations. The dilemma of time reversible microscopic physical laws and the irreversible nature of thermodynamical equations are re-examined from this point of view. The subjective character of statistical mechanics, i.e. with respect to the theoretical formulation relative to a given level of description, is reconsidered as well. A complex symmetric ansatz, incorporating both time reversible and time irreversible evolutions charts the evolution of the basic laws of nature and reveals novel orders of organization. Examples are drawn from the self-organizational behaviour of complex biological systems as well as background dependent relativistic structures including Einstein's laws of relativity and the perihelion movement of Mercury. A possible solution to the above mentioned conundrum is provided for, as a consequence of a specific informity rule in combination with a Gödelian like decoherence code protection. The theory comprises an interesting cosmological scenario in concert with the second law.

1.1 Introduction

The most recognized dilemma in the theoretical description of physical events is the problem related to irreversible behaviour and the associated time asymmetry of entropic increase. In this appraisal lies the more fundamental re-interpretation of thermodynamics from the viewpoint of statistical mechanics, the choice of initial probability distributions as well as the emergence of temporal asymmetry from

E.J. Brändas (✉)
Department of Quantum Chemistry, Uppsala University, Box 518 SE-751 20, Uppsala, Sweden
e-mail: Erkki.Brandas@kvac.uu.se

P.E. Hoggan et al. (eds.), *Advances in the Theory of Quantum Systems in Chemistry and Physics*, Progress in Theoretical Chemistry and Physics 22, DOI 10.1007/978-94-007-2076-3_1, © Springer Science+Business Media B.V. 2012

perfectly time symmetric microscopic dynamics. This mystery, moreover, carries over to the cosmological picture, which, regardless of the materialization of modern Big Bang models, is far from adequately resolved [1].

One radical solution to this puzzle has been offered by I. Prigogine in his theory of time irreversibility, see e.g. [2] and references therein. Without taking recourse to any course graining he took irreversibility to be a fundamental fact due to dynamics alone. However, in speaking of intrinsic irreversibility he did attract challenging criticism from candid scientists and philosophers alike. The key to realize the Prigogine causal sub-dynamics and to understand the reduction of macroscopic laws to microscopic ones lies in a mathematical ingredient, the star-unitary transformation, which attempts to obtain a symmetry breaking time evolution of the original probability distribution via apposite semi-group selections. It is important to recognize that the Liouville formalism applies to both classical and quantum mechanical formulations and furthermore that the emergence of irreversibility rests in the coupling between the dynamics of the open dissipative system and entropic evolution via an explicitly given Lyapunov function.

Although it may be too early to fully evaluate the vision and foresight of Ilya Prigogine, there have appeared over time various fundamental and also critical objections to his programme of open dissipative systems. With some risk of oversimplification one can say that a scientist and a philosopher disagree by and large in that the former nurtures concept unification, while the latter on the contrary espouses concept differentiation. Parallel derivations and analogous interpretations may be construed as unification in one domain and conflation in the other. Both viewpoints are indisputably important if properly balanced.

Neglecting philosophical critique we bring attention to an alternative derivation of subdynamics that has its roots in quantum mechanics, viz. the utilization of the dilation group through a mathematical theorem due to Balslev and Combes [3]. The reformulation of the Nakajima-Zwanzig Generalized Master Equation [4, 5] within a retarded-advanced formulation made it possible to evaluate the relevant residue contributions of the projective decompositions of the appropriate resolvent, i.e. the non-hermitean collision operator etc., via the proper analytic continuation explicitly defined via the aforementioned theorem [6]. This derives the dissipativity condition for quantum mechanical systems with an absolutely continuous spectrum (the situation is a bit more complex in the classical formulation). Another essential difference, comparing the Prigogine causal dynamics [7] with the present development, see e.g. Refs. [6,8], is that the retarded-advanced dynamics allows conversion into contracted semigroups with the positivity preserving condition (probabilistic interpretation) relaxed [9]. The latter step is important since it carries with it an inevitable objective loss of information. Furthermore we have demonstrated that the present representation via Bloch thermalization empowers microscopic self-organization through integrated quantum-thermal correlations [10,11]. The resulting *Coherent Dissipative Structure, CDS*, provides a rich variation of timescales as well as being code protected against decoherence, see more details below and also reference [12].

With this idea as background we will develop the non-probabilistic formulation further and generalize its framework incorporating complex biological systems and most importantly providing an alternative formulation of special and general relativity. We will demonstrate its significance as regards time irreversibility, biological organization and the Einstein laws of general relativity. In addition we will demonstrate its expediency and accuracy by determining the perihelion motion of Mercury concluding with a possible cosmologic scenario extending and going beyond popular big bang – inflation type settings particularly demonstrating that cosmic memory loss provides cosmic sensorship and an objective platform for the second law.

1.2   Time Evolution, Partitioning Technique and Associated Dynamics

We will first demonstrate the subtleties involved in the derivations of the proper dynamical equations. Although the presentation below can, without problem, be extended to a Liouville formulation we will, for simplicity review the "time reversible" case of the Schrödinger equation based on the self-adjoint Hamiltonian H. Thus we write the following causal expressions ($h = 2\pi$) of the time-independent and time-dependent Schrödinger Equation assuming the existence of an absolutely continuous spectrum σ_{AC}

$$(E - H)\psi(E) = 0; \quad \left(i\frac{\partial}{\partial t} - H\right)\psi(t) = 0; \quad \psi(t) = e^{-iHt}\psi(0) \qquad (1.1)$$

In passing we adopt the traditional definition of the spectrum, σ, of a general unbounded (closable) operator H, defined in a complete separable Hilbert space, characterized according to the standard decomposition theorem, i.e. σ_P, the pure point-, σ_{AC} the absolutely continuous- and σ_{SC}, the singularly continuous part. Note that molecular Hamiltonians do not contain σ_{SC}, so this option is not discussed here. In Eq. 1.1 above we will particularly investigate the situation when the energy E belongs to the continuum, i.e. $E \in \sigma_{AC}$, $\psi(E)$, $\psi(t)$ are the time independent and time dependent wave functions respectively and t is the time parameter. The two equations above are connected through the Fourier-Laplace transform and we will explain this formulation in more detail below. Although this outline has been presented several times by the author, one needs to redevelop some of the main equations for impending use and conclusions.

It follows that in order to guarantee the existence of the transforms, one has to define integration paths along suitable complex contours $C^{\pm}$ by the lines in the upper/lower complex plane via $C^{\pm} : (\pm id - \infty \to \pm id + \infty)$ where $d > 0$ may be arbitrary small. Using the well known Heaviside and Dirac delta functions $\theta(t)$ and

$\delta(t)$, respectively, we can separate out positive and negative times (with respect to an arbitrary chosen time $t = 0$) as related with the relevant contours $C^{\pm}$ according to

$$G^{\pm}(t) = \pm(-i)\,\theta(\pm t)e^{-iHt}; \quad G(z) = (z-H)^{-1} \tag{1.2}$$

where the retarded-advanced propagators $G^{\pm}(t)$ and the resolvents $G(z^{\pm})$; $z^{\pm} = \mathcal{R}z \pm i\mathcal{I}z$ are connected through ($\mathcal{R}z$, $\mathcal{I}z$ are the real and imaginary parts of $z = E + i\varepsilon$)

$$G^{\pm}(t) = \frac{1}{2\pi}\int_{C^{\pm}} G(z)e^{-izt}dz; \quad G(z) = \int_{-\infty}^{+\infty} G^{\pm}(t)e^{izt}dt$$

$$\{t > 0; \, \mathcal{I}z > 0\}$$
$$\{t < 0; \, \mathcal{I}z < 0\} \tag{1.3}$$

The Fourier-Laplace transform Eq. 1.2 exists under quite general conditions by e.g. closing the contours $C^{\pm}$ in the lower and upper complex planes respectively. More details regarding the specific choice of contours in actual cases can be found in Refs. [13–15] and references therein. The formal retarded-advanced formulation corresponding to Eq. 1.1 including the memory terms follows from

$$\left(i\frac{\partial}{\partial t} - H\right)G^{\pm}(t) = \delta(t); \quad \psi^{\pm}(t) = \pm iG^{\pm}(t)\psi^{\pm}(0)$$

$$\left(i\frac{\partial}{\partial t} - H\right)\psi^{\pm}(t) = \pm i\delta(t)\psi^{\pm}(0) \tag{1.4}$$

and

$$(z-H)G(z) = I; \psi^{\pm}(z) = \pm iG(z)\psi^{\pm}(0)$$
$$(z-H)\psi^{\pm}(z) = \pm i\psi^{\pm}(0) \tag{1.5}$$

It is usual to normalize the time dependent wavefunction $\psi^{\pm}(t)$, which means that $\psi(z)$ obtained from partitioning technique as a result is not. Since we are primarily interested in the case $E \in \sigma_{AC'}$ we will take the limits $\mathcal{I}(z) \to \pm 0$, obtaining the dispersion relations

$$G(E+i\varepsilon) = \lim_{\varepsilon \to \pm 0}(E+i\varepsilon - H)^{-1} = \mathcal{P}(E-H)^{-1} \pm (-i)\pi\delta(E-H) \tag{1.6}$$

where $\mathcal{P}$ is the principal value of the integral.

The goal is now to evaluate full time dependence from available knowledge of the wave function φ at time $t = 0$, for simplicity we assume that the limits $t \to \pm 0$ are the same, although this is not a necessary condition in general. Making the choice $O = |\phi\rangle\langle\phi|\phi\rangle^{-1}\langle\phi|; \phi = \varphi(0)$, one obtains (the subspace, defined by the projector O can easily be extended to additional dimensions)

$$\psi(0) = \psi^{+}(0) = \psi^{-}(0); \quad O\psi(0) = \varphi(0); \quad P\psi(0) = \kappa(0); \quad O + P = I \tag{1.7}$$

Using familiar operator relations of the time dependent partitioning technique, see again e.g. Ref. [15] for a recent review, we obtain

$$O(z-H)^{-1} = O(z-O\mathcal{H}(z)O)^{-1}(I+HT(z))$$
$$\mathcal{H}(z) = H+HT(z)H; T(z) = P(z-PHP)^{-1}P \qquad (1.8)$$

As we have pointed out at several instances the present equations are essentially analogous to the development of suitable master equations in statistical mechanics [4–7], where the "wavefunction" here plays the role of suitable probability distributions. Note for instance the similarity between the reduced resolvent, based on $\mathcal{H}(z)$, and the collision operator of the Prigogine subdynamics. The eigenvalues of the latter define the spectral contributions corresponding to the projector that defines the map of an arbitrary initial distribution onto a kinetic space obeying semigroup evolution laws, for more details we refer to Ref. [6] and the following section.

Rewriting the inhomogeneous version of the Schrödinger equation, where the boldface wave vectors below signify added dimensions, one obtains (the poles of the first line of Eq. 1.8 correspond to eigenvalues below)

$$(z-H)\Psi(z) = O(z-\mathcal{H}(z))O\phi \qquad (1.9)$$

From Eq. 1.9 the formulas of the time-dependent partitioning technique follows straightforwardly

$$\varphi^{\pm}(t) = \pm\frac{i}{2\pi}\int_{C^{\pm}} O(z-H)^{-1}\psi(0)e^{-izt}dz = \pm\frac{i}{2\pi}\int_{C^{\pm}}\varphi(z)e^{-izt}dz \qquad (1.10)$$

where

$$\varphi(z) = O(z-H)^{-1}\psi(0) = O(z-O\mathcal{H}(z)O)^{-1}(\varphi(0)+HT(z)\kappa(0)) \qquad (1.11)$$

The equations of motion, restricted to subspace O, is directly obtained from the convolution theorem of the Fourier-Laplace transform, i.e.

$$\left(i\frac{\partial}{\partial t} - OHO\right)\varphi^{\pm}(t) = \pm i\delta(t)\varphi(0)$$

$$+OHP\left\{\int_{0}^{t}(G_P(t-\tau)PH\varphi(\tau))^{\pm}d\tau \pm iG_P^{\pm}(t)\kappa(0)\right\}$$

$$(1.12)$$

with

$$G_P^{\pm}(t) = \pm(-i)\theta(\pm t)e^{-iPHPt} = \frac{1}{2\pi}\int_{C^{\pm}}T(z)e^{-izt}dz \qquad (1.13)$$

and

$$\int_0^t (G_P(t-\tau)PH\varphi(\tau))^\pm d\tau = \pm\frac{i}{2\pi}\int_{C^\pm} T(z)H\varphi(z)e^{-izt}dz \quad (1.14)$$

The first term on the right in Eq. 1.12 yields the description of the amplitude $\varphi^\pm$ evolving according to the Hamiltonian OHO. Furthermore the second term depends on all times between 0 and t, while the last term evolves the unknown part κ at $t=0$ completing the memory at initial time. Sofar no approximations have been set and no loss of information acquiesced.

Introducing the auxiliary operator $G_L^\pm(t)$, through

$$O\mathcal{H}(z^\pm)O = \int_{-\infty}^{+\infty} G_L^\pm(t)e^{iz^\pm t}dt$$

and the non-local operator $G^\pm(t;0)$ via

$$G^\pm(t;0)\varphi(0) = \int_0^t (G_L(t-\tau)\varphi(\tau))^\pm d\tau = \pm\frac{i}{2\pi}\int_{C^\pm} O\mathcal{H}(z^\pm)O\varphi(z)e^{-izt}dz \quad (1.15)$$

one gets the more compact expression

$$\left(i\frac{\partial}{\partial t}\varphi^\pm(t) - G^\pm(t;0)\right)\varphi(0) = \pm i\left[\delta(t)\varphi(0) + OHPG_P^\pm(t)\kappa(0)\right] \quad (1.16)$$

Note that analogous evolution formulas hold within the subspace P, i.e. with O and P interchanged. Although any localized wave packet under free evolution disperses, it is however traditionally recognized that the complete formulation of an elementary scattering set-up describes a time symmetric process provided the generator of the evolution commutes with the time reversal operator and the time-dependent equation imparts time symmetric boundary conditions. Compare for instance analogous discussions in connection with the electromagnetic field, obeying Maxwell's equations, via retarded-, advanced- or symmetric potentials. Although time symmetric equations may exhibit un-symmetric solutions via specific initial conditions the fundamental point here concerns the "master evolution equation" itself. Hence, as already pointed out, we re-emphasize that no approximations have been admitted and consequently time evolution proceeds without loss of information.

At this junction it is common to discuss various short time expansions and/or long time situations, i.e. to consider partitions of relevant time scales. We will principally mention two interdependent scales, i.e. a global relaxation time τ_{rel} and a local collision time τ_c. For instance, during τ_c the amplitude φ is not supposed to alter very much. Hence one approximates the convolution in Eq. 1.16 for positive times, i.e.

$$G^+(\tau_c;0) \approx OHO - i\tau_c\{OHPHO + 1/2!(-i\tau_c)OH(PH)^2O\cdots\} \tag{1.17}$$

The relaxation time τ_{rel} obtains from time independent partitioning technique. Accordingly starting from Eqs. 1.5 and 1.6 one attains the limits

$$(E - H)\psi^\pm(E) = \pm i\psi(0)$$

$$\psi^\pm(E) = \pm i \lim_{\varepsilon \to \pm 0} (E + i\varepsilon - H)^{-1}\psi(0) = \lim_{\varepsilon \to \pm 0} \int_{-\infty}^{+\infty} \psi^\pm(t)e^{i(E+i\varepsilon)t} dt \tag{1.18}$$

Incidentally we recover the stationary wave, $\psi_c(E)$, via the causal propagator $G_c(t)$

$$\psi_c(t) = G_c(t)\psi(0) = \psi^+(t) + \psi^-(t); \, G_c(t) = e^{-iHt}; \tag{1.19}$$

and formally

$$\psi_c(E) = \lim_{\varepsilon \to +0} \int_{-\infty}^{+\infty} \psi^+(t)e^{i(E+i\varepsilon)t} dt + \lim_{\varepsilon \to -0} \int_{-\infty}^{+\infty} \psi^-(t)e^{i(E+i\varepsilon)t} dt = \int_{-\infty}^{+\infty} \psi_c(t)e^{iEt} dt.$$

Since the relaxation or life time, τ_{rel}, is directly related to a "hidden" complex resonance eigenvalue of Eq. 1.9, or pole of Eq. 1.8, we need to derive and investigate the dispersion relation for the reduced resolvent $T(z)$, i.e.

$$\lim_{\varepsilon \to \pm 0} T(E + i\varepsilon) = \lim_{\varepsilon \to \pm 0} (E + i\varepsilon - PHP)^{-1}$$

$$= \mathcal{P}(E - PHP)^{-1} \pm (-i)\pi\delta(E - PHP) \tag{1.20}$$

For instance, in the none-degenerate case one finds by getting the real and imaginary parts of $f(z)$

$$f(z) = \langle \phi|H(z)|\phi\rangle \tag{1.21}$$

and

$$f^\pm(E) = f_R(E) \pm (-i)f_I(E) \tag{1.22}$$

with

$$f_R(E) = E + \langle\phi|H\mathcal{P}(E - PHP)^{-1}H|\phi\rangle$$
$$f_I(E) = \pi\langle\phi|H\delta(E - PHP)H|\phi\rangle \geq 0 \tag{1.23}$$

In the limit $\varepsilon \to \pm 0$, assuming full information for simplicity at the initial time, $t = 0$, i.e. $\kappa(0) = 0; \phi = \varphi(0) = \psi(0)$, Eq. 1.9 yields

$$(E - H)\Psi^\pm(E) = \pm f_I(E)\varphi(0) \tag{1.24}$$

keeping in mind that $z = E - i\varepsilon$ from Eq. 1.23

$$E = f_R(E); \varepsilon(E) = f_1(E) \tag{1.25}$$

Note that Eq. 1.25 only gives the resonance approximately. The exact complex resonance eigenvalue (if it exists, see more on this below) has to be found by analytical continuation, Eq. 1.9, by e.g. successive iterations until convergence for each individual resonance eigenvalue. Hence in the first iteration, one gets the lifetime τ_{rel} given by

$$f_1(E) = \varepsilon(E) = \{2\tau_{rel}(E)\}^{-1} \tag{1.26}$$

To find an "uncertainty-like" relation between the two time scales we combine the expansion Eq. 1.17 with

$$O\mathcal{H}(z)O = OHO + \frac{1}{z}\left\{ OHPHO + \frac{1}{z}OH(PH)^2O\cdots \right\}$$

and

$$O\mathcal{H}(E \pm i0)O = OHO + OHP\,\mathcal{P}\,(E - PHP)^{-1}PHO$$
$$\pm(-i)\pi OHP\delta(E - PHP)PHO$$

obtaining

$$\tau_c\sigma^2(E) = \varepsilon(E) = \frac{1}{2\,\tau_{rel}} \tag{1.27}$$

with σ^2 being the variance at $E_0 = \langle\varphi(0)|H|\varphi(0)\rangle$, i.e.

$$\sigma^2 = \langle\varphi(0)|HPH|\varphi(0)\rangle = \langle\varphi(0)|(H^2 - \langle\varphi(0)|H|\varphi(0)\rangle^2)|\varphi(0)\rangle \tag{1.28}$$

The relations above contain many approximations, e.g. break-up of convolutions, truncation of various expansions etc., not to mention the assumed existence of a rigorous analytical continuation into higher order Riemann sheets of the complex energy plane. Since the original Schrödinger equation, or Liouville equation, as pointed out, is time reversible and rests on a unitary time evolution, it is obvious that the present lifetime analysis is contradictory. This is a well-known fact amongst the practitioners in the field; nevertheless, a lot of physical and chemical interpretations have been made and found to be meaningful portrayals of fundamental experimental situations. How come that this still appears to work satisfactorily? In the next section we will examine the reasons why, as well as develop the necessary mathematical machinery to rigorously extend resolvent- and propagator domains and examine its evolutionary consequences.

The object of our description is twofold: first to show that analytic continuation into the complex plane can be rigorously carried out and secondly to examine the end result for the associated time evolution. Not only will we find that time symmetry is by necessity broken, but also that novel complex structures appear with fundamental consequences for the validity of the second law as well as giving further guidelines on a proper self-referential approach to the theory of gravity.

1.3 Non-self-Adjoint Problems and Dissipative Dynamics

Staying within our original focus in the introduction, we recapitulate the current dilemma, i.e. how to connect, if possible, the exact microscopic time reversible dynamics, see above, with a time irreversible macroscopic entropic formulation without making use of any approximations whatsoever. In the present setting one must ascertain precisely what it takes to go from a stationary- to a quasi-stationary scenario. To begin with, we return to the practical problem of extracting meaningful life times out of the exact dynamics presented above and in particular to dwell on the consequences if any.

The understanding, interpretation and practical tools to approach the problem of resonances states in quantum chemistry and molecular physics are basically very well studied. Generally one has either (i) *concentrated on the properties of the stationary time-independent scattering solution* (ii) *attempted to extract the Gamow wave by analytic continuation and/or* (iii) *considered the time-dependent problem via a suitably prepared reference function or wave-packet*. In each case the analysis prompts different explanations, numerical techniques and understanding, see e.g. Ref. [15] for a review and more details.

In order to appreciate the significance of this situation, we will portray one of the most significant and successful approaches to quasi-stationary unstable quantum states by re-connecting with the previously mentioned theorem due to Balslev and Combes [3]. The authors derived general spectral theorems of many-body Schrödinger operators, employing rigorous mathematical properties of so-called dilatation analytic interactions (with the absence of singularly continuous spectra). The possibility to "move" or rotate the absolutely continuous spectrum, σ_{AC}, appealed almost instantly and was right away exploited in a variety of quantum theoretical applications in both quantum chemistry and nuclear physics [16].

The principle idea stems from a suitable change, or scaling, of all the coordinates in the second order partial differential equation (Schrödinger equation), which if allowing a complex scale factor, permits outgoing growing exponential solutions, so-called Gamow waves, to be treated via stable numerical methods without being forced to leave Hilbert space. Although this trick admits standard usage of alleged $\mathcal{L}^2$ techniques, there is a price, i.e. the emergence of non-self-adjoint operators which brings about a lot of important consequences to be summarized further below.

The strategy is best illustrated by considering a typical matrix element of a general quantum mechanical operator $W(r)$ over the basis functions, $\varphi(r)$ and $\phi(r)$, where we write $r = r_1, r_2, \ldots r_N$; assuming $3N$ fermionic degrees of freedom.

Employing the scaling $r' = \eta^{3N}r; \eta = e^{i\vartheta}$ (or $\eta = |\eta|e^{i\vartheta}$), where the phase $\vartheta \le \vartheta_0$ for some ϑ_0 that in general depends on the operator, one finds straightforwardly

$$\int \varphi^*(r)W(r)\phi(r)dr = \int \varphi^*(r'^*)W(r')\phi(r')dr' \qquad (1.29)$$

or in terms Dirac bra-kets (with $\varphi^*(\eta^*) = \varphi(\eta)$)

$$\langle \varphi|W|\varphi\rangle = \langle \varphi(\eta^*)|W(\eta)|\phi(\eta)\rangle \qquad (1.30)$$

We assume that the operator $W(r)$ as well as $\varphi(r)$ and $\phi(r)$ are properly defined for the scaling process to be justified. For simplicity we also take the interval of the radial components of r to be $(0,\infty)$. Note that Eq. 1.29 contains the requirement that the matrix element should be analytic in the parameter η, demanding the complex conjugate of η in the "bra" side of Eq. 1.30. This is the reason why many complex scaling treatments in quantum chemistry are implemented using complex symmetric forms.

In order to appreciate the fine points in this analysis, we therefore return to the domain issues, i.e. how to define the operator and the basis functions so that the scaling operation above becomes meaningful. Following Balslev and Combes [3], we introduce the N-body (molecular) Hamiltonian as $H = T + V$, where T is the kinetic energy operator and V is the (dilatation analytic) interaction potential (expressed as sum of two-body potentials V_{ij} bounded relative $T_{ij} = \Delta_{ij}$, where the indices i and j refers to particles i and j respectively). As a first crucial point we realize that the complex scaling transformation is unbounded, which necessitates a restriction of the domain of H; note that H is normally bounded from below. Hence we need to specify the domain $\mathcal{D}(H)$ of H as

$$\mathcal{D}(H) = \{\Phi \in \hbar, H\Phi \in \hbar\} \qquad (1.31)$$

where $\hbar$ denotes the well-known Hilbert space. The essential property of a dilatation analytic operator is that each individual pair potential of the interaction V is bounded relative the corresponding part of the kinetic energy. Hence the unboundedness is due to the latter i.e. $\mathcal{D}(H) = \mathcal{D}(T)$. With these preliminaries one can prove that the scaling operator $U(\vartheta) = \exp(iA\theta)$ is unitary for real ϑ and generated by

$$A = \frac{1}{2}\sum_{k=1}^{k=N}[\boldsymbol{p}_k\boldsymbol{x}_k + \boldsymbol{x}_k\boldsymbol{p}_k] \qquad (1.32)$$

where $\boldsymbol{x}_k$ and $\boldsymbol{p}_k$ are coordinate and momentum vectors of the particle k. As a result we get

$$U(e^{\vartheta})\Phi(r) = \exp(iA\theta)\Phi(r) = e^{\frac{3N\vartheta}{2}}\Phi(e^{\vartheta}r) \qquad (1.33)$$

and with $\vartheta \to i\vartheta; \eta = e^{i\vartheta}$, or more generally $\eta = |\eta|e^{i\vartheta}$ we write

$$H(\eta) = U(\eta)H(1)U^{-1}(\eta) = \eta^{-2}T(1) + V(\eta) \qquad (1.34)$$

At this stage it is crucial to emphasize that the formal expression Eq. 1.34 must be obtained in two steps due to the unboundedness of T and the complex scaling transformation. First we introduce $\Omega = \{\eta, |\arg(\eta)| \leq \vartheta_0\}$ in agreement what has been said above, then decompose Ω in its upper and lower parts, partitioned by the real axis R, where $\Omega = \Omega^+ \cup \Omega^- \cup R$ and $R = R^+ \cup R^- \cup \{0\}$. To avoid problems we will exclude the point $\{0\}$. The first step consists of the real scaling, i.e. $\eta \in R^+$, which corresponds to a unitary transformation, followed by an analytical continuation to $\eta \in \Omega^+$, corresponding to a similarity, non-unitary operation. Since this is an important point we will consider the scaling operator $U(\eta); \eta \in \Omega^+$ more exactly by bringing in the dense subset $\mathcal{N}(\Omega)$

$$\mathcal{N}(\Omega) = \{\Phi, \Phi \in h; \ H(\eta)\Phi \in h; \ U(\eta) \in h; \ \eta \in \Omega\} \qquad (1.35)$$

as the well-known Nelson's class of dilatation analytic vectors [17] more specifically defined as follows. A vector $\phi \in \mathcal{D}(A)$ is an analytic vector of A if the series expansion of $e^{A\vartheta}\phi$ has a positive radius of absolute convergence, i.e.

$$\sum_{n=0}^{\infty} \frac{\|A^n \phi\|}{n!} \vartheta^n < \infty$$

for some $\vartheta > 0$. For our purpose, to be explained below, we introduce the Hilbert (or Banach) space norm

$$\sup_{\eta \in \Omega} \|U(\eta)\phi\| < \infty; \quad \int_{-\vartheta_0}^{\vartheta_0} \|U(\eta)\phi\|_{L^2}^2 \, d\vartheta = \|\phi\|_{\mathcal{N}(\vartheta_0)}^2 \qquad (1.36)$$

To include the kinetic energy operator in our discussion it is natural to request that the first and second partial derivatives should also satisfy Eq. 1.36 hence introducing the spaces $\mathcal{N}_{\vartheta_0}^{(i)}$ with $i = 0, 1, 2$ analogously.

With these preliminaries we can now make the precise definition of the self-adjoint analytic family $H(\eta)$ as

$$\begin{cases} H(\eta) = U(\eta)H(\eta)U^{-1}(\eta); \quad \mathcal{D}(UHU^{-1}) = \mathcal{N}_{\vartheta_0}^2 \\ H(\eta) = \eta^{-2}T(1) + V(\eta); \quad \mathcal{D}(UHU^{-1}) \to \mathcal{D}(T) \end{cases} \qquad (1.37)$$

Recapitulating, the first step consists of restricting the Hilbert space to a smaller domain $\mathcal{N}_{\vartheta_0}^{(2)}$ for which the scaling U is defined for all complex η values with its arguments smaller in absolute value than ϑ_0. The second step, after the parameter ϑ has been made complex ($\vartheta \to i\vartheta$), consists of completing the Nelson class of dilation analytic vectors to the domain of H or in this case T. Here this means convergence with respect to the standard L^2 norm (for both the functions and its first and second partial derivatives).

To appreciate the reason for our painstaking carefulness at this particular stage we come back to our frequent references to the fundamental dilemma expressed above and in the introduction. First the theorem of Balslev and Combes provides us with a rigorous path into the second Riemann sheet of the complex energy plane. The factor, η^{-2}, appearing in front of the kinetic energy operator T, see Eqs. 1.34 and 1.37, has a simple and natural effect. It means that the absolutely continuous spectrum of $H(\eta)$ is rotated in the complex plane with a phase angle equal to -2ϑ. In this process complex resonance eigenvalues become "exposed" in agreement with the aforesaid generalized mathematical spectral theorem [3]. However, there is a small price to be paid, viz, in the process of the analytic continuation the two steps mentioned above entails a small inevitable loss of information represented by the restrictions necessary for the definition of the whole analytic family of the operators $H(\eta)$. As we will see this will have consequences both for the entropic as well as the temporal evolution. These results, all the same, guarantee that the approximations made in the previous section could be meaningful despite our words of warning.

There are in effect two principal consequences that we will examine. Firstly the spectral generalization [3] in terms of appearing complex poles of the actual resolvent, with the complex part interpreted essentially as the reciprocal life-time of the state and secondly the dynamical outcome regarding the time evolution, i.e. the conversion of an isometry to a contractive semigroup [18].

To appreciate the first generalization, i.e. modifying the projection operator formulations of Sect. 1.2, the following construal is supplied

$$T(\eta;z) = P(\eta)(z - P(\eta)U(\eta)H(1)U^{-1}(\eta)P(\eta))^{-1}P(\eta)$$

$$O(\eta) = |\phi(\eta)\rangle\langle\phi(\eta^*)\,|; \, P(\eta) = I - O(\eta)$$

$$\langle\phi(\eta^*)|\phi\,(\eta)\rangle = \langle\phi(1)|\phi(1)\rangle = 1 \tag{1.38}$$

Apart from giving the impression of being a rather formal extension, there are two important points to consider. First the projectors are oblique, i.e. idempotent

$$O^2(\eta) = O(\eta)$$

but not self-adjoint

$$O^\dagger(\eta) = O(\eta^*) \neq O(\eta)$$

Furthermore the present bi-orthogonal construction authorize non-probabilistic formulations allowing e.g. the possibility of zero norms, *viz.* from Eq. 1.9 one may encounter, starting with a none-degenerate eigenvalue, that

$$\langle\Psi(\eta^*;z^*)|\Psi(\eta;z)\rangle = 1 + \Delta(\eta;z) = 1 - f'(z) = 0 \tag{1.39}$$

The emerging singularity is associated with a degeneracy of so-called Jordan-block type, an abysmal situation in matrix theory; see e.g. Ref. [18] and references therein. In our case, as we will see, this will actually be a "blessing in disguise." In passing we note that the observed loss of information carries an unexpected increase of entropy. The occurrence of Jordan blocks, see more below about associated spectral degeneracies and their interpretations, implies that full information as to the given state becomes uncertain at the "bifurcation point", with an associated entropic increase as a result.

The second consequence regards the dynamics. As already pointed out the step-wise approach is of basic relevance for the use of dilatation analytic Hamiltonians as generators of contractive semigroups. The problem of comparing classical and quantum dynamics and the appropriate choice of so-called Lyapunov converters were examined in some detail in Ref. [8]. Briefly we will review the implication as follows. Consider an isometric semigroup, cf. the causal propagator in (1.19), $G(t); t \geq 0$, defined on some Hilbert space h. If there exists a contractive semigroup $S(t); t \geq 0$ and a densely defined closed invertible linear operator Λ, with the domain $\mathcal{D}(\Lambda)$ and range $\mathcal{R}(\Lambda)$ both dense in h, such that

$$S(t) = \Lambda G(t)\Lambda^{-1}; t \geq 0 \tag{1.40}$$

on a dense linear subset of h, then Λ is called a Lyapunov converter. A necessary condition for the existence of Λ for a given $G(t) = e^{-iHt}$ is that the generator H has a non-void absolutely continuous spectral part, i.e. $\sigma_{AC} \neq \varnothing$. In view of what has been said above it is natural to ask whether $H(\eta)$ generates a contractive semi-group, i.e. that (note that the $+$-sign in S^+ is not a "dagger")

$$S^+(t,\eta) = U(\eta)G(t)U^{-1}(\eta) = e^{-iH(\eta)t}; t \geq 0 \tag{1.41}$$

This is indeed true for many types of potentials, but unfortunately not for the case of the attractive Coulomb interaction. Although the Balslev-Combes theorem for dilation analytic Hamiltonians guarantee that the modified spectrum lies on the real axis (bound states) and in a subset of the closed lower complex halfplane, a further requirement (using the Hille-Yosida theorem) is that the numerical range must also be contained in the lower part of the complex energy plane. In addition the $1/r$ potential is problematic both at the origin and at infinity; see e.g. Ref. [19] for a detailed treatment of resonance trajectories and spectral concentration for a short-range perturbation resting on a Coulomb background. Here a resonance in the continuous spectrum carries typical ground-state properties [20] and allows for complex curve crossings (Jordan blocks) [21]. Since the long-range Coulomb part in a many-body system will be screened by the other particles the anomalies of the Coulomb problem should not be crucial with respect to the isometric-contractive semi-group conversion in realistic physical systems. For additional discussions on this point, involving a slightly more general definition in terms of quasi-isometries see Ref. [8].

Summarizing; the loss of information, i.e. restricting the full unitary time evolution to an isometry (weak convergence) before the conversion via a suitable Lyapunov converter to a contractive semi-group (strong convergence) is an objective process in contrast to the subjective preparation of any initial state involving various levels of course graining. It is also important to realize that the completion of a dense subset of Hilbert space with respect to the appropriate norm gives different limits depending on whether it is carried out before or after the conversion, hence we will speak of an *informity rule*, i.e. a certain natural loss of information, which is compatible with broken temporal symmetry.

Finally, on account of the impairment of information loss, it is all the same important to mention that, in the classical as well as in the quantum case, it is not enough to conclude that the mere existence of a Lyapunov converter explains or derives time irreversibility and, in the Liouville formulation, guarantees the approach to equilibrium [8]. In the next section we will concentrate on the degenerate state before moving on to the relativistic situation looking for the only remaining explanation of irreversibility in terms of a formulation involving a spatio-temporal dependent background.

1.4   The Jordan Block and the Coherent Dissipative Ensemble

As already mentioned the *informity rule* prompts several consequences one being the emergence of so-called Jordan blocks or exceptional points. Although belonging to standard practise in linear algebra formulations we will proffer some extra time to this concept. In addition to demonstrate its simple nature we will also establish a simple complex symmetric form not previously obtained, see e.g. Refs. [11, 14, 21, 22]. Let us start with the 2×2 case, where it is easy to demonstrate that the Jordan canonical form J and the complex symmetric form Q are unitarily connected through the transformation B, i.e.

$$Q = B^{-1} JB = B^{\dagger} JB \qquad (1.42)$$

where

$$Q = \frac{1}{2} \begin{pmatrix} 1 & -i \\ -i & -1 \end{pmatrix}; J = \begin{pmatrix} 0 & 1 \\ 0 & 0 \end{pmatrix}; B = \frac{1}{\sqrt{2}} \begin{pmatrix} 1 & i \\ 1 & -i \end{pmatrix} \qquad (1.43)$$

We note that the squares of Q and J are zero, yet the rank is one. Although these Jordan forms do not appear in conventional quantum mechanical energy variation calculations they are not uncommon in extended formulations. For various examples of the latter instigated in quantum physical situations, we refer to [11] and also to applications of a new reformulation of the celebrated Gödel(s) theorem(s) in terms of exceptional points, Ref. [12]. Note that the alternative formulation in terms of an antisymmetric construction is not anti-hermitean as wrongly indicated in Ref. [12].

As demonstrated, complex symmetric forms are naturally exploited in quantum chemistry and molecular physics and therefore we need to extend Eqs. 1.42 and 1.43 to general $n \times n$ matrices. The mathematical theorem that a triangular matrix is similar to a complex symmetric form goes back to Gantmacher [23], but the explicit form to be used here was first derived by Reid and Brändas [21], see also [22]. Since every matrix, with distinct eigenvalues, can be brought to diagonal form, the critical situation under study obtains from the general canonical form $J_n(\lambda) = \lambda\mathbf{1} + J_n(0)$ where $\mathbf{1}$ is the n-dimensional unit matrix and λ the n-fold degenerate eigenvalue

$$J_n(0) = \begin{pmatrix} 0 & 1 & 0 & \cdot & \cdot & 0 \\ 0 & 0 & 1 & \cdot & \cdot & 0 \\ 0 & 0 & 0 & 1 & \cdot & 0 \\ \cdot & \cdot & \cdot & \cdot & \cdot & \cdot \\ 0 & 0 & \cdot & \cdot & 0 & 1 \\ 0 & 0 & \cdot & \cdot & 0 & 0 \end{pmatrix} \tag{1.44}$$

The explicit complex symmetric representation thus becomes, [21,22]

$$Q = B^{-1}J_nB; Q_{kl} = \exp\left\{\frac{i\pi}{n}(k+l-2)\right\}\left(\delta_{kl} - \frac{1}{n}\right) \tag{1.45}$$

with the similarity (also unitary) transformation matrix B given by

$$B = \frac{1}{\sqrt{n}}\begin{pmatrix} 1 & \omega & \omega^2 & \cdot & \omega^{n-1} \\ 1 & \omega^3 & \omega^6 & \cdot & \omega^{3(n-1)} \\ \cdot & \cdot & & \cdot & \cdot \\ \cdot & \cdot & & \cdot & \cdot \\ 1 & \omega^{2n-1} & \omega^{2(2n-1)} & \cdot & \omega^{(n-1)(2n-1)} \end{pmatrix}; \; \omega = e^{\frac{i\pi}{n}} \tag{1.46}$$

Generalizations to various powers of J_n can easily be found by elementary means [11,14,22]. Incidentally we note an interesting factorization property of the columns of B to be further discussed below.

Revisiting the subdynamics formulation referred to earlier, we replace the Schrödinger equation by the Liouville equation ($h = 2\pi$)

$$i\frac{\partial\rho}{\partial t} = \hat{L}\rho; \hat{L}|\cdot\rangle\langle\cdot| = H|\cdot\rangle\langle\cdot| - |\cdot\rangle\langle\cdot|H \tag{1.47}$$

where ρ is the density matrix (an analogous equation for the classical case appears with the commutator above being substituted with the Poisson bracket). Thermalization, on the other hand, obtains through the Bloch equation

$$\frac{\partial\rho}{\partial\beta} = \hat{L}_B\rho; \hat{L}_B|\cdot\rangle\langle\cdot| = \frac{1}{2}\{H|\cdot\rangle\langle\cdot| + |\cdot\rangle\langle\cdot|H\} \tag{1.48}$$

with the temperature parameter $\beta = (k_B T)^{-1}$, and T the absolute temperature. The difference between the properties of the energy superoperator $\hat{L}_B$ and the Liouvillian $\hat{L}$ generate non-trivial analytic extensions, a somewhat technical yet straightforward procedure [6,11,16].

To exemplify the generality of our formulation we will consider $M = N/2$ bosons (or N fermions) described by a set of $n \geq N/2$ localized pair functions or geminals $\boldsymbol{h} = h_1, h_2, \ldots h_n$ obtained from appropriate pairing of one-particle basis spin functions. For simplicity we will briefly overlook the fermionic level. With this somewhat imprecise model we will demonstrate its portrayal of various interesting phenomena via the density operator

$$\Gamma = \rho = \sum_{k,l}^{n} |h_k\rangle \rho_{kl} \langle h_l|; \quad Tr\{\rho\} = \frac{N}{2} \tag{1.49}$$

to be further examined below, for more details see also Ref. [11]. A general quantum statistical argument illustrates the model and its quantum content. The matrix elements ρ_{kl} define probabilities for finding particles at site or state k and transition probabilities for "particles to go" from state k to l. Hence the matrix ρ has the elements

$$\rho_{kk} = p; \quad \rho_{kl} = p(1-p); \quad k \neq l; \quad p = \frac{N}{2n} \tag{1.50}$$

The associated secular equation reveals a non-degenerate large eigenvalue $\lambda_L = np - (n-1)p^2$ and a small $(n-1)$-degenerate $\lambda_S = p^2$. As a result the density operator now writes

$$\Gamma = \rho = \lambda_L |g_1\rangle \langle g_1| + \lambda_S \sum_{k,l=1}^{n} |h_k\rangle \left(\delta_{kl} - \frac{1}{n} \right) \langle h_l| \tag{1.51}$$

or using the transformation $|\boldsymbol{h}\rangle \boldsymbol{B} = |\boldsymbol{g}\rangle = |g_1, g_2, \ldots g_n\rangle$, Eq. 1.46, a compact diagonal representation for the degenerate part obtains as

$$\Gamma = \rho = \lambda_L |g_1\rangle \langle g_1| + \lambda_S \sum_{k=2}^{n} |g_k\rangle \langle g_k| \tag{1.52}$$

Although the model has a quantum probabilistic origin it is important to note that we do not have to incorporate any approximations at this stage. For instance, one may consider the reduction of a many-body fermionic pure state to an N-representable two-matrix, with the latter effectively mimicking Eq. 1.52. Since the density matrix in Eq. 1.52, through its relation to Coleman's extreme case [24], see also [10], is essentially N-representable, one might, via appropriate projections, completely recover the proper information corresponding to the partitioning procedure of Sect. 1.2.

To proceed we will concentrate on the Bloch equation, i.e. opt for the integration of thermal- and quantum fluctuations via appropriate incorporation of the temperature T. This is basically a complicated problem, since we are contending with non-equilibrium systems, yet one might all at once consider constructive interaction from the environment on our open system. In a few words an open or dissipative system is: *a system in which there exists a flow of entropy due to exchange of energy or information with the environment.* Additionally we will append, see below, specifications for a so-called *coherent-dissipative structure.* In passing we stress that we do not apply the thermodynamic limit, unless explicitly employed, and that no subjective loss of knowledge is at all conceded.

Introducing the Lyapunov converter, here the complex scaling operation with the bi-orthogonal complex symmetric form to be signified by the complex conjugate in the bra-position in Eq. 1.52, we will assign to our model a complex energy $\mathcal{E}_k = E_k - i\varepsilon_k$ to every site described by the basis function h_k. The total energy expression is given by

$$\mathcal{E} = Tr\{H_2\Gamma\} \tag{1.53}$$

where H_2 is the reduced Hamiltonian of the ensemble, see e.g. [11]. Without restriction we can put $\mathcal{E} = 0$. The formal solution of Eq. 1.49 is

$$\Gamma_T = e^{-\beta \hat{L}_B}\Gamma$$

or using the standard factorization property of the exponential superoperator

$$e^{-\beta \hat{L}_B}\Gamma = \lambda_L \sum_{k,l}^{n} |h_k\rangle e^{i\beta \frac{1}{2}(\varepsilon_k+\varepsilon_l)} \langle h_l| + \lambda_S \sum_{k,l}^{n} |h_k\rangle e^{i\beta \frac{1}{2}(\varepsilon_k+\varepsilon_l)} \left(\delta_{kl} - \frac{1}{n}\right) \langle h_l| \tag{1.54}$$

As can be proven the assumption that the real part of the energies E_k for each site can be set equal to zero is commensurate with $\mathcal{E} = 0$. Taking advantage of the usual relation between the imaginary part of the energy and the time scale

$$\varepsilon_k = \frac{h}{4\pi\tau_k}$$

one can, from the examination of a simple thermal scattering process [11] derive the following "quantization condition", see also derivations in conclusion

$$\beta\varepsilon_l = 2\pi\frac{l-1}{n}; \quad l = 1,2,\ldots n \tag{1.55}$$

From (1.55) we realize that the thermalized matrix in Eq. 1.54 assumes the Jordan form

$$\Gamma_T = \lambda_L \sum_{k,l}^{n} |h_k\rangle e^{i\frac{\pi}{n}(k+l-2)} \langle h_l| + \lambda_S \sum_{k,l}^{n} |h_k\rangle e^{i\frac{\pi}{n}(k+l-2)} \left(\delta_{kl} - \frac{1}{n}\right) \langle h_l| \tag{1.56}$$

or introducing the canonical basis $|h\rangle B^{-1} = |f\rangle = |f_1, f_2, \ldots f_n\rangle$ we obtain the Dunford formula

$$\Gamma_T = \lambda_L J^{(n-1)} + \lambda_S J; \quad J = \sum_{k=1}^{n-1} |f_k\rangle \langle f_{k+1}|; \quad J^{(n-1)} = |f_1\rangle \langle f_n| \tag{1.57}$$

From Eq. 1.55 we conclude that the present spatio-temporal structure is very special acquiring a prolonged survival time or time scale as well as optimum spatial properties. This defines a so-called *coherent dissipative system, CDS*, by requiring additionally (to that of a dissipative system) that *(a) they are created or destroyed by integrated quantum- and thermal correlations* $(T \neq 0)$, *(b) they exchange energy and information with an entangled environment and (c) they can not have a size smaller than a critical one.* Accurate dynamical evolution of such systems leads to non-exponential decay and the law of microscopic self-organization. We have furthermore considered unambiguous conditions where matching patterns survive as well as examined unpredicted organizations and associated emergent properties. Specifically we have studied anomalies of proton transport in water, ionic conductance of molten salts, conjectures regarding long-range proton correlations in DNA and further, quantum correlation effects in high-TC cuprates, see Ref. [11] for a recent review and discussion.

The spatio-temporal structure, *CDS*, has been developed as a replica for the characterization of a *living system* [12]. This portrait must account for, in addition to the *metabolic process*, the *genetic function* and *homeostasis*, the protracted survival of autonomic meta-codes, which assign mappings between genotypic and phenotypic spaces. Briefly, the modus operandi relate to the transformation B, Eq. 1.46, which, as emphasized, demonstrates interesting factorizations indicating that various groups of sites (localizations) are to be correlated. Selecting particular values of n, we will find all the factors appearing in the vectors of B, the importance of which follows, since it diagonalizes Γ, while B^{-1} puts $\tilde{\Gamma}$ in canonical form. A simple example of the divisor property is illustrated by $n = 6$ below, showing $\sqrt{6}B$ with the first column removed:

$$
\begin{matrix}
 & & (2) & & \\
 & (3) & & (3) & \\
(6) & & (2) & & (6) \\
 & (3) & & (3) & \\
 & & (2) & &
\end{matrix}
\tag{1.58}
$$

The present attribute suggests various modes of encoding for the autonomic assembly involving microscopic self-organization, see e.g. [11, 12] for more details. Note also that the genetic- or more generally, the evolution program is stored via the transformation B, while the "transfer" is mediated through B^{-1}. If this model is correct, there is no specific origin of life, since self-referential laws are intrinsic to the evolution and a fundamental property of our universe.

Summarizing this section, we have found that *the informity rule* and the quantization condition Eq. 1.55 lead to entropic growth as the system in a well-defined quantum state with the entropy $S = 0$, will increase to $S = k_B \ln\{n\}$ in the *CDS* structure without any subjective approximations as e.g. statistical deductions based on the pre-assumed level of descriptions being invoked. Still, we have not derived or even explained the origin of time irreversibility.

1.5 General Relativity and the Global Superposition Principle

It is an interesting observation that the degenerate description above with respect to an instruction-based evolution relates developmental and building matters with functional issues. This leads us to general questions like the origin of life and associated cosmological evolution, the theme of the last two sections.

In order to place our theoretical framework in the most general position with respect to our contradiction, i.e. positing irreversible thermodynamics versus reversible dynamics, we need to direct our focus towards a relativistically invariant (or covariant in the general case) theory. It would perhaps be more pertinent to use the word consistent, since our more general complex symmetric framework, allows broken symmetry solutions, while relevant symmetries are properly embedded if necessary. In a previous communication [12], see also [25], we have derived *a Global Superposition Principle* that applies to both classical and quantum mechanical interpretations.

We will not re-derive this formulation, see e.g. [12] and references therein, except make a brief outline of the main results. Rather than going through the construction through apposite complex symmetric forms, we will here proceed directly via the observation that the classical-quantum equations of relativity cf. the Klein-Gordon equation, is a quadratic form in the actual observables. Considering the non-positive square root from the simple ansatz of the Hamiltonian $\mathcal{H}$ below

$$H^2 = m_0^2 \begin{pmatrix} 1 & 0 \\ 0 & 1 \end{pmatrix}; \quad H^{\pm} = \begin{pmatrix} m_0 & 0 \\ 0 & \pm m_0 \end{pmatrix}$$

$$H^- = \mathcal{H} = \begin{pmatrix} m_0 & 0 \\ 0 & -m_0 \end{pmatrix} = \begin{pmatrix} m & -iv \\ -iv & -m \end{pmatrix} \tag{1.59}$$

where Eq. 1.59 stipulates that $m_0^2 = m^2 - v^2$ with the obvious identifications $v = p/c$; p (not be confused with the probability p of the previous section) the momentum of a particle of mass m, relative to an inertial system, where m_0 is the (non-zero) rest mass, and finally with c the velocity of light. We have chosen the "negative" square root in the complex symmetric form above in order to supply an appropriate frame for our non-stationary description. Note also that velocities and momenta should in general be represented as operators-vectors.

From the secular equation, based on the complex symmetric matrix $\mathcal{H}$, we obtain the eigenvalues $\lambda_{\pm} = \pm m_0$ via the Klein-Gordon-like equation.

$$\lambda^2 = m_0^2 = m^2 - v^2 = m^2 - p^2 c^{-2}. \tag{1.60}$$

It is natural to introduce the parameter $\beta = p/mc = $ ("classical particles") $= v/c$, with v the velocity of the particle, with Eq. 1.60 yielding the well-known formula

$$m = \frac{m_0}{\sqrt{1 - v^2/c^2}} = \frac{m_0}{\sqrt{1 - \beta^2}}. \tag{1.61}$$

Note that the entities appearing in the matrix, Eq. 1.59, are generally operators, which compel a formal reading of the relations Eqs. 1.60 and 1.61. Nevertheless, as we have seen above, we have the machinery to define rather general operator domains and ranges, which should go with any relevant application. Hence, irrespective of whether we subscribe to a classical and/or a quantum mechanical description, our scheme begets biorthogonality, i.e. introducing the general ket-vectors $|\,m\rangle\,;|\,\overline{m}\rangle$ and $|\,m_0\rangle\,;|\,\overline{m}_0\rangle$ corresponding to the "abstract states" of a particle m and its antiparticle $\overline{m}$ respectively and similarly for m_0 and $\overline{m}_0$, the rest mass in the interaction free case. The biorthogonal construction ascends from

$$|\,m_0\rangle = c_1|\,m\rangle + c_2|\,\overline{m}\rangle\,;\ \lambda_+ = m_0$$

$$|\,\overline{m}_0\rangle = -c_2|\,m\rangle + c_1|\,\overline{m}\rangle\,;\ \lambda_- = -m_0$$

$$|\,m\rangle = c_1|m\,_0\rangle - c_2|\overline{m}\,_0\rangle$$

$$|\,\overline{m}\rangle = c_2|m\,_0\rangle + c_1|\overline{m}\,_0\rangle$$

(1.62)

$$c_1 = \sqrt{\frac{1+X}{2X}};\ \ c_2 = -i\sqrt{\frac{1-X}{2X}};\ \ X = \sqrt{1-\beta^2};\ \ c_1^2 + c_2^2 = 1.$$

To sum up we note that the present level of formulation does not distinguish between classical- and quantum mechanics. A further characteristic reveals biorthogonality implying that the coefficients c_i are not to be associated with a probability interpretation, since they obey the rule $c_1^2 + c_2^2 = 1$. As emphasized, the operators in Eq. 1.63 are principally non-self-adjoint and non-normal and hence they might not commute with each other as well as their own adjoint. The order appearing in the resulting operator relations therefore has to be respected.

Another actuality arises from the general ket-dependence on the energy and the momentum, while the conjugate problem, see below, depends entirely on time and position. Returning to Eqs. 1.59 and 1.60, introducing the well known operator identifications, ($h = 2\pi$ and ∇ the gradient operator)

$$E_{\mathrm{op}} = i\frac{\partial}{\partial t};\ \mathbf{p} = -i\nabla$$

(1.63)

we identify, from Eq. 1.60, the explicit connection between the present Klein-Gordon-like equation and Maxwell's equations for vacuum, i.e. by setting the determinant of $\boldsymbol{H}$ in Eqs. 1.59 and 1.64 below, equal to zero

$$\begin{vmatrix} i\dfrac{\partial}{\partial t} & -i\boldsymbol{p}c \\[2mm] -i\boldsymbol{p}c & -i\dfrac{\partial}{\partial t} \end{vmatrix} = \frac{\partial^2}{\partial t^2} - c^2\nabla^2.$$

(1.64)

We can now formally introduce the conjugate operators

$$\tau = T_{\mathrm{op}} = -i\frac{\partial}{\partial E};\ \mathbf{x} = i\nabla_p$$

(1.65)

and

$$\mathcal{H}_{conj} = \begin{pmatrix} c\tau & -i\mathbf{x} \\ -i\mathbf{x} & -c\tau \end{pmatrix} \tag{1.66}$$

from which follows trivially that the eigenvalues τ_0, and $x_0 = \sqrt{\mathbf{x}_0 \cdot \mathbf{x}_0}$, obtains in analogy with Eqs. 1.59–1.61, using $\mathbf{x} = \upsilon\tau$ in Eq. (1.66) i.e.

$$\tau = \frac{\tau_0}{\sqrt{1 - \upsilon^2/c^2}} = \frac{\tau_0}{\sqrt{1 - \beta^2}} \tag{1.67}$$

and

$$x = \frac{x_0}{\sqrt{1 - \beta^2}}. \tag{1.68}$$

In passing we note, despite working within a general complex symmetric framework including biorthogonality, that we have recovered the well-known Lorentz invariance and that our universal superposition principle applies irrespective of whether we uphold classical wave propagation, quantum mechanical matter waves or classical point particles. The present formulation also carries a very important message with respect to the theme of our present investigation. The conjugate pair description unequivocally couples the time direction with the selected particle excitation. Nevertheless the present equations, whether interpreted classically or quantum mechanically, are not protected against decoherence towards a time reversible formulation and hence we have not yet succeeded in explaining our original dilemma. We will not generalize the discussion here to more general invariances with regard to charge conjugation and the strong parity concept (*CPT*), compare also related issues in connection with Kramer's degeneracy, but it can be done.

To incorporate gravitational interactions in our conjugate pair operator framework we will attach, to our model in the basis $|m, \overline{m}\rangle$, the interaction

$$m\kappa(r) = m\mu/r; \; \mu = \frac{G \cdot M}{c^2} \tag{1.69}$$

resulting in the modified Hamiltonian (operator) matrix $(m_0 \neq 0)$

$$\mathcal{H} = \begin{pmatrix} m(1 - \kappa(r)) & -i\upsilon \\ -i\upsilon & -m(1 - \kappa(r)) \end{pmatrix} \tag{1.70}$$

with μ the gravitational radius, G the gravitational constant, M a spherically symmetric non-rotating mass distribution (which does not change sign when $m \to -m$) and $v = p/c$ as in Eq. 1.59. For more details on the fundamental nature of M, and the emergence of black hole like objects see further below and Refs. [11, 26]. Note that the operator, $\kappa(r) \geq 0$, depends on the coordinate (or in general operator) r of the particle m with "origin" at the center of mass of M. However, $\mathbf{x}$ and τ are subject to the adjoint description to be addressed below, cf. e.g. Eqs. 1.65 and 1.68 and hence we will, all things considered, return to curved space-time geometry indicative of classical theories.

In analogy, with Eqs. 1.60–1.63 we obtain

$$\lambda^2 = m^2 (1 - \kappa(r))^2 - p^2/c^2$$
$$\lambda_\pm = \pm m_0(1 - \kappa(r)); \; v = p/c \tag{1.71}$$

with the eigenvalues $\lambda_\pm$ scaled so that m_0 is consistent with the special theory, i.e.

$$m_0^2 = m^2 - p^2/((1 - \kappa(r))^2 c^2)$$
$$\lambda_\pm/(1 - \kappa(r)) = \pm m_0 = \pm\sqrt{m^2 - p^2/((1 - \kappa(r))^2 c^2)}$$
$$m = m_0/\sqrt{1 - \beta'^2}; \quad \beta' \le 1; \; 1 > \kappa(r)$$
$$\beta' = p/(mc(1 - \kappa(r))) = v/(c(1 - \kappa(r))). \tag{1.72}$$

The intimate relation to special relativity is displayed by rewriting the operator matrix (note that $\kappa(r) < 1/2$ rather than $\kappa(r) < 1$, see below)

$$\mathcal{H}_{sp} = \begin{pmatrix} m & -ip'/c \\ -ip'/c & -m \end{pmatrix}; \quad p' = p(1 - \kappa(r))^{-1}. \tag{1.73}$$

Assuming that the gravitational source is a spherical black hole-like object at rest, i.e. that the angular momentum of the non-zero rest-mass particle is a constant of motion one obtains

$$mvr = m\mu c \tag{1.74}$$

or

$$v = \kappa(r)c = \mu c/r \tag{1.75}$$

Note that the simple relations (1.74 and 1.75) above are a bit more involved than what gives the impression. Although an appropriate rotation by the polar angle φ commutes with the operators m, p and r, it is important to remember that m has a nonzero eigenvalue, m_0, cf. Eq. 1.84 below, which defines the velocity v according to Eq. 1.75. Note also that the end result will agree with the covariant form given by the Schwarzschild gauge as will be examined in what follows.

Above we have postulated the limit velocity c at the limiting distance given by the gravitational radius μ. Eq. 1.75 serves as a boundary condition for the operator matrix model. With the replacement $v/c = \kappa(r)$, our complex symmetric representation, reads (given that $\kappa(r) < \frac{1}{2}$, see below)

$$\mathcal{H} = m \begin{pmatrix} (1 - \kappa(r)) & -i\kappa(r) \\ -i\kappa(r) & -(1 - \kappa(r)) \end{pmatrix} \to m \begin{pmatrix} \sqrt{(1 - 2\kappa(r))} & 0 \\ 0 & -\sqrt{(1 - 2\kappa(r))} \end{pmatrix}. \tag{1.76}$$

While m is not determined by Eq. 1.76, the quotient m/m_0 (non-zero rest-mass) follows from Eqs. 1.71 and 1.72 and the eigenvalues of (1.76), i.e.

$$\frac{m}{m_0} = \frac{1 - \kappa(r)}{\sqrt{1 - 2\kappa(r)}} \tag{1.77}$$

Equations 1.61, 1.72 and 1.76 invoke the scalings $m\sqrt{1-\beta^2}$, $m\sqrt{1-\beta'^2}$, and $m\sqrt{1-2\kappa(r)}$ respectively. The singularity in Eq. 1.77 at $r = 2\mu$ or $\kappa(r) = \frac{1}{2}$, corresponds to a degeneracy of the matrix $\mathcal{H} = \mathcal{H}_{\text{deg}}$ consistent with Eqs. 1.70, 1.74 and 1.76, i.e.

$$\mathcal{H}_{\text{deg}} = m \begin{pmatrix} \frac{1}{2} & -i\frac{1}{2} \\ -i\frac{1}{2} & -\frac{1}{2} \end{pmatrix} \rightarrow \begin{pmatrix} 0 & m \\ 0 & 0 \end{pmatrix} \tag{1.78}$$

where the transformation to classical canonical form is accomplished by the unitary transformation

$$|m_0\rangle \rightarrow |0\rangle = \frac{1}{\sqrt{2}}|m\rangle - i\frac{1}{\sqrt{2}}|\bar{m}\rangle ;$$

$$|\bar{m}_0\rangle \rightarrow |\bar{0}\rangle = \frac{1}{\sqrt{2}}|m\rangle + i\frac{1}{\sqrt{2}}|\bar{m}\rangle . \tag{1.79}$$

It is remarkable that the exceptional point Eq. 1.79 corresponds to the celebrated Laplace-Schwarzschild radius $r = 2\mu = R_{LS}$ (given that M is confined inside a sphere with radius R_{LS}). Note that the present result is a universal property of the present formulation in contrast to the classical "Schwartzschild singularity", which depends on the choice of coordinate system. Stated in a different way decoherence to classical reality might take place for $0 < \kappa(r) < \frac{1}{2}$ whilst potential quantum like structures appears inside R_{LS} for $\frac{1}{2} \leq \kappa(r) < 1$.

The significance of the point $\kappa(r) = \frac{1}{2}$ has been recently discussed in connection with Gödel's celebrated incompleteness theorem, [12], illustrating the self-referential character of the description. Hence self-referentiability serves as code protection against decoherence. On the other hand it carries a slight complication in that we need to discuss the case $m_0 = 0$ separately, i.e. applying the present theory to the particles of light or photons. This will lead to predictable incongruities in the operator-conjugate operator structure in comparison with particles with $m_0 \neq 0$. Consistency demands a separate gravitational law for zero rest-mass particles. By means of the specific notation $\kappa_0(r) = G_0 \cdot M/(c^2 r)$, rewriting Eqs. 1.70–1.72 for particles with $m_0 = 0$, the first inconsistency reveals that $\mathcal{H}$ is singular, i.e. cannot be diagonalised since

$$m(1 - \kappa_0(r)) = p/c \tag{1.80}$$

where $\kappa_0(r)$ is to be uniquely defined below, and

$$\mathcal{H} = \begin{pmatrix} p/c & -ip/c \\ -ip/c & -p/c \end{pmatrix} \rightarrow \begin{pmatrix} 0 & 2p/c \\ 0 & 0 \end{pmatrix} \tag{1.81}$$

where one employs the same unitary transformation, see Eq. 1.79. In contradistinction to $m_0 \neq 0$ we note that Eqs. 1.80 and 1.81 display, for all r, the triangular form or in other words an eigenvalue degeneracy with a Segrè characteristic equal to 2. Consistency therefore adds the requirement, see Eq. 1.82 below, in order for Eq. 1.81 to be commensurate with the boundary condition Eq. 1.75 and the eigenvalue relations Eqs. 1.76–1.78. The necessary condition to be added is zero average momentum; see Eq. 1.80, at the Schwarzschild radius $r = 2\mu = R_{LS}$. Hence the stipulation of $\bar{p} = 0$ at $\kappa(r) = 1/2$ leads to $G_0 = 2G$ or

$$\kappa_0(r) = 2\kappa(r) \tag{1.82}$$

From Eq. 1.82 follows Einstein's laws of light deflection, the gravitational redshift and the time delay [12].

The difference in the limiting behavior at $r = 2\mu = R_{LS}$, for the cases $m_0 \neq 0$ and $m_0 = 0$, motivates a separate examination of the conjugate operator formulation. Generalizing Eqs. 1.63–1.66 to

$$d\mathcal{H}_{\text{conj}} = \begin{pmatrix} cds & 0 \\ 0 & -cds \end{pmatrix} = \begin{pmatrix} cAd\tau & -iBd\mathbf{x} \\ -iBd\mathbf{x} & -cAd\tau \end{pmatrix} \tag{1.83}$$

where the modified conjugate operators are introduced via

$$E_{\text{op}} \sqrt{1 - 2\kappa(r)} = E_s = i\frac{\partial}{\partial s}$$

$$s = -i\frac{\partial}{\partial E_s} \tag{1.84}$$

and with $A(r)$ and $B(r)$ to be decided below, it was proven in Ref. [12] that

$$d\mathcal{H}_{\text{conj}} = \begin{pmatrix} cdt(1 - 2\kappa(r))^{1/2} & -idr(1 - 2\kappa(r))^{-1/2} \\ -idr(1 - 2\kappa(r))^{-1/2} & -cdt(1 - 2\kappa(r))^{1/2} \end{pmatrix} \tag{1.85}$$

One note that Eq. 1.85 is decoherence protected for the case $m_0 = 0$ since $d\mathcal{H}_{\text{conj}}$ is non-diagonal with the determinant equal to zero for all values of r. For $m_0 \neq 0$ the only singular point is at $r = R_{LS}$. We also stress that the operator E_s replaces the classical notion of rest-mass. Thus we obtain an invariant $|d\mathcal{H}_{\text{conj}}| = -c^2 ds^2$, which is zero (null-vector) for photons, and the line element expression (in the spherical case)

$$-c^2 ds^2 = -c^2 dt^2 (1 - 2\kappa(r)) + dr^2 (1 - 2\kappa(r))^{-1} \tag{1.86}$$

Again we note that the present complex symmetric ansatz has generated a covariant transformation, compatible with the classical Schwartzschild gauge but, as has been stated recurrently, we have not yet proven or explained the origin of time irreversibility.

Before ending this section on relativity theory, we reflect on a remark made by Löwdin [27] regarding the perihelion motion of Mercury. Describing a gravitational approach within the construction of special relativity, he demonstrated that the perihelion moved but that the effect was only half the correct value. The problem here is the fundamental inconsistency between the force-, momentum and the energy laws, while the discrepancy for so-called normal distances are almost impossible to observe directly since $(1 - \kappa(r)) \approx 1$. However using the present method to the classical constant of motion

$$E_{tot} = m(1 - \kappa(r)) \tag{1.87}$$

we construe from Eq. 1.77 that E_{tot} is singular at $\kappa(r) = 1/2$. Hence one needs to include a perturbation consistent with the boundary condition Eq. 1.75, i.e.

$$\Delta E_{tot} = -m\kappa(r). \tag{1.88}$$

With this addition the correct value from general relativity theory is recovered for the Mercury perihelion move for each rotation of the standard polar angle. Still, although classical descriptions are essentially right outside the Schwartzschild domain, it is nevertheless important to point out that the self-referential property of the gravitational interaction, resulting in the modified Eqs. 1.87 and 1.88, impinges on the selected time direction through a uniquely given background dependence. In the final section we will relate this choice to a cosmological setting that demonstrates the uni-directedness of evolution, code protection against decoherence, cosmic sensorship and the absolute validity of the second law.

1.6 Cosmological Scenarios and Conclusions

In order to connect the present formulation with the cosmological problem, we need to base our development on the key quantity, viz. the emergence of the black-hole object or M characterized as a spherical mass distribution, which does not change sign when $m \to -m$, see also previous discussion after Eq. 1.70. As stated in connection with the density matrix in Eq. 1.52, there is a close relation between the simple form derived and Yang's famous concept of ODLRO, Off-Diagonal Long-Range Order, [28], and Coleman's notion of an extreme state [24]. For more details we refer to [10, 11, 16] and the appendix of Ref. [26] as well as references therein. We will demonstrate that our black hole type entity is analogous to the organization of ODLRO. In addition we will extend the discussion to a Kerr-type [29] rotating black hole characterized by its mass and total angular momentum.

Let us begin with the analogy of ODLRO, i.e. the consideration of a finite number of fermion- or particle-anti-particle pairs in a vacuum (or particle-like environment, cf. Cooper pairs in a superconductor) written as

$$|0\rangle \wedge |\bar{0}\rangle = i|m\rangle \wedge |\bar{m}\rangle \tag{1.89}$$

using Eqs. 1.76, 1.79, 1.81 and 1.84. From the knowledge of the (second order reduced) density matrix the energy W (not be confused with the operator in Eq. 1.29) obtains, cf. Eq. 1.53

$$W = \mathrm{Tr}\{H_{12}\Gamma^{(2)}\} \tag{1.90}$$

where H_{12} is a general two-body potential between the constituents, i.e. the particle-anti-particle combinations based on the pairing Eq. 1.89. We have also explicitly denoted the density matrix as $\Gamma^{(2)}$ to demonstrate its connection with the Coleman extreme case. Nevertheless the difference between $\Gamma^{(2)}$ here and Γ in the Sect. 1.4 above is minor, save a missing tail contribution that will not play any role here (this information can in principle be retrieved so it will not cause additional information loss). In what remains we can use the basic definitions and their transformation properties as presented. The following simplifications and relations are easy to derive, i.e.

$$W = \lambda_L \langle g_1|H_{12}|g_1\rangle + \lambda_S \sum_{k=2}^{n} \langle g_k|H_{12}|g_k\rangle \tag{1.91}$$

For an "adequately" localised basis $\boldsymbol{h}$, i.e.

$$\langle h_k|H_{12}|h_l\rangle = \langle h_k|H_{12}|h_k\rangle \, \delta_{kl} \tag{1.92}$$

one gets for large n

$$W \approx \frac{N}{2}w$$

$$w = \frac{1}{n}\sum_{k=1}^{n} \langle h_k|H_{12}|h_k\rangle \tag{1.93}$$

On the other hand assuming that the pairing is established by one original interaction with all matrix elements $\langle h_k|H_{12}|h_l\rangle = w_{LS}$, where $w_{LS} < 0$ is independent on the sites k and l, we obtain an enormous energy stabilization. In passing this is indeed a realistic supposition in connection with ODLRO leading to e.g. coherent phenomena like e.g. superconductivity. Here this is a somewhat difficult proposition, since we are not able to describe any localisation centres inside the fundamental radius (cf. the discussion related to Eq. 1.76 and the appearance of the Schwarzschild radius). We also have the problem of defining space and time and their conjugate operators m_0 and p. Fortunately we have, as will be seen below, the possibility to characterize the black hole entity in terms of its mass and angular momentum. The formula derived in Ref. [26] writes

$$W = \lambda_L \langle g_1 | H_{12} | g_1 \rangle = \frac{N}{2} n w_{LS} \qquad (1.94)$$

Although the formula Eq. 1.94 usually obtains from a description of highly correlated fermionic pairs in an apt spatio-temporal-independent background, we find a macroscopically large energy lowering, where W above is the product of the mass $M \propto N/2$ and the number, n, of rotational degrees of freedom, i.e. with the rotational quantum number $J \approx n/2$. Hence we have obtained, as a result of a unique fundamental interaction, the possible existence of a quantum mechanical version of a (fermionic-quasibosonic) black hole. Incidentally, this could not be a charged black hole, since we will always have an equal amount of particle-antiparticle pairs in the "condensate", see e.g. Eqs. 1.79 and 1.89. This realization will also take care of the comment after Eq. 1.70: "a mass M not changing sign under $m \rightarrow -m$". Another interesting observation concerns the analysis of the rotational spectrum of the "black hole" in particular its J_z components and "figure axis rotations".

A small lesson in quantum chemistry reveals that rotational spectra in non-relativistic quantum mechanics is very simple yet very informative as regards molecular types and their transitions, heat capacities etc. For e.g. a symmetric rotor one obtains trivially in the rigid case

$$E(J, M_J, K) = BJ(J+1) + (A-B)K^2 \qquad (1.95)$$

with M_J and K specifying the components of the angular momentum along the laboratory- and the principal molecular axis, respectively, and the rotational constants given by

$$A = \frac{h}{8\pi^2 I_{\parallel}}; \; B = \frac{h}{8\pi^2 I_{\perp}} \qquad (1.96)$$

where $I_{\perp}, I_{\parallel}$ are the moments of inertia orthogonal and parallel, respectively, to the internal figure axis. For an oblate object, i.e. $A < B$, one finds that the energy decreases with K^2 and for $K \approx J$ the object rotates mostly around the figure axis. The degeneracy is $2(2J+1)$ except when $K = 0$. On the other hand when $K = 0$ the molecule rotates mostly around an axis perpendicular to the symmetry axis. For $A = B$, i.e. the spherical rotor, the degeneracy is $(2J+1)^2$ which significantly influences its physical properties.

With this simple analogy we will draw some conclusions regarding our black-hole-like entity obtained above. Comparing the Kerr- [29] and the Schwarzschild metric one distinguishes two physical surfaces in the former where the metric either changes sign or becomes singular (spherical coordinate system), i.e.

$$R_{Ki} = \frac{1}{2} \left\{ R_{LS} + \sqrt{R_{LS}^2 - 4\frac{K^2}{M^2 c^2}} \right\}$$

$$R_{Ko} = \frac{1}{2} \left\{ R_{LS} + \sqrt{R_{LS}^2 - 4\frac{K^2 \cos^2 \theta}{M^2 c^2}} \right\} \qquad (1.97)$$

where R_{Ki} are the inner surface that corresponds to the event horizon, R_{Ko} the outer one that touches the inner surface at the poles of the rotation axis, θ the colatitude and K the angular momentum along the entity's principle axis, cf. the analysis of the symmetric rotor. The space in between is called the ergosphere. It is easy to show that $K/Mc \leq \mu; \mu = M/Gc^2$ so that $\min\{R_{Ki}\} = \frac{1}{2}R_{LS} = \mu$, with the ring singularity shrinking to a point in the Schwarzschild case, albeit the quantum analogue of the cosmic sensorship hypothesis of Penrose, see e.g. [30] holds.

Rather than discussing the significance of these quantities in the classical theory, we will go back to the fermionic-quasibosonic black-hole system discussed above, remembering that the energy stabilization Eq. 1.94, emerged from rotational degrees of freedom only. In order to discuss the exchange of matter and energy with the system, mimicked as a gigantic non-elastic resonance scattering process, we will first consider n degrees of freedom modelling baryonic matter waves being correlated on a so-called relaxation time scale τ_{rel}. The scale corresponds to the average lifetime of the "scattering process" and depends generally on the type of particles or properties of the units being represented, cf. the Einstein relation in physical chemistry, which connect transport displacements with the diffusion constant D. Furthermore we will define a spherically averaged total reaction cross section denoted by σ_{tot}. This area, cf. the surface of the event horizon, should be consistent with the physical parameters of the model so that on average we will detect one particle or degree of freedom in the differential solid-angle element $d\Omega$ during the limit time scale τ_{lim} given by Heisenberg's uncertainty relation, i.e. $\tau_{lim} = h/(2\pi k_B T)$. Here k_B is Boltzmann's constant and T the absolute temperature. Again we emphasize that our goal is not to determine cross section data or evaluating lifetimes, reaction rates etc. Instead we want to identify consistent relations between temperatures, sizes of the dissipative structure, time scales, etc. and to utilize this information as input to our quantum statistical equations.

With the input above we find the consistency relations between incident and scattered fluxes. The incident flux, i.e. the number of particles or degrees of freedom per unit area and time is

$$N_{inc} = \frac{n}{\sigma_{tot}\tau_{rel}} \tag{1.98}$$

Our model defines the number, $N_s d\Omega$, of particles scattered into $d\Omega$ per unit time as

$$N_s d\Omega = \frac{d\Omega}{\tau_{lim}} = \frac{2\pi k_B T}{h} d\Omega \tag{1.99}$$

Hence we find the consistent relations from total cross section given by

$$\sigma_{tot} = \int \sigma(\Omega) d\Omega = \int \frac{N_s}{N_{inc}} d\Omega \tag{1.100}$$

From Eqs. 1.98 to 1.100 we get

$$n = \frac{8\pi^2 k_B T}{h} \tau_{rel} \tag{1.101}$$

Since σ_{tot} cancels out in (1.100), Eq. 1.101 yields the number of correlated degrees of freedom depending uniquely on the temperature and the relaxation time. This consistency relation will be used as input to the thermalization formula, see Eq. 1.54, i.e. we want to study the behavior of the matrix element $\hat{\rho}_{kl} = |h_k\rangle \langle h_l^*|$ yielding (remember the degenerate energy is set to zero)

$$e^{-\frac{\beta}{2}H_2} \hat{\rho}_{kl} e^{-\frac{\beta}{2}H_2} = e^{i\frac{\beta}{2}(\varepsilon_k + \varepsilon_l)} \hat{\rho}_{kl} \qquad (1.102)$$

In an interaction free environment the "black-hole rotor" exhibits a very large degeneracy for each choice of J (and K). The interesting quantum number is K since it describes the shapes, i.e. "perpendicular rotation", or spherical shape, for $K = 0$ and an oblate, more distorted one, as K grows in magnitude. For $K = \pm J$ the disc-shape is maximum. For a classical oblate top this corresponds to the lowest energy. However, in our present case we have no electromagnetic fields present, i.e. no dipoles, quadrupoles etc., which means the absence of standard rotational selection rules. It is reasonable to assume that every incoming degree of freedom adds a constant amount of inertia to the system implying that we can organize the rotational "black-hole" excitation spectrum of the collective cluster of particles/degrees of freedom harmonically, i.e. with the distance between the levels being equal, i.e. $h(2\pi \tau_{rel})^{-1}$. Hence the $(l-1)$:th level is given by the angular frequency τ_l^{-1}, uniquely specified by the harmonic spectrum. Hence one obtains $\tau_{rel} = (l-1)\tau_l$; $l = 2,3,\ldots n$, and from Eq. 1.55 (note a mistake in the second line of Eq. (A7) of Ref. [11] even if the end result is correct)

$$\frac{1}{2}\beta(\varepsilon_k + \varepsilon_l) = \frac{h}{8\pi k_B T}\left\{\frac{1}{\tau_k} + \frac{1}{\tau_l}\right\} = \frac{h}{8\pi k_B T\, \tau_{rel}}\{k+l-2\}$$

$$= \frac{\pi}{n}(k+l-2); \quad k,l = 1,2,3,\ldots n-1 \qquad (1.103)$$

With the insertion of this result in Eqs. 1.54 and 1.57 follows, i.e.

$$\Gamma_T^{(2)} = \lambda_L J^{(n-1)} + \lambda_S J; \quad J = \sum_{k=1}^{n-1} |f_k\rangle\langle f_{k+1}|$$

However the Jordan form above displays peculiar properties. From the thermalized density matrix above we find that the energy, see the analogue of Eq. 1.90, is zero, yet we have an extremely large degeneracy with Segrè characteristic n. Since "crossing states" refer to rotational degrees of freedom, we see that the first term above, with the largest weight, concerns transitions between $K = -J$ and $K = J$, while the second term displays all the other ones from $K = -J$ via $K = -J+1, -J+2, \ldots J-1, J$. The eigenvalue of $\Gamma_T^{(2)}$ is zero corresponding $K = 0$. Since the direction of the flux into the black hole is arbitrary, the orthogonal projection to the (arbitrary) symmetry axis must also be zero. Hence our generalized quantum state is characterized by the quantum numbers $2J+1 = n$; $M_J = 0$; $K = 0$, i.e. with an enormous rotational energy (large n) but with the rotational z-component in an arbitrary direction in space equal to zero! Although we are advocating a quantum mechanical description it is reasonable to expect rotational effects similar to frame-dragging, see e.g. Ref. [31]. Hence from the perspective of an observer outside the black hole one would experience its effects all around us, since $M_J = 0$, i.e. we would infer to have our cosmological horizon limited by the event horizon or in other words the cosmological or particle horizon equal to the event horizon. In this model it is obvious that the big bang singularity disperses as an illusion, yet effectively persistent

Summarizing the scattering model: we have an incoming flux of matter and radiation attracted to the "black hole", exciting the "black-hole" from a lower state to a higher state, or to a white-hole-like object, getting rid of matter and radiation, leaving the "target" in a CDS-like state, characterized by its mass M and angular momentum J. The lower state has low entropy, $S \approx 0$, (the ground state has of course zero entropy), while the "excited state" (or rather the CDS structure) + ejected matter and energy has high entropy. Note that our CDS object is not really a black or a white hole but rather close quantum analogues.

From the viewpoint of an expelled observer the cosmological- and the event horizon coalesce and hence the process may be considered as an "implosion in reverse". This connects the process with inflationary scenarios based on negative energy pressure. One may offer several consequences and speculations from the model; see more below, including possible solutions to the "Newton bucket paradox". Also, observing that the normalised eigenvalue, λ_L/n, occurring in Eqs. 1.29–1.52, assumes its maximum value 0.25 for $p = 0.5$, indicates that about 25% of the energy content is expelled as mostly matter-antimatter constituents outside the cosmological horizon, invisible (dark) matter, while the remainder consists of essentially "dark energy" as well as some radiation and baryonic matter. Although the relative amount of the latter is difficult to estimate one deduce from Eq. 1.97 that it might be proportional to some suitable average of $(K/J)^2$ and thus very small.

The present quantum model of the universe is the very simplest to say the least. It describes a scattering target with only rotational degrees of freedom, however with an almost infinite cross section obtained from a high rotational energy state, with the projection of the angular momentum in any arbitrary direction equal to zero. Even if simplifying assumptions have been made the theory incorporates elemental covariance. However, more importantly, there is a self-referential component in the formulation that generates a new kind of organization, which is protected against decoherence to time reversible dynamics. This prompts several immediate consequences, e.g. obvious candidates for dark energy (rotating black-white-hole-systems), dark matter (outside the particle horizon), the explanation of the galaxy rotation problem and with the concept of mass related to rotation (negative or imaginary). One might also speculate whether appearing radiation concentrations along certain directions and the subsistence of acoustic peaks in the harmonic analysis of the cosmic microwave background may be established in our model, since temperature variations are coupled to n or J via Eq. 1.101.

Irrespective of the soundness of these inferences, it is clear that the overall model of our universe by way of a gigantic scattering process has certain affinities with the steady state model, while rotational excitations of the black-hole-system may mimic a big-bang-like scenario. The capture of baryonic matter by the black-hole (scattering) system appears to lower the entropy outside the event horizon, while the forward evolution simultaneously increases it via information loss. Furthermore the "excitation process" leads to a Coherent-Dissipative Structure, *CDS*, with initially low entropy for the target, or $S \approx 0$ for the "ground state" rising to $S = k_B \ln\{n\}$. In addition the "course of action" adds the entropy of the inflated energy. Thus the scattering process incorporates *the informity rule* and the Hawking radiation (see [12] for a derivation within the present framework), which guarantees that the second law holds absolutely. Since the *CDS* structure is code protected from decohering to a time reversible state, our formulation is fundamentally time irreversible, in concert with Prigogine's contention. Reflecting on the time-dependent partitioning technique, it follows that the unknown part $\kappa(0)$, see Eq. 1.7, must always be nonzero, although its presence, unlike unitary evolution, if present, will gradually vanish under proper contractive evolution. Since the equations in Sect. 1.2 live in symbiosis with time reversible dynamics there are no contradictions at the level of a fixed spatio-temporal background as long as one realizes that proper cosmological evolution, by necessity, links the evolution through an appropriate formulation of the spatio-temporal dependent metric causing cosmic amnesia. The indirect example of parity violations in weak interactions, which if the CPT theorem is valid, e.g. seem to support fundamental time irreversible evolution of our universe.

Acknowledgements The present results have been presented at QSCP XV held at Magdalene College, Cambridge, England, August 31- September 5, 2010. The author thanks the organisers of QSCP XV, in particular the Chair Prof. Philip E. Hoggan for an excellent programme and organization as well as generous hospitality.

References

1. L. Sklar, *Physics and Chance*. (Cambridge, University Press, 1993).
2. I. Prigogine, *The End of Certainty. Time, Chaos and the New Laws of Nature*. The Free Press, Simon & Schuster Inc., New York (1996).
3. E. Balslev, J. M. Combes, Commun. Math. Phys. **22**, 280 (1971).
4. S. Nakajima, Progr. Theoret. Phys. **20**, 948 (1958).
5. R. Zwanzig, J. Chem. Phys. **33**, 1338 (1960).
6. Ch. Obcemea, E. Brändas, Ann. Phys. **151**, 383 (1983).
7. I. Prigogine, Physica, **A263**, 528 (1999).
8. J. Kumicák, E. Brändas, Int. J. Quant. Chem. **32**, 669 (1987); *ibid.* **46**, 391 (1993).
9. E. J. Brändas, I. E. Antoniou, Int. J. Quant. Chem. **46**, 419 (1993).
10. E. Brändas, Chatzidimitriou-Dreismann, Int. J. Quant. Chem. **40**, 649 (1991).

11. E. Brändas, *Complex Symmetry, Jordan Blocks and Microscopic Selforganization: An examination of the limits of quantum theory based on nonself-adjoint extensions with illustrations from chemistry and physics.* in Molecular Self-Organization in Micro-, Nano, and Macro-dimensions: From Molecules to Water, to Nanoparticles, DNA and Proteins. Eds. N. Russo, V. Antonchenko and E. Kryachko; NATO Science for Peace and Security Series A: Chemistry and Biology, Springer Science+Business Media B.V., Dordrecht, 49-87 (2009).

12. E. Brändas, Int. J. Quant. Chem. **111**, 215 (2011); *ibid.* **111**, 1321 (2011).

13. E. Engdahl, E. Brändas, M. Rittby, N. Elander, J. Math. Phys. **27**, 2629 (1986).

14. E. Brändas, *Resonances and Dilatation Analyticity in Liouville Space.* Adv. Chem. Phys. **99**, 211 (1997).

15. E. Brändas, Mol. Phys. **108**, 3259 (2010).

16. E. Brändas, N. Elander Eds., *Resonances-The Unifying Route Towards the Formulation of Dynamical Processes – Foundations and Applications in Nuclear, Atomic and Molecular Physics,* Springer Verlag, Berlin, *Lecture Notes in Physics,* **325**, pp. 1-564 (1989).

17. E. Nelson, Ann. Math. **70**, 572, (1959).

18. E. Brändas, P. Froelich, M. Hehenberger, Int. J. Quant. Chem. **14**,419 (1978).

19. E. Engdahl, E. Brändas, M. Rittby, N. Elander, Phys. Rev. **A 37**, 3777 (1988).

20. M. A. Natiello, E. J. Brändas, A. R. Engelmann, Int. J. Quant. Chem. **S21**, 555 (1987).

21. C. E. Reid, E. Brändas, Lecture Notes in Physics **325**, 476 (1989).

22. E. Brändas, Int. J. Quant. Chem. **109**, 2860 (2009).

23. F. R. Gantmacher, *The Theory of Matrices,* Chelsea, New York, Vols I, II (1959).

24. A. J. Coleman, Rev. Mod. Phys. **35**, 668 (1963).

25. E. Brändas, Adv. Quant. Chem. **54**, 115 (2008).

26. E. J. Brändas, *The Equivalence Principle from a Quantum Mechanical Perspective,* in Frontiers in Quantum Systems in Chemistry and Physics, eds. P. Piecuch, J. Maruani, G. Delgado-Barrio and S. Wilson, Springer Verlag, Vol.**19**, 73 (2009).

27. P.-O. Löwdin, *Some Comments on the Foundations of Physics,* World Scientific, Singapore, (1998).

28. C. N. Yang, Rev. Mod. Phys. **34**, 694 (1962).

29. R. P. Kerr, Phys. Rev. Lett. **11**, 237 (1963).

30. R. Penrose, *Cycles of Time,* The Bodly Head Random House, London SW1V 2SA (2010).

31. H. Pfister, *General Relativity and Gravitation* **39** (11), 1735 (2007). Company, Vols. I, II 1959).

Part II
Model Atoms

Chapter 2
Spatially-Dependent-Mass Schrödinger Equations with Morse Oscillator Eigenvalues: Isospectral Potentials and Factorization Operators

G. Ovando, J.J. Peña, and J. Morales

Abstract In this work an algorithm based on the point canonical transformation method to convert any general second order differential equation of Sturm-Liouville type into a Schrödinger-like equation is applied to the position-dependent mass Schrödinger equation (PDMSE). This algorithm is next applied to find potentials isospectral to Morse potential and associated to different position-dependent mass distributions in the PDMSE. Factorization of worked PDMSE are also obtained.

2.1 Introduction

Quantum systems with a position-dependent mass have attracted attention in recent years due to their relevance in describing the features of many microstructures of current interest as for example the determination of physical properties in quantum wells and quantum dots [1], quantum liquids [2], nuclei [3], ^{3}He and metal type clusters [4] and graded alloys [5]. At this regard, the different analytical and algebraic approaches used in the study of systems with constant mass such as the Lie algebraic techniques [6], kinetic energy approach [7], factorization method [8] and supersymmetric quantum mechanics [9], among many others, have been extended to the treatment of the position-dependent mass Schrödinger equation (PDMSE). In the case of the supersymmetric treatment applied to the PDMSE, the point canonical transformation method (PCTM) has been used extensively [10–15] in the mapping of the nonconstant mass Schrödinger equation into a standard one. In a similar way, in this work we propose an algorithm to transform a general second order differential equation of Sturm-Liouville type, into a standard Schrödinger-like equation.

G. Ovando (✉) · J.J. Peña · J. Morales
Area de Física Atómica Molecular Aplicada, CBI Universidad Autónoma
Metropolitana-Azcapotzalco Av. San Pablo 180, Col. Reynosa-Tamps.
02200 México, D. F. Mexico
e-mail: gaoz@correo.azc.uam.mx; jjpg@correo.azc.uam.mx; jmr@correo.azc.uam.mx

P.E. Hoggan et al. (eds.), *Advances in the Theory of Quantum Systems in Chemistry and Physics*, Progress in Theoretical Chemistry and Physics 22,
DOI 10.1007/978-94-007-2076-3_2, © Springer Science+Business Media B.V. 2012

This approach is applied to the PDMSE with the aim to find the isospectral potentials associated to different position-dependent mass distributions (PDMD). Although the proposal is general as it is depicted in Sect. 2.2, we consider explicitly in Sect. 2.3 the useful application which is the Morse potential model for the standard Schrodinger equation. Its consequences for different PDMD in the PDMSE are analyzed, which are the corresponding isospectral potentials, wavefunctions and factorization operators as it is indicated in each studied case. Beyond the case considered, the proposal is general and the algorithm can be extended to find exactly solvable PDMSE that fulfill another specific effective potential models along with different PDMD that could be useful in the quantum treatment of different systems in material science and condensed matter physics where spatially-dependent mass is a point to consider.

2.2 PCTM Applied to the Position-Dependent Mass Schrödinger Equation

The one-dimensional position-dependent mass Schrodinger equation is given by [16]

$$\frac{1}{2m(x)}\psi_n''(x) + \left(\frac{1}{2m(x)}\right)' \psi_n'(x) + (E_n - V(x))\psi_n(x) = 0, \tag{2.1}$$

where the prime denotes derivative respect to the argument, E_n is the energy spectra and $\hbar = 1$. The point canonical transformation method applied to the above equation has the purpose of reducing it to a constant mass Schrödinger-like equation, solutions of which are usually easier to find than for the PDMSE. Furthermore, due to the fact that Eq. 2.1 has the form

$$P(x)f_n''(x) + Q(x)f_n'(x) + R(x)f_n = 0 \tag{2.2}$$

provided that $P(x) = \frac{1}{2m(x)}$, $Q(x) = (\frac{1}{2m(x)})'$ and $R(x) = E_n - V(x)$, we can extend the algorithm given by Peña et al. [17] to eliminate the $P(x)$ coefficient. For this purpose we consider the variable change

$$x = F(u) = g^{-1}(u), \tag{2.3}$$

with

$$g(x) = \int^x \sqrt{2m(t)}dt, \tag{2.4}$$

which imply

$$\left(\frac{1}{2m(x)}\right)' \frac{d}{dx} = \frac{d\ln\left(\frac{1}{2m(F(u))}\right)}{du}\frac{d}{du} \tag{2.5}$$

and

$$\frac{1}{2m(x)}\frac{d^2}{dx^2} = \frac{d^2}{du^2} + \frac{d\ln\sqrt{(2m(F(u)))}}{du}\frac{d}{du}. \tag{2.6}$$

Thus, Eq. 2.1 can now be transformed into a constant mass Schrödinger-like equation

$$-\varphi_n''(u) + V_{\mathit{eff}}(u)\varphi_n(u) = E_n\phi_n(u), \tag{2.7}$$

where $\varphi_n(u)$ are the corresponding normalized eigenfunctions

$$\varphi_n(u) = \psi_n(F(u))\exp\left[\int^u W(t)dt\right], \tag{2.8}$$

and $V_{\mathit{eff}}(u)$ is the *effective potential* given by the Riccati-type equation

$$V_{\mathit{eff}}(u) = V(F(u)) + W^2(u) + W'(u), \tag{2.9}$$

with

$$W(u) = \frac{d}{du}\ln(2m(F(u)))^{-\frac{1}{4}}, \tag{2.10}$$

such that the position-dependent mass distribution in terms of $W(u)$ is

$$m(x) = \frac{1}{2}e^{-4\int^{g(x)} W(t)dt}. \tag{2.11}$$

That is, the PDMSE with potential

$$V(x) = V_{\mathit{eff}}(g(x)) - \left(W^2(g(x)) + W'(g(x))\right), \tag{2.12}$$

has eigenvalues E_n and normalized eigenfunctions

$$\psi_n(x) = \varphi_n(g(x))\exp\left[-\int^{g(x)} W(t)dt\right]. \tag{2.13}$$

Therefore, the applicability of the above algorithm gives rise to various possibilities depending on the choice of the different elements that characterize the Riccati-type Eq. 2.9: an ansatz for the equivalent of the Witten superpotential $W(u)$, a starter $V(F(u))$, or a proposal for the effective potential $V_{\mathit{eff}}(u)$. In the next paragraph, as a useful application of the proposed approach we are going to consider this last choice in the specific case of the Morse potential model for the constant mass Schrödinger equation.

2.3 PDMSE with Morse Potential Eigenvalues

To obtain exactly solvable isospectral potentials on the position-dependent mass problem, let us consider the Morse potential model in the standard Schrodinger-like relationship of Eq. 2.7 that means to take as effective potential [9, 18]

$$V_{eff}(u) = A^2 + B^2 e^{-2\alpha u} - 2B\left(A + \frac{\alpha}{2}\right)e^{-\alpha u}, \qquad (2.14)$$

where A, B, α are parameters usually related by condition $A + \frac{\alpha}{2} = B = \sqrt{D}$, being D the potential's well depth. In this case, the wavefunctions are

$$\varphi_n(u) = N_n \left(\frac{2B}{\alpha}e^{-\alpha u}\right)^{s-n} e^{-\frac{B}{\alpha}e^{-\alpha u}} L_n^{2s-2n}\left(\frac{2B}{\alpha}e^{-\alpha u}\right), \qquad (2.15)$$

where $s = \frac{A}{\alpha}$ and $N_n = \sqrt{\frac{2\alpha(s-n)\Gamma(n+1)}{\Gamma(2s-n+1)}}$ is the normalization constant. Consequently, the former potential in the PDMSE becomes

$$V(x) = \left[A^2 + B^2 e^{-2\alpha u} - 2B\left(A + \frac{\alpha}{2}\right)e^{-\alpha u} - (W^2(u) + W'(u))\right]_{u=g(x)}, \qquad (2.16)$$

with Morse energy spectra $E_n = A^2 - (A - n\alpha)^2$, and according to Eqs. 2.11 and 2.13, wavefunctions

$$\psi_n(x) = \varphi_n(g(x))\sqrt{g'(x)}. \qquad (2.17)$$

Besides, due to the fact that constant mass Schrodinger-like equation is factorized in this case as

$$\left[\left(-\frac{d}{du} - A + Be^{-\alpha u}\right)\left(\frac{d}{du} - A + Be^{-\alpha u}\right)\right]\varphi_n(u) = E_n\varphi_n(u), \qquad (2.18)$$

where $\frac{d}{du}$ is given from Eqs. 2.3 and 2.4 by

$$\frac{d}{du} = (2m(x))^{-1/2}\frac{d}{dx}, \qquad (2.19)$$

and from Eqs. 2.8 and 2.11

$$\varphi_n(u) = (2m(x))^{-1/4}\psi_n(x), \qquad (2.20)$$

the factorization of PDMSE given in Eq. 2.1 becomes

$$A^+(x)A^-(x)\psi_n(x) = E_n\psi_n(x), \qquad (2.21)$$

where the operators are

$$A^{\pm}(x) = \mp \left((2m(x))^{-1/2} \frac{d}{dx} + W(g(x)) \right) - A + Be^{-\alpha g(x)}, \qquad (2.22)$$

satisfying the commutation relation

$$[A^{-}(x), A^{+}(x)] = -2\alpha Be^{-\alpha g(x)}. \qquad (2.23)$$

At this point, it is worth to notice that $A^{\pm}(x)$ are equivalent to the Darboux transform that relates the partner Hamiltonians H^{+} and H^{-} by means of

$$H^{\pm} = H^{\mp} \pm 2\alpha Be^{-\alpha g(x)}. \qquad (2.24)$$

Let us now show the usefulness of the above results by considering a selection of position-dependent mass distributions leading to new isospectral potentials with Morse-type eigenvalues.

2.3.1 Inverse Squared Position-Dependent Mass Distribution

We have taken a mass function in the form $2m(x) = (\beta x)^{-2}$ where hereafter β is a parameter for the mass function. From Eqs. 2.4 and 2.10 one leads to $g(x) = \ln x^{1/\beta}$ and $W(g(x)) = \beta/2$. Consequently, the exactly solvable former potential in the PDMSE becomes in this case

$$V(x) = A^2 + B^2 x^{-2\alpha/\beta} - 2B \left(A + \frac{\alpha}{2} \right) x^{-\alpha/\beta} - \frac{1}{4}\beta^2, \qquad (2.25)$$

having eigenfunctions

$$\psi_n(x) = N_n (\beta x)^{-1/2} \left(\frac{2B}{\alpha} x^{-\alpha/\beta} \right)^{s-n} e^{-\frac{B}{\alpha} x^{-\alpha/\beta}} L_n^{2s-2n} \left(\frac{2B}{\alpha} x^{-\alpha/\beta} \right). \qquad (2.26)$$

and Morse energy eigenvalues. In order to show the principal features of the generated solution we have taken the values $\alpha = 1, A = 5/2, B = 3$ in the Morse Potential throughout the paper, which correspond to a case of only three eigenstates (Fig. 2.1).

The potential of Eq. 2.25 is factorized by means of

$$A^{\pm}(x) = \mp \left(\beta x \frac{d}{dx} + \frac{1}{2}\beta \right) - A + Bx^{-\alpha/\beta}. \qquad (2.27)$$

Fig. 2.1 The principal
features of the isospectral
exactly solvable potential
$V(x)$ given in Eq. 2.25 along
with their eigenfunctions and
energy spectra, as well as the
position-dependent mass
distribution. The mass
parameter is $\beta = 1/2$

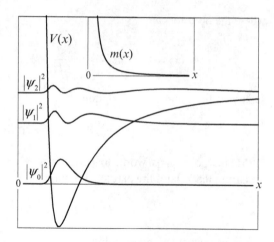

2.3.2 A First Rational Position-Dependent Mass Distribution

For the $2m(x) = (1+\beta^2 x^2)^{-1}$ position-dependent mass distribution one has $g(x) = \frac{1}{\beta}\operatorname{arcsinh}(\beta x)$ and

$$W(g(x)) = \frac{\beta^2}{2}\frac{x}{\sqrt{1+\beta^2 x^2}}. \tag{2.28}$$

Thus, the exactly solvable isospectral potential in the corresponding PDMSE is (Fig. 2.2)

$$V(x) = A^2 + B^2 \left(\sqrt{1+\beta^2 x^2} - \beta x\right)^{2\alpha/\beta} - 2B\left(A+\frac{\alpha}{2}\right)$$

$$\times \left(\sqrt{1+\beta^2 x^2} - \beta x\right)^{\alpha/\beta} - \frac{\beta^2}{4}\frac{\beta^2 x^2 + 2}{\beta^2 x^2 + 1}, \tag{2.29}$$

with eigenfunctions

$$\psi_n(x) = N_n(1+\beta^2 x^2)^{-1/4}\left(\frac{2B}{\alpha}(\sqrt{1+\beta^2 x^2} - \beta x)^{\alpha/\beta}\right)^{s-n}$$

$$\times e^{-\frac{B}{\alpha}(\sqrt{1+\beta^2 x^2} - \beta x)^{\alpha/\beta}} L_n^{2s-2n}\left(\frac{2B}{\alpha}(\sqrt{1+\beta^2 x^2} - \beta x)^{\alpha/\beta}\right). \tag{2.30}$$

The factorization operators for the PDMSE with potentials given in Eq. 2.29 are

$$A^{\pm} = \mp\left((1+\beta^2 x^2)^{1/2}\frac{d}{dx} + \frac{\beta^2}{2}\frac{x}{\sqrt{1+\beta^2 x^2}}\right) - A + B\left(\sqrt{1+\beta^2 x^2} - \beta x\right)^{\alpha/\beta}. \tag{2.31}$$

Fig. 2.2 Similarly to the
above case, the exactly
solvable potential of Eq. 2.29,
their eigenfunctions and
energy spectra, as well as the
corresponding
position-dependent mass
distribution. The mass
parameter is $\beta = 1$

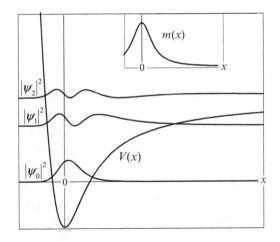

2.3.3 Exponential Position-Dependent Mass Distribution

Let us consider now $2m(x) = (1 + e^{\beta x})^2$ which, after taking properly the integration constant gives $g(x) = \frac{1}{\beta}(e^{\beta x} + \beta x - \beta)$ and

$$W(g(x)) = -\frac{\beta e^{\beta x}}{2(e^{\beta x} + 1)^2}. \tag{2.32}$$

In consequence, for this exponential position-dependent mass distribution, the corresponding isospectral potential is given by

$$V(x) = A^2 + B^2 e^{-2\frac{\alpha}{\beta}(e^{\beta x} + \beta x - \beta)} - 2B\left(A + \frac{\alpha}{2}\right)$$
$$\times e^{-\frac{\alpha}{\beta}(e^{\beta x} + \beta x - \beta)} - \frac{\beta^2}{4} e^{\beta x} \frac{3e^{\beta x} - 2}{(e^{\beta x} + 1)^4}, \tag{2.33}$$

with eigenfunctions

$$\psi_n(x) = N_n(1 + e^{\beta x})^{1/2}\left(\frac{2B}{\alpha}e^{-\frac{\alpha}{\beta}}\left(e^{\beta x} + \beta x - \beta\right)\right)^{s-n}$$
$$\times e^{-\frac{B}{\alpha}e^{-\frac{\alpha}{\beta}}\left(e^{\beta x} + \beta x - \beta\right)} L_n^{2s-2n}\left(\frac{2B}{\alpha}e^{-\frac{\alpha}{\beta}(e^{\beta x} + \beta x - \beta)}\right), \tag{2.34}$$

which are shown in Fig. 2.3. Similarly to the above cases, the factorization operators are now given by

$$A^\pm = \mp\left(\frac{1}{1 + e^{\beta x}}\frac{d}{dx} - \frac{\beta e^{\beta x}}{2(e^{\beta x} + 1)^2}\right) - A + Be^{-\frac{\alpha}{\beta}(e^{\beta x} + \beta x - \beta)}. \tag{2.35}$$

Fig. 2.3 The exactly solvable
potential given in Eq. 2.33 for
a exponential-type
position-dependent mass.
The mass parameter is $\beta = 1$

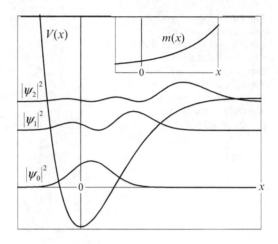

2.3.4 Step-Like Position-Dependent Mass Distribution

As it can be seen at the right down corner of Fig. 2.4, the position-dependent
mass distribution

$$2m(x) = \frac{\exp(2\beta x)}{\cosh^2(\beta x)}, \tag{2.36}$$

behaves as a step-like function. It is possible to define a point transformation with
range in the interval $(-x_e, \infty)$

$$g(x) = \frac{1}{\beta}\left(\ln\left(\frac{\exp(2\beta x)+1}{\exp(\beta x_e)}\right)\right). \tag{2.37}$$

With this definition one has $W(g(x)) = -\beta \exp(-2\beta x)/2.$ and the isospectral
singular potential of the PDMSE will be

$$V(x) = A^2 + B^2 \frac{e^{-2\alpha(x-x_e)}}{(2\cosh\beta x)^{2\alpha/\beta}} - 2B\left(A + \frac{\alpha}{2}\right)$$

$$\times \frac{e^{-\alpha(x-x_e)}}{(2\cosh\beta x)^{\alpha/\beta}} - \frac{1}{2}\beta^2 e^{-2\beta x}\left(1 + \frac{3}{2}e^{-2\beta x}\right), \tag{2.38}$$

with eigenfunctions

$$\psi_n(x) = N_n \frac{e^{\beta x/2}}{\cosh^{1/2}\beta x}\left(\frac{2B}{\alpha}\frac{e^{-\alpha(x-x_e)}}{(2\cosh\beta x)^{\alpha/\beta}}\right)^{s-n}$$

$$\times e^{-\frac{B}{\alpha}\frac{e^{-\alpha(x-x_e)}}{(2\cosh\beta x)^{\alpha/\beta}}} L_n^{2s-2n}\left(\frac{2B}{\alpha}\frac{e^{-\alpha(x-x_e)}}{(2\cosh\beta x)^{\alpha/\beta}}\right), \tag{2.39}$$

Fig. 2.4 Case of a
exponential type position
dependent mass that tends
to zero in the negative x
direction showing that the
potential diverges towards
minus infinity

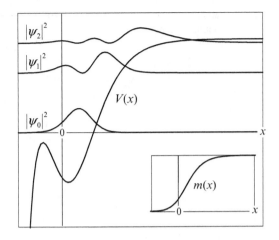

Figure 2.4 shows that in this case $m(x)$ decrease faster than $e^{-4\beta|x|}$ in the $-x$ axis and
the potential breaks towards $-\infty$. The corresponding factorization of the PDMSE
comes from the operators

$$A^{\pm} = \mp \left(e^{-\beta x}\cosh\beta x \frac{d}{dx} - \frac{\beta}{2}e^{-2\beta x} \right) - A + B\frac{e^{-\alpha(x-x_e)}}{(2\cosh\beta x)^{\alpha/\beta}}. \tag{2.40}$$

2.3.5 Powered Position-Dependent Mass Distribution

In this case one has the positive function $2m(x) = x^{2\beta}$ which leads to

$$g(x) = \frac{x^{\beta+1}}{\beta+1}, \tag{2.41}$$

where the value of β will assure that the interval $(-\infty, \infty)$ be the range of this
function. Then $W(g(x)) = -\beta x^{-\beta-1}/2$ and consequently, for the powered $m(x)$
the isospectral potential becomes

$$V(x) = A^2 + B^2 e^{-\frac{2\alpha}{\beta+1}x^{\beta+1}} - 2B\left(A+\frac{\alpha}{2}\right)e^{-\frac{\alpha}{\beta+1}x^{\beta+1}} - \frac{1}{4x^{2\beta+2}}\beta(3\beta+2), \tag{2.42}$$

with eigenfunctions

$$\psi_n(x) = N_n x^{\beta/2}\left(\frac{2B}{\alpha}e^{-\frac{\alpha}{\beta+1}x^{\beta+1}}\right)^{s-n}e^{-\frac{B}{\alpha}e^{-\frac{\alpha}{\beta+1}x^{\beta+1}}}L_n^{2s-2n}\left(\frac{2B}{\alpha}e^{-\frac{\alpha}{\beta+1}x^{\beta+1}}\right), \tag{2.43}$$

Fig. 2.5 The potential and
the eigenfunctions in the case
of the mass $2m(x) = x^{2\beta}$

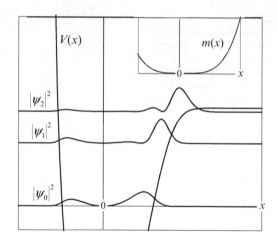

and factorization operators

$$A^{\pm} = \mp\left(x^{-\beta}\frac{d}{dx} - \frac{\beta}{2}x^{-\beta-1}\right) - A + Be^{-\frac{\alpha}{\beta+1}x^{\beta+1}}. \tag{2.44}$$

Figure 2.5 displays the potential of Eq. 2.43 and their eigenfunctions considering $\beta = 2$. We remark that if $m(x)$ reaches the zero value it causes the break of the potential toward $-\infty$.

2.3.6 Increasingly Mass Values

We have selected $2m(x) = \cos^{-2}(\beta x)$ to show a case where the mass values increase quickly. We have

$$g(x) = \frac{1}{\beta}\ln\left(\frac{1+\sin(\beta x)}{\cos(\beta x)}\right). \tag{2.45}$$

Which leads to $W(g(x)) = -\beta\sin(\beta x)/2$, the isospectral potential is

$$V(x) = A^2 + B^2\left(\frac{\cos\beta x}{1+\sin\beta x}\right)^{2\alpha/\beta} - 2B\left(A+\frac{\alpha}{2}\right)$$

$$\times\left(\frac{\cos\beta x}{1+\sin\beta x}\right)^{\alpha/\beta} - \frac{\beta^2}{4}(1-3\cos^2\beta x). \tag{2.46}$$

Fig. 2.6 The potential and
the eigenfunctions in the case
of the position dependent
mass $2m(x) = \cos^{-2}(\beta x)$

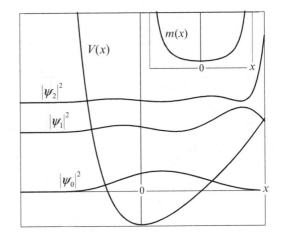

The factorization operators are

$$A^{\pm} = \mp \left(\cos(\beta x)\frac{d}{dx} - \frac{1}{2}\beta\sin\beta x\right) - A + B\left(\frac{\cos\beta x}{1+\sin\beta x}\right)^{\alpha/\beta}. \qquad (2.47)$$

And the eigenfunctions

$$\psi_n(x) = \frac{N_n}{\cos^{1/2}\beta x}\left(\frac{2B}{\alpha}\left(\frac{\cos\beta x}{1+\sin\beta x}\right)^{\alpha/\beta}\right)^{s-n}$$

$$\times e^{-\frac{B}{\alpha}\left(\frac{\cos\beta x}{1+\sin\beta x}\right)^{\alpha/\beta}} L_n^{2s-2n}\left(\frac{2B}{\alpha}\left(\frac{\cos\beta x}{1+\sin\beta x}\right)^{\alpha/\beta}\right), \qquad (2.48)$$

whose behavior is shown in Fig. 2.6. It should be pointed out, that wavefunctions
$\varphi_n(u)$ should tend to zero in the frontier faster than $(2m(x))^{-1/4}$ to assure that $\psi_n(x)$
satisfy the boundary conditions. This is not the case for $\psi_2(x)$.

2.3.7 A Second Rational-Type Position-Dependent Mass Distribution

The last two cases we show are missing of the anomalies studied in the three
previous cases, both are of rational type. Let us consider first (Fig. 2.7)

$$m(x) = \frac{1}{2\beta^2}\left(\frac{x^2+\beta}{x^2+1}\right)^2 \quad \beta \neq 0, \qquad (2.49)$$

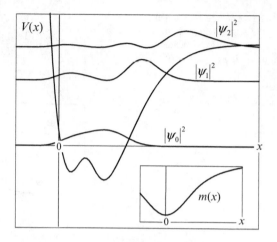

which has been used as model of solvability [19–21]. In this case one has $g(x) = (x + (\beta - 1) \arctan x)/\beta$, $W(g(x)) = \beta(\beta - 1)x/(x^2 + \beta)^2$, and the corresponding exactly solvable isospectral potential is given by

$$V(x) = A^2 + B^2 e^{-\frac{2\alpha}{\beta}(x + (\beta - 1) \arctan x)}$$

$$-2B\left(A + \frac{\alpha}{2}\right) e^{\frac{\alpha}{\beta}(x + (\beta - 1) \arctan x)}$$

$$-\frac{\beta^2(\beta - 1)}{(x^2 + \beta)^4}(\beta + 2x^2\beta - 4x^2 - 3x^4), \tag{2.50}$$

with eigenfunctions

$$\psi_n(x) = N_n \frac{1}{\sqrt{\beta}} \sqrt{\frac{x^2 + \beta}{x^2 + 1}} \left(\frac{2B}{\alpha} e^{-\frac{\alpha}{\beta}(x + (\beta - 1) \arctan x)}\right)^{s-n}$$

$$\times e^{-\frac{B}{\alpha} e^{-\frac{\alpha}{\beta}(x + (\beta - 1) \arctan x)}} L_n^{2s-2n}\left(\frac{2B}{\alpha} e^{-\frac{\alpha}{\beta}(x + (\beta - 1) \arctan x)}\right), \tag{2.51}$$

and factorization operators

$$A^{\pm} = \mp\left(\beta \frac{x^2 + 1}{x^2 + \beta} \frac{d}{dx} + \frac{\beta(\beta - 1)x}{(x^2 + \beta)^2}\right) - A + B e^{-\frac{\alpha}{\beta}(x + (\beta - 1) \arctan x)}. \tag{2.52}$$

2.3.8 A Third Rational-Type Position-Dependent Mass Distribution

In this case is taken the mass function

$$m(x) = \frac{1}{2(\beta x + 1)^2}, \quad x > -\frac{1}{\beta}, \beta \neq 0. \tag{2.53}$$

Which, in the case $\beta = \alpha$ has been used to identify a Coulomb type potential; see reference [22]. The change of variable is $g(x) = (\ln(\beta x + 1))/\beta$, which will have the range $(-\infty, \infty)$ provided $x > -\frac{1}{\beta}$ is taken. Thus $W(g(x)) = \beta/2$ allow to calculate the potential

$$V(x) = A^2 + B^2(\beta x + 1)^{-\frac{2\alpha}{\beta}} - 2B\left(A + \frac{\alpha}{2}\right)(\beta x + 1)^{-\frac{\alpha}{\beta}} - \frac{1}{4}\beta^2, \tag{2.54}$$

the eigenfunctions

$$\psi_n(x) = N_n \frac{1}{\sqrt{\beta x + 1}} \left(\frac{2B}{\alpha}(\beta x + 1)^{-\frac{\alpha}{\beta}}\right)^{s-n}$$
$$\times e^{-\frac{B}{\alpha}(\beta x + 1)^{-\frac{\alpha}{\beta}}} L_n^{2s-2n}\left(\frac{2B}{\alpha}(\beta x + 1)^{-\frac{\alpha}{\beta}}\right), \tag{2.55}$$

and the factorization operators

$$A^{\pm} = \mp\left((\beta x + 1)\frac{d}{dx} + \frac{1}{2}\beta\right) - A + B(\beta x + 1)^{-\frac{\alpha}{\beta}}. \tag{2.56}$$

In Fig. 2.8 is displayed the potential of Eq. 2.54 and their eigenfunctions considering $\beta = 1/3$.

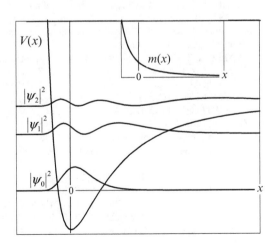

Fig. 2.8 Exactly solvable potential for the position dependent mass distribution of Eq. 2.53

2.4 Concluding Remarks

The purpose of this work has been to find exactly solvable potentials for Schrödinger equations with a position-dependent mass distribution. In the search of such a goal, it was proposed an approach to find the isospectral potentials that correspond to a specific position-dependent mass distributions and effective potential. This approach is based on a point canonical and a gauge transformation applied to a general PDMSE in order to convert it into a standard Schrödinger-like equation. As a useful application of the method we have considered the special case of the Morse potential model as effective potential in order to find the exactly solvable isospectral potentials and their factorization operators in the case of some models of PDMD. That is, in all cases considered as examples the potentials are isospectral allowing Morse eigenvalues. Furthermore, it has been pointed out that the proposal is general and can be used in the search of those exactly solvable isospectral potentials involved in a PDMSE for other effective potential models and PDMD, which in turn could be useful in the quantum chemical treatment in materials science.

Acknowledgments This work was partially supported by the projects UAM-A-CBI-2232001, 2232004 and 2232006 México. We gratefully acknowledge the SNI-Conacyt-Mexico for the distinction of our membership and the stipend received.

References

1. P. Harrison, *Quantum Wells, Wires and Dots*, Wiley, New York (2000)
2. F. Arias et al., Phys. Rev. B **50**(1997) 4248
3. P. Ring, P. Schuck, *The Nuclear Many-Body Problem*, Springer-Verlag, New York (1980)
4. M.Barranco, M. Pi, S. M. Gatica, E. S. Hernandez, J. Navarro, Phys. Rev. B **56**(1997) 8997
5. C. Weisbuch, B. Vinter, *Quantum Semiconductor Heterostructure*, Academic Press, New York (1993) and references therein
6. G. Levai, J. Phys. A: Math. Gen. **27**(1994) 3809
7. B. Bagchi, P. Gorain, C. Quesne, R. Roychoudhury, Mod. Phys, Lett. A **19**(2004) 2765
8. B. Gönül, B. Gönül, D. Tutcu, O. Özer, Modern Phys, Lett. A **17**(2002) 2057
9. F. Cooper, A. Khare, U. Sukhatme, *Supersymmetry in Quantum Mechanics*, World Scientific Publishing Co., Singapore (2001)
10. O. Mustafa, S. H. Mazharimousavi, J. Phys. A: Math. Gen. **39**(2006) 10537
11. C. Tezcan, R. Sever, J. Math. Chem. **42**(2007) 387
12. K. Bencheikh, S. Berkane, S. Bouizane, J. Phys. A: Math. Gen. **37**(2004) 10719
13. L. Jiang, L. Yi, L. C. S. Jia, Phys. Lett. A **345**(2005) 279
14. J. Yu, J. Dong, J. Phys. Lett. A **325**(2004) 194
15. G. Chen, Z. D. Chen, Phys. Lett. A **331**(2004) 312
16. J. M. Lévy-Leblond, Phys. Rev. A, **52**(1995) 1845
17. J. J. Peña, J. Morales, E. Zamora-Gallardo, J. García-Ravelo, Int. J. Quant. Chem. **100**(2004) 957
18. J. López-Bonilla, V. Gaftoi, G. Ovando, South East Asian J. Math. & Math. Sci. **4**(2005) 61
19. B. Roy, P. Roy, Phys. Lett. A **340**(2005) 70
20. R. Koc, H. Tütüncüler, Ann. Phys. (Leipzig)**12**(2003) 684
21. M. Solimannejad, S.K. Moayedi, M. Tavakoli., Int. J. Quant. Chem **106**(2006) 1027
22. C. Pacheco-García, J. García-Ravelo, J. Morales, J.J. Peña, Int. J. Quant. Chem. **110**(2010) 2880

Chapter 3
Relativistic Theory of Cooperative Muon – γ-Nuclear Processes: Negative Muon Capture and Metastable Nucleus Discharge

Alexander V. Glushkov, Olga Yu. Khetselius, and Andrey A. Svinarenko

Abstract We present a new consistent energy approach to calculation of the cross-section for the negative muon capture by an atom, based on the relativistic many-body perturbation (PT) theory. The calculation results for cross-section of the μ^- capture by He atom are listed. It is presented a generalized energy approach in the relativistic theory of discharge of a metastable nucleus with emission of γ quantum and further muon conversion, which initiates this discharge. The numerical calculation of the corresponding probabilities is firstly carried out for the scandium nucleus ($A = 49, N = 21$) with using the Dirac-Woods-Saxon model. The theoretical and experimental studying the muon-γ-nuclear interaction effects opens prospects for nuclear quantum optics, probing the structural features of a nucleus and muon spectroscopy in atomic and molecular physics.

3.1 Introduction

Methods for influencing the radioactive decay rate have been sought from early years of Nuclear physics. Nuclear transmutation (i.e. change in the nuclear charge) induced by nuclear reactions are often accompanied by a redistribution of the electrons, muons (mesons in the hadronic atoms) around the final transmuted nucleus. Muonic atoms have always been useful tools for nuclear spectroscopy employing atomic-physics techniques. Muonic atoms also play an important role

A.V. Glushkov (✉)
Odessa University OSENU, P.O.Box 24a, Odessa-9, SE-65009, Ukraine

Institute for Spectroscopy (ISAN), Russian Academy of Sciences, Troitsk-Moscow, 142090, Russia
e-mail: glushkov@paco.net

O.Yu. Khetselius • A.A. Svinarenko
Odessa University OSENU, P.O.Box 24a, Odessa-9, SE-65009, Ukraine
e-mail: nuclei08@mail.ru; quantsvi@mail.ru

P.E. Hoggan et al. (eds.), *Advances in the Theory of Quantum Systems in Chemistry and Physics*, Progress in Theoretical Chemistry and Physics 22, DOI 10.1007/978-94-007-2076-3_3, © Springer Science+Business Media B.V. 2012

as catalysts for nuclear fusion. It should be also mentioned the growing importance of muon spectroscopy in molecular physics. Electrons, muons (other particles such as kaons, pions etc.) originally in the ground state of the target atom can be excited reversibly either to the bound or continuum states. The elementary cooperative e-,n-β-,γ-nuclear processes in atoms and molecules were considered in the pioneering papers by Migdal (1941), Levinger (1953), Schwartz (1953), Carlson et al. (1968), Kaplan et al. (1973–1975), Gol'dansky-Letokhov-Ivanov (1974–1981), Freedman (1974), Law-Campbell (1975), Martin-Cohen (1975), Isozumi et al. (1977), Mukouama et al. (1978), Batkin-Smirnov (1980), Law-Suzuki (1982), Intemann (1983), Wauters-Vaeck (1997) et al. [1–16]. In this context, the known Mössbauer, Szilard-Chalmers and other cooperative effects should be mentioned [9]. The elementary cooperative electron α-nuclear processes were considered in the papers by Levinger (1953), Hansen (1974), Watson (1975), Law (1977), Anholt-Amundsen (1982), Mukoyama-Ito (1988) et al. [10, 17–22]. The consistent relativistic theory of the cooperative electron γ-nuclear processes in atoms and molecules is developed in Refs. [19–22]. With appearance of the intensive neutron pencils, laser sources studying the $\gamma-\mu$-nuclear interactions is of a great importance [4–6, 16]. The rapid progress in laser technology even opens prospects for nuclear quantum optics via direct laser-nucleus coupling [19–26]. It is known that a negative muon μ^- captured by a metastable nucleus may accelerate a discharge of the latter by many orders of magnitude [18–22]. A principal possibility of storage of the significant quantities of the metastable nuclei in the nuclear technology processes and their concentrating by chemical and laser methods leads to problem of governing their decay velocity [5, 19, 22]. The μ-atom system differs advantageously of the usual atom; the relation r_n/r_a (r_n is a radius of a nucleus and r_a is a radius of an atom) can vary in the wide limits in dependence upon the nuclear charge. Because of the large muon mass and the correspondingly small Bohr radius, the muonic wave function has a large overlap with the nucleus and thus effectively probes its structural features, such as finite size, deformation, polarization etc. For a certain relation between the energy range of the nuclear and muonic levels a discharge of the metastable nucleus may be followed by the ejection of a muon, which may then participate in the discharge of other nuclei. The estimates of probabilities for discharge of a nucleus with emission of γ quantum and further muon or electron conversion are presented in ref. [19, 22]. Despite the relatively long history, studying processes of the muon-atom and muon-nucleus interactions hitherto remains very actual and complicated problem. One could remind the known difficulties of the corresponding experiment. On the other hand, theoretical estimates in different models differ significantly [1–4, 22]. According to Mann & Rose, the μ capture occurs mainly at the energies of E $\sim$ 10 κeV, but according to Baker, muons survive till thermal energies [2]. In many papers different authors predicted the μ capture energies in the range from a few dozens to thousands eV. The standard theoretical approach to problem bases on the known Born approximation with the plane or disturbed wave functions and the hydrogen-like functions for the discrete states. In papers by Vogel et al. and Leon-Miller the well-known Fermi-Teller model is used (the atomic electrons are treated as an electron gas and a muon is classically described) [2–4]. In paper by Cherepkov and Chernysheva [2] the Hartree-Fock (HF) method is used to calculate the cross-sections of the capture, elastic and inelastic scattering of the negative μ on the He atom. In recent years more advanced approaches using the fermion molecular dynamics method are used to solve the scattering and capture problem [4, 5]. The Kravtsov-Mikhailov model [4] describes transition of a muon from the excited muonic hydrogen to the helium based on quasimolecular concept. The series of papers by Ponomarev et al. on treating the muonic nuclear catalysis use ideas of Alvarets et al. [5]. More sophisticated methods of the relativistic (QED) PT should be used for correct treating the muon capture effects by multielectron atoms (nuclei). We present here a new consistent energy approach to calculation of the cross-section of the negative muon capture by atoms, using relativistic many-body PT [27–35]. The numerical results calculationals for the cross-section of the μ^- capture by the He atom are listed. Relativistic theory of discharge of a metastable nucleus with emission of γ quantum and further μ^- conversion is presented. The numerical calculation is firstly carried out for scandium nucleus ($A = 49$, $N = 21$) with using the Dirac-Woods-Saxon model.

3.2 Relativistic Energy Approach to the Muon-Atom Interaction

3.2.1 General Formalism

In atomic theory, a convenient field procedure is known for calculating the energy shifts ΔE of the degenerate states. Secular matrix M diagonalization is used. In constructing M, the Gell-Mann and Low adiabatic formula for ΔE is used. A similar approach, using this formula with the QED scattering matrix, is applicable in the relativistic theory [27–31]. In contrast to the non-relativistic case, the secular matrix elements are already complex in the PT second order (first order of the inter-electron interaction). Their imaginary parts relate to radiation decay (transition) probability. The total energy shift of the state is usually presented as follows:

$$\Delta E = \text{Re}\Delta E + i\,\text{Im}\Delta E, \qquad (3.1a)$$

$$\text{Im}\,\Delta E = -\Gamma/2, \qquad (3.1b)$$

where Γ is interpreted as the level width, and the decay possibility $P = \Gamma$. The whole calculation of energies and decay probabilities of a non-degenerate excited state is reduced to calculation and diagonalization of the complex matrix M. To start with the Gell-Mann and Low formula it is necessary to choose the PT zero-order approximation. Usually, the one-electron Hamiltonian is used, with a central potential that can be treated as a bare potential in the formally exact QED PT. There are many well-known attempts to find the fundamental optimization principle for construction of the bare one-electron Hamiltonian (for free atom or atom in a field) or (what is the same) for the set of the one-quasiparticle (QP) functions, which represent such a Hamiltonian [27, 36–42].

Here we consider closed electron shell atoms (ions). For example, the ground state $1s^2$ of the He atom or He-like ion. Note that we operate in the relativistic approximation, though the non-relativistic approach is suitable for light atoms (H or He). As the bare potential, one usually includes the electric nuclear potential V_N and some parameterized screening potential V_C. The parameters of the bare potential may be chosen to generate the accurate eigen-energies of all two-QP states. In the PT second order the energy shift is expressed in terms of the two-QP matrix elements [27–30]:

$$V(1,2;4,3) = \sqrt{(2j_1+1)(2j_2+1)(2j_3+1)(2j_4+1)}(-1)^{j_1+j_2+j_3+j_4+m_1+m_2}$$

$$\times \sum_{\lambda,\nu}(-1)^\nu \begin{bmatrix} j_1.....j_3...\lambda \\ m_1.-m_3..\nu \end{bmatrix}\begin{bmatrix} j_2.....j_4...\lambda \\ m_2.-m_4..\nu \end{bmatrix}Q_\lambda^{Qul} \quad (3.2)$$

Here Q_λ^{Qul} is corresponding to the Coulomb inter-particle interaction:

$$Q_\lambda^{Qul} = \{R_\lambda(1243)S_\lambda(1243)+R_\lambda(\tilde{1}24\tilde{3})S_\lambda(\tilde{1}24\tilde{3})$$

$$+R_\lambda(1\tilde{2}4\tilde{3})S_\lambda(1\tilde{2}4\tilde{3})+R_\lambda(\tilde{1}\tilde{2}4\tilde{3})S_\lambda(\tilde{1}\tilde{2}4\tilde{3})\}, \quad (3.3)$$

where $R_\lambda(1,2;4,3)$ is the radial integral of the Coulomb inter-QP interaction with large radial Dirac components; the tilde denotes a small Dirac component; S_λ is the angular multiplier (see details in Refs. [27–35]). To calculate all necessary matrix elements one must have the 1QP relativistic functions. Further we briefly outline the main idea using, as an example, the negative muon capture by He atom:

$$\left((1s)^2[J_iM_i], \varepsilon_{in}^\mu\right) \rightarrow \left(1s\varepsilon l, \varepsilon_{nl}^\mu\right).$$

Here J_i is the total angular moment of the initial target state; indices ε_{in}^μ and ε_{fk}^μ are the incident and discrete state energies, respectively to the incident and captured muons. Further it is convenient to use the second quantization representation. In particular, the initial state of the system "atom plus free muon" can be written as $a_{in}^{+\mu}\Phi_0$ state. The final state is that of an atom with the discrete state electron, removed electron and captured muon; in further $|I>$ represents one-particle (1QP) state, and $|F>$ represents the three-quasiparticle (3QP) state. The imaginary (scattering) part of the energy shift $Im\Delta E$ in the atomic PT second order (fourth order of the QED PT) is as follows [27, 31]:

$$Im\Delta E = \pi G(\varepsilon_{iv},\varepsilon_{ie},\varepsilon_{in}^\mu,\varepsilon_{fk}^\mu), \quad (3.4)$$

where indices e, v are corresponding to atomic electrons and G is a definite squired combination of the two-QP matrix elements (2). The value $\sigma = -2Im\Delta E$ represents the capture cross-section if the incident muon eigen-function is normalized by the unit flow condition. The different normalization conditions are used for the incident and captured state QP wave functions. The details of the whole numerical procedure of calculation of the cross-sections can be found in Refs. [27–35].

3.2.2 The Dirac-Kohn-Sham Relativistic Wave Functions

Usually, a multielectron atom is defined by a relativistic Dirac Hamiltonian a.u.:

$$H = \sum_i h(r_i) + \sum_{i>j} V(r_i r_j). \tag{3.5}$$

Here, $h(r)$ is one-particle Dirac Hamiltonian for electron in a field of the finite size nucleus and V is potential of the inter-electron interaction. In order to take into account the retarding effect and magnetic interaction in the lowest order on parameter α^2 (α is the fine structure constant) one could write [27, 28]:

$$V(r_i r_j) = exp(i\omega_{ij} r_{ij}) \cdot \frac{(1 - \alpha_i \alpha_j)}{r_{ij}}, \tag{3.6}$$

where ω_{ij} is the transition frequency; α_i, α_j are the Dirac matrices. The Dirac equation potential includes the electric and polarization potentials of a nucleus and exchange-correlation potential. The Kohn-Sham (KS) exchange potential is [36]:

$$V_X^{KS}(r) = -(1/\pi)[3\pi^2 \rho(r)]^{1/3}. \tag{3.7}$$

In the local density approximation the relativistic potential is as follows:

$$V_X[\rho(r), r] = \frac{\delta E_X[\rho(r)]}{\delta \rho(r)}, \tag{3.8}$$

where $E_X[\rho(r)]$ is the exchange energy of the multielectron system corresponding to the homogeneous density $\rho(r)$, which is obtained from a Hamiltonian having a transverse vector potential describing the photons. In this theory the exchange potential is [37]:

$$V_X[\rho(r), r] = V_X^{KS}(r) \cdot \left\{ \frac{3}{2} ln \frac{[\beta + (\beta^2 + 1)^{1/2}]}{\beta(\beta^2 + 1)^{1/2}} - \frac{1}{2} \right\}, \tag{3.9}$$

where

$$\beta = [3\pi^2 \rho(r)]^{1/3}/c. \tag{3.10}$$

The corresponding correlation functional is [20, 37]:

$$V_C[\rho(r), r] = -0.0333 \cdot b \cdot ln[1 + 18.3768 \cdot \rho(r)^{1/3}], \qquad (3.11)$$

where b is the optimization parameter (see details in Refs. [20, 27, 32]). Earlier it has been shown [27–32] that an adequate description of the atomic characteristics requires using an optimized base of the wave functions. In Ref. [27] a new ab initio optimization procedure is proposed. It is reduced to minimization of the gauge dependent multielectron contribution $Im\delta E_{ninv}$ of the lowest QED PT corrections to the radiation widths of atomic levels. In the fourth order of QED PT (the second order of the atomic PT) there appear the diagrams, whose contribution to the $Im\delta E_{ninv}$ accounts for correlation effects. This contribution is determined by the electromagnetic potential gauge (the gauge dependent contribution). All the gauge dependent terms are multielectron by their nature (the known example of the gauge dependence is difference of the oscillator strength values calculated with using the "length" and "velocity" transition operator forms). The dependent contribution to imaginary part of the electron energy is obtained after involved calculation, as follows [27]:

$$
\begin{aligned}
Im\delta E_{ninv}(\alpha - s|b) = -C\frac{e^2}{4\pi} &\iiiint dr_1 dr_2 dr_3 dr_4 \sum_{n>f, m\leq f} \left(\frac{1}{\omega_{mn} + \omega_{\alpha s}} + \frac{1}{\omega_{mn} - \omega_{\alpha s}}\right) \\
&\times \Psi_\alpha^+(r_1)\Psi_m^+(r_2)\Psi_s^+(r_4)\Psi_n^+(r_3) \cdot [(1 - \alpha_1\alpha_2)/r_{12}] \\
&\cdot \{[\alpha_3\alpha_4 - (\alpha_3 n_{34})(\alpha_4 n_{34})]/r_{34} \times sin[\omega_{\alpha n}(r_{12} + r_{34})] \\
&+ [1 + (\alpha_3 n_{34})(\alpha_4 n_{34})]\omega_{\alpha n}cos[\omega_{\alpha n}(r_{12} + r_{34})]\} \\
&\times \Psi_m(r_3)\Psi_\alpha(r_4)\Psi_n(r_2)\Psi_s(r_1). \qquad (3.12)
\end{aligned}
$$

Here, C is the gauge constant, f is the boundary of the closed shells; $n \geq f$ indicating the vacant band and the upper continuum electron states; $m \leq f$ indicates the finite number of states in the atomic core and the states of a negative continuum (accounting for the electron vacuum polarization). The minimization of the functional $Im\delta E_{ninv}$ leads to the Dirac-Kohn-Sham-like equations for the electron density that are numerically solved. Finally an optimal set of the 1QP functions results. In concrete calculation it is sufficient to use the simplified procedure, which is reduced to the functional minimization using the variation of the correlation potential parameter b in Eq. 3.11 [20, 32]. The Dirac equations for the radial functions F and G (the large and small Dirac components respectively) are:

$$\frac{\partial F}{\partial r} + (1 + \chi)\frac{F}{r} - (\varepsilon + m - V)G = 0, \qquad (3.13a)$$

$$\frac{\partial G}{\partial r} + (1 - \chi)\frac{G}{r} + (\varepsilon - m - V)F = 0, \qquad (3.13b)$$

where χ is the quantum number. At large χ the functions F and G vary rapidly at the origin. This creates difficulties in numerical integration of the equations in the region $r \to 0$. To prevent the integration step from becoming too small it is convenient to turn to new functions isolating the main power dependence: $f = Fr^{1-|\chi|}$, $g = Gr^{1-|\chi|}$, $\gamma = (\chi^2 - (\alpha Z)^2)^{1/2}$. The Dirac equations are transformed as follows:

$$f' = -(\chi + |\chi|)f/r - \alpha ZVg - (\alpha ZE_{n\chi} + 2/\alpha Z)g, \tag{3.14a}$$

$$g' = (\chi - |\chi|)g/r - \alpha ZVf + \alpha ZE_{n\chi}f. \tag{3.14b}$$

Here $E_{n\chi}$ is one-electron energy without the rest energy. The boundary values are defined by the first terms of the Taylor expansion:

$$g = (V(0) - E_{n\chi})\,r\alpha Z/(2\chi + 1); \quad f = 1 \tag{3.15a}$$

$$f = (V(0) - E_{n\chi} - 2/\alpha^2 Z^2)\,\alpha Z; \quad g = 1. \tag{3.15b}$$

The condition $f, g \to 0$ at $r \to \infty$ determines the $E_{n\chi}$ state quantified energies. Normalization of radial functions f and g give behaviour for large r as follows:

$$g_x(r) \to r^{-1}[(E+1)/E]^{1/2}\sin(pr + \delta_\chi), \tag{3.16a}$$

$$f_x(r) \to r^{-1}(\chi/|\chi|)[(E-1)/E]^{1/2}\cos(pr + \delta_\chi). \tag{3.16b}$$

The system of Eqs. 3.14a and 3.14b are numerically solved by the Runge-Kutta method ('Superatom' package is used [19–22, 27–35]). Our theory takes into account the nuclear (finite size etc.) and radiative effects if necessary (heavy isotope) etc. (see details in Refs. [33, 34, 43, 44]).

3.2.3 Capture of Negative Muons by Helium Atom

The results of calculation of the cross-section for the negative muon capture by atom of He are shown in Figs. 3.1–3.4. The scheme includes $2 \cdot 10^3$ points till distance $25a_B$ (a_B is the Bohr radius). The main contribution to the capture cross-section is provided by transitions with the moment $l = 0 - 3$.

First we studied the behaviour of curves of the μ^- capture cross-section in dependence on the principal quantum number n with summation on the orbital moments l for several values of the muon initial energy. In whole our curves are lying a little higher than the corresponding curves of Refs. [1–3]. The analysis shows that for the incident μ energies 16 and 50 eV the capture cross-section begins to decrease for all n with growth of the l number ($l > 10$). The states with large l

Fig. 3.1 The calculated dependences of the Auger capture cross-section (*solid line* – E = 50 eV; *dotted line* – E = 20 eV) on orbital number *l* for different *n* values for incident μ⁻ energies 20, 50 eV

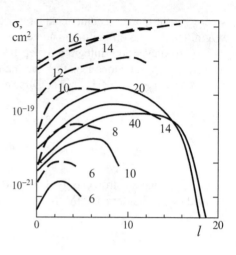

Fig. 3.2 The capture cross-sections in dependence on the orbital number *l* after summation on the *n* number (digits in figure – the muon energies in eV)

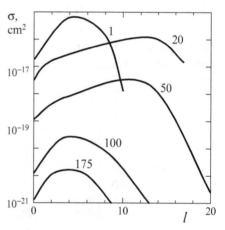

for the muon energies (lower or higher in comparison with the atomic ionization potential value) are populated less probably than in a case of the μ⁻ energy of the ionization potential order. In Fig. 3.1 we present the calculated dependences of the Auger capture cross-section on the orbital number *l* for different *n* values for the incident μ energies of 20 and 50 eV. In Fig. 3.2 we present the calculated capture cross-section in the dependence on the *l* number after summation on *n*. In Fig. 3.3 we present the Auger capture cross-sections in dependence on the principal quantum number after summation on all orbital moment values. In Fig. 3.4 we present the calculated total capture cross-section in terms of energy (with summation on all *n,l*): our data (the Auger capture cross-section) – curve 7 (elastic and inelastic scattering cross-sections) – curves 2,3. In figure we also present the results by Copenman and Rogova in the Born approximation with using the hydrogen-like wave functions (curve 5) and the HF data by Cherepkov-Chernysheva (curve 1), the inelastic scattering cross-section by Rosenberg (curve 4), the transport cross-section

Fig. 3.3 The calculated
Auger capture cross-section
in dependence upon the
principal quantum number n
after summation on all orbital
moment values for different
muon energies (the digits in
figure – the muon energies
in eV)

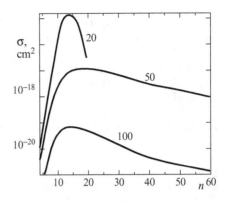

Fig. 3.4 Total cross-section
of μ^- capture in dependence
on an energy: our data (the
Auger capture cross-section)
– curve 7 (elastic and
inelastic scattering
cross-sections) – curves 2,3;
cross-section of capture by
Copenman and Rogova
(curve 5); the HaF data by
Cherepkov-Chernysheva –
curve 1; inelastic scattering
cross-section by Rosenberg –
curve 4; the transport
cross-section – x

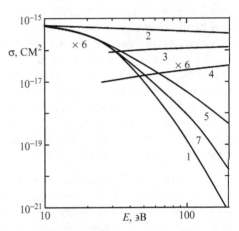

(x symbol) [2,3,32]. The analysis of the results shows that our data are in physically reasonable agreement with the HF data by Cherepkov-Chernysheva and Rosenberg. But, there is an essential difference of the Mann-Rose and Bayer data [1–3]. The relativistic corrections were to found to be small here, but a calculation for a heavy systems (atoms, nuclei) requires a proper treatment for both relativistic and correlation effects.

3.3   Relativistic Theory of Metastable Nucleus Discharge During Negative Muon Capture

3.3.1   General Formalism

For simplicity, we consider the model of a nucleus as the 1-QP system [19, 22]. Further we suppose that a nucleus consists off a twice-magic core and a single

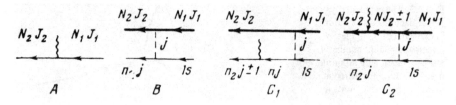

Fig. 3.5 Feynman diagrams corresponding to different channels of a decay (see text)

proton and single muon, which move in the nuclear core field. The proton and muon interact through the Coulomb potential. This interaction will be accounted for in the first order of the atomic PT or in the second order of the QED PT [22]. Surely a majority of the known excited nuclear states have the multi-body character and it is hardly possible to describe their structure within the one-QP model [45–47]. Nevertheless, the studied effects of the muon-proton interaction are not covered by the one-particle model. It is possible to consider a dynamical interaction of two particles through the core too. It accounts for finite core mass. However, this interaction may decrease a multipolarity of the nuclear transition only by unity. Strongly forbidden transitions of high multipolarity are of a great interest.

We calculate the decay probabilities to different channels of the system, which consists of the proton (in an excited state $\Phi_{N_1 J_1}$) and muon (in the ground state Ψ_{1s}^{μ}). Three channels should be taken into account:

(i) a radiative purely nuclear 2^j-poled transition (probability P_1);
(ii) a non-radiative decay, when proton transits to the ground state and the muon leaves a nucleus with energy: $E = \Delta E_{N_1 J_1}^p - E_{\mu}^i$; $\Delta E_{N_1 J_1}^p$ is the energy of nuclear transition; E_{μ}^i is a bond energy of muon in the $1s$ state (probability P_2);
(iii) transition of a proton to the ground state with excitation of muon and emission of γ-quantum with energy $\hbar\omega = \Delta E_{N_1 J_1}^p - \Delta E_{nl}^{\mu}$ (probability P_3).

Feynman diagrams, corresponding to different muon-nuclear decay channels, are shown in Fig. 3.5. Diagram A (Fig. 3.1) corresponds to the first channel (i), diagram B – to the second channel (ii) and diagrams C_1 and C_2 – to the third channel (iii).

The thin line on the diagrams (Fig. 3.5) corresponds to the muon state, the bold line – to the proton state. The initial and final states of proton and muon are designated by indices on the lines. The dashed line with the index j means the Coulomb interaction between muon and proton with an exchange of the 2^j-pole quanta. The wavy line corresponds to operator of the radiative dipole transition. This effect is due to the muon-proton interaction. The diagram A (Fig. 3.1) has the zeroth order on the muon-proton interaction; other diagrams (Fig. 3.1) are first order. The probability of purely radiative nuclear 2^j – pole transition is defined by convention ($r_n = 510^{-13}$ cm) [45]:

$$P_1 = 2 \cdot 10^{20} \cdot \frac{j+1}{j[(2j+1)!!]^2} \left(\frac{3}{j+3}\right)^2 \left(\frac{\Delta E[MeV]}{40}\right)^{2j+1} \tag{3.17}$$

(standard notation). Diagrams C_1 and C_2 (Fig. 3.5) account for an effect of the QP interaction on the initial state. Surely there are other versions of these diagrams, but their contribution to probabilities of the decay processes is significantly less than contributions of the diagrams C_1 and C_2 [19, 22].

Within the relativistic energy approach [21, 22, 26] the total probability is divided into the sum of the partial contributions, connected with decay to the definite final states of a system. These contributions are equal to the corresponding transition probabilities (P_i). E.g., if $\Delta E^P_{N_1 J_1} > E^i_\mu$ the probability determination reduces to relativistic calculation of probability for two-QP system autoionization decay. An imaginary excited state energy in the lowest QED PT order is written: [21, 22]:

$$\mathrm{Im}\Delta E = e^2 \mathrm{Im} i \lim \int \int d^4 x_1 d^4 x_2 \exp\left[\gamma(t_1 + t_2)\right] \{D\left(r_{c_1 t_1}, r_{c_2 t_2}\right) \cdot$$

$$< \Phi_I | (j_{cv}(x_1) j_{cv}(x_2)) | \Phi_I > + D\left(r_{p_1 t_1}, r_{p_2 t_2}\right)$$

$$< \Phi_I | (j_{pv}(x_1) j_{pv}(x_2)) | \Phi_I > + + D\left(r_{\mu_1 t_1}, r_{\mu_2 t_2}\right)$$

$$< \Phi_I | (j_{\mu v}(x_1) j_{\mu v}(x_2)) | \Phi_I > \tag{3.18}$$

Here $D(r_1 t_1, r_2 t_2)$ is the photon propagator; $j_{cv}, j_{pv}, j_{\mu v}$ are the 4-dimensional components of a current operator for the particles: core, proton, muon; $x = (r_c, r_p, r_\mu, t)$ is the four-dimensional space-time coordinate of the particles, respectively; γ is an adiabatic parameter.

Further one should use the exact electrodynamic expression for the photon propagator. Below we are limited by the lowest order of the QED PT, i.e. the next QED corrections to $\mathrm{Im}\Delta E$ will not be considered. Finally, the imaginary part of energy of the excited state can be represented as a sum of the corresponding QP contributions [22]:

$$\mathrm{Im}\Delta E = \mathrm{Im}E_c + \mathrm{Im}E_p + \mathrm{Im}E_\mu,$$

$$\mathrm{Im}E_a = -Z^2_a/4\pi \sum_F \int \int dr_{c1} dr_{c2} \int \int dr_{p1} dr_{p2} \int \int dr_{\mu1} dr_{\mu2}$$

$$\Phi^*_i(1)\Phi^*_F(2) T_a(1,2)\Phi_F(1)\Phi_1(2),$$

$$T_a(1,2) = \frac{\exp(w_{IF} r_{a12})}{r_{a12}}(1 - \alpha_1 \alpha_2), \tag{3.19}$$

Here $r_{a12} = |r_{a1} - r_{a2}|$; Φ_c, Φ_p, Φ_e are the second quantization operators of field of the core particles, proton and muon. Sum on F means the summation on the final states of a system.

Consider a case, for excitation energy of a nucleus ΔE^P of more than the ionization energy of the muon-atomic system E^i_μ. The value $P_3 \ll P_2$, as P_3 has an additional small parameter upon interaction with an electromagnetic field. Calculation of the probability P_2 can be reduced to calculation of probability of the autoionization

Table 3.1 Probabilities
(s^{-1}) of the radiative 2^j-pole
nuclear transition P_1, muon
conversion P_2^μ and electron
conversion P_2^e for $\Delta E^p \Delta \approx \Delta E_i^\mu$
(unit of energy 1 MeV, $Z = 20$)

J	P_1	P_2^μ	P_2^e
1	$4.710^{14}E^3$	1.310^{16}	$1.710^{10}E^{-0,5}$
2	$5.210^9 E^5$	1.710^{16}	$5.810^5 E^{0,5}$
3	$3.910^4 E^7$	3.310^{15}	$9.210° E^{1,5}$
4	$2.610^{-1}E^9$	4.910^{14}	$8.410^{-5}E^{2,5}$
5	$4.210^{-7}E^{11}$	5.710^{13}	$5.410^{-10}E^{3,5}$
6	$2.910^{-12}E^{13}$	5.610^{12}	$2.310^{-15}E^{4,5}$

decay of the two-QP system state, i.e. $P_2 = 2\mathrm{Im}\Delta E/\hbar$, where $\mathrm{Im}\Delta E$ is defined by
Eq. 3.19. In Table 3.1 we present the values of probabilities of the electron (P_2^e) and
muon (P_2^μ) conversion and 2^j-pole (P_1) nuclear transition (for Ca; $Z = 20$).

The results show that the following relationships between corresponding proba-
bilities hold: $P_2^\mu \ll P_1 \gg P_2^e$. The relation $P_2^\mu/P_1 > P_2^\mu/P_2^e$ increases very quickly
with growth of the transition multipolarity. Indeed, it is qualitatively corresponding
to the estimates [19]. For example, for $J = 1$ the estimate by Letokhov-Ivanov gives
the following values: $P_1 = 4,010^{14}E^3, P_2^\mu = 10^{16}, P_2^e = 1,310^{10}E^{-0,5}$. So, when
$\Delta E^p > E_\mu^i$, a reaction of the periodic muon capture by the metastable nucleus occurs
with further muon conversion. The opposite case $\Delta E^p < E_\mu^i$ corresponds to muon
capture in the lowest $1s$ state (the resonant effect and a nucleus is a trap for muons)
[22]. Further we present numerical results of different decay probabilities for given
states of the nucleus $^{49}_{21}Sc_{28}$ with using two versions of the model nuclear potentials,
including the well-known Dirac-Woods-Saxon model [45].

3.3.2 Numerical Calculation for the Nucleus $^{49}_{21}Sc_{28}$

The nucleus of $^{49}_{21}Sc_{28}$ contains a single proton above the twice magic core $^{49}_{20}Ca_{28}$.
The scheme of the corresponding energy levels for this nucleus is presented in
Fig. 3.6. The transitions of proton and muon on the first and second stages are noted
by the solid and dotted lines.

The level $p_{1/2}$ is connected with the ground level $f_{7/2}$ by the E4 transition
and with the low-lying level $p_{3/2}$ by the E2 transition. The levels $p_{3/2}$ and $f_{7/2}$
are connected by the E2 transition. One could consider also magnetic transitions
between these levels. The life-time for the isolated nucleus in the excited state is of
order 10^{-11} s. We use two different approaches to modelling the proton motion in a
field of a nuclear core. The first model corresponds to the well-known Dirac-Wood-
Saxon model [45]. Another approach uses the Bloumkvist-Wahlborn potential [46]
(see also Refs. [22,47]). In the latter, a proton moves in an effective field of the core:

$$V - 25 \cdot f(l,j) \cdot V'/r \qquad (3.20)$$

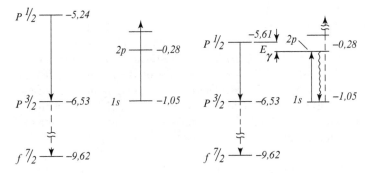

Fig. 3.6 Schemes of energy levels of proton (the *left part* of the figure) and muon (the *right part* of the figure) in $^{49}_{21}Sc_{28}$. Transitions of proton and muon on the first and second stages are noted by the *solid* and *dotted* lines

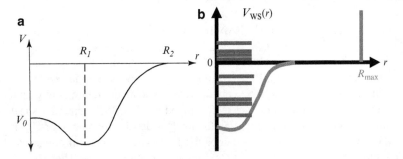

Fig. 3.7 Nuclear potential: (**a**) – the Bloumkvist-Wahlborn potential; (**b**) – the Woods-Saxon potential (see text)

where the self-nuclear part of the interaction V is as follows (Fig. 3.7a):

$$V_0 - a[r^4/4 - r^3(R_1 + R_2)/3 + r^2 R_1 R_2/2], \dots r < R_2$$
$$0, \dots r > R_2 \qquad (3.21)$$

This potential is more suitable in the numerical calculation because it does not lead to divergence (under $r \to 0$) of the spin-orbit interaction $-25f(l, j)V'/r$.

So, it differs advantageously from the well-known Woods-Saxon potential (see below). The electric core potential is given as potential of the charged sphere (the upper sign is for proton and the down sign – for muon):

$$U(r) = \pm Ze^2 \left\{ \begin{array}{l} 3/(2R) - r^2/(2R)^3, r < R \\ 1/r, \dots r > R \end{array} \right\} \qquad (3.22)$$

The parameters are calculated from the fitting condition for the experimental and theoretical energies of the ground and first excited states. For the values $V_0 = -47.6$, $R_1 == 2$, $R_2 = 7.65$, $R = 4.75$ it has been obtained for the proton states: $E(f7/2) = -9.62$, $E(p3/2) = -6.53$, $E(p1/2) = -5.24$ and for the muon states: $E(1s) = -1.05$, $E(2s) = -0.272 E(2p) = -0.281$ (the units of energy 1 MeV and length units 10^{-13} cm are used). To calculate the corresponding integrals in the formula (32) (33), we use the technique, described in Refs. [27–29]. In the relativistic Dirac-Woods-Saxon model the wave functions are defined by solution of the Dirac equations with the Woods-Saxon potential (see below):

$$[\alpha \cdot p + V_{WS}(r) + \beta(M + S_{WS}(r)] \psi^0_{nkm} = \varepsilon^0_{nkm} \psi^0_{nkm} \tag{3.23}$$

$$\{[\varepsilon^0_{nkm}, \psi^0_{nkm}(r,s,t)]; \varepsilon^0_{nkm} >< 0; n = 0,1,2,...;k = \pm1,\pm2,...\}$$

The Woods-Saxon potential is defined as follows (see Fig. 3.7b):

$$V_{WS}(r) = \left\{ \begin{array}{l} V_0[1 + \exp\{(r - R_0)/a_0\}]^{-1}, r < R_{max} \\ \infty, ..., r > R_{max} \end{array} \right\} \tag{3.24}$$

where the parameters V_0, a_0, R_0 are fitting using the levels energies as above said. It should be noted that in the last years the relativistic mean field model with using the Dirac-Woods-Saxon orbital set is developed too [21, 46]. Generally speaking, here any relativistic mean field model, the nuclear density functional or HF theory with the density dependent forces can be used [45–48]. We present numerical data (calculation with two nuclear potentials) for the scandium nucleus. The probabilities of the muon-atomic decay (in s^{-1}) for a most interesting nuclear transitions are:

(i) for the Bloumkvist-Wahlborn potential [22]:

$$P_2(p_{1/2}-p_{3/2}) = 3,93 \cdot 10^{15}, P_2(p_{1/2}-f_{7/2}) = 3,15 \cdot 10^{12},$$
$$P_2(p_{3/2}-f_{7/2}) = 8,83 \cdot 10^{14},$$

(ii) for the Woods-Saxon potential:

$$P_2(p_{1/2}-p_{3/2}) = 3.87 \cdot 10^{15}, P_2(p_{1/2}-f_{7/2}) = 3,09 \cdot 10^{12},$$
$$P_2(p_{3/2}-f_{7/2}) = 8.75 \cdot 10^{14}.$$

For both potentials the presented values are significantly higher that the corresponding non-relativistic estimates of Refs. [19,32]. For example, according to [19]: $P_2(p_{1/2}-p_{3/2}) = 3.3 \cdot 10^{15}$. If a muon-atomic system is in the initial state $p_{1/2}$, than the cascade discharge occurs with an ejection of the muon on the first stage and the γ quantum emission on the second stage.

To consider a case when the second channel is closed and the third channel is opened, let us suppose that $E^P(p_{1/2})-E^P(p_{3/2}) = 0.92$ MeV (Fig. 3.6). The nuclear transition energy is not sufficient to provide transition of the muon to the continuum state. However, it is sufficient for the excitation to the $2p$ state. It is important to note here that this energy is not lying in the resonant range. The diagram C_1 (see Fig. 3.5) describes the proton transition $p_{1/2}-p_{3/2}$ with the virtual excitation of muon to states of the series nd and with emitting γ quantum of the following energy:

$$hw = E^p(p_{1/2}) + E^m(1s) - E^p(p_{3/2}) - E^m(2p).$$

Further the dipole transition $2p - 1s$ can occur. The calculated value for the probability of this transition is $P_3 = 1.9 \cdot 10^{13}\,\text{s}^{-1}$. It should be noted that the value P_3 is more than the probability value for the radiation transition $p_{1/2}-p_{3/2}$ and the probability value for the transition $p_{1/2}-f_{7/2}$. The transition $p_{3/2}-f_{7/2}$ occurs during $\sim 10^{-15}\,\text{s}$ without emission, but with the ejection of a muon.

3.4 Concluding Remarks and Future Perspectives

We have presented a new relativistic approach to calculation of the cross-section of the negative μ capture by atoms and a consistent relativistic theory of discharge of a metastable nucleus with emission of γ quantum and further muon conversion, which initiates this discharge. The approaches are based upon the relativistic many-body PT theory, energy approach and the shell nuclear models. The calculation results are presented for the μ^--He system and Sc nucleus. The experimental possibilities of search of the metastable nucleus discharge effect have been discussed in Refs. [2–5, 22, 48] and require a choice of the special type nuclei (a target). Probability of the μ^- capture by the excited nucleus must be comparable or being more than a probability of the capture by other (non-excited) nuclei of a target. As result, the target must be prepared as the excited nuclei concentrate with the minimal size of order or more than the free μ running length l in relation to a capture by a nucleus. The condition for fewest excited nuclei in a target is $N_{min} > l^3 n_0$, where n_0 is a density of the target atoms. For initial μ slowing to energies of $0.1–0.3\,\text{MeV}$, the free running length is $\sim 0.1\,\text{cm}$. The required number of metastable nuclei is $N_{min} > 10^{19}$ for the density $n_0 = 10^{22}\,\text{cm}^{-3}$. The radioactivity of such a target is $R = N_{min}/T$, where T is the decay time. For example, for $T = 100$ days one can get the estimate $R \sim 10^3\text{Ci}(1\text{Ci} = 3.7 \cdot 10^{10}$ decays per sec).

In conclusion, note that further development of electron-μ-nuclear spectroscopy of atoms (nuclei) is of a great theoretical and practical interest. The development of new approaches [2–6, 21–23] to studying the cooperative e-,μ – γ-nuclear processes promises the rapid progress in our understanding of the nuclear decay. Such an approach is useful, providing perspective for developing advanced nuclear models, search for new cooperative effects on the boundary of atomic and nuclear physics, carrying out new methods for treating (by muonic chemistry tools) the spatial structure of molecular orbitals, studying the chemical bond nature and checking various models in quantum chemistry and solids physics [3–8, 18–23, 49].

Finally, availability of more exact data on the μ-nucleus and μ-atom interactions is important in astrophysics, studying substance transformation in the Universe, testing the Standard Model etc. [6, 21, 49, 50]]. e-μ-γ-nuclear spectroscopy of atoms (molecules) opens new prospects in combining nuclear physics and quantum chemistry (atomic physics). These possibilities are strengthened by quickly developed nuclear quantum optics [19–26]. Superintense lasers (raser or even graser) field may provide a definite measurement of the change in the dynamics of the nuclear processes, including the muon capture (and/or γ-, β-, α-decay) [18–26, 48, 50].

Acknowledgements The authors thank Dr. P. E. Hoggan for his invitation to the workshop QSCP-XV (Cambridge, UK). Authors are grateful to Drs. E. P. Ivanova, L. N. Ivanov, V. S. Letokhov, W. Kohn, E. Brandas, J. Maruani, S. Wilson, I. Kaplan, S. Malinovskaya, A. Theo-philou for useful comments. The support of the Institute for Spectroscopy (Russian Academy of Sciences, Troitsk, Russia), Universities of Geneva and Frieburg (Switzerland, Germany) is acknowledged.

References

1. Tiomno JJ, Weller JA (1949) Rev Mod Phys 21:153; Zaretsky DF, Novikov V (1960) Nucl Phys 14:540; Foldi LL, Walecka JD (1964) Nuovo Cim 34:1026
2. Baker GA (1960) Phys Rev 117:1130; Mann R, Rose M (1961) Phys Rev 121:293; Leon M, Miller J (1977) Nucl Phys A282:461; Cherepkov N, Chernysheva L (1980) J Nucl Phys 32:709; Cohen J, Leon M (1985) Phys Rev Lett 55:52; Cohen J (2004) Phys Rev A 69:022501
3. Haff P, Tombrello T (1974) Ann Phys 86:178; Vogel P, Haff P, Akylas V, Winther A (1975) Nucl Phys A 254:445; Vogel P, Winther A, Akylas V (1977) Phys Lett B70:39
4. Naumannm RA, Schmidt G, Knight JD, Mausner LF, Orth CJ, Schillaci ME (1980) Phys Rev A 21:639; Auerbach N, Klein A (1984) Nucl Phys A 422:480; Kolbe E, Langanke K, Vogel P (2006) Phys Rev C 62:055502
5. Ponomarev L, Fiorentini G (1987) Muon Cat Fus 1:3; Ponomarev L, Gerstain S, Petrov Y (1990) Phys -Usp 160:3; Kravtsov A, Mikhailov A (1994) Phys Rev A49:3566
6. Dykhne A, Yudin G (1996) Sudden perturbation and quantum evolution. UFN, Moscow; Lauss B (2009) Nucl Phys A 827:401; Gazit D (2009) Nucl Phys A 827:408
7. Kaplan IG, Smutny VN (1988) Adv Quantum Chem 19:289; Kaplan IG, Markin AP (1977) Reports of the USSR Acad Sci 232:319; (1975) JETP 69:9
8. Takahashi K, Boyd R, Mathews G, Yokoi K (1987) Phys Rev C 36:1522; Mikheev V, Morozov V, Morozova N (2008) Phys Part Nucl Lett 5:371; Vysotskii VI (1998) Phys Rev C 58:337
9. Mössbauer RM (1958) Z Phys A: Hadrons Nuclei 151:124; Szilard L, Chalmers T (1934) Nature (London) 134:462
10. Migdal AB (1941) J Phys USSR 4:449; Levinger JS (1953) Phys Rev 90:11
11. Cioccheti G, Molinari A (1965) Nuovo Cim 40:69; Carlson T, Nestor CW, Tucker TC, Malik FB (1968) Phys Rev 169:27
12. Amudsen P, Barker PH (1994) Phys Rev C 50:2466; Anholt R, Amundsen P (1982) Phys Rev A 25:169
13. Wolfgang RL, Anderson R, Dodson RW (1956) J Chem Phys 24:16; Martin RL, Cohen JS (1985) Phys Lett A 110: 95
14. Hansen JS (1974) Phys Rev A 9:40; Law J (1977) Nucl Phys A 286:339; (1980) Can J Phys 58:504; Law J, Campbell JL (1982) Phys Rev C 25:514
15. Mukoyama T, Ito S (1988) Phys Lett A 131:182; Mukoyama T, Shimizu S (1978) J Phys G: Nucl Part 4:1509
16. Kienle P (1993) Phys Scripta 46:81; Wauters L, Vaeck N (1996) Phys Rev C 53:497; Wauters L, Vaeck N, Godefroid M, van der Hart H, Demeur M (1997) J Phys B30:4569

17. Glushkov AV, Makarov I, Nikiforova E, Pravdin M, Sleptsov I (1995) Astroparticle Phys 4:15; Glushkov AV, Dedenko L, Pravdin M, Sleptsov I (2004) JETP 99:123
18. Baldwin GG, Salem JC, Gol'dansky VI (1981) Rev Mod Phys 53:687; Letokhov VS (1979) In: Prokhorov AM, Letokhov VS (eds) Application of lasers in atomic, molecular and nuclear physics. Nauka, Moscow, p 412
19. Letokhov VS, Gol'dansky VI (1974) JETP 67:513; Ivanov LN, Letokhov VS (1976) JETP 70:19; Ivanov LN, Letokhov VS (1985) Com Mod Phys D 4:169; Glushkov AV, Ivanov LN, Letokhov VS (1991) Nuclear quantum optics, Preprint of Inst. for Spectroscopy of USSR Academy of Sciences (ISAN), N5 Troitsk, pp 1–18
20. Glushkov AV, Khetselius OYu, Lovett L (2010) In: Piecuch P, Maruani J, Delgado-Barrio G, Wilson S (eds) Advances in the theory of atomic and molecular systems dynamics, spectroscopy, clusters, and nanostructures. Progress in theoretical chemistry and physics, vol 20. Springer, Berlin, pp 125–172
21. Glushkov AV, Khetselius OYu, Malinovskaya SV (2008) In: Wilson S, Grout PJ, Maruani J, Delgado-Barrio G, Piecuch P (eds) Frontiers in quantum systems in chemistry and physics, progress in theoretical chemistry and physics, vol 18. Springer, Berlin, p 523; (2008) Eur Phys J ST 160:195; (2008) Mol Phys 106:1257
22. Glushkov AV, Malinovskaya S, Vitavetskaya L, Dubrovskaya Yu (2006) In: Julien J-P, Maruani J, Mayou D, Wilson S, Delgado-Barrio G (eds) Recent advances in theoretical physics and chemistry systems, progress in theoretical chemistry and physics, vol 15. Springer, Berlin, pp 301–318; Glushkov A, Malinovskaya S, Gurnitskaya E, Khetselius O (2006) J Phys CS 35:426
23. Shahbaz A, Müller C, Bürvenich TJ, Keitel CH (2009) Nucl Phys A 821:106; Müller C, Di Piazza A, Shahbaz A, Bürvenich TJ, Evers J, Hatsagortsyan HZ, Keitel CH (2008) Laser Physics 11:175; Shahbaz A, Müller C, Staudt A, Burnevich TJ, Keitel CH (2007) Phys Rev Lett 98:263901; Burnevich TJ, Evers J, Keitel CH (2006) Phys Rev C 74:044601; (2006) Phys Rev Lett 96:142501
24. Romanovsky M (1998) Laser Phys 1:17; Harston MR, Caroll JJ (2004) Laser Phys 7:1452; Tkalya EV (2007) Phys Rev A 75:022509
25. Ahmad I, Dunfird R, Esbensen H, Gemmell DS, Kanter EP, Run U, Siuthwirth SH (2000) Phys Rev C 61:051304; Kishimoto S, Yoda Y, Kobayashi Y, Kitao S, Haruki R, Masuda R, Seto M (2006) Phys Rev C 74:031301
26. Olariu S, Sinor T, Collins C (1994) Phys Rev B 50:616; Glushkov AV, Ivanov LN (1991) Preprint of Institute for Spectroscopy of USSR Academy of Sciences (ISAN), N-2AS; Glushkov AV, Svinarenko A (2010) Sensor Electron Microsyst Technol 1:13
27. Glushkov AV, Ivanov LN, Ivanova EP (1986) Autoionization phenomena in atoms. Moscow University Press, Moscow; Glushkov AV, Ivanov LN (1992) Phys Lett A 170:33; Glushkov AV (1992) JETP Lett 55:97
28. Ivanov LN, Ivanova EP, Knight L (1993) Phys Rev A 48:4365; Ivanova EP, Ivanov LN, Aglitsky EV (1988) Phys Rep 166:315; Ivanova EP, Ivanov LN (1996) JETP 83:258; Ivanova EP, Grant IP (1998) J Phys B 31:2871
29. Ivanova EP, Ivanov LN, Glushkov AV, Kramida A (1985) Phys Scripta 32:512; Glushkov AV, Ivanova EP (1986) J Quant Spectr Rad Transfer 36:127
30. Glushkov AV, Khetselius OYu, Loboda AV, Svinarenko AA (2008) In: Wilson S, Grout PJ, Maruani J, Delgado-Barrio G, Piecuch P (eds) Frontiers in quantum systems in chemistry and physics, progress in theoretical chemistry and physics, vol 18. Springer, Berlin, p 541; Glushkov AV, Loboda AV, Gurnitskaya EP, Svinarenko AA (2009) Phys Scripta T 134:137740
31. Glushkov AV, Malinovskaya SV, Loboda AV, Gurnitskaya EP, Korchevsky DA (2005) J Phys: Conf Ser 11:188; Glushkov AV, Ambrosov SV, Loboda AV, Gurnitskaya EP, Prepelitsa GP (2005) Int J Quantum Chem 104:562; Glushkov A, Malinovskaya S, Prepelitsa G, Ignatenko V (2004) J Phys CS 11:199
32. Glushkov AV (1989) Opt Spectr (USSR) 66:31; (1991) 70:952; (1991) 71:395; (1992) 72:55; (1992) 72:542; (1992) Russ J Phys Chem 66:589; (1992) 66:1259; (1992) Russ J Struct Chem 32:11; (1993) 34:3

33. Glushkov AV, Khetselius OYu, Gurnitskaya EP, Loboda AV, Florko TA, Sukharev DE, Lovett L (2008) In: Wilson S, Grout PJ, Maruani J, Delgado-Barrio G, Piecuch P (eds) Frontiers in quantum systems in chemistry and physics, progress in theoretical chemistry and physics, vol 18. Springer, Berlin, p 505

34. Glushkov AV, Ambrosov SV, Loboda AV, Gurnitskaya EP, Khetselius OYu (2006) In: Julien J-P, Maruani J, Mayou D, Wilson S, Delgado-Barrio G (eds) Recent advances in theoretical physics and chemistry systems, progress in theoretical chemistry and physics, vol 15. Springer, Berlin, p 285

35. Glushkov A, Malinovskaya S, Loboda A, Prepelitsa G (2006) J Phys: Conf Ser 35:420; Glushkov AV, Malinovskaya SV, Chernyakova Yu G, Svinarenko AA (2004) Int J Quantum Chem 99:889; (2005) Int J Quantum Chem 104:496

36. Kohn W, Sham LJ (1964) Phys Rev A 140:1133; Hohenberg P, Kohn W (1964) Phys Rev B 136:864; Feller D, Davidson ER (1981) J Chem Phys 74:3977

37. Gunnarsson O, Lundqvist B (1976) Phys Rev B 13:4274; Das M, Ramana M, Rajagopal A (1980) Phys Rev 22:9; Schmid R, Engel E, Dreizler RM (1995) Phys Rev C 52:164; Dreizler RM, Gross EKU (1990) Density FT. Springer, Berlin

38. Hehenberger M, McIntosh H, Brändas E (1974) Phys Rev A 10:1494; Glushkov AV, Ivanov LN (1993) J Phys B: At Mol Opt Phys 26:L379; Glushkov AV, Ambrosov S, Ignatenko A, Korchevsky D (2004) Int J Quantum Chem 99:936

39. Grant IP (2007) Relativistic quantum theory of atoms and molecules, theory and computation, vol 40. Springer, Berlin, p 587

40. Dyall KG, Faegri K Jr (2007) Introduction to relativistic quantum theory. OUP, Oxford

41. Glushkov AV (2008) Relativistic quantum theory. Quantum mechanics of atomic systems. Astroprint, Odessa; Rusov VD, Tarasov V, Litvinov D (2008) Theory of beta processes in georeactor. URS, Moscow

42. Wilson S (2007) In: Maruani J, Lahmar S, Wilson S, Delgado-Barrio G (eds) Recent advances in theoretical physics and chemistry systems, progress in theoretical chemistry and physics, vol 16. Springer, Berlin, p 11

43. Mohr P (1993) Atom Dat Nucl Dat Tables 24:453; Santos J, Parenre F, Boucard S, Indelicato P, Desclaux J (2005) Phys Rev A 71:032501; Sapirstein J, Cheng KT (2005) Phys Rev A 71:022503

44. Safranova UI, Safranova MS, Johnson W (2005) Phys Rev A71:052506; Shabaev V, Tupitsyn I, Pachucki K, Plunien G, Yerokhin VA (2005) Phys Rev A 72:062105; Dzuba V, Flambaum V, Safranova M (2006) Phys Rev A73:022112

45. Bohr O, Motelsson B (1971) Structure of atomic nucleus. Plenum, New York; Ring P, Schuck P (2000) The nuclear many-body problem. Springer, Heidelberg; (2010) In: Yang F, Hamilton JH (eds) Fundamentals of nuclear models. World Scientific, Singapore

46. Serot B, Walecka J (1986) Adv Nucl Phys 16:1; Bender M, Heenen P, Reinhard P (2003) Rev Mod Phys 75:121; Bloumkvist J, Wahlborn S (1960) Ark Fysic 16:545

47. Glushkov AV (2007) In: Krewald S, Machner H (eds) Meson-nucleon physics and the structure of the nucleon. IKP, Juelich. SLAC eConf C070910 Menlo Park 2:111; Glushkov AV, Khetselius OYu, Loboda AV, Malinovskaya SV (2007) ibid 2:118; Glushkov A, Lovett L, Khetselius O, Loboda A, Dubrovskaya Yu, Gurnitskaya E (2009) Int J Modern Phys A Part, Fields, Nucl Phys 24:611

48. Glushkov AV, Rusov VD, Ambrosov S, Loboda A (2003) In: Fazio G, Hanappe F (eds) New projects and new lines of research in nuclear phys. World Scientific, Singapore, p 126; Zagrebaev V, Oganessian Yu, Itkis M, Greiner W (2006) Phys Rev C 73:031602

49. Behr J, Gwinner G (2009) J Phys G: Nucl Part 36:033101; Kettner K, Becker H, Strieder F, Rolfs C (2006) J Phys G: Nucl Part 32:489; Godovikov S (2001) RAN Izv Ser Phys 65:1063; Zinner N (2007) Nucl Phys A 781:81

50. Glushkov AV, Malinovskaya SV (2003) In: Fazio G, Hanappe F (eds) New projects and new lines of research in nuclear phys. World Scientific, Singapore, p 242; Glushkov A (2005) In: Grzonka D, Czyzykiewicz R, Oelert W, Rozek T, Winter P (eds) Low energy antiproton Phys. AIP, Melville, NY, 796:206

Part III
Atoms and Molecules with Exponential-Type Orbitals

Chapter 4
Two-Range Addition Theorem for Coulomb Sturmians

Daniel H. Gebremedhin and Charles A. Weatherford

Abstract A new compact two-range addition theorem for Coulomb Sturmians is presented. This theorem has been derived by breaking up the exponential-type orbitals into convenient elementary functions: the Yukawa potential $(e^{-\alpha r}/r)$ and "evenly-loaded solid harmonics," $(r^{2\nu+l}Y_l^m(\hat{r})$ for which translation formulas are available. The resulting two-range translation formula for the exponential orbital is presented and used to construct a new addition theorem for the Coulomb Sturmians.

4.1 Introduction

Exponential type orbitals (ETO) are the natural choice for the basis set in atomic and molecular electronic structure calculations. The resulting multicenter integrals (for molecules) are notoriously difficult to accurately evaluate. One of the fundamental reasons is the lack of a compact and rapidly convergent addition theorem (ADT) for ETOs.

ADTs for many of the exponential-type functions have been extensively studied and are generally given in non-terminating two-range forms, which, according to Weniger [1], result from the lack of analyticity at the origin. For non-exponential functions such as the regular solid spherical harmonics [1], terminating single-range addition theorems are available. The irregular solid harmonics have a double-range non-terminating addition theorem [1]. For function $f(\mathbf{r} + \mathbf{r}')$ where $\mathbf{r}$ and $\mathbf{r}'$ are vectors in real three dimensional space, the ADTs expand f in terms of two functions with arguments $\mathbf{r}_<$ and $\mathbf{r}_>$ such that $\mathbf{r}_<$ ($\mathbf{r}_>$) is which of $\mathbf{r}$ and $\mathbf{r}'$ that is lesser (greater) in magnitude. Direct numerical tests of such translations (see for

D.H. Gebremedhin (✉) • C.A. Weatherford
Department of Physics, Florida A & M University, FL 32307
e-mail: daniel1.gebremedhin@famu.edu; charles.weatherford@famu.edu

P.E. Hoggan et al. (eds.), *Advances in the Theory of Quantum Systems in Chemistry and Physics*, Progress in Theoretical Chemistry and Physics 22,
DOI 10.1007/978-94-007-2076-3_4, © Springer Science+Business Media B.V. 2012

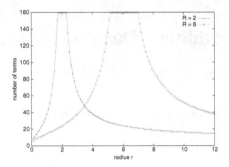

Fig. 4.1 Number of terms needed for Laplace expansion in order to calculate $|\mathbf{R}+\mathbf{r}|^{-1}$ to within 10^{-12} accuracy. Two cases are shown where one of the two radii, R, is fixed to $R = 2$ & $R = 6$. Number of maximum terms was set to be 160. We can see that convergence improves as the radial separation of the two vectors increases (Angular variables were chosen at random and held constant)

example Eq. 4.10 where $\mathbf{r} = \mathbf{r}_> + \mathbf{r}_<$) reveal that the convergence of the expansions deteriorate as $|r_> - r_<|$ decreases and vice versa (See Fig. 4.1). This behavior is also exhibited by multicenter integrals. ADTs are mainly used as a tool for separation of variables in integrals whenever the integrands involve two or more centers – the ADTs converge better if the relative radial separation of the centers is relatively large.

Coulomb Sturmians (CSs) are an exponential-type complete set of basis functions which satisfy a Sturm–Liouville equation [2]. The main objective of the present work is to derive an ADT for the Slater-type orbitals (STOs), which are the fundamental ETO, and thereby for the CSs, which are a linear combination of STOs. The expression for the two-center overlap integral is then worked out for the CSs as an illustration and numerical results and conclusions are presented.

4.2 Translation of Coulomb Sturmians

For a position vector $\mathbf{r}$ in three dimensional coordinate space, the following notation is used:

$$\mathbf{r} \equiv \mathbf{r}_1 + \mathbf{r}_2, \qquad r \equiv |\mathbf{r}|, \qquad r_1 < r_2, \qquad \hat{r} \equiv \frac{\mathbf{r}}{r}. \tag{4.1}$$

The main topic of the paper is to give a compact addition theorem for the CS:

$$\Psi_{nlm}(\alpha, \mathbf{r}) = N_{nl}(2\alpha)^{3/2}(2\alpha r)^l e^{-\alpha r} L_{n-l-1}^{2l+1}(2\alpha r) Y_l^m(\hat{r}) \tag{4.2}$$

where Y is spherical harmonics and L is associated Laguerre polynomial explicitly given by:

$$L_n^k(x) = \sum_{\mu=0}^{n} \sigma_{\mu,n}^k x^\mu \tag{4.3}$$

with the coefficients, σ, and normalization constants, N:

$$\sigma^k_{\mu,n} = \frac{(-1)^\mu (n+k)!}{(n-\mu)!(k+\mu)!\mu!} \qquad N_{nl} = \sqrt{\frac{(n-l-1)!}{2n(n+l)!}} \tag{4.4}$$

where $x!$ is factorial function of x. From Eqs. 4.2 and 4.3, we can see that the CSs are finite superposition of Slater-type orbitals (STOs) and it is for these STOs that we will work out the addition theorems. The unnormalized STO's are divided into two types according to their powers of r and are defined as:0

$$\Phi_{vlm}(\alpha,\mathbf{r}) \equiv \frac{e^{-\alpha r}}{r}\mathscr{Y}_{vlm}(\mathbf{r}) \quad \& \quad \bar{\Phi}_{vlm}(\alpha,\mathbf{r}) \equiv \frac{e^{-\alpha r}}{r}\bar{\mathscr{Y}}_{vlm}(\mathbf{r}) \tag{4.5}$$

where $\mathscr{Y}$ and $\bar{\mathscr{Y}}$ are termed (by us), "evenly- and, oddly-loaded spherical harmonics" and are respectively

$$\mathscr{Y}_{vlm}(\mathbf{r}) \equiv r^{2v+l}Y^m_l(\hat{r}) \quad \& \quad \bar{\mathscr{Y}}_{vlm}(\mathbf{r}) \equiv r^{2v+l+1}Y^m_l(\hat{r}). \tag{4.6}$$

Here, v, unlike n, l & m, is not a quantum number but a dummy non-negative integer. We will now derive a two-range ADT for the two types of orbitals defined in Eq. 4.5 by further breaking them up into convenient components. Two-range ADTs are more naturally expressed in terms of bipolar spherical harmonics, X, defined as [3]:

$$X^{LM}_{l_1 l_2}(\hat{r}_1, \hat{r}_2) = \sum_{m_1=-l_1}^{l_1} C^{LM}_{l_1 m_1 l_2 (M-m_1)} Y^{m_1}_{l_1}(\hat{r}_1) Y^{M-m_1}_{l_2}(\hat{r}_2) \tag{4.7}$$

where C is Clebsch-Gordan coefficient. Coordinate interchange in the X's results in sign change according to:

$$X^{LM}_{l_1 l_2}(\hat{r}_1, \hat{r}_2) = (-1)^{l_1+l_2-L} X^{LM}_{l_2 l_1}(\hat{r}_2, \hat{r}_1). \tag{4.8}$$

One of the basic ADTs we will use is to that of Yukawa potential. The equation is commonly known as Gegenbauer addition theorem and is given by:

$$\frac{e^{-\alpha r}}{r} = 4\pi \sum_{l=0}^{\infty} \omega_l f_l(\alpha, r_1, r_2) X^{00}_{ll}(\hat{r}_1, \hat{r}_2) \tag{4.9}$$

where $\omega_l \equiv \sqrt{2l+1}$, $f_l(\alpha, r_1, r_2) \equiv \alpha \hat{i}_l(\alpha r_1)\hat{k}_l(\alpha r_2)$ and $\hat{i}$ & $\hat{k}$ are spherical modified Bessel functions of the first and second kind respectively [4].

This is the fundamental expansion we are going to build upon and in our opinion, this is the best that can be done in terms of an ADT for the function $\frac{e^{-\alpha r}}{r}$ (Yukawa potential). The famous Laplace expansion (Coulomb potential addition theorem) has a similar form to Eq. 4.9:

$$\frac{1}{r} = \frac{4\pi}{r_2} \sum_{l=0}^{\infty} \frac{1}{\omega_l} \left(\frac{r_1}{r_2}\right)^l X_{ll}^{00}(\hat{r}_1, \hat{r}_2). \tag{4.10}$$

The evenly-loaded solid spherical harmonics have a terminating two-range ADT given by:

$$\mathscr{Y}_{vlm}(\mathbf{r}) = \gamma_{vl} r_2^{2v+l} \sum_{l_1=0}^{v+l} \sum_{l_2=|l_1-l|}^{l_{2max}} \mathscr{C}_{l_1 l_2}^{l} X_{l_1 l_2}^{lm}(\hat{r}_1, \hat{r}_2) \sum_{q=l_1,2}^{2v+l-l_2} \left(\frac{r_1}{r_2}\right)^q \Bigg/ Q_{vlq}^{l_1 l_2} \tag{4.11}$$

where

- $l_{2max} \equiv min(l_1 + l, 2v + l - l_1)$ with $(l_1 + l_2 + l)$ even
- $\gamma_{vl} \equiv 4\pi(2v)!!(2v + 2l + 1)!!$
- $\mathscr{C}_{l_1 l_2}^{l} \equiv \sqrt{\frac{(2l_1+1)(2l_2+1)}{4\pi(2l+1)}} C_{l_1 0 l_2 0}^{l0}$
- $Q_{vlq}^{l_1 l_2} \equiv (2v + l + l_2 - q + 1)!!(2v + l - q - l_2)!!(q + l_1 + 1)!!(q - l_1)!!$

and $x!!$ is double factorial function of x. A direct product of Eqs. 4.9 and 4.11 gives:

$$\Phi_{vlm}(\alpha, \mathbf{r}) = 4\pi \gamma_{vl} r_2^{2v+l} \sum_{l_0=0}^{\infty} \omega_{l_0} f_{l_0}(\alpha, r_1, r_2) X_{l_0 l_0}^{00}(\hat{r}_1, \hat{r}_2)$$

$$\sum_{l_1=0}^{v+l} \sum_{l_2=|l_1-l|}^{l_{2max}} \mathscr{C}_{l_1 l_2}^{l} X_{l_1 l_2}^{lm}(\hat{r}_1, \hat{r}_2) \sum_{q=l_1,2}^{2v+l-l_2} \left(\frac{r_1}{r_2}\right)^q \Bigg/ Q_{vlq}^{l_1 l_2} \tag{4.12}$$

The above equation can be further simplified to give the ADT for the Φ's:

$$\Phi_{vlm}(\alpha, \mathbf{r}) = 4\pi \gamma_{vl} r_2^{2v+l} \sum_{l_0=0}^{\infty} \omega_{l_0} f_{l_0}(\alpha, r_1, r_2) \sum_{l_1=0}^{v+l} \sum_{l_2=|l_1-l|}^{l_{2max}} \mathscr{C}_{l_1 l_2}^{l}$$

$$\sum_{q=l_1,2}^{2v+l-l_2} \left(\frac{r_1}{r_2}\right)^q \Bigg/ Q_{vlq}^{l_1 l_2} \sum_{l_1'=|l_0-l_1|}^{l_0+l_1} \sum_{l_2'=l_{2min}'}^{l_{2max}'} B_{l_0 l_1 l_2}^{l_1' l_2' l} X_{l_1' l_2'}^{lm}(\hat{r}_1, \hat{r}_2) \tag{4.13}$$

where for the sum on l_1', we have an additional symmetry requirement that $(l_0 + l_1 + l_1')$ be even. $l_{2min}' \equiv max(|l_2 - l_0|, |l - l_1'|)$, $l_{2max}' \equiv min(l_2 + l_0, l + l_1')$ with $(l_0 + l_2 + l_2')$ even. B is a coefficient for combination of two bipolar harmonics given as:

$$B_{l_0 l_1 l_2}^{l_1' l_2' l} = \frac{(-1)^{l_0 + l_2 + l_1' + l}}{4\pi} \sqrt{(2l_0+1)(2l_1+1)(2l_2+1)} C_{l_0 0 l_1 0}^{l_1' 0} C_{l_0 0 l_2 0}^{l_2' 0} \begin{Bmatrix} l_2' & l_2 & l_0 \\ l_1 & l_1' & l \end{Bmatrix} \tag{4.14}$$

where the term in bracket is a 6j symbol. Equation 4.12 is a sometimes useful form although (4.13) is more compact.

To derive an ADT for the $\bar{\Phi}$'s in a similar fashion, we might want to look for an expansion for the $\bar{\mathscr{Y}}$'s. Such a translation, given below, exists and is a structurally similar expression to that of the Eq. 4.11, but for one major difference – it has an open sum.

$$\bar{\mathscr{Y}}_{vlm}(\mathbf{r}) = \bar{\gamma}_{vl} r_2^{2v+l+1} \sum_{l_2=\alpha_1=|l_2-l|}^{\infty} \sum_{l_2=\alpha_1=|l_2-l|}^{l_2+l} \bar{\mathscr{C}}_{l_1l_2}^l \left(\frac{r_1}{r_2}\right)^{l_1} X_{l_1l_2}^{lm}(\hat{r}_1,\hat{r}_2) \sum_{q=0,2}^{2v+l+l_2-l_1+2} \left(\frac{r_1}{r_2}\right)^q \Big/ \bar{Q}_{vlq}^{l_1l_2}$$

$$(4.15)$$

where

- $\bar{\gamma}_{vl} \equiv 4\pi(-1)^l (2v+1)!!(2v+2l+2)!!$

- $\bar{\mathscr{C}}_{l_1l_2}^l \equiv (-1)^{l_1+l_2} \sqrt{\frac{(2l_1+1)(2l_2+1)}{4\pi(2l+1)}} C_{l_10l_20}^{l0}$

- $\bar{Q}_{vlq}^{l_1l_2} \equiv (2v+l+l_2-q-l_1+2)!!(2v+l-q-l_1-l_2+1)!!(q+2l_1+1)!!q!!$

Employing the above equation would clearly result in double open sums on the ADT for the $\bar{\Phi}$, which would render it to be computationally very costly. Instead, we rearrange the second equation in (4.5) as $\bar{\Phi}_{vlm}(\alpha,\mathbf{r}) \equiv e^{-\alpha r}\mathscr{Y}_{vlm}(\mathbf{r})$, and seek an ADT for the exponential function. This is elegantly derived by a simple parametric differentiation with respect to α on both sides of Eq. 4.9. Using the recurrence relation of the spherical modified Bessel functions and their derivatives, and after some simplification, we get the equation given below for the first time here:

$$e^{-\alpha r} = 4\pi \sum_{l=0}^{\infty} \omega_l \left[f_l^{(+)}(\alpha,r_1,r_2) - f_l^{(-)}(\alpha,r_1,r_2) \right] X_{ll}^{00}(\hat{r}_1,\hat{r}_2) \qquad (4.16)$$

where $f_l^{(+)}(\alpha,r_1,r_2) \equiv (\alpha r_2)\hat{i}_l(\alpha r_1)\hat{k}_{l+1}(\alpha r_2)$ and $f_l^{(-)}(\alpha,r_1,r_2) \equiv (\alpha r_1)\hat{i}_{l-1}(\alpha r_1)$ $\hat{k}_l(\alpha r_2)$. Comparison of Eqs. 4.16 and 4.9 reveals that a mere replacement of $f_{l_0}(\alpha,r_1,r_2)$ by $\left[f_{l_0}^{(+)}(\alpha,r_1,r_2) - f_{l_0}^{(-)}(\alpha,r_1,r_2) \right]$ in Eq. 4.13 gives the required ADT for the $\bar{\Phi}$'s. We find it notationally convenient to express this as:

$$\bar{\Phi}_{vlm}(\alpha,\mathbf{r}) = \Phi_{vlm}^{(+)}(\alpha,\mathbf{r}) - \Phi_{vlm}^{(-)}(\alpha,\mathbf{r}) \qquad (4.17)$$

where, this time, the expressions for each of $\Phi_{vlm}^{(+)}(\alpha,\mathbf{r})$ and $\Phi_{vlm}^{(-)}(\alpha,\mathbf{r})$ are exactly similar to the right hand side of (4.13), except for the replacement of $f_{l_0}(\alpha,r_1,r_2)$ by $f_{l_0}^{(+)}(\alpha,r_1,r_2)$ and $f_{l_0}^{(-)}(\alpha,r_1,r_2)$ respectively. We have managed to expand the ETO $e^{-\alpha r}\mathscr{Y}_{vlm}(\mathbf{r})$ with only one open sum and this is the main result reported in this work.

Finally, expressing the Ψ's in terms of the STOs Φ and $\bar{\Phi}$ completes the addition theorem for the CS functions:

$$\Psi_{nlm}(\alpha,\mathbf{r}) = N_{nl}(2\alpha)^{l+3/2}\left[\sum_{v=0}^{\lfloor(n-l-1)/2\rfloor}\sigma_{2v,n-l-1}^{2l+1}(2\alpha)^{2v}\bar{\Phi}_{vlm}(\alpha,\mathbf{r})\right.$$

$$\left.+\sum_{v=1}^{\lfloor(n-l)/2\rfloor}\sigma_{2v-1,n-l-1}^{2l+1}(2\alpha)^{2v-1}\Phi_{vlm}(\alpha,\mathbf{r})\right] \quad (4.18)$$

where $\lfloor x \rfloor$ is floor function of x. The above ADT has been constructed by breaking up the CS in such a way that, only the "best available" expansions are used. As far as the two-range (with point-wise convergence) translation of the CSs for physically useful internuclear radial distances is concerned, we posit that our formula is the best that can be done in terms of accuracy and compactness. The striking similarity of the forms of Φ, $\Phi^{(+)}$ & $\Phi^{(-)}$ is an added convenience for their applications in multicenter molecular integrals and related studies.

For the sake of completeness, multiplying the above equation by $\frac{n}{\alpha}$ and switching Φ & $\bar{\Phi}$ results in an ADT for the reduced CS, $\mathbf{S}$, defined as:

$$\mathbf{S}_{nlm}(\alpha,\mathbf{r}) = \frac{n}{\alpha r}\Psi_{nlm}(\alpha,\mathbf{r}) \quad (4.19)$$

They satisfy the orthonormality relation:

$$\langle\mathbf{S}_{n'l'm'}(\alpha,\mathbf{r})|\Psi_{nlm}(\alpha,\mathbf{r})\rangle = \delta_{n'n}\delta_{l'l}\delta_{m'm}. \quad (4.20)$$

4.3 Two-Center Overlap Integrals – Numerical Example

4.3.1 Calculations

We have applied the ADT (Eq. 4.18) to the calculation of a two-center overlap integral of CSs. To this end, let us introduce a new notation, so that, $0 \leq r_A \leq r_B \leq \infty$ and $\mathbf{A} \equiv \mathbf{r} + \mathbf{r}_A$ with analogous definition for vector $\mathbf{B}$. We would also interchangeably use $j \equiv (nlm)_j \equiv n_jl_jm_j$. Then, the two-center overlap integral between two CS is:

$$V_{jk}(\alpha,\beta,\mathbf{r}_A,\mathbf{r}_B) \equiv \int \Psi_j^*(\alpha,\mathbf{A})\Psi_k(\beta,\mathbf{B})\,d^3r. \quad (4.21)$$

In passing, we comment that neither of the two translation vectors for the two centers in the above integral can be safely set to zero due to the irregularity of the $\hat{k}$'s at the origin.

V is first expressed in terms of nine different overlap integrals between the three expansions Φ, $\Phi^{(+)}$ and $\Phi^{(-)}$. The resulting overlap integrals are further divided over three ranges in radius: namely, $[0, r_A]$, $[r_A, r_B]$ & $[r_B, \infty)$. Hence, a total of $3 \times 9 = 27$ unique integrals has to be calculated. All of the 27 integrals will have one open sum. The basic integrals needed are

$$\Lambda^N_{l_1 l_2}(\alpha, \beta, r_1, r_2) \equiv \int_{r_1}^{r_2} \hat{i}_{l_1}(\alpha, r) \hat{i}_{l_2}(\beta, r) r^N \, dr$$

$$\Omega^N_{l_1 l_2}(\alpha, \beta, r_1, r_2) \equiv \int_{r_1}^{r_2} \hat{i}_{l_1}(\alpha, r) \hat{k}_{l_2}(\beta, r) r^N \, dr$$

$$\Gamma^N_{l_1 l_2}(\alpha, \beta, r_1, r_2) \equiv \int_{r_1}^{r_2} \hat{k}_{l_1}(\alpha, r) \hat{k}_{l_2}(\beta, r) r^N \, dr$$

$$\Theta^{LML'M'}_{l l_1 l' l'_1 l_c}(\hat{r}_A, \hat{r}_B, \hat{r}_C) \equiv \int X^{*LM}_{l l_1}(\hat{r}, \hat{r}_A) X^{L'M'}_{l' l'_1}(\hat{r}, \hat{r}_B) X^{00}_{l_c l_c}(\hat{r}, \hat{r}_C) \, d\hat{r}. \qquad (4.22)$$

The importance of accurately calculating the above four integrals cannot be overstated. The accuracy of the first three integrals is heavily dependent upon a judicious choice of a representation for the spherical modified Bessel function among the ones that are available on the literature. When the upper integration limit, r_2, is infinite (this only occurs on the third equation of (4.22)), we used the following terminating sum representation of the $\hat{k}$'s:

$$\hat{k}_l(x) = \frac{e^{-x}}{x} \sum_{j=0}^{l} \frac{(l+j)!}{(2j)!!(l-j)!} x^{-j} \qquad (4.23)$$

This leads to the exponential integral and an efficient numerical algorithm has already been provided for them [5]. When r_2 is finite, however, the exponential function presents a numerically unstable expression for the respective integrals, so we rather have to use a purely polynomial representation of the two Bessel functions. Such an expansion, a highly convergent one, already exists for the $\hat{i}$'s:

$$\hat{i}_l(x) = x^l \sum_{j=0}^{\infty} \frac{x^{2j}}{(2j)!!(2l+2j+1)!!} \qquad (4.24)$$

And for the $\hat{k}$'s, the following identity is useful

$$\hat{k}_l(x) = (-1)^l [\hat{i}_{-(l+1)}(x) - \hat{i}_l(x)] \qquad x \neq 0 \qquad (4.25)$$

In this way, all of the above four integrals can be numerically evaluated accurately even for very high quantum numbers. To give an example, we give below the resulting formula for one of the 27 integrals:

$$\int d\hat{r} \int_0^{r_A} dr\, r^2 \Phi_j^*(\alpha, \mathbf{A})\Phi_k(\beta, \mathbf{B}) = (4\pi)^2 (\alpha\beta)\, \gamma_{(vl)_j}\gamma_{(vl)_k}\, r_A^{2\nu_j + l_j}\, r_B^{2\nu_k + l_k}$$

$$\times \sum_{l_{0j}=0}^{\infty} \omega_{l_{0j}}\hat{k}_{l_{0j}}(\alpha r_A)\sum_{l_{1j}}\sum_{l_{2j}}\mathscr{C}_{(l_1 l_2)_j}^{l_j}\sum_{q_j}\left[\mathcal{Q}_{(vlq)_j}^{(l_1 l_2)_j} r_A^{q_j}\right]^{-1}$$

$$\times \sum_{l_1' l_2'}\sum B_{(l_0 l_1 l_2)_j}^{(l_1' l_2' l)_j}\sum_{l_{1k}}\sum_{l_{0k}=|l_{1k}-l_{1j}'|}^{l_{1k}+l_{1j}'}\omega_{l_{0k}}\hat{k}_{l_{0k}}(\beta r_B)$$

$$\times \sum_{l_{2k}}\mathscr{C}_{(l_1 l_2)_k}^{l_k}\Theta_{(l_1' l_2')_j(l_1 l_2)_k l_{0k}}^{(lm)_j(lm)_k}(\hat{r}_A, \hat{r}_B, \hat{r}_B)$$

$$\times \sum_{q_k}\Lambda_{l_{0j} l_{0k}}^{q_j + q_k + 2}(\alpha, \beta, 0, r_A)\bigg/\left[\mathcal{Q}_{(vlq)_k}^{(l_1 l_2)_k} r_B^{q_k}\right] \quad (4.26)$$

The sum on l_{0k} has an additional symmetry requirement that $(l_{1k} + l_{1j}' + l_{0k})$ be even. The summation limits for the rest of the sums can be inferred from Eqs. 4.12 and/or 4.13. Note that the form of Eq. 4.13 was used for the bra side & Eq. 4.12 was used for the ket side of the above integral, so that we could easily recognize where one of the open sums of the orbitals terminates. Thanks to the similarities of the ADTs for both ETO's, the remaining 26 integrals fall into the same pattern and can be worked out rather straightforwardly.

4.3.2 Numerical Example

To give a numerical example, let $\alpha = \beta = 1.0$, $r_A = (1.0, \frac{2\pi}{5}, \frac{\pi}{6})$ & $r_B = (2.0, \frac{3\pi}{5}, \frac{7\pi}{6})$. Table 4.1 shows the convergence of the integral V for $j = (3,2,1)$ & $k = (2,1,0)$. ε determines the size of the added term on the open sum compared to one. For a variable x_j that accumulates on the open sums, the code fragment would look like:

$$x_j \leftarrow x_j + t$$

$$\text{if } (|t/x_j| \leq \varepsilon) \text{ then exit (open loop)}$$

This way of numerically terminating the open loops gives us better control of the accuracy used to evaluate the integrals than, for example, terminating all 27 integrals on a fixed number of the loop count. δ is the relative error with respect to the last row and it is interesting to note how consistent it is with the anticipated accuracy, ε. Table 4.2 shows values of the same integral for different quantum numbers for $\varepsilon = 1.0 \times 10^{-16}$. A fortran 95 serial code has been written for numerically calculating V and the CPU times shown are from a 4 GB, 2.6 GHz Intel Core 2 Duo macbook pro laptop.

Table 4.1 Relative Convergence for $V_{(3,2,1)(2,1,0)}(1.0,1.0,\mathbf{r}_A,\mathbf{r}_B)$

Real part	Imaginary part	δ_{real}	δ_{imag}	ε^a	CPU time(sec)
−0.59873949437907303	0.34568240825421820	1.06	1.06	1.0E-01	2.032E-03
−0.29515476803702206	0.17040768477877541	1.40E-02	1.40E-02	1.0E-02	0.202
−0.29103936033692773	0.16803165303530088	1.78E-04	1.78E-04	1.0E-03	0.315
−0.29101610142056444	0.16801822449367737	2.58E-04	2.58E-04	1.0E-04	0.463
−0.29109857012531110	0.16806583782256260	2.58E-05	2.58E-05	1.0E-05	0.633
−0.29109047967150004	0.16806116679687697	1.97E-06	1.97E-06	1.0E-06	0.879
−0.29109112968151762	0.16806154208033555	2.66E-07	2.66E-07	1.0E-07	1.160
−0.29109104169200689	0.16806149127956788	3.64E-08	3.64E-08	1.0E-08	1.501
−0.29109105247787853	0.16806149750679378	6.64E-10	6.64E-10	1.0E-09	1.893
−0.29109105233626992	0.16806149742503609	1.77E-10	1.77E-10	1.0E-10	2.364
−0.29109105227864435	0.16806149739176596	2.09E-11	2.09E-11	1.0E-11	2.849
−0.29109105228522064	0.16806149739556273	1.67E-12	1.67E-12	1.0E-12	3.447
−0.29109105228465826	0.16806149739523804	2.62E-13	2.62E-13	1.0E-13	4.126
−0.29109105228474391	0.16806149739528750	3.26E-14	3.27E-14	1.0E-14	4.825
−0.29109105228473481	0.16806149739528223	1.34E-15	1.32E-15	1.0E-15	5.699
−0.29109105228473442	0.16806149739528201	0.00E+00	0.00E+00	1.0E-16	6.553

[a] Strictly speaking, ε was calculated with respect to the absolute value, but in this particular case, the real and imaginary part happened to be of the same order of magnitude, hence, both converged at a similar rate

Table 4.2 Numerical values for $V_{jk}(1.0,1.0,\mathbf{r}_A,\mathbf{r}_B)$ for $\varepsilon = 1.0 \times 10^{-16}$

$(nlm)_j$	$(nlm)_k$	Real part	Imaginary part	CPU time(sec)
1, 0, 0	1, 0, 0	0.34850947857504333	2.23342113801633579E-22	0.200
2, 0, 0	2, 0, 0	0.61735964776149899	5.49395794007445780E-19	1.053
2, 0, 0	2, 1, 0	0.11077236162560267	−1.39457735123553977E-20	1.527
2, 1, 0	2, 0, 0	−0.11077236162559734	−1.25792472295851408E-18	1.774
2, 1, 0	2, 1, 0	0.40809499372155028	1.68240824753530884E-20	2.542
2, 1, 1	2, 1, 0	0.11289934515219595	−6.51824673149524469E-02	2.650
2, 1, 0	2, 1, 1	0.11289934515219550	6.51824673149525441E-02	2.661
3, 0, 0	3, 0, 0	0.57895248073483607	2.14237420820933209E-19	6.506
3, 1, 0	3, 0, 0	−4.84528149260349261E-02	−1.98399696407086199E-18	10.265
3, 0, 0	3, 1, 0	4.84528149260701202E-02	1.65984230188867265E-18	7.900
3, 2, 1	3, 2, 1	0.26286186227813135	−5.23498551681719911E-17	14.322
3, 2, 0	3, 2, 0	0.33380515163527102	3.27871802188479513E-18	14.505
3, 2, −1	3, 2, −1	0.26286186227813230	5.13533725770497534E-17	14.344
3, 2, 1	2, 1, 0	−0.29109105228473442	0.16806149739528201	6.557

4.4   One-Center Overlap Integrals

The integral V can also be done by inserting a decomposition of unity (using Dirac's notation),

$$1 = \sum_i |\Psi_i(\beta,\mathbf{r})\rangle\langle S_i(\beta,\mathbf{r})| \tag{4.27}$$

so that, V is expressed in terms of two one-center overlap integrals as:

$$V_{jk}(\alpha,\beta,\mathbf{r}_A,\mathbf{r}_B) = \sum_i \langle \Psi_j(\alpha,\mathbf{A})|\Psi_i(\beta,\mathbf{r})\rangle\langle \mathbf{S}_i(\beta,\mathbf{r})|\Psi_k(\beta,\mathbf{B})\rangle. \tag{4.28}$$

All such one-center overlap integrals, will terminate using our new ADT. We have checked that the above equation numerically converges to the results shown on Table 4.2 as the size of the expansion increases. The second integral in Eq. 4.28 is particularly familiar – it is the famous Shibuya–Wulfman integral [2,6,7], which has a finite form in terms of the 9j symbols [7,8]. We not only have come up with a new way of calculating it, but also generalized it to all four combinations of one-center overlaps between Ψ & $\mathbf{S}$ and different screening parameters.

4.5 Conclusions

A new efficient two-range, point-wise convergent ADT for the CSs and the STOs, that is attractive both in notation and convergence, has been given. The numerical application given are the basic one- and two-center overlap integrals, which demonstrates its value as a first choice ADT for STOs in any relevant applications.

For the more complicated three- and four-center integrals, direct substitution of the Laplace expansion (Eq. 4.10), will clearly result in a multiple open sum expressions such that, numerical calculations will be computationally too slow to be of any realistic usage. However, the Poisson equation technique [9, 10] will avoid several of the open sums and allow for the efficient calculation of the three- and four-center integrals. Using these techniques, the calculation of the three- and four-center integrals is under investigation.

Acknowledgements This work was supported by the National Science Foundation of the United States under the CREST program's Florida A&M University Center for Astrophysical Science and Technology.

References

1. Weniger EJ (2000) Addition theorems as three-dimensional Taylor expansions. Int J Quantum Chem 76:280–295
2. Avery JA, Avery JO (2006) Generalized sturmians and atomic spectra. World Scientific, Singapore
3. Varshalovich DA, Moskalev AN, Khersonskii VK (1988) Quantum theory of angular momentum. World Scientific, Singapore
4. Arfken GB, Weber HJ (2005) Mathematical methods for physicists. Elsevier Academic, San Diego
5. Press WH, Teukolsky SA, Vetterling WT, Flannery BP (1992) Numerical recipes in Fortran. Cambridge University Press, New York

6. Red E, Weatherford CA (2004) Derivation of a general formula for the Shibuya–Wulfman matrix. Int J Quantum Chem 100:208–213
7. Judd BR (1975) Angular momentum theory of diatomic molecules. Academic, New York
8. Aquilanti V, Caligiana A (2002) Sturmian approach to one-electron many-center systems: integrals and iteration schemes. Chem Phys Lett 366:157–164
9. Weatherford CA, Red E, Hoggan PE (2005) Solution of Poisson's equation using spectral forms. Mol Phys 103:15–16, 2169–2172
10. Weatherford CA, Red E, Joseph D, Hoggan P (2006) Poisson's equation solution of Coulomb integrals in atoms and molecules. Mol Phys 104:9, 1385–1389

Chapter 5
Why Specific ETOs are Advantageous for NMR and Molecular Interactions

Philip E. Hoggan and Ahmed Boufergyène

Abstract This paper advocates use of the atomic orbitals which have direct physical interpretation, i.e., Coulomb Sturmians and hydrogen-like orbitals. They are exponential type orbitals (ETOs). Their radial nodes are shown to be essential in obtaining accurate local energy for Quantum Monte Carlo, molecular interactions a nuclear and shielding tensors for NMR work. The NMR work builds on a 2003 French PhD and many numerical results were published by 2007. The improvements in this paper are noteworthy, the key being the actual basis function choice. Until 2008, their products on different atoms were difficult to manipulate for the evaluation of two-electron integrals. Coulomb resolutions provide an excellent approximation that reduces these integrals to a sum of one-electron overlap-like integral products that each involve orbitals on at most two centers. Such two-center integrals are separable in prolate spheroidal co-ordinates. They are thus readily evaluated. Only these integrals need to be re-evaluated to change basis functions. In this paper, a review of more recent applications to ETOs of a particularly convenient Coulomb resolution in QMC work is illustrated.

5.1 Introduction

The criteria for choice between gaussian and exponential basis sets for molecules do not seem obvious at present. In fact, it appears to be constructive to regard them as being complementary, depending on the specific physical property required from molecular electronic structure calculations.

P.E. Hoggan (✉)
LASMEA, CNRS and Université Blaise Pascal, Aubière, France
e-mail: pehoggan@yahoo.com

A. Boufergyène
University of Alberta, Edmonton, Canada
e-mail: ahmed.bouferguene@ualberta.ca

P.E. Hoggan et al. (eds.), *Advances in the Theory of Quantum Systems in Chemistry and Physics*, Progress in Theoretical Chemistry and Physics 22, DOI 10.1007/978-94-007-2076-3_5, © Springer Science+Business Media B.V. 2012

The present work describes a breakthrough in two-electron integral calculations, as a result of Coulomb operator resolutions. This is particularly significant in that it eliminates the arduous orbital translations which were necessary until now for exponential type orbitals. The bottleneck has been eliminated from evaluation of three- and four- center integrals over Slater type orbitals and related basis functions.

The two-center integrals are replaced by sums of overlap-like one-electron integrals. This implies a speed-up for all basis sets, including gaussians. The improvement is most spectacular for exponential type orbitals. A change of basis set is also facilitated as only these one-electron integrals need to be changed. The gaussian and exponential type orbital basis sets are, therefore interchangeable in a given program. The timings of exponential type orbital calculations are no longer significantly greater than for a gaussian basis, when a given accuracy is sought for molecular electronic properties.

Atomic orbitals are physically meaningful one-electron atom eigenfunctions for the Schrödinger equation. This gives well-known analytical expressions: hydrogen-like orbitals.

Boundary conditions allow the principal quantum number n to be identified as the order of the polynomial factor in the radial variable. It must therefore be positive and finite. It is also defined such that $n - l - l$ is greater than or equal to 0. This gives the number of zeros of the polynomial (radial nodes). Here, $l = 0$ or a positive integer, which defines the angular factor of the orbital. (i.e., a spherical harmonic, or, more rarely, its Cartesian equivalent) The number n gives the energy of the one-electron atomic bound states. Frequently, basis set studies focus on the radial factor.

Certain physical properties, such as NMR shielding tensor calculations directly involve the nuclear cusp and correct treatment of radial nodes, which indicates that basis sets such as Coulomb Sturmians are better suited to their evaluation than gaussians [4, 16, 33].

There is also evidence to suggest that CI expansions converge in smaller exponential basis sets compared to gaussians [22, 72]. Benchmark overlap similarity work is available [5, 12]

5.2   Wave-Function Quality

To test wave-function quality, the following quantity must be smooth.

It is, to varying degrees, in different basis sets.

$$-1/2 \frac{\nabla \rho(r)}{\rho(r)}$$

Atomic positions must give cusps. The importance for stable and accurate kinetic energy terms, particularly in DFT.

Much molecular quantum chemistry is carried out using gaussian basis sets and they are indeed convenient and lead to rapid calculations. The essential advantage was the simple product theorem for gaussians on two different atomic centers. This allows all the two-electron integrals, including three- and four- center terms to be expressed as single-center two-electron integrals. The corresponding relationship for exponential type orbitals generally led to infinite sums and the time required, particularly for four-center integrals could become prohibitive.

Recent work by Gill has, nevertheless, been used to speed up all three and four center integral evaluation, regardless of basis using the resolution of the Coulomb operator [27, 35, 66]. This reduces the three- and four- center two-electron integrals to a sum of products of overlap-like (one-electron) integrals, basically two-centered. This algorithm was coded in a Slater type orbital (STO) basis within the framework of the STOP package [8] (in fortran) during summer 2008. Note, however, that other exponential or gaussian basis sets can readily be used. The set of one-electron overlap-like auxiliary integrals is the only calculation that needs to be re-done. They may be re-evaluated for the basis set that the user selects for a given application. This procedure makes the approach highly versatile. A modular or object oriented program is available to do this efficiently [35, 47, 48].

The layout is as follows: the review begins with a brief recap of basis sets and programming strategy in the next two sections. Atom pairs are the physical entity used for integral evaluation, both in the Poisson equation technique and the Coulomb resolution. A case study of molecular interaction by Quantum Monte Carlo simulation, using ETOs is followed by accurate NMR chemical shift evaluation.

5.3 Basis Sets

Although the majority of electronic quantum chemistry uses gaussian expansions of atomic orbitals [10, 11], the present work uses exponential type orbital (ETO) basis sets which satisfy Kato's conditions for atomic orbitals: they possess a cusp at the nucleus and decay exponentially at long distances from it [39]. It updates work since 1970 and detailed elsewhere [3, 6, 15, 18, 28, 31, 42, 51, 52, 57, 60, 62].

Two types of ETO are considered here: Slater type orbitals (STOs) [58, 59] and Coulomb Sturmians and their generalisation, which may be written as a finite combination thereof [69]. Otherwise, STOs may be treated as multiple zeta basis functions in a similar way to the approach used with gaussian functions.

Many exponential type functions exist [69]. Preferential use of Sturmian and related functions with similar radial nodes is discussed [35].

Coulomb Sturmians have the advantage of constituting a complete set without continuum states because they are eigenfunctions of a Sturm-Liouville equation involving the nuclear attraction potential i.e., the differential equation below.

$$\nabla_{\mathbf{r}}^2 \mathscr{S}_{nl}^m(\beta, \mathbf{r}) = \left[\beta^2 - \frac{2\beta n}{r} \right] \mathscr{S}_{nl}^m(\beta, \mathbf{r}).$$

The exponential factor of Coulomb Sturmians; $e^{-\beta r}$ has an arbitrary screening parameter β. In the special case when $\beta = \zeta/n$ with n the principal quantum number and ζ the Slater exponent, we obtain hydrogen-like functions, which do not span the same space and require inclusion of continuum states to form a complete set [69]. Hydrogen-like functions are, however well known as atomic orbitals: the radial factor contains the associated Laguerre polynomial of order $2l+1$ with suffix $n-l-1$ and the exponential $e^{-\zeta r/n}$ as indicated above. The angular factor is just a spherical harmonic of order l. These functions are ortho-normal. The optimal values of the β parameters may be determined analytically by setting up secular equations which make use of the fact that the Sturmian eigenfunctions also orthogonalise the nuclear attraction potential, as developed by Avery [2].

$$ \int \mathscr{S}_{nl}^m(r,\theta,\phi)\,\mathscr{S}_{n'l'}^{m'}(r,\theta,\phi)\,\frac{dr}{r} = \delta_{nn'll'mm'}. $$

These functions are further generalised by varying α from the Coulomb Sturmian value of 1. In such a case, the basis remains ortho-normal and othogonalises a/r^α. This eliminates the r^2 term, arising for quadrupole moments when $\alpha = -2$, thus confirming the very recent numerics by Guseinov's group [29]. Similarly, it would be expected that $\alpha = -1$ ETOs would constitute the optimal basis for magnetic dipole integrals of NMR shielding. Furthermore, a negative value of α will not modify the number of radial nodes: the functions will simply breath.

Recently, a physical interpretation of α was given by Guseinov. It is shown that the Lorentz friction of electrodynamics gives rise to an additional potential term in the Schrödinger equation for atoms. Interestingly, this term is zero when $\alpha = 1$, so that this special case reduces to the Coulomb potential and the Sturm Liouville equation defining Coulomb Sturmians is simply obtained. Otherwise, an additional term, depending on orbital angular momentum, represents the 'drag' on the electron by moving within the field of the nucleus (Guseinov submitted to JTCC 2010).

Definitions: the generalised exponential functions constitute finite complete orthonormal sets. Their expression is as follows:

$$ \chi_{nlm}(\mathbf{r}) = \left[\frac{(-1)}{\sqrt{2n}}\right]^\alpha N_{nl}L_{n-l-1-\alpha}^{2l+2-\alpha}(2\zeta r)\,r^l e^{-\zeta r}Y_l^m(\theta,\phi) \qquad (5.1) $$

Here, N is the normalisation constant previously obtained for Coulomb Sturmians, L is the associated Laguerre polynomial of order $2l+2-\alpha$ with suffix $n-l-1-\alpha$ (recall that $\alpha = 1$ defines the Coulomb Sturmians.

Define a variable including the screening constant:

$$ x = 2\zeta r $$

Subsequently, rewriting the norm as $N(\alpha)_{nl}$ and introducing $p = 2l+2-\alpha$ and $q = n+l+1-\alpha$ gives the simplified expression for the generalised orthonormal basis sets of ETO, used by Guseinov.

$$ \chi_{nlm}(\mathbf{x}) = N(\alpha)_{nl}L_q^p(x)\,r^l e^{-x/2}Y_l^m(\theta,\phi) \qquad (5.2) $$

In past applications, no obvious advantage has been evidenced for the functions with negative α indices over the well-known Slater type orbitals, Coulomb Sturmians ($\alpha = 1$) or Shull–Loewdin functions ($\alpha = 0$). In fact, the infinite series arising when Hartree–Fock two-electron integrals that do not possess closed forms (three and four center terms) are evaluated converge much more slowly when the negative α functions are used. This has recently also proven to be the case of a set of electric field integrals [73]. This paper records the precedent of electric quadrupole integrals, already published by Guseinov's colleagues, where the negative α basis converges as well as (if not better than) the STO [29] and presents a new application to the dipole integrals in the NMR experimental setup.

The investigations are extended to comparisons with previous work on the nuclear dipole integrals that are so important to the evaluation of nuclear shielding tensors and NMR chemical shifts. Furthermore, these nuclear magnetic dipole integrals are closely related to the one-electron nuclear attraction integral, required in all Hartree–Fock and DFT work.

In the case of electric quadrupole integrals, accounting for the Lorenz drag and the r^2 orthogonalisation in these integrals, favor the $\alpha = -2$ ETO basis functions, as predicted graphically from numerical results in [29]. In an analagous manner, the dipole term in the perturbative treatment of NMR Shielding tensors favors the use of $\alpha = -1$ ETO eigenfunction basis sets, as illustrated in the Results tables of the present work.

Alternative ETOs would be Slater type orbitals and B-functions with their simple Fourier transforms. Strictly, they should be combined as linear combinations to form hydrogen-like or, better, Sturmian basis sets prior to use.

STOs allow us to use routines from the STOP package [8] directly, whereas Coulomb Sturmians still require some coding. The relationship to STOs is used to carry out calculations over Coulomb Sturmian basis with STOP until the complete Sturmian code is available. The present state-of-the-art algorithms require at most twice as long per integral than GTO codes but the CI converges with fewer functions and the integrals may be evaluated after gaussian expansion or expressed as overlaps to obtain speed up [68].

After a suitably accurate electron density has been obtained for the optimized geometry over a Coulomb Sturmian basis set, the second order perturbation defining the nuclear shielding tensor should be evaluated in a Coupled perturbed Hartree Fock scheme.

The integrals involved may conveniently be evaluated using B-functions with linear combinations giving the Coulomb Sturmians.

$$\mathscr{S}_{nl}^{m}(r) = (2\alpha)^{3/2} \frac{2^{2l+1}}{(2l+1)!!} \sum_{l=0}^{n-l-1} \frac{(-n+l+1)_t \, (n+l+1)_t}{t! \, (l+3/2)_t} B_{t+1,l}^{m}(r)$$

The techniques exploit properties of Fourier transforms of the integrand.

Note that either HF or DFT can serve as zero order for the present nuclear shielding tensor calculation over ETOs.

A full ab initio B-function code including nuclear shielding tensor work is expected to be complete shortly.

Some tests show that Slater type orbitals (STO) or B-functions (BTO) are less adequate basis functions that Coulomb Sturmians, because only the Sturmians possess the correct nuclear cusp and radial behavior.

5.4   Two-Center Integrals and Inter-Molecular Interactions

Asymptotic Coulomb and exchange integral decay is $1/R$ and exponentially.

The short range form involves powers of R as a factor of exponentials [50].

The total energy obtained for the isolated H_2 molecule in a quadruple-s basis is -1.1284436 Ha as compared to a Hartree–Fock limit estimate of -1.1336296 Ha. Nevertheless, the Van der Waals well, observed at 6.4 a.u. with a depth of 0.057 kcal/mol is quite reasonably reproduced [32].

Dimer geometry: rectangular and planar. Distance between two hydrogen atoms of neighboring molecules: 6 a.u. Largest two-center integral between molecules: $4.162864 \cdot 10^{-5}$. (Note that this alone justifies the expression dimer-the geometry corresponds to two almost completely separate molecules, however, the method is applicable in any geometry).

5.4.1   Intermolecular Interaction for ETOs: A Case Study

Exponentially decaying orbital are required for accurate representation of the atom-atom interactions involved in molecular adsorption on a solid surface. The present application involves CO adsorption on copper. This modifies the carbon partial charge so that it becomes the seat of nucleophilic attack.

The plane wave basis used as the basis for 2-D periodic solid wave-functions is approximated by localised B-splines, that can also describe the exponential type orbitals of the molecules.

The Cu (001) surface is exposed. This truncation of the bulk lattice, as well as adsorption, leads to drastic changes in electronic correlation. They are not adequately taken into account by density-functional theory (DFT). A method is required that gives almost all the electronic correlation. The ideal choice is the quantum Monte Carlo (QMC) approach. In variational quantum Monte Carlo (VMC) correlation is taken into account by using a trial many-electron wave function that is an explicit function of inter-particle distances. Free parameters in the trial wave function are optimised by minimising the energy expectation value in accordance with the variational principle. The trial wave functions that used in this work are of Slater–Jastrow form, consisting of Slater determinants of orbitals taken from Hartree–Fock or DFT codes, multiplied by a so-called Jastrow factor that includes electron pair and three-body (two-electron and nucleus) terms.

A second, more accurate step, takes the optimised Jastrow factor as data and carries out a diffusion quantum Monte Carlo (DMC) calculation, based on transforming the time-dependent Schrödinger equation to a diffusion problem in imaginary time. An ensemble in configuration space is propagated to obtain a highly accurate ground state.

Carbon monoxide adsorption was considered in a preliminary DMC study showing preferential bridged adsorption.

The QMC calculations will be carried out using CASINO. This shows good linear scaling up to at least 100,000 processors.

5.5 Methods

5.5.1 Defining the Model System

Physically, a slab of copper is defined and oriented to expose the (100) surface, perpendicular to the z-axis. This surface was shown to be active towards oxidation and CO adsorption in previous DFT (Density Functional Theory) work on the molecule-metal surface interface, at the Pascal Institute [43, 46]. It was also shown to re-arrange its geometry to a certain extent to minimise energy and this nano-structuring was intimately related to the presence of the adsorbed oxygen or CO whilst generating specific adsorption sites [30].

In this work, the copper lattice is truncated by a planar surface initially. The wave-function is generated on a plane-wave basis, in the first Brillouin zone of reciprocal space, using the Monkhorst-Pack grid of k-points. The PBE GGA functional is chosen (the Perdew Burke Ernzerhof functional, from 1996 works within the Generalised Gradient Approximation (GGA) where the energy is a functional of the density and its gradients [45], because of its non-empirical nature, thus enabling ab initio DFT calculations to be carried out. In order to reduce the number of active electrons (to 54 per super-cell, for a two-mesh thick slab), the data-bank of pseudo-potentials pre-calculated over grids of k-points from the Fritz Haber Institute is used [30]. An argon core is used for copper, restricting the active electrons to 11 per atom.

5.5.2 Generating the Trial Wave-Function

The trial wave-function is generated as plane-wave output from a DFT-PBE calculation carried out using the ABINIT freeware (see www.abinit.org) [34].

5.5.3 Variation Monte Carlo

This input is then used for the CASINO programme which is a Quantum Monte Carlo code adapted to solid state work. It is converted into a B-spline basis, which has negligible influence on the accuracy and speeds up the Quantum Monte Carlo calculations considerably. These are cubic splines limited to radius 2 (i.e., $f(u) = 0$ if $r = |u| \geq 2$) of the form:

$$f(u) = 1 - 3/2u^2 + 3/4|u|^3. \quad r \geq 0, r \leq 1 \tag{5.3}$$

$$f(u) = 1/4(2 - |u|)^3. \quad r \geq 1, r \leq 2 \tag{5.4}$$

for all three axes. See [1].

A preliminary Variational Monte Carlo (VMC) calculation is carried out in order to generate several thousand configurations (instantaneous points in the direct space of the electrons) e.g., 5,000–10,000 configurations/core. VMC is driven by energy minimisation.

Electron correlation is introduced via a Jastrow factor which can be optimised by Variational Monte Carlo methods.

This optimisation procedure generates a correlation.data file containing the optimised numerical parameters for the electron–electron and electron pair-nuclear contributions to the Jastrow factor.

A final VMC calculation generates the initial configurations required for the Diffusion Monte Carlo step (DMC) also 5,000–10,000 per core, typically.

5.5.4 Diffusion Monte Carlo

Linear scaling tests and memory requirement (memory sharing of a little less than 2Mo by four cores for the present application) adapting CASINO specifically on Bluegene were completed successfully.

5.6 Application

5.6.1 Adsorption of Carbon Monoxide on Copper Surfaces

Quantum Monte Carlo (QMC) simulations are used to describe small molecules adsorbed on a Cu(001) surface. The surface presents 2-D periodicity and the molecules interact with both the surface and each other.

Carbon monoxide (CO) adsorption was considered. The strong bonding within this molecule is weakened by interactions with the surface, making the molecule easier to attack by nucleophilic molecules. The carbon monoxide molecule is polarised by its adsorbtion on the surface (gaseous CO has a negatively charged carbon, whereas once adsorbed it can present the typical partial positive charge of a carbonyl carbon which is the site of nucleophilic attack). Investigation of this phenomenon is followed by a study of reaction dynamics with the hydride as model nucleophile before protonation.

Removal of toxic carbon monoxide molecules is a typical depollution reaction, of interest in catalytic exhausts. Furthermore, the reaction with water is of industrial importance in producing clean fuels (hydrogen gas) in a sustainable process., whereas that with hydrogen, via hydride attack and addition of the $H+$ counter-ion produces formaldehyde. Formaldehyde, i.e., H_2CO is a valuable molecule for organic synthesis and solvation.

5.6.2 Adsorbed CO with Ionised H_2: Hydride Model Nucleophile

To follow the model reaction, with adsorbed CO under nucleophilic attack from a hydride (H–) ion, the set of geometries for CO approaching a Cu(100) surface is obtained by DFT plane-wave calculations using the PBE functional.

Initially, the interaction is repulsive: the partial negative charge on carbon closer to the surface than the oxygen repels mobile electrons, causing repulsion with positive charges induced at the metal surface. Closer to the surface, the carbon acquires a partial positive charge and the system reaches an equilibrium, after some electron transfer to the metal (about 0.1e) and a slight stabilisation occurs.

As seen from previous work, hydrogen is dissociated at the surface and remains available by diffusion [30, 46].

Examining the electron distribution indicates that the gas-phase mechanism is of free-radical type: $H. + CO \rightarrow HCO$. and, since the hydrogen radical comes from (e.g., photo-induced) homolytic fission of H_2, the reaction terminates by radical addition:

$$HCO. + H. \rightarrow H_2CO$$

A similar reaction is already the subject of study by kinetic Monte Carlo techniques (kMC) on Ni(100), where it has been observed experimentally. Here, the aim is selectively to produce formaldehyde in an addition reaction.

This work is conducted with A Bouferguène, using a model surface and kMC techniques to study the kinetics of the following reaction (manuscript in press: Int. J. Mod. Phys. C 2011):

$$H_2 \rightarrow H + H + CO \rightarrow HCHO.$$

Note, however, that the H_2 bond is known to participate in ring-systems, in which the energy difference between atom-centered radical and ionic species is zero. A good DMC value for the total H_2 energy is -1.07757 a.u. [64].

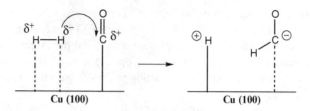

Fig. 5.1 Stretched surface H_2 giving hydride ion as model nucleophile to attack adsorbed CO

Charge transfer to the Cu(100) surface is necessary to polarise the CO molecule, adsorb it and give it the "organic" carbonyl carbon, with partial positive charge. Energy criteria suggest a *concerted* mechanism, with the H_2 bond stretching and also being polarised, in the vicinity of the co-adsorbed CO. This will inevitably lead to partial charges, since the energetic cost of forming a hydride ion is prohibitive. Nevertheless, this species is a model nucleophile and it is to be expected that a hydride species, with partial negative charge would rapidly form the $HCO-$ intermediate. Since this intermediate is so much stabler on the Cu(100) surface than HCO., it is assumed that the rate limiting step is the initial concerted bond-stretch and charge re-arrangement (shown in the scheme and Fig. 5.1 below):

$$H_2 + CO \rightarrow H^+ - -H^- + C^{\delta+}O \rightarrow HCO^- + H+$$

All the molecules involved have near-Hartree–Fock-limit wave functions available from the STOP (QCPE 667) [8] code using basis sets due to I. Ema [23], which are triple-zeta STOs with one d and one f orbital on O and C. The works on water and CO_2 are referenced below, and those on diatomics are unpublished tests [14, 54]. An extensive QMC study showing improvements obtained using STO was recently carried out on Bluegene calculators [63].

The geometry of reaction intermediates and transition states, provided by DFT/plane-wave calculation may be used as input geometry to QMC/blip basis set calculations.

Preliminary results have successfully shown the initial repulsion and de-stabilisation of CO adsorbed on the Cu (100) surface. Skin depth effects are limited to a two-layer copper slab, with periodic boundary conditions in two dimensions. The CO molecule transfers electronic charge to the surface and subsequently presents a partial positive charge on the carbon atom, near the copper surface. This has then be shown to provide a site for nucleophilic attack by the hydride ion (H–). This adsorbed reaction has been compared to its gas-phase counterpart and shown to have a lower activation barrier. All these systems require a DMC fixed node calculation to obtain sufficiently low QMC variance energies to argue reliably the case for a surface catalyst effect. Other nucleophiles are now being considered.

As a result, we advocate the use of plane-wave or blip [1] basis sets because they are easy to manipulate numerically and increase in number to improve accuracy. They are exponential type atomic orbitals with complex exponents. They may be used to represent the analytical orbitals over a numerical grid to arbitrary accuracy.

5.6.3 ABINIT Adsorbtion Calculations

Complete CO adsorption calculations were carried out using a slab with six layers of copper at the bulk geometry, oriented to expose the Cu(001) surface perpendicular to the z-axis, and periodic within the slab. Co-ordinates of the molecule were allowed to vary around a distance of 6 a.u. from the C and O atoms to the surface (i.e., little overlap). This slab size has been shown to be the smallest that adequately converges the physical properties under investigation in DFT-PBE studies. CO coverage was one molecule per face.

The molecules were systematically polarised by the presence of the surface, with the bond length increasing by about one third during geometry optimisations.

Copper atoms are modeled using argon core Troullier-Martins pseudopotentials generated using the utility provided by FHI Berlin [46].

The other atoms also use this type of pseudopotential for the 1s cores.

The system size is thus about $N = 4 \times 11 + 2 = 46$ in all for the CO+H_2 adsorbed system.

An $8 \times 8 \times 1$ Monkhorst-Pack grid converges the surface-formation and adsorption energies to better than micro Hartree accuracy in DFT work. In QMC work, the cells need to be limited, to a single **k** point, or at most a $2 \times 2 \times 1$ grid. This limitation is not too severe for the properties sought. A high kinetic-energy cutoff of 60 a.u. is required for micro Hartree accuracy.

These DFT calculations take 20–30 core hours on a local PC grid (32 cores).

DFT is inadequate because the optimal orientation and adsorption energy for CO vary with the choice of (exchange-correlation) functional. In particular, there is no guarantee of which leads to an improvement, LDA included.

A 2-D periodic system in a plane-wave basis is adopted. This approach can be initialised with a trial wave function generated by the ABINIT code. Some test calculations on CO have been successfully completed, suggesting a very stringent error margin should be used (corresponding to interactions involving a pair of weak bonds at a total of 0.2–0.4 eV). A tolerable error margin (about 5% of the smallest interactions involved) for this is 15 meV.

5.6.4 The Need for QMC and Its Originality

This rather modest number of carefully selected electrons allows an accurate wave-function representation of the complex system involved, thus lending itself to the use of QMC to account accurately for electron correlation. To our knowledge, this is the first time that QMC will have been used to study a reaction taking place on a surface.

Since the software shows linear scaling within the memory requirements of Bluegene, runs on 2048 cores have become routine.

5.6.5   Preliminary Results

In fact, preliminary QMC runs have shown that the "atop" mode of mono-dentate adsorption is at 1.43 eV and the bi-dentate, bridged state at 1.55 eV. These appear to be significant differences and the values should be distinguished, both led to dipole reversal for CO (cf plane-wave DFT calculations).

Energetics:

HCO. (gas phase) $-22.022551238664 +/- 0.0075780$ a.u.

CO (gas phase) $-21.489665250511 +/- 0.0003418$ a.u.

Gas phase barrier: 86.2 kJ/mol.

Note that this assumes a collision occurs in the gas phase. The collision dynamics has not been investigated here. In the adsorbed phase, the molecules may be physisorbed and consequently fixed in the vicinity of the surface. Reagent molecule co-adsorption is considered, in which the model nucleophile and CO are adsorbed at the surface and close to *each other*. This is by no means unlikely, since CO tends to adsorb strongly at temperatures up to a few hundred K, whereas the hydrogen atoms diffuse readily. See [30].Below the DMC standard error (se) are considered.

Clean 2-D periodic slab exposing Cu(100) two unit mesh thick: -204.8894 se 0.00187.
Slab with co-adsorbed hydrogen and CO: $-227.440789886210 +/- 0.006013$
Stabilisation, compared to HCO.+H. (gas phase): 75.5 kJ/mol.
Barrier height, compared to $H_2 + CO$ (gas phase): 40.0 kJ/mol (min).

Therefore, some evidence has been accumulated for the Cu(100) surface stabilising activated species and thus acting as a heterogeneous catalyst. The key issue is that gas phase CO is inert towards nucleophiles (but may react by a radical pathway) and that the rate limiting step appears to be concerted H_2 bond stretch, accompanying the hydride attack on the adsorbed CO carbon, that is closer to the surface than the oxygen and thereby acquires a partial positive charge.

These steps are all reversible and more evidence is required regarding the transition state energy and structures.

Therefore, more extensive work is required, both on setting up the pseudo-potentials and above all on DMC simulations for Cu(100) slabs as large as resources will allow and carried out for the time required to obtain variance compatible with differentiating these values that are only about 0.15 eV apart. This would appear to be the smallest significant energy difference in the process, as the addition of hydrogen-bearing molecules will result in weak interactions of the same order of magnitude.

The implementation of plane-wave and blip basis sets in the CASINO program simply requires it to read output from, e.g., ABINIT. Hence ABINIT is the obvious choice for these applications.

5.7 Numerical Methods and Algorithms Used

Two QMC methods are used: variational quantum Monte Carlo and diffusion quantum Monte Carlo (VMC and DMC). In the VMC method, the expectation value of a quantum mechanical operator is taken with respect to a trial many-body wave function. The integration is performed using a Monte Carlo method, with the Metropolis algorithm being used to sample electronic configurations distributed as the square of the trial wave function. Free parameters in the trial wave function can be optimised by minimising the energy expectation value in accordance with the variational principle. In the DMC method the ground-state component of the trial wave function is projected out by solving the Schrödinger equation (SE) in imaginary time. This is accomplished by noting that the imaginary-time SE is a diffusion equation in the $3N$-dimensional space of electron coordinates, with the potential energy acting as a source/sink term. The imaginary-time SE can therefore be solved by a combination of diffusion and branching/dying processes. The introduction of importance sampling using the trial wave function transforms the problem into one involving drift as well as diffusion, but greatly reduces the population fluctuations due to the branching/dying process. The Fermionic antisymmetry of the wave function has to be maintained by constraining the nodal surface to equal that of the trial wave function. Reference [24] provides an overview of the VMC and DMC methods.

The VMC and DMC methods are implemented in the CASINO code [44] (version 3.0), with which we intend to carry out this project. CASINO is written in Fortran 95 and parallelised using MPI with a master/slave program model. CASINO has been in existence for more than thirteen years and has been used on a wide variety of high-performance computer platforms. There are 360 registered users of the code. CASINO requires only the MPI library.

In these calculations, our trial wave function will be of Slater–Jastrow form. The Slater determinants will contain orbitals taken from density functional theory (DFT) calculations. The Jastrow factor is an explicit function of electron–electron distance, enabling a highly accurate and compact description of electron correlation. The Jastrow factor consists of polynomial expansions in electron–electron separation, electron–nucleus separation, in which the polynomial expansion coefficients are optimisable parameters [21]. These parameters were determined by minimising the VMC energy.

The DMC method will then be used to determine a highly accurate value for the ground-state energy. The computational effort required by the DMC calculations is very much greater than that required by the preliminary DFT or VMC calculations, dominating contribution to the computational effort involved.

By calculating DMC energies for various DFT-generated atomic configurations, the height of the energy barriers were obtained, for a model reaction taking place on the copper surface. The physics underpinning the chemical process of catalysis was described.

The initial interaction energy from DFT work is 0.1 a.u., i.e., 2.7 eV, but differences as low as 0.1–0.2 eV may be significant, corresponding to weak (hydrogen bonding or polarisation) interactions. This is the value retained as threshold for significant energy differences to be resolved by QMC.

The statistical error bar Δ on the QMC total energy must be small compared with the energy difference to be resolved. Assuming the cost of the equilibration phase of a QMC calculation is negligible, the statistical error bar falls off as $1/\sqrt{T}$, where T is the computational effort in core-hours. The computational cost to achieve a given error bar increases as N^3 [24]. Hence the error bar in eV can be estimated as

$$\Delta = c\sqrt{N^3/T}, \tag{5.5}$$

The pre-factor c in Eq. 5.5 was determined from DMC studies of copper clusters [37], giving $c = 0.00126$.

The orbitals were initialised using DFT in a plane-wave basis, then re-represent them in a blip (B-spline) basis for the QMC calculations, in order to improve the system-size scaling of QMC [1]. The blip coefficients and plane-wave coefficients are related by a Fourier transform. The number of blip coefficients is usually somewhat larger than the number of non-zero plane-wave coefficients, in order to make the blip grid finer in real space.

5.7.1 The NMR Nuclear Shielding Tensor

More complete work is referred to here and the present description is a brief summary [19, 20, 38]. In NMR, the nuclear shielding tensor is a second order perturbation energy correction, for derivatives with respect to the nuclear dipole moment and the external field.

The perturbed Fock matrix element when including the effect of the external field contains both one and two electron terms. In this example, we focus on the one electron terms.

The purpose of the present section is to give a case study of one of the contributing energy integrals involving the dipole $1/r_N^3$ operator.

In the applied magnetic field, the question of gauge invariance must be resolved. A method of circumventing the problem was devised by Ditchfield using the London GIAO [41]. These Gauge Including Atomic Orbitals reduce to STO for zero field and contain the required phase factor otherwise [19, 20, 67].

The integrals were evaluated for GTO at zero field and nuclear shielding tensor or chemical shifts have been available since Gaussian 72 based on this pioneering work [31] and distributed to academics by QCPE. It is nevertheless important that users input the appropriate structure in order to obtain accurate chemical shifts corresponding to the species studied and note that for work in solution (or in solids) some structural changes may occur.

Define the nuclear shielding tensor as a second order energy perturbation:

$$\sigma_{\alpha\beta}^N = \left[\frac{\partial^2 \langle 0| \mathscr{H}(\boldsymbol{\mu}_N, \mathbf{B}_0)|0\rangle}{\partial \mu_{N,\alpha} \partial B_{0,\beta}} \right]_{\mu_N = 0,\, \mathbf{B}_0 = 0} \tag{5.6}$$

with μ_N the nuclear dipole moment of nucleus N and $\mathbf{B}_0$ the external field. $|0\rangle$ is a closed shell ground state Slater determinant. α and β stand for cartesian coordinates.

A coupled Hartree–Fock treatment of the above equation leads to [19, 20, 61]:

$$\sigma_{\alpha\beta}^N = Tr \left[P_\beta^{(0,1)} h_\alpha^{(1,0)} + P^{(0)} h_{\alpha\beta}^{(1,1)} \right] \tag{5.7}$$

where $P^{(0)}$ and $P_\beta^{(0,1)}$ are the density matrix of zero order and first order with respect to the external magnetic field, $h_\alpha^{(1,0)}$ is the core Hamiltonian of the first order with respect to nuclear dipole moment and $h_{\alpha\beta}^{(1,1)}$ is the second order one-electron Hamiltonian with respect to the nuclear moment μ_α and the external field B_β. The non-zero orders in (5.7) involve integrals which are absent from ab initio Hartree–Fock calculations. In this work, we focus our attention on integrals involving $1/r_N^3$ in their operator. These integrals appearing in the second order expression for the approximate perturbed Hamiltonian:

$$h_{\mu\nu,\alpha\beta}^{(1,1)} = \frac{\mu_0}{4\pi} \frac{e^2}{2m_e} \left\langle \chi_\mu \left| \frac{\mathbf{r}_\nu \cdot \mathbf{r}_N \delta_{\alpha\beta} - r_{\nu,\alpha} r_{N,\beta}}{r_N^3} + \frac{(\mathbf{R}_{\mu\nu} \wedge \mathbf{r})_\beta (\mathbf{r}_N \wedge \nabla)_\alpha}{r_N^3} \right| \chi_\nu \right\rangle \tag{5.8}$$

The integral which we have chosen to investigate in detail within the Fourier transform approach, is the three-center one electron integral:

$$I = \left\langle \chi_\mu \left| \frac{\mathbf{r}_\nu \cdot \mathbf{r}_N \delta_{\alpha\beta} - r_{\nu,\alpha} r_{N,\beta}}{r_N^3} \right| \chi_\nu \right\rangle \tag{5.9}$$

here $\mathbf{r}_N$ is the instantaneous position of the electron with respect to the nuclei N.

Analytical treatment: [67]

The algorithm is available in Fortran, within the STOP (Slater Type Orbital Package) set of programs, at the coupled perturbed Hartree–Fock level with the ETOs expanded in Slater type orbitals.

DFT coding proves more accurate for NMR chemical shifts because it accounts for the majority of the electron correlation energy. In this case, the ETOs are fitted to large Gaussian expansions, following the algorithm in [49] and Gaussian 03 is subsequently used.

5.7.1.1 Application

The ^{15}N chemical shifts measured for a set of benzothiazoles are evaluated with the above expressions. These molecules possess a ring nitrogen and have been studied previously in our group [67]. The measurements were made in natural

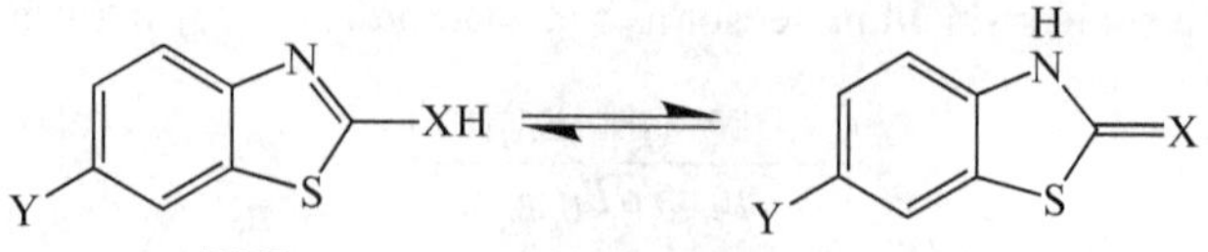

XBT   PXBT

Fig. 5.2 Tautomerism protonating the nitrogen investigated by NMR

Table 5.1 Numerical values of chemical shifts for tautomer structures

Molecule	Substituent	a	b	c
BT:benzothiazole:	No X	−72.5	−71.8	−61.4
OBT:	X=O	−238.8	−238.9	−133.3
OHOBT:	X=O; Y=OH	−240.4	−239.9	−135.3
ABT:	X=NH	−153.1	−152.1	−131.6
OHABT:	X=NH; Y=OH	−153.1	−153.6	−132.3
MBT:	X=S	−199.6	−199.9	−79.9
OHMBT:	X=S; Y=OH	−205.5	−205.5	−83.2
MTB:	X=N(CH$_3$)CONHCH$_3$	−124.0	−125.4	−141.0

abundance. The intensity of signals due to the nitrogen must be amplified by a 2-D NMR technique involving cross-polarization to benefit from the intensity of proton resonances coupled to that of the ^{15}N in the molecule.

The in vivo NMR benefited from measurements by B. Combourieu: these molecules are metabolized by bacteria and researchers in the group try to follow the pathway by NMR. Since such studies are very difficult to do, we tried calculating some chemical shifts accurately from structures to assign them (see [67]).

The Y substituent, generally a hydroxide was found to be in the position indicated (for mechanistic reasons, it is the only accessible and stabilised position for ring hydroxylation which has been found to take place in vivo after experiments in our group).

In solution, these molecules undergo a tautomeric equilibrium reaction transferring a proton towards this nitrogen as shown in the figure below (also used for nomenclature; P=protonated on resonating nitrogen) (Fig. 5.2).

NDDO-PM3 fitted STO molecular-site calculations on unprotonated tautomers (b). When the Gaussian03/PBE 6−311++G(2d,p) calculation (c) differs substantially from the measured value (a) (ppm/CH_3NO_2) that the resonating nitrogen is mostly protonated. This serves as a guideline for ab initio structures studied for these equilibria (Table 5.1).

The above results prompted use of a structure, protonated on the resonating N, (denoted P) to obtain the zero-order wave-function, in all cases apart from benzothiazole (BT) and ABT. Below, the same cases are treated in the DFT work.

Note that the basis sets including hydrogen-like orbitals perform better than the STO basis sets that in turn improve upon dense-core Gaussian basis sets [6−311++G(2d,p)].

Table 5.2 Numerical values for DFT calculations

Molecule	Substituent	a–c	a–b	a–d	a–e	a–f
BT:benzothiazole:	No X	1.3	8.3	11.1	1.2	2.8
POBT:	X=O	4.6	11.7	20.0	3.8	5.3
POHOBT:	X=O; Y=OH	4.5	7.4	14.9	2.9	5.2
ABT:	X=NH	1.1	3.8	21.5	0.9	3.1
OHABT:	X=NH; Y=OH	4.5	10.1	20.8	2.8	6.1
PMBT:	X=S	3.0	11.2	21.2	2.1	4.5
POHMBT:	X=S; Y=OH	2.5	10.1	18.8	1.7	4.8

Basis sets augmented with hydrogen-like orbitals are within 5 ppm of the experimental values (measured within 2 ppm) for the discrete solvated model. This model explicitly includes several deuterated methanol molecules to cater for the specific hydrogen bonding interactions.

Next, examining the generalised basis sets, compare b, c, d, and f with the measured chemical shifts and evaluate the difference in ppm. Differences between calculated and observed ^{15}N chemical shifts for commercial benzothiazoles and some metabolites (in ppm).

a-Measured values with respect to nitromethane standard in deuterated methanol solvent (B. Combourieu in [67]) error bars of 2 ppm.

b-Coupled perturbed STO.

c-Gaussian [25] with hydrogen-like AOs (c.f. Coulomb Sturmians $\alpha = 1$).

d-Gaussian [25].

e-Generalised ETO $\alpha = -1$.

f-Generalised ETO $\alpha = -2$.

Note. b through f involve solvation models, detailed below (Table 5.2).

a Measured chemical shift for ring nitrogen.

b STO: DFT PBE $6-311++G(2d,p)$ calculations with two discrete CD_3OD molecules on OH, NH, and SH (one on N, O, S) for minimal total energy.

c Gaussian 2003 as (b) with hydrogen-like orbital DFT PBE aug-6.311G**(2d, p) calculations.

d Gaussian 2003 DFT PBE $6-311++G(2d,p)$ calculations.

The content of this table is original and based on the previous work of the author [35] i.e., geometries are re-optimized from the co-ordinates of [35].

5.8 Conclusions

The scientific promise in studying a reaction at the copper (001) surface that removes toxic carbon monoxide using water as reagent and produces hydrogen as clean fuel is significant in providing the theory for a model catalyst which should be helpful in designing its industrial counterpart.

For NMR work, it is essential to use a basis set which comprises orbitals with the correct nuclear cusp behavior. This implies a non-zero value of the function at the origin for spherically symmetric cases and satisfying Kato's conditions. Hydrogen-like atomic orbital basis sets therefore perform better than Slater type orbitals which are an improvement upon Gaussians.

The NDDO-PM3 molecular site approach has the advantage of rapidity. Calculations take about a minute instead of 50–75 h on the IBM-44P-270. They cannot be systematically improved, however once the site Slater exponents have been fitted. Note that the 2s Slater exponent fluctuates wildly in fits, providing further evidence that shielding must be of the form (2-r) for the 2s ETO.

Fundamental work on orbital translation (see previous contribution) is also in progress to speed up these calculations within the test-bed of the STOP programmes [53].

The interplay of these discrete molecule solvent models and accurate in vivo NMR measurements is satisfactory, in that the structures postulated give calculated chemical shifts to similar accuracy as obtained for experimental values (on the order 2 ppm). It should be stressed that energy minimization in this case does evidence directional hydrogen bonds but can lead to several possible solvent geometries.

It is a remarkable gain in simplicity that the Coulomb operator resolution [66] now enables the exponential type orbital translations to be completely avoided, although some mathematical structure has been emerging in the BCLFs used to translate Slater type orbitals [74] and even more in the Shibuya–Wulfman matrix used to translate Coulomb Sturmians.

References

1. Alfè D, Gillan MJ (2004) Phys Rev B 70:161101
2. Avery J, Avery J (2007) Generalised sturmians and atomic spectra. World Scientific, Singapore
3. Baerends EJ, Ellis DE, Ros P (1973) Chem Phys 2:41
4. Berlu L (2003) PhD thesis, Université Blaise Pascal, Clermont Ferrand, France, Directed by P. E. Hoggan only.
5. Berlu L, Hoggan PE (2003) J Theor Comput Chem 2:147
6. Bonaccorsi R, Scrocco E, Tomasi J (1970) J Chem Phys 52:5270
7. BouferguèneA (1992) PhD thesis, Nancy I University, France
8. (a) Bouferguène A, Fares M, Hoggan PE (1996) Stater type orbital package for general molecular electronic structure calculations ab initio. Int J Quantum Chem 57:801
   (b) Bouferguène A, Hoggan PE (1996) STOP slater type orbital package program description. Q.C.P.E. Quart Bull 16:1
9. Bouferguène A, Rinaldi DL (1994) Int J Quantum Chem 50:21
10. Boys SF (1950) Proc Roy Soc [London] A 200:542
11. Boys SF, Cook GB, Reeves CM, Shavitt I (1956) Automated molecular electronic structure calculations. Nature 178:1207
12. Carbó R, Leyda L, Arnau M (1980) Int J Quantum Chem 17:1185
13. Cesco JC, Pérez JE, Denner CC, Giubergia GO, Rosso AE (2005) Appl Num Math 55(2):173 and references therein
14. Chuluunbaatar O, Joulakian B (2010) Three center continuum wave-function: application to first ionisation of molecular orbitals of CO_2 by electron impact. J Phys B 43:155201
15. Clementi E, Raimondi DL (1963) J Chem Phys 38:2686

16. Cohen AJ, Handy NC (2002) J Chem Phys 117:1470
Watson MA, Cohen AJ, Handy NC (2003) J Chem Phys 119:6475
Watson MA, Cohen AJ, Handy NC, Helgaker T (2004) J Chem Phys 120:7252
17. Condon EU, Shortley GH (1978) The theory of atomic spectra. Cambridge University Press, Cambridge, UK, p 48
18. Csizmadia IG, Harrison MC, Moskowitz JW, Seung S, Sutcliffe BT, Barnett MP (1963) POLYATOM: Program set for non-empirical molecular calculations. Massachusetts Institute of Technology Cambridge, 02139 Massachusetts. QCPE No 11, Programme 47 and Barnett MP Rev Mod Phys 35:571
19. Ditchfield R (1972) J Chem Phys 56:5688
20. Ditchfield R (1974) Mol Phys 27:789
21. Drummond ND, Towler MD, Needs RJ (2004) Phys Rev B 70:235119
22. Fernández Rico J, López R, Aguado A, Ema I, Ramírez G (1998) J Comput Chem 19(11):1284
23. Fernández Rico J, López R, Ramírez G, Ema I, Ludeña EV (2004) J Comput Chem 25:1355, and online archive
24. Foulkes, WMC, Mitas L, Needs RJ, Rajagopal G (2001) Rev Mod Phys 73:33
25. Frisch MJ (2004) Gaussian 03, Revision C.02. Gaussian, Inc., Wallingford
26. Gaunt JA (1929) Phil Trans R Soc A 228:151
27. Gill PMW, Gilbert ATB (2009) Resolutions of the Coulomb Operator. II The Laguerre Generator. Chem Phys (Accepted, 2008) 130(23)
28. Glushov VN, Wilson S (2002) Int J Quantum Chem 89:237, (2004) 99:903, (2007) 107:in press. Adv Quantum Chem (2001) 39:123
29. Guseinov I, Seckin Gorgun N (2010) J Mol Model 17(16):1517–1524
30. Hamouda AB, Absi N, Hoggan PE, Pimpinelli A (2008) Phys Rev B 77:245430
31. Hehre WJ, Lathan WA, Ditchfield R, Newton MD, Pople JA (1973) GAUSSIAN 70: Ab initio SCF-MO calculations on organic molecules QCPE 11, Programme number 236
32. Hinde RJ (2008) Six dimensional potential energy surface of H_2–H_2. J Chem Phys 128 (15):154308
33. Hoggan PE (2004) Int J Quantum Chem 100:218
34. Hoggan PE (2007) Trial wavefunctions for Quantum Monte Carlo simlations over ETOs. AIP Proc 963(2):193–197
35. Hoggan PE (2010) Four-center Slater type orbital molecular integrals without orbital translations. Int J Quantum Chem 110:98–103
36. Homeier HHH, Joachim Weniger E, Steinborn EO (1992) Programs for the evaluation of overlap integrals with B functions. Comput Phys Commun 72:269–287; Homeier HHH, Steinborn EO (1993) Programs for the evaluation of nuclear attraction integrals with B functions. Comput Phys Commun 77:135–151
37. Hsing, C-R, Wei C-M, Drummond ND, Needs RJ (2009) Phys Rev B 79:245401
38. Joudieh N (1998) PhD thesis, Faculté des Sciences de l'Université de Rouen, Rouen, France
39. Kato T (1957) Commun Pure Appl Math 10:151
40. Lee RM et al arXiv:1006.1798
41. London F (1937) J Phys Radium 8:397
42. McLean AD, Yoshimine M, Lengsfield BH, Bagus PS, Liu B (1991) ALCHEMY II. IBM research, Yorktown Heights, MOTECC 91
43. MSc Thesis. Clermont, 2010
44. Needs RJ, Towler MD, Drummond ND, López Ríos P (2010) J Phys:Condens Matter 22:023201; CASINO website: www.tcm.phy.cam.ac.uk/~mdt26/casino2.html. Accessed 10 Feb 2010
45. Perdew J, Burke K, Ernzerhof M (1996) Phys Rev Lett 77:3865
46. PhD Thesis. Clermont, 2009. Available in pdf at www.inist.fr
47. Pinchon D, Hoggan PE (2007) J Phys A 40:1597
48. Pinchon D, Hoggan PE (2007) Int J Quantum Chem 107:2186
49. Pinchon D, Hoggan PE (2009) Int J Quantum Chem 109:135
50. Podolanski J (1931) Ann Phys 402(7):868

51. Pople JA, Beveridge DL (1970) Approximate molecular orbital theory. McGraw Hill, New York
52. Pritchard HO (2001) J Mol Graph Model 19:623
53. Red E, Weatherford CA (2004) Int J Quantum Chem 100:204
54. Reinhardt P, Hoggan PE (2009) Cusps and derivatives for H_2O wave functions using Hartree-Fock Slater code: a density study. Int J Quantum Chem 109:3191–3198
55. Roothaan CCJ (1951) J Chem Phys 19:1445
56. Shao Y, White CA, Head-Gordon M (2001) J Chem Phys 114 6572
57. Shavitt I (1963) In: Alder B, Fernbach S, Rotenberg M (eds) Methods in computational physics, vol 2. Academic, New York, p. 15
58. Slater JC (1930) Phys Rev 36:57
59. Slater JC (1932) Phys Rev 42:33
60. Smith SJ, Sutcliffe BT (1997) In: Lipkowtz KB, Boyd BD (eds) The development of computational chemistry in the United Kingdom in Reviews in computational chemistry. VCH Academic Publishers, New York
61. Stevens RM, Lipscomb, WN (1964) J Chem Phys 40:2238–2247
62. Stevens RM (1970) The POLYCAL program. J Chem Phys 52:1397
63. Toulouse, J, Hoggan PE, Reinhardt P, Caffarel M, Umrigar CJ (2011) PTCP, B22
64. Toulouse J, Hoggan PE, Reinhardt P Progress in QMC, ACS (in press 2011)
65. Tully JC (1973) J hem Phys 58:1396
66. Vagranov SA, Gilbert ATB, Duplaxes E, Gill PMW (2008) J Chem Phys 128:201104
67. Vieille L, Berlu L, Combourieu B, Hoggan PE (2002) J Theor Comput Chem 1(2):295
68. Weatherford CA, Red E, Joseph D, Hoggan PE (2006) Mol Phys 104:1385
69. Weniger EJ (1985) J Math Phys 26:276
70. Weniger EJ, Steinborn EO (1982) Comput Phys Commun 25:149
71. Weniger EJ, Steinborn EO (1989) J Math Phys 30(4):774
72. Werner H-J, Knowles PJ, Lindh R, Manby FR et al (2011) MOLPRO, version 2006.1 a package of ab initio programs. www.molpro.net . Accessed 6 Aug 2011
73. Zaim N (2010) Poster presentation at QSCP XV Cambridge
74. Sidi A, Hoggan PE (2011) Int J Pure Appl Math 71:481–498

Chapter 6
Progress in Hylleraas-CI Calculations on Boron

María Belén RUIZ

Abstract Preliminary results on Hylleraas-Configuration Interaction calculations of the boron atom ground state are presented. The wave function consists of a 954 term Configuration Interaction part and 192 configurations including all interelectronic distances. An energy value of -24.64815076 a.u. was been obtained with a minimal orbital basis [4s3p2d]. Calculations with more configurations are in progress. Correct description of the electronic cusp is important, as discussed and the most recent benchmark calculations in the field are concisely reviewed. The computational techniques for matrix element evaluation are described. Those employed for the B atom can be readily used for C and N atoms, and further for the highly accurate calculation of the nonrelativistic energy of second row elements.

6.1 Introduction

After Egil Hylleraas [26] proposed (in 1929) the He atom wave function, including the interelectronic coordinate r_{12}, in the following decades one of the goals has been to improve it and extend it to determine energies and properties of ground and excited states of light atoms and small molecules with the highest possible accuracy. Hylleraas pointed out: *"It is then quite clear for me, that here if the extreme limit of capability in these calculations would be achieved, the exact treatment of more difficult systems would be hopeless. (...) I have finally succeeded to find a method, that for helium leads to the desired results and hopefully will prove to be fruitful*

M.B. RUIZ (✉)
Department of Theoretical Chemistry of the Friedrich-Alexander-University, Erlangen-Nürnberg, Egerlandstraße 3, D-91058, Erlangen, Germany
e-mail: Belen.Ruiz@chemie.uni-erlangen.de

P.E. Hoggan et al. (eds.), *Advances in the Theory of Quantum Systems in Chemistry and Physics*, Progress in Theoretical Chemistry and Physics 22, DOI 10.1007/978-94-007-2076-3_6, © Springer Science+Business Media B.V. 2012

also for other problems." In this work we shall show that the Hylleraas method can be successfully extended to the calculations of atoms with five and more electrons. Concretely, this can be done using a variant of the Hylleraas method, the Hylleraas-Configuration Interaction method [59, 60, 71], which is a hybrid of two methods: that Hylleraas and the Configuration Interaction method (CI).

The CI wave function, which is of great importance in quantum mechanical calculations of the electron structure of atoms and molecules, has well-known shortcomings, like the fact that the wave function does not fulfill the electronic cusp condition [29]:

$$\left(\frac{1}{\Psi}\frac{\partial\Psi}{\partial r_{ij}}\right)_{r_{ij}=0} = \frac{1}{2}. \tag{6.1}$$

This is because the CI wave function does not contain linear odd powers of the interelectronic coordinate r_{ij}. Note that the CI wave function does contain terms r_{ij}^2, r_{ij}^4, $\cdots r_{ij}^{2n}$ which are formed by combination of angular orbitals p, d, f of the one-electron basis, for instance:

$p(1)p(2) \equiv s(1)s(2)r_{12}^2$ [17, 61]. Although the presence of the even powers r_{ij}^{2n} explains the accuracy of the full-CI wave functions, its form is responsible for the extremely slow convergence of the CI method to the exact solution. Odd powers of r_{ij} are significant energetically [69]. They are equivalent to an infinite expansion of one-electron orbitals. It can be demonstrated using the addition theorem of spherical harmonics that:

$$s(1)s(2)r_{12} \equiv s(1)s(2) + p(1)p(2) + d(1)d(2) + f(1)f(2) + \cdots \tag{6.2}$$

In the CI wave function higher and higher angular terms included attempt to represent the term r_{ij} c.f. Taylor expansion [69]. The energy improvement when increasing the quantum number l obeys an asymptotic formula proportional to $(l+1/2)^{-4}$ for two-electron systems [24], and in general for a larger number of electrons [35,56]. Therefore, the correct description of the electron cusp is of utmost importance for the convergence of the energy by the different methods and, therefore the feasibility of achieving highly accurate energy results with existing computer resources and in a reasonable computer time.

The improvement of Hylleraas-type wave functions has continued over the years, represented among others by calculations of James and Coolidge [27], Frankowski and Pekeris [20]. The past two decades saw a 'Renaissance' of Hylleraas-type methods, with numerous benchmark calculations on light atoms and small molecules. This progress has been accelerated by the recent developments in computer technology, the use of high precision arithmetic [3], and the appearance of algebraic programs like Maple [37], which permits one to work with complicated algebraic expressions with no human error. Nowadays the field of highly accurate calculations is in continuous development. For example, we briefly mention here the latest benchmark calculations on light atoms and small molecules.

For the He atom and two-electron systems the Hylleraas wave function has been recently improved including logarithmic terms by Schwartz and Nakatsuji et al.

The nonrelativistic energies obtained for the ground [42,57] and excited states [43] have more than 40 decimal digits of accuracy (these results are essentially exact for practical purposes, beyond what can be obtained experimentally). Note that highly accurate computational data is used together with experimental data, to determine fundamental physical constants, e.g., the fine structure constant α, or the lamb shift in He atom (CODATA) [41]. Also highly accurate energy results are needed e.g., to calculate electron affinities and ionization potentials. A beautiful example is the calculation of the ionization energy of lithium atom [30], where the leading terms are the nonrelativistic energies. Data for highly ionized levels of atoms and ions is very scarce. The calculation of anions is specially difficult because the odd electron is weakly bound. Other possible applications of highly accurate wave functions are the calculation of the energy levels of atoms and small molecules confined in cavities [2], interesting for their application to several problems of modern Physics and Chemistry, and the calculation of the transition probabilities during β-decay of light atoms [21], relevant in Nuclear Physics and Nuclear Medicine.

Recently, impressive calculations using Hylleraas wave functions have been done for the H_2 molecule by the Hylleraas method [44, 63], the Iterative Complement Iteration (ICI) [36], and Explicitly Correlated Gaussian (ECG) [12] methods. Few molecules have yet been calculated using Hylleraas-type wave functions: HeH^+ and some other species [72] using the Hylleraas method, the helium dimer He_2 interaction energy [46] and the ground state of the BH molecule [7], both using the ECG method.

The Hy-CI method was applied in 1976 to LiH molecule by Clary [14] using elliptical STOs. For two-center molecules the three-electron and four-electron integrals occurring in the Hy-CI method have been developed by Budzinski [8]. Clementi et al. extended the Hy-CI method to molecules using Gaussian orbitals [45], and applied it to the calculation of H_3. For a review on molecular methods using Slater orbitals and their history, see Refs. [25,51].

The electronic structure of light atoms e.g., Li and Be have been investigated using the Hylleraas, Hy-CI and ECG methods by several authors. For instance, the Li atom has been recently determined beyond nanohartree accuracy ($>1.10^{-9}$ a.u.) [47] using the Hylleraas approach[1] and the Hy-CI methods [64]. The best CI calculation on the Li atom using Slater orbitals needs 2.6 million configurations. The Be atom is currently the subject of investigations, now better than one tenth of a microhartree accuracy ($<0.1 \cdot 10^{-6}$ a.u.) has been achieved using the ECG [66] and the Hy-CI methods [65]. This quest for high-accuracy requires a profound

---

[1]In the Hylleraas (Hy) method all r_{ij} may be included simultaneously into a configuration. This method converges faster to the exact solution, but it has been applied only for $N \leq 4$ electrons with the restriction of single and double-linked products of r_{ij} [10]. The integrals for Hy double-linked wave functions are very complex [31]. For a $N \geq 5$ electrons the Hy method would lead to the many-electron integrals, which are yet to be solved in general.

knowledge in the construction of wave functions, which will be helpful to calculate with relative ease larger systems.

In the years 2009, 2010 two highly accurate calculations on the B atom have been reported using the CI and ECG methods. A huge selected CI calculation with L-S eigenfunctions, carried out by Bunge [1, 9] with the following set of orbitals [24s23p22d21f20g19h18i17k16l15m14n13o12 q11r10t9u8v7w6x5y4z] and about 12 million selected configurations, has led to the energy -24.6538373 a.u. which is $\approx 1 \cdot 10^{-4}$ a.u. accurate. The immense dimension of the matrix is treated using a "CI by parts" dividing the space into subspaces. The CI calculations were extremely fast, of the order of hours/day.

The second calculation on B atom by Adamowicz et al. [7] was done using the ECG method with 2000 configurations, and led to the energy -24.65384 a.u. ($< 0.1 \cdot 10^{-4}$ a.u.). The reported computational time was immense, about 1 year of continuous calculation using parallelization and energy gradients, due to the large number of non-linear parameters (30,000) to be optimized.

For comparison, using the r12-MR-CI method the value -24.65379 a.u., (also $< 0.1 \cdot 10^{-4}$ a.u. accurate) [23] was obtained with a very large basis of Gaussian orbitals. The ab initio result from the Diffusion Monte Carlo method is $-24.65357(3)$ a.u. [6]. The estimated nonrelativistic energy using theoretical and experimental data is -24.65391 a.u. [13]. The mentioned calculations are less accurate than a microhartree. The relativistic energy value including mass and Darwin corrections is estimated to be: -24.659758 a.u.

The carbon atom has been recently calculated employing the ECG method thanks to the use of analytical energy gradients [58]. Although neon atom was calculated by Clary and Handy using the Hy-CI method [15], for N and larger atoms the most accurate calculations have been obtained using the R12 [34] and F12 methods [32].

In this paper we present some preliminary Hy-CI calculations on the B atom using all interelectronic distances and only some of all possible configurations. In the next sections the techniques used for the calculation of the matrix elements, construction of the configurations and exponent optimization are treated. These techniques can be readily extended to the C and N atoms. We shall apply the Hy-CI method to the B atom and show that the preliminary calculations are very promising, because a small number of configurations is needed to obtain an accurate energy result compared with other methods which need millions of configurations. Work is in progress to add all types of configurations into the wave function.

6.2 Theory

The Hylleraas-Configuration Interaction wave function [59, 60, 71] may be written:

$$\Psi = \sum_{p=1}^{N} C_p \Phi_p, \qquad \Phi_p = \hat{O}(\hat{L}^2) \mathscr{A} \phi_p \chi, \qquad (6.3)$$

where Φ_p are symmetry adapted configurations, N is the number of configurations and the constants C_p are determined variationally. The operator $\hat{O}(\hat{L}^2)$ projects over the appropriate space, so that every configuration is an eigenfunction of the square of the angular momentum operator $\hat{L}^2$. $\mathscr{A}$ is the antisymmetrization operator, and χ is the spin eigenfunction. In the case of the B atom:

$$\chi = [(\alpha\beta - \beta\alpha)(\alpha\beta - \beta\alpha)\alpha]. \tag{6.4}$$

The spatial part of the basis functions are Hartree products of Slater orbitals:

$$\phi_p = r_{ij}^{\nu} \prod_{k=1}^{n} \phi_k(r_k, \theta_k, \varphi_k). \tag{6.5}$$

The terms r_{ij}^{ν} are effectively reduced to $\nu = 0,1$ since all higher terms can be expressed as a product of r_{ij} times a polynomial in r_i, r_j and angular functions.

The basis functions ϕ_p, are products of s-, p- and d-Slater orbitals. For simplicity, we do not consider higher angular momentum orbitals. These can be added later if necessary to obtain higher accuracy. We use unnormalized complex Slater orbitals for which the exponents are adjustable parameters, defined as:

$$\phi(\mathbf{r}) = r^{n-1}e^{-\alpha r}Y_l^m(\theta, \phi). \tag{6.6}$$

The spherical harmonics with Condon and Shortley phase [16, p. 52] are given by:

$$Y_l^m(\theta, \phi) = (-1)^m \left[\frac{2l+1}{4\pi} \frac{(l-m)!}{(l+m)!} \right]^{1/2} P_l^m(\cos\theta)e^{im\phi}, \tag{6.7}$$

where $P_l^m(\cos\theta)$ are the associated Legendre functions. The spherical harmonics and associated Legendre functions used throughout this work are written explicitly in [67, p. 14]. Using the complex spherical harmonics we have constructed a set of configurations of P symmetry, which have shown to be important in CI calculations [53]. The ground state configuration is $ssssp$ and the configurations following an energetic order are $ssppp$, $sspds$, $sssspd$, $ssddp$. Additional configurations rotating the orbitals are $ppssp$, $psspp$, $spdss$, $pdsss$, $ddssp$, $dssdp$. Combinations of inner-shell configurations $S(pp)$, $S(dd)$ with $P(sp)$ and $P(pd)$ lead to the configurations $ppppp$, $ppddp$, $ddppp$, $ppspd$, $ddspd$, and $ddddp$, which may be important at the micro- and nano-hartree level of accuracy. The number of configurations grows very fast when adding the ten r_{ij} factors.

The Hamiltonian in Hylleraas coordinates may be written [50][2]:

$$
\begin{aligned}
\hat{H} = {} & -\frac{1}{2}\sum_{i=1}^{n}\frac{\partial^2}{\partial r_i^2} - \sum_{i=1}^{n}\frac{1}{r_i}\frac{\partial}{\partial r_i} - \sum_{i=1}^{n}\frac{Z}{r_i} - \sum_{i<j}^{n}\frac{\partial^2}{\partial r_{ij}^2} - \sum_{i<j}^{n}\frac{2}{r_{ij}}\frac{\partial}{\partial r_{ij}} + \sum_{i<j}^{n}\frac{1}{r_{ij}} \\
& -\frac{1}{2}\sum_{i\neq j}^{n}\frac{r_i^2+r_{ij}^2-r_j^2}{r_i r_{ij}}\frac{\partial^2}{\partial r_i \partial r_{ij}} - \frac{1}{2}\sum_{i\neq j}^{n}\sum_{k>j}^{n}\frac{r_{ij}^2+r_{ik}^2-r_{jk}^2}{r_{ij}r_{ik}}\frac{\partial^2}{\partial r_{ij}\partial r_{ik}} \\
& -\frac{1}{2}\sum_{i=1}^{n}\frac{1}{r_i^2}\frac{\partial^2}{\partial \theta_i^2} - \frac{1}{2}\sum_{i=1}^{n}\frac{1}{r_i^2\sin^2\theta_i}\frac{\partial^2}{\partial \varphi_i^2} - \frac{1}{2}\sum_{i=1}^{n}\frac{\cot\theta_i}{r_i^2}\frac{\partial}{\partial \theta_i} \\
& -\sum_{i\neq j}^{n}\left(\frac{r_j}{r_i r_{ij}}\frac{\cos\theta_j}{\sin\theta_i} + \frac{1}{2}\cot\theta_i\frac{r_{ij}^2-r_i^2-r_j^2}{r_i^2 r_{ij}}\right)\frac{\partial^2}{\partial \theta_i \partial r_{ij}} \\
& -\sum_{i\neq j}^{n}\frac{r_j}{r_i r_{ij}}\frac{\sin\theta_j}{\sin\theta_i}\sin(\varphi_i-\varphi_j)\frac{\partial^2}{\partial \varphi_i \partial r_{ij}}.
\end{aligned}
\tag{6.8}
$$

The kinetic energy operator has been separated into several radial and angular parts. For any atomic number two- and three-electron kinetic-energy integrals occur. This operator has the advantage that for the case of three-electron kinetic integrals the expansion of r_{ij} into $r_<$ and $r_>$ is avoided, and therefore no three-electron auxiliary W integrals are needed. This fact saves not only calculations but also memory space. Only the easily computed two-electron auxiliary integrals $V(n,m;\alpha,\beta)$ are needed.

The angular momentum operator can be extracted from Eq. 6.8:

$$
\sum_{i=1}^{n}\frac{1}{r_i^2}\hat{L}_i^2 = -\frac{1}{2}\sum_{i=1}^{n}\frac{1}{r_i^2}\frac{\partial^2}{\partial \theta_i^2} - \frac{1}{2}\sum_{i=1}^{n}\frac{1}{r_i^2\sin^2\theta_i}\frac{\partial^2}{\partial \varphi_i^2} - \frac{1}{2}\sum_{i=1}^{n}\frac{\cot\theta_i}{r_i^2}\frac{\partial}{\partial \theta_i},
\tag{6.9}
$$

and its eigenvalue equation used:

$$
L_i^2\phi_i = l_i(l_i+1)\phi_i,
\tag{6.10}
$$

with l_i the angular quantum number of the orbital ϕ_i. In the case of Hy-CI wave functions the term $\partial^2/(\partial r_{ij}\partial r_{ik})$ containing derivatives with respect to two r_{ij} vanishes. The Be wave function used by Kleindienst et al. [10] contained this term in the so-called double-linked Hy-CI calculations.

From the variational principle one obtains the matrix eigenvalue problem:

$$
(\mathbf{H}-E\mathbf{S})\mathbf{C}=\mathbf{0},
\tag{6.11}
$$

---

[2]This formula has been derived independently by the author and by Barrois et al. [4, 40] (some angular terms are written differently but they are equivalent, the derivations are different). A similar formula was also proposed by Walsh and Borowitz [68], but it was incomplete.

where the matrix elements are:

$$H_{kl} = \int \Phi_k H \Phi_l d\tau, \qquad S_{kl} = \int \Phi_k \Phi_l d\tau. \qquad (6.12)$$

The occurring integrals in Hy-CI calculations are: two-electron type[3]:

$$\langle r_{12} \rangle, \qquad \langle r_{12}^2 \rangle, \qquad \left\langle \frac{1}{r_{12}} \right\rangle,$$

$$\langle r_{12} \rangle \langle r_{34} \rangle, \qquad \langle r_{12} \rangle \left\langle \frac{1}{r_{34}} \right\rangle, \qquad (6.13)$$

three-electron type:

$$\left\langle r_{12} r_{13} \right\rangle, \qquad \left\langle r_{12}^2 r_{13} \right\rangle, \qquad \left\langle \frac{r_{12}}{r_{13}} \right\rangle, \qquad \left\langle \frac{r_{12} r_{13}}{r_{23}} \right\rangle \qquad (6.14)$$

the first three cases after direct integration over both one r_{ij} and one electron are reduced to a linear combination of two-electron integrals [48]. Only the so-called triangle integral $\langle r_{12} r_{13}/r_{23} \rangle$ has not been treated by us and we use a very efficient subroutine from Sims and Hagstrom [62]. The four-electron integrals occurring for any atomic number are of three types:

$$\left\langle \frac{r_{12} r_{13}}{r_{14}} \right\rangle, \qquad \left\langle \frac{r_{12} r_{13}}{r_{34}} \right\rangle \qquad \left\langle \frac{r_{12} r_{34}}{r_{23}} \right\rangle. \qquad (6.15)$$

All four-electron integrals have been evaluated using the method of direct integration over one interelectronic coordinate and one electron [49] leading to linear combinations of three-electron ones. And these lead to linear combinations of two-electron ones. The two-electron integrals are calculated in terms of the auxiliary two-electron integrals $V(m,n;\alpha,\beta)$, defined:

$$V(m,n;\alpha,\beta) = \int_0^\infty r_1^m e^{-\alpha r_1} dr_1 \int_{r_1}^\infty r_2^n e^{-\beta r_2} dr_2 , \qquad (6.16)$$

For a review on the calculation of all cases of $V(m,n;\alpha,\beta)$ integrals, see Ref. [48] and references therein. The subroutine for the triangle integral $\langle r_{12} r_{13}/r_{23} \rangle$ does require three-electron auxiliary integrals $W(f,g,h;\alpha,\beta,\gamma)$. In our program the W integrals are computed directly when needed (without constructing tables) using a very fast and stable subroutine from Sims and Hagstrom described in Ref. [62].

---

[3]The notation e.g., $\langle r_{12} r_{13}/r_{14} \rangle$ represents the integral where the left and right hand orbitals of electrons 1, 2, 3 and 4 are involved: $\langle \phi(\mathbf{r}_1)\phi(\mathbf{r}_2)\phi(\mathbf{r}_3)\phi(\mathbf{r}_4)|r_{12}r_{13}/r_{14}|\phi(\mathbf{r}_1)\phi(\mathbf{r}_2)\phi(\mathbf{r}_3)\phi(\mathbf{r}_4) \rangle$. The indices of the actual integrals can be interchanged to write them in these forms.

Therefore only two-electron integrals, as in the case of the CI method, and triangle integrals have to be computed. This fact will be extremely helpful when extending the application of Hy-CI method to larger systems. In our code, the same computer memory is needed for CI and Hy-CI calculations. Note, that in the Hylleraas-CI method for any electron number no higher order integrals than four-electron ones appear.

6.3 Computational Aspects

We have developed a Hy-CI computer program for the B atom in Fortran 90, working with quadruple precision (about 30 decimal digits of accuracy), which permits us to calculate more than 5,000 configurations in sequential form, using only a two-processor work station. We are convinced that the structure of this program is a very good approach to a future general Hy-CI program. The program is based on a CI program for B atom with L-S configurations of Slater-type orbitals [53]. The structure of the new Hy-CI program part is based on the recognition of whether an interelectronic distance r_{ij}, and which interelectronic distance r_{ij}, appears on the right hand side of a matrix element. We have set the antisymmetrization operator (note that it is quasi-idempotent) on the left hand side of the matrix elements, and we perform a loop over all generated permutations on the left. Note that CI configurations do not contain any r_{ij}-term, this is taken into account globally in the program. Due to the permutations generated by the antisymmetrization operator we may find any r_{ij} (or none) on the left hand side of the matrix elements, the program performs a loop with if-statements over all possible r_{ij} (in case of B there are 10) on the left. Once the r_{ij} that appear on left and right parts of the matrix elements are identified, the indices of the r_{ij} together with the type of operator lead to connections of indices which are translated into graphs and every graph corresponds to a given type of integral or product of integrals, which are computed by calls to the corresponding functions. The matrix elements have been derived from the Hamiltonian in Hylleraas coordinates Eq. 6.8. The program is divided into two major subroutines, one which computes the one-electron operators (overlap, kinetic and potential energy operators) leading to one- two- and three-electron integrals and the other which computes the two-electron repulsion integrals which lead to products of 2 two-electron integrals, to three-electron integrals, products of three- and two-electron integrals, and to four-electron integrals. There are $N!$ permutations, N being the number of electrons. In addition, N must be multiplied by the number of primitives of the spin eigenfunction. The spin functions are also computed using loops over the number of spin primitives.

Note that the program permutes the interelectronic distances (antisymmetrization) and during the construction of the matrix elements identifies which interelectronic distance acts on the left and on the right, calling the appropriate integral subroutine. As in our CI program, we use an automatic generation of the permutations of 2–6 electrons for atoms, ions, and molecules [52], combined with the necessary spin functions, performing also the "spin integration" [11]. The extension of the Hy-CI computer program from 6 to 7,8 electron systems needs only minor changes in the structure of the program.

In atom/ions with $N = 5$ due to the presence of angular orbitals in the wave function, $L - S$ there are many kinds of configurations to be constructed, whose contribution to the nonrelativistic energy is important. In particular, the ground states of B and C atoms are of P-symmetry. The Hy-CI program is general for any type of orbital, although in Hy-CI we use unnormalized s-, p- and d-orbitals.

The Hy-CI program is still under development and future steps will be to add several subroutines to account for configurations containing simultaneously angular orbitals ($l_i > 0$) and interelectronic coordinates r_{ij}. These contributions originate from terms in the Hamiltonian Eq. 6.8 containing $\partial^2/\partial\theta_i\partial r_{ij}$ and $\partial^2/\partial\phi_i\partial r_{ij}$. This work is in progress.

6.4 Calculations

We have performed preliminary calculations on the ground ^{2}P state of boron atom, using a set of s-, p-, and d-Slater orbitals. Higher angular momentum orbitals could also be used. We plan, in future, to add f-orbitals. We have used a set of three exponents, considering double occupancy of the orbitals, which were optimized for the Hy-CI wave function: $\alpha = 5.47575$, $\beta = 1.469125$ and $\gamma = 1.17595$. These exponents are used for all configurations. Further optimization of the orbital exponents using more configurations is planned.

In order to observe the energy effect of configurations containing r_{ij}, we show in Table 6.1, the energy contribution in μ-hartrees of the first Hy-CI configurations $1s1s2s2s2pr_{ij}$ and $1s2s2s3s2pr_{ij}$. The correlated configurations of s-type have a predominant role, they pick up most correlation energy. The r_{12} configuration improves the energy by about 100 millihartrees, followed by r_{13} with about 50 millihartrees contribution. The next configurations are r_{34} and r_{35} with contributions larger than 1 millihartree. Note that due to the symmetry and double occupancy of the orbitals, some configurations are equivalent and therefore do not both separately contribute to the energy, for instance $E(1s2s2s2s2pr_{13}) = E(1s2s2s2s2pr_{14})$. Here, a second type of configuration $1s2s2s3s2p$ where all electrons are different is used. Then all configurations including r_{ij} contribute.

The structure of the calculations is the following: first a selected CI calculation (configurations which contributed less than $1 \cdot 10^{-6}$ a.u. have been eliminated), then configurations containing r_{ij} are added. For a basis[4] $n = 4$ and some configurations from $n = 5$ we have obtained the energy value -24.641886656 a.u. using 954 configurations. This result is better than that of Schaefer and Harris -24.639194 a.u. [55] who also employed a selected CI with Slater orbitals. Our CI result agrees with

---

[4] $n = 3$ stands for the orbital set: $1s, 2s, 3s, 2p, 3p, 3d$ and the notation $[3s2p1d]$.

Table 6.1 Effect on the energy of the first Hy-CI configurations in the calculations on the ^{2}P ground state of boron atom

Conf.	Wave function	E(a.u.)	Virial	E_{diff} in μh
1	1s1s2s2s2p	−24.3945 2842	−1.88614	
2	1s1s2s2s2pr_{12}	−24.4937 3920	−1.94976	−99210.8
3	1s1s2s2s2pr_{13}	−24.5398 5132	−1.97943	−46112.1
4	1s1s2s2s2pr_{14}	−24.5398 5132	−1.97943	0.0
5	1s1s2s2p2sr_{15}	−24.5398 7159	−1.97948	−20.3
6	1s1s2s2s2pr_{23}	−24.5398 7159	−1.97948	0.0
7	1s1s2s2s2pr_{24}	−24.5398 7159	−1.97948	0.0
8	1s1s2s2p2sr_{25}	−24.5398 7159	−1.97948	0.0
9	1s1s2s2s2pr_{34}	−24.5659 5970	−1.97844	−26088.1
10	1s1s2s2p2sr_{35}	−24.5660 0326	−1.97830	−43.6
11	1s1s2p2s2sr_{45}	−24.5660 0326	−1.97830	0.0
12	1s2s2s3s3s2p	−24.5701 2584	−1.99796	−4122.6
13	1s2s2s3s3s2pr_{12}	−24.5704 1459	−1.99878	−288.7
14	1s2s2s3s3s2pr_{13}	−24.5705 3553	−1.99881	−120.9
15	1s2s2s3s3s2pr_{14}	−24.5705 7212	−1.99888	−36.6
16	1s2s2s3p2sr_{15}	−24.5707 0461	−1.99890	−132.5
17	1s2s2s3s2pr_{23}	−24.5715 7357	−1.99880	−868.9
18	1s2s2s3s3s2pr_{24}	−24.5716 0635	−1.99883	−32.8
19	1s2s2s3p2sr_{25}	−24.5716 1502	−1.99882	−8.7
20	1s2s2s3s3s2pr_{34}	−24.5729 1963	−2.00094	−1304.6
21	1s2s2s3p2sr_{35}	−24.5740 3579	−2.00050	−1116.2
22	1s2s2p3s2sr_{45}	−24.5740 4953	−2.00049	−13.7

Orbital exponents K-shell: 4.914, L-shell: $s_L = p_L = 1.48025$ and $s_{L'} = p_{L'} = 1.206375$. Configurations 4,6–8,11 are equivalent to previous configurations and do not contribute to the energy

the value of Froese Fischer −24.64046977 a.u. for $n = 5$ using the MC-HF method [19]. More details of the CI calculation can be found in Ref. [53].

Configurations including r_{ij} terms have been added to the CI wave function. These first Hy-CI type configurations are *ssssp*, or its permutations, with a basis [4s4p] orbitals. In Table 6.2 we show intermediate Hy-CI results on the ground state of B atom which are very promising: using 1146 configurations and a small basis [4s3p2d] we obtained an energy of −24.64814427 a.u. We can compare this result with the CI ones using Slater and Gaussian orbitals, respectively, using the same computer code, computer and basis set, see Table 6.3. The CI with Slater orbitals and 2066 configurations leads to −24.6410472 a.u., and the CI using Gaussian orbitals with comparable basis set cc-pTVZ needs 21.5 million configurations yielding −24.645585 a.u. [38]. The comparison shows that the Hy-CI wave function converges much faster. We can also compare with the CI calculation of Almora and Bunge [1] with 13 million configurations leading to −24.6538 a.u. According to the pattern of convergence the Hy-CI needs a small number of configurations. This is due to the inclusion of r_{ij} into the wave function. Comparing with the ECG calculation of Bubin et al. [7], our Hy-CI calculations are performed in a reasonable amount of computer time, because no continuous optimizations are needed, other

Table 6.2 Hylleraas-Configuration Interaction (Hy-CI) calculations on the ^2P ground state of boron atom using a wave function with s-, p- and d-orbitals. Configurations are filtered and kept when their contribution $> 1.0 \times 10^{-6}$ a.u

Conf.	Wave function	N	N_{tot}	E(a.u.)	E_{diff} in μh
ssssp	1:5s 1:5s 1:5s 1:5s 2:5p	205	205	−24.5521 6835	
sssps	1:5s 1:5s 1:5s 2:5p 1:5s	127	332	−24.5552 1161	−3043
spsss	1:5s 2:5p 1:5s 1:5s 1:5s	74	406	−24.5556 4742	−436
ssppp	1:5s 1:5s 2:5p 2:5p 2:5p	149	555	−24.5918 5267	−36205
ppssp	2:5p 2:5p 1:5s 1:5s 2:5p	72	627	−24.6148 1880	−22966
sppsp	1:5s 2:5p 2:5p 1:5s 2:5p	28	655	−24.6167 9454	−1976
ppppp	2:5p 2:5p 2:5p 2:5p 2:5p	2	657	−24.6177 2748	−933
ssspd	1:5s 1:5s 1:5s 2:5p 3:5d	121	778	−24.6309 4026	−13213
sssdp	1:5s 1:5s 1:5s 3:5d 2:5p	36	814	−24.6377 6821	−6828
sspds	1:5s 1:5s 2:5p 3:5d 1:5s	26	840	−24.6378 2704	−59
pdsss	2:5p 3:5d 1:5s 1:5s 1:5s	32	872	−24.6384 3562	−609
spsds	1:5s 2:5p 1:5s 3:5d 1:5s	19	891	−24.6390 1796	−582
sdssp	1:5s 3:5s 1:5s 1:5s 2:5p	8	899	−24.6390 3370	−16
ssddp	1:5s 1:5s 3:5d 3:5d 2:5p	10	909	−24.6411 3144	−2098
sspdd	1:5s 1:5s 2:5p 3:5d 3:5d	5	914	−24.6411 5589	−24
sdsdp	1:5s 3:5d 1:5s 3:5d 2:5p	17	931	−24.6414 2490	−269
sppdp	1:4s 2:4p 2:4p 3:4d 2:4p	2	933	−24.6414 4000	−15
spppd	1:4s 2:4p 2:4p 2:4p 3:4d	2	935	−24.6414 5975	−20
ppsdp	2:4p 2:4p 1:4s 3:4d 2:4p	1	936	−24.6414 6453	−5
ppspd	2:4p 2:4p 1:4s 2:4p 3:4d	8	944	−24.6417 8394	−319
pppdd	2:4p 2:4p 2:4p 3:4d 3:4d	1	945	−24.6417 9721	−14
ddppp	3:4d 3:4d 2:4p 2:4p 2:4p	3	948	−24.6418 8434	−87
ppddp	2:4d 2:4d 3:4p 3:4p 2:4p	3	951	−24.6418 8635	−2
ddddp	3:4d 3:4d 3:4d 3:4d 2:4p	3	954	−24.6418 8666	0
ssssp r12	1s 1s 2s 2s 2p	1	955	−24.6464 6510	−4578
ssssp r13	1s 1s 2s 2s 2p	1	956	−24.6467 5150	−286
ssssp r15	1s 1s 2s 2s 2p	1	957	−24.6467 5158	0
ssssp r34	1s 1s 2s 2s 2p	1	958	−24.6474 4083	−689
sssps r35	1s 1s 2s 2s 2p	1	959	−24.6474 4415	−3
ssssp $\sum r_{ij}$	1:3s 1:3s 2:3s 2:3s 2:3p	199	1158	−24.6481 4522	−703

Orbital exponents K-shell: $s_K = p_K = 5.47575$, L-shell: $s_L = p_L = 1.469125$ and $s_{L'} = p_{L'} = 1.17595$. The virial factor 2.0001 is almost constant during the calculation. The notation 1:5s means groups of configurations builded with 1s,2s,3s,4s,5s orbitals

than an initial optimization of the orbital exponents to treat ground or excited state using a smaller basis set.

Our next steps are to add more configurations and to select them using the Brown formula [5]. Now the selection has been done comparing the energy difference between the N and N-1 dimensional basis calculations. Other possible improvements will be to include f-orbitals at the end of the wave function expansion, and to use different exponents for different orbitals. For very large wave function expansions major changes in the program are needed, e.g., parallelization. Since the number of configurations increases rapidly we have used symmetry adapted

Table 6.3 Comparison of variational upper bonds to the energy of the ^2P ground state of boron atom calculated by different methods

Method	Orb.	References	Basis set	Confs.[c]	Energy (a.u.)
CI	STO	Weiss (1969) [70]	[4s3p2d]	35	−24.5975 0
FCI	GTO	Mayer[a] (2009) [38]	[6s5p2d] aug-cc-pCVDZ[d]	2.8 mill.	−24.6375 7357
CI	STO	Schaefer and Harris (1968) [55]	[10s7p4d] AOs	91	−24.6378 3865
CI	STO	Schaefer and Harris (1968) [55]	[10s7p4d] AOs	180	−24.6391 94
FCI	STO	Ruiz et al. (2009) [53]	[5s4p3d]	3,957	−24.6401 3999
VMC	STO	Gálvez et al. (2005) [22]	[2s2p1d]		−24.6450 2(6)
FCI	GTO	Mayer[a] (2009) [38]	[6s5p3d1f] aug-cc-pCVTZ[d]	21.7 mill.	−24.6455 8509
Hy-CI	STO	This work	[5s4p3d]	1,146	−24.6481 4427
CI(SDTQ)	STO	Sasaki and Yoshimine (1974) [54]	[8s7p6d5f4g3h2i]	798	−24.6500
MR-CI	GTO	Feller and Davidson (1988) [18]	[23s12p10d4f2g]		−24.6511
MC-HF	Num.	Jönsson and Froese Fischer (1994) [28]	[8s7p6d5f4g3h2i1k]	7,096	−24.6510 09
MC-CI	GTO	Meyer et al. (1995) [39]	[15s10p7d6f4g]	1.5 mill.	−24.6518 1
MR-CI	Num.	Jönsson and Froese Fischer (1994) [28]	[8s7p6d5f4g3h2i1k]	32,456	−24.6527 25
DMC	STO	Brown et al. (2007) [6]	$n \le 7, 1 \le 5$[b]		−24.65357(3)
r_{12}-MR-CI	GTO	Gdanitz (1998) [23]	[18s13p2d1f]		−24.65379
FCI	STO	Almora and Bunge[a] (2010) [1]	$n \le 17, 1 \le 15$	7.2 mill.	−24.6538 00034
ECG	GTO	Komasa[a] (2009) [33]		2,048	−24.65381
ECG	GTO	Bubin et al. (2009) [7]		2,000	−24.65384
Estimated		Almora and Bunge (2010) [1]			−24.653862(2)
Estimated exact		Chakravorty et al. (1993) [13]			−24.65391

[a]Personal communication

[b]Notation orbitals i.e., here $n = 3$, $l = 2$ means the set of orbitals $1s, 2s, 2p, 3s, 3p, 3d$

[c]Lenght of the wave function expansion. No. of configurations in case of STOs, with GTO No. Slater determinants

[d]core/val. functions cc-pCVTZ added

configurations or functions (SAF) which are linear combinations of five-electron basis functions being eigenfunctions of the L-operator. The use of SAF saves memory space and computer time.

The program will be used for the calculation of the final state probabilities ^{10}B in ground and various excited states necessary to obtain the probabilities of formation of the He, Li ions during the boron nuclear reaction in the Boron Neutron Capture Therapy (BNCT), and to the study of the β-decay process of B atom to C^+ ion.

6.5 Conclusions

We have done and are currently extending the first up-to-date genuine Hy-CI calculations for the boron atom, showing that such calculations are possible. The methods, techniques and programs developed here can be adapted in a straightforward fashion to treat the C and N atoms and further to the second row elements of the periodic table. Therefore, from the computational point of view, calculations on the second row of elements are possible using the Hy-CI method. From the theoretical point of view this is also the case, since all necessary integrals have been analytically evaluated for atoms with $N \geq 5$.

In view of the short expansion of the Hy-CI wave function compared with other methods it is shown that the Hy-CI method is a very powerful one. When all kinds of configuration including all the different interelectronic distances are added to the present wave function, the final results are expected to be improved.

Acknowledgements I would like to thank Philip E. Hoggan for the invitation to contribute to these Proceedings. I am indebted to James Sims for interesting discussions about the Hy-CI method, and for providing highly accurate results of three- and four-electron integrals to test the program code. The high precision Vkl, Condon and Shortley coefficients and triangle integral programs of James Sims and Stanley Hagstrom are greatly acknowledged. I would like to thank very much Carlos Bunge for helpful advice on the CI method. It is a pleasant duty to acknowledge Peter Otto for advising in efficient Fortran programming and for supporting this project. Finally, I am indebted to the anonymous Referee for the careful revision, valuable comments and insights.

References

1. Almora-Díaz CX, Bunge CF (2010) Int J Quantum Chem 10:2982–2988
2. Aquino N, Garza J, Flores-Riveros A, Rivas-Silva JF, Sen KD (2006) J Chem Phys 124:054311
3. Bailey DH High-precision software directory. Available at: http://crd.lbl.gov/dhbailey/mpdist/ Accessed 10 Aug 2011
4. Barrois R, Lüchow A, Kleindienst H (1997) Int J Quantum Chem 62:77
5. Brown RE (1967) PhD. thesis, Department of Chemistry, Indiana University, USA
6. Brown MD, Trail JR, López Ríos P, Needs RJ (2007) J Chem Phys 126:224110
7. Bubin S, Stanke M, Adamowicz L (2009) J Chem Phys 131:044128
8. Budzinski J (2004) Int J Quantum Chem 97:832
9. Bunge CF (2010) Theor Chem Acc 126:139

10. Büsse G, Kleindienst H, Lüchow A (1998) Int J Quantum Chem 66:241
11. Cencek W, Rychlewski J (1993) J Chem Phys 98:1252
12. Cencek W, Szalewicz K (2008) Int J Quantum Chem 108:2191
13. Chakravorty SJ, Gwaltney SR, Davidson ER, Parpia FA, Fischer CF (1993) Phys Rev A 47:3649
14. Clary DC (1977) Mol Phys 34:793
15. Clary DC, Handy NC (1976) Phys Rev A 14:1607
16. Condon EU, Shortley GH (1967) The theory of atomic spectra. Cambridge University Press, Cambridge, MA
17. Drake GWF (1999) Phys Scripta T83:83
18. Feller D, Davidson ER (1988) J Chem Phys 88:7580
19. Fischer CF Personal communication
20. Frankowski K, Pekeris CL (1966) Phys Rev 146:46
21. Frolov AM, Ruiz MB (2010) Phys Rev A 82:042511
22. Gálvez FJ, Buendía E, Sarsa A (2005) J Chem Phys 122:154307
23. Gdanitz RJ (1998) J Chem Phys 109:9795
24. Hill RN (1985) J Chem Phys 83:1173
25. Hoggan PE, Ruiz MB, Özdogan T (2011) Molecular integrals over slater-type orbitals. From pioneers to recent progress. In: Putz MV (ed) Quantum frontiers of atoms and molecules. Nova Publishing Inc., New York, pp 61–89
26. Hylleraas EA (1929) Z Phys 54:347
27. James HM, Coolidge AS (1936) Phys Rev 49:688
28. Jönsson P, Froese Fischer C (1994) Phys Rev A 50:3080
29. Kato T (1957) Commun Pure Appl Math 10:151
30. King FW (1999) Adv Mol Opt Phys 40:57
31. kleindienst H, Büsse G, Lüchow A (1995) Int J Quantum Chem 53:575
32. Klopper W, Bachorz RA, Tew DP, Hättig C (2010) Phys Rev A 81:022503
33. Komasa J personal communication
34. Kutzelnigg W (1985) Theor Chim Acta 86:445
35. Kutzelnigg W, Morgan JD III (1992) J Chem Phys 96:4484
36. Kurokawa Y, Nakashima H, Nakatsuji H (2005) Phys Rev A 72:062502
37. MAPLE 9 Release by Waterloo Maple Inc. Copyright 2003
38. Mayer I (1991) Personal communication. HONDO-8, from MOTECC-91, contributed and documented by M. Dupuis and A. Farazdel, IBM Corporation Center for Scientific & Engineering Computations, Kingston, NY. Knowles PJ, Handy NC Chem Phys Lett 111:315 (1984); Comput Phys Commun 54:75 (1989)
39. Meyer H, Müller T, Schweig A (1995) Chem Phys 191:213
40. Merckens H-P (2002) Eigenwertberechnungen an angeregten ^1S-Zuständen des Berylliumatoms, Thesis, Düsseldorf, Germany
41. Mohr PJ, Taylor BN, Newell DB (2008) CODATA Recommended values of the fundamental physical constants: 2006. Rev Mod Phys 80:633–730
42. Nakashima H, Nakatsuji H (2008) J Chem Phys 128:154107
43. Nakashima H, Hijikata Y, Nakatsuji H (2008) J Chem Phys 128:154108
44. Pachucki K (2010) Phys Rev A 82:032509
45. Preiskorn A, Frey D, Lie GC, Clementi E (1991) In: Clementi E (ed) Modern techniques in computational chemistry: MOTECC-91. ESCOM Science Publishers, Leiden, Chapter 13
46. Przybytek M, Cencek W, Komasa J, Lach G, Jeziorski B, Szalewicz K (2010) Phys Rev Lett 104:183003
47. Puchalski M, Kedziera D, Pachucki K (2009) Phys Rev A 80:032521
48. Ruiz MB (2009) J Math Chem 46:24
49. Ruiz MB (2009) J Math Chem 46:1322
50. Ruiz MB (2005) Int J Quantum Chem 100:246

51. Ruiz MB, Peuker K (2008) In: Ozdogan T, Ruiz MB Recent advances in computational chemistry: molecular integrals over slater orbitals. Transworld Research Network, Kerala, pp 99–144
52. Ruiz MB, Rojas M (2003) Comput Methods Sci Technol 9(1–2):101
53. Ruiz MB, Rojas M, Chicón G, Otto P (2010) Int J Quantum Chem 111, 1921 (2011)
54. Sasaki F, Yoshimine M (1974) Phys Rev A 9:17
55. Schaefer HF, Harris FE (1968) Phys Rev 167:67
56. Schwartz C (1962) Phys Rev 126:1015
57. Schwartz C (2006) Int J Mod Phys E15:877; e-Print arXiv:physics/0208004; Updated results in e-Print math-phys/0605018
58. Sharkey KL, Bubin S, Adamowicz L (2010) J Chem Phys 132:184106
59. Sims JS, Hagstrom SA (1971) J Chem Phys 55:4699
60. Sims JS, Hagstrom SA (1971) Phys Rev A 4:908
61. Sims JS, Hagstrom SA (2002) Int J Quantum Chem 90:1600
62. Sims JS, Hagstrom SA (2004) J Phys B: At Mol Opt Phys 37:1519
63. Sims JS, Hagstrom SA (2006) J Chem Phys 124:094101
64. Sims JS, Hagstrom SA (2009) Phys Rev A 80:052507
65. Sims JS, Hagstrom SA (2011) Phys Rev A 83:032512
66. Stanke M, Komasa J, Bubin S, Adamowitz L (2009) Phys Rev A 80:022514
67. Stevenson R (1965) Multiplet structure of atoms and molecules. W.B. Saunders Company, Philadelphia & London
68. Walsh P, Borowitz S (1960) Phys Rev 119:1274
69. Weiss AW (1961) Phys Rev 122:1826
70. Weiss AW (1969) Phys Rev 188:119
71. Woznicki W (1971) In: Jucys A (ed) Theory of electronic shells in atoms and molecules. Mintis, Vilnius, p 103
72. Zhou BL, Zhu JM, Yan ZC (2006) Phys Rev A 73:064503

Chapter 7
Structural and Electronic Properties of Po under Hydrostatic Pressure

A. Rubio-Ponce, J. Morales, and D. Olguín

Abstract Although Polonium (Po) is the only element of the periodic table that has a simple cubic (sc) crystal structure at ambient pressure, it is one of the least-studied elements, and its phase diagram is still unknown. Thus, with the aim to contribute to the study of the Po phase diagram, we present in this work theoretical calculations focused on determining the structural and electronic properties of Po under hydrostatic pressure. For that purpose, our theoretical study considers the sc structure as well as the hypothetical rhombohedral (r), body-centered cubic (bcc) and face-centered cubic (fcc) crystal structures. The calculations were performed using the full potential linearized augmented plane wave (FLAPW) method by using the local density approximation (LDA) for the exchange-correlation energy and by including the spin-orbit coupling to take into account relativistic effects. The total energy results were fitted to the third-order Birch-Murnaghan equation of state (EOS) for pressures up to 100 GPa. Consequently, the energy results, along with the enthalpy findings, indicate that phase transitions follow the sequence sc → r → bcc → fcc at pressure values of 2.47, 5.75, and 70.27 GPa, respectively. These results are consistent with the fact that the hydrostatic pressure induces a change in the atomic distance and in the orbital hybridization that leads to different crystal structures.

A. Rubio-Ponce (✉) • J. Morales
Departamento de Ciencias Básicas, Universidad Autónoma Metropolitana-Azcapotzalco,
Av. San Pablo 180, 02200 México, D.F., MÉXICO
e-mail: arp@correo.azc.uam.mx; jmr@correo.azc.uam.mx

D. Olguín
Departamento de Física, Centro de Investigación y de Estudios Avanzados del Instituto
Politécnico Nacional, A.P. 14740, C.P. 07300 México, D.F. MÉXICO
e-mail: daniel@fis.cinvestav.mx

P.E. Hoggan et al. (eds.), *Advances in the Theory of Quantum Systems in Chemistry and Physics*, Progress in Theoretical Chemistry and Physics 22, DOI 10.1007/978-94-007-2076-3_7, © Springer Science+Business Media B.V. 2012

7.1 Introduction

Polonium (Po) is a radioactive element that was discovered in 1898 by Marie
Skłodowska-Curie and Pierre Curie. Po is used in brushes to remove dust from
photographic films and to avoid charge static accumulation produced by several
processes, such as the rolling of paper, wire, and sheet metal. In addition, Po has
been alloyed with beryllium to be used as a neutron source. All these and other
applications depend on Po's structural properties. Po is the only element of the
periodic table that adopts the simple cubic (sc) structure at ambient pressure (a few
other elements such as Ca-III and As-II present the sc, but only at high pressure
[1]), and this structure has a low atomic packing factor and is rare in nature. The
first experimental studies of Po's crystal structure, by using electron diffraction,
were reported in 1936 by Rollier et al. [2]. Several years later, Beamer and Maxwell
[3, 4] and Sando and Lange [5] reported on their X-ray diffraction experiments on
metallic Po. From these reports, we know that Po exhibits two structural phases: the
α phase (α-Po), which has the sc structure $[O_h^1(Pm3m)]$, $a = 3.345(2)$ Å [4], and
the β phase (β-Po), stable above $77(9)°$C, which has the rhombohedral (r) structure
$[D_{3d}^5(R3m)]$, $a = 3.359(1)$ Å, and $\alpha = 98.22(5)°$.

   To date, Po is one of the least-studied elements, and its phase diagram is
unknown. Po is located in the VIA column of the periodic table and has an atomic
number equal to 84 and the following electronic distribution: [Xe] $4f^{14}5d^{10}6s^26p^4$.
The first theoretical study on Po was focused on explaining its existence as the only
element of the periodic table with the sc structure. Using the pseudopotential band
method, Kraig et al. [6] showed that the sc structure has the lowest total energy
and, therefore, that this structure is preferred by Po instead of the hypothetical
face-centered cubic structure (fcc) or body-centered cubic structure (bcc). They
argued that the large s-p splitting would produce the stable sc structure. In addition,
Lach-hab et al. [7] performed a structural study for Po using the tight-binding (TB)
approach; the parameters were fitted to the semirelativistic results obtained from the
full-potential linearized augmented plane wave (FLAPW) method. They also found
that the most stable structure is the sc; their calculations were done both with and
without spin-orbit (SO) interaction because for solids with a large atomic number,
such as lanthanides, actinides, and elements of layer six, the relativistic effects are
very important. In fact, Min et al. [8] found that the sc structure of Po is due to
strong SO interaction. Legut et al. [9, 10] using the FLAPW method within the local
density approximation (LDA) and the gradient generalized approximation (GGA),
including relativistic effects and analyzing the impact of different relativistic terms
on the structural properties, concluded that the stability of the sc structure of Po
is due to the relativistic mass-velocity and Darwin terms. They also predicted that
the sc structure of Po becomes unstable when the hydrostatic pressure is increased
above 3 GPa.

   With the aim of contributing to the study of the Po phase diagram, we discuss in
this paper the structural and electronic properties of Po under hydrostatic pressure.
We consider the sc structure and the r, bcc, and fcc as hypothetical crystal structures

for pressures up to 100 GPa. The rest of the paper is organized as follows: Sect. 7.2 summarizes the computational details. Section 7.3 shows our results. Finally, Sect. 7.4 shows our conclusions.

7.2 Computational Details

The structural and electronic properties of Po have been obtained by using the FLAPW method as implemented in the WIEN2k package [11]. In our study, we have used the LDA correction for the exchange-correlation part of the total energy [12], and we have included the SO coupling as well. To minimize the total energy, we have used the next variational parameters, the plane-wave expansion parameter $R_{mt}K_{max} = 9.0$ (where R_{mt} is the muffin tin radius and K_{max} is the plane wave cutoff), while the energy cutoff parameter was $G_{max} = 14$. The convergence criterion corresponds to the energy difference less than 10^{-5} Ry in all cases.

In our study, we have considered the sc crystal structure ($Pm3m = O_h^1$, space group 221) and three other hypothetical crystal structures: the r ($R\bar{3}m = D_{3d}^5$, space group 166), the bcc ($Im3m = O_h^9$, space group 229), and the fcc ($Fm3m = O_h^5$, space group 225). In the full, irreducible part of the first Brillouin zone (FBZ), the number of k-points used was 816 for sc, fcc, and bcc, and for the r structure, the number of k-points used was 2,736.

7.3 Results

The structural bulk modulus and lattice parameters are obtained by fitting the total energy of each primitive cell to the third-order Birch-Murnaghan equation of state (EOS) [13], which reads

$$P = \frac{3}{2}B_0 \left(\eta^{\frac{7}{3}} - \eta^{\frac{5}{3}}\right)\left(1 - \frac{3}{4}(4 - B_0')\left(\eta^{\frac{2}{3}} - 1\right)\right), \qquad (7.1)$$

where $\eta = (V_0/V)$, V_0 is the volume at zero pressure. Using as a reference the total energy of the sc phase, Fig. 7.1 shows the total energy differences (ΔE) for the proposed hypothetical structures without considering SO coupling. As can be observed, the sc phase has the lowest energy. With respect to the sc curve, the minimum of the energy curve for the hypothetical crystal structures r, bcc, and fcc is located at 12.667 meV, 198.194 meV, and 278.236 meV, respectively.

Similarly to the above mentioned case, Fig. 7.2 displays the ΔE curves obtained, including the SO interaction. As previously mentioned, the sc phase has the lowest energy, but the energy difference between the sc and r structures decreases by 4.3% (12.123 meV), whereas between sc and bcc, the energy difference decreases by 38.1% (122.614 meV) and between sc and fcc around the 15.5% (235.038 meV).

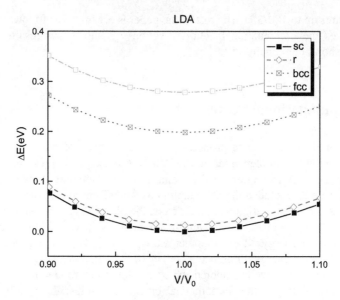

Fig. 7.1 (Color online) Total energy difference of α-Po as compared with the proposed hypothetical structures. The calculations were performed within the LDA correction, and here no SO coupling is considered

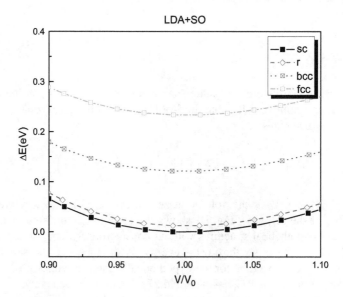

Fig. 7.2 (Color online) Total energy difference for α-Po as compared with the proposed hypothetical structures obtained by including SO coupling in the calculations

Table 7.1 Experimental and theoretical data found in the literature for the lattice parameter (a), bulk moduli (B_0) and pressure derivative bulk modulus (B'_0) of α-Po

	a(Å)	B_0 (GPa)	B'_0	Ref.
Exp.	3.345(2)			[4]
	3.359(1)			[5]
LDA	3.28	56		[6]
	3.34	44		[7]
	3.277	57.1		[10]
	3.335	39.45	4.89	[14]
	3.272	59.93	4.32	This work
LDA + SO	3.334	42.3		[10]
	3.323	47.35	4.59	This work
GGA	3.356	46.9		[10]
GGA + SO	3.34			[8]
	3.411	40.1		[10]
TB	3.26	59		[7]
TB +SO	3.29	51		[7]

From these graphs, we can conclude that the SO interaction is not necessary to stabilize the α-Po phase. Our calculations confirm the interpretation of Legut et al. [9] that the scalar relativistic terms are sufficient to stabilize the simple cubic phase.

Our calculated structural parameters for Po are in good agreement with known experimental and recent theoretical data [4–10,14,15]; see Table 7.1. By considering SO coupling, for α-Po, our calculated bulk modulus (B_0) and first derivative of the bulk modulus (B'_0) are 47.35 GPa and 4.59, respectively, while the lattice parameter is $a = 3.323$ Å. On the other hand, the corresponding values without SO coupling are $B_0 = 59.93$ GPa, $B'_0 = 4.32$, and $a = 3.272$ Å. That is, the bulk modulus, which is related to the lattice constant value, decreases when SO is included. As can be seen, the calculated values for the lattice parameter, with the exception of those calculated using GGA and GGA + SO [10], are smaller than the experimental data. Unfortunately, there are no experimental data for the elastic properties, bulk, and Shear and Young's modulus to compare to our findings. Lach-hab et al. [7] claim that the experimental bulk modulus of sc Po is 26 GPa; however, they do not provide any reference for it. On the other hand, our bulk modulus result is consistent with other theoretical calculations.

In order to analyze the influence of the SO interaction on the electronic properties of Po, we compare the calculated electronic band structure with and without SO coupling. Figure 7.3 shows the Po electronic band structure along high-symmetry directions in the FBZ of the sc structure. The energy values have been uniformly shifted to set the Fermi level (E_F) at zero energy. Solid lines show the calculations including SO coupling, and dashed lines show the calculations without SO. In that figure, it is easy to see the effect of SO; when SO interaction is included, some valence bands do not cross the E_F, these bands have mainly the $6p$ character. Another interesting effect is the splitting of bands placed around -30 eV. These bands give origin to two new bands with a gap of 3.4 eV; the bands are shown at the

Fig. 7.3 (Color online)
Electronic band structure for
α-Po. Here we show our
results with SO coupling
(*solid line*) and without SO
coupling (*dashed line*)

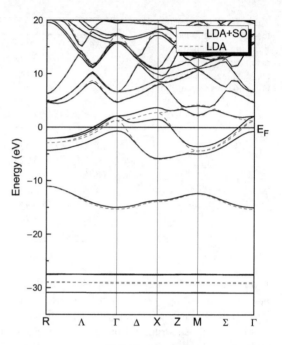

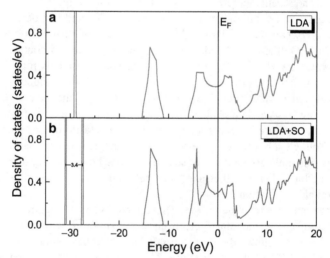

Fig. 7.4 (Color online) Comparison of the calculated DOS without SO coupling (*upper panel*)
and including the SO coupling (*lower panel*). Note the induced band splitting for the lower states
(around −30.0 eV) due to the SO coupling, which is around 3.4 eV

bottom of Fig. 7.3. These results are also shown in Fig. 7.4, in which we compare
the calculated density of states (DOS) for both theoretical calculations, with and
without SO coupling.

Fig. 7.5 (Color online) Total and partial contributions to the calculated DOS. The splitting of the lower bands occurs on the *d* orbital states, *panel d*

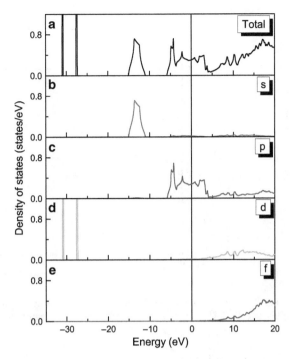

The partial contributions of the *s*, *p*, *d*, and *f* states of Po to the total DOS, including the SO coupling, are shown in Fig. 7.5. Figure 7.5a shows the total DOS. For energies around -12 eV, Fig. 7.5b shows that the main contribution to these energies is from the 6*s* orbitals. Due to the SO coupling, the band width of these states decreases from 4.40 to 4.04 eV. Figure 7.5c presents the contribution of the 6*p* orbitals, and these states are located around the E_F. It appears that the SO coupling does not change the shape of *s-p* hybridization. Consequently, we conclude that Po can be classified as a *p*-type element. The states around -30 eV are derived mainly from the 5*d* orbitals, and are shown in Fig. 7.5d. Above the E_F we found the 4*f* orbitals, as it is shown in Fig. 7.5e.

Finally, with the aim of finding possible phase transitions between the different crystal structures considered in this work, we have calculated the enthalpy ($H = E + pV$) relative to the sc structure in the range from 0 to 100 GPa. According to our results, the phase transition sequence is given by sc → r → bcc → fcc, that corresponds to pressure values of 2.47, 5.75, and 70.27 GPa, respectively (see Fig. 7.6). These values are in agreement with the results of Legut et al. [10]. They claim that when the pressure increases, Po loses its sc structure and becomes rhombohedral in the interval of 1–3 GPa. It turns out that, to the best of our knowledge, experimental information on phase transition as function of the pressure is not available.

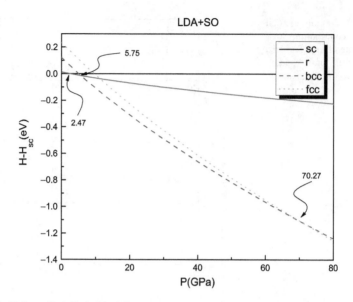

Fig. 7.6 (Color online) Enthalpy difference vs. pressure for the different Po structures. The cross line at 2.47, 5.75, and 70.27 GPa indicates the phase transition

7.4 Conclusions

From total energy calculations, we have found that α-Po has the lowest energy with respect to r, bcc, and fcc structures. Our calculated lattice parameter, $a = 3.323$ Å, has been found to be in good agreement with recent theoretical data. The bulk modulus B_0 and its first pressure derivative B_0' were found to be equal to 47.35 GPa and 4.59, respectively. Our results show that both LDA and LDA + SO calculations predict that, in comparison with the r, bcc, or fcc structures, Po has the sc structure as the stable phase. In addition, we can conclude that the SO coupling is not necessary to stabilize the α-Po structure. However, we found that the SO coupling should be included to properly describe the Po electronic band structure. Finally, according to our results, it becomes clear that there is a sequence for the phase transition as a function of the pressure: sc $\rightarrow$ r $\rightarrow$ bcc $\rightarrow$ fcc, from low to high pressure.

Acknowledgements This work was partially supported by the projects UAM-A-CBI-2232004 and 2232013 México. We gratefully acknowledge the SNI-Conacyt-México for the distinction of our membership and the stipend received.

References

1. McMahon MI, Nelmes RJ (2006) Chem Soc Rev 35:943
2. Rollier MA, Hendricks SB, Maxwell LR (1936) J Chem Phys 4:648

3. Beamer WH, Maxwell CR (1946) J Chem Phys 14:569
4. Beamer WH, Maxwell CR (1949) J Chem Phys 17:1293
5. De Sando RJ, Lange RC (1966) J Inorg Nucl Chem 28:1837
6. Kraig RE, Roundy D, Cohen ML (2004) Solid State Commun 129:411
7. Lach-hab M, Akdim M, Papaconstantopoulos DA, Mehl MJ, Bernstein N (2004) J Phys Chem Solids 65:1837
8. Min I, Shim JH, Park MS, Kim K, Kwon SK, Youn SJ (2006) Phys Rev B 73:132102
9. Legut D, Friák M, Šob M (2007) Phys Rev Lett 99:016402
10. Legut D, Friák M, Šob M (2010) Phys Rev B 81:214118
11. Blaha P, Schwarz K-H, Madsen GKH, Kvasnicka D, Luitz J (2001) WIEN2K, an augmented plane wave plus local orbitals program for calculating crystal properties. Vienna University of Technology, Vienna
12. Perdew JP, Wang Y (1992) Phys Rev B 45:13244
13. Birch F (1947) Phys Rev 71:809
14. Verstraete MJ (2010) Phys Rev Lett 104:035501
15. Lohr LL (1987) Inorg Chem 26:2005

Chapter 8
Complexity Analysis of the Hydrogenic Spectrum in Strong Fields

R. González-Férez, J.S. Dehesa, and K.D. Sen

Abstract The hydrogenic spectrum in strong parallel magnetic and electric fields is studied here. The statistical shape complexity measure is used for the identification and characterization of the most distinctive spectroscopic features of complex systems caused by slowly varying perturbations, the so-called avoided crossings. This is illustrated for some pairs of hydrogenic levels in presence of strong magnetic and electric fields. It is found that the LMC shape complexities of the two states involved in an avoided crossing are mutual mirror images with respect to the axis at the critical magnetic field strength giving rise to the avoided crossing. This is a clear manifestation of the information-theoretic character exchange taking place between the two quantum states involved in the crossing.

8.1 Introduction

The avoided crossings between a pair of neighboring levels with the same symmetry [1] are the most distinctive non-linear signatures of the spectra of numerous physical systems which play a significant role from both scientific and technological points of view. They are characteristic of the chaotic region of the spectrum of

R. González-Férez
Instituto 'Carlos I' de Física Teórica y Computacional and Departamento de Física Atómica, Molecular y Nuclear, Universidad de Granada, E-18071 Granada, Spain
e-mail: rogonzal@ugr.es

J.S. Dehesa (✉)
Instituto 'Carlos I' de Física Teórica y Computacional and Departamento de Física Atómica, Molecular y Nuclear, Universidad de Granada, E-18071 Granada, Spain
e-mail: dehesa@ugr.es

K.D. Sen
School of Chemistry, University of Hyderabad, Hyderabad 500046, India
e-mail: sensc@uohyd.ernet.in

P.E. Hoggan et al. (eds.), *Advances in the Theory of Quantum Systems in Chemistry and Physics*, Progress in Theoretical Chemistry and Physics 22,
DOI 10.1007/978-94-007-2076-3_8, © Springer Science+Business Media B.V. 2012

the hydrogen atom in strong electric and magnetic fields. They also appear in the electronic spectra of anisotropic two-electron quantum dots as the condition defining the spatial confinement changes from a nano-ring to nano-wire [2]. In the rovibrational spectra of polar dimers in homogeneous static electric fields, the avoided crossings lead to strongly distorted and asymmetric molecular states including well pronounced localization effects of their probability density [3]. Recently, a Stückelberg interferometer with ultracold Cs_2 molecules has been developed by means of weak avoided crossings, which are used as beam splitter for molecular states, providing a full control on the interferometer dynamics [4]. An efficient experimental scheme based on Zeeman tuning of molecular states towards the avoided level crossing has been devised to transfer molecules within the neighboring levels [5].

An avoided crossing is characterized by (a) an energy repulsion, and (b) an *information-theoretic character* exchange between the associated pair of quantum mechanical states. The former provides a mechanism for state reordering with energy under the slowly varying external perturbation (e.g. when a magnetic field changes adiabatically). The latter can be identified and characterized by means of information-theoretic measures of the probability densities corresponding to the two states involved in the crossing. This has been shown by means of the Shannon entropy [6, 7] and the Fisher information [8], which manifest sharp extrema and crossing behaviors, respectively, through this highly irregular region.

The complexity measures (See, e.g. [9–14]) have been recently shown to describe and characterize the non-uniformity of the quantum-mechanical probability density of the energy levels of physical systems (so, the internal disorder of the system at those energies) in a better way that a single information-theoretic element such as the Shannon entropy and the Fisher information. This is because they are defined in terms of the product of two complementary information-theoretic measures representing the global and local characteristics of the probability distribution. Consequently, such complexity measures may be used to disentangle and highlight the irregularities of the probability densities most distinctively. Moreover, they reflect the intuitive notion of disorder in the sense that they vanish for the two extreme probability distributions associated to perfect order (Dirac-like delta distribution) and maximum disorder (uniform distribution). Among all the composite complexity measures, the so-called LMC (López-Ruiz-Mancini-Calbet) shape complexity [9, 10, 15] occupies a special position because it satisfies the invariance properties under coordinate scaling, translation and replication. The LMC shape complexity of the energy level characterized by the normalized-to-unity quantum-mechanical density $\rho(\mathbf{r})$ is defined by

$$C \equiv C[\rho] \equiv D[\rho] \exp(S[\rho]) \tag{8.1}$$

where

$$S[\rho] = -\int \rho(\mathbf{r}) \log \rho(\mathbf{r}) d\mathbf{r} \tag{8.2}$$

denotes de Shannon entropy of the system [16], which measures the bulk spreading of the charge cloud, and

$$D[\rho] = \int [\rho(\mathbf{r})]^2 d\mathbf{r} \qquad (8.3)$$

is the averaging density or disequilibrium of the system which gives the quadratic distance to the equilibrium or most probable state [17–19], at times experimentaly amenable [20].

In this paper we study the avoided crossing phenomenon by means of the shape complexity of the two quantum states involved in it. We note here that this physical quantity is determined not by the energy eigenvalue but by the eigenfunction $\psi(\mathbf{r})$ of the Hamiltonian since $\rho(\mathbf{r}) = |\psi(\mathbf{r})|^2$. So, we are not so much interested in the repulsion of the two energy levels (although we will show it for completeness) but rather in locating the information-theoretic character exchange between the corresponding quantum eigenfunctions which is revealed by the LMC shape complexity. This will be illustrated by the detailed numerical analysis of the avoided crossings of the $(3p_0, 3d_0)$ and $(5p_0, 5d_0)$ of hydrogen atom in the presence of a combination of intense parallel electric and magnetic fields.

We believe that the present analysis of the avoided crossings of levels through the statistical complexity measure, C, introduces a novel theoretical tool based on the electron density integrals exclusively, rather than those which are usually derived from the quantum mechanical expectation values. The density based analysis offers the future possibilities of evaluating such crossings under the influence of the spatial confinement by the changes in the electron probability distribution brought about, e.g., by an additional hard or soft external potential, including the ultra-high magnetic fields.

8.2 Methodology for Hydrogenic Applications

In this section, we consider as a prototype system the hydrogen atom exposed to a combination of parallel magnetic and electric fields, and numerically analyze some avoided crossings appearing in the corresponding spectrum. Atomic units will be used throughout, unless stated otherwise. The Hamiltonian of the hydrogen atom in the presence of uniform magnetic $\mathbf{B}$ and electric $\mathbf{F}$ fields, both oriented along the z-axis, is given by

$$H = -\frac{1}{2r^2}\frac{\partial}{\partial r}r^2\frac{\partial}{\partial r} + \frac{1}{2r^2}\mathbf{L}^2(\theta,\phi) - \frac{1}{r} + \frac{B^2}{2}r^2\sin^2\theta + Fr\cos\theta, \qquad (8.4)$$

where spherical coordinates have been used and $\mathbf{L}^2(\theta,\phi)$ denotes the squared angular momentum. The magnetic field strength B is measured in units of $B_0 = \frac{2\alpha^2 m_e^2 c^2}{\hbar e} \approx 4.701 \cdot 10^5\, T$, and the electric field strength F in units of $F_0 = \frac{\alpha^3 m_e^2 c^3}{e\hbar} \approx 5.142 \cdot 10^{11} V/m$. Note that we have neglected the relativistic corrections [21], and the spin-orbit coupling [22]. Besides, the motion of the nucleus is not explicitly considered because in strong magnetic fields its effect may be accounted for by a constant shift in the energy which is exact for parallel fields [23]. In addition we have disregarded in the Hamiltonian (8.4) the paramagnetic term $B \cdot L_z$, where L_z is the z-component of the angular momentum.

In the presence of a magnetic field, parallel to the z-axis, the magnetic quantum number and the z-axis parity are good quantum numbers, while in an additional electric field only the azimuthal symmetry remains. Nevertheless, we will label the field-dressed states with the corresponding field-free hydrogenic quantum numbers n, l and m for reasons of simplicity, and to facilitate the physical discussion later on. Furthermore, only $m = 0$ states are considered. The computational approach to solve the non-integrable two-dimensional equation of motion associated to the Hamiltonian (8.4), is based in a combination of the finite element method for the radial coordinate and the discrete variable technique for the angular one, together with a Krylov type diagonalization technique [24].

8.3   Main Results

In this section, we interpret the numerical results on the avoided crossings between the pairs of hydrogenic states $(3p_0, 3d_0)$ and $(5p_0, 5d_0)$ in the presence of intense electric and magnetic fields. We first consider the pair of states $3p_0$ and $3d_0$ of the hydrogen atom. Figures 8.1a and b show the corresponding ionization energies and shape complexities (SCs, in short), respectively, as the magnetic field strength varies

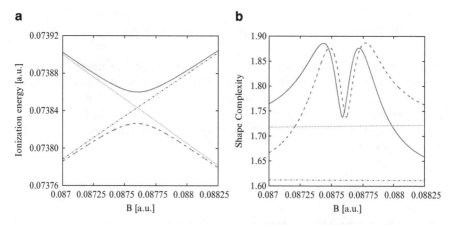

Fig. 8.1 Color online. The ionization energies (**a**) and shape complexities (**b**) of the states $3p_0$ (*dashed line*) and $3d_0$ (*solid line*), of the hydrogen atom in parallel electric and magnetic fields as a function of the magnetic field strength. The electric field was fixed to $F = 1.946 \times 10^{-6}$ a.u. Besides the results for vanishing electric field have been also included for the $3p_0$ (*dotted-dashed line*) and $3d_0$ (*dotted line*) levels

over the range 0.087 a.u. $- 0.08825$ a.u. and the electric field strength $F = 1.946 \times 10^{-6}$ a.u. For the sake of completeness, the vanishing electric field results have been also included. For $F = 0$, these states have different z-parity, and their energies cross for a certain magnetic field strength B_c (see Fig. 8.1a). Indeed, the energy of the $3d_0$ level monotonically decreases as the magnetic field is enhanced, while the behavior of the $3p_0$ energy is just the opposite. These two levels do not interact between each other, and the SCs keep constant in the magnetic field range (see Fig. 8.1b). Their difference $\Delta C^s = C^s_{3p_0} - C^s_{3d_0}$ is always negative, which indicates that the electronic cloud of the $3p_0$ state is more confined.

The presence of the electric field breaks the degeneracy of this pair, hence they have the same symmetry and due to the Wigner-non-crossing rule [1] an avoided crossing is formed between them. Their ionization energies present the typical behavior of this non-linear phenomenon. They approach each other with increasing magnetic field, until they come close and strongly interact, splitting apart thereafter. The minimal energetic spacing $\Delta E = |E_{3p_0} - E_{3d_0}| = 3.35 \times 10^{-5}$ a.u. occurs at the field strength $B_c = 8.760038 \times 10^{-2}$ a.u. Far away from the strongly interacting region the behavior of the energies resemble the $F = 0$ results.

Before taking up the specific discussion on our results, let us highlight that the existence of extremum points in the SC is a consequence of the confinement effects brought about by the presence of external fields. This is in line with the recent finding of Patil et al. [25], who show for some specific constrained Coulomb potentials that the simplest composite uncertainty measure, the Heisenberg uncertainty product, of the electron density presents an extremum located at a critical position which scales as the reciprocal value of the potential strength.

We shall now analyze in detail the interesting structure observed for the SCs of the two states involved in this avoided crossing. Both quantities as a function of B have similar behaviors. They present a double-hump structure with a mirror symmetry with respect to (the axis located at) the critical value B_c. Even more, they are very different from their electric-field-free values. Indeed both quantities are no more constant but they show up two maxima, what illustrates the strong interaction taking place between these levels in this irregular region. The SCs increase as the magnetic field strength is augmented, pass through a first maximum, and with a further enhancement of the magnetic field they achieve a pronounced minimum, followed by a second maximum, and finally they decrease thereafter. In particular, their difference, $\Delta C^s = C^s_{3p_0} - C^s_{3d_0}$, changes its sign several times and there are three B values at which both SCs are equal. Indeed, the SCs cross at $B_c = 8.760044 \times 10^{-2}$ a.u., being their difference $\Delta C^s = 6.29 \times 10^{-6}$ a.u., which is very close to the critical B-value for the minimal energetic spacing. The SC minimal values are equal for both states, $C^s_{3p_0} = C^s_{3d_0} = 1.7380$, and are located at symmetric positions with respect to the critical magnetic field value B_c, i.e., they shifted to the left $B = B_c + 1.569 \times 10^{-5}$ a.u. and to the right $B = B_c - 1.566 \times 10^{-5}$ a.u. for the $3p_0$ and $3d_0$ states, respectively. The first hump of $C^s_{3d_0}$ and the second one of $C^s_{3p_0}$, also have a very similar value ($C^s_{3d_0} = 1.8856$, $C^s_{3p_0} = 1.8867$), and are shifted to the left by 1.6706×10^{-4} a.u. and to the right by 1.6632×10^{-4} a.u., respectively.

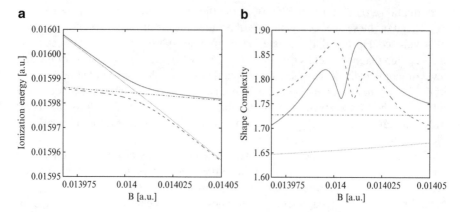

Fig. 8.2 Color online. The ionization energies (**a**) and shape complexities (**b**) of the states $5p_0$ (*dashed line*) and $5d_0$ (*solid line*), of the hydrogen atom in parallel electric and magnetic fields as a function of the magnetic field strength. The electric field was fixed to $F = 10^{-7}$ a.u. Besides the results for vanishing electric field have been also included for the $5p_0$ (*dotted-dashed line*) and $5d_0$ (*dotted line*) levels

In a similar way, the second maxima of $C^s_{3p_0}$ and the first one of $C^s_{3d_0}$ are located at $B = B_c + 1.1591 \times 10^{-4}$ a.u. and $B = B_C - 1.1566 \times 10^{-4}$ a.u., respectively, with $C^s_{3p_0} = 1.8766$ and $C^s_{3d_0} = 1.8762$.

Analogous studies were done for other pairs of hydrogenic states with higher principal quantum numbers, and similar results were obtained. Let us now discuss the avoided crossing between the levels $5p_0$ and $5d_0$ of the hydrogen atom in parallel magnetic and electric fields. The corresponding ionization energies and SCs are presented in Figs. 8.2a and b, respectively, as a function of the magnetic field strength, within the range 0.013975 a.u. $\leq B \leq 0.01405$ a.u. and with $F = 10^{-7}$ a.u. The electric field-free results are also shown in the same figures. Even such a weak electric field breaks the degeneracy of this pair, and a very narrow avoided crossing is formed. The minimal energetic spacing between the levels is $\Delta E = |E_{5p_0} - E_{5d_0}| = 7.054 \times 10^{-6}$ a.u. at the critical field value $B_c = 1.400888 \times 10^{-2}$ a.u. Both energies approach the corresponding electric field-free results as we move away from the non-linear region.

For $F = 0$, the SCs of both states smoothly increase as the magnetic field is enhanced, showing that the confinement of the electronic clouds does not experiment large variations over the considered range of magnetic field strengths. The SC difference $\Delta C^s = C^s_{5p_0} - C^s_{5d_0}$ is always positive, indicating the higher confinement of the $5p_0$ state. Again, the electric field induces drastic changes on the SCs. As in the previous example, they are characterized by a double-hump structure and the specular symmetry with respect to the critical value $B_c = 1.40088 \times 10^{-2}$ a.u., where the SC difference is $\Delta C^s = 1.08 \times 10^{-4}$ a.u. As the magnetic field is tuned through this highly irregular region, ΔC^s changes its sign three times and finally, for $B > B_c$ we encounter that $\Delta C^s > 0$. These features indicate that together with a strong coupling and interaction between both states, the electronic clouds get significantly affected in the avoided crossing region and the $5d_0$ level is now confined more strongly.

8.4 Conclusions

The phenomenon of avoided crossing between a pair of quantum states is characterized by the repulsion of the two energy eigenvalues and the character exchange of the corresponding eigenfunctions. In this paper we emphasize the latter characterization as revealed by the statistical shape complexity. This has been done by numerically analyzing two avoided crossings appearing in the spectrum of the hydrogen atom in the presence of a combination of intense parallel electric and magnetic fields. The main conclusions derived in this work can be summarized as follows. First, the shape complexities of the quantum-mechanical states involved in an avoided crossing (a) show up a double-hump structure because of the confinement provoked by the external fields and (b) present a mirror symmetry with respect to the axis located at the critical magnetic field strength which has given rise to this highly irregular phenomenon. Moreover, the state with stronger (weaker) confinement get less (more) confined in going through that region. This confinement exchange of the two energy levels involved in the avoided crossing seems to indicate that in order to understand the enormous complexity of the whole spectrum of a hydrogenic system in presence of external fields, it is useful to study the statistical complexity measure of the quantum wavefunctions rather than the redistribution of the corresponding eigenenergies. In this sense, it would be very instructive to analyze the LMC shape complexity of the eigenfunctions in both the position and momentum space representations. Second, the minimal energetic spacing of the levels occurs at the same magnetic field strength at which the two hydrogenic states involved in the avoided crossing have equal shape complexities.

Following an analysis similar to that performed by Patil et al. [25] for composite information measures it can be shown that under external fields the shape complexity is expressed as a *product* of the functions of two independent parameters $s_1 = BZ^{-2}$ and $s_2 = FZ^{-3}$. Such a study reveals the presence of extrema in the shape complexity as a function of the two external fields. The specific nature and the number of extrema can be obtained only if the functional form is known for a given electronic state which is not obtained through the dimensional analysis. The accurate numerical results presented here should be useful in the exploration of the analytic form of shape complexity of the np_0 and nd_0 states in presence of the external fields. We note here that while the general *product* form of shape complexity is independent of the relative orientation of the two fields, the specific form could depend on orientation. In view of this, it would be interesting to carry out numerical investigations similar to those presented here for a few arbitrarily fixed orientations of the external fields.

Acknowledgements Financial support by the Spanish project FIS2008–02380 (MICINN) as well as the Grants FQM-2445 and FQM–4643 (Junta de Andalucía) is gratefully appreciated. RGF and JSD belong to the Andalusian research group FQM-207. KDS acknowledges the financial grant under the J.C. Bose National Fellowship, Department of Science and Technology, New Delhi.

References

1. Von Neumann J, Wigner E (1929) Phys Z 30:467
2. Drouvelis PS, Schmelcher P, Diakonos FK (2003) Europhys Lett 64:232
3. González-Férez R, Schmelcher P (2005) Europhys Lett 72:555
4. Mark M, Kraemer T, Waldburger P, Herbig J, Chin C, Nägerl H-C, Grimm R (2007) Phys Rev Lett 99:113201
5. Lang F, Straten PVD, Brandsttter B, Thalhammer G, Winkler K, Julienne PS, Grimm R, Hecker Denschlag J (2008) Nature Phys 4:223
6. González-Férez R, Dehesa JS (2003) Chem Phys Lett 373:615
7. González-Férez R, Dehesa JS (2003) Phys Rev Lett 91:113001
8. González-Férez R, Dehesa JS (2005) Eur Phys J D 32:39
9. López-Ruiz R, Mancini HL, Calbet X (1995) Phys Lett A 209:321; Pipek J, Varga I, Nagy T (1990) Int J Quantum Chem 37:529; Pipek J, Varga I (1992) Phys Rev A 46:3148; Pipek J, Varga I (2002) Phys Rev E 68:026202. The structural entropy S_{str} introduced independently as a localization quantity characteristic of the decay of the distribution function is related to the shape complexity as $\ln C$
10. Catalán RG, Garay J, López-Ruiz R (2002) Phys Rev E 66:011102
11. Angulo JC, Antolín J (2008) J Chem Phys 128:164109
12. Antolín J, Angulo JC (2009) Int J Quantum Chem 109:586
13. López-Rosa S, Angulo JC, Antolín J (2009) Phys A 388:2081
14. Shiner JS, Davison M, Landsberg PT (1999) Phys Rev E 59:1459
15. López-Ruiz R (2005) Biophys Chem 115:215
16. Shannon CE (1948) Bell Syst Tech 27:379; ibid. 27:623 (1948)
17. Onicescu O (1966) CR Acad Sci Paris A 263:25
18. Hall MJW (1999) Phys Rev A 59:2602
19. Anteneodo C, Plastino AR (1996) Phys Lett A 223:348; Martin MT, Plastino A, Roso OA Phys Lett A 311:126 (2003); Pennini F, Plastino A Phys Lett A 365:262 (2007)
20. Hyman AS, Yaniger SI, Liebman JL (1978) Int J Quantum Chem 19:757
21. Chen Z, Goldman SP (1993) Phys Rev A 48:1107
22. Garstang RH (1977) Rep Prog Phys 40:105
23. Pavlov-Verevkin VB, Zhilinskii BI (1980) Phys Lett 75A:279; Phys Lett 78A:244 (1980)
24. Schweizer W, Fassbinder P, González-Férez R, Braun M, Kulla S, Stehele M (1999) J Comput Appl Math 109:95
25. Patil SH, Sen KD, Watson NA, Montgomery HE (2007) J Phys B 40:2147

Part IV
Density-Oriented Methods

Chapter 9
Atomic Density Functions: Atomic Physics Calculations Analyzed with Methods from Quantum Chemistry

Alex Borgoo, Michel R. Godefroid, and Paul Geerlings

Abstract This contribution reviews a selection of findings on atomic density functions and discusses ways for reading chemical information from them. First an expression for the density function for atoms in the multi-configuration Hartree–Fock scheme is established. The spherical harmonic content of the density function and ways to restore the spherical symmetry in a general open-shell case are treated. The evaluation of the density function is illustrated in a few examples. In the second part of the paper, atomic density functions are analyzed using quantum similarity measures. The comparison of atomic density functions is shown to be useful to obtain physical and chemical information. Finally, concepts from information theory are introduced and adopted for the comparison of density functions. In particular, based on the Kullback–Leibler form, a functional is constructed that reveals the periodicity in Mendeleev's table. Finally a quantum similarity measure is constructed, based on the integrand of the Kullback–Leibler expression and the periodicity is regained in a different way.

9.1 Introduction

Density Functional Theory (DFT) plays a prominent role in present day investigations of the electronic structure of atoms and molecules. Within DFT the electron density function plays a central role as it caries all the information to describe the investigated system. The idea that all physical and chemical information is contained

A. Borgoo (✉) • P. Geerlings
Department of Chemistry, Vrije Universiteit Brussel, Pleinlaan 2, 1050, Brussels
e-mail: aborgoo@vub.ac.be; pgeerlin@vub.ac.be

M.R. Godefroid
Service de Chimie quantique et Photophysique CP160/09, Université Libre de Bruxelles,
50, av. F.D. Roosevelt, B-1050 Bruxelles
e-mail: mrgodef@ulb.ac.be

P.E. Hoggan et al. (eds.), *Advances in the Theory of Quantum Systems in Chemistry and Physics*, Progress in Theoretical Chemistry and Physics 22,
DOI 10.1007/978-94-007-2076-3_9, © Springer Science+Business Media B.V. 2012

in the density incited the present authors to try and recover some of this information. Although the proof of the Hohenberg–Kohn theorems [1], which guarantee the presence of all physical information in the density function, is generally said to be disarmingly simple, it does not provide a method to get to the relevant information. The continued search for improved energy functionals is the most evident example to illustrate the challenge researchers are confronted with.

In recent years, the present authors have developed an interest in obtaining chemical information from atonic density functions. The application of concepts from quantum chemistry shows that some particular aspects of physical and chemical interest can be read from the density functions. In particular the comparison of density functions using quantum similarity measures or functionals from information theory plays an important role. The original goal of the work was to find a way of regaining the periodicity in Mendeleev's table through the comparison of density functions.

The purpose of this contribution is to give an overview of the results which center around the atomic density function and the recovery of the periodicity. Since all the calculations are based on atomic density functions, it is appropriate to revisit the construction of these densities in some depth. First a workable definition of the density function is established in the framework of the multi-configuration Hartree–Fock method (MCHF) [2] and the spherical harmonic content of the density function is discussed. A spherical density function is established in a natural way, by using spherical tensor operators. The proposed expression can be evaluated for any multi-configuration state function corresponding to an atom in a particular well-defined state and a recently developed extension of the MCHF code [3] is used for that purpose. Three illustrative examples are given. In the next section relativistic density functions for the relativistic Dirac–Hartree–Fock method [4] are defined. The latter will be used for a thorough analysis of the influence of relativistic effects on electron density functions later on in this paper.

The analysis of atomic density functions can be furthered by comparing them in pairs. Specifically, the use of quantum similarity measures and indices as defined by Carbó [5] has shown that particular influences on the density functions can be estimated in this way. Here this feature is demonstrated by reviewing three case studies: (1) the *LS*-term dependence of Hartree–Fock densities, (2) the comparison of atoms throughout the periodic table [6], and (3) the quantitative evaluation of the influence of relativistic effects, via a comparison of non-relativistic Hartree–Fock densities with Dirac–Hartree–Fock relativistic densities [7].

In the final part of this contribution, information theory is introduced. After a brief revision of the relevant concepts, a functional based on the Kullback–Leibler measure [8] is constructed for the investigation of atomic density functions throughout Mendeleev's table. Since the quantum similarity does not reveal the expected periodic patterns, it is significant to show that it is actually possible to regain the periodicity by constructing an appropriate functional [6]. By considering the integrand of the Kullback–Leibler measure and comparing it locally for two atoms, a quantum similarity measure can be constructed which does reveal periodic patterns [7].

9.2   The Multi-configuration Many-Electron Wave Function

The total energy which results from the Hartree–Fock equations is due to electrons moving independently in an averaged, central potential. Any improvements to the energy (still in the context of the non-relativistic Schrödinger equation) are said to be due to correlation effects as a direct consequence of the electron–electron interaction. In fact it is common to define the correlation energy as

$$E_{\text{CORR}} = E_{\text{EXACT}} - E_{\text{HF}},\tag{9.1}$$

where E_{EXACT} is the exact non-relativistic energy and E_{HF} the energy due to the solutions of the corresponding Hartree–Fock equations.

A particularly natural way to improve the Hartree–Fock energy – i.e., include correlation energy – is by departing from the single-configuration approximation. In Hartree–Fock calculations a rigid orbital picture is assumed, where electrons have a fixed place in a given electron configuration. By allowing electrons to occupy different orbitals and allowing several electron configurations, the variational approach can be applied on a significantly larger set of trial wave functions.

In the multi-configuration Hartree–Fock (MCHF) approach, the N-electron wave function $\Psi_{\alpha LSM_LM_S}$ is a linear combination of M configuration state functions (CSFs) $\Phi_{\alpha_i LSM_LM_S}$ which are eigenfunctions of the total angular momentum L^2, the spin momentum S^2 and their projections L_z and S_z, with eigenvalues $\hbar^2 L(L+1)$, $\hbar^2 S(S+1)$, $\hbar M_L$ and $\hbar M_S$, respectively

$$\Psi_{\alpha LSM_LM_S}(\mathbf{x}_1,\cdots\mathbf{x}_N) = \sum_{i=1}^{M} c_i\,\Phi(\alpha_i LSM_LM_S;\mathbf{x}_1,\cdots\mathbf{x}_N),\tag{9.2}$$

with

$$\sum_{i=1}^{M} |c_i|^2 = 1.\tag{9.3}$$

The mixing coefficients $\{c_i\}$ and the radial functions $\{R_{n_il_i}(r)\}$, constituting the one-electron basis, are solutions of the multi-configuration Hartree–Fock method in the MCHF approach. For a given set of orbitals, the mixing coefficients may also be the solution of the configuration interaction (CI) problem. The relativistic corrections can be taken into account by diagonalizing the Breit–Pauli Hamiltonian [9] in the LSJ-coupled CSF basis to get the intermediate coupling eigenvectors

$$\Psi_{\alpha JM}(\mathbf{x}_1,\cdots\mathbf{x}_N) = \sum_{i=1}^{M} a_i\,\Phi(\alpha_i L_i S_i JM;\mathbf{x}_1,\cdots\mathbf{x}_N),\tag{9.4}$$

with

$$\sum_{i=1}^{M} |a_i|^2 = 1.\tag{9.5}$$

In the MCHF approach, the trial wave functions are of the form (9.2) and the energy expression for a given state becomes

$$E(\alpha LS) = \sum_i^M \sum_j^M c_i^* c_j H_{ij},$$ (9.6)

where

$$H_{ij} \equiv \langle \Phi_{\alpha_i LSM_LM_S} | H | \Phi_{\alpha_j LSM_LM_S} \rangle = \sum_{ab} q_{ab}^{ij} I_{ab} + \sum_{abcd;k} v_{abcd;k}^{ij} R^k(ab,cd),$$ (9.7)

where $a \equiv (n_a l_a)$ and $b \equiv (n_b l_b)$ and the sums run over all occupied orbitals in the respective configuration i and j.

In the MCHF context the variational principle is applied to the energy functional in Eq. 9.6. A stationary solution is obtained by minimizing the energy with respect to variations in the radial wave functions $P_{nl}(r) \equiv rR_{nl}(r)$ satisfying the orthonormality conditions

$$N_{nl,n'l} \equiv \int_0^\infty P_{nl}(r) P_{n'l}(r) \, dr = \delta_{nn'}$$ (9.8)

and

$$\sum_{i=1}^M c_i^2 = 1.$$ (9.9)

The energy expression (9.6) can be written in matrix notation

$$E(\alpha LS) = \mathbf{C}^t \mathbf{H} \mathbf{C},$$ (9.10)

where $\mathbf{H} \equiv (H_{ij})$, $\mathbf{C}$ is the column matrix of the expansion coefficients and $\mathbf{C}^t$ its transpose.

This gives the eigenvalue problem for the expansion coefficients

$$\mathbf{H}\mathbf{C} = \mathbf{C}E.$$ (9.11)

To obtain a self consistent field solution to the MCHF problem, two optimizations need to be performed i.e., one for the variation of the one-electron radial orbitals $\{P_{nl}(r)\}$ in the wave function and one for the expansion coefficients. This can be done by consecutively iterating, first the orbital optimization followed by the coefficient optimization.

9.3   On the Symmetry of the Density Function

In the first part of this paragraph we review the study of the spherical harmonic content of the density function for atoms in a well defined state (9.2) or (9.4). The spherical density functions, which reveal the familiar shell structure, are discussed and illustrative examples are given.

9.3.1 The Non-Spherical Density Function

The so-called "generalized density function" [10] or the "first order reduced density matrix" [11] is a special case of the reduced density matrix [10, 12]

$$\gamma_1(\mathbf{x}_1,\mathbf{x}'_1) = N \int \Psi(\mathbf{x}_1,\mathbf{x}_2,\dots,\mathbf{x}_N)\, \Psi^*(\mathbf{x}'_1,\mathbf{x}_2,\dots,\mathbf{x}_N)\, d\mathbf{x}_2\dots d\mathbf{x}_N, \qquad (9.12)$$

where $\Psi(\mathbf{x}_1,\mathbf{x}_2,\dots,\mathbf{x}_N)$ is the total wave function of an N electron system and $\Psi^*(\mathbf{x}_1,\mathbf{x}_2,\dots,\mathbf{x}_N)$ is its complex conjugate. The spin-less total electron density function $\rho(\mathbf{r})$ is defined as the first order reduced density matrix, integrated over the spin and evaluated for $\mathbf{x}_1 = \mathbf{x}'_1$

$$\rho(\mathbf{r}_1) = \int \gamma_1(\mathbf{x}_1,\mathbf{x}_1)\, d\sigma_1. \qquad (9.13)$$

This electron density function is normalized to the number of electrons of the system

$$\int \rho(\mathbf{r})\, d\mathbf{r} = \int \rho(\mathbf{r})\, r^2 \sin\vartheta\, dr d\vartheta d\varphi = N. \qquad (9.14)$$

As discussed in [11], the single particle density function can be calculated by evaluating the expectation value of the $\delta(\mathbf{r})$ operator,

$$\rho(\mathbf{r}) = \int \Psi(\mathbf{x}_1,\mathbf{x}_2,\dots,\mathbf{x}_N)\, \delta(\mathbf{r})\, \Psi^*(\mathbf{x}_1,\mathbf{x}_2,\dots,\mathbf{x}_N)\, d\mathbf{x}_1 d\mathbf{x}_2\dots d\mathbf{x}_N, \qquad (9.15)$$

where $\delta(\mathbf{r})$ probes the presence of electrons at a particular point in space and can be written as the one-electron first-quantization operator

$$\delta(\mathbf{r}) = \sum_{i=1}^{N} \delta(\mathbf{r}-\mathbf{r}_i). \qquad (9.16)$$

The exact spin-less total electron density function (9.15) evaluated for an eigenstate with well-defined quantum numbers (LSM_LM_S)

$$\rho(\mathbf{r})^{LSM_LM_S} = \sum_{lm} Y_{lm}(\vartheta,\varphi)\frac{1}{r^2}\, \langle \Psi_{\alpha LSM_LM_S}| \sum_{i=1}^{N} \delta(r-r_i)\, Y_{lm}^*(\vartheta_i,\varphi_i)|\Psi_{\alpha LSM_LM_S}\rangle, \qquad (9.17)$$

becomes

$$\rho(\mathbf{r})^{LSM_LM_S} = \sum_{l=0}^{L} \rho(r)_{2l}^{LSM_LM_S} Y_{2l\,0}(\vartheta,\varphi), \qquad (9.18)$$

where

$$\rho(r)_{2l}^{LSM_LM_S} = \frac{1}{r^2}(-1)^{L-M_L}\begin{pmatrix} L & 2l & L \\ -M_L & 0 & M_L \end{pmatrix}$$

$$\times\langle\Psi_{\alpha LS}\|\sum_{i=1}^{N}\delta(r-r_i)\,Y_{2l}^*(\vartheta_i,\varphi_i)\|\Psi_{\alpha LS}\rangle. \qquad (9.19)$$

This result, which can be found by applying the Wigner–Eckart theorem [13], recovers Fertig and Kohn's analysis [14] for the density corresponding to a well-defined (LSM_LM_S) eigenstate of the Schrödinger equation. The spherical harmonic components in the density are limited to l-even contributions, because the bra and the ket need to be of the same parity $\pi = (-1)^{\Sigma_i l_i}$.

In this paper, the authors observed that the self-consistent field densities obtained via the Hartree and Hartree–Fock methods generally violate the specific finite spherical harmonic content of $\rho(\mathbf{r})^{LSM_LM_S}$. They also mention that this exact form can be obtained by spherically averaging the effective potential, yielding single-particle states with good angular momentum quantum numbers. The atomic structure software package ATSP2K [2] applies this approach, as was done in the original atomic Hartree–Fock theory [15–17]. This implies two things: (1) the density function $\rho(\mathbf{r})^{LSM_LM_S}$ calculated from any multiconfiguration wave function of the form (9.2), is not a priori spherically symmetric, (2) this density function will contain all spherical harmonic components (up to $2L$) as long as the one-electron orbital active set spanning the configuration space is l-rich enough.

The density function can also be expressed in second quantization [10]. Introducing the notation $q \equiv n_q l_q m_{l_q} m_{s_q}$ for spin-orbitals, expression (9.12) becomes

$$\gamma_1(\mathbf{x}_1,\mathbf{x}'_1) = \sum_{pq}D_{pq}\,\psi_p^*(\mathbf{x}'_1)\psi_q(\mathbf{x}_1), \qquad (9.20)$$

where D_{pq} are elements of the density matrix which are given by

$$D_{pq} \equiv \langle\Psi|a_p^\dagger a_q|\Psi\rangle. \qquad (9.21)$$

The sum in Eq. 9.20 runs over all possible pairs of quartets of quantum numbers p and q. The spin-less density function (9.13) calculated from $\rho(\mathbf{r}) = \langle\Psi|\hat{\delta}(\mathbf{r})|\Psi\rangle$, using the second quantized form of the operator

$$\hat{\delta}(\mathbf{r}) \equiv \sum_{pq}a_p^\dagger a_q\,\delta_{m_{s_p},m_{s_q}}\langle\psi_p(\mathbf{r}')|\frac{1}{r^2\sin\vartheta}\,\delta(r-r')\,\delta(\vartheta-\vartheta')\,\delta(\varphi-\varphi')|\psi_q(\mathbf{r}')\rangle$$

$$= \sum_{pq}a_p^\dagger a_q\,\delta_{m_{s_p},m_{s_q}}R_{n_p l_p}^*(r)Y_{l_p m_{l_p}}^*(\vartheta,\varphi)R_{n_q l_q}(r)Y_{l_q m_{l_q}}(\vartheta,\varphi), \qquad (9.22)$$

yields

$$\rho(\mathbf{r}) = \sum_{pq} D_{pq} \, \delta_{m_{s_p}, m_{s_q}} R^*_{n_p l_p}(r) Y^*_{l_p m_{l_p}}(\vartheta, \varphi) R_{n_q l_q}(r) Y_{l_q m_{l_q}}(\vartheta, \varphi). \tag{9.23}$$

To illustrate the spherical harmonics content of the density in the Hartree–Fock approximation, consider the atomic term $1s^2 2p^2(\,^3P)3d\,^4F$ for which the $(M_L, M_S) = (+3, +3/2)$ subspace reduces to a single Slater determinant

$$\Psi_{\alpha L S M_L M_S} = \Phi(1s^2 2p^2(\,^3P)3d\,^4F_{+3, +3/2}) = |1s\overline{1s}2p_{+1}2p_0 3d_{+2}|. \tag{9.24}$$

When evaluating (9.23), all non-zero D_{pq}-values appear on the diagonal ($p = q$), yielding

$$\rho(\mathbf{r})^{^4F_{+3,+3/2}} = |\psi_{1s}(\mathbf{r})|^2 + |\psi_{\overline{1s}}(\mathbf{r})|^2 + |\psi_{2p_{+1}}(\mathbf{r})|^2 + |\psi_{2p_0}(\mathbf{r})|^2 + |\psi_{3d_{+2}}(\mathbf{r})|^2. \tag{9.25}$$

This density has a clear non-spherical angular dependence. However, referring to [18]

$$W^{\|}_{JM}(\vartheta) \equiv |Y_{JM}(\vartheta, \varphi)|^2 = \sum_{n=0}^{J} b_n(J, M) \, P_{2n}(\cos \vartheta) = \sum_{n=0}^{J} b'_n(J, M) \, Y_{2n\,0}(\vartheta, \varphi) \tag{9.26}$$

one recovers the even Legendre polynomial content of the density, although not reaching the $(2L = 6)$ limit $Y_{6\,0}(\vartheta, \varphi)$ of the exact density (9.18). This limit will be attained when extending the one-electron orbital active set to higher angular momentum values for building a correlated wave function.

As discussed in detail in [3], contributions to the density function corresponding to $(p \neq q)$ can appear through off-diagonal matrix elements in the CSF basis. These contributions will be present for parity conserving single electron excitations of the type $|l_1 q\rangle \to |l_2 q\rangle$.

The "offending" spherical harmonic contributions described by Fertig and Kohn [14] do not occur in the MCHF calculation of the density function, whatever the maximum l-value of the orbital active space [3].

9.3.2 The Spherical Density Function

A *spherically* symmetric density function can be obtained for an arbitrary CSF $\Phi_{\alpha L S M_L M_S}$ by averaging the $(2L+1)(2S+1)$ magnetic components of the spin-less density function

$$\rho(\mathbf{r})^{LS} \equiv \frac{1}{(2L+1)(2S+1)} \sum_{M_L M_S} \rho(\mathbf{r})^{L S M_L M_S}, \tag{9.27}$$

where $\rho(\mathbf{r})^{LSM_LM_S}$ is constructed according to Eq. 9.23

$$\rho(\mathbf{r})^{LSM_LM_S} = \sum_{pq} \langle \Phi_{\alpha LSM_LM_S} | a_p^\dagger a_q | \Phi_{\alpha LSM_LM_S} \rangle \, \delta_{m_{s_p}, m_{s_q}} \, \psi_p^*(\mathbf{r}) \psi_q(\mathbf{r}). \qquad (9.28)$$

Applying Eqs. 9.27 and 9.28 for the atomic term $1s^2 2p^2(^3P)3d\,^4F$ considered in the previous section, we simply get

$$\rho(\mathbf{r})^{4F} = \frac{1}{4\pi r^2} \left\{ 2P_{1s}^2(r) + 2P_{2p}^2(r) + P_{3d}^2(r) \right\}, \qquad (9.29)$$

which is, *in contrast* to Eq. 9.25, obviously spherically symmetric. The sum over (M_L, M_S) performed in (9.27) guarantees, for any nl-subshell, the presence of all necessary components $\{Y_{lm_l} \mid m_l = -l, \ldots +l\}$ with the same weight factor, which permits the application of Unsöld's theorem [19]

$$\sum_{m_l=-l}^{+l} |Y_{lm_l}(\vartheta, \varphi)|^2 = \frac{2l+1}{4\pi} \qquad (9.30)$$

and yields the spherical symmetry. This result is valid for any single CSF

$$\rho(\mathbf{r})^{LS} = \frac{1}{4\pi r^2} \sum_{nl} q_{nl} P_{nl}^2(r), \qquad (9.31)$$

where q_{nl} is the occupation number of nl-subshell. Its sphericity explicitly appears by rewriting (9.31) as

$$\rho(\mathbf{r}) = \rho(r) \, |Y_{00}(\vartheta, \varphi)|^2 = \frac{D(r)}{r^2} \, |Y_{00}(\vartheta, \varphi)|^2, \qquad (9.32)$$

with

$$\rho(r) \equiv \frac{1}{r^2} \sum_{nl} q_{nl} P_{nl}^2(r), \qquad (9.33)$$

and

$$D(r) \equiv r^2 \rho(r) = \sum_{nl} q_{nl} P_{nl}^2(r) = \sum_{nl} q_{nl} \, r^2 R_{nl}^2(r). \qquad (9.34)$$

The *radial distribution* function $D(r)$ represents the probability of finding an electron between the distances r and $r + dr$ from the nucleus, regardless of directionThis radial density function reveals the atomic shell structure when plotted as function of r. Its integration over r gives the total number of electrons of the system

$$\int_0^\infty D(r)\,dr = \int_0^\infty r^2 \rho(r)\,dr = \sum_{nl} q_{nl} = N. \qquad (9.35)$$

Where above the spherical symmetry of the average density (9.27) is demonstrated for a single CSF thanks to Unsöld's theorem, it can be demonstrated in the general case by combining (9.27), (9.18) and the 3-j sum rule [13]

$$\sum_{M_L}(-1)^{L-M_L} \begin{pmatrix} L & k & L \\ -M_L & 0 & M_L \end{pmatrix} = (2k+1)^{1/2}\,\delta_{k,0} \tag{9.36}$$

for each $k = 2l$ contribution (9.19). However, the radial density $\rho(r)$ will be more complicated than (9.33), involving mixed contributions of the type $P_{n'l}(r)P_{nl}(r) = r^2R_{n'l}(r)R_{nl}(r)$, as developed below.

Instead of obtaining a spherically symmetric density function by averaging the magnetic components $\rho(\mathbf{r})^{LSM_LM_S}$ through Eq. 9.27, one can build a radial density operator associated to the function (9.34) which is spin- and angular-independent, i.e., independent of the spin (σ) and angular (ϑ, φ) variables. Adopting the methodology used by Helgaker et al. [11] for defining the spin-less density operator, we write a general first quantization spin-free *radial* operator

$$f = \sum_{i=1}^{N} f(r_i) \tag{9.37}$$

in second quantization as

$$\hat{f} = \sum_{pq} f_{pq}\,a_p^\dagger a_q, \tag{9.38}$$

where f_{pq} is the one-electron integral

$$f_{pq} = \int \psi_p^*(\mathbf{x})f(r)\psi_q(\mathbf{x})r^2 \sin\vartheta\, dr d\vartheta d\varphi d\sigma. \tag{9.39}$$

Applying this formalism to the radial density operator

$$\delta(r) \equiv \sum_{i=1}^{N} \delta(r - r_i), \tag{9.40}$$

and using the spin-orbital factorization for both p and q quartets, we obtain the second quantization form

$$\hat{\delta}(r) = \sum_{pq} d_{pq}(r)\,a_p^\dagger a_q, \tag{9.41}$$

with

$$d_{pq}(r) = \delta_{l_p l_q}\,\delta_{m_{l_p} m_{l_q}}\,\delta_{m_{s_p} m_{s_q}}\, R_{n_p l_p}^*(r)R_{n_q l_q}(r)r^2, \tag{9.42}$$

where the Kronecker delta arises from the orthonormality property of the spherical harmonics and spin functions. With real radial one-electron functions, the operator (9.41) becomes

$$\hat{\delta}(r) = \sum_{n',l',m'_l,m'_s,n,l,m_l,m_s,} \delta_{l'l}\, \delta_{m'_l m_l}\, \delta_{m'_s m_s}\, a^{\dagger}_{n'l'm'_l m'_s} a_{nlm_l m_s} R_{n'l'}(r)R_{nl}(r)r^2 \quad (9.43)$$

$$= \sum_{n',nl,m_l,m_s} a^{\dagger}_{n'lm_l m_s} a_{nlm_l m_s} R_{n'l}(r)R_{nl}(r)r^2. \quad (9.44)$$

Its expectation value provides the radial density function $D(r) = r^2\rho(r) = 4\pi r^2\rho(\mathbf{r})$ defined by (9.32) and (9.34).

Building the coupled tensor of ranks (00) from the $[2(2l+1)]$ components of the creation and annihilation operators [20]

$$\left(a^{\dagger}_{n'l}a_{nl}\right)^{(00)}_{00} = -\frac{1}{\sqrt{2(2l+1)}} \sum_{m_l m_s} a^{\dagger}_{n'lm_l m_s} a_{nlm_l m_s}, \quad (9.45)$$

the operator (9.43) becomes

$$\hat{\delta}(r) = -\sum_{l} \sqrt{2(2l+1)} \sum_{n',n} \left(a^{\dagger}_{n'l}a_{nl}\right)^{(00)}_{00} R_{n'l}(r)R_{nl}(r)r^2. \quad (9.46)$$

The expectation value of this operator provides the spherical density function for any atomic state. Note that, in contrast to (9.28), the tensorial ranks (00) garantee the diagonal character in L,S,M_L and M_S, thanks to Wigner–Eckart theorem

$$\langle \alpha LSM_LM_S|T^{(00)}_{00}|\alpha'L'S'M'_LM'_S\rangle = (-1)^{L+S-M_L-M_S}$$

$$\times \begin{pmatrix} L & 0 & L' \\ -M_L & 0 & M'_L \end{pmatrix} \begin{pmatrix} S & 0 & S' \\ -M_S & 0 & M'_S \end{pmatrix}$$

$$\times \langle \alpha LS\|T^{(00)}\|\alpha'L'S'\rangle. \quad (9.47)$$

Moreover, the M_L/M_S independence emerges from the special $3j$-symbol

$$\begin{pmatrix} j & 0 & j' \\ -m_j & 0 & m'_j \end{pmatrix} = (-1)^{j-m}(2j+1)^{-1/2}\delta_{jj'}\delta_{m_j m'_j}. \quad (9.48)$$

In other words, where the non-spherical components are washed out by the averaging process (9.27), they simply do not exist and will never appear for the density calculated from (9.46), for any (M_L,M_S) magnetic component.

The radial distribution function $D(r) \equiv r^2 \rho(r)$ can be calculated from the expectation value of the operator (9.46), using the wave function (9.2) or (9.4). In the most general case (expansion (9.4)), using the $(LS)J$-coupled form of the excitation operator,

$$\left(a_{n'l}^{\dagger}a_{nl}\right)_0^{(00)0} = \left(a_{n'l}^{\dagger}a_{nl}\right)_{00}^{(00)},$$ (9.49)

one obtains

$$\langle\Psi_{\alpha JM}|\hat{\delta}(r)|\Psi_{\alpha JM}\rangle = (-1)^{J-M}\begin{pmatrix} J & 0 & J \\ -M & 0 & M \end{pmatrix}\langle\Psi_{\alpha J}\|\widehat{F}_{\rho}^{(00)0}\|\Psi_{\alpha J}\rangle$$ (9.50)

with

$$\widehat{F}_{\rho,0}^{(00)0} = -\sum_{l=1}\sqrt{2(2l+1)}\sum_{n,n'}\left(a_{n'l}^{\dagger}a_{nl}\right)_0^{(00)0} I_{\rho}\left(n'l,nl\right),$$ (9.51)

and

$$I_{\rho}\left(n'l,nl\right)(r) \equiv R_{n'l}(r)R_{nl}(r)r^2.$$ (9.52)

The diagonal reduced matrix element (RME) evaluated with the Breit–Pauli eigenvector (9.4) has the following form

$$\langle\Psi_{\alpha J}\|\widehat{F}_{\rho}^{(00)0}\|\Psi_{\alpha J}\rangle = \sum_{i,j}a_i^* a_j \langle\Phi(\alpha_i L_i S_i J)\|\widehat{F}_{\rho}^{(00)0}\|\Phi(\alpha_j L_j S_j J)\rangle$$ (9.53)

where the RME in the $(LS)J$ coupled basis reduces to

$$\langle\Phi(\alpha_i L_i S_i JM)\|\widehat{F}_{\rho}^{(00)0}\|\Phi(\alpha_j L_j S_j JM)\rangle$$

$$= \sqrt{\frac{2J+1}{(2L_i+1)(2S_i+1)}}\delta_{L_i,L_j}\delta_{S_i,S_j}$$

$$\times\langle\Phi(\alpha_i L_i S_i)\|\widehat{F}_{\rho}^{(00)}\|\Phi(\alpha_j L_j S_j)\rangle$$ (9.54)

and

$$\widehat{F}_{\rho,00}^{(00)} = -\sum_{l=1}\sqrt{2(2l+1)}\sum_{n,n'}\left(a_{n'l}^{\dagger}a_{nl}\right)_{00}^{(00)} I_{\rho}\left(n'l,nl\right).$$ (9.55)

From a comparison of the operator (9.55) with the non-relativistic one-body Hamiltonian operator (see Eq. A5 of [21]), one observes that the angular coefficients of the radial functions $I_{\rho}\left(n'l,nl\right)(r)$ are identical to those of the one-electron Hamiltonian radial integrals $I_{n'l,nl}$, as anticipated from McWeeny analysis [10]. These angular coefficients can be derived by working out the matrix elements of a one–particle scalar operator $\widehat{F}_{\rho}^{(00)}$ between configuration state functions with u open shells, as explicitly derived by Gaigalas et al. [22].

9.4   Three Tangible Examples

First the evaluation of the density function and the influence of correlation effects
is illustrated by plotting in Fig. 9.1 the radial density distribution $D(r) = r^2\rho(r)$ for
a CAS-MCHF wave function of the beryllium ground state (Be $1s^2 2s^2$ ^1S), using
a $n = 9$ orbital active set. In the same figure, the Hartree–Fock radial density is
compared with the one obtained with two correlation models: (1) the $n = 2$ CAS-
MCHF expansion, largely dominated by the near-degeneracy mixing associated to
the Layzer complex $1s^2\{2s^2 + 2p^2\}$ and (2) the $n = 9$ CAS-MCHF. From the plotted
results we notice that the density of the $n = 2$ calculation already contains the
major correlation effects, compared to the $n = 9$ calculation. Indeed, the density
does not seem to change a lot by going from the $n = 2$ to the $n = 9$ orbital basis,
the valence double excitation $1s^2 2p^2$ contributing for 9.7% of the wave function.
From the energy point of view however, this observation is somewhat surprising
(see Table 9.1): the correlation energy associated to the $n = 2$ CAS-MCHF solution
"only" represents 47% of the $n = 9$ correlation energy.

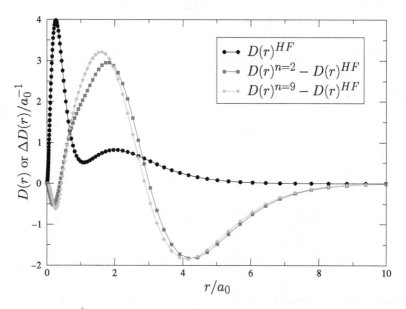

Fig. 9.1 Density of ^1S Be ground state for different CAS-MCHF wave function as compared to
Hartree–Fock (HF). Density differences have been scaled by a factor 100

Table 9.1 Total energy for
the ground state of Be with
different correlation models

Model	Energy (a.u.)	Correlation energy (a.u.)
HF	$-14.573\,023$	
$n = 2$-CAS	$-14.616\,856$	$E_{corr}^{n=2} = E^{n=2} - E^{HF} = 0.043\,832$
$n = 9$-CAS	$-14.667\,013$	$E_{corr}^{n=9} = E^{n=9} - E^{HF} = 0.093\,986$

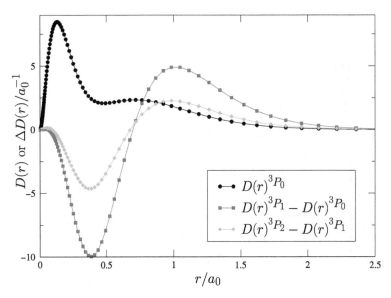

Fig. 9.2 Comparison of the $^3P_0$, $^3P_1$ and $^3P_2$ radial density functions for O^{4+}. Density differences have been scaled by a factor 10,000

Table 9.2 Total energy for 2s2p $^3P^o_J$ fine-structure levels of O^{4+}

Model	Energy (a.u.)	Energy difference (a.u.)
$1s^2 2s2p\ ^3P^o_0$	−68.032 086	
$^3P^o_1$	−68.031 473	$E(\,^3P_1 - {}^3P_0) = 0.000\ 613$
$^3P^o_2$	−68.030 102	$E(\,^3P_2 - {}^3P_1) = 0.001\ 370$

As a second example, we illustrate the influence of relativistic effects – in the Breit–Pauli approximation – on the density function of the Be-like O^{4+} atom, by comparing the densities of the fine-structure states $1s^2 2s2p\ ^3P^o_0$, $^3P^o_1$ and $^3P^o_2$. From the plots in Fig. 9.2 and the data given in Table 9.2 we observe that the largest energy difference corresponds to the largest difference in density function. More bound is the level, higher is the electron density in the inner region. The influence of relativistic effects on the density function will be discussed thoroughly below.

Finally a third example is given that is relevant when studying the electron affinities, as it is often interesting to investigate the differential correlation effects between the negative ion and the neutral system [23]. Figure 9.3 displays the density functions $D(r)$ of both the [Ne]$3s^2 3p^4\ ^3P$ ground state of neutral Sulphur (S) and the [Ne]$3s^2 3p^5\ ^2P^o$ ground state of the negative ion S^-, evaluated with elaborate correlation models [23], together with their difference $\Delta D(r)$. The latter integrates to unity and reveals that the "extra" electron resides in the valence area.

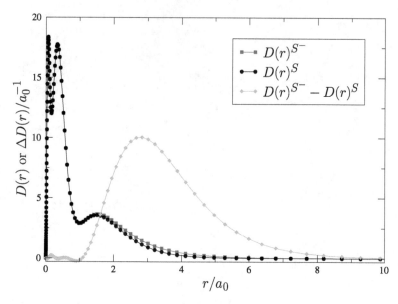

Fig. 9.3 S and S⁻ density functions. Density differences have been scaled by a factor 30

9.5 Relativistic Density Functions

9.5.1 Relativistic Multi-configuration Wave Functions

In the relativistic scheme, the atomic wave function is, in the most general case, a combination of configuration state functions

$$|\pi J M_J\rangle = \sum_v c_v |v\pi J M_J\rangle, \tag{9.56}$$

eigenfunction of the inversion operator I, the total angular momentum $\mathbf{J}^2$ and its projection J_z. v denotes all the necessary information for specifying the relativistic configuration unambiguously. The CSFs are built on the one-electron Dirac four-spinor

$$\psi_i(r) = \frac{1}{r} \begin{pmatrix} P_i(r)\chi_{\kappa_i}^{\mu_i}(\Omega) \\ iQ_i(r)\chi_{-\kappa_i}^{\mu_i}(\Omega) \end{pmatrix}, \tag{9.57}$$

where $\chi_{\kappa_i}^{\mu_i}(\Omega)$ is a two-dimensional vector harmonic. It has the property that $K\psi_i(r) = \kappa\psi_i(r)$ where $K = \beta(\sigma \cdot \mathbf{L} + 1)$. The large $\{P(r)\}$ and small $\{Q(r)\}$ components are solutions of a set of coupled integro-differential equations (9.62) [24]. The mixing coefficients $\{c_v\}$ are obtained by diagonalizing the matrix of the no-pair Hamiltonian containing the magnetic and retardation terms

[25]. The two coupled variational problems are solved iteratively. For a complete discussion on relativistic atomic structure we refer to [26].

It is to be noted that the relativistic scheme rapidly becomes more complicated than the corresponding non-relativistic one. For example, if the ground term of Carbon atom is described, in the non-relativistic one-configuration Hartree–Fock approximation, by a single CSF $|1s^2 2s^2 2p^2\ {}^3P\rangle$, the relativistic equivalent implies the specification of the J-value. For $J = 0$ corresponding to the ground level of Carbon, the following two-configuration description becomes necessary

$$|\text{``}1s^2 2s^2 2p^2\text{''}(J=0)\rangle = c_1|1s^2 2s^2 (2p^*)^2 (J=0)\rangle + c_2|1s^2 2s^2 2p^2 (J=0)\rangle, \quad (9.58)$$

implicitly taking into account the relativistic mixing of the two LS-terms (^1S and ^3P) arising from the $2p^2$ configuration and belonging to the $J = 0$ subspace. p^* and p in expression (9.58) correspond to the j-values, $j = 1/2$ ($\kappa = +1$) and $j = 3/2$ ($\kappa = -2$), respectively.

9.5.2 Multi-configuration Dirac–Hartree–Fock Equations

For the calculations of relativistic density functions we used a multi-configuration Dirac–Fock approach (MCDF), which can be thought of as a relativistic version of the MCHF method. The MCDF approach implemented in the MDF/GME program [4, 27] calculates approximate solutions to the Dirac equation with the effective Dirac–Breit Hamiltonian [27]

$$H^{DB}(\mathbf{r}_1, \mathbf{r}_2, \ldots, \mathbf{r}_N) = \sum_{i=1}^{N} h^D(\mathbf{r}_i) + \sum_{i<j} h^B(\mathbf{r}_i, \mathbf{r}_j), \quad (9.59)$$

with

$$h^D(\mathbf{r}_i) = c\,\alpha \cdot \mathbf{p} + c^2(\beta - 1)V_i(r) \quad (9.60)$$

and

$$h^B(\mathbf{r}_i, \mathbf{r}_j) = \frac{1}{r_{ij}} - \frac{\alpha_i \cdot \alpha_j}{r_{ij}} \cos(\omega_{ij} r_{ij}) + (\alpha \cdot \nabla)_i (\alpha \cdot \nabla)_j \frac{\cos(\omega_{ij} r_{ij}) - 1}{\omega_{ij}^2 r_{ij}}. \quad (9.61)$$

The total Hamiltonian contains three types of contributions: the one-electron Dirac Hamiltonion, the Coulomb repulsion and the Breit interaction. These contributions, which appear in the energy expression, give rise to radial integrals, which need to be calculated for the two component wave function. We will simply state the MCDF equations, which can be obtained by applying the variational principle to the energy expression, for variations in the expansion coefficients c_i in Eq. 9.56 and both the large and the small components of the radial wave function. The coefficients c_i can be determined from the diagonalization of a Hamiltonian matrix and the radial components can be optimized by solving the coupled integro-differential equations, here given for the orbital A

$$
\begin{bmatrix} \frac{d}{dr} + \frac{\kappa_A}{r} & -\frac{2}{\alpha} + \alpha V_A(r) \\ -\frac{2}{\alpha} + \alpha V_A(r) & \frac{d}{dr} + \frac{\kappa_A}{r} \end{bmatrix} \begin{pmatrix} P_A(r) \\ Q_A(r) \end{pmatrix} = \alpha \sum_B \varepsilon_{AB} \begin{pmatrix} Q_B(r) \\ -P_B(r) \end{pmatrix} + \begin{pmatrix} X_{Q_A}(r) \\ -X_{P_A}(r) \end{pmatrix},
$$
(9.62)

where the summation over B contains only the contributions for $\kappa_A = \kappa_B$, where V_A is the sum of the nuclear and the direct Coulomb potentials and where X_{P_A} contains all the two electron integrals, except the instantaneous direct Coulomb repulsion. With the Lagrangian parameters ε_{AB} the orthonormality constraint

$$
\int \{P_A(r)P_B(r) + Q_A(r)Q_B(r)\} \, dr = \delta_{\kappa_A \kappa_B} \delta_{n_A n_B},
$$
(9.63)

is enforced.

9.5.3 Relativistic Density Functions

For the purpose of quantifying the relativistic effects on the electron density functions, which are discussed later on in this contribution, we evaluate Dirac–Fock density functions, using a point nucleus approximation. In this section we describe how density functions can be obtained.

By averaging the sublevel densities

$$
\rho(\mathbf{r}) = \frac{1}{(2J+1)} \sum_{M_J=-J}^{+J} \rho^{JM_J}(\mathbf{r}),
$$
(9.64)

the total electron density becomes spherical for any open-shell system, as found in the non-relativistic scheme (see Sect. 9.3.2) and can be calculated from

$$
\rho(r) = \frac{1}{4\pi} \sum_{n\kappa} \frac{P_{n\kappa}^2(r) + Q_{n\kappa}^2(r)}{r^2} q_{n\kappa},
$$
(9.65)

where $q_{n\kappa}$ is the occupation number of the relativistic subshell $(n\kappa)$. Expression (9.65) can be considered as the relativistic version of Eq. 9.27.

9.6 Analyzing Atomic Densities: Concepts from Quantum Chemistry

9.6.1 The Shape Function

In the context of information theory (cf. Sect. 9.8) the shape function, defined as the density per particle

$$\sigma(\mathbf{r}) = \frac{\rho(\mathbf{r})}{N}, \tag{9.66}$$

where N is the number of electrons, given by

$$N = \int \rho(\mathbf{r})\,d\mathbf{r}, \tag{9.67}$$

plays a role as carrier of information. In particular, the shape function is employed as a probability distribution which describes an atom or a molecule.

The shape function first came to the scene of quantum chemistry in 1983 with the work of Parr and Bartolotti [28]. Although the shape function had appeared before in another context, it is due to Parr and Bartolotti's work that the quantity owes its name. Just as for the density function, the shape function can be shown to determine the external potential $v(\mathbf{r})$ and the number of electrons N [29] and so completely determines the system. The Kato cusp condition [30] leads to the nuclear charges and the relationship between the logarithmic derivative of the shape function and the ionization potential determines the number of electrons [29]. On this basis a Wilson-like argument has been constructed [31] (similar to the original DFT), confirming the shape function "as carrier of information" [32]. Relationships between the shape function and concepts from conceptual DFT were established. In [33] a slightly different perspective is given on the fundamental nature of the shape function.

9.6.2 Quantum Similarity

In chemistry the similarity of molecules plays a central role. Indeed, comparable molecules, usually molecules with a similar shape, are expected to show similar chemical properties and reactivity patterns. Specifically, there chemical behavior is expected to be similar [34]. The concept of functional groups is extensively used in organic chemistry [35], through which certain properties are transferable (to a certain extent) from one molecule to another, and the intense QSAR investigations in pharmaceutical chemistry [36] are illustrations of the attempts to master and exploit similarity in structure, physicochemical properties and reactivity of molecular systems.

Remarkably, the question of quantifying similarity within a quantum mechanical framework has been addressed relatively late, in the early 1980s. The pioneering work of Carbó and co-workers [5,37] led to a series of quantum similarity measures (QSM) and indices (QSI). These were essentially based on the electron density distribution of the two quantum objects (in casu molecules) to be compared. The link between similarity analysis and DFT [38, 39] built on the electron density as the basic carrier of information, and pervading quantum chemical literature at that time, is striking.

The last 15 years witnessed a multitude of studies on various aspects of quantum similarity of molecules (the use of different separation operators [37], the replacement of the density by more appropriate reactivity oriented functions [38, 39] within the context of conceptual DFT [40], the treatment of enantiomers [31, 41–43]). With the exception of two papers by Carbó and co-workers, the study of isolated atoms remained surprisingly unexplored. In the first paper [44] atomic self-similarity was studied, whereas the second one [45] contains a relatively short study on atomic and nuclear similarity, leading to the conclusion that atoms bear the highest resemblance to their neighbors in the Periodic Table.

The work discussed below is situated in the context of a mathematically rigorous theory of quantum similarity measures (QSM) and quantum similarity indices (QSI) as developed by Carbó [5, 37]. Following Carbó, we define the similarity of two atoms (a and b) as a QSM

$$Z_{ab}(\Omega) = \int \rho_a(\mathbf{r})\,\Omega(\mathbf{r},\mathbf{r}')\,\rho_b(\mathbf{r}')\,d\mathbf{r}\,d\mathbf{r}', \qquad (9.68)$$

where $\Omega(\mathbf{r}_1,\mathbf{r}_2)$ is a positive definite operator. Renormalization to

$$SI_\Omega = \frac{Z_{ab}(\Omega)}{\sqrt{Z_{aa}(\Omega)}\sqrt{Z_{bb}(\Omega)}}, \qquad (9.69)$$

yields a QSI SI_Ω with values comprised between 0 and 1.

The two most successful choices for the separation operator $\Omega(\mathbf{r},\mathbf{r}')$ are the Dirac-delta $\delta(\mathbf{r},\mathbf{r}')$ and the Coulomb repulsion $\frac{1}{|\mathbf{r}-\mathbf{r}'|}$. The first is known to reflect comparison of geometrical shape of molecules, whereas the second is said to reflect the charge concentrations [44].

9.7 Analyzing Atomic Densities: Some Examples

In the previous sections the calculation of the density function was discussed and a methodology for comparing them was introduced. In this section a quantitative analysis of atomic density functions is made. It seemed interesting to employ concepts from molecular similarity studies (cf. Sect. 9.6.2), as well developed in quantum chemistry and chemical reactivity studies. Here molecular quantum similarity measures will be applied in a straightforward fashion to investigate (1) the *LS*-dependence of the electron density function in a Hartree–Fock approximation, (2) the density functions of the atoms in their ground state, throughout the periodic table, based on the density function alone, and (3) a quantization of relativistic effects by comparing density functions from Hartree–Fock and Dirac–Hartree–Fock models.

9.7.1 On the LS-term Dependence of Atomic Electron Density Functions

From the developments on the density function in Sect. 9.3.2 it is clear that the *LS*-dependent restricted Hartree–Fock approximation yields *LS*-dependent Hartree–Fock equations for atoms with open sub shells. In Hartree–Fock this dependence can be traced back to the term-dependency of the coulomb interaction. In the single incomplete shell case, corresponding to the ground state configurations we are interested in for the present study, the term dependency is usually fairly small. Froese-Fischer compared mean radii of the radial functions [17]. The differences in $\langle r \rangle$ for the outer orbitals between the values obtained from a Hartree–Fock calculation on the lowest term and those for the average energy of the configuration are of the order of 1–5%. Although the *LS*-dependency does not show up explicitly in the direct and exchange potentials of the closed-subshell radial Hartree–Fock equations, the closed-subshell radial functions are ultimately *LS*-dependent through the coupling between the orbitals in the HF equations to be solved in the iterative procedure, but these relaxation effects turn out to be even smaller.

As explicitly indicated through (9.31) the density built from the one-electron radial functions could therefore be *LS*-dependent but this issue has not yet been investigated quantitatively. Combining the term-dependent densities $\rho_A = \rho_{\alpha_A L_A S_A}$ and $\rho_B = \rho_{\alpha_B L_B S_B}$ for the same atom in the same electronic ground state configuration, but possibly different states (and adopting the Dirac δ-function for Ω) for evaluating the quantum similarity measure Z_{AB} of (9.68), the similarity matrix can be constructed according to (9.69). Its matrix elements have been estimated for the np^2 configuration of Carbon ($n = 2$) and Silicon ($n = 3$) [6]. As expected, the deviation of the off-diagonal elements from 1 is very small, the HF orbitals for the different terms ^3P, ^1D and ^1S being highly similar, although not identical.

9.7.2 A Study of the Periodic Table

As a first step to the recovery of the periodic patterns in Mendeleev's table, Carbó's quantum similarity index (9.69) was used, with the Dirac-δ as separation operator. In this case the expression (9.69) reduces to an expression for shape functions (9.66).

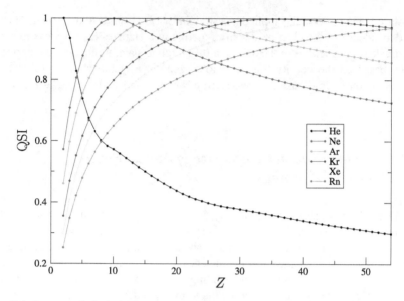

Fig. 9.4 Quantum similarity indices for noble gases, using the Dirac-delta function as separation operator

For the evaluation of the QSI in expression (9.69), we used atomic density functions of atoms in their ground state e.g., corresponding to the lowest energy term. As elaborated in Sect. 9.3.2 the involvement of all degenerate magnetic components allows to construct a spherical density function. For the density functions in this study, we limited ourselves to the Hartree–Fock approximation where no correlation effects are involved and the state functions are built with one CSF.

In Fig. 9.4 we extract, as a case study from the complete atom QSI-matrix, the relevant information for the noble gases. Here the similarities were calculated using the Dirac delta function as separation operator. From these data it is clear that the similarity indices are higher, the closer the atoms are in the periodic table (smallest ΔZ, Z being the atomic number). The tendency noticed by Robert and Carbó in [45] is regained in the present study at a more sophisticated level. It can hence be concluded that the QSI involving $\rho(\mathbf{r})$ and evaluated with $\delta(\mathbf{r} - \mathbf{r}')$ as separation operator Ω, does not generate periodicity.

The discussion of the work on the retrieval of periodicity is continued below in Sects. 9.9.1 and 9.9.2, where concepts from information theory are employed to construct a functional which quantifies the difference between two density functions in a different way.

9.7.3 On the Influence of Relativistic Effects

In this section we investigate the importance of relativistic effects for the electron density functions of atoms. From the relativistic effects on total energies one can infer these effects have implications for the electron densities. The effect of relativity on atomic wave function has been studied in the pioneering work of Burke and Grant [46] who presented graphs and tables to show the order of magnitude of corrections to the hydrogenic charge distributions for $Z = 80$. The relative changes in the binding energies and expectation values of r due to relativistic effects are known from the comparison of the results obtained by solving both the Schrödinger and Dirac equations for the same Coulomb potential. The contraction of the ns-orbitals is a well known example of these relativistic effects. But as pointed out by Desclaux in his *"Tour historique"* [47], for a many-electron system, the self-consistent field effects change this simple picture quite significantly. Indeed, contrary to the single electron solution of the Dirac equation showing mainly the mass variation with velocity, a Dirac–Fock calculation includes the changes in the spatial charge distribution of the electrons induced by the self-consistent field.

We first illustrate the difference of the radial density functions $D(r)$ defined as (see also expression (9.34))

$$D(r) \equiv 4\pi r^2 \rho(\mathbf{r}), \tag{9.70}$$

calculated in the Hartree–Fock (HF) and Dirac–Fock (DF) approximations for the ground state $6p^2\,^3P_0$ of Pb I ($Z = 82$) according to Eqs. 9.65 and 9.70, respectively. These are plotted in Fig. 9.5, which shows the global relativistic contraction of the shell structure.

Employing the framework of QSI to compare non-relativistic Hartree–Fock electron density functions $\rho^{HF}(\mathbf{r})$ with relativistic Dirac–Fock electron density functions $\rho^{DF}(\mathbf{r})$ for a given atom, the influence of relativistic effects on the total density functions of atoms can be quantified via the QSI defined as

$$Z_{HF,\,DF}(\delta) = \int \rho^{HF}(\mathbf{r})\,\delta(\mathbf{r} - \mathbf{r}')\rho^{DF}(\mathbf{r}')\,d\mathbf{r}\,d\mathbf{r}' \tag{9.71}$$

$$SI_\delta = \frac{Z_{HF,\,DF}(\delta)}{\sqrt{Z_{HF,\,DF}(\delta)}\sqrt{Z_{HF,\,DF}(\delta)}}, \tag{9.72}$$

where δ is the Dirac-δ operator.

In Fig. 9.6 we supply the QSI between atomic densities obtained from numerical Hartree–Fock calculation and those obtained from numerical Dirac–Fock calculations, for all atoms of the periodic table.

The results show practically no relativistic effects on the electron densities for the first periods, the influence becoming comparatively large for heavy atoms. To illustrate the evolution through the table the numerical results of the carbon group elements are highlighted in the graph in Fig. 9.6. From the graph it is also noticeable that the relativistic effects rapidly gain importance for atoms heavier than Pb ($Z = 86$).

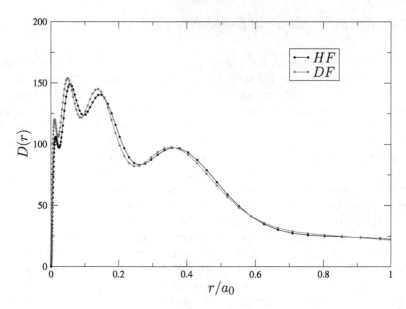

Fig. 9.5 DF and HF density distributions $D(r) = 4\pi r^2 \rho(\mathbf{r})$ for the neutral Pb atom ($Z = 82$). The contraction of the first shells is clearly visible

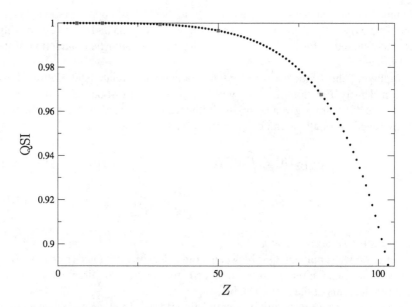

Fig. 9.6 Similarity of non-relativistic Hartree–Fock with relativistic Dirac–Fock atomic density functions with highlighted results for the C group atoms (C, Si, Ge, Sn, Pb)

9.8 Analyzing Atomic Densities: Concepts from Information Theory

Nowadays density functional theory (DFT) is the most widely used tool in quantum chemistry. Its relatively low computational cost and the attractive way in which chemical reactivity can be investigated made it a good alternative to traditional wave function based approaches. DFT is based on the Hohenberg–Kohn theorems [1]. In other words an atom's or a molecule's energy – and in fact any other physical property – can be determined by evaluating a density functional. However, the construction of some functionals corresponding to certain physical property, has proven very difficult. Moreover, to the present day no general and systematic way for constructing such functionals has been established. Although energy functionals, which are accurate enough for numerous practical purposes, have been around for some time now, the complicated rationale and the everlasting search for even more accurate energy functionals are proof of the difficulties encountered when constructing such functionals. In the domain of conceptual DFT, where chemical reactivity is investigated, a scheme for the construction of functionals, based on derivatives of the energy with respect to the number of electrons and/or the external potential, has proven very successful [40, 48]. Inspiration for the construction of chemically interesting functionals has also come from information theory and statistical mathematics. The functionals used for analyzing probability distributions have been successfully applied to investigate electron density functions of atoms and molecules. In this chapter we introduce those functionals and discuss several studies where they have been applied to construct chemically interesting functionals.

Shannon is generally recognized as one of the founding fathers of information theory. He defined a measure for the amount of information in a message and based on that, he developed a mathematical theory of communication. His theory of communication is concerned with the amount of information in a message rather than the information itself or the semantics. It is based on the idea that – from the physical point of view – the message itself is irrelevant, but its size is an objective quantity. Shannon saw his measure of information as a measure of uncertainty and referred to it as an entropy. Since Shannon's seminal publication in 1948 [49], information theory became a very useful quantitative theory for dealing with problems of transmission of information and his ideas found many applications in a remarkable number of scientific fields. The fundamental character of information, as defined by Shannon, is strengthened by the work of Jaynes [50, 51], who showed that it is possible to develop a statistical mechanics on the basis of the principle of maximum entropy.

In the literature the terms entropy and information are frequently interchanged. Arih Ben-Naim, the author of *"Farewell to Entropy: Statistical Thermodynamics Based on Information"* [52] insists on going one step further and motivates "not only to use the principle of maximum entropy in predicting the probability distribution [which is used in statistical physics], but to replace altogether the concept of entropy with the more suitable information." In his opinion "this would replace an essentially meaningless term [entropy] with an actual objective, interpretable physical quantity [information]." We do not intend to participate in this discussion at this time, since the present chapter is not concerned with the development of information theory itself, but rather with an investigation of the applicability of some concepts, borrowed from information theory, in a quantum chemical context. The interested reader can find an abundance of treatments on information theory itself and its applications to statistical physics and thermodynamics in the literature.

Information theoretical concepts found their way into chemistry during the seventies. They were introduced to investigate experimental and computed energy distributions from molecular collision experiments. The purpose of the information theoretical approach was to measure the significance of theoretical models and conversely to decide which parameters should be investigated to gain the best insight into the actual distribution. For an overview of this approach to molecular reaction dynamics, we refer to Levine's work [53]. Although the investigated energy distributions have little relation with electronic wave functions and density functions, the same ideas and concepts found their way to quantum chemistry and the chemical reactivity studies which are an important study field of it. Most probably this is stimulated by the fact that atoms and molecules can be described by their density function, which is ultimately a probability distribution. The first applications of information theoretical concepts in quantum chemistry, can be found in the literature of the early eighties. The pioneering work of Sears et al. [54] quickly lead to more novel ideas and publications. Since then, many applications of information theoretical concepts to investigate wave functions and density functions, have been reported. In [55] Gadre gives a detailed review of the original ideas behind and the literature on "Information Theoretical Approaches to Quantum Chemistry." To motivate our work in this field we paraphrase the author's concluding sentence:

> Thus it is felt that the information theoretical principles will continue to serve as powerful guidelines for predictive and interpretive purposes in quantum chemistry.

The initial idea in our approach was to construct a density functional, which reveals chemical and physical properties of atoms, since the periodicity of the Table is one of the most important and basic cornerstones of chemistry. Its recovery on the basis of the electron density alone can be considered a significant result. In an information theoretical context, the periodicity revealing functional can be interpreted as a quantification of the amount of information in a given atom's density function, missing from the density function of the noble gas atom which precedes it in the periodic table. The results indicate that information theory offers a method for the construction of density functionals with chemical interest and based on this we continued along the same lines and investigated if more chemically interesting information functionals could be constructed.

In the same spirit, the concept of complexity has been taken under consideration for the investigation of electron density functions. Complexity has appeared in many fields of scientific inquiry e.g., physics, statistics, biology, computer science and economics [56]. At present there does not exist a general definition which quantifies complexity, however several attempts have been made. For us, one of these stands out due to its functional form and its link with information theory.

Throughout this chapter it becomes clear that different information and complexity measures can be used to distinguish electron density functions. Their evaluation and interpretation for atomic and molecular density functions gradually gives us a better understanding of how the density function carries physical and chemical information. *This exploration of the density function using information measures teaches us to read this information.*

Before going into more details about our research several concepts should be formally introduced. For our research, which deals with applications of functional measures to atomic and molecular density functions, a brief discussion of these measures should suffice. The theoretical sections are followed by an in depth discussion of our results. In the concluding section we formulate general remarks and some perspectives.

9.8.1   Shannon's Measure: An Axiomatic Definition

In 1948 Shannon constructed his information measure – also referred to as "entropy" – for probability distributions according to a set of characterizing axioms [49]. A subsequent work showed that, to obtain the desired characterization, Shannon's original four axioms should be completed with a fifth one [57]. Different equivalent sets of axioms exist which yield Shannon's information measure. The original axioms, with the necessary fifth, can be found in [58]. Here we state the set of axioms described by Kinchin [55,59].

For a stochastic event with a set of n possible outcomes (called the event space) $\{A_1, A_2, \ldots, A_n\}$ where the associated probability distribution $P = \{P_1, P_2, \ldots, P_n\}$ with $P_i \geq 0$ for all i and $\sum_{i=1}^{n} P_i = 1$, the measure S should satisfy:

1. The entropy functional S is a continuous functional of P
2. The entropy is maximal when P is the uniform distribution i.e., $P_i = 1/n$
3. The entropy of independent schemes are additive i.e., $S(P_A + P_B) = S(P_A) + S(P_B)$ (a weaker condition for dependent schemes exists)
4. Adding any number of impossible events to the event space does not change the entropy i.e., $S(P_1, P_2, \ldots, P_n, 0, 0, \ldots, 0) = S(P_1, P_2, \ldots, P_n)$.

It can be proven [59] that these axioms suffice to uniquely characterize Shannon's entropy functional

$$S = -k \sum_i P_i \log P_i, \qquad (9.73)$$

with k a positive constant. The sum runs over the event space i.e., the entire probability distribution. In physics, expression (9.73) also defines the entropy of a given macro-state, where the sum runs over all micro-states and where P_i is the probability corresponding to the i-th micro-state. The uniform distribution possesses the largest entropy indicating that the measure can be considered as a measure of randomness or uncertainty, or alternatively, it indicates the presence of information.

When Shannon made the straightforward generalization for continuous probability distributions $P(x)$

$$S[P(x)] = -k \int P(x) \log P(x) \, dx, \tag{9.74}$$

he noticed that the obtained functional depends on the choice of the coordinates. This is easily demonstrated for an arbitrary coordinate transformation $y = g(x)$, by employing the transformation rule for the probability distribution $p(x)$

$$q(y) = p(x)J^{-1} \tag{9.75}$$

and the integrandum

$$dy = J dx, \tag{9.76}$$

where J is the Jacobian of the coordinate transformation and J^{-1} its inverse. The entropy hence becomes

$$\int q(y) \log(q(y)) \, dy = \int p(x) \log(p(x)J^{-1}) \, dx, \tag{9.77}$$

where the residual J^{-1} inhibits the invariance of the entropy. Although Shannon's definition lacks invariance and although it is not always positive, it generally performs very well. Moreover, its fundamental character is emphasized by Jaynes's maximum entropy principle, which permits the construction of statistical physics, based on the concept of information [50,51]. In the last decade several investigations of the Shannon entropy in a quantum chemical context have been reported. Those relevant to our research are discussed in more detail below.

9.8.2 Kullback–Leibler Missing Information

Kullback–Leibler's information deficiency was introduced in 1951 as a generalization of Shannon's information entropy [8]. For a continuous probability distribution $P(x)$, relative to the reference distribution $P_0(x)$, it is given by

$$\Delta S[P(x)|P_0(x)] = \int P(x) \log \frac{P(x)}{P_0(x)} \, dx. \tag{9.78}$$

As can easily be seen from expression (9.77), the introduction of a reference probability distribution $P_0(x)$ yields a measure independent of the choice of the coordinate system. The Kullback–Leibler functional quantifies the amount of information which discriminates $P(x)$ from $P_0(x)$. In other words, it quantifies the distinguishability of the two probability distributions. Sometimes it can be useful to see $\Delta S[P(x)|P_0(x)]$ as the distance in information from P_0 to P, although strictly speaking the lack of symmetry under exchange of $P(x)$ and $P_0(x)$ makes it a directed divergence.

Kullback–Leibler's measure is an attractive quantity from a conceptual and formal point of view. It satisfies the important properties positivity, additivity, invariance, respectively:

1. $\Delta S[P(x)|P_0(x)] \geq 0$
2. $\Delta S[P(x,y)|P_0(x,y)] = \Delta S[P(x)|P_0(x)] + \Delta S[P(y)|P_0(y)]$ for independent events i.e., $P(x,y) = P(x)P(y)$
3. $\Delta S[P(y)|P_0(y)] = \Delta S[P(x)|P_0(x)]$ if $y = f(x)$.

Besides the lack of symmetry, the Kullback–Leibler functional has other formal limitations e.g., it is not bound, nor is it always well defined. In [60] the lack of these properties was addressed and the Jensen–Shannon divergence was introduced as a symmetrized version of Kullback–Leibler's functional. In [61] the Jensen–Shannon distribution was first proposed as a measure of distinguishability of two quantum states. Chatzisavvas et al. investigated the quantity for atomic density functions [62].

For our investigation of atomic and molecular density functions, as carrier of physical and chemical information, we constructed functionals based on the definition of information measures. In Sect. 9.9.1 below, the research is discussed in depth.

9.9 Examples from Information Theory

9.9.1 Reading Chemical Information from the Atomic Density Functions

This section contains a detailed description of our research on the recovery of the periodicity of Mendeleev's Table. The novelty in this study is that we managed to generate the chemical periodicity of Mendeleev's table in a natural way, by constructing and evaluating a density functional. As discussed before in Sect. 9.7.2, the comparison of atomic density functions on the basis of a quantum similarity index (using the $\delta(\mathbf{r}_1 - \mathbf{r}_2)$ operator), masks the periodic patterns in Mendeleev's table. On the other hand, the importance of the periodicity, as one of the workhorses in chemistry, can hardly be underestimated. Due to the Hohenberg-Kohn theorems, the electron density can be considered as the basic carrier of information, although, for many properties it is unknown how to extract the relevant information from the density function. This prompted us to investigate whether the information measures, which gained a widespread attention by the quantum chemical community, could be used to help extract chemical information from atomic density functions in general and help to regain chemical periodicity in particular.

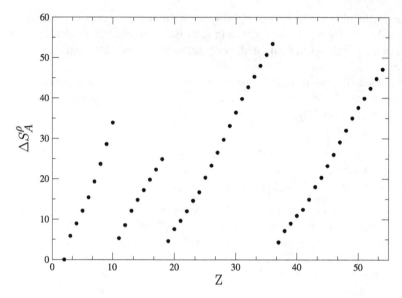

Fig. 9.7 Kullback–Leibler information (9.79) versus Z for atomic densities with the noble gas of the previous row as reference

Tempted by the interpretation of the Kullback–Leibler expression (9.78) as a tool to distinguish two probability distributions, the possibility of using it to compare atomic density functions is explored. To make a physically motivated choice of the reference density $P_0(x)$ we consider the construction of Sanderson's electronegativity scale [63], which is based on the compactness of the electron cloud. Sanderson introduced a hypothetical noble gas atom with an average density scaled by the number of electrons. This gives us the argument to use renormalized noble gas densities as reference in expression (9.78). This gives us the quantity

$$\Delta S_A^\rho \equiv \Delta S[\rho_A(\mathbf{r})|\rho_0(\mathbf{r})] = \int \rho_A(\mathbf{r}) \log \frac{\sigma_A(\mathbf{r})}{\sigma_0(\mathbf{r})} \, d\mathbf{r}, \qquad (9.79)$$

where $\rho_A(\mathbf{r})$ and $\sigma_A(\mathbf{r})$ are the density and shape function of the investigated system and $\sigma_0(\mathbf{r})$ the shape function of the noble gas atom preceding atom A in Mendeleev's table. The evaluation of this expression for atoms He through Xe shows a clear periodic pattern, as can be seen in Fig. 9.7.

Reducing the above expression to one that is based on shape functions only, leads to

$$\Delta S_A^\sigma \equiv \Delta S[\sigma_A(\mathbf{r})|\sigma_0(\mathbf{r})] = \int \sigma_A(\mathbf{r}) \log \frac{\sigma_A(\mathbf{r})}{\sigma_0(\mathbf{r})} \, d\mathbf{r} \qquad (9.80)$$

and its evolution is shown in Fig. 9.8. The periodicity is clearly present and this with the fact that the distance between points in a given period is decreasing gradually from first to fourth row is in agreement with the evolution of many

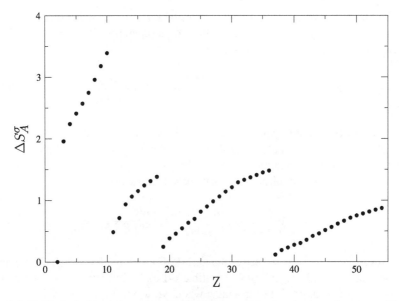

Fig. 9.8 Kullback–Leibler information (Eq. 9.80) versus Z for atomic shape functions with the noble gas of the previous row as reference

chemical properties throughout the periodic table. One hereby regains one of the basic characteristics of the Periodic Table namely that the evolution in (many) properties through a given period slows down when going down in the Table. The decrease in slope of the four curves is a further illustration.

9.9.2 Information Theoretical QSI

Continuing the search for periodic patterns based on similarity measures, as introduced in Sect. 9.7 and motivated by the results obtained in an information theoretical framework in Sect. 9.9.1, we will now combine the ideas from both efforts and construct an information theoretical similarity measure.

For the construction of the functional in the above section, the choice to set the reference (the prior) density to that of a hypothetical noble gas atom, in analogy to Sanderson's electronegativity scale, was motivated and the particular choice lead to results which could be interpreted chemically. Following these findings one can see that it would be interesting to compare the information entropy, evaluated locally as

$$\Delta S_A^\rho(\mathbf{r}) \equiv \rho_A(\mathbf{r}) \log \frac{\rho_A(\mathbf{r})}{\frac{N_A}{N_0}\rho_0(\mathbf{r})}, \tag{9.81}$$

for two atoms by use of a QSM, which can be constructed straightforwardly, by considering the overlap integral (with Dirac-δ as separation operator) of the local information entropies of two atoms A and B

$$Z_{AB}(\delta) = \int \rho_A(\mathbf{r}) \log \frac{\rho_A(\mathbf{r})}{\frac{N_A}{N_0}\rho_0(\mathbf{r})} \rho_B(\mathbf{r}) \log \frac{\rho_B(\mathbf{r})}{\frac{N_B}{N_{0'}}\rho_{0'}(\mathbf{r})} \, d\mathbf{r}. \qquad (9.82)$$

A QSI can be defined by normalizing the QSM as before, via expression (9.69). The QSM and the normalized QSI give a quantitative way of studying the resemblance in the information carried by the valence electrons of two atoms. The obtained QSI trivially simplifies to a shape based expression

$$SI_{(\delta)} = \frac{\int \Delta S_A^\sigma(\mathbf{r}) \Delta S_B^\sigma(\mathbf{r}) d\mathbf{r}}{\sqrt{\int \Delta S_A^\sigma(\mathbf{r}) \Delta S_A^\sigma(\mathbf{r}) d\mathbf{r}} \sqrt{\int \Delta S_B^\sigma(\mathbf{r}) \Delta S_B^\sigma(\mathbf{r}) d\mathbf{r}}}. \qquad (9.83)$$

To illustrate the results we select the QSI (9.83) with the top atoms of each column as prior. Formulated in terms of Kullback–Leibler information discrimination the following is evaluated. For instance, when we want to investigate the distance of the atoms Al, Si, S and Cl from the N-column (group Va), we consider the information theory based QSI in expression (9.83), where the reference densities ρ_0 and $\rho_{0'}$ are set to ρ_N, ρ_A to ρ_{Al}, ρ_{Si}, ρ_P, etc., respectively and ρ_B to ρ_P, i.e., we compare the information contained in the shape function of N to determine that of P, with its information on the shape function of Al, Si, S, Cl. Due to the construction a 1. is yielded for the element P and the other values for the elements to the left and to the right of the N-column decrease, as shown in Fig. 9.9. This pattern is followed for the periods 3 up to 6, taking As, Sb and Bi as reference, with decreasing difference along a given period in accordance with the results above. Note that the difference from 1. remains small, due to the effect of the renormalization used to obtain the QSI.

9.10 General Conclusion

Results on the investigation of atomic density functions are reviewed. First, ways for calculating the density of atoms in a well-defined state are discussed, with particular attention for the spherical symmetry. It follows that the density function of an arbitrary open shell atom is not a priori spherically symmetric. A workable definition for density functions within the multi-configuration Hartree–Fock framework is established. By evaluating the obtained definition, particular influences on the density function are illustrated. A brief overview of the calculation of density functions within the relativistic Dirac–Hartree–Fock scheme is given as well.

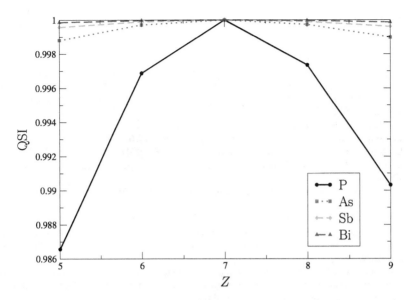

Fig. 9.9 Results of the information theory based QSI with the atom on top of the column as prior. The symbol in the legend indicates the period of the investigated atom and the nuclear charge Z-axis indicates the column of the investigated atom (For example Ga can be found as a square $Z = 5$)

After discussing the definition of atomic density functions, quantum similarity measures are introduced and three case studies illustrate that specific influences on the density function of electron correlation and relativity can be quantified in this way. Although no periodic patterns were found in Mendeleev's table, the methodology is particularly successful for quantifying the influence of relativistic effects on the density function.

In the final part the application of concepts from information theory is reviewed. After covering the necessary theoretical background a particular form of the Kullback–Liebler information measure is adopted and employed to define a functional for the investigation of density functions throughout Mendeleev's Table. The evaluation of the constructed functional reveals clear periodic patterns, which are even further improved when the shape function is employed instead of the density functions. These results clearly demonstrate that it is possible to retrieve chemically interesting information from the density function. Moreover the results indicate that the shape function further simplifies the density function without loosing essential information. The latter point of view is extensively treated in [64], where the authors elaborately discuss "information carriers" such as the wave function, the reduced density matrix, the electron density function and the shape function.

References

1. Hohenberg P, Kohn W (1964) Phys Rev B 136(3B):B864
2. Fischer CF, Tachiev G, Gaigalas G, Godefroid MR (2007) Comput Phys Commun 176(8):559
3. Borgoo A, Scharf O, Gaigalas G, Godefroid M (2010) Comput Phys Commun 181(2):426. doi:10.1016/j.cpc.2009.10.014. URL http://www.sciencedirect.com/science/article/B6TJ5-4XG3SF0-1/2/d040eb496c97b1d109b779bede692437
4. Desclaux JP (1975) Comput Phys Commun 9(1):31
5. Carbó R, Leyda L, Arnau M (1980) Int J Quantum Chem 17(6):1185
6. Borgoo A, Godefroid M, Sen KD, De Proft F, Geerlings P (2004) Chem Phys Lett 399(4-6):363
7. Borgoo A, Godefroid M, Indelicato P, De Proft F, Geerlings P (2007) J Chem Phys 126(4):044102
8. Kullback S, Leibler RA (1951) Ann Math Stat 22(1):79. URL http://www.jstor.org/stable/2236703
9. Hibbert A, Glass R, Fisher CF (1991) Comput Phys Commun 64(3):455
10. McWeeny R (1992) Methods of molecular quantum mechanics. Academic, London
11. Helgaker T, Jorgensen P, Olsen J (2000) Molecular electronic structure theory. Wiley, Chichester
12. Davidson ER (1976) Reduced density matrices in quantum mechanics. Academic, New York
13. Cowan R (1981) The theory of atomic structure and spectra. Los Alamos Series in Basic and Aplied Sciences. University of California Press
14. Fertig HA, Kohn W (2000) Phys Rev A 62(5):052511
15. Slater J (1930) Phys Rev 35:210
16. Hartree D (1957) The calculation of atomic structures. Wiley, New York
17. Fischer CF (1977) The Hartree-Fock method for atoms. A numerical approach. Wiley, New York
18. Varshalovich D, Moskalev A, Khersonskii V (1988) Quantum theory of angular momentum. World Scientific, Singapore
19. Unsöld A (1927) Ann Phys 82:355
20. Judd B (1967) Second quantization and atomic spectroscopy. The Johns Hopkins Press, Baltimore
21. Olsen J, Godefroid MR, Jönsson P, Malmqvist PA, Fischer CF (1995) Phys Rev E 52(4):4499. doi:10.1103/PhysRevE.52.4499
22. Gaigalas G, Fritzsche S, Grant, IP (2001) Comput Phys Commun 139(3):263
23. Carette T, Drag C, Scharf O, Blondel C, Delsart C, Froese Fischer C, Godefroid M (2010) Phys Rev A 81:042522. doi:10.1103/PhysRevA.81.042522
24. Indelicato P (1995) Phys Rev A 51(2):1132
25. Indelicato P (1996) Phys Rev Lett 77(16):3323
26. Grant I (1996) In: Drake G (ed) Atom, molecular ant optical physics. AIP, New York
27. Desclaux J (1993) In: Clementi E (ed) Methods and techniques in computational chemistry - vol. A: small systems of METTEC, Stef, Cagliari, p. 253
28. Parr RG, Bartolotti LJ (1983) J Phys Chem 87(15):2810
29. Ayers PW (2000) Proc Natl Acad Sci U S A 97(5):1959
30. Kato T (1957) Commun Pure Appl Math 10(2):151
31. Geerlings P, Boon, G, Van Alsenoy C, De Proft F (2005) Int J Quantum Chem 101(6):722
32. Geerlings P, De Proft F, Ayers P (2007) In: Toro Labbé A (ed) Theoretical aspects of chemical reactivity, Elsevier, Amsterdam
33. Chattaraj P (ed) (2009) Chemical reactivity theory; a density functional view. CRC/Taylor & Francis Group, Boca Raton
34. Rouvray D (1995) In: Sen K (ed) Topics in current chemistry, vol 173. Springer, Berlin/New York, p. 2
35. Patai S (ed) (1992) The chemistry of functional groups. Interscience Pubishers, London

36. Bultinck P, De Winter H, Langeneaker W (2003) Computation medicinal chemistry for drug discovery. Decker Inc., New York
37. Bultinck P, Girones X, Carbó-Dorca R (2005) Rev Comput Chem 21:127
38. Boon G, De Proft F, Langenaeker W, Geerlings P (1998) Chem Phys Lett 295(1–2):122
39. Boon G, Langenaeker W, De Proft F, De Winter H, Tollenaere JP, Geerlings P (2001) J Phys Chem A 105(38):8805
40. Geerlings P, De Proft F, Langenaeker W (2003) Chem. Rev. 103(5):1793
41. Boon G, Van Alsenoy C, De Proft F, Bultinck P, Geerlings P (2003) J Phys Chem A 107(50):11120
42. Janssens S, Boon G, Geerlings P (2006) J Phys Chem A 110(29):9267
43. Janssens S, Van Alsenoy C, Geerlings P (2007) J Phys Chem A 111(16):3143
44. Sola M, Mestres J, Oliva JM, Duran M, Carbó R (1996) Int J Quantum Chem 58(4):361
45. Robert D, Carbó-Dorca R (2000) Int J Quantum Chem 77(3):685
46. Burke VM, Grant IP (1967) Proc Phys Soc Lond 90(2):297
47. Desclaux JP (2002) The relativistic electronic structure theory book. Theoretical and computational chemistry, vol 11. Elsevier, Amsterdam
48. Geerlings P, De Proft F (2008) Phys Chem Chem Phys 10(21):3028
49. Shannon S (1948) Bell Syst Tech 27:379
50. Jaynes ET (1957) Phys Rev 106(4):620
51. Jaynes ET (1957) Phys Rev 108(2):171
52. Ben-Naim A (2008) Farewell to entropy: statistical thermodynamics based on information. World Scientific Publishing Co. Pte. Ltd., Singapore
53. Levine RD (1978) Annu Rev Phys Chem 29:59
54. Sears SB, Parr RG, Dinur U (1980) Israel J Chem 19(1–4):165
55. Gadre S (2002) In: Sen K (ed) Reviews of modern quantum chemistry, vol 1. World Scientific Publishing Co., Singapore, pp. 108–147
56. Catalan RG, Garay J, Lopez-Ruiz R (2002) Phys Rev E 66(1):011102
57. Mathai A (1975) Basic concepts in information theory and statistics axiomatic foundations and applications. Wiley Eastern, New Delhi
58. Ash R (1967) Information theory. Interscience Publishers, New York
59. Kinchin A (1957) Mathematical foundations of information theory. Dover, New York
60. Lin J (1991) IEEE Trans Inf Theory 37(1):145
61. Majtey A, Lamberti PW, Martin MT, Plastino A (2005) Eur Phys J D 32(3):413
62. Chatzisavvas KC, Moustakidis CC, Panos CP (2005) J Chem Phys 123(17):174111
63. Sanderson RT (1951) Science 114:670
64. Geerlings P, Borgoo A (2011) Phys Chem Chem Phys 13(3):911

Chapter 10
Understanding Maximum Probability Domains with Simple Models

Osvaldo Mafra Lopes Jr., Benoît Braïda, Mauro Causà, and Andreas Savin

Abstract The paper presents maximum probability domains (MPDs). These are regions of the three dimensional space for which the probability to find a given number of electrons is maximal. In order to clarify issues hidden by numerical uncertainties, some simple models are used. They show that MPDs reproduce features which one would expect using chemical intuition. For a given number of electrons, there can be several solutions, corresponding to different chemical situations (e.g. different bonds). Some of them can be equivalent, by symmetry. Symmetry can produce, however, alternative solutions. The models show that MPDs do not exactly partition space, and they can also be formed by disjoint subdomains. Finally, an example shows that a partition of space, as provided by loge theory, can lead to situations difficult to deal with, not present for MPDs.

10.1 Introduction

In the last few years a method was explored which allows to analyze electronic wave functions by describing the regions of space for which the probability to find a given number of electrons, v, is maximal [1–11]. When $v = 2$, it relates to Lewis' concept of electron pairs and provides thus a connection between quantum mechanics and the traditional way of thinking of chemists. This paper summarizes

O. Mafra Lopes Jr. • B. Braïda • A. Savin (✉)
Laboratoire de Chimie Théorique, CNRS and UPMC Univ Paris 6, 4 place Jussieu,
75252 Paris, France
e-mail: osvaldo.mafra@gmail.com; braida@lct.jussieu.fr; andreas.savin@lct.jussieu.fr

M. Causà
Dip. di Chimica, Universita' di Napoli Federico II, Via Cintia, 80126 Napoli, Italy
e-mail: mauro.causa@unina.it

P.E. Hoggan et al. (eds.), *Advances in the Theory of Quantum Systems in Chemistry and Physics*, Progress in Theoretical Chemistry and Physics 22,
DOI 10.1007/978-94-007-2076-3_10, © Springer Science+Business Media B.V. 2012

its main features, and illustrates them by using simple quantum mechanical models. These allow to clarify some features which may be blurred by numerical issues in realistic situations.

10.2   Method

10.2.1   Maximal Probability Domains

For a system in the state described by the wave function Ψ, the probability to find ν and only ν electrons out of N in a three-dimensional region Ω is given by

$$p_\nu(\Omega) = \binom{N}{\nu} \int_\Omega dx_1...dx_\nu \int_{\bar{\Omega}} dx_{\nu+1}...dx_N |\Psi(x_1,...,x_N)|^2 \tag{10.1}$$

where $\bar{\Omega}$ is the complement of Ω, $R^3 \setminus \Omega$, and the binomial coefficient is added to take into account electron indistinguishability. A maximal probability domain, MPD, is a region of space maximizing $p_\nu(\Omega)$. It will be denoted by Ω_ν.

Please notice that p_ν is not restricted to ground states, and that Ω can be formed of disjoint subdomains.

The computation of p_ν, Eq. 10.1, is less difficult as it may seem, at least for certain forms of the wave function. In particular, for a single Slater determinant, one first computes the overlaps of all occupied orbitals over the regions Ω,

$$S_{ij}(\Omega) = \int_\Omega \phi_i(x)\phi_j(x)dx \tag{10.2}$$

Next, the eigenvalues of the matrix with elements S_{ij} are obtained. From them, the probabilities are quickly computed for all ν, with a recursive formula [3].

The presently running programs use a grid of small cubes. To represent a spatial domain Ω, a collection of such cubes is used. The procedure to optimize Ω is the following. We first start by guessing a domain, either in a trivial way, like using a larger cube (union of small unit cubes) located in the region of interest, or by using ELF basins [12, 13], produced, e.g., with the TopMod program [14]. In the present version of the MPD program, two optimization algorithms are available. In one of them, small cubes are randomly added or deleted, the step being accepted when the probability of the new domain is increased. This algorithm was already used and described in more detail in Ref. [15]. The other algorithm which can be used, takes advantage of the availability of shape derivatives, as described in Ref. [3]. The derivative indicates whether one should add or delete small cubes on the surface of the domain in order to increase the probability.

Programs to provide MPDs now available can use single determinant wave functions from calculations produced by the Gaussian suite of programs [16] for molecules, or by Crystal-98 [17] for periodic systems. MPDs can be produced also with correlated wave functions, via a Quantum Monte Carlo program, cf. [10, 15]. Probabilities can be computed for multi-determinant wave functions [18], but the optimization of Ω is not implemented yet.

10.2.2   Similarities and Differences

For a well-localized pair of electrons, MPDs provide regions of space which resemble the regions where orbitals can be localized. Notice, however, that localized orbitals extend to infinity, while the MPDs extend over a given region of space. In this respect, they resemble the basins showing up in the Quantum Theory of Atoms in Molecules [19], or in the Electron Localization Function (ELF) approach [12,13]. For single Slater determinants, in the ideal limiting case of strictly localized (non-overlapping) orbitals, the localization domain of the orbitals, the ELF basins, and the MPDs become identical [20].

The expression for the probability p_ν may remind of the ν-particle reduced density matrices. The latter are obtained, however, by integrating $x_{\nu+1},...,x_N$ over the whole space, not just over $\bar{\Omega}$. In particular, $p_1(\Omega)$ is not equal to the integral of the one-particle density over the region Ω,

$$\int_\Omega \rho(x_1)dx_1 = N \int_\Omega dx_1 \int dx_2...dx_N |\Psi(x_1,...,x_N)|^2 \qquad (10.3)$$

The latter is, in fact, not the probability to find one particle in Ω, but the average number of particles in Ω [1], the population of Ω. For example, for the dissociated hydrogen molecule in its electronic ground state, the probability to find one electron in the half-space containing one of the protons is equal to 1, as is the population. For the ionic excited state, $H^+...H^- \leftrightarrow H^-...H^+$, the probability to find one electron in one half-space is zero, while the population is still one.

With population analysis, and also with the valence bond approach, a reference space of atomic functions is defined. Such a space is absent when defining the MPDs as the search is carried out in three-dimensional space.

MPDs remind of Daudel's loges [21,22] which also use $p_\nu(\Omega)$. There, the idea is to partition molecular space into domains, called "loges", and look for all different possibilities to distribute electrons into them. After some initial trials it was decided to minimize the missing information function,

$$H(x_1,..,x_M) = -\sum_k P(x_k)\ln(P(x_k)) \qquad (10.4)$$

as the criterion to determine the partition. In the definition of H, an event x_k is a given distribution of electrons in the loges, $P(x_k)$ is the probability to have such a distribution, and the sum goes over all possible distribution of electrons in the loges.[1]

10.2.3 Models

We will treat below some non-interacting particle models for which the exact solution of the Schrödinger equation is known, because it is possible to compute $p_V(\Omega)$ for these systems with arbitrary accuracy, e.g., with Mathematica [23]. More details about the models can be found in the appendix.

In order to have a significance for chemistry, we will assume that particles are fermions. Thus, although non-interacting, the particles are not independent, as they have to obey the Pauli principle. Please notice that this holds for same-spin particles, while particles with opposite spin are independent in these models.

In the following, we will sometimes consider situations where only particles of a given spin are present. For N particles of one spin, the probabilities can be related to those for N pairs of fermions of opposite spin. This can be easily seen, by writing the wave function as a product of a Slater determinant for α spin with one for β spin (see, e.g. [24]). This product yields the same expectation values as the Slater determinant written with all spin-orbitals, for both spins. Equation 10.1 yields for the probability of finding v_α and v_β electrons in Ω the product of the probabilities computed for each spin individually, for v_α, and v_β, respectively. When we consider the restricted Hartree-Fock closed shell form, we obtain the same terms for each spin. Thus, one finds the same Ω when maximizing the probability of finding v_α and v_β electrons as when one maximizes the probability of finding v_α or v_β electrons only.

10.3 Results

10.3.1 Experience with MPDs

Up to now, experience has shown that MPD correspond to regions to which a chemist would associate bonds, or cores, or lone pairs (see, e.g. [10]). In this respect, when a single Lewis picture is sufficient, Ω_Vs resemble images produced by other tools, e.g., ELF basins [12, 13]. For example, for the MgO crystal, when

---

[1]The basis of the logarithm is arbitrary; we have chosen in this paper the natural logarithm, and not the binary logarithm, as usually done.

Fig. 10.1 Ω_{10} solutions for the MgO crystal: Mg^{2+}, *top*, and O^{2-}, *bottom*

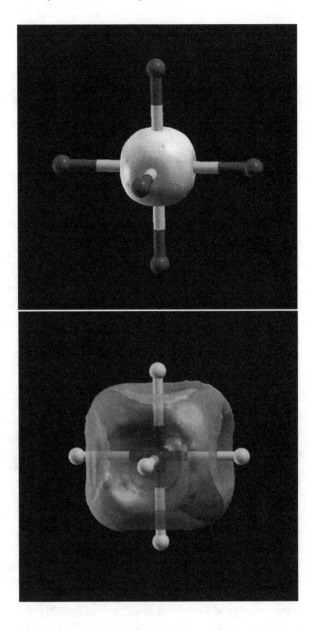

maximizing p_{10}, it yields two solutions, one corresponding to Mg^{2+} and another to O^{2-}, see Fig. 10.1. In fact, it was noticed that the results are slightly closer to chemical intuition than with ELF. For example, with MPDs, the population of atomic shells is closer to the integer numbers one intuitively expects than it is with ELF basins [1].

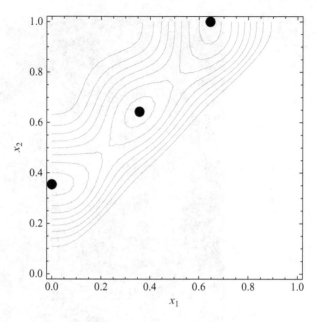

Fig. 10.2 $p_1([x_1,x_2])$ for three same-spin fermions in a box. The *dots mark* the maxima

10.3.2 MPDs Are not Unique

In general, there can be several MPDs, Ω_v, for a given v. We just mentioned (cf. Fig. 10.1) that maximizing p_{10} can yield two physically relevant solutions, one corresponding to Mg^{2+}, and another corresponding to O^{2-}. This is physically motivated, and is not a limitation of MPDs. From the practical viewpoint, it means that the program searching for Ω_v can yield several solutions, typically by using different starting guesses.

An accurate calculation showing multiple solutions can be produced for three same-spin particles in a box of unit length. The probability to find one particle in the interval between x_1 and x_2, $p([x_1,x_2])$, is shown in Fig. 10.2. There are two maxima corresponding to symmetrically arranged Ω, given by the intervals $[0,0.35]$, at the left of the box, and $[0.65,1]$, at the right of the box. There is a third Ω between them. It is no surprise that the values for p_1 are different for the Ω_1 at the borders of the box (≈ 0.84) and in the center of the box (≈ 0.75). The lower value for the central Ω_1 can be understood by the existence of two penetrable walls for this MPD, while there is only one for the terminal Ω_1s.

The example above also shows that there are solutions which are equivalent, the Ω_1 corresponding to the left is equivalent to that on the right border of the box. The origin of this equivalence, is the symmetry of the box. It reminds of the equivalence of localized orbitals [25]: a symmetry operation can transform one localized orbital into another. In a molecule, such equivalent MPDs are common. For example, in

Fig. 10.3 The two lone pair Ω_2s in the H_2O molecule

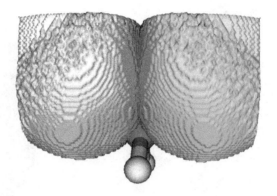

the water molecule, there is a MPD corresponding to one of the lone pairs, that is equivalent to another one, corresponding to the other lone pair, cf. Fig. 10.3.

However, symmetry operations can also produce physically reasonable, equivalent solutions, but not transform one chemically relevant unit (like a bond) into another. In certain situations, there may be an infinite number of equivalent solutions. Take, for example, three same-spin particles in a ring. There is a solution corresponding to θ between 0 and $\approx 2\pi/3$. Of course, there are two more solutions, one corresponding to roughly $[2\pi/3, 4\pi/3]$, and another one for roughly $[4\pi/3, 2\pi]$. These are produced from the first domain by threefold rotations. However, the ring is invariant to rotations by an arbitrary angle. Thus, besides the chemically understandable existence of threefold solutions, one can find an infinity of Ω_1, as produced by a rotation by an arbitrary angle: $p_1 \approx 0.68$ does not change when the lower and the upper limit of the interval is displaced by the same arbitrary constant.

A similar result is obtained for acetylene. There is a solution corresponding to one of the banana bonds, see Fig. 10.4. There are two more MPDs, corresponding to the other two banana bonds. One can generate, however, infinitely many new Ω_1, by rotating the previous Ω_1s around the internuclear axis. Please notice that the three Ω produced by the rotation about a threefold axis are essentially non-overlapping. One can see the rotation by an arbitrary angle as the arbitrariness in the choice of the set of three banana bonds.

Sometimes, there only is a finite number of supplementary solutions dictated by symmetry. Let us consider the case of three protons at infinite separation, occupying the vertices of an equilateral triangle. Let us put two electrons of the same spin into this system. One of the degenerate wave functions of the system is given by the Slater determinant:

$$\Phi_a = \left| \frac{1}{\sqrt{3}}(\chi_a + \chi_b + \chi_c) \frac{1}{\sqrt{6}}(2\chi_a - \chi_b - \chi_c) \right| \tag{10.5}$$

It can also be written as:

$$\Phi_a = \frac{1}{\sqrt{2}} \left(|\chi_a \chi_b| + |\chi_a \chi_c| \right) \tag{10.6}$$

Fig. 10.4 Two views of an Ω_2 corresponding to one of the three banana bonds in acetylene

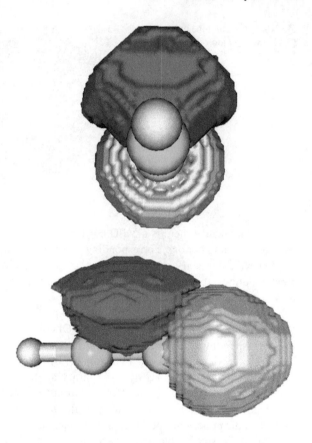

From the latter formulation, one can immediately see that for this wave function, one has $p_1 = 1$ when Ω_1 contains either the space associated to one proton, H_a, or that for two of them, H_b and H_c, cf. Fig. 10.5. One has a system $H...H_2^+$. This separation can be done, of course, also by isolating H_b, or H_c. These solutions can be obtained by a rotation along the threefold axis. Notice that there are three solutions, but as there are only two electrons of same spin these solutions overlap significantly, and do not form a partition of space, as it is given by $H...H_2^+$. Symmetry thus provides alternative chemically equivalent solutions.

A similar situation has been noticed for the Si_2H_2 molecule, in D_{2h} symmetry (bent acetylene structure) [10]. One also obtains MPDs corresponding to banana bonds. By the bending, however, the C_∞ axis which was present in acetylene is not present anymore, so that an arbitrary rotation around the Si-Si axis does not produce an equivalent solution. The molecule still has inversion symmetry. By inversion, the arrangement of the three banana bond like MPDs, in the $\triangle$ arrangement, are transformed into one having a $\triangledown$ arrangement.

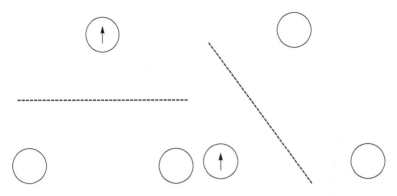

Fig. 10.5 Two choices leading to Ω_1 with $p_1 = 1$ for three protons and two same-spin electrons

10.3.3   MPDs do not Always Provide an Exact Partition of Space

In the ground state of two non-interacting same-spin fermions in a one-dimensional box, Ω_1 corresponds to exactly one half of the box. The two solutions for Ω_1 thus provide a partition of the box. In general, however, MPDs do not necessarily provide an exact partition of space. For example, small overlaps were noticed in some molecular calculations (CH_5^+ or FHF^- [10]). However, the discretization by small cubes, and numerical noise left a question mark with this statement. In order to clarify whether MPDs necessarily provide a partition of space, we consider now a model, for which we can state with certitude that MPDs do not provide a partition of space. We choose again the example of the three non-interacting fermions of the same spin in a one-dimensional box. The Ω_1s are given by: $0 \leq x \leq 0.3547$, $0.3572 \leq x \leq 0.6428$, and $0.6453 \leq x \leq 1$. Small regions $0.3547 < x < 0.3572$ and $0.6428 < x < 0.6453$ have not been attributed. Similarly, for the three electrons on a ring, Ω_1 is not delimited by 0 and exactly $2\pi/3$; the upper limit is $\approx 2.06 < 2\pi/3$. With the present accuracy of our programs, such small effects could not yet be detected with certainty in molecular or crystalline systems.

10.3.4   MPDs can be Disjoint in Space

We have seen that for the ground state of two same-spin non-interacting fermions in a box, the Ω_1 correspond to half of the box, with $p_1 \approx 0.86$. In the excited states, this is not the case anymore. For example, if one fermion is excited from the $n = 2$ to the $n = 3$ state, there is a solution for Ω_1, $x \in [0.32, 0.62]$ with $p_1 \approx 0.88$. There is another solution for Ω_1, with the same p_1, for the rest of the box, $x \in [0, 0.32] \cup [0.62, 1]$. For it, a description with a single basin is not possible. This is a situation that does not correspond to the classical image of localized electrons, but shows nevertheless a high probability. Such a situation can appear in resonating systems (when several Lewis structures are needed).

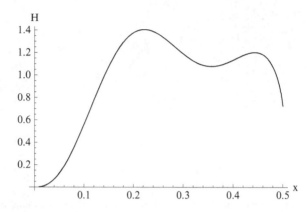

Fig. 10.6 Missing information function H for loges defined by the intervals $[0,x]$, $(x, 1-x)$, $[1-x, 1]$, as a function of x

10.3.5 MPDs and Loges

In this section we will highlight some differences between MPDs and loges. Let us consider again the system of three same-spin non-interacting particles in a box. We define the loges by the intervals $[0,x]$, $(x, 1-x)$, $[1-x, 1]$. The missing information function H, Eq. 10.4, depends on x, see Fig. 10.6. H presents a minimum, for $x \approx 0.3565$, close to the value obtained by maximizing p_1. This is a local minimum, and there are two other, lower minima. The first one, at $x = 0$ is trivial. It corresponds to making the border loges vanish, keeping only the central loge. It is a minimum of H, as the probability of finding all three electrons in the whole box is equal to 1. This situation can be easily identified and discarded. There is another minimum, however, for $x = 1/2$ taking a lower value than for the physically interesting minimum. It is the case where the central box vanishes, and only the distributions where there is no electron in the central box contribute to H. Please notice that taking $x < 1/2$, but close to this value, the central loge has not vanished yet, but H takes a low value. Thus, it is not easy to detect such a situation when minimizing H.

10.4 Conclusions

In order to better understand the maximum probability domains (the region of space maximizing the probability to find a given number of electrons in it), we studied some simple model systems.

Electronic systems well described by a single Lewis structure produce MPDs which correspond roughly to a partition of space which permits an association to bonds, lone pairs, or cores. However, our models have shown that this partition is not strict. A further point is that symmetry can produce alternative solutions.

The model calculations have proven that MPDs formed by disjoint subdomains can exist.

It was shown that using the missing information function, as it is done in loge theory, can produce unwanted results.

10.5   Appendix: Detailed Description of the Models

Let us now describe the simple models used in this paper. The first one is that of non-interacting particles in a one-dimensional box. We will choose the box to have unit length, and the orbitals are given by:

$$\varphi_n(x) = \sqrt{2}\sin(n\pi x) \tag{10.7}$$

The overlaps for Ω defined by the interval $[x_1, x_2]$ is given by

$$S_{ij}([x_1, x_2]) = \int_{x_1}^{x_2} \varphi_i(x)\varphi_j(x)dx \tag{10.8}$$

and easily computed and diagonalized, yielding the eigenvalues λ_i. For two fermions of the same spin the probability to find one electron in this interval is given by [3]

$$p_1([x_1, x_2]) = \lambda_1 + \lambda_2 - 2\lambda_1\lambda_2 \tag{10.9}$$

A slightly more complicated expression is obtained for three particles of the same spin.

For particles in a ring, the orbitals are given by

$$\varphi_k(\theta) = \frac{1}{\sqrt{2\pi}}e^{ik\theta} \tag{10.10}$$

for θ between 0 and 2π.

For three infinitely distant hydrogen atoms, with the nuclei on the vertices of an equilateral triangle, the orbitals are:

$$\varphi_1 = \frac{1}{\sqrt{3}}(\chi_a + \chi_b + \chi_c)$$

$$\varphi_2 = \frac{1}{\sqrt{6}}(2\chi_a - \chi_b - \chi_c)$$

$$\varphi_3 = \frac{1}{\sqrt{6}}(-\chi_a + 2\chi_b - \chi_c)$$

$$\varphi_4 = \frac{1}{\sqrt{6}}(-\chi_a - \chi_b + 2\chi_b) \tag{10.11}$$

where only two of $\varphi_2, \varphi_3, \varphi_4$ are linearly independent. Please notice that $S(\Omega)$ is simplified as all products between the functions centered on different atoms are vanishing.

References

1. Savin A (2002) In: Sen KD (ed) Reviews of modern quantum chemistry: a celebration of the contributions of Robert G. Parr. World Scientific, Singapore, p 43
2. Chamorro E, Fuentealba P, Savin A (2003) J Comp Chem 24:496
3. Cancès E, Keriven R, Lodier F, Savin A (2004) Theor Chem Acc 111:373
4. Gallegos A, Carbó-Dorca R, Lodier F, Cancès E, Savin A (2005) J Comp Chem 26:455
5. Francisco E, Martín Pendás A, Blanco MA (2007) J Chem Phys 126(9):094102
6. Martín Pendás A, Francisco E, Blanco MA (2007) J Chem Phys 127:144103
7. Martín Pendás A, Francisco E, Blanco MA (2007) Chem Phys Lett 437:287
8. Martín Pendás A, Francisco E, Blanco MA (2007) Phys Chem Chem Phys 9:1087
9. Martín Pendás A, Francisco E, Blanco MA (2007) J Phys Chem A 111:1084
10. Scemama A, Caffarel M, Savin A (2007) J Comp Chem 28:442
11. Francisco E, Martín Pendás A, Blanco MA (2009) J Chem Phys 131:124, 125
12. Silvi B, Savin A (1994) Nature 371:683
13. Becke A, Edgecombe KE (1992) J Chem Phys 92:5397
14. Noury S, Krokidis X, Fuster F, Silvi B (1999) Comput Chem 23:597
15. Scemama A (2005) J Theor Comp Chem 4:397
16. Frisch MJ, Trucks GW, Schlegel HB, Scuseria GE, Robb MA, Cheeseman JR, Scalmani G, Barone V, Mennucci B, Petersson GA, Nakatsuji H, Caricato M, Li X, Hratchian HP, Izmaylov AF, Bloino J, Zheng G, Sonnenberg JL, Hada M, Ehara M, Toyota K, Fukuda R, Hasegawa J, Ishida M, Nakajima T, Honda Y, Kitao O, Nakai H, Vreven T, Montgomery JA Jr, Peralta JE, Ogliaro F, Bearpark M, Heyd JJ, Brothers E, Kudin KN, Staroverov VN, Kobayashi R, Normand J, Raghavachari K, Rendell A, Burant JC, Iyengar SS, Tomasi J, Cossi M, Rega N, Millam JM, Klene M, Knox JE, Cross JB, Bakken V, Adamo C, Jaramillo J, Gomperts R, Stratmann RE, Yazyev O, Austin AJ, Cammi R, Pomelli C, Ochterski JW, Martin RL, Morokuma K, Zakrzewski VG, Voth GA, Salvador P, Dannenberg JJ, Dapprich S, Daniels AD, Farkas Ö, Foresman JB, Ortiz JV, Cioslowski J, Fox DJ (2009) Gaussian 09 revision A.1. Gaussian Inc., Wallingford CT
17. Saunders VR, Dovesi R, Roetti C, Orlando R, Harrison NM, Zicovich-Wilson CM (1998) CRYSTAL98 (CRYSTAL98 user's manual). University of Torino, Torino
18. Francisco E, Pendás AM, Blanco MA (2008) Comput Phys Commun 178:621
19. Bader RFW (1990) Atoms in molecules: a quantum theory. Oxford University Press, Oxford
20. Savin A (2005) J Chem Sci 117:473
21. Daudel R (1953) CR Acad Sci France 237:691
22. Aslangul C, Constanciel R, Daudel R, Kottis P (1972) Adv Quantum Chem 6:93
23. Wolfram S (2008) Mathematica edition: version 7.0. Wolfram Research, Inc., Champaign
24. Wigner E (1934) Phys Rev 46:1002
25. Lennard-Jones J, Pople JA (1950) Proc Roy Soc A 202:166

Chapter 11
Density Scaling for Excited States

Á. Nagy

Abstract The theory for a single excited state based on Kato's theorem is revisited. Density scaling proposed by Chan and Handy is used to construct a Kohn-Sham scheme with a scaled density. It is shown that there exists a value of the scaling factor for which the correlation energy disappears. Generalized OPM and KLI methods incorporating correlation are proposed. A ζKLI method as simple as the original KLI method is presented for excited states.

11.1 Introduction

Density functional theory [1] in its original form is a ground-state theory which is valid for the lowest-energy state in each symmetry class [2, 3]. It was, however, applied to excited states as well, starting with the transition state method of Slater [4]. The first rigorous generalization for excited states was proposed by Theophilou [5]. The variational principle for excited states was studied by Perdew and Levy [6] and Lieb [7]. Formalisms for excited states were also provided by Fritsche [8] and English et al. [9]. Gross, Oliveira and Kohn [10] worked out the theory of unequally weighted ensembles of excited states. Relativistic generalization of this formalism was also done [11]. A theory of excited states was presented utilizing Görling-Levy perturbation theory [12, 13] and a quasi-local-density approximation [14] was proposed. Coordinate scaling and adiabatic connection were studied [15, 16]. (For reviews of excited-state theories see [17, 18].) The optimized potential method was generalized to ensembles [19–24].

The ensemble theory has the disadvantage that one has to calculate all the ensemble energies lying under the given ensemble energy to obtain the desired

Á. Nagy (✉)
Department of Theoretical Physics, University of Debrecen, H–4010 Debrecen, Hungary
e-mail: anagy@phys.unideb.hu

P.E. Hoggan et al. (eds.), *Advances in the Theory of Quantum Systems in Chemistry and Physics*, Progress in Theoretical Chemistry and Physics 22,
DOI 10.1007/978-94-007-2076-3_11, © Springer Science+Business Media B.V. 2012

excitation energy. It is especially inconvenient to use it if one is interested in highly excited states. That is why it is important to extend density functional theory to a single excited state. Two theories for a single excited state [25, 26, 28, 29] exist. A non-variational theory [25–27] based on Kato's theorem and a variational density functional theory [28, 29]. In this paper the non-variational theory is extended and combined with density scaling [30, 31].

Density scaling was proposed by Chan and Handy [32]. In density scaling the density $n(\mathbf{r})$ is changed to $\zeta n(\mathbf{r})$. It is shown that there exist a value of the scaling factor for which the correlation energy disappears. The optimized potential method (OPM) [60] and the Krieger-Li-Iafrate (KLI) [61] approach are generalized to incorporate correlation. In this paper only a non-degenerate excited state is treated.

Section 11.2 presents the non-variational theory. In Sect. 11.3 Kohn-Sham-like equations are obtained through adiabatic connection. Density scaling is applied to obtain a generalized Kohn-Sham scheme in Sect. 11.4. The optimized potential and the KLI methods are generalized in Sect. 11.5. The last section is devoted to illustrative examples and discussion.

We mention that there are other noteworthy single excited-state theories: the stationary-principle theory of Görling [33], the formalism of Sahni [34] or the local scaling approach of Ludena and Kryachko [35]. Beyond the time-independent theories mentioned above the time-dependent density functional theory [36] provides an alternative (see e.g. [37]).

11.2 Non-variational Theory for a Single Excited State

The consequence of the Hohenberg-Kohn theorem is that the ground-state electron density determines all molecular properties. E. Bright Wilson [38] noticed that Kato's theorem [39, 40] provides an explicit procedure for constructing the Hamiltonian of a Coulomb system from the electron density:

$$Z_\beta = -\frac{1}{2\bar{n}(r)} \left. \frac{\partial \bar{n}(r)}{\partial r} \right|_{r=R_\beta}. \tag{11.1}$$

Here $\bar{n}$ denotes the angular average of the density n and the right-hand side is evaluated at the position of nucleus β. From Eq. 11.1, the cusps of the density show us where the nuclei are (R_β) and what are their atomic numbers Z_β. The integral of the density provides the number of electrons:

$$N = \int n(\mathbf{r}) d\mathbf{r}. \tag{11.2}$$

Therefore one can readily obtain the Hamiltonian from the density and then determine every property of the system by solving the Schrödinger equation. One has to emphasize, however, that this argument holds only for Coulomb systems. By contrast, the density functional theory formulated by Hohenberg and Kohn is valid for any local external potential.

Kato's theorem is valid not only for the ground state but also for the excited states. Consequently, if the density n_i of the i-th excited state is known, the Hamiltonian $\hat{H}$ is also in principle known and its eigenvalue problem

$$\hat{H}\Psi_k = E_k\Psi_k \qquad (k = 0, 1, \ldots, i, \ldots) \tag{11.3}$$

can be solved, where

$$\hat{H} = \hat{T} + \hat{V} + \hat{V}_{ee}. \tag{11.4}$$

$$\hat{T} = \sum_{j=1}^{N}\left(-\frac{1}{2}\nabla_j^2\right), \tag{11.5}$$

$$\hat{V}_{ee} = \sum_{k=1}^{N-1}\sum_{j=k+1}^{N}\frac{1}{|\mathbf{r}_k - \mathbf{r}_j|} \tag{11.6}$$

and

$$\hat{V} = \sum_{k=1}^{N}\sum_{J=1}^{M}\frac{-Z_J}{|\mathbf{r}_k - \mathbf{R}_J|} \tag{11.7}$$

are the kinetic energy, the electron-electron energy and the electron-nuclear energy operators, respectively.

We emphasize that there are certain special cases, where Eq. 11.1 does not provide the atomic number. The simplest example is the $2p$ excited state of the hydrogen atom, where the density

$$n_{2p}(r) = cr^2 e^{-Zr} \tag{11.8}$$

and the derivative of the density are zero at the nucleus. Though Kato's theorem (11.1) is valid in this case, too, it does not provide the atomic number. Similar cases occur in other highly excited atoms, ions or molecules, for which the spherical average of the derivative of the wave function is zero at a nucleus, that is where we have no s-electrons.

Pack and Brown [41] derived cusp relations for the wave functions of these systems. The corresponding cusp relations for the density [42,43] were also derived. Let us define

$$\eta_l(r) = \frac{\bar{n}(r)}{r^{2l}}, \tag{11.9}$$

where l is the smallest integer for which η_l is not zero at the nucleus. The corresponding cusp relations for the density are

$$\frac{\partial \eta^l(r)}{\partial r}\bigg|_{r=0} = -\frac{2Z}{l+1}\eta^l(0).$$ (11.10)

For the example of a one-electron atom in the $2p$ state, Eq. 11.9 leads to

$$\eta_{2p}(r) = \frac{n_{2p}}{r^{2l}} = ce^{-Zr}$$ (11.11)

and the cusp relation has the form:

$$-2Z\eta_{2p}(0) = 2\eta'_{2p}(0).$$ (11.12)

So we can again readily obtain the atomic number from the electron density. Other useful cusp relations have also been derived [44, 45]. There are several other works concerning the cusp of the density [46–55].

11.3   Kohn-Sham-Like Equations

Making use of adiabatic connection [2, 56] Kohn-Sham-like equations can be derived. We suppose the existence of a continuous path between the interacting and the non-interacting systems. The density n_i of the i-th electron state is the same along the path.

$$\hat{H}_i^\alpha \Psi_k^\alpha = E_k^\alpha \Psi_k^\alpha,$$ (11.13)

where

$$\hat{H}_i^\alpha = \hat{T} + \alpha \hat{V}_{ee} + \hat{V}_i^\alpha.$$ (11.14)

The subscript i denotes that the density of the given excited state is supposed to be the same for any value of the coupling constant α. $\alpha = 1$ corresponds to the fully interacting case, while $\alpha = 0$ gives the non-interacting system:

$$\hat{H}_i^0 \Psi_k^0 = E_k^0 \Psi_k^0.$$ (11.15)

For $\alpha = 1$ the Hamiltonian $\hat{H}_i^\alpha$ is independent of i. For any other value of α the 'adiabatic' Hamiltonian depends on i and we have different Hamiltonians for different excited states. Thus the non-interacting Hamiltonian ($\alpha = 0$) is different for different excited states. If there are several 'external' potentials $V^{\alpha=0}$ leading to the same density n_i we select that potential for which the non-interacting kinetic energy is closest to the interacting one. It is important to emphasize that this procedure only works for the Coulomb case.

To apply the Kohn-Sham-like equations (11.15) one has to find an approximation to the potential of the non-interacting system. The optimized potential method [60] can be generalized for a single excited state, too. It was shown [25] that because the energy is stationary at the true wave function, the energy is stationary at the true potential. This is the consequence of the well-known fact that, when the energy is considered to be a functional of the wave function, the only stationary points of $E[\Psi]$ are those associated with the eigenvalues/eigenvectors of the Hamiltonian

$$\frac{\delta E}{\delta \Psi_k} = 0 \qquad (k = 1, ..., i, ...). \tag{11.16}$$

The density n_i of a given excited state determines the Hamiltonian and via adiabatic connection the non-interacting effective potential $V_i^{\alpha=0}$. Therefore we can consider the total energy as a functional of the non-interacting effective potential:

$$E[\Psi_i] = E[\Psi_i[V_i^0]]. \tag{11.17}$$

Utilizing Eq. 11.16 we obtain

$$\frac{\delta E}{\delta V_i^0} = \int \frac{\delta E}{\delta \Psi_i} \frac{\delta \Psi_i}{\delta V_i^0} + c.c. = 0. \tag{11.18}$$

So an optimized effective potential can be found for the given excited state. The KLI approximation to the optimized effective potential can also be derived straightforwardly [25].

Exchange identities utilizing the principle of adiabatic connection and coordinate scaling and a generalized 'Koopmans' theorem' were derived and the excited-state effective potential was constructed [57]. The differential virial theorem was also derived for a single excited state [58].

11.4   Density Scaling for a Single Excited State

Now, the density scaling is applied to obtain a Kohn-Sham scheme with a scaled density. Here, we suppose a non-degenerate excited state. Extension to degenerate excited states will be detailed elsewhere. Consider a non-interacting system with excited state density $n_{\zeta i}(\mathbf{r}) = n_i(\mathbf{r})/\zeta$, where $\zeta = N/N_\zeta$ is a positive number. In the present theory we suppose that ζ is larger but close to 1. If the original real system has N-electrons

$$\int n_i(\mathbf{r})d\mathbf{r} = N, \tag{11.19}$$

the Kohn-Sham system with the scaled density has N_ζ-electrons:

$$\int n_{\zeta i}(\mathbf{r})d\mathbf{r} = N_\zeta. \tag{11.20}$$

N is always integer, but N_ζ is generally non-integer. Therefore, the Kohn-Sham-like equations will differ from the ones corresponding to the N-electron Kohn-Sham system (11.15). To construct another Kohn-Sham system we define the density

$$n_{\zeta i} = (1-q)n_i + qn_{ion}, \tag{11.21}$$

where

$$q = N - N_\zeta = N(1 - 1/\zeta). \tag{11.22}$$

We consider only that case for which q is a small positive number: $q \ll 1$. The second term in Eq. 11.21 corresponds to the density of the ion that is obtained after ionization. It is the ground state of the non-interacting $N-1$-electron system if we consider an excitation where the electron is excited from the highest occupied level to a highest state. It can be an excited state of the non-interacting $N-1$-electron system in other cases. The Kohn-Sham system is a non-interacting system with the scaled density $n_{\zeta i}$. The non-interacting kinetic energy can be constructed from the non-interacting wave function Ψ_i^0 of the original 'Kohn-Sham' system (Eq. 11.15) and the non-interacting wave function Ψ_{ion}^0 of the ion:

$$T_\zeta^0[n_i] = \zeta\left[(1-q)\langle\Psi_i^0|\hat{T}|\Psi_i^0\rangle + q\langle\Psi_{ion}^0|\hat{T}^{ion}|\Psi_{ion}^0\rangle\right]. \tag{11.23}$$

The Kohn-Sham equations with the scaled density have the form

$$\left[-\frac{1}{2}\nabla^2 + v_{\zeta KSi}(\mathbf{r})\right]u_{\zeta i,j} = \varepsilon_{\zeta i,j}u_{\zeta i,j}, \tag{11.24}$$

where the scaled Kohn-Sham density has the form

$$n_{\zeta i}(\mathbf{r}) = \sum_j^M \lambda_{\zeta i,j}|u_{\zeta i,j}(\mathbf{r})|^2. \tag{11.25}$$

$\lambda_{\zeta i,j}$ are the occupation numbers and M is the largest integer for which $\lambda_{\zeta i,j} \neq 0$. $u_{\zeta i,j}$ are the scaled Kohn-Sham orbitals. The non-interacting kinetic energy takes the form

$$T_\zeta^0[n_i] = \zeta\sum_j^M \lambda_{\zeta i,j}\langle u_{\zeta i,j}| - \frac{1}{2}\nabla^2|u_{\zeta i,j}\rangle. \tag{11.26}$$

In the original theory the exchange-correlation energy $E_{xci}[n_i]$ is defined by the total energy expression

$$E[n_i] = T^0[n_i] + J[n_i] + E_{xci}[n_i] + \int n_i(\mathbf{r})v(\mathbf{r})d\mathbf{r}, \qquad (11.27)$$

where

$$J[n_i] = \frac{1}{2}\int \frac{n_i(\mathbf{r}_1)n_i(\mathbf{r}_2)}{|\mathbf{r}_1 - \mathbf{r}_2|}d\mathbf{r}_1 d\mathbf{r}_2 \qquad (11.28)$$

is the classical Coulomb energy and v is the external potential.

Similarly, in this Kohn-Sham theory with the scaled density the exchange-correlation energy $E_{\zeta xc}[n_i]$ is defined by the total energy expression

$$E[n_i] = T^0_{\zeta}[n_i] + J[n_i] + E_{\zeta xci}[n_i] + \int n_i(\mathbf{r})v(\mathbf{r})d\mathbf{r}. \qquad (11.29)$$

A comparison of Eqs. 11.27 and 11.29 leads to the important relation

$$T^0[n_i] + E_{xci}[n_i] = T^0_{\zeta}[n_i] + E_{\zeta xci}[n_i]. \qquad (11.30)$$

Obviously, $T^0_{\zeta=1} = T^0$. The relationship between T^0_{ζ} and T^0 derived by Chan and Handy [32] is valid for the excited state

$$T_{\zeta}[n_i]^0 = \zeta T^0[n_{\zeta i}]. \qquad (11.31)$$

Now, we define the correlation energy as follows:

$$E_{\zeta ci}[n_i] = \langle \Psi_i | \hat{T} + \hat{V}_{ee} | \Psi_i \rangle$$
$$- [(1-q)\langle \Psi^0_i | \zeta \hat{T} + \zeta^2 \hat{V}_{ee} | \Psi^0_i \rangle + q\langle \Psi^0_{ion} | \zeta \hat{T}^{ion} + \zeta^2 \hat{V}^{ion}_{ee} | \Psi^0_{ion} \rangle]. \qquad (11.32)$$

Theorem 11.1. *There exists a parameter ζ_i for which the correlation energy disappears: $E_{\zeta ci} = 0$.*

Proof. If $\zeta_i = 1$, $E_{\zeta ci}$ is equal to the correlation energy E_{ci} of the original theory. Consider a small change in ζ_i and notice that from the definition (11.32) follows that $E_{\zeta ci}$ is almost quadratic in ζ. $E_{\zeta ci} = 0$ means that

$$\zeta^2[(1-q)\langle \Psi^0_i | \hat{V}_{ee} | \Psi^0_i \rangle + q\langle \Psi^0_{ion} | \hat{V}^{ion}_{ee} | \Psi^0_{ion} \rangle] + \zeta[(1-q)\langle \Psi^0_i | \hat{T} | \Psi^0_i \rangle$$
$$+ q\langle \Psi^0_{ion} | \hat{T}^{ion} | \Psi^0_{ion} \rangle] - \langle \Psi_i | \hat{T} + \hat{V}_{ee} | \Psi_i \rangle = 0. \qquad (11.33)$$

This equation has solutions as

$$[(1-q)\langle\Psi_i^0|\hat{T}|\Psi_i^0\rangle + q\langle\Psi_{ion}^0|\hat{T}^{ion}|\Psi_{ion}^0\rangle]^2 + 4[(1-q)\langle\Psi_i^0|\hat{V}_{ee}|\Psi_i^0\rangle$$
$$+ q\langle\Psi_{ion}^0|\hat{V}_{ee}^{ion}|\Psi_{ion}^0\rangle]\langle\Psi_i|\hat{T} + \hat{V}_{ee}|\Psi_i\rangle > 0. \tag{11.34}$$

Consequently, there exist a value of ζ for which $E_{\zeta ci} = 0$. Note that Eq. 11.33 has two solutions, however, the other solution is not close to 1. Moreover it can even be negative and thus physically not acceptable.

11.5   The ζOPM and ζKLI Methods for a Single Excited State

In order to perform calculations one needs explicit expressions for the functionals. In the ground-state theory, exchange can be treated exactly (or very accurately) via the optimized potential method [60] (or KLI method [61]). Now, these methods are combined with density scaling for a single excited state.

In the optimized potential method the following problem is solved: find the potential such that when it is given a small variation, the energy of the system remains stationary:

$$\frac{\delta E}{\delta V} = 0. \tag{11.35}$$

From the fact the energy is stationary at the true wave function follows that the energy is stationary at the true potential. It is well-known that considering the energy as a functional of the wave function $E[\Psi]$, the eigenvalues of the Hamiltonian are stationary points of E

$$\frac{\delta E}{\delta \Psi_k} = 0 \qquad (k = 1, ..., i, ...), \tag{11.36}$$

and only the eigenvalues are stationary points.

As we emphasized above from the density of a given excited state n_i one can obtain the Hamiltonian, the eigenvalues and eigenfunctions and through adiabatic connection the Kohn-Sham potential $V_i^{\alpha=0}$ and certainly the solution of the Kohn-Sham equations leads to the density n_i:

$$n_i \to \hat{H} \to E_k, \Psi_k \quad (k = 1, ..., i, ...) \to v_{iKS} \to n_i. \tag{11.37}$$

For the scaled density we also have

$$n_{\zeta i} \to \hat{H} \to E_k, \Psi_k \quad (k = 1, ..., i, ...) \to v_{\zeta iKS} \to n_{\zeta i}. \tag{11.38}$$

Thus, we can consider the total energy as a functional of the Kohn-Sham potential:

$$E[\Psi_i] = E[\Psi_i[v_{\zeta iKS}]]. \tag{11.39}$$

Making use of Eq. 11.16 we obtain

$$\frac{\delta E}{\delta v_{\zeta i}^0} = \int \frac{\delta E}{\delta \Psi_i} \frac{\delta \Psi_i'}{\delta v_{\zeta i}^0} + c.c. = 0. \tag{11.40}$$

Thus, from the fact that the energy is stationary at the true wave function follows that the energy is stationary at the true potential. (We mention in passing that there is a condition of this sort in the Levy-Nagy theory [28]. It is also one of the key results in the potential functional theory of Yang et al. [59]).

However, as the energy is only stationary and not minimum at the true density it is difficult to find adequate approximations. The Kohn-Sham wave function should be orthogonal to the exact Kohn-Sham wave function(s) of the lower state(s). Since the exact Kohn-Sham wave functions are not known, one is satisfied if approximate orthogonality with respect to the approximate lower Kohn-Sham wave function(s) is assured.

In the ground-state theory exchange can be treated exactly via the optimized potential method [60]. This method has been generalized for excited states [19, 25] and extension for the scaled density is straightforward. To find the optimized potential is very tedious even in the ground-state. However, Krieger, Li and Iafrate [61] introduced a very accurate approximation. This method can be readily generalized to excited states [19,25]. An extension to the scaled density is presented here using an alternative derivation of the KLI approximation [62].

Both the OPM and KLI methods can be applied when the total energy is known as a functional of the one-electron orbitals. Let us consider the exchange-only case. The exchange energy is known. The expression is the same as the Hartree-Fock one, but contains the Kohn-Sham orbitals. As we have just shown above for a certain critical value of scaling factor ζ the correlation energy disappears. Using the stationary principle the first variation of the total energy with respect to the one-electron orbitals leads to Hartree-Fock-like equations:

$$-\frac{1}{2}\nabla^2 \psi_{\zeta i,j}(\mathbf{r}) + \left(v(\mathbf{r}) + v_{\zeta Ji}(\mathbf{r})\right)\psi_{\zeta i,j}(\mathbf{r}) - \int d\mathbf{r}' w_{\zeta i}(\mathbf{r},\mathbf{r}')\psi_{\zeta i,j}(\mathbf{r}') = \varepsilon_{\zeta i,j}\psi_{\zeta i,j}(\mathbf{r}), \tag{11.41}$$

where v is the external potential and $v_{\zeta Ji}$ is the classical Coulomb potential:

$$v_{\zeta Ji}(\mathbf{r}) = \int d\mathbf{r}' n_i(\mathbf{r}')/|\mathbf{r} - \mathbf{r}'|. \tag{11.42}$$

194

Á. Nagy

The total electron density n_i can be expressed with the Hartree-Fock like spinorbitals $\psi_{\zeta i,j}$:

$$n_i(\mathbf{r}) = \zeta \sum_j \lambda_{\zeta i,j} |\psi_{\zeta i,j}(\mathbf{r})|^2, \tag{11.43}$$

while the exchange kernel $w_{\zeta i}(\mathbf{r},\mathbf{r}')$ takes the form

$$w_{\zeta i}(\mathbf{r},\mathbf{r}') = \sum_j \lambda_{\zeta i,j} \psi^*_{\zeta i,j}(\mathbf{r}') \psi_{\zeta i,j}(\mathbf{r}) / |\mathbf{r}-\mathbf{r}'|. \tag{11.44}$$

After multiplying Eq. 11.41 by $\zeta \lambda_{\zeta i,j} \psi^*_{\zeta i,j}$ and summing for all orbitals we obtain

$$-\frac{1}{2} \sum_j \lambda_{\zeta i,j} \psi^*_{\zeta i,j} \nabla^2 \psi_{\zeta i,j} + (v+v_{\zeta Ji}+v_{\zeta Si})n_i = \zeta \sum_j \lambda_{\zeta i,j} \varepsilon_{\zeta i,j} |\psi_{\zeta i,j}|^2 ; \tag{11.45}$$

where $v_{\zeta Si}$ is the Slater potential:

$$v_{\zeta Si}(\mathbf{r}) = \frac{\zeta}{n_i(\mathbf{r})} \sum_j \psi^*_{\zeta i,j}(\mathbf{r}) v_{\zeta x i,j}(\mathbf{r}) \psi_{\zeta i,j}(\mathbf{r}) \tag{11.46}$$

and $v_{\zeta x i,j}$ is the Hartree-Fock-like exchange potential

$$v_{\zeta x i,j}(\mathbf{r}) \psi_{\zeta i,j}(\mathbf{r}) = -\int d\mathbf{r}' w(\mathbf{r},\mathbf{r}') \psi_{\zeta i,j}(\mathbf{r}'), \tag{11.47}$$

Now the Kohn-Sham equations (11.24) are multiplied by $\zeta \lambda_{\zeta i,j} u^*_{\zeta i,j}$ and summed for all orbitals

$$-\frac{1}{2} \sum_j \lambda_{\zeta i,j} u^*_{\zeta i,j} \nabla^2 u_{\zeta i,j} + v_{\zeta KSi} n_i = \zeta \sum_j \lambda_{\zeta i,j} \varepsilon_{\zeta i,j} |u_{\zeta i,j}|^2 . \tag{11.48}$$

Now we consider the case when both Eqs. 11.43 and 11.25 provide the same excited-state density n_i. Moreover, it is supposed that the Hartree-Fock-like orbitals $\psi_{\zeta i,j}$ can be approximated by the scaled Kohn-Sham orbitals $u_{\zeta i,j}$. Then comparing Eqs. 11.45 and 11.48 we obtain the generalized ζKLI approximation for the Kohn-Sham potential:

$$v_{\zeta KSi} = v+v_{\zeta Ji}+v_{\zeta xi} , \tag{11.49}$$

where

$$v_{\zeta xi} = v_{\zeta Si} + \frac{\zeta}{n_i} \sum_j \langle u_{\zeta i,j} | \lambda_{\zeta i,j} v_{\zeta xi} - v_{\zeta xi,j} | u_{\zeta i,j} \rangle |u_{\zeta i,j}|^2 , \tag{11.50}$$

Table 11.1 The values of ζ_c and q_c for which the ζKLI and experimental total energies are equal for the ground and some excited states of the Li atom

Configuration	ζ_c	q_c
$1s^2 2s$	1.00961	0.02856
$1s^2 2p$	1.00939	0.02791
$1s^2 3s$	1.00871	0.02590
$1s^2 3p$	1.00881	0.02621

Table 11.2 The values of ζ_c and q_c for which the ζKLI and experimental total energies are equal for the ground and some excited states of the Na atom

Configuration	ζ_c	q_c
$1s^2 2s^2 2p^6 3s$	1.00407	0.04454
$1s^2 2s^2 2p^6 3p$	1.00398	0.04364
$1s^2 2s^2 2p^6 4s$	1.00393	0.04308
$1s^2 2s^2 2p^6 4p$	1.00392	0.04295

For $\zeta = 1$ Eq. 11.49 gives the original KLI exchange potential. As we used Hartree-Fock like expression we obtained only the exchange. The results above are valid for any value of ζ. We search that ζ_c for which the correlation energy disappears. For that value of ζ the ζKLI method provides a very simple approximation that includes correlation.

11.6 Illustrative Examples and Discussion

As an illustration the ground- and some excited state-energies of Li and Na atoms have been calculated using the ζKLI method. The value ζ_c for which the ζKLI and experimental energies [63] are equal has been determined. Tables 11.1 and 11.2 present these values ζ_c. $\zeta > 1$ means that the scaled number of electrons N_ζ is smaller than the true electron number N. The difference $q_c = N - N_\zeta$ is also shown in the Tables. The values of ζ_c and q_c for the ground- and excited states are not the same but their difference is small.

The ζKLI method is as simple as the original KLI method. But it contains correlation as well. The ζKLI method is not exact, because $E_{\zeta_c,i}[n_i] = 0$ is valid only for a single density and the functional derivative, that is, the correlation potential is not zero.

Acknowledgements This work is supported by the TAMOP 4.2.1/B-09/1/KONV-2010-0007 project. The project is co-financed by the European Union and the European Social Fund. Grant OTKA No. K67923 is also gratefully acknowledged.

References

1. Hohenberg P, Kohn W (1864) Phys Rev 136:B864
2. Gunnarsson O, Lundqvist BI (1976) Phys Rev B 13:4274; Gunnarsson O, Jonson M, Lundqvist BI (1979) Phys Rev B 20:3136
3. von Barth U (1979) Phys Rev A 20:1693

196

Á. Nagy

4. Slater JC (1974) Quantum theory of molecules and solids, vol 4. McGraw-Hill, New York
5. Theophilou AK (1978) J Phys C 12:5419
6. Perdew JP, Levy M (1985) Phys Rev B 31:6264; Levy M, Perdew JP (1985) In: Dreizler RM, da Providencia J (eds) Density functional methods in physics. NATO ASI Series B, vol 123. Plenum, New York, p 11
7. Lieb EH (1985) In: Dreizler RM, da Providencia J (eds) Density functional methods in physics. NATO ASI Series B, vol 123. Plenum, New York, p 31
8. Fritsche L (1986) Phys Rev B 33:3976; (1987) Int J Quantum Chem 21:15
9. English H, Fieseler H, Haufe A (1988) Phys Rev A 37:4570
10. Gross EKU, Oliveira LN, Kohn W (1988) Phys Rev A 37:2805, 2809, 2821
11. Nagy Á (1994) Phys Rev A 49:3074
12. Görling A (1996) Phys Rev A 54:3912
13. Görling A, Levy M (1995) Int J Quantum Chem S 29:93; (1993) Phys Rev B 47:13105
14. Kohn W (1986) Phys Rev A 34:737; (1995) Nagy Int J Quantum Chem S 29:297
15. Nagy Á (1995) Int J Quantum Chem 56:225
16. Nagy Á (1996) J Phys B 29:389
17. Nagy Á (1997) Adv Quant Chem 29:159
18. Singh R, Deb BD (1999) Phys Rep 311:47
19. Nagy Á (1998) Int J Quantum Chem 69:247
20. Gidopoulos NI, Papakonstantinou P, Gross EKU (2001) Phys Rev Lett 88:033003
21. Nagy Á (2001) J Phys B 34:2363
22. Tasnádi F, Nagy Á (2003) Int J Quantum Chem 92:234
23. Tasnádi F, Nagy Á (2003) J Phys B 36:4073
24. Tasnádi F, Nagy Á (2003) J Chem Phys 119:4141
25. Nagy Á (1998) Int J Quantum Chem 70:681
26. Nagy Á (1999) In: Gonis A, Kioussis N, Ciftan M (eds) Electron correlations and materials properties. Kluwer, New York, p 451
27. Ayers PW, Levy M (2009) Phys Rev A 80:012508
28. Levy M, Nagy Á (1999) Phys Rev Lett 83:4631
29. Nagy Á, Levy M (2001) Phys Rev A 63:2502
30. Nagy Á (2005) Chem Phys Lett 411:492
31. Nagy Á (2005) J Chem Phys 123:044105
32. Chan GK-L, Handy NC (1999) Phys Rev A 59:2670
33. Görling A (1999) Phys Rev A 59:3359; (2000) Phys Rev Lett 85:4229
34. Sahni V, Massa L, Singh R, Slamet M (2001) Phys Rev Lett 87:113002; Sahni V, Pan XY (2003) Phys Rev Lett 90:123001
35. Kryachko ES, Ludena EV, Koga T (1992) J Math Chem 11:325
36. Gross EKU, Dobson JF, Petersilka M (1996) In: Nalewajski R (ed) Density functional theory, Topics in Current Chemistry, vol 181. Springer-Verlag, Heidelberg, p 81
37. Marques MAL, Ullrich CA, Nogueira F, Rubio A, Burke K, Gross EKU (2006) Time-deendent density functional theory. Series Lecture Notes in Physics, vol 706, and references therein. Springer-Verlag, Heidelberg
38. Handy NC (1996) In: Bicout D, Field M (eds) Quantum mechanical simulation methods for studying biological systems. Springer-Verlag, Heidelberg, p 1
39. Kato T (1957) Commun Pure Appl Math 10:151
40. Steiner E (1963) J Chem Phys 39:2365; March NH (1975) Self-consistent fields in atoms. Pergamon, Oxford
41. Pack RT, Brown WB (1966) J Chem Phys 45:556
42. Nagy Á, Sen KD (2000) J Phys B 33:1745
43. Ayers PW (2000) Proc Natl Acad Sci 97:1959
44. Nagy Á, Sen KD (2000) Chem Phys Lett 332:154
45. Nagy Á, Sen KD (2001) J Chem Phys 115:6300
46. Gálvez FJ, Porras J, Angulo JC, Dehesa JS (1988) J Phys B 21:L271

47. Angulo JC, Dehesa JS, Gálvez FJ (1990) Phys Rev A 42:641; erratum (1991) Phys Rev A 43:4069
48. Angulo JC, Dehesa JS (1991) Phys Rev A 44:1516
49. Gálvez FJ, Porras J (1991) Phys Rev A 44:144
50. Porras J, Gálvez FJ (1992) Phys Rev A 46:105
51. Esquivel RO, Chen J, Stott MJ, Sagar RP, Smith Jr VH (1993) Phys Rev A 47:936
52. Esquivel RO, Sagar RP, Smith Jr VH, Chen J, Stott MJ (1993) Phys Rev A 47:4735
53. Dehesa JS, Koga T, Romera E (1994) Phys Rev A 49:4255
54. Koga T (1997) Theor Chim Acta 95:113; Koga T, Matsuyama H (1997) Theor Chim Acta 98:129
55. Angulo JC, Koga T, Romera E, Dehesa JS (2000) THEOCHEM 501–502:177
56. Langreth DC, Perdew JP (1977) Phys Rev B 15:2884; Harris J, Jones RO (1974) J Phys F 4:1170; Harris J (1984) Phys Rev A 29:1648
57. Nagy Á (2001) In: Maruani J, Minot C, McWeeny R, Smeyers YG, Wilson S, (eds) New trends in quantum systems in chemistry and physics. Progress in theoretical chemistry and physics, vols 1, 6. Kluwer, Dordrecht, p 13; Nagy Á (2001) Adv Quant Chem 39:35
58. Nagy Á (2002) In: Barone V, Bencini A, Fantucci P (eds) Recent advances in computational chemistry. Recent advances in the density functional methods, Part III. World Scientific, Singapore, p 247
59. Yang WT, Ayers PW, Wu Q (2004) Phys Rev Lett 92:146404
60. Sharp RT, Horton GK (1953) Phys Rev A 30:317; Aashamar K, Luke TM, Talman JD (1978) At Data Nucl Data Tables 22:443
61. Krieger JB, Li Y, Iafrate GJ (1992) Phys Rev A 45:101; (1992) Phys Rev A 46:5453
62. Nagy Á (1997) Phys Rev A 55:3465
63. Moore CE (1949) Atomic energy levels NBS circular No 467. US Govt. Printing Office, Washington, DC

Chapter 12
Finite Element Method in Density Functional Theory Electronic Structure Calculations

Jiří Vackář, Ondřej Čertík, Robert Cimrman, Matyáš Novák, Ondřej Šipr, and Jiří Plešek

Abstract We summarize an ab-initio real-space approach to electronic structure calculations based on the finite-element method. This approach brings a new quality to solving Kohn Sham equations, calculating electronic states, total energy, Hellmann–Feynman forces and material properties particularly for non-crystalline, non-periodic structures. Precise, fully non-local, environment-reflecting real-space ab-initio pseudopotentials increase the efficiency by treating the core-electrons separately, without imposing any kind of frozen-core approximation. Contrary to the variety of well established k-space methods that are based on Bloch's theorem

J. Vackář (✉) • O. Šipr
Institute of Physics, Academy of Sciences of the Czech Republic
e-mail: vackar@fzu.cz; sipr@fzu.cz

O. Čertík
Institute of Physics, Academy of Sciences of the Czech Republic,
Na Slovance 2, 182 21 Praha 8, Czech Republic

Charles University in Prague, Faculty of Mathematics and Physics, Ke Karlovu 3,
121 16 Praha 2, Czech Republic

University of Nevada, Reno, USA
e-mail: ondrej@certik.cz

R. Cimrman
University of West Bohemia, New Technologies Research Centre, Univerzitní 8,
306 14 Plzeň, Czech Republic
e-mail: cimrman3@students.zcu.cz

M. Novák
Charles University in Prague, Faculty of Mathematics and Physics, Ke Karlovu 3,
121 16 Praha 2, Czech Republic
e-mail: info@czechmodels.cz

J. Plešek
Institute of Thermomechanics, Academy of Sciences of the Czech Republic,
Dolejškova 1402/5, 182 00 Praha 8, Czech Republic
e-mail: plesek@it.cas.cz

P.E. Hoggan et al. (eds.), *Advances in the Theory of Quantum Systems in Chemistry and Physics*, Progress in Theoretical Chemistry and Physics 22,
DOI 10.1007/978-94-007-2076-3_12, © Springer Science+Business Media B.V. 2012

and applicable to periodic structures, we don't assume periodicity in any respect. The main asset of the present approach is the efficient combination of an excellent convergence control of standard, universal basis of industrially proved finite-element method and high precision of ab-initio pseudopotentials with applicability not restricted to periodic environment.

12.1 Introduction

For understanding and predicting material properties such as density, elasticity, magnetization or hardness from first principles quantum mechanical calculations, a reliable and efficient tool for electronic structure calculations is necessary. The reciprocal space methods, to which most attention has been dedicated so far, are very powerful and sophisticated but by their nature are suitable mostly for crystals. For systems without translational symmetry such as metallic clusters, defects, quantum dots, adsorbates and nanocrystals, use of real-space methods is more promising.

We introduce new ab-initio real-space method based on (1) density functional theory, (2) finite element method, and (3) environment-reflecting pseudopotentials. It opens various ways for further development and applications: restricted periodic boundary conditions in a desired sub-region or in a requisite direction (e.g. for non-periodic objects with bonds to periodic surroundings), adaptive finite-element mesh and basis playing the role of variational parameters (hp-adaptivity) and various approaches to Hellman-Feynman forces and sensitivity analysis for structural optimizations and molecular dynamics.

The present method focuses on solving Kohn Sham equations and calculating electronic states, total energy and material properties of non-crystalline, non-periodic structures. Contrary to the variety of well established k-space methods that are based on Bloch's theorem and applicable to periodic structures, we don't assume periodicity in any respect. Precise ab-initio environment-reflecting pseudopotentials proven within the plane wave approach are connected with real space finite-element basis in the present approach. The main expected asset of the present approach is the combination of efficiency and high precision of ab-initio pseudopotentials with universal applicability, universal basis and excellent convergence control of finite-element method not restricted to periodic environment.

In the next sections, we give a general overview how the Density Functional Theory is applied to electronic structure calculations within the framework of the finite-element method. We show how to incorporate pseudopotentials into the equations, explaining some technical difficulties that had to be solved and sorting all the ideas out and presenting them in a fashion applicable to our problem.

12.2 Density Functional Theory and Pseudopotentials

Using the approach described e.g. in [3, 6–8, 10] and [1], i.e. making use of the Hohenberg-Kohn theorem and applying the ab-initio pseudopotential approach, the many-particle Schödinger equation is decomposed into the Kohn–Sham equations

$$\left(-\frac{1}{2}\nabla^2 + V_H(\mathbf{r}) + V_{xc}(\mathbf{r}) + \hat{V}(\mathbf{r})\right)\psi_i(\mathbf{r}) = \varepsilon_i\psi(\mathbf{r}) \tag{12.1}$$

which yield the orbitals ψ_i that reproduce the density $n(\mathbf{r})$ of the original interacting system. The core electrons, separated from valence electrons, are represented by a non-local Hermitian operator $\hat{V}$ together with nuclear charge. The density is formed by the sum over the occupied single-electron states in a system of N electron

$$n(\mathbf{r}) = \sum_i^N |\psi_i(\mathbf{r})|^2 \tag{12.2}$$

and V_H is the electrostatic potential obtained as a solution to the Poisson equation

$$V_H(\mathbf{r}) = \frac{1}{2}\int\frac{n(\mathbf{r}')}{|\mathbf{r}-\mathbf{r}'|}\mathrm{d}^3r'. \tag{12.3}$$

All the non-electrostatic interactions are represented by the exchange-correlation potential term $V_{xc}(\mathbf{r}) = \delta E_{xc}[n]/\delta n(\mathbf{r})$, where E_{xc} is the exchange and correlation energy.

Kohn–Sham equations are solved within the iterative scheme described e.g. in [4] and [9].

12.2.1 Semilocal and Separable Potentials

The pseudopotential having the form

$$\hat{V} = V_{loc}(\rho) + \sum_{lm}|lm\rangle V_l(\rho)\langle lm|, \tag{12.4}$$

usually denoted as semilocal $l-$dependent, is a general hermitian operator in the spherically symmetric problem (i.e. $\hat{V} = R^{-1}\hat{V}R$) and it is radially local. This form is general, i.e. any such operator can be written in the form (12.4). Equivalently, in the $\mathbf{r}$ representation:

$$V(\mathbf{r},\mathbf{r}') = \langle\mathbf{r}|\hat{V}|\mathbf{r}'\rangle = V_{loc}(\rho)\delta(\mathbf{r}-\mathbf{r}') + \frac{\delta(\rho-\rho')}{\rho^2}\sum_{lm}Y_{lm}(\hat{\mathbf{r}})V_l(\rho)Y_{lm}^*(\hat{\mathbf{r}}')$$

The first term doesn't cause a problem. Let's denote the second term (which is semilocal) simply by v:

$$v = \sum_{lm}|lm\rangle V_l(\rho)\langle lm|$$

Let's choose a complete but otherwise arbitrary set of functions $|\phi_i\rangle$ (they contain both a radial and an angular dependence) and define a matrix U by the equation

$$\sum_j U_{ij} \langle \phi_j | v | \phi_k \rangle = \delta_{ik}$$

then ($|\psi\rangle = |\phi_k\rangle \alpha_k$):

$$v|\psi\rangle = \sum_{ik} v|\phi_i\rangle \delta_{ik}\alpha_k = \sum_{ijk} v|\phi_i\rangle U_{ij} \langle \phi_j|v|\phi_k\rangle \alpha_k = \sum_{ij} v|\phi_i\rangle U_{ij} \langle \phi_j|v|\psi\rangle$$

So any Hermitian operator (including v) can be transformed exactly into the following form

$$v = \sum_{ij} v|\phi_i\rangle U_{ij} \langle \phi_j|v \qquad (12.5)$$

We diagonalize the matrix U by choosing such functions $|\bar{\phi}_i\rangle$ for which the matrix $\langle \bar{\phi}_j|v|\bar{\phi}_k\rangle$ (and hence the corresponding matrix U) is equal to identity. We can find such functions for example using the Gram–Schmidt orthogonalization procedure on $|\phi_i\rangle$ with a norm $\langle f|v|g\rangle$ (for functions f and g), more on that later. Then

$$v = \sum_i v|\bar{\phi}_i\rangle \frac{1}{\langle \bar{\phi}_i|v|\bar{\phi}_i\rangle} \langle \bar{\phi}_i|v = \sum_i v|\bar{\phi}_i\rangle \langle \bar{\phi}_i|v \qquad (12.6)$$

We could take any $|\phi_i\rangle$ and orthogonalize them. But because we have v in the form of (12.4), we will be using $|\phi_i\rangle$ in the form $|\phi_i\rangle = |R_{nl}\rangle |lm\rangle$, because it turns out we will only need to orthogonalize the radial parts. The first term in (12.6) then corresponds to the KB [5] potential. Taking more terms leads to more accurate results without ghost states.

Let's look at the orthogonalization. We start with the wavefunctions:

$$|\phi_i\rangle = |R_{nl}\rangle |lm\rangle$$

where $R_{nl}(\rho) = \langle \rho|R_{nl}\rangle$ and i goes over all possible triplets (nlm).

We can also relate the i and n, l, m using this formula

$$i_{nlm} = \sum_{k=1}^{n-1} k^2 + \left(\sum_{k=0}^{l-1} (2k+1) \right) + (l+m+1) = \frac{(n-1)n(2n-1)}{6} + l(l+1) + m + 1$$

The operator v acts on these $|\phi_i\rangle$ like this

$$\langle \mathbf{r}|v|\phi_i\rangle = \langle \mathbf{r}|v|R_{nl}\rangle |lm\rangle = \langle \hat{\mathbf{r}}| \langle \rho|V_l(\rho)|R_{nl}\rangle |lm\rangle = V_l(\rho)R_{nl}(\rho)Y_{lm}(\hat{\mathbf{r}})$$

Now we need to construct new orthogonal set of functions $|\bar{\phi}_i\rangle$ satisfying

$$\langle \bar{\phi}_i|v|\bar{\phi}_j\rangle = \delta_{ij}$$

This can be done using several methods, we chose the Gram–Schmidt orthogonalization procedure, which works according to the following scheme:

$$|\tilde{\phi}_1\rangle = \mathbb{1}\frac{1}{\sqrt{\langle\phi_1|v|\phi_1\rangle}}|\phi_1\rangle; \qquad\qquad |\bar{\phi}_1\rangle = \frac{1}{\sqrt{\langle\tilde{\phi}_1|v|\tilde{\phi}_1\rangle}}|\tilde{\phi}_1\rangle$$

$$|\tilde{\phi}_2\rangle = (\mathbb{1}-|\bar{\phi}_1\rangle\langle\bar{\phi}_1|v)\frac{1}{\sqrt{\langle\phi_2|v|\phi_2\rangle}}|\phi_2\rangle; \qquad |\bar{\phi}_2\rangle = \frac{1}{\sqrt{\langle\tilde{\phi}_2|v|\tilde{\phi}_2\rangle}}|\tilde{\phi}_2\rangle$$

$$|\tilde{\phi}_3\rangle = (\mathbb{1}-|\bar{\phi}_1\rangle\langle\bar{\phi}_1|v-|\bar{\phi}_2\rangle\langle\bar{\phi}_2|v)\frac{1}{\sqrt{\langle\phi_3|v|\phi_3\rangle}}|\phi_3\rangle; \quad |\bar{\phi}_3\rangle = \frac{1}{\sqrt{\langle\tilde{\phi}_3|v|\tilde{\phi}_3\rangle}}|\tilde{\phi}_3\rangle$$

$$\dots$$

We can verify by a direct calculation that this procedure ensures

$$\langle\bar{\phi}_i|v|\bar{\phi}_j\rangle = \delta_{ij}$$

It may be useful to compute the normalization factors explicitly:

$$\langle\tilde{\phi}_1|v|\tilde{\phi}_1\rangle = 1$$

$$\langle\tilde{\phi}_2|v|\tilde{\phi}_2\rangle = 1 - \frac{\langle\phi_2|v|\bar{\phi}_1\rangle\langle\bar{\phi}_1|v|\phi_2\rangle}{\langle\phi_2|v|\phi_2\rangle}$$

$$\langle\tilde{\phi}_3|v|\tilde{\phi}_3\rangle = 1 - \frac{\langle\phi_3|v|\bar{\phi}_1\rangle\langle\bar{\phi}_1|v|\phi_3\rangle + \langle\phi_3|v|\bar{\phi}_2\rangle\langle\bar{\phi}_2|v|\phi_3\rangle}{\langle\phi_3|v|\phi_3\rangle}$$

$$\dots$$

we can also write down a first few orthogonal vectors explicitly:

$$|\bar{\phi}_1\rangle = \frac{|\phi_1\rangle}{\sqrt{\langle\phi_1|v|\phi_1\rangle}}$$

$$|\bar{\phi}_2\rangle = \frac{|\phi_2\rangle\langle\phi_1|v|\phi_1\rangle - |\phi_1\rangle\langle\phi_1|v|\phi_2\rangle}{\sqrt{(\langle\phi_1|v|\phi_1\rangle\langle\phi_2|v|\phi_2\rangle - \langle\phi_2|v|\phi_1\rangle\langle\phi_1|v|\phi_2\rangle)\langle\phi_1|v|\phi_1\rangle\langle\phi_2|v|\phi_2\rangle}}$$

Now the crucial observation is

$$\langle lm|\langle R_{nl}|v|R_{n'l'}\rangle|l'm'\rangle = \langle R_{nl}|V_l(\rho)|R_{n'l'}\rangle\,\delta_{ll'}\,\delta_{mm'}$$

which means that $\langle\phi_i|v|\phi_j\rangle = 0$ if $|\phi_i\rangle$ and $|\phi_j\rangle$ have different l or m. In other words $|\phi_i\rangle$ and $|\phi_j\rangle$ for different $|lm\rangle$ are already orthogonal. Thus the G-S orthogonalization procedure only makes the R_{nl} orthogonal for the same $|lm\rangle$. To get explicit expressions for $|\bar{\phi}_i\rangle$, we simply use the formulas above and get:

$$|\phi_i\rangle = |R_{nl}\rangle\,|lm\rangle \quad \rightarrow \quad |\bar{\phi}_i\rangle = |\bar{R}_{nl}\rangle\,|lm\rangle$$

where we have constructed new $|\bar{R}_{nl}\rangle$ from original $|R_{nl}\rangle$:

$$|\bar{R}_{10}\rangle = \frac{|R_{10}\rangle}{\sqrt{\langle R_{10}|V_0|R_{10}\rangle}}$$

$$|\bar{R}_{20}\rangle = \frac{|R_{20}\rangle - |\bar{R}_{10}\rangle\,\langle \bar{R}_{10}|V_0|R_{20}\rangle}{\sqrt{\cdots}}$$

$$|\bar{R}_{21}\rangle = \frac{|R_{21}\rangle}{\sqrt{\langle R_{21}|V_1|R_{21}\rangle}}$$

$$|\bar{R}_{30}\rangle = \frac{|R_{30}\rangle - |\bar{R}_{10}\rangle\,\langle \bar{R}_{10}|V_0|R_{30}\rangle - |\bar{R}_{20}\rangle\,\langle \bar{R}_{20}|V_0|R_{30}\rangle}{\sqrt{\cdots}}$$

$$|\bar{R}_{31}\rangle = \frac{|R_{31}\rangle - |\bar{R}_{21}\rangle\,\langle \bar{R}_{21}|V_1|R_{31}\rangle}{\sqrt{\cdots}}$$

$$|\bar{R}_{32}\rangle = \frac{|R_{32}\rangle}{\sqrt{\langle R_{32}|V_1|R_{32}\rangle}}$$

$$|\bar{R}_{40}\rangle = \frac{|R_{40}\rangle - |\bar{R}_{10}\rangle\,\langle \bar{R}_{10}|V_0|R_{40}\rangle - |\bar{R}_{20}\rangle\,\langle \bar{R}_{20}|V_0|R_{40}\rangle - |\bar{R}_{30}\rangle\,\langle \bar{R}_{30}|V_0|R_{40}\rangle}{\sqrt{\cdots}}$$

$$|\bar{R}_{41}\rangle = \frac{|R_{41}\rangle - |\bar{R}_{21}\rangle\,\langle \bar{R}_{21}|V_1|R_{41}\rangle - |\bar{R}_{31}\rangle\,\langle \bar{R}_{31}|V_1|R_{41}\rangle}{\sqrt{\cdots}}$$

$$\cdots$$

We have constructed new $|\bar{R}_{nl}\rangle$ from $|R_{nl}\rangle$ which obey

$$\langle \bar{R}_{nl}|V_l|\bar{R}_{n'l}\rangle = \delta_{nn'} \tag{12.7}$$

so for every V_l, we construct $|\bar{R}_{nl}\rangle$ for $n = l+1,\, l+2, \cdots$. Let's continue:

$$v\,|\bar{\phi}_i\rangle = V_l(\rho)\,|\bar{R}_{nl}\rangle\,|lm\rangle$$

and finally we arrive at the separable form of the l dependent pseudopotential

$$v = \sum_i v\,|\bar{\phi}_i\rangle\,\langle \bar{\phi}_i|\,v = \sum_{l,i} V_l(\rho)\,|\bar{R}_{nl}\rangle\,|lm\rangle\,\langle lm|\,\langle \bar{R}_{nl}|V_l(\rho) \tag{12.8}$$

To have some explicit formula, let's write how the separable potential acts on a wavefunction:

$$
\begin{aligned}
(v\psi)(\mathbf{r}) = \langle \mathbf{r}|v|\psi\rangle &= \sum_i \langle \hat{\mathbf{r}}|lm\rangle \langle \rho|V_l(\rho)|\bar{R}_{nl}\rangle \langle \bar{R}_{nl}|V_l(\rho)\langle lm|\psi\rangle \\
&= \sum_i Y_{lm}(\hat{\mathbf{r}})\bar{R}_{nl}(\rho)V_l(\rho) \int \bar{R}_{nl}(\rho')V_l(\rho') \int Y_{lm}^*(\hat{\mathbf{r}}')\psi(\mathbf{r}')\,d\Omega'\,\rho'^2 d\rho' \\
&= \sum_i Y_{lm}(\hat{\mathbf{r}})\bar{R}_{nl}(\rho)V_l(\rho) \int \bar{R}_{nl}(\rho')V_l(\rho')Y_{lm}^*(\hat{\mathbf{r}}')\psi(\mathbf{r}')\,d^3r' \qquad (12.9)
\end{aligned}
$$

denoting $\hat{\mathbf{r}}$ and ρ the angular and radial component of $\mathbf{r}$.

To have some insight on what we are actually doing: we are making the local potential V_l nonlocal using:

$$
V_l = \sum_{n=l+1}^{\infty} V_l|\bar{R}_{nl}\rangle \langle \bar{R}_{nl}|V_l \qquad (12.10)
$$

where

$$
\langle \bar{R}_{nl}|V_l|\bar{R}_{n'l}\rangle = \delta_{nn'}
$$

or in $\mathbf{r}$ representation:

$$
V_l(\rho)\psi(\rho\hat{\mathbf{r}}) = \sum_n V_l(\rho)\bar{R}_{nl}(\rho) \int \bar{R}_{nl}(\rho')V_l(\rho')\psi(\rho'\hat{\mathbf{r}})\rho'^2 d\rho'
$$

which is useful when computing integrals of this type

$$
V_{ij} = \int \phi_i(\rho)V_l\phi_j(\rho)\rho^2 d^3\rho = \langle i|V_l|j\rangle = \sum_n \langle i|V_l|\bar{R}_{nl}\rangle \langle \bar{R}_{nl}|V_l|j\rangle
$$

$$
\langle i|V_l|\bar{R}_{nl}\rangle = \int \phi_i(\rho)V_l(\rho)\bar{R}_{nl}(\rho)\rho^2 d\rho
$$

because the integral on the left hand side actually represents N^2 integrals, where N is the number of basis vectors $|\phi_i\rangle$. The sum on the right hand side however only represents $K \cdot N$ integrals, where K is the number of terms taken into account in (12.10). Of course taking only finite number of terms in (12.10) is only an approximation to $\hat{V}_l$. In our case, we don't need these 1D integrals (which can be easily computed directly, because V_l is local and the basis functions ϕ_i are nonzero only around a node in the mesh, which means that the matrix V_{ij} is sparse), but 3D integrals, where angular parts of V are nonlocal and radial part is local (so the matrix V_{ij} is dense), so the above procedure is the only way how to proceed, because we decompose the matrix V_{ij} into the sum of matrices in the form $p_i p_j^*$, which can easily be handled and solved.

The scheme for the separation described above works for any functions $R_{nl}(\rho)$. Because of the form of the expansion (12.10) however, we will use R_{nl} from one atomic calculation. We need to approximate V_l by as few terms as possible, so imagine how the $V_l(\rho)$ acts on the lowest radial function in the l subspace, which is $|R_{l+1;l}\rangle$ and we see that all the terms in (12.10) except the first one $V_l|\bar{R}_{l+1;l}\rangle\langle\bar{R}_{l+1;l}|V_l$ give zero, because they are orthogonal to $|R_{l+1;l}\rangle$. For the function $|R_{l+2;l}\rangle$ all the terms except the first two are zero, because $\langle\bar{R}_{nl}|V_0|R_{l+2;l}\rangle \neq 0$ only for $n = l+1$ or $n = l+2$ (because the vectors $|R_{l+1;l}\rangle$ and $|R_{l+2;l}\rangle$ span the same subspace as $|\bar{R}_{l+1;l}\rangle$ and $|\bar{R}_{l+2;l}\rangle$ and using (12.7)). For functions, which are a little different from all $|R_{nl}\rangle$ ($n > l$), we won't generally get precise results taking any (finite) number of terms in (12.10), but the higher terms should give smaller and smaller corrections.

The described method is general, the only drawback is that if we don't take functions $|R_{nl}\rangle$ which are similar to the solution, we need to take a more terms in (12.10), resulting in more matrices of the form $p_i p_j^*$.

12.2.2 Environment-Reflecting (Environment-Adaptive, All-Electron) Pseudopotentials

The difference with respect to what has been described above consists in the atomic calculation: instead of the asymptotically vanishing wave functions, the boundary conditions reflecting the actual charge distribution in a given material are taken into account. The initial step of *all-electron pseudopotential* generating procedure provides an all-electron atomic calculation where the valence states are treated selfconsistently with the calculation in a given surrounding. The charge density corresponding to valence radial atomic wavefunctions $R_{E,l}(r)$ matches the partial charge density in a solid by the logarithmic derivative at the cut-off radius R_C,

$$\frac{d}{dr}\ln\left(r^2\left|R^{at}_{Eval,l,l}(r)\right|^2\right)\bigg|_{r=R_C} = \frac{d}{dr}\ln\left(r^2\rho_l^{sps}(r)\right)\bigg|_{r=R_C}, \qquad (12.11)$$

where the partial charge density in a solid $\rho_l^{sps}(r)$ is evaluated by summing over all occupied states,

$$\rho_l^{sps}(r) = \sum_{\mathbf{k},n}\sum_{m=-l}^{l}\frac{1}{4\pi r^2}\int_{SPH}d\Omega d\Omega'$$

$$\psi_{\mathbf{k},n}^*(r\hat{\mathbf{n}})Y_{lm}(\hat{\mathbf{n}})Y_{lm}^*(\hat{\mathbf{n}}')\psi_{\mathbf{k},n}(r\hat{\mathbf{n}}'), \qquad (12.12)$$

where $\psi_{\mathbf{k},n}$ denotes the crystal pseudo-wavefunction with $\mathbf{k}$ as the vector of the first Brillouin zone and a band index n. For a non-periodic environment, the sum over $\mathbf{k}, n$ is simply replaced by the sum over occupied states indexed in any appropriate way.

Fig. 12.1 DFT
selfoconsistent scheme, with
environment-adaptive
pseudopotentials

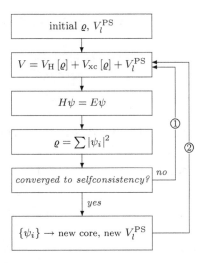

The boundary condition of Eq. 12.11 replaces the standard condition for the wavefunctions to be normalizable and determines the eigenvalue E_l. The normalization condition for the valence atomic-like radial wavefunctions is

$$\int_0^{R_C} \left| R_{E_{\mathrm{val},l,l}}^{\mathrm{at}}(r) \right|^2 r^2 dr = \int_0^{R_C} \rho_l^{\mathrm{sps}}(r) r^2 dr. \qquad (12.13)$$

Using the results of the selfconsistent atomic-like calculation with the boundary conditions of the solid, we apply the phase-shift technique for the construction of pseudopotentials. By varying the screened pseudopotential, this technique minimizes a functional assembled from a set of conditions to be satisfied by the pseudopotential. The common condition requiring the continuous augmentation of the pseudo-wave-function at the cut-off radius R_C is transformed into a condition for the generalized phase-shift that depends on the energy monotonously, which ensures the numerical stability and gives the name to the technique.

Each component $V_l^{\mathrm{scr}}(r)$ of the screened pseudopotential is constructed in such a way that the following conditions for $V_l^{\mathrm{scr}}(r)$ and for each corresponding pair of pseudo-wave-functions (in the case of two energy windows) are fulfilled.

1. At $r = R_C$ the potential $V_l^{\mathrm{scr}}(r)$ matches the all-electron potential $V^{\mathrm{at}}(r)$ up to the second derivative
2. At $r = R_C$ the radial pseudo-wavefunctions $R_{E_{n,l,l}}^{\mathrm{ps}}(r)$ match the corresponding atomic-like radial functions by their values and first derivatives
3. The correct energy derivative of the pseudo-wavefunction is ensured by the norm-conserving condition

More about the all-electron pseudopotential technique can be found in [10] (Fig. 12.1).

12.3 Finite Element Method

This chapter explains FEM and gives concrete formulas which are needed in the calculation.

12.3.1 Weak Formulation of the Schrödinger Equation

One particle Schrödinger equation is

$$\left(-\frac{\hbar^2}{2m}\nabla^2 + V\right)\psi = E\psi.$$

We multiply both sides by a test function v

$$-\left(\frac{\hbar^2}{2m}\nabla^2\psi\right)v = (E-V)\psi v,$$

and integrate over the whole volume we are interested in

$$\int -\left(\frac{\hbar^2}{2m}\nabla^2\psi\right)v\,dV = \int (E-V)\psi v\,dV, \tag{12.14}$$

and using the vector identity

$$-\left(\nabla^2\psi\right)v = \nabla\psi\cdot\nabla v - \nabla\cdot((\nabla\psi)v),$$

we rewrite the left hand side of (12.14)

$$\int \frac{\hbar^2}{2m}\nabla\psi\cdot\nabla v\,dV = \int (E-V)\psi v\,dV + \int \frac{\hbar^2}{2m}\nabla\cdot((\nabla\psi)v)\,dV,$$

now we apply Gauss Theorem

$$\int \frac{\hbar^2}{2m}\nabla\psi\cdot\nabla v\,dV = \int (E-V)\psi v\,dV + \oint \frac{\hbar^2}{2m}(\nabla\psi)v\cdot\mathbf{n}\,dS,$$

and rewriting $\nabla\psi\cdot\mathbf{n}\equiv\frac{d\psi}{dn}$

$$\int \frac{\hbar^2}{2m}\nabla\psi\cdot\nabla v\,dV + \int vV\psi\,dV = \int E\psi v\,dV + \oint \frac{\hbar^2}{2m}\frac{d\psi}{dn}v\,dS, \tag{12.15}$$

which is the weak formulation. The problem reads: find a function ψ such that (12.15) holds for every v.

The boundary condition of all functions is that they disappear at the boundary (infinity) and have zero normal derivative there. That follows from the physical requirement of normalizability of the wave function.

12.3.2 Finite Elements

We choose a basis ϕ_i and substitute ϕ_i for v and expand $\psi = \sum q_j \phi_j$

$$\left(\int \frac{\hbar^2}{2m} \nabla \phi_j \cdot \nabla \phi_i \, dV + \int \phi_i V \phi_j \, dV \right) q_j = \left(\int E \phi_j \phi_i \, dV \right) q_j + \oint \frac{\hbar^2}{2m} \frac{d\psi}{dn} \phi_i \, dS,$$

(12.16)

which can be written in a matrix form

$$(K_{ij} + V_{ij}) q_j = E M_{ij} q_j + F_i,$$

where

$$V_{ij} = \int \phi_i V \phi_j \, dV,$$

$$M_{ij} = \int \phi_i \phi_j \, dV,$$

$$K_{ij} = \frac{\hbar^2}{2m} \int \nabla \phi_i \cdot \nabla \phi_j \, dV,$$

$$F_i = \frac{\hbar^2}{2m} \oint \frac{d\psi}{dn} \phi_i \, dS.$$

Usually we set $F_i = 0$.

We decompose the domain into elements and compute the integrals as the sum over elements. For example:

$$K_{ij} = \sum_{E \in elements} K_{ij}^E$$

where K_{ij}^E is the integral over one element only

$$K_{ij}^E = \int \frac{\hbar^2}{2m} \nabla \phi_j \cdot \nabla \phi_i \, dV^E \approx \sum_{q=0}^{N_q-1} \frac{\hbar^2}{2m} \nabla \phi_i(x_q) \cdot \nabla \phi_j(x_q) w_q |\det J(\hat{x}_q)|.$$

The integral is computed numerically using a Gauss integration: x_q are Gauss points (there are N_q of them), w_q is the weight of each point, and the Jacobian $|\det J(\hat{x}_q)|$ is there because we are actually computing the integral on the reference element instead in the real space.

The surface integrals are computed similarly, but in our case they are all zero, as the normal derivative of the wave function is zero at the boundary.

The actual assembly is a little more complex (you need to loop over all elements, calculate the integrals above and put the numbers to the correct place in the global matrix), but the above sketch should make the main idea clear.

As to the concrete form (shape) of the basis (shapefunctions) – it is well-known that quadratic elements are generally more precise than linear, and that cubes are generally more precise compared to tetrahedra. However, in practise, that needs to be tried on a case by case basis, also depending on the mesh. Our code can work with both linear and quadratic elements and with tetrahedra, cubes, prisms and pyramids.

12.3.3   Pseudopotentials Formulation

There are no problems with other matrix elements in (12.16) except

$$V_{ij} = \int \phi_i(\mathbf{r})V\phi_j(\mathbf{r})\mathrm{d}^3 r = \int \langle i|\mathbf{r}\rangle \langle \mathbf{r}|\hat{V}|j\rangle \mathrm{d}^3 r = \langle i|\hat{V}|j\rangle$$

where

$$\hat{V} = V_{loc}(\rho) + \sum_{nlm} V_l(\rho)|\bar{R}_{nl}\rangle |lm\rangle \langle lm| \langle \bar{R}_{nl}| V_l(\rho)$$

so

$$V_{ij} = \langle i|V_{loc}(\rho)|j\rangle + \langle i|\sum_{nlm} V_l(\rho)|\bar{R}_{nl}\rangle |lm\rangle \langle lm| \langle \bar{R}_{nl}| V_l(\rho)|j\rangle = V_{ij}^{loc} + \sum_{nlm} p_i p_j^*$$

where the complex vector p_i is given by

$$p_i^{(nlm)} = \langle i|lm\rangle V_l(\rho)|\bar{R}_{nl}\rangle = \int \langle i|\mathbf{r}\rangle \langle \hat{\mathbf{r}}|lm\rangle \langle \rho|V_l(\rho)|\bar{R}_{nl}\rangle \mathrm{d}^3 r$$

$$= \int \phi_i(\mathbf{r})Y_{lm}(\hat{\mathbf{r}})V_l(\rho)\bar{R}_{nl}(\rho)\mathrm{d}^3 r$$

and

$$V_{ij}^{loc} = \int \phi_i(\mathbf{r})V_{loc}(\rho)\phi_j(\mathbf{r})\mathrm{d}^3 r$$

and $Y_{lm}(\hat{\mathbf{r}})$, $\bar{R}_{nl}(\rho)$ and $V_l(\rho)$ are given functions. Noticing that

$$\sum_m p_i p_j^* = \int \phi_i(\mathbf{r})Y_{lm}(\hat{\mathbf{r}})V_l(\rho)\bar{R}_{nl}(\rho)\mathrm{d}^3 r \int \phi_j(\mathbf{r}')Y_{lm}^*(\hat{\mathbf{r}}')V_l(\rho')\bar{R}_{nl}(\rho')\mathrm{d}^3 r'$$

$$= \int\int Y_{lm}(\hat{\mathbf{r}})Y_{lm}^*(\hat{\mathbf{r}}')V_l(\rho)\bar{R}_{nl}(\rho)\phi_i(\mathbf{r})\phi_j(\mathbf{r}')V_l(\rho')\bar{R}_{nl}(\rho')\mathrm{d}^3 r\mathrm{d}^3 r'$$

and so we get

$$\sum_m p_i p_j^* = \int \int \frac{4\pi}{2l+1} P_l(\hat{\mathbf{r}} \cdot \hat{\mathbf{r}}') V_l(\rho) \bar{R}_{nl}(\rho) \phi_i(\mathbf{r}) \phi_j(\mathbf{r}') V_l(\rho') \bar{R}_{nl}(\rho') \mathrm{d}^3 r \mathrm{d}^3 r'$$

which is a real number, thus $\sum_{nlm} p_i p_j^*$ is also a real number, which means that we can calculate with only the real parts of the matrix $p_i p_j^*$, because the imaginary parts cancels out in the result:

$$\sum_{nlm} p_i p_j^* = \Re \left(\sum_{nlm} p_i p_j^* \right) = \sum_{nlm} \Re(p_i p_j^*).$$

12.3.4 Separable Potential – More General View

In Sect. 12.2.1 we described a particular way of converting the potential into the separable form, using a direct product of V-orthogonalized atomic radial functions and spherical harmonics as V-orthonormal basis. In real calculations, we never use the complete infinite basis. Therefore, creating a separable potential form always represents some kind of linearization. This fact is independent of the previous potential generating procedure: It concerns not only pseudopotentials, but also e.g. a Projector Augmented Waves (PAW) [2]. In this section we present a more general view on the separable potential and we discuss how to make the linearization more precise via improving the finite V-orthonormal basis (consisting of just a few terms, in practice).

Separable potential with incomplete radial basis is, in fact, a projection of V to the subspace generated by chosen radial projector functions. For each dimension of the subspace there is an energy window, i.e. some neighborhood of the energy for which the corresponding radial projector function has been generated, where the potential acts on a trial function with a given accuracy. Our aim is to enhance the precision by extending the energy window.

We enrich the radial basis with energy derivatives of radial functions

$$\dot{R}_{El}(\rho) = \frac{\partial}{\partial E} R_{E,l}, \tag{12.17}$$

similarly as in the Full-Potential Linearized Augmented Plane Wave (FLAPW) method. In the case of environment-reflecting pseudopotentials, it requires to calculate the energy derivative in the neighborhood of the condition for the logarithmic derivative of Eq. 12.11. The radial basis, therefore, is obtained by means of V_l-orthogonalization procedure of the set of pairs $\{R_{nl}, \dot{R}_{El}\}$.

Let's now consider an alternative notation for separable pseudopotential

$$V = \sum_i c_i |\xi_i\rangle c_i \langle\xi_i| \tag{12.18}$$

In our previously used notation, the coefficients c_i are hidden in the more general matrix U (see Eq. 12.5):

$$U_{i,i} = c_i^2 \qquad U_{i,j} = 0 \text{ for } i \neq j \tag{12.19}$$

The matrix U was eliminated from equations (see the text before Eq. 12.6) by using Gram–Schmidt orthonormalization. This orthonormalization can be performed in $\mathbb{R}$ supposing that the operator V is positive definite and that the norm $\langle\xi_i|V|\xi_j\rangle$ exists. If we keep the norm of radial projectors fixed and just V-ortogonalize the basis (and so diagonalize U), we obtain the solution also for non-positive definite pseudopotential. Using the coefficients

$$d_i = U_{i,i} = \langle\alpha_i^l|V_l|\alpha_i^l\rangle^{-1} \tag{12.20}$$

as variation coefficients and optimizing them to match the all-electron calculation via the logarithmic derivatives, we get an additional tool for improving accuracy.

Using the basis α_i^l derived from $\{R_{nl}, \dot{R}_{El}\}$ with rigid norm, we can write the operator V_l as

$$V_l = \sum_i \left(\left(V_l\alpha_i^l\right) \otimes Y_{lm}\right) d_i^l \left(\left(V_l\alpha_i^l\right) \otimes Y_{lm}\right)^+ \tag{12.21}$$

and reintroduce Eq. 12.9 in the form

$$\langle\psi|V_l\psi'\rangle = \tag{12.22}$$

$$\sum_l \left\langle\psi\left|\left(V_l \otimes \sum_{m=-l}^{l} Y_{lm}Y_{lm}^+\right)\psi'\right\rangle = \tag{12.23}$$

$$\left\langle\psi\left|\sum_l \left(\sum_i \left(V_l\alpha_i^l\right) d_i^l \left(V_l\alpha_i^l\right)^+ \otimes \sum_{m=-l}^{l} Y_{lm}Y_{lm}^+\right)\psi'\right\rangle = \tag{12.24}$$

$$\sum_{l,i,m} d_i^l \left\langle\psi\left|\left(\left(V_l\alpha_i^l\right) d_i^l \left(V_l\alpha_i^l\right)^+ \otimes Y_{lm}Y_{lm}^+\right)\psi'\right\rangle = \tag{12.25}$$

$$\sum_{l,i,m} d_i^l \left\langle\psi\left|\left(V_l\alpha_i^l \otimes Y_{lm}\right)\left(\left(V_l\alpha_i^l\right)^+ \otimes Y_{lm}^+\right)\psi'\right\rangle = \tag{12.26}$$

$$\sum_{l,i,m} d_i^l \left\langle\left(V_l\alpha_i^l \otimes Y_{lm}\right)^+\psi\left|\left(V_l\alpha_i^l \otimes Y_{lm}\right)^+\psi'\right\rangle \tag{12.27}$$

By that, we get a general separable potential form defined by coefficients d_i^l and by vectors ϑ_{ilm}

$$\vartheta_{ilm} = V_l \alpha_i^l \otimes Y_{lm} \tag{12.28}$$

in the form

$$V_l = \sum_{i,l,m} d_i^l \, \vartheta_{ilm} (\vartheta_{ilm})^+ \tag{12.29}$$

For finite-element discretization we have just to perform numeric integration of

$$\int\limits_{element} V_l \alpha_i^l Y_{lm} \phi_j \, dV = \iiint\limits_{element} V_l(r)\alpha_i^l(r)Y_{lm}(\theta,\varphi)\phi_j(r,\theta,\varphi) \, dr \, d\theta \, d\varphi \tag{12.30}$$

for each α_i^l over each element ϕ_j of finite-element basis in spherical coordinates with respect to the center of each atom.

12.3.5   Eigenvalue Problem

A correction of the matrix $A^{m \times m}$ in a form $A + UU^+$, where U is a matrix of the shape $n \times m$, is usually denoted as rank-n-update.

Applying the previously derived relations, we can write the Kohn–Sham equations as a sum of sparse matrices with the same pattern plus the rank-n-update (omitting the variation coefficients d_{ni}^l for the moment)

$$\left(K + V_h + V_{xc} + V_{loc} + \sum_{n=0}^{N} \sum_{l=0}^{\max(l) \atop (n)} \sum_{i=0}^{\max(i) \atop (l,n)} \sum_{m=-l}^{l} \vartheta_{ni \atop lm} (\vartheta_{ni \atop lm})^+ \right) \bar{q} = \varepsilon M \bar{q} \tag{12.31}$$

and so, in the matrix notation of rank-n-update:

$$\left(K_V + UU^+ \right) \bar{q} = \lambda M \bar{q} \tag{12.32}$$

To employ preconditioning for solving the rank-n-update eigenvalue problem, we can take advantage of the knowledge how the rank-n-update has been obtained. Taking into account that

$$V_l = \sum_i d_i^l (V\alpha_i^l)(V\alpha_i^l)^+ \tag{12.33}$$

V_l^{-1} can be constructed as

$$V_l^{-1} = \sum_i d_i^l \alpha_i^l (\alpha_i^l)^+ \tag{12.34}$$

This fact can be easily shown from the orthogonality of the basis. For a vector $u = \sum_j a_j \alpha_j^l$ it holds

$$\sum_i \left(d_i^l \alpha_i^l \left(\alpha_i^l \right)^+ \right) V_l u = \sum_i \left(d_i^l \alpha_i^l \left(\alpha_i^l \right)^+ \right) \sum_j a_j V_l \alpha_j^l$$

$$= \sum_{i,j} \left(d_i^l \alpha_i^l \left(\alpha_i^l \right)^+ \right) a_j V_l \alpha_j^l = \sum_{i,j} d_i^l \alpha_i^l \left(a_j \left\langle \alpha_i^l | V_l | \alpha_j^l \right\rangle \right)$$

$$= \sum_j d_j^l \alpha_j^l \frac{a_j}{d_j^l} = \sum_j a_j \alpha_j^l = u \tag{12.35}$$

Since V_l are short-range potentials, ϑ_{nilm} have non-zero values only in those finite elements located near enough to the corresponding atomic center. This fact can be employed for optimization. The vectors can be treated as sparse by means of the pairs (index, value). Since each set of vectors have the same pattern, this form can be made even more efficient by means of the structure (index, {values}). By choosing convenient order of finite elements according to the pertinence to atomic centers, additional optimization could be performed, but, in this case, the benefit would hardly balance the higher computational costs in finite-element mesh generation.

12.3.5.1 Real Spherical Harmonics

Since the angular basis of Y_{lm} serves just for projecting to l-subspaces and since for any kind of atomic-like radial calculations (including the Environment-Reflecting Pseudopotentials and Projector Augmented Waves) the wavefunctions and all the quantities can be considered real, we can reach considerable simplification by using real spherical harmonics $Y_{lm}^{\mathbb{R}}$ instead of Y_{lm} so that all the vectors for rank-n update also remain real:

$$Y_{\ell m}^{\mathbb{R}} = \begin{cases} Y_\ell^0 & \text{if } m = 0 \\ \frac{1}{\sqrt{2}} \left(Y_\ell^m + (-1)^m Y_\ell^{-m} \right) = \sqrt{2} N_{(\ell,m)} P_\ell^m(\cos\theta) \cos m\varphi & \text{if } m > 0 \\ \frac{1}{i\sqrt{2}} \left(Y_\ell^{-m} - (-1)^m Y_\ell^m \right) = \sqrt{2} N_{(\ell,m)} P_\ell^{-m}(\cos\theta) \sin m\varphi & \text{if } m < 0. \end{cases} \tag{12.36}$$

where P_l^m are associated Legendere polynomials.

For the real separable potentials U, it holds $U^+ = U^T$, which brings considerable simplification at some points (Figs. 12.2, 12.3 and 12.4).

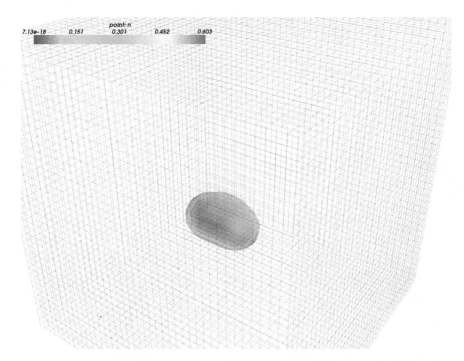

Fig. 12.2 Charge density of two-atom nitrogen molecule in finite elements; let's note that the error is $O(h^6)$, where h is the mesh spacing

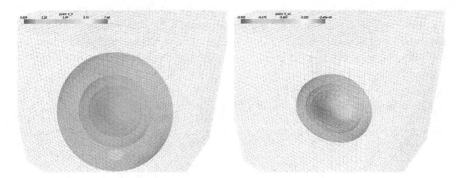

Fig. 12.3 Hartree and XC potentials

12.4 Conclusion

We introduced a method to solve Kohn Sham equations and to calculate electronic states and other properties of non-crystalline, non-periodic structures with fully non-local real-space environment-reflecting ab-initio pseudopotentials using finite elements, together with some preliminary results of our program. We believe this

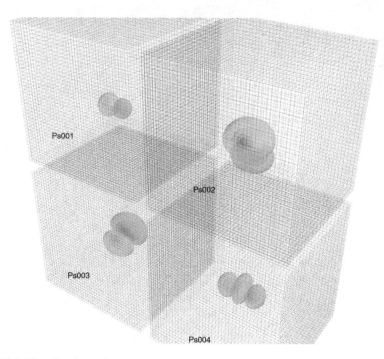

Fig. 12.4 Wave functions of two-atom nitrogen molecule

new approach could yield some new interesting results in electronic structure calculations. The main asset of the present approach is the combination of efficiency and high precision of ab-initio pseudopotentials with universal applicability, universal basis and excellent convergence control of finite-element method not restricted to periodic environment.

Acknowledgements We acknowledge support provided by the Grant Agency of the Czech Republic, under Grant No. 101/09/1630 (the first ant the last authors) and P108-11-0853 (the third and fifth authors). The work of the second author was partially supported by the LC06040 research center project.

References

1. Blochl PE (1990) Generalized separable potentials for electronic-structure calculations. Phys Rev B 41(8):5414–5416
2. Blochl PE (1994) Projector augmented-wave method. Phys Rev B 50:17953–17979
3. Dreizler RM, Gross EKU (1990) Density functional theory. Springer, Berlin
4. Eyert V (1996) A comparative study on methods for convergence acceleration of iterative vector sequences. J Comput Phys 124:271–285
5. Kleinman L, Bylander DM (1982) Efficacious form for model pseudopotentials. Phys Rev Lett 48:1425–1428

6. Parr RG, Weitao Y (1994) Density-functional theory of atoms and molecules. Oxford University Press, New York
7. Pickett WE (1989) Pseudopotential methods in condensed matter applications. Comput Phys Rep 9:115–198
8. Richard MM (2005) Electronic structure: basic theory and practical methods. Cambridge University Press, Cambridge, MA
9. Srivastava GP (1984) Broyden's method for self-consistent field convergence acceleration. J Phys A 17:L317–L321
10. Vackář J, Šimůnek A (2003) Adaptability and accuracy of all-electron pseudopotentials. Phys Rev B 67:125113

Chapter 13
Shifts in Excitation Energies Induced by Hydrogen Bonding: A Comparison of the Embedding and Supermolecular Time-Dependent Density Functional Theory Calculations with the Equation-of-Motion Coupled-Cluster Results

Georgios Fradelos, Jesse J. Lutz, Tomasz A. Wesołowski, Piotr Piecuch, and Marta Włoch

Abstract Shifts in the $\pi \rightarrow \pi^*$ excitation energy of the cis-7-hydroxyquinoline chromophore induced by hydrogen bonding with small molecules, obtained with the frozen-density embedding theory (FDET), are compared with the results of the high-level equation-of-motion coupled-cluster (EOMCC) calculations with singles, doubles, and noniterative triples, which provide the reference ab initio data, the supermolecular time-dependent density functional theory (TDDFT) calculations, and the available experimental data. It is demonstrated that the spectral shifts resulting from the FDET calculations employing nonrelaxed environment densities and their EOMCC counterparts are in excellent agreement with one another, whereas the analogous shifts obtained with the supermolecular TDDFT approach do not agree with the EOMCC reference data. Among the discussed issues are the effects of higher-order correlations on the excitation energies and complexation-induced excitation energy shifts resulting from the EOMCC calculations, and the choice of the approximants that represent the nonadditive kinetic energy contributions to the embedding potential of FDET.

G. Fradelos • T.A. Wesołowski (✉)
Département de Chimie Physique, Université de Genève, 30, quai Ernest-Ansermet,
CH-1211 Genève 4, Switzerland
e-mail: georgios.fradelos@unige.ch; tomasz.wesolowski@unige.ch

J.J. Lutz • P. Piecuch (✉)
Department of Chemistry, Michigan State University, East Lansing, Michigan 48824, USA
e-mail: lutzjess@chemistry.msu.edu; piecuch@chemistry.msu.edu

M. Włoch
Department of Chemistry, Michigan Technological University, Houghton, Michigan 49931, USA
e-mail: wloch@mtu.edu

P.E. Hoggan et al. (eds.), *Advances in the Theory of Quantum Systems in Chemistry and Physics*, Progress in Theoretical Chemistry and Physics 22, DOI 10.1007/978-94-007-2076-3_13, © Springer Science+Business Media B.V. 2012

13.1   Introduction

Noncovalent interactions, such as hydrogen bonds, can qualitatively affect the electronic structure and properties of molecules embedded in molecular environments. Among such properties, electronic excitation energies are of particular interest because of the wide use of organic chromophores as probes in various environments [1–4]. Typically, hydrogen bonding results in shifts in the positions of absorption and emission bands anywhere between a few hundred and about $3,000\,cm^{-1}$ [5]. This means that if one is to use computer modeling for interpretation of experimental data, the errors of the calculated shifts must be very small, on the order of $100\,cm^{-1}$ or less.

Unfortunately, the brute force application of the conventional supermolecular approach to evaluate the excitation energy shifts induced by hydrogen bonding with environment molecules, in which one determines the shift as a difference of two large numbers representing the excitation energy for a given electronic transition in the complex and the analogous excitation energy characterizing the isolated chromophore, encounters several challenges. For example, the supermolecular approach relies on the ability of a given electronic structure method to provide an accurate and well-balanced description of excitation energies in systems that have different sizes, which in the specific case of spectral shifts induced by complexation are the total system consisting of the chromophore and environment molecules and the system representing the isolated chromophore. In great many cases, particularly when the low-level quantum-chemistry methods are exploited and larger molecules are examined, this condition is difficult to satisfy. Ab initio methods based on the equation-of-motion (EOM) [6–10] or linear-response [11–16] coupled-cluster (CC) [17–22] theories (cf. Refs. [23–25] for selected reviews), including, among several schemes proposed to date, the basic EOMCC approach with singles and doubles (EOMCCSD) [7–9] and the so-called δ-CR-EOMCC$(2,3)$ method, which represents a suitably modified variant of the completely renormalized (CR) EOMCC theory with singles, doubles, and noniterative triples based on the CR-CC$(2,3)$ [26–28] and CR-EOMCC$(2,3)$ [29–31] approximations and which is used in the present work to provide the reference data, or the related EOMCCSD$(2)_T$ [32] and EOMCCSD$(\tilde{T})$ [33] approaches, satisfy this condition, since they provide an accurate and systematically improvable description of the electronic excitations in molecular systems and satisfy the key property of size-intensivity [16, 34], but their applicability is limited to smaller molecular problems due to the CPU steps that typically scale as $\mathcal{N}^6 - \mathcal{N}^7$ with the system size $\mathcal{N}$. Significant advances have been made in recent years toward extending the EOMCC and response CC methods to larger molecules through code parallelization and the use of local correlation methodologies [35–38], combined, in analogy to QM/MM techniques, with molecular mechanics [39–43], but none of the resulting approaches is as practical, as far as computer costs are concerned, as methods based on the time-dependent density functional theory (TDDFT) [44]. Unfortunately, the existing

TDDFT approaches, although applicable to large molecular systems due to low computer costs, are not always accurate enough to guarantee a robust description of the complexation-induced spectral shifts in weakly bound complexes when the supermolecular approach is employed due to their difficulties with describing dispersion and charge-transfer interactions, and other intrinsic errors. It is, for example, unclear if the existing practical implementations of the TDDFT methodology satisfy the aforementioned condition of size intensivity, which is critical for obtaining a well-balanced description of electronic excitations in molecules that differ in size, and accurate values of the spectral shifts. In a size-intensive approach, the vertical excitation energy of a noninteracting system $A + B$, in which fragment A is excited, is the same as that obtained for the isolated system A. The aforementioned EOMCCSD, δ-CR-EOMCC(2,3), EOMCCSD(2)$_T$, and EOMCCSD($\tilde{T}$) methods are size intensive, but, based on the results presented in this study, the widely used TDDFT approximations may violate this property.

Methods employing the embedding strategy, including those based on the frozen-density embedding theory (FDET) [45–48] that interests us in this work, provide an alternative to the supermolecular approach for evaluating the excitation energy shifts. In all embedding methods, of both empirical (e.g., QM/MM) and FDET types, the effect of the environment on the molecular properties of interest is not treated explicitly, but, rather, through the use of the suitably designed embedding potential. Thus, instead of solving the electronic Schrödinger equation for the total $(N_A + N_B)$-electron system AB consisting of the N_A-electron molecule A and N_B-electron environment B, i.e.,

$$\hat{H}^{(AB)}|\Psi^{(AB)}\rangle = E^{(AB)}|\Psi^{(AB)}\rangle, \tag{13.1}$$

where $\hat{H}^{(AB)}$ is the Hamiltonian of the total system AB, and the electronic Schrödinger equation for the isolated molecule A,

$$\hat{H}^{(A)}|\Psi^{(A)}\rangle = E^{(A)}|\Psi^{(A)}\rangle, \tag{13.2}$$

where $\hat{H}^{(A)}$ is the Hamiltonian of molecule A in the absence of environment, and then calculating the shift in an observable of interest associated with an operator $\hat{O}$ by forming the difference of the expectation values of $\hat{O}$ computed for systems AB and A that have different numbers of electrons,

$$\Delta\langle\hat{O}\rangle = \langle\Psi^{(AB)}|\hat{O}|\Psi^{(AB)}\rangle - \langle\Psi^{(A)}|\hat{O}|\Psi^{(A)}\rangle, \tag{13.3}$$

as one does in supermolecular calculations, one solves two eigenvalue problems characterized by the same number of electrons, namely, Eq. 13.2 and

$$\left[\hat{H}^{(A)} + \hat{V}_{emb}^{(A)}\right]|\Psi_{emb}^{(A)}\rangle = E_{emb}^{(A)}|\Psi_{emb}^{(A)}\rangle, \tag{13.4}$$

where $|\Psi_{emb}^{(A)}\rangle$ is the auxiliary N_A-electron wavefunction describing the effective state of molecule A in the presence of environment B and $\hat{V}_{emb}^{(A)} = \sum_{i=1}^{N_A} v_{emb}(\mathbf{r}_i)$ is the suitable embedding operator defined in terms of the effective one-electron potential $v_{emb}(\mathbf{r})$. As a result, the complexation-induced shift in an observable represented by the $\hat{O}$ operator is evaluated in the embedding strategy as

$$\Delta\langle\hat{O}\rangle = \left\langle \Psi_{emb}^{(A)} | \hat{O} | \Psi_{emb}^{(A)} \right\rangle - \left\langle \Psi^{(A)} | \hat{O} | \Psi^{(A)} \right\rangle, \qquad (13.5)$$

i.e., by using the many-electron wavefunctions $|\Psi^{(A)}\rangle$ and $|\Psi_{emb}^{(A)}\rangle$ that represent two different physical states of system A corresponding to the same number of electrons, the state of the isolated molecule A and the state of A embedded in the environment B. This has two advantages over the conventional supermolecular approach. First, the embedding strategy does not require the explicit consideration of the total $(N_A + N_B)$-electron system consisting of the complex of the molecule of interest and its environment, which leads to a significant cost reduction in the computer effort, particularly in applications involving larger environments. Indeed, it has already been demonstrated that FDET can be applied in large-scale multi-scale molecular simulations [46, 49–51]. Second, by determining the complexation-induced shift $\Delta\langle\hat{O}\rangle$ using the wavefunctions $|\Psi^{(A)}\rangle$ and $|\Psi_{emb}^{(A)}\rangle$ that correspond to the same number of electrons, the errors due to approximations used to solve the N_A-electron problems represented by Eqs. 13.2 and 13.4 largely cancel out as we do not have to be concerned about the possible dependence of the error in the calculated $\Delta\langle\hat{O}\rangle$ value on the system size. In the calculations of the shifts in excitation energies using the size-intensive EOMCC methods, one does not have to worry about it either, but the supermolecular EOMCC approach requires an explicit consideration of the $(N_A + N_B)$-electron system consisting of the chromophore and environment, which may lead to a significant cost increase when N_B is larger. By representing systems A and AB as two quantum-mechanical states of A with a fixed number of electrons N_A that correspond to the state of the isolated molecule A and the state of A perturbed by the environment B, we can effectively enforce the condition of size intensivity of excitation energies in the embedding approaches.

Clearly, the accuracy of the complexation-induced shifts obtained in the embedding calculations largely depends on the quality of the embedding operator $\hat{V}_{emb}^{(A)}$. Thus, when compared with the supermolecular approach, the challenge is moved from assuring the cancellation of errors in approximate solutions of two Schrödinger equations for systems that differ in the number of electrons to developing the appropriate form of $\hat{V}_{emb}^{(A)}$ that can accurately describe the state of the chromophore in the weakly bound complex with environment. According to the FDET formalism [45–48], the embedding operator $\hat{V}_{emb}^{(A)}$ can be represented in terms of a local potential $v_{emb}(\mathbf{r})$ (orbital-free embedding potential), which is determined by the pair of electron densities, ρ_A describing the embedded system A and constructed using the $|\Psi_{emb}^{(A)}\rangle$ wavefunction, and ρ_B representing the electron density of the environment B. Unfortunately, except for a small number of analytically solvable problems [52], the precise dependence of $v_{emb}(\mathbf{r})$ on ρ_A and ρ_B is not known. Only its electrostatic component is known exactly. The nonelectrostatic component, which arises from the nonadditivity of the density functionals for the exchange-correlation and kinetic energies, must be approximated or reconstructed, either analytically (if possible) [52] or numerically [53–55]. In the case of hydrogen-bonded environments that interest us in this work, the electrostatic component of the exact embedding potential is expected to dominate and the overall accuracy of the environment-induced changes of the electronic structure of embedded species should be quite high. Indeed, a number of earlier studies [56, 57] demonstrate that the currently known approximants to the relevant functional representations of $v_{emb}(\mathbf{r})$ in terms of ρ_A and ρ_B are adequate. Still, it is instructive to examine how the FDET results depend on the approximants that are used to represent the nonelectrostatic components of $v_{emb}(\mathbf{r})$, such as the nonadditive kinetic energy potential.

Our previous studies of the methodological and practical aspects of the FDET approach have largely focused on the analytically solvable model systems and direct comparisons with experiment. The exactly solvable model systems are certainly important, since they may provide useful information about the analytic dependence of the embedding potential $v_{emb}(\mathbf{r})$ on densities ρ_A and ρ_B, but one has to keep in mind that real molecular systems may be quite different from such models. A comparison with the experimental data is clearly the final goal of any computational technique, and the FDET approach is no different in this regard, but we also have to remember that experiments have their own error bars and their interpretation may require the incorporation of physical effects that are not included in the purely electronic structure calculations. For these reasons, the present study chooses an alternative way of examining the performance of the FDET methodology in which we make a direct comparison of the benchmark results obtained in the high-level, wavefunction-based EOMCC calculations, using the aforementioned size-intensive modification of the CR-EOMCC(2,3) method [29–31], designated as δ-CR-EOMCC(2,3), with those produced by the embedding-theory-based FDET approach [45–48] and the supermolecular TDDFT methodology. In order to address this objective, we first obtain the δ-CR-EOMCC(2,3)-based shifts in the vertical excitation energy corresponding to the $\pi \rightarrow \pi^*$ transition in the cis-7-hydroxyquinoline (cis-7HQ) chromophore, induced by formation of hydrogen-bonded complexes with eight different environments defined by the water, ammonia, methanol, and formic acid molecules, and their aggregates consisting of up to three molecules, for which, as demonstrated in this study, reliable EOMCC data can be obtained and which were previously examined using the laser resonant two-photon UV spectroscopy [5, 57]. Then, we use the resulting EOMCC reference shift values to assess the quality of the analogous spectral shifts obtained in the FDET and supermolecular TDDFT calculations. By having access to the highly accurate reference EOMCC data, we can explore various aspects of the FDET methodology and approximations imposed within. One such aspect is the possible dependence of the shifts in the excitation energy of cis-7HQ induced by the complexation with hydrogen-bonded molecules on the approximations used for one

of the nonelectrostatic components of the embedding potential $v_{emb}(\mathbf{r})$ resulting
from the nonadditivity of density functionals for the kinetic energy. We examine this
issue by performing two sets of the FDET calculations, one in which the nonadditive
kinetic energy potential entering the definition of $v_{emb}(\mathbf{r})$ is approximated using
the generalized gradient approximation (GGA97) [58], and another one in which
the nonadditive kinetic energy component of $v_{emb}(\mathbf{r})$ is approximated with the
help of the recently developed NDSD approximant [59] that incorporates the exact
conditions relevant for the proper behavior of this component of $v_{emb}(\mathbf{r})$ in the
vicinity of nuclei. Other aspects of the FDET considerations, such as the role of
approximations that are used to define the exchange-correlation potentials in the
FDET (and supermolecular TDDFT) calculations, the usefulness of the monomer vs
supermolecular basis expansions in the FDET calculations, the basis set dependence
of the FDET results, and the effect of the form of the electronic density of the
environment on the FDET results, will be discussed elsewhere [60].

Although our main goal is a direct comparison of the embedding-theory-
based FDET and supermolecular TDDFT results with the high-level ab initio
EOMCC data, which demonstrates the advantages of the FDET approach over the
supermolecular TDDFT methodology in a realistic application, a comparison of the
calculated shifts with the corresponding experimental data [5, 57] is discussed as
well. The gas-phase complexes of the cis-7HQ chromophore are of great interest,
since some of them, particularly the larger ones, can be regarded as models of
proton-transferring systems in biomolecular systems [1, 61].

13.2   Methods

This section describes the electronic structure methods used in the present study.
Since the supermolecular TDDFT approach is an established methodology, we focus
on the FDET and EOMCC schemes exploited in our calculations.

13.2.1   Frozen-Density Embedding Theory

The FDET formalism [45–48, 52, 56, 59] (cf., also, Refs. [62–66]), provides
basic equations for the variational treatment of a quantum-mechanical subsystem
embedded in a given electronic density. In order to introduce the FDET-based
computational methods for describing the molecular system A embedded in the
environment B created by some other molecule(s), one introduces two types of
electronic densities to represent the total system AB. The first one is the density
of the subsystem A defined by embedded molecule(s), $\rho_A(\mathbf{r})$, which is typically
represented using one the following auxiliary quantities: (1) the occupied orbitals
of a noninteracting reference system $\{\phi_i^{(A)}(\mathbf{r}),\ i=1,\ldots,N_A\}$ [45], (2) the occupied

and unoccupied orbitals of a noninteracting reference system [56], (3) the interacting wavefunction [47], or (4) the one-particle density matrix [48]. The second one is the density of the subsystem B describing the environment, $\rho_B(\mathbf{r})$, which is fixed for a given electronic problem ("frozen density"). The optimum density $\rho_A(\mathbf{r})$ of the system A embedded in the environment B, represented by the fixed density $\rho_B(\mathbf{r})$ satisfying

$$\int \rho_B(\mathbf{r})d\mathbf{r} = N_B, \tag{13.6}$$

is obtained by performing the following constrained search:

$$E_{emb}^{(A)}[\rho_B] = \min_{\rho \geq \rho_B} E_{HK}[\rho] = \min_{\rho_A} E_{HK}[\rho_A + \rho_B], \tag{13.7}$$

subject to the conditions

$$\int \rho(\mathbf{r})d\mathbf{r} = N_A + N_B \tag{13.8}$$

and

$$\int \rho_A(\mathbf{r})d\mathbf{r} = N_A, \tag{13.9}$$

where $E_{HK}[\rho]$ in Eq. 13.7 is the usual Hohenberg-Kohn energy functional.

In practice, the search for the optimum density ρ_A, defined by Eq. 13.7, is performed by exploiting the Kohn-Sham formulation [67] of DFT [68] to solve Eq. 13.4, in which $\hat{H}^{(A)}$ is the environment-free Hamiltonian of the isolated system A and $\hat{V}_{emb}^{(A)} = \sum_{i=1}^{N_A} v_{emb}(\mathbf{r}_i)$ is the potential energy operator describing the effect of environment B on system A, where $\hat{v}_{emb}(\mathbf{r})$ has the form of a local, orbital-free, embedding potential $v_{emb}^{eff}(\mathbf{r})$, determined by the pair of densities $\rho_A(\mathbf{r})$ and $\rho_B(\mathbf{r})$ and designated by $v_{emb}^{eff}[\rho_A, \rho_B; \mathbf{r}]$. As shown in our earlier work [45, 47, 48], the relationship between $v_{emb}^{eff}[\rho_A, \rho_B; \mathbf{r}]$ and densities $\rho_A(\mathbf{r})$ and $\rho_B(\mathbf{r})$ depends on the quantum-mechanical descriptors that are used as the auxiliary quantities for representing $\rho_A(\mathbf{r})$. If we use the orbitals of a noninteracting reference system, the wavefunction of the full configuration interaction (CI) form, or the one-particle density matrix as the descriptors to define $\rho_A(\mathbf{r})$, the local, orbital-free, embedding potential $v_{emb}^{eff}[\rho_A, \rho_B; \mathbf{r}]$ can be given the following form:

$$v_{emb}^{eff}[\rho_A, \rho_B; \mathbf{r}] = v_{ext}^B(\mathbf{r}) + \int \frac{\rho_B(\mathbf{r}')}{|\mathbf{r}' - \mathbf{r}|} d\mathbf{r}' + v_{xc}^{nad}[\rho_A, \rho_B](\mathbf{r}) + v_t^{nad}[\rho_A, \rho_B](\mathbf{r}), \tag{13.10}$$

where

$$v_{xc}^{nad}[\rho_A, \rho_B](\mathbf{r}) = \left. \frac{\delta E_{xc}[\rho]}{\delta \rho} \right|_{\rho=\rho_A+\rho_B} - \left. \frac{\delta E_{xc}[\rho]}{\delta \rho} \right|_{\rho=\rho_A} \tag{13.11}$$

and

$$v_t^{nad}[\rho_A, \rho_B](\mathbf{r}) = \left.\frac{\delta T_s[\rho]}{\delta \rho}\right|_{\rho=\rho_A+\rho_B} - \left.\frac{\delta T_s[\rho]}{\delta \rho}\right|_{\rho=\rho_A}. \tag{13.12}$$

As one can see, $v_{emb}^{eff}[\rho_A, \rho_B; \mathbf{r}]$ involves the external and Coulomb potentials due to the environment B, and the $v_{xc}^{nad}[\rho_A, \rho_B](\mathbf{r})$ and $v_t^{nad}[\rho_A, \rho_B](\mathbf{r})$ components that arise from the nonadditivities of the exchange-correlation and kinetic energy functionals of the Kohn-Sham DFT, $E_{xc}[\rho]$ and $T_s[\rho]$, respectively.

Once $v_{emb}^{eff}[\rho_A, \rho_B; \mathbf{r}]$ is defined, as in Eq. 13.10, and if we use a noninteracting reference system to conduct the constrained search given by Eq. 13.7, the orbitals $\phi_i^{(A)}$, $i = 1, \ldots, N_A$, of the system A embedded in the environment B are determined by solving the Kohn-Sham-like equations (cf. Eqs. 20 and 21 in Ref. [45])

$$\left[-\frac{1}{2}\nabla^2 + v_{KS}^{eff}[\rho_A; \mathbf{r}] + v_{emb}^{eff}[\rho_A, \rho_B; \mathbf{r}]\right] \phi_i^{(A)} = \varepsilon_i^{(A)} \phi_i^{(A)}, \tag{13.13}$$

where $v_{KS}^{eff}[\rho_A; \mathbf{r}]$ is the usual expression for the potential of the Kohn-Sham DFT for the isolated system A. After obtaining the orbitals $\phi_i^{(A)}$ and the corresponding orbital energies $\varepsilon_i^{(A)}$, we calculate the ground- and excited-state energies and properties other than energy, which in this case describe the system A embedded in the environment B, in a usual manner, using standard algorithms of DFT or TDDFT.

It may be worth mentioning that there are two other approaches related to FDET that aim at the description of a system consisting of subsystems, including the situation of a molecule embedded in an environment which interests us here, namely, the subsystem formulation of DFT (SDFT) [69, 70] and the recently developed partition DFT (PDFT) [71]. There are, however, differences between the SDFT and PDFT methods on the one hand and the FDET formalism on the other hand. Indeed, in the exact limit, both SDFT and PDFT lead to the exact ground-state electronic density and energy of the total system under investigation, providing an alternative to the conventional supermolecular Kohn-Sham framework. This should be contrasted with FDET, which does not target the exact ground-state electronic density of the total system AB, but, rather, the density of subsystem A that minimizes the Hohenberg-Kohn energy functional of the total system, $E_{HK}[\rho_A + \rho_B]$, using a fixed form of the environment density ρ_B given in advance in the presence of constraints, as in Eqs. 13.6–13.9. Thus, FDET may lead to the same total ground-state density as SDFT, Kohn-Sham DFT, or PDFT, but only when the specific set of additional assumptions and constraints is employed [46]. Otherwise, it can only give the upper bound to the exact ground-state energy of the total system AB, $E_{emb}^{(A)}[\rho_B] \geq E^{(AB)}$ (see Refs. [45–48, 60] and references cited therein for further information).

The effectiveness of FDET methods based on Eq. 13.13, with $v_{emb}^{eff}[\rho_A, \rho_B; \mathbf{r}]$ determined using Eq. 13.10, in the calculations of changes in the electronic structure due to the interactions between the embedded system and its environment was demonstrated in a number of applications, including excitation energies [56,57,72], ESR hyperfine coupling constants [73, 74], ligand-field splittings of f-levels in lanthanide impurities [75], NMR shieldings [76], and multipole moments and frequency dependent polarizabilities [72]. Based on these positive experiences, the FDET approach is expected to provide accurate values of the complexation-induced shifts in the vertical excitation energy corresponding to the $\pi \rightarrow \pi^*$ transition in the cis-7HQ chromophore, and the present study demonstrates that this is indeed the case by comparing the FDET and δ-CR-EOMCC(2,3)-based EOMCC results.

13.2.2 Equation-of-Motion Coupled-Cluster Calculations

The basic idea of the EOMCC theory is the following wavefunction ansatz for the excited-state wavefunctions $|\Psi_\mu\rangle$ [6–9] (cf. Refs. [10, 23, 25, 30] for reviews):

$$|\Psi_\mu\rangle = R_\mu |\Psi_0\rangle, \tag{13.14}$$

where the linear excitation operator R_μ generates $|\Psi_\mu\rangle$ from the CC ground state

$$|\Psi_0\rangle = e^T |\Phi\rangle, \tag{13.15}$$

with T representing the cluster operator and $|\Phi\rangle$ the reference determinant [in all of the EOMCC calculations discussed in this work, the restricted Hartree-Fock (RHF) configuration]. Throughout this paper, we use a convention in which we allow index μ in Eq. 13.14 to become zero by defining the $\mu = 0$ operator R_μ as a unit operator, $R_{\mu=0} = \mathbf{1}$, so that we can incorporate the ground- and excited-state cases corresponding to $\mu = 0$ and $\mu > 0$, respectively, within a single set of formulas.

The cluster operator T in Eq. 13.15 is typically obtained by truncating the corresponding many-body expansion

$$T = \sum_{n=1}^{N} T_n, \tag{13.16}$$

where

$$T_n = \sum_{i_1 < \cdots < i_n, a_1 < \cdots < a_n} t_{a_1 \ldots a_n}^{i_1 \ldots i_n} a^{a_1} \cdots a^{a_n} a_{i_n} \cdots a_{i_1} \tag{13.17}$$

is the n-body component of T and N is the number of correlated electrons in a system, at some, preferably low, excitation level $M < N$, and by solving the nonlinear system of equations for cluster amplitudes $t_{a_1 \ldots a_n}^{i_1 \ldots i_n}$ with $n \leq M$, which define the truncated form of T, designated as $T^{(M)}$, that results from projecting the electronic Schrödinger equation on the excited determinants $|\Phi_{i_1 \ldots i_n}^{a_1 \ldots a_n}\rangle$ which correspond to the many-body components T_n included in $T^{(M)}$. Once T and the corresponding ground-state CC energy E_0 are determined, one obtains the many-body components

$$R_{\mu,n} = \sum_{i_1 < \cdots < i_n, a_1 < \cdots < a_n} r_{\mu,a_1 \ldots a_n}^{i_1 \ldots i_n} \, a^{a_1} \cdots a^{a_n} a_{i_n} \cdots a_{i_1} \tag{13.18}$$

of the linear excitation operator

$$R_\mu = r_{\mu,0} \, \mathbf{1} + \sum_{n=0}^{N} R_{\mu,n}, \tag{13.19}$$

which is usually truncated at the same excitation level M as T, and the corresponding vertical excitation energies

$$\omega_\mu = E_\mu - E_0, \tag{13.20}$$

by diagonalizing the similarity-transformed Hamiltonian $\bar{H}^{(M)} = e^{-T^{(M)}} H e^{T^{(M)}}$ in the subspace of the N-electron Hilbert space spanned by the excited determinants $|\Phi_{i_1 \ldots i_n}^{a_1 \ldots a_n}\rangle$ that correspond to the many-body components $R_{\mu,n}$ included in R_μ.

The basic EOMCCSD approach [7–9], in which $M = 2$, so that $T \approx T^{(2)} = T_1 + T_2$ and $R_\mu \approx R_\mu^{(2)} = r_{\mu,0} \, \mathbf{1} + R_{\mu,1} + R_{\mu,2}$ (in general, $R_\mu^{(M)}$ designates the R_μ operator truncated at the M-body component $R_{\mu,M}$), in which one diagonalizes the similarity-transformed Hamiltonian of CCSD, $\bar{H}^{(2)} = e^{-T^{(2)}} H e^{T^{(2)}}$, in the space spanned by singly and doubly excited determinants, $|\Phi_i^a\rangle$ and $|\Phi_{ij}^{ab}\rangle$, respectively, and its linear-response CCSD counterpart [15,16] have been successful in describing excited states dominated by one-electron transitions, but this success does not extend to the more complicated excited states, such as those characterized by a significant two-electron excitation nature (cf. Refs. [29–31, 77–83] for examples). There also are cases of excited states dominated by one-electron transitions, particularly when larger molecular systems are examined, where the EOMCCSD theory level is not sufficiently accurate, producing errors in the computed excitation energies on the order of 0.3–0.5 eV [84, 85]. Thus, particularly in the context of this study, where the molecular systems of interest are not small and where we expect the EOMCC theory to provide accurate reference data for the FDET and TDDFT calculations of the relatively small spectral shifts induced by the formation of weakly bound complexes, it is important to examine if the EOMCC results used by us as a reference are reasonably well converged with respect to the truncations in the T and R_μ operators. Ideally, one would like to perform the full EOMCCSDT (EOMCC with singles, doubles, and triples) calculations [86–88] and compare them with the corresponding EOMCCSD results to examine this. Unfortunately, it is not possible to carry out the full EOMCCSDT calculations for the cis-7HQ system and its complexes investigated in this work due to a steep increase of the CPU time and storage requirements characterizing the EOMCCSDT approach that scale as $n_o^3 n_u^5$ and $\sim n_o^3 n_u^3$ with the numbers of occupied and unoccupied orbitals, n_o and n_u, respectively, as compared to the $n_o^2 n_u^4$ CPU time and $\sim n_o^2 n_u^2$ storage requirements of EOMCCSD. One has to resort to an approximate treatment of triple excitations in the EOMCC theory that replaces the prohibitively expensive iterative CPU steps of full EOMCCSDT that scale as $\mathcal{N}^8$ with the system size $\mathcal{N}$ to the more manageable, $\mathcal{N}^6 - \mathcal{N}^7$, steps.

A large number of approximate EOMCCSDT approaches and their linear-response analogs have been developed to date [29–33, 77–79, 81, 82, 86, 87, 89–96]. The noniterative EOMCC methods, in which one adds corrections due to triples to the EOMCCSD energies, such as, for example, EOM-CCSD(2)$_T$ [32], CCSDR3 [92, 93], EOMCCSD($\tilde{T}$) [33], CR-EOMCCSD(T) [77, 78, 81, 82], N-EOMCCSD(T) [96], and CR-EOMCC(2,3) [29–31], are particularly promising, since they represent computational black boxes similar to those of the widely used ground-state CCSD(T) approach [97] and its CCSD[T] predecessor [98], or their CR-CCSD(T) [77, 78, 99, 100], CR-CC(2,3) [26–28], CCSD(2)$_T$ [90, 101–103], and CCSD(T)$_\Lambda$ [104–106] analogs (cf. Refs. [107, 108] for related work). All of the above methods greatly reduce the computer costs of full EOMCCSDT calculations, while improving the EOMCCSD results. For example, the aforementioned EOM-CCSD(2)$_T$, CCSDR3, EOMCCSD($\tilde{T}$), CR-EOMCCSD(T), N-EOMCCSD(T), and CR-EOMCC(2,3) approaches are characterized by the iterative $n_o^2 n_u^4$ steps of EOM-CCSD and the noniterative $n_o^3 n_u^4$ steps needed to construct the triples corrections to the EOMCCSD energies, while eliminating the need for storing the $\sim n_o^3 n_u^3$ triply excited amplitudes defining the T and R_μ operators. This makes these methods applicable to much larger problems than those that can be handled by full EOMCCSDT, including the complexes of cis-7HQ examined in the present study.

As mentioned in the Introduction, our focus in this work is on the size-intensive modification of the CR-EOMCC(2,3) method of Refs. [29–31], defining the δ-CR-EOMCC(2,3) approach. The CR-EOMCC(2,3) scheme and the underlying ground-state CR-CC(2,3) approximation [26–28] are examples of the renormalized CC/EOMCC schemes, which are based on the idea of adding the *a posteriori*, noniterative, and state-specific corrections δ_μ due to higher-order excitations, neglected in the conventional CC/EOMCC calculations defined by some truncation level M, such as CCSD or EOMCCSD, to the corresponding CC/EOMCC energies. The formal basis for deriving the computationally tractable expressions for corrections δ_μ is provided by the moment expansions which describe the differences between the full CI and CC/EOMCC energies [26, 27, 77–79, 99, 100, 109–112]. If we are interested in correcting the energies $E_\mu^{(M)}$ obtained in the CC/EOMCC calculations truncated at M-tuple excitations, the CC/EOMCC moments that enter the expressions for the corresponding corrections δ_μ are defined as projections of the CC/EOMCC equations with T and R_μ truncated at the M-body components T_M and $R_{\mu,M}$, respectively, on the excited determinants $|\Phi_{i_1 \ldots i_n}^{a_1 \ldots a_n}\rangle$ with $n > M$ that are disregarded in the conventional CC/EOMCC calculations, i.e.,

$$\mathfrak{M}_{\mu, a_1 \ldots a_n}^{i_1 \ldots i_n}(M) = \left\langle \Phi_{i_1 \ldots i_n}^{a_1 \ldots a_n} \left| \left(\bar{H}^{(M)} R_\mu^{(M)} \right) \right| \Phi \right\rangle. \tag{13.21}$$

All of the resulting moment expansions of the corrections δ_μ that define the differences between the full CI energies E_μ and the corresponding CC/EOMCC energies $E_\mu^{(M)}$, developed to date [26, 27, 77–79, 99, 100, 109–112], can be written as

$$\delta_\mu \equiv E_\mu - E_\mu^{(M)} = \sum_{n=M+1}^{N_{\mu,M}} \sum_{i_1 < \cdots < i_n, a_1 < \cdots < a_n} \ell_{\mu,i_1\ldots i_n}^{a_1\ldots a_n} \, \mathfrak{M}_{\mu,a_1\ldots a_n}^{i_1\ldots i_n}(M), \qquad (13.22)$$

where $N_{\mu,M}$ represents the highest value of the many-body rank n for which $\mathfrak{M}_{\mu,a_1\ldots a_n}^{i_1\ldots i_n}(M)$ is still nonzero (in the CCSD/EOMCCSD $M = 2$ case, $N_{\mu,M}$ is 6 when $\mu = 0$ and 8 when $\mu > 0$). The only essential difference between various approximations based on Eq. 13.22 is in the way one handles the $\ell_{\mu,i_1\ldots i_n}^{a_1\ldots a_n}$ coefficients. Equation 13.22 can be obtained by considering the asymmetric energy expression [26, 27, 77–79, 100, 109] $\tilde{E}_\mu = \langle \tilde{\Psi}_\mu | H R_\mu^{(M)} e^{T^{(M)}} | \Phi \rangle / \langle \tilde{\Psi}_\mu | R_\mu^{(M)} e^{T^{(M)}} | \Phi \rangle$, which gives the exact, full CI, energy E_μ, independent of the truncation level M in $T^{(M)}$ and $R_\mu^{(M)}$, if $\langle \tilde{\Psi}_\mu |$ is the full CI bra state $\langle \Psi_\mu |$. When $\langle \Psi_\mu |$ is represented as $\langle \Phi | \mathcal{L}_\mu e^{-T^{(M)}}$, where $\mathcal{L}_\mu$ is a suitably defined deexcitation operator satisfying $\langle \Phi | \mathcal{L}_\mu R_\mu^{(M)} | \Phi \rangle = 1$, the $\ell_{\mu,i_1\ldots i_n}^{a_1\ldots a_n}$ coefficients in Eq. 13.22 have a meaning of amplitudes defining $\mathcal{L}_\mu$ [26, 27]. In that case, the asymmetric energy expression $\tilde{E}_\mu$ gives the formula for the exact energy in the form $E_\mu = \langle \Phi | \mathcal{L}_\mu \bar{H}^{(M)} R_\mu^{(M)} | \Phi \rangle$, which becomes equivalent to the conventional CC energy functional [25] (used, for example, to derive the CCSD(T)$_\Lambda$ approach [104–106]) when $\mu = 0$ and $M = N$, and which directly leads to Eq. 13.22 for the δ_μ correction after subtracting the CC/EOMCC energy $E_\mu^{(M)}$ from E_μ and performing straightforward analysis [26, 27].

In the specific case of the CR-EOMCC(2,3) approach that interests us here, which corresponds to setting M in Eq. 13.22 at 2 and considering only the $n = 3$ term in the resulting moment expansion for δ_μ, one calculates the energies of the ground and excited states as

$$E_\mu = E_\mu^{(\text{CCSD})} + \sum_{i<j<k,a<b<c} \ell_{\mu,ijk}^{abc} \, \mathfrak{M}_{\mu,abc}^{ijk}(2), \qquad (13.23)$$

where $E_\mu^{(\text{CCSD})} \equiv E_\mu^{(2)}$ are the CCSD ($\mu = 0$) and EOMCCSD ($\mu > 0$) energies, $\mathfrak{M}_{\mu,abc}^{ijk}(2)$ are the moments of the CCSD/EOMCCSD equations corresponding to triple excitations, which are defined by Eq. 13.21 in which $M = 2$, and $\ell_{\mu,ijk}^{abc}$ are the deexcitation amplitudes that one can calculate using the quasi-perturbative expressions presented in Refs. [29, 31]. The $\ell_{\mu,ijk}^{abc}$ amplitudes used in the CR-EOMCC(2,3) considerations are expressed in terms of the one- and two-body components of the deexcitation operator defining the left eigenstate of EOM-CCSD [9], and the one-body, two-body, and – in the full implementation of CR-EOMCC(2,3) defining variant D of it designated as CR-EOMCC(2,3),D – selected three-body components of the similarity-transformed Hamiltonian of CCSD, $\bar{H}^{(2)}$.

In particular, the one-, two-, and three-body components of $\bar{H}^{(2)}$ enter the Epstein-Nesbet-like denominator for triple excitations which defines the $\ell^{abc}_{\mu,ijk}$ amplitudes in the CR-EOMCC(2,3),D approach. In variant A of CR-EOMCC(2,3), abbreviated as CR-EOMCC(2,3),A and equivalent to the EOM-CC(2)PT(2) method of Ref. [90], one replaces the Epstein-Nesbet-like denominator defining the $\ell^{abc}_{\mu,ijk}$ amplitudes, which in variant D of CR-EOMCC(2,3) is calculated as $[\omega^{(CCSD)}_{\mu} - (\langle\Phi^{abc}_{ijk}|\bar{H}^{(2)}_1|\Phi^{abc}_{ijk}\rangle + \langle\Phi^{abc}_{ijk}|\bar{H}^{(2)}_2|\Phi^{abc}_{ijk}\rangle + \langle\Phi^{abc}_{ijk}|\bar{H}^{(2)}_3|\Phi^{abc}_{ijk}\rangle)]$, where $\omega^{(CCSD)}_{\mu}$ is the EOMCCSD excitation energy and $\bar{H}^{(2)}_n$ is the n-body component of $\bar{H}^{(2)}$, by the simplified form of it which represents the Møller-Plesset-like denominator for triple excitations, $[\omega^{(CCSD)}_{\mu} - (\varepsilon_a + \varepsilon_b + \varepsilon_c - \varepsilon_i - \varepsilon_j - \varepsilon_k)]$. The differences between variants A and D are substantial, in favor of CR-EOMCC(2,3),D, when the excited states of interest are dominated by two-electron transitions. When the excited states of interest are dominated by one-electron transitions, as is the case in the present work, where we examine the $\pi \to \pi^*$ excitations in cis-7HQ and its complexes, the CR-EOMCC(2,3),A and CR-EOMCC(2,3),D approaches provide similar results (cf. Sect. 13.3.1). We refer the reader to the original Refs. [29, 31] for further details of the CR-EOMCC(2,3) approach and its variants A–D.

We now explain how to obtain the desired size-intensive δ-CR-EOMCC(2,3) results within the CR-EOMCC(2,3) framework. As demonstrated in Refs. [31, 32], although the ground-state CR-CC(2,3),D method and its CR-CC(2,3),A counterpart, which is equivalent to the CCSD(2)$_T$ approach of Ref. [101], are size extensive, being perfectly suited for examining weakly bound complexes involving larger molecules [113, 114], such as those studied in this work, their excited state CR-EOMCC(2,3),D and CR-EOMCC(2,3),A [or EOM-CC(2)PT(2)] analogs violate the property of size intensivity discussed in the Introduction and satisfied by EOMCCSD [16, 34]. Although the departure from strict size intensivity in the CR-EOMCC calculations of vertical and adiabatic excitation energies is in many cases of minor significance when compared to other sources of errors [82], this may be a more serious issue when examining the shifts in the excitation energy due to formation of weakly bound complexes. The lack of size intensivity of the CR-EOMCC(2,3) and EOM-CC(2)PT(2) approaches can be traced back to the presence of the contribution

$$\beta_\mu = \sum_{i<j<k,a<b<c} \left(r_{\mu,0}\ell^{abc}_{\mu,ijk} - \ell^{abc}_{0,ijk} \right) \mathfrak{M}^{ijk}_{0,abc}(2) \qquad (13.24)$$

in the corresponding vertical excitation energies

$$\omega^{(CR\text{-}EOMCC(2,3))}_{\mu} = E^{(CR\text{-}EOMCC(2,3))}_{\mu} - E^{(CR\text{-}CC(2,3))}_0, \qquad (13.25)$$

which is, as shown in Ref. [32], size extensive (cf. the $E^{(T0)}_p - E^{(T0)}_0$ term in Eq. 17 of Ref. [32]; see, also, Ref. [31] for additional remarks). Indeed, using the above equations for the CR-EOMCC(2,3) energies, particularly Eq. 13.23, we can decompose the CR-EOMCC(2,3) excitation energy $\omega_\mu^{(\text{CR-EOMCC}(2,3))}$ and its EOM-CC(2)PT(2) analog as follows [31, 32]:

$$\omega_\mu^{(\text{CR-EOMCC}(2,3))} = \omega_\mu^{(\text{CCSD})} + \alpha_\mu + \beta_\mu. \tag{13.26}$$

Here, $\omega_\mu^{(\text{CCSD})}$ is the vertical excitation energy of EOMCCSD,

$$\alpha_\mu = \sum_{i<j<k,a<b<c} \ell_{\mu,ijk}^{abc}\, \widetilde{\mathfrak{M}}_{\mu,abc}^{ijk}(2), \tag{13.27}$$

with $\widetilde{\mathfrak{M}}_{\mu,abc}^{ijk}(2) = \langle \Phi_{ijk}^{abc}|\bar{H}^{(2)}(R_{\mu,1}+R_{\mu,2})|\Phi\rangle$ representing the contribution to the triply excited moment $\mathfrak{M}_{\mu,abc}^{ijk}(2)$ of EOMCCSD due to the one- and two-body components of $R_\mu^{(2)}$, and β_μ is defined by Eq. 13.24. Since the EOMCCSD approach is size intensive and, as shown in Ref. [32], the α_μ term, Eq. 13.27, is size intensive as well (cf. the $E_p^{(T1)} + E_p^{(T2)}$ contribution in Eq. 17 of Ref. [32]), the $[\omega_\mu^{(\text{CCSD})} + \alpha_\mu(2,3)]$ part of the CR-EOMCC(2,3) excitation energy $\omega_\mu^{(\text{CR-EOMCC}(2,3))}$ is a size-intensive quantity. Unfortunately, the β_μ term defined by Eq. 13.24, being a size-extensive contribution, grows with the system size [31, 32], destroying the size intensivity of $\omega_\mu^{(\text{CR-EOMCC}(2,3))}$. In order to address this concern, in this work we have implemented the rigorously size-intensive variant of CR-EOMCC(2,3), designated as δ-CR-EOMCC(2,3), by neglecting the problematic β_μ term in Eq. 13.26 and redefining the vertical excitation energy in the following manner [31, 32]:

$$\omega_\mu^{(\delta\text{-CR-EOMCC}(2,3))} = \omega_\mu^{(\text{CCSD})} + \alpha_\mu, \tag{13.28}$$

with α_μ given by Eq. 13.27. The δ-CR-EOMCC(2,3) method provides a size-intensive description of the excitation energies and, by defining the total energy E_μ of a given electronic state μ as a sum of the size-extensive ground-state CR-CC(2,3) energy and size-intensive excitation energy $\omega_\mu^{(\delta\text{-CR-EOMCC}(2,3))}$, Eq. 13.28, so that

$$
\begin{aligned}
E_\mu &= E_0^{(\text{CR-CC}(2,3))} + \omega_\mu^{(\delta\text{-CR-EOMCC}(2,3))} \\
&= E_\mu^{(\text{CCSD})} + \sum_{i<j<k,a<b<c} \ell_{0,ijk}^{abc}\, \mathfrak{M}_{0,abc}^{ijk}(2) + \sum_{i<j<k,a<b<c} \ell_{\mu,ijk}^{abc}\, \widetilde{\mathfrak{M}}_{\mu,abc}^{ijk}(2),
\end{aligned}
$$

$$\tag{13.29}$$

the size-extensive description of state μ, assuming that the electronic excitation in AB is localized on either A or in B, but not on both fragments simultaneously (cf. Refs. [16, 34, 82] for a detailed analysis). As in the case of CR-EOMCC(2,3), we can distinguish between the full variant D of $\omega_\mu^{(\delta\text{-CR-EOMCC}(2,3))}$, designated as δ-CR-EOMCC(2,3),D, and its various approximations, including variant A. The δ-CR-EOMCC(2,3),A method is equivalent to the EOMCCSD(2)$_T$ approach of

Ref. [32] and, if we limit ourselves to the vertical excitation energies only, to the EOMCCSD($\tilde{\text{T}}$) approach of Ref. [33]. The latter connection is worth commenting on. The EOMCCSD($\tilde{\text{T}}$) method of Ref. [33] and its various EOMCCSD(T)-like analogs [33, 91] are based on the idea of directly correcting the EOMCCSD excitation energies rather than determining the total energies first and forming the excitation energies afterwards, as described above. This has an advantage in the fact that the resulting triples corrections, such as those defining EOMCCSD($\tilde{\text{T}}$), are immediately size intensive. Unfortunately, they are not robust enough in applications involving excited-state potential energy surfaces along bond breaking coordinates (see, e.g., Ref. [81]). The CR-EOMCC(2,3) approach, which is used to design the δ-CR-EOMCC(2,3) method, and its CR-EOMCCSD(T) predecessor [77, 78, 81], as well as their EOM-CC(2)PT(2)-based analogs [32], in which one corrects the total CCSD and EOMCCSD energies first and computes the excitation energies later, are very useful in calculations of ground- and excited-state potential energy surfaces (see Refs. [78, 81]; cf., also, Ref. [32]), but one has to take extra steps to eliminate terms violating the property of size intensivity that cannot be ignored in applications reported in this work. As shown in Sect. 13.3.1, the size-intensivity-corrected δ-CR-EOMCC(2,3),A and δ-CR-EOMCC(2,3),D methods provide very similar $\pi \rightarrow \pi^*$ excitation energies in the cis-7HQ chromophore and its complexes, which also are in good agreement with the experimental data reported in Refs. [5, 57].

13.2.3 The Remaining Computational Details

In order to examine the performance of the FDET approach and to demonstrate its advantages when compared with the supermolecular TDDFT calculations, both benchmarked against the high-level EOMCC data, we have investigated the shifts $\Delta\omega_{\pi\rightarrow\pi^*}$ in the vertical excitation energy $\omega_{\pi\rightarrow\pi^*}$ corresponding to the lowest $\pi \rightarrow \pi^*$ transition in the cis-7HQ chromophore induced by the formation of hydrogen-bonded complexes shown in Fig. 13.1. The eight complexes considered in this study, which were examined experimentally using the laser resonant two-photon UV spectroscopy [5, 57], include the cis-7HQ$\cdots B$ systems, where B represents one of the following environments: a single water molecule, a single ammonia molecule, a water dimer, a single molecule of methanol, a single molecule of formic acid, a trimer consisting of ammonia and two water molecules, a trimer consisting of ammonia, water, and ammonia, and a trimer consisting of two ammonia and one water molecules (cf. Fig. 13.1). For each cis-7HQ$\cdots B$ complex and for each electronic structure approach used in this study, the corresponding environment-induced shift $\Delta\omega_{\pi\rightarrow\pi^*}$ was determined as a difference between the value of $\omega_{\pi\rightarrow\pi^*}$ characterizing the complex and that obtained for the isolated cis-7HQ molecule. The relevant nuclear geometries of the cis-7HQ$\cdots B$ and cis-7HQ systems were optimized in the second-order Møller-Plesset perturbation theory (MP2) [115] calculations employing the aug-cc-pVTZ basis set [116, 117], using the analytic MP2 gradients available in Gaussian(R) 03 [118]. As in all other post-Hartree-Fock

Fig. 13.1 The
hydrogen-bonded complexes
of the cis-7HQ molecule
examined in the present study

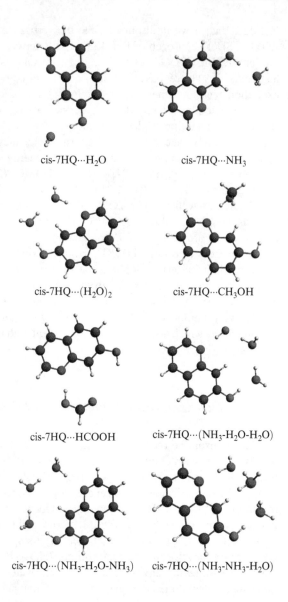

cis-7HQ···H$_2$O   cis-7HQ···NH$_3$

cis-7HQ···(H$_2$O)$_2$ cis-7HQ···CH$_3$OH

cis-7HQ···HCOOH   cis-7HQ···(NH$_3$-H$_2$O-H$_2$O)

cis-7HQ···(NH$_3$-H$_2$O-NH$_3$)   cis-7HQ···(NH$_3$-NH$_3$-H$_2$O)

wavefunction calculations discussed in this paper, the core molecular orbitals (MOs)
correlating with the 1s shells of the C, N, and O atoms were frozen in these
optimizations.

Once the nuclear geometries of the cis-7HQ and cis-7HQ···B systems were obtained, we performed the desired FDET and supermolecular TDDFT and EOMCC
calculations of the vertical excitation energies $\Delta\omega_{\pi\to\pi^*}$ and the complexation-
induced shifts $\Delta\omega_{\pi\to\pi^*}$. First, in order to establish the reference EOMCC values

of the environment-induced shifts $\Delta\omega_{\pi\to\pi^*}$, we carried out a series of EOMCCSD calculations for the cis-7HQ, 7HQ$\cdots$H$_2$O, and 7HQ$\cdots$NH$_3$ systems using five different basis sets, including 6-31+G(d) [119–121], 6-311+G(d) [121, 122], aug-cc-pVDZ [116, 117], and the [5s3p2d/3s2p] basis of Sadlej [123], designated as POL, followed by the complete set of EOMCCSD and δ-CR-EOMCC(2,3) computations using the largest basis sets we could afford for all eight cis-7HQ$\cdots$B complexes investigated in this work, which were 6-311+G(d) in the EOMCCSD case and 6-31+G(d) in the case of the δ-CR-EOMCC(2,3) approach. The main objective of these initial calculations was to determine the stability of the final EOMCC values of the $\Delta\omega_{\pi\to\pi^*}$ shifts recommended for the use in benchmarking the FDET and supermolecular TDDFT data with respect to the basis set choice and the role of the higher-order correlation effects neglected in EOMCCSD, but included in δ-CR-EOMCC(2,3). All of the EOMCC calculations reported in this work were carried out with the programs developed at Michigan State University described, for example, in Refs. [26, 31, 81, 82], that form part of the GAMESS package [124, 125]. In order to obtain the final δ-CR-EOMCC(2,3) results, as defined by Eqs. 13.28 and 13.29, we modified the previously developed [26, 31] CR-CC(2,3)/CR-EOMCC(2,3) GAMESS routines in a suitable manner. The corresponding ground-state CCSD calculations, which precede the determination of the left CCSD and right and left EOMCCSD eigenstates that enter the formulas for the triples corrections of δ-CR-EOMCC(2,3) and the steps needed to compute the triples corrections of the ground-state CR-CC(2,3) and excited-state CR-EOMCC(2,3) and δ-CR-EOMCC(2,3) approaches, were performed with the routines described in Ref. [126], which form part of GAMESS as well. The RHF orbitals were employed throughout and, as pointed out above, the core MOs that correlate with the 1s shells of the nonhydrogen atoms were frozen in the CCSD, EOMCCSD, and δ-CR-EOMCC(2,3) calculations. The CCSD/EOMCCSD energies were converged to 10^{-7} Hartree. Further details of the EOMCC computer codes and algorithms exploited in this work can be found in Refs. [26, 31, 81, 82].

Once the reference EOMCC data were established, we moved to the FDET and supermolecular TDDFT calculations, which were performed using the linear-response TDDFT routines available in the ADF2009.01 code [127]. The FDET calculations followed the protocol described in Ref. [56], in which the occupied and unoccupied orbitals of the embedded chromophore that are obtained by solving the Kohn-Sham-like system defined by Eq. 13.13 are subsequently used within the linear-response TDDFT framework [44] to obtain excitation energies. All of the FDET and supermolecular TDDFT calculations were performed using the STO ATZ2P basis set [127], which is a STO-type triple-zeta basis with two sets of polarization functions, augmented with one set of diffuse s-STO and p-STO functions. As shown in a separate study [60], the results of the FDET and supermolecular TDDFT calculations using the STO ATZ2P basis set can be viewed as converged with respect to the basis set choice. Because of the small energy differences that define the spectral shifts examined in this work, we used tight convergence criteria when solving the Kohn-Sham and linear-response TDDFT equations (10^{-10} Hartree).

The environment density ρ_B used in the FDET calculations reported in this work was nonrelaxed, i.e., we used the ground-state electronic density of the environment obtained by solving the conventional Kohn-Sham equations for the environment molecules in the absence of the chromophore. Moreover, we exploited the so-called monomer-expansion FDET technique, in which the orbitals of the chromophore A embedded in B and the corresponding density ρ_A are represented using the atomic centers of A, whereas the environment density ρ_B and the corresponding orbitals of B are represented using the atomic centers of B (see Refs. [58, 128] for further information). The monomer-expansion FDET technique using nonrelaxed ρ_B, which is the recommended variant of FDET for the type of applications reported in this work, relies on the approximation, referred to as the Neglect of Dynamic Response of the Environment (NDRE), in which we assume that the dynamic response of the whole system AB to the process of electronic excitation is limited to chromophore A and that the coupling between the excitations in the embedded system and in its environment can be neglected. The NDRE approximation and the monomer-expansion-based FDET approach that results from it are very effective in eliminating spurious electronic excitations involving the environment [50]. The effect of the relaxation of the electronic density of the environment ρ_B in the presence of the chromophore A and the usefulness of the monomer vs supermolecular basis expansions in the FDET calculations will be discussed in a separate study [60].

In both the FDET and supermolecular TDDFT calculations, we used the SAOP scheme [129] to approximate the relevant exchange-correlation potential contributions. To examine the possible dependence of the complexation-induced shifts in the excitation energy of the cis-7HQ chromophore on the approximations exploited for one of the nonelectrostatic components of the embedding potential $v_{emb}^{eff}[\rho_A, \rho_B; \mathbf{r}]$, Eq. 13.10, used in the FDET calculations, resulting from the nonadditivity of density functionals for the kinetic energy, the nonadditive kinetic energy potential $v_t^{nad}[\rho_A, \rho_B](\mathbf{r})$, Eq. 13.12, that forms part of $v_{emb}^{eff}[\rho_A, \rho_B; \mathbf{r}]$ was determined using two different approximations, namely, GGA97 [58] and NDSD [59]. Let us recall that the latter approximant incorporates the exact conditions that are relevant for the proper behavior of $v_t^{nad}[\rho_A, \rho_B](\mathbf{r})$ in the vicinity of nuclei. The nonadditive exchange-correlation component of $v_{emb}^{eff}[\rho_A, \rho_B; \mathbf{r}]$, represented by the $v_{xc}^{nad}[\rho_A, \rho_B](\mathbf{r})$ potential, Eq. 13.11, was approximated using the Perdew-Wang (PW91) functional [130]. Other treatments of the exchange-correlation contributions in FDET and supermolecular TDDFT calculations for the cis-7HQ system and its hydrogen-bonded complexes will be examined elsewhere [60], where we will show that the spectral shifts obtained with the nonrelaxed FDET approach are almost insensitive to the choice of functionals used in the calculations.

13.3 Results and Discussion

The results of our FDET, supermolecular TDDFT, and EOMCC calculations for the shifts in the vertical excitation energy $\omega_{\pi \to \pi^*}$ corresponding to the lowest $\pi \to \pi^*$ transition in the cis-7HQ chromophore induced by the formation of the eight complexes shown in Fig. 13.1 are summarized in Tables 13.1 and 13.2, and Fig. 13.2. We begin our discussion with the analysis of the EOMCCSD and δ-CR-EOMCC(2,3) calculations aimed at establishing the reference EOMCC values.

13.3.1 Reference EOMCC Results

In order to determine the level of EOMCC theory that would be suitable for serving as a reference for the FDET and supermolecular TDDFT calculations, we first examine the dependence of the environment-induced shifts $\Delta\omega_{\pi\to\pi^*}$ resulting from the EOMCCSD calculations on the basis set. We first compare the EOMCCSD results for the two smallest complexes, $7HQ\cdots H_2O$ and $7HQ\cdots NH_3$, for which we could afford the largest number of computations, including the 6-31+G(d), 6-311+G(d), aug-cc-pVDZ, and POL basis sets. The results in Table 13.1 indicate that although the vertical excitation energies $\omega_{\pi\to\pi^*}$ in the bare cis-7HQ system and its complexes with the water and ammonia molecules vary with the basis set (for the four basis sets tested here by as much as about $600\,\text{cm}^{-1}$), the environment-induced shifts $\Delta\omega_{\pi\to\pi^*}$ are almost insensitive to the basis set choice. Although we could not perform a similarly thorough analysis for the remaining six complexes due to prohibitive computer costs of the EOMCCSD calculations with the aug-cc-pVDZ and POL basis sets, we were able to obtain the EOMCCSD $\omega_{\pi\to\pi^*}$ and $\Delta\omega_{\pi\to\pi^*}$ values for all eight complexes examined in this study using the 6-31+G(d) and 6-311+G(d) bases. As shown in Table 13.1, the differences between the EOMCCSD/6-31+G(d) and EOMCCSD/6-311+G(d) values of the $\Delta\omega_{\pi\to\pi^*}$ shifts remain small for all complexes of interest, ranging from $43\,\text{cm}^{-1}$ in the $7HQ\cdots NH_3$ case to $43\,\text{cm}^{-1}$ in the case of $7HQ\cdots(H_2O)_2$, or 1–3%. We conclude that the choice of the basis set, although important for obtaining the converged $\omega_{\pi\to\pi^*}$ values, is of almost no importance when the environment-induced shifts $\Delta\omega_{\pi\to\pi^*}$ are considered.

Although the EOMCCSD approach is known to provide an accurate description of excited states dominated by one-electron transitions, such as the $\pi\to\pi^*$ transition in cis-7HQ and its complexes, there have been cases of similar states reported in the literature, where the EOMCCSD level has not been sufficient to obtain high-quality results [84, 85]. Moreover, the small energy differences defining the environment-induced shifts $\Delta\omega_{\pi\to\pi^*}$ may be sensitive to the higher-order correlation effects neglected in the EOMCCSD calculations. For these reasons, we also examined the effect of triple excitations on the $\omega_{\pi\to\pi^*}$ and $\Delta\omega_{\pi\to\pi^*}$ values by performing the δ-CR-EOMCC(2,3) calculations with the 6-31+G(d) basis set. As shown in Table 13.1, triple excitations have a significant effect on the vertical excitation energies $\omega_{\pi\to\pi^*}$, reducing the 4,000–5,000 cm^{-1} differences between the EOMCCSD and experimental data to no more than about $800\,\text{cm}^{-1}$, when the δ-CR-EOMCC(2,3),A/6-31+G(d) calculations are performed, and no more than about $500\,\text{cm}^{-1}$ when the δ-CR-EOMCC(2,3),D/6-31+G(d) approach is employed, while bringing the $\Delta\omega_{\pi\to\pi^*}$ values closer to the experimentally observed shifts

Table 13.1 The vertical excitation energies $\omega_{\pi\to\pi^*}$ and the environment-induced shifts $\Delta\omega_{\pi\to\pi^*}$ (in cm^{-1}) obtained with the EOMCCSD/6-31+G(d), EOMCCSD/6-311+G(d), δ-CR-EOMCC(2,3), δ-CR-EOMCC(2,3),A/6-31+G(d), and δ-CR-EOMCC(2,3),D/6-31+G(d) approaches, and their composite EOMCC,A and EOMCC,D analogs defined by Eq. 13.30, corresponding to the lowest $\pi\to\pi^*$ transition in the cis-7HQ chromophore and its various complexes

Environment	EOMCCSD/ 6-31+G(d)	EOMCCSD/ 6-311+G(d)	EOMCCSD/ aug-cc-pVDZ	EOMCCSD/ POL	δ-CR-EOMCCSD(2,3)/ 6-31+G(d)	δ-CR-EOMCC(2,3),A/ 6-31+G(d)	δ-CR-EOMCC(2,3),D/ 6-31+G(d)	EOMCC,A[a]	EOMCC,D[b]	Exp.[c]
$\omega_{\pi\to\pi^*}$										
None	35171	35046	34707	34596		31103	30711	30977	30586	30830
H_2O	34643	34500	34182	34077		30558	30199	30415	30056	30240
NH_3	34396	34263	33923	33819		30291	29922	30157	29788	29925
$2H_2O$	33867	33699				29700	29378	29532	29210	29193
CH_3OH	34830	34695				30717	30428	30582	30293	30363
$HCOOH$	34505	34371				30368	30056	30235	29922	29816
NH_3–H_2O–H_2O	33381	33218				29171	28863	29008	28701	28340
NH_3–H_2O–NH_3	33542	33385				29355	29036	29197	28879	28694
NH_3–NH_3–H_2O	33302	33136				29088	28812	28922	28646	28348
$\Delta\omega_{\pi\to\pi^*}$										
H_2O	−528	−546	−525	−519		−544	−512	−562	−530	−590
NH_3	−775	−783	−784	−777		−812	−789	−820	−797	−905
$2H_2O$	−1304	−1347				−1403	−1333	−1446	−1376	−1637
CH_3OH	−341	−351				−386	−283	−396	−292	−467
$HCOOH$	−666	−675				−734	−655	−743	−664	−1014
NH_3–H_2O–H_2O	−1790	−1828				−1932	−1847	−1969	−1885	−2490
NH_3–H_2O–NH_3	−1629	−1661				−1748	−1675	−1780	−1707	−2136
NH_3–NH_3–H_2O	−1869	−1910				−2014	−1899	−2055	−1940	−2482

[a] Defined by Eq. 13.30, in which variant A of CR-EOMCC(2,3) is employed

[b] Defined by Eq. 13.30, in which variant D of CR-EOMCC(2,3) is employed

[c] Obtained with the laser resonant two-photon ionization UV spectroscopy by examining the shifts in the origin of the $S_0 \to S_1$ absorption band [5]

Table 13.2 A comparison of the environment-induced shifts $\Delta\omega_{\pi\to\pi^*}$ (in cm^{-1}) of the vertical excitation energy corresponding to the lowest $\pi\to\pi^*$ transition in the cis-7HQ chromophore that result from the monomer-expansion-based FDET calculations employing the nonrelaxed environment densities ρ_B, the PW91 approximant for the nonadditive exchange-correlation contribution $v_{xc}^{nad}[\rho_A,\rho_B](\mathbf{r})$ to the embedding potential $v_{emb}^{eff}[\rho_A,\rho_B;\mathbf{r}]$, and the GGA97 and NDSD approximants for the nonadditive kinetic energy contribution $v_t^{nad}[\rho_A,\rho_B](\mathbf{r})$ to $v_{emb}^{eff}[\rho_A,\rho_B;\mathbf{r}]$, with the results of the supermolecular TDDFT calculations (all using the SAOP approximant for the exchange-correlation potential and the STO ATZ2P basis set) and the reference EOMCC,A data

| | Supermolecular | | FDET | |
Environment	EOMCC,A	TDDFT	GGA97	NDSD
H$_2$O	−562	−944	−645	−669
NH$_3$	−820	−1222	−816	−849
2H$_2$O	−1446	−2280	−1624	−1648
CH$_3$OH	−396	−805	−454	−439
HCOOH	−743	−1569	−972	−952
NH$_3$–H$_2$O–H$_2$O	−1969	−2838	−1863	−1876
NH$_3$–H$_2$O–NH$_3$	−1780	−2594	−1791	−1811
NH$_3$–NH$_3$–H$_2$O	−2055	−2899	−1890	−1931
Aver. dev. from EOMCC,A	0	−673	−36	−51
Aver. abs. dev. from EOMCC,A	0	673	104	105

when compared with the EOMCCSD data. Although the differences between the δ-CR-EOMCC$(2,3)$ and EOMCCSD values of the environment-induced shifts $\Delta\omega_{\pi\to\pi^*}$ resulting from the calculations with the 6-31+G(d) basis set do not exceed 15–16% of the EOMCCSD values, triples corrections improve the EOMCCSD results and, as such, are useful for generating the reference EOMCC data.

It would be great if we could perform the δ-CR-EOMCC$(2,3)$ calculations using basis sets larger than 6-31+G(d), such as 6-311+G(d), but the hydrogen-bonded complexes of cis-7HQ examined in this study are too large for performing such calculations on our computers. In the absence of the δ-CR-EOMCC$(2,3)$ larger basis set data and considering the fact that the triples corrections to the $\Delta\omega_{\pi\to\pi^*}$ shifts are relatively small when compared to their EOMCCSD values, we have decided to combine the EOMCCSD/6-311+G(d) results with the triples corrections to the EOMCCSD energies extracted from the δ-CR-EOMCC$(2,3),X$/6-31+G(d) ($X=$ A, D) calculations, using the formula

$$\omega_{\pi\to\pi^*}(\text{EOMCC},X) = \omega_{\pi\to\pi^*}(\text{EOMCCSD/6-311+G(d)})$$
$$+\,[\omega_{\pi\to\pi^*}(\delta\text{-CR-EOMCC}(2,3),X/\text{6-31+G(d)})$$
$$-\,\omega_{\pi\to\pi^*}(\text{EOMCCSD/6-31+G(d)})]\,, \qquad (13.30)$$

where $X=$ A or D. As shown in Table 13.1, the resulting composite EOMCC,A and EOMCC,D approaches provide vertical excitation energies $\omega_{\pi\to\pi^*}$ that are in excellent agreement with the experimental excitation energies, while offering further improvements in the environment-induced shifts $\Delta\omega_{\pi\to\pi^*}$ when compared with the EOMCCSD/6-311+G(d) and δ-CR-EOMCC$(2,3)$/6-31+G(d) calculations. Indeed,

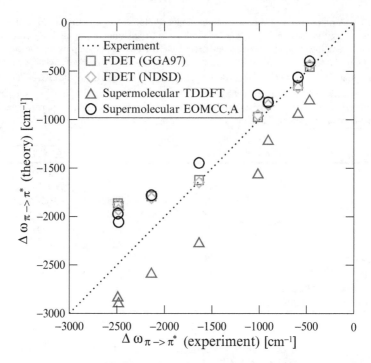

Fig. 13.2 A comparison of the environment-induced shifts $\Delta\omega_{\pi\to\pi^*}$ of the vertical excitation energy corresponding to the lowest $\pi \to \pi^*$ transition in the cis-7HQ chromophore and resulting from the monomer-expansion-based FDET/GGA97 and FDET/NDSD calculations with nonrelaxed-ρ_B, and the supermolecular TDDFT calculations (all using the STO ATZ2P basis set), with the reference supermolecular EOMCC,A data and the experimental shifts in the origin of the $S_0 \to S_1$ absorption band induced by complex formation [5]

the EOMCC,A approach, which adds the triples correction extracted from the δ-CR-EOMCC(2,3),A/6-31+G(d) calculation to the EOMCCSD/6-311+G(d) energy, gives errors in the calculated excitation energies $\omega_{\pi\to\pi^*}$ relative to experiment that range between $147\,\mathrm{cm}^{-1}$ in the case of the bare cis-7HQ system and $668\,\mathrm{cm}^{-1}$ in the case of the $7\mathrm{HQ}\cdots(\mathrm{NH_3\text{-}H_2O\text{-}H_2O})$ complex, never exceeding 2% of the experimental excitation energies. The EOMCC,D approach, which adds the triples correction obtained in the δ-CR-EOMCC(2,3),D/6-31+G(d) calculation to the EOMCCSD/6-311+G(d) energy, gives errors in the calculated $\omega_{\pi\to\pi^*}$ values relative to experiment that range between $17\,\mathrm{cm}^{-1}$ in the case of the $7\mathrm{HQ}\cdots(\mathrm{H_2O})_2$ complex and $361\,\mathrm{cm}^{-1}$ for $7\mathrm{HQ}\cdots(\mathrm{NH_3\text{-}H_2O\text{-}H_2O})$, or no more than 1% of the experimental values. These results should be compared to the much larger differences between the EOMCCSD/6-311+G(d) and experimental excitation energies that range between 14 and 17%. The complexation-induced spectral shifts $\Delta\omega_{\pi\to\pi^*}$ resulting from the EOMCC,A and EOMCC,D calculations agree with their experimental counterparts to within 5–27% or 15% on average in the case of EOMCC,A and 10–37% or 22% on average in the EOMCC,D case. In other words, the EOMCC,D approach, while bringing the excitation energies $\omega_{\pi \to \pi^*}$ to a much closer agreement with experiment than the EOMCCSD/6-311+G(d) calculations, does not offer improvements in the calculated $\Delta \omega_{\pi \to \pi^*}$ values. The composite EOMCC,A approach provides additional small improvements in the calculated $\Delta \omega_{\pi \to \pi^*}$ shifts, reducing the 7–33% errors relative to experiment resulting from the EOMCCSD/6-311+G(d) calculations to 5–27%. Based on these observations, we consider the EOMCC,A values of the spectral shifts $\Delta \omega_{\pi \to \pi^*}$ as the theoretical reference values for assessing the quality of the FDET and supermolecular TDDFT calculations, although the use of EOMCC,D would not change any of our main conclusions. Clearly, a comparison of the purely electronic EOMCC and experimental data discussed above has limitations, since we would have to investigate the effect of nuclear motion on the EOMCCSD and δ-CR-EOMCC$(2,3)$ excitation energies and use basis sets larger than 6-31+G(d) in the δ-CR-EOMCC$(2,3)$ calculations to make more definitive statements. We believe, however, that the EOMCC,A results obtained by combining the EOMCCSD/6-311+G(d) excitation energies with the triples corrections extracted from the δ-CR-EOMCC$(2,3)$,A/6-31+G(d) calculations, as in Eq. 13.30, are of sufficiently high quality to allow us to assess the quality of the FDET and supermolecular TDDFT results in applications involving the environment-induced spectral shifts in complexes of cis-7HQ, which are discussed next.

13.3.2 A Comparison of the Excitation Energy Shifts From the FDET and Supermolecular TDDFT Calculations with the Reference EOMCC Data

In agreement with the experimental data reported in Ref. [5], the excitation energy shifts $\Delta \omega_{\pi \to \pi^*}$ for the hydrogen-bonded $7HQ \cdots B$ complexes investigated in this work resulting from the EOMCC calculations are always negative and the magnitude of $\Delta \omega_{\pi \to \pi^*}$ correlates, to a large extent, with the size of the hydrogen-bonded environment B in the cis-$7HQ \cdots B$ complex (see Table 13.1). In particular, according to the reference EOMCC,A calculations, the shifts in the vertical excitation energy $\omega_{\pi \to \pi^*}$ corresponding to the lowest $\pi \to \pi^*$ transition in the cis-7HQ chromophore vary from $(-396)–(-820) \text{cm}^{-1}$ in the case of the smaller $7HQ \cdots B$ complexes involving the CH_3OH, H_2O, $HCOOH$, and NH_3 monomers, through $-1,446 \text{cm}^{-1}$ in the case of the $7HQ \cdots (H_2O)_2$ complex, to $(-1,780)–(-2,055) \text{cm}^{-1}$ in the case of the largest $7HQ \cdots B$ systems involving the $(NH_3 - H_2O - NH_3)$, $(NH_3 - H_2O - H_2O)$, and $(NH_3 - NH_3 - H_2O)$ trimers. It is interesting to examine how well the FDET and supermolecular TDDFT calculations reproduce these data.

As shown in Table 13.2 and Fig. 13.2, the overall agreement of the monomer-expansion-based FDET/ATZ2P data employing the nonrelaxed environment densities ρ_B with the reference EOMCC,A results is excellent, independent of the approximant used to determine the nonadditive kinetic energy potential $v_t^{nad}[\rho_A, \rho_B](\mathbf{r})$

that forms part of the embedding potential $v_{emb}^{eff}[\rho_A, \rho_B; \mathbf{r}]$. The absolute values of the deviations between the $\Delta\omega_{\pi\to\pi^*}$ values resulting from the FDET calculations employing the GGA97 approximant to represent $v_t^{nad}[\rho_A, \rho_B](\mathbf{r})$ and the corresponding supermolecular EOMCC,A calculations range from $4\,cm^{-1}$ in the case of the $7HQ\cdots NH_3$ complex, where the EOMCC,A shift is $-820\,cm^{-1}$, to $229\,cm^{-1}$ in the case of the $7HQ\cdots HCOOH$ system, where the EOMCC,A result for $\Delta\omega_{\pi\to\pi^*}$ is $-743\,cm^{-1}$. The mean signed and unsigned errors in the FDET results for the environment-induced shifts $\Delta\omega_{\pi\to\pi^*}$ obtained with the $v_t^{nad}[\rho_A, \rho_B](\mathbf{r})$ potential approximated with GGA97 relative to the EOMCC,A reference data are -36 and $104\,cm^{-1}$, respectively, or 11%, if we average the individual relative errors. The use of the NDSD approximant to represent $v_t^{nad}[\rho_A, \rho_B](\mathbf{r})$ has virtually no effect on the FDET $\Delta\omega_{\pi\to\pi^*}$ values, producing errors relative to EOMCC,A that range, in absolute value, between $29\,cm^{-1}$ for the $7HQ\cdots NH_3$ complex and $209\,cm^{-1}$ for the $7HQ\cdots HCOOH$ system, or the mean signed and unsigned errors of -51 and $105\,cm^{-1}$, respectively (again, 11%, if we average the individual relative errors). Based on the analysis of the EOMCCSD and δ-CR-EOMCC$(2,3)$ calculations presented in Sect. 13.3.1, the deviations between the nonrelaxed, monomer-expansion-based FDET/ATZ2P and reference EOMCC,A data shown in Table 13.2 and Fig. 13.2 are well within the accuracy of the EOMCC calculations, independent of the approximant used to represent the nonadditive kinetic energy potential $v_t^{nad}[\rho_A, \rho_B](\mathbf{r})$.

The performance of the FDET method has to be contrasted with the supermolecular TDDFT calculations, which are a lot less accurate than their FDET counterparts, when both types of calculations are compared with the reference EOMCC,A $\Delta\omega_{\pi\to\pi^*}$ values. Indeed, as shown in Table 13.2 (see, also, Fig. 13.2), the differences between the $\Delta\omega_{\pi\to\pi^*}$ shift values obtained in the supermolecular TDDFT calculations using the STO ATZ2P basis set and their reference EOMCC,A counterparts range, in absolute value, from $382\,cm^{-1}$ in the case of the $7HQ\cdots H_2O$ complex, where the EOMCC,A shift is $-562\,cm^{-1}$, to $869\,cm^{-1}$ in the case of the $7HQ\cdots(NH_3\text{-}H_2O\text{-}H_2O)$ system, where the EOMCC,A $\Delta\omega_{\pi\to\pi^*}$ value is $-1,969\,cm^{-1}$. The mean unsigned error in the supermolecular TDDFT/ATZ2P values of the $\Delta\omega_{\pi\to\pi^*}$ shifts relative to EOMCC,A is $673\,cm^{-1}$ or, if we average the individual relative errors, 65%. Obviously, these are much larger differences when compared with the corresponding nonrelaxed, monomer-expansion-based FDET calculations that give the 4–$229\,cm^{-1}$ deviations from the EOMCC,A data when the GGA97 approximant is used to represent the nonadditive kinetic energy potential $v_t^{nad}[\rho_A, \rho_B](\mathbf{r})$ and the 29–$209\,cm^{-1}$ deviations from the EOMCC,A $\Delta\omega_{\pi\to\pi^*}$ values when the NDSD approximant is employed to construct $v_t^{nad}[\rho_A, \rho_B](\mathbf{r})$. The 65% average relative error characterizing the $\Delta\omega_{\pi\to\pi^*}$ shift values resulting from the supermolecular TDDFT/ATZ2P calculations is six times larger than the analogous error characterizing the FDET calculations. Based on the analysis of the EOMCC calculations presented in Sect. 13.3.1, the differences between the supermolecular TDDFT and reference EOMCC,A data are well outside the accuracy of the EOMCC calculations for the $\Delta\omega_{\pi\to\pi^*}$ shifts, indicating the poor performance of the supermolecular TDDFT approach. Unlike in the FDET case, the differences between the supermolecular TDDFT and EOMCC,A $\Delta\omega_{\pi\to\pi^*}$ values increase with the size of the environment bound to the cis-7HQ chromophore. This indicates the difficulties with obtaining the balanced description of excitation energies in systems that have different sizes in the supermolecular TDDFT calculations, which are not present in the FDET and EOMCC calculations. Based on our numerical results, it is not entirely unlikely that the supermolecular TDDFT approach is not size intensive. All of the EOMCC approximations employed in this work, including the composite EOMCC,A approach, are rigorously size intensive. The FDET methodology offers a size intensive description of the complexation-induced spectral shifts by design.

13.3.3   A Comparison of the FDET and Supermolecular TDDFT Excitation Energy Shifts with the Experimental Data

Finally, it is instructive to comment on the quality of the shifts resulting from the FDET and supermolecular TDDFT calculations in the context of the available experimental information [5]. In analogy to the EOMCC,A results discussed in Sect. 13.3.1, a comparison of the purely electronic FDET or supermolecular TDDFT and experimental data has limitations, since one cannot measure vertical excitation energies obtained in the FDET and supermolecular TDDFT calculations in a direct manner. The experimental shifts reported in Ref. [5] correspond to the complexation-induced shifts in the position of the 0^0_0 $\pi\to\pi^*$ absorption band in the cis-7HQ chromophore. Thus, although the experimental shifts obtained in Ref. [5] are closely related to the theoretical shifts obtained in this study, the two types of quantities differ because of the following factors: (1) the geometry relaxation in the excited states of the cis-7HQ and cis-7HQ$\cdots B$ systems when compared to the corresponding ground electronic states, and (2) the MP2/aug-cc-pVTZ geometries of cis-7HQ and its complexes employed in this work, although probably reasonable, are not the experimental geometries. All of these factors certainly contribute to the deviations between the theoretical shifts calculated in this study and their experimental counterparts reported in Ref. [5]. On the other hand, the careful EOMCC calculations reported in this work which, as pointed out in Sect. 13.3.1, closely follow the experimental excitation energies corresponding to the lowest $\pi\to\pi^*$ transition in the cis-7HQ and cis-7HQ$\cdots B$ systems, particularly when the EOMCC,A and EOMCC,D approaches corrected for triple excitations are employed, indicate that the above factors, although important, lead to a relatively small overall effect. It is, therefore, interesting to compare our FDET and supermolecular TDDFT results for the excitation energy shifts $\Delta\omega_{\pi\to\pi^*}$, given in Table 13.2, with the experimentally derived shifts reported in Ref. [5] and listed in Table 13.1.

This comparison is shown in Fig. 13.2. As one can see by inspecting Tables 13.1 and 13.2, and Fig. 13.2, the $\Delta\omega_{\pi\to\pi^*}$ values obtained in the nonrelaxed, monomer-expansion-based FDET/ATZ2P calculations are in good agreement with the shifts in the experimental UV absorption spectra. The mean unsigned error in the $\Delta\omega_{\pi\to\pi^*}$ values resulting from the FDET calculations employing the GGA97 approximant to represent the nonadditive kinetic energy potential $v_t^{nad}[\rho_A,\rho_B](\mathbf{r})$, relative to the spectral shifts observed in experiment, is $222\,\mathrm{cm}^{-1}$, in excellent agreement with the EOMCC,A approach which gives $244\,\mathrm{cm}^{-1}$. The analogous mean unsigned error resulting from the FDET calculations employing the NDSD approximant to represent $v_t^{nad}[\rho_A,\rho_B](\mathbf{r})$ is very similar ($216\,\mathrm{cm}^{-1}$). For comparison, the mean unsigned error in the $\Delta\omega_{\pi\to\pi^*}$ values resulting from the supermolecular TDDFT/ATZ2P calculations, relative to the experimental spectral shifts, is twice as large ($429\,\mathrm{cm}^{-1}$), demonstrating once again the advantages of the embedding vs supermolecular strategy within the TDDFT framework. It is very encouraging to observe that the FDET approach, which can be applied to large molecular systems, is capable of providing shifts in the excitation energy corresponding to the lowest $\pi\to\pi^*$ transition in the cis-7HQ system due to formation of hydrogen-bonded complexes that can compete with the results of the considerably more expensive EOMCC calculations and that are a lot better than the supermolecular TDDFT results.

13.4   Summary and Concluding Remarks

We used the embedding FDET approach to determine the shifts in the excitation energy corresponding to the lowest $\pi\to\pi^*$ transition in cis-7-hydroxyquinoline (cis-7HQ), induced by the formation of hydrogen-bonded complexes of cis-7HQ with a number of small molecules, and compared the resulting shift values with the reference EOMCC data and the analogous shifts obtained in the conventional supermolecular TDDFT calculations. The main difference between the embedding strategy exploited in the FDET formalism and the conventional supermolecular approach is in the fact that in the former case one evaluates the excitation energy shifts induced by the interactions of the chromophore with its molecular environment as the differences of the excitation energies of the same many-electron system, representing the chromophore fragment with two different effective potentials, whereas in the latter case one has to perform calculations for two systems that differ in the number of electrons, the complex formed by the chromophore and its molecular environment and the isolated chromophore.

By considering eight complexes of cis-7HQ with up to three small hydrogen-bonded molecules, we demonstrated that the spectral shifts resulting from the FDET calculations with the nonrelaxed environment densities are in excellent agreement with the reference EOMCC data obtained in the size-intensive EOMCC calculations with singles, doubles, and noniterative triples, whereas the analogous shifts obtained with the supermolecular TDDFT approach are far from those obtained with EOMCC. The nonrelaxed FDET calculations provide shifts that agree with their EOMCC analogs to within about $100\,\mathrm{cm}^{-1}$ or 10% on average, where the absolute values of the excitation energy shifts in the complexes of cis-7HQ examined in this study resulting from the EOMCC calculations range between about 500 and 2,000 cm^{-1}. As shown in the present study, the accuracy of the FDET shift calculations employing nonrelaxed environment densities is on the order of the accuracy of the high-level EOMCC calculations. This should be contrasted with the excitation energy shifts obtained with the supermolecular TDDFT approach, which differ from the reference EOMCC values by about 700 cm^{-1} or 65% on average and which are well outside the accuracy of the EOMCC calculations. We demonstrated that none of the above findings are affected by the type of approximant used to represent the non-additive kinetic energy potential $v_t^{nad}[\rho_A, \rho_B](\mathbf{r})$ that forms part of the local, orbital-free, embedding potential $v_{emb}^{eff}[\rho_A, \rho_B; \mathbf{r}]$ employed in the FDET calculations. Two such approximants, GGA97 and NDSD, were examined, giving the virtually identical FDET spectral shift values.

Although the main focus of the present study was the comparison of the FDET and supermolecular TDDFT results for the complexation-induced shifts in the excitation energy corresponding to the lowest $\pi \to \pi^*$ transition in cis-7HQ with the EOMCC data, we also compared the FDET, supermolecular TDDFT, and reference EOMCC shift values with the experimental shifts reported in Ref. [5]. Although such a comparison has limitations due to the neglect of the effect of nuclear motion on photoabsorption spectra in purely electronic calculations performed in this work, the spectral shifts obtained with the FDET approach using nonrelaxed environment densities and those obtained with the EOMCC methodology agree with the experimental shifts quite well, whereas the supermolecular TDDFT calculations produce very large errors. This reinforces the superiority of the FDET strategy when compared with the conventional supermolecular TDDFT approach in applications involving complexation-induced spectral shifts.

Acknowledgements This work has been supported by the U.S. Department of Energy (Grant No. DE-FG02-01ER15228; P.P.) and Fonds National Suisse de la Recherche Scientifique (Grant No. 200020-134791; T.A.W.).

References

1. Tanner C, Manca C, Leutwyler S (2003) Science 302:1736
2. Bruhwiler D, Calzaferri G, Torres T, Ramm JH, Gartmann N, Dieu LQ, Lopez-Duarte I, Martinez-Diaz MV (2009) J Mater Chem 19:8040
3. Hernandez FE, Yu S, Garcia M, Campiglia AD (2005) J Phys Chem B 109:9499
4. Goldberg JM, Batjargal S, Petersson EJ (2010) J Am Chem Soc 132:14719
5. Thut M, Tanner C, Steinlin A, Leutwyler S (2008) J Phys Chem A 112:5566
6. Emrich K (1981) Nucl Phys A 351:379
7. Geertsen J, Rittby M, Bartlett RJ (1989) Chem Phys Lett 164:57
8. Comeau DC, Bartlett RJ (1993) Chem Phys Lett 207:414
9. Stanton JF, Bartlett RJ (1993) J Chem Phys 98:7029
10. Piecuch P, Bartlett RJ (1999) Adv Quantum Chem 34:295
11. Monkhorst H (1977) Int J Quantum Chem Symp 11:421
12. Dalgaard E, Monkhorst H (1983) Phys Rev A 28:1217

13. Mukherjee D, Mukherjee PK (1979) Chem Phys 39:325
14. Takahashi M, Paldus J (1986) J Chem Phys 85:1486
15. Koch H, Jørgensen P (1990) J Chem Phys 93:3333
16. Koch H, Jensen HJA, Jørgensen P, Helgaker T (1990) J Chem Phys 93:3345
17. Coester F (1958) Nucl Phys 7:421
18. Coester F, Kümmel H (1960) Nucl Phys 17:477
19. Čížek J (1966) J Chem Phys 45:4256
20. Čížek J (1969) Adv Chem Phys 14:35
21. Čížek J, Paldus J (1971) Int J Quantum Chem 5:359
22. Paldus J, Shavitt I, Čížek J (1972) Phys Rev A 5:50
23. Gauss J (1998) In: Schleyer PVR, Allinger NL, Clark T, Gasteiger J, Kollman PA, Schaefer HF III, Schreiner PR (eds) Encyclopedia of computational chemistry, vol 1. Wiley, Chichester, pp 615–636
24. Paldus J, Li X (1999) Adv Chem Phys 110:1
25. Bartlett RJ, Musiał M (2007) Rev Mod Phys 79:291
26. Piecuch P, Włoch M (2005) J Chem Phys 123:224105
27. Piecuch P, Włoch M, Gour JR, Kinal A (2006) Chem Phys Lett 418:467
28. Włoch M, Gour JR, Piecuch P (2007) J Phys Chem A 111:11359
29. Włoch M, Lodriguito MD, Piecuch P, Gour JR (2006) Mol Phys 104:2149
30. Piecuch P, Włoch M, Lodriguito M, Gour JR (2006) In: Wilson S, Julien JP, Maruani J, Brändas E, Delgado-Barrio G (eds) Progress in theoretical chemistry and physics, vol 15. Springer, Dordrecht, pp 45–106
31. Piecuch P, Gour JR, Włoch M (2009) Int J Quantum Chem 109:3268
32. Shiozaki T, Hirao K, Hirata S (2007) J Chem Phys 126:244106
33. Watts JD, Bartlett RJ (1996) Chem Phys Lett 258:581
34. Meissner L, Bartlett RJ (1995) J Chem Phys 102:7490
35. Korona T, Werner HJ (2003) J Chem Phys 118:3006
36. Korona T, Schütz M (2006) J Chem Phys 125:104106
37. Kats D, Korona T, Schütz M (2007) J Chem Phys 127:064107
38. Crawford TD, King RA (2002) Chem Phys Lett 366:611
39. Fan PD, Valiev M, Kowalski K (2008) Chem Phys Lett 458:205
40. Valiev M, Kowalski K (2006) J Chem Phys 125:211101
41. Valiev M, Kowalski K (2006) J Phys Chem A 110:13106
42. Kowalski K, Valiev M (2008) J Phys Chem A 112:5538
43. Epifanovsky E, Kowalski K, Fan PD, Valiev M, Matsika S, Krylov AI (2008) J Phys Chem A 112:9983
44. Casida ME (1995) In: Chong DP (ed) Recent advances in density-functional methods, part-i. World Scientific, Singapore, pp 155–192
45. Wesołowski TA, Warshel A (1993) J Phys Chem 97:8050
46. Wesołowski TA (2006) In: Leszczyński J (ed) Computational chemistry: reviews of current trends, vol 10. World Scientific, Singapore, pp 1–82
47. Wesolowski TA (2008) Phys Rev A 77:012504
48. Pernal K, Wesolowski TA (2009) Int J Quantum Chem 109:2520
49. Wesolowski TA, Warshel A (1994) J Phys Chem 98:5183
50. Neugebauer J, Louwerse MJ, Baerends EJ, Wesolowski TA (2005) J Chem Phys 122:094115
51. Kaminski JW, Gusarov S, Kovalenko A, Wesolowski TA (2010) J Phys Chem A 114:6082
52. Savin A, Wesolowski TA (2009) In: Piecuch P, Maruani J, Delgado-Barrio G, Wilson S (eds) Progress in theoretical chemistry and physics, vol 19. Springer, Dordrecht, pp 327–339
53. Roncero O, de Lara-Castells M, Villarreal P, Flores F, Ortega J, Paniagua M, Aguado A (2008) J Chem Phys 129:184104
54. Fux S, Jacob C, Neugebauer J, Visscher L, Reiher M (2010) J Chem Phys 132:164101
55. Goodpaster JD, Ananth N, Manby FR, Miller TF III (2010) J Chem Phys 133:084103
56. Wesolowski T (2004) J Am Chem Soc 126:11444
57. Fradelos G, Kaminski JW, Wesolowski TA, Leutwyler S (2009) J Phys Chem A 19:9766

58. Wesolowski TA, Chermette H, Weber J (1996) J Chem Phys 105:9182
59. Lastra JMG, Kaminski JW, Wesolowski TA (2008) J Chem Phys 129:074107
60. Fradelos G, Lutz JJ, Wesołowski TA, Piecuch P, Włoch M (2011) J Chem Theory Comput 7:1647
61. Domcke W, Sobolewski AL (2003) Science 302:1963
62. Stefanovich EV, Truong TN (1996) J Chem Phys 104:2946
63. Govind N, Wang YA, Carter EA (1999) J Chem Phys 110:7677
64. Neugebauer J, Jacob CR, Wesolowski TA, Baerends EJ (2005) J Phys Chem A 109:7805
65. Hodak M, Lu W, Bernholc J (2008) J Chem Phys 128:014101
66. Gomes ASP, Jacob CR, Visscher L (2008) Phys Chem Chem Phys 10:5353
67. Kohn W, Sham LJ (1965) Phys Rev 140:A1133
68. Hohenberg P, Kohn W (1964) Phys Rev 136:B864
69. Cortona P (1991) Phys Rev B 44:8454
70. Senatore G, Subbaswamy K (1986) Phys Rev B 34:5754
71. Elliott P, Cohen MH, Wasserman A, Burke K (2009) J Chem Theory Comput 5:827
72. Jacob CJ, Neugebauer J, Jensen L, Visscher L (2006) Phys Chem Chem Phys 8:2349
73. Wesolowski TA (1999) Chem Phys Lett 311:87
74. Neugebauer J, Louwerse MJ, Belanzoni P, Wesolowski TA, Baerends EJ (2005) J Chem Phys 123:114101
75. Zbiri M, Atanasov M, Daul C, Garcia-Lastra JM, Wesolowski TA (2004) Chem Phys Lett 397:441
76. Jacob CR, Visscher L (2006) J Chem Phys 125:194104
77. Piecuch P, Kowalski K, Pimienta ISO, McGuire MJ (2002) Int Rev Phys Chem 21:527
78. Piecuch P, Kowalski K, Pimienta ISO, Fan PD, Lodriguito M, McGuire MJ, Kucharski SA, Kuś T, Musiał M (2004) Theor Chem Acc 112:349
79. Kowalski K, Piecuch P (2001) J Chem Phys 115:2966
80. Kowalski K, Piecuch P (2002) J Chem Phys 116:7411
81. Kowalski K, Piecuch P (2004) J Chem Phys 120:1715
82. Włoch M, Gour JR, Kowalski K, Piecuch P (2005) J Chem Phys 122:214107
83. Kowalski K, Hirata S, Włoch M, Piecuch P, Windus TL (2005) J Chem Phys 123:074319
84. Coussan S, Ferro Y, Trivella A, Roubin P, Wieczorek R, Manca C, Piecuch P, Kowal-ski K, Włoch M, Kucharski SA, Musiał M (2006) J Phys Chem A 110:3920
85. Kowalski K, Krishnamoorthy S, Villa O, Hammond JR, Govind N (2010) J Chem Phys 132:154103
86. Kowalski K, Piecuch P (2001) J Chem Phys 115:643
87. Kowalski K, Piecuch P (2001) Chem Phys Lett 347:237
88. Kucharski SA, Włoch M, Musiał M, Bartlett RJ (2001) J Chem Phys 115:8263
89. Kowalski K, Piecuch P (2000) J Chem Phys 113:8490
90. Hirata S, Nooijen M, Grabowski I, Bartlett RJ (2001) J Chem Phys 114:3919. 115, 3967 (2001) [Erratum]
91. Watts JD, Bartlett RJ (1995) Chem Phys Lett 233:81
92. Christiansen O, Koch H, Jørgensen P (1996) J Chem Phys 105:1451
93. Christiansen O, Koch H, Jørgensen P, Olsen J (1996) Chem Phys Lett 256:185
94. Koch H, Christiansen O, Jørgensen P, Olsen J (1995) Chem Phys Lett 244:75
95. Christiansen O, Koch H, Jørgensen P (1995) J Chem Phys 103:7429
96. Kowalski K (2009) J Chem Phys 130:194110
97. Raghavachari K, Trucks GW, Pople JA, Head-Gordon M (1989) Chem Phys Lett 102:479
98. Urban M, Noga J, Cole SJ, Bartlett RJ (1985) J Chem Phys 83:4041
99. Piecuch P, Kowalski K (2000) In: Leszczyński J (ed) Computational chemistry: reviews of current trends, vol 5. World Scientific, Singapore, pp 1–104
100. Kowalski K, Piecuch P (2000) J Chem Phys 113:18
101. Hirata S, Fan PD, Auer AA, Nooijen M, Piecuch P (2004) J Chem Phys 121:12197
102. Gwaltney SR, Head-Gordon M (2000) Chem Phys Lett 323:21
103. Gwaltney SR, Head-Gordon M (2001) J Chem Phys 115:2014

104. Kucharski SA, Bartlett RJ (1998) J Chem Phys 108:5243
105. Taube AG, Bartlett RJ (2008) J Chem Phys 128:044110
106. Taube AG, Bartlett RJ (2008) J Chem Phys 128:044111
107. Stanton JF (1997) Chem Phys Lett 281:130
108. Crawford TD, Stanton JF (1998) Int J Quant Chem 70:601
109. Kowalski K, Piecuch P (2005) J Chem Phys 122:074107
110. Piecuch P, Kowalski K, Fan PD, Pimienta ISO (2003) In: Maruani J, Lefebvre R, Brändas E
     (eds) Progress in theoretical chemistry and physics, vol 12. Kluwer, Dordrecht, pp 119–206
111. Piecuch P, Włoch M, Varandas AJC (2007) In: Lahmar S, Maruani J, Wilson S,
     Delgado-Barrio G (eds) Progress in theoretical chemistry and physics, vol 16. Springer,
     Dordrecht, pp 63–121
112. Lodriguito M, Piecuch P (2008) In: Wilson S, Grout P, Maruani J, Delgado-Barrio G,
     Piecuch P (eds) Progress in theoretical chemistry and physics, vol 18. Springer, Dordrecht,
     pp 67–174
113. Li W, Gour JR, Piecuch P, Li S (2009) J Chem Phys 131:114109
114. Li W, Piecuch P (2010) J Phys Chem A 114:6721
115. Møller C, Plesset MS (1934) Phys Rev 46:618
116. Dunning TH Jr (1989) J Chem Phys 90:1007
117. Kendall RA, Dunning TH Jr, Harrison RJ (1992) J Chem Phys 96:6796
118. Frisch MJ, Trucks GW, Schlegel HB, Scuseria GE, Robb MA, Cheeseman JR,
     Montgomery JA Jr, Vreven T, Kudin KN, Burant JC et al (2003) Gaussian 03, revision B.03.
     Gaussian, Inc, Pittsburgh PA
119. Hehre WJ, Ditchfield R, Pople JA (1972) J Chem Phys 56:2257
120. Hariharan PC, Pople JA (1973) Theor Chim Acta 28:213
121. Clark T, Chandrasekhar J, Spitznagel GW, Schleyer PVR (1983) J Comput Chem 4:294
122. Krishnan R, Binkley JS, Seeger R, Pople JA (1980) J Chem Phys 72:650
123. Sadlej AJ (1988) Coll Czech Chem Commun 53:1995
124. Schmidt MW, Baldridge KK, Boatz JA, Elbert ST, Gordon MS, Jensen JH, Koseki S,
     Matsunaga N, Nguyen KA, Su SJ, Windus TL, Dupuis M, Montgomery JA (1993) J Comput
     Chem 14:1347
125. Schmidt MW, Baldridge KK, Boatz JA, Elbert ST, Gordon MS, Jensen JH, Koseki S,
     Matsunaga N, Nguyen KA, Su SJ, Windus TL, Dupuis M, Montgomery JA (1993) J Comput
     Chem 14:1347
126. Piecuch P, Kucharski SA, Kowalski K, Musiał M (2002) Comp Phys Commun 149:71
127. ADF2009 suite of programs. Theoretical Chemistry Department, Vrije Universiteit, Amster-
     dam. http://www.scm.com
128. Wesolowski TA, Weber J (1996) Chem Phys Lett 248:71
129. Gritsenko OV, Schipper PRT, Baerends EJ (1999) Chem Phys Lett 302:199
130. Perdew JP, Chevary JA, Vosko SH, Jackson KA, Pederson MR, Singh DJ, Fiolhais C (1993)
     Phys Rev B 48:4978

Chapter 14
Multiparticle Distribution of Fermi Gas System in Any Dimension

Shigenori Tanaka

Abstract The multiparticle distribution functions and density matrices for ideal Fermi gas system in the ground state are calculated for any spatial dimension. The results are expressed as determinant forms, in which a correlation kernel plays a vital role. The expression obtained for the one-dimensional Fermi gas is essentially equivalent to that observed for the eigenvalue distribution of random unitary matrices, and thus to that conjectured for the distribution of the non-trivial zeros of the Riemann zeta function. Their implications are discussed briefly.

14.1 Introduction

The distribution and correlation functions of fermion systems have long been studied theoretically since the birth of quantum mechanics. In 1933, Wigner and Seitz [1] derived an expression for the pair distribution function in the three-dimensional free fermion system, which provided a deep insight into the structure of interparticle correlations in Fermi systems. Most of theoreticians and experimenters, however, are usually interested only in the pair correlations, probably because detailed investigations concerning the higher-order, many-body correlations are formidable tasks in experiments, theories and simulations. As seen below in this article, we nevertheless find that systematic studies on many-body distribution functions sometimes yield useful and interesting information about the equivalence of underlying mathematical structures between the Fermi and other seemingly different systems.

We hereafter consider the noninteracting, ideal Fermi gas system in any dimension d with the particle number N and the volume V (taken to be unity for notational

S. Tanaka (✉)
Department of Computational Science, Graduate School of System Informatics,
Kobe University, Japan
e-mail: tanaka2@kobe-u.ac.jp

P.E. Hoggan et al. (eds.), *Advances in the Theory of Quantum Systems in Chemistry and Physics*, Progress in Theoretical Chemistry and Physics 22,
DOI 10.1007/978-94-007-2076-3_14, © Springer Science+Business Media B.V. 2012

simplicity) in the ground state (at zero temperature). We investigate the n-particle distribution function defined by

$$g^{(n)}\left(\mathbf{R}_1, \mathbf{R}_2, \cdots, \mathbf{R}_n\right) = \left\langle \frac{1}{N^n} \sum_{i_1, i_2, \cdots, i_n=1}^{N} \delta\left(\mathbf{R}_1 - \mathbf{r}_{i_1}\right) \delta\left(\mathbf{R}_2 - \mathbf{r}_{i_2}\right) \cdots \times \delta\left(\mathbf{R}_n - \mathbf{r}_{i_n}\right) \right\rangle$$

$$(14.1)$$

in the limit of $N \to \infty$. Here $\mathbf{r}_i$ refers to the spatial coordinate of the i-th Fermi particle and $\delta(\mathbf{r})$ represents the Dirac delta function; $\langle \rangle$ means the statistical average over the ground-state ensemble. In the following we consider the case that all the coordinates $\mathbf{R}_j$ ($1 \leq j \leq n \leq N$) in the distribution function are different from each other. Further, for simplicity, we confine ourselves to the case of the spin-polarized (ferromagnetic) system, while the extension to the spin-unpolarized (paramagnetic) system is straightforward.

The expression for the pair ($n = 2$) distribution function in three ($d = 3$) dimension is well known [1, 2]. However, the general one for any n and d is much less known. Interestingly, the distribution of Fermi particles in one ($d = 1$) dimension has a mathematical structure similar to those found for the eigenvalues of the random matrices [3–5] and for the zeros of the Riemann zeta function [6,7], as shown below. In the following Sects. 14.2 and 14.3, explicit expressions for the pair and ternary distribution functions of the ideal Fermi gas system in any dimension are derived. We then find an expression for the n-particle distribution function as a determinant form in Sect. 14.4. Another representation for the multiparticle distribution for finite N is given in terms of density matrix in Sect. 14.5. The explicit formula for correlation kernel which plays an essential role for the description of the multiparticle correlations in the Fermi system is derived in Sect. 14.6. The relationship with the theories for the random matrices and the Riemann zeta function is addressed in Sect. 14.7.

14.2 Pair Distribution Function

First, we calculate the pair distribution function,

$$g^{(2)}\left(\mathbf{R}_1, \mathbf{R}_2\right) = \left\langle \frac{1}{N^2} \sum_{i,j=1}^{N} \delta\left(\mathbf{R}_1 - \mathbf{r}_i\right) \delta\left(\mathbf{R}_2 - \mathbf{r}_j\right) \right\rangle, \qquad (14.2)$$

for the ideal Fermi gas. Here, introducing $\mathbf{r}_A = \mathbf{R}_1 - \mathbf{R}_2$ and $\mathbf{r}_B = \mathbf{R}_2 - \mathbf{R}_1$, we make a symmetrization as

$$\delta\left(\mathbf{R}_1 - \mathbf{r}_i\right) \delta\left(\mathbf{R}_2 - \mathbf{r}_j\right) \longrightarrow \delta\left(\mathbf{r}_A + \mathbf{r}_j - \mathbf{r}_i\right) \delta\left(\mathbf{r}_B + \mathbf{r}_i - \mathbf{r}_j\right). \qquad (14.3)$$

A symmetrized distribution function is then written as

$$f^{(2)}(\mathbf{r}_A, \mathbf{r}_B) = \left\langle \frac{1}{N^2} \sum_{i,j=1}^{N} \delta(\mathbf{r}_A + \mathbf{r}_j - \mathbf{r}_i)\, \delta(\mathbf{r}_B + \mathbf{r}_i - \mathbf{r}_j) \right\rangle. \tag{14.4}$$

Fourier transformation from the variables $\mathbf{r}_A$ and $\mathbf{r}_B$ to $\mathbf{k}_1$ and $\mathbf{k}_2$ gives a two-body structure factor as

$$S^{(2)}(\mathbf{k}_1, \mathbf{k}_2) = \int d\mathbf{r}_A \int d\mathbf{r}_B \exp(-i\mathbf{k}_1 \cdot \mathbf{r}_A - i\mathbf{k}_2 \cdot \mathbf{r}_B)\, f^{(2)}(\mathbf{r}_A, \mathbf{r}_B)$$

$$= \frac{1}{N^2} \left\langle \sum_{i,j=1}^{N} \exp[-i\mathbf{k}_1 \cdot (\mathbf{r}_i - \mathbf{r}_j) - i\mathbf{k}_2 \cdot (\mathbf{r}_j - \mathbf{r}_i)] \right\rangle$$

$$= \frac{1}{N^2} \left\langle \sum_{i=1}^{N} \exp[-i(\mathbf{k}_1 - \mathbf{k}_2) \cdot \mathbf{r}_i] \sum_{j=1}^{N} \exp[-i(\mathbf{k}_2 - \mathbf{k}_1) \cdot \mathbf{r}_j] \right\rangle. \tag{14.5}$$

In order to calculate the structure factor in the wavenumber space, we consider a density fluctuation operator [2],

$$\rho(\mathbf{r}) = \sum_{i=1}^{N} \delta(\mathbf{r} - \mathbf{r}_i), \tag{14.6}$$

and its Fourier transform,

$$\rho_{\mathbf{k}} = \int d\mathbf{r} \exp(-i\mathbf{k} \cdot \mathbf{r})\, \rho(\mathbf{r}) = \sum_{i=1}^{N} \exp(-i\mathbf{k} \cdot \mathbf{r}_i), \tag{14.7}$$

with a relation,

$$\rho(\mathbf{r}) = \sum_{\mathbf{k}} \rho_{\mathbf{k}} \exp(i\mathbf{k} \cdot \mathbf{r}). \tag{14.8}$$

Here, for the dimension d, the summation over $\mathbf{k}$ is represented by

$$\sum_{\mathbf{k}} = \frac{1}{(2\pi)^d} \int d\mathbf{k}, \tag{14.9}$$

and the volume V is taken to be unity in this article. The two-body structure factor is thus given by

$$S^{(2)}(\mathbf{k}_1, \mathbf{k}_2) = \frac{1}{N^2} \langle \rho_{\mathbf{k}_1 - \mathbf{k}_2} \rho_{\mathbf{k}_2 - \mathbf{k}_1} \rangle. \tag{14.10}$$

To proceed further, we consider the representation of second quantization [2]. The density fluctuation operator is expressed in terms of the creation operator $c_{\mathbf{k}}^{\dagger}$ and the annihilation operator $c_{\mathbf{k}}$ as

$$\rho_{\mathbf{k}} = \sum_{\mathbf{p}} c_{\mathbf{p}-\mathbf{k}}^{\dagger} c_{\mathbf{p}}, \tag{14.11}$$

where these operators satisfy the following anticommutation relations:

$$\left\{ c_{\mathbf{k}}, c_{\mathbf{p}}^{\dagger} \right\} = c_{\mathbf{k}} c_{\mathbf{p}}^{\dagger} + c_{\mathbf{p}}^{\dagger} c_{\mathbf{k}} = \delta_{\mathbf{k},\mathbf{p}}, \tag{14.12}$$

$$\left\{ c_{\mathbf{k}}, c_{\mathbf{p}} \right\} = 0, \tag{14.13}$$

$$\left\{ c_{\mathbf{k}}^{\dagger}, c_{\mathbf{p}}^{\dagger} \right\} = 0, \tag{14.14}$$

with the spin variables being omitted. The product of the density fluctuation operators in Eq. 14.10 is then given by

$$
\begin{aligned}
\rho_{\mathbf{k}_1-\mathbf{k}_2} \rho_{\mathbf{k}_2-\mathbf{k}_1} &= \sum_{\mathbf{p}} c_{\mathbf{p}-\mathbf{k}_1+\mathbf{k}_2}^{\dagger} c_{\mathbf{p}} \sum_{\mathbf{q}} c_{\mathbf{q}+\mathbf{k}_1-\mathbf{k}_2}^{\dagger} c_{\mathbf{q}} \\
&= -\sum_{\mathbf{p},\mathbf{q}} c_{\mathbf{p}-\mathbf{k}_1+\mathbf{k}_2}^{\dagger} c_{\mathbf{q}+\mathbf{k}_1-\mathbf{k}_2}^{\dagger} c_{\mathbf{p}} c_{\mathbf{q}} + \sum_{\mathbf{q}} c_{\mathbf{q}}^{\dagger} c_{\mathbf{q}}.
\end{aligned}
\tag{14.15}
$$

Accounting for the ground state of the noninteracting fermion system, the occupation number density in the wavenumber space,

$$n_{\mathbf{k}} = \left\langle c_{\mathbf{k}}^{\dagger} c_{\mathbf{k}} \right\rangle, \tag{14.16}$$

is expressed as

$$n_{\mathbf{k}} = \theta \left(k_{\mathrm{F}} - |\mathbf{k}| \right), \tag{14.17}$$

using the step function, $\theta(x)$, where k_{F} refers to the Fermi wavenumber. This occupation number density satisfies the normalization condition,

$$\sum_{\mathbf{k}} n_{\mathbf{k}} = \langle \rho_0 \rangle = N. \tag{14.18}$$

We thus have the two-body structure factor as

$$S^{(2)} \left(\mathbf{k}_1, \mathbf{k}_2 \right) = -\frac{1}{N^2} \left\langle \sum_{\mathbf{p},\mathbf{q}} c_{\mathbf{p}-\mathbf{k}_1+\mathbf{k}_2}^{\dagger} c_{\mathbf{q}+\mathbf{k}_1-\mathbf{k}_2}^{\dagger} c_{\mathbf{p}} c_{\mathbf{q}} \right\rangle + \frac{1}{N}. \tag{14.19}$$

We first consider the case of $\mathbf{k}_1 \neq \mathbf{k}_2$. The first term on the right-hand side in Eq. 14.19 survives when $\mathbf{p} = \mathbf{q} + \mathbf{k}_1 - \mathbf{k}_2$, and we thus find

$$S^{(2)}(\mathbf{k}_1, \mathbf{k}_2) = -\frac{1}{N^2} \sum_{\mathbf{p}} n_{\mathbf{p}+\mathbf{k}_1} n_{\mathbf{p}+\mathbf{k}_2} + \frac{1}{N}. \qquad (14.20)$$

On the other hand, when $\mathbf{k}_1 = \mathbf{k}_2$, $S^{(2)}(\mathbf{k}_1, \mathbf{k}_2) = 1$. Hence, the structure factor is expressed by

$$S^{(2)}(\mathbf{k}_1, \mathbf{k}_2) = \frac{1}{N}\left(1 - \frac{1}{N}\sum_{\mathbf{p}} n_{\mathbf{p}+\mathbf{k}_1} n_{\mathbf{p}+\mathbf{k}_2}\right)(1 - \delta_{\mathbf{k}_1,\mathbf{k}_2}) + \delta_{\mathbf{k}_1,\mathbf{k}_2}. \qquad (14.21)$$

To make the inverse Fourier transformation of Eq. 14.21, we note a convolution formula for the following Fourier-transform pair functions,

$$F_i(\mathbf{k}) = \int d\mathbf{r} \exp(-i\mathbf{k} \cdot \mathbf{r}) f_i(\mathbf{r}) \qquad (14.22)$$

and

$$f_i(\mathbf{r}) = \sum_{\mathbf{k}} \exp(i\mathbf{k} \cdot \mathbf{r}) F_i(\mathbf{k}). \qquad (14.23)$$

A formula,

$$\sum_{\mathbf{k}_1,\mathbf{k}_2,\cdots,\mathbf{k}_n} \exp(i\mathbf{k}_1 \cdot \mathbf{r}_1 + i\mathbf{k}_2 \cdot \mathbf{r}_2 + \cdots + i\mathbf{k}_n \cdot \mathbf{r}_n)$$

$$\times \sum_{\mathbf{k}_0} F_1(\mathbf{k}_0 + \mathbf{k}_1) F_2(\mathbf{k}_0 + \mathbf{k}_2) \cdots \times F_n(\mathbf{k}_0 + \mathbf{k}_n)$$

$$= \delta(\mathbf{r}_1 + \mathbf{r}_2 + \cdots + \mathbf{r}_n) f_1(\mathbf{r}_1) f_2(\mathbf{r}_2) \cdots \times f_n(\mathbf{r}_n), \qquad (14.24)$$

can be proved easily. We then obtain an explicit form for the symmetrized distribution function as

$$f^{(2)}(\mathbf{r}_A, \mathbf{r}_B) = \sum_{\mathbf{k}_1} \sum_{\mathbf{k}_2} \exp(i\mathbf{k}_1 \cdot \mathbf{r}_A + i\mathbf{k}_2 \cdot \mathbf{r}_B) S^{(2)}(\mathbf{k}_1, \mathbf{k}_2)$$

$$= \left[1 - \frac{1}{N^2}\tilde{n}(\mathbf{r}_A)\tilde{n}(\mathbf{r}_B)\right]\delta(\mathbf{r}_A + \mathbf{r}_B) + \frac{1}{N}\delta(\mathbf{r}_A)\delta(\mathbf{r}_B), \qquad (14.25)$$

where $\tilde{n}(\mathbf{r})$ is given by

$$\tilde{n}(\mathbf{r}) = \sum_{\mathbf{k}} n_{\mathbf{k}} \exp(i\mathbf{k} \cdot \mathbf{r}). \qquad (14.26)$$

Here, in passing, we calculate an explicit form of $\tilde{n}(\mathbf{r})$ for the three dimensional case ($d = 3$). The Fermi wavenumber and $\tilde{n}(\mathbf{r})$ are then given by

$$k_F = (6\pi^2 N)^{1/3} \qquad (14.27)$$

and

$$\tilde{n}(\mathbf{r}) = \frac{1}{(2\pi)^3} \int d\mathbf{k}\exp(i\mathbf{k}\cdot\mathbf{r})n_\mathbf{k}$$

$$= \frac{3N}{(k_F r)^3}(\sin k_F r - k_F r\cos k_F r). \qquad (14.28)$$

By defining the correlation kernel as

$$K(\mathbf{r}) = \frac{1}{N}\tilde{n}(\mathbf{r}) = \frac{1}{N}\sum_\mathbf{k}\exp(i\mathbf{k}\cdot\mathbf{r})n_\mathbf{k}, \qquad (14.29)$$

we find

$$K(\mathbf{r}) = \frac{3}{(k_F r)^3}(\sin k_F r - k_F r\cos k_F r) = \frac{3}{k_F r}j_1(k_F r), \qquad (14.30)$$

where $j_1(z)$ is the spherical Bessel function of order one.

Thus, we obtain the expression for the symmetrized distribution function as

$$f^{(2)}(\mathbf{r}_A,\mathbf{r}_B) = [1 - K(\mathbf{r}_A)K(\mathbf{r}_B)]\delta(\mathbf{r}_A+\mathbf{r}_B)+\frac{1}{N}\delta(\mathbf{r}_A)\delta(\mathbf{r}_B). \qquad (14.31)$$

The second term on the right-hand side in this expression refers to a singular part which represents the correlation of a particle with itself; it should therefore be neglected when $\mathbf{R}_1 \neq \mathbf{R}_2$ and $N \to \infty$. Further, eliminating the $\delta(\mathbf{r}_A+\mathbf{r}_B)$ factor which comes from the symmetrization (Eq. 14.3) for $\mathbf{r}_A = \mathbf{R}_1 - \mathbf{R}_2$ and $\mathbf{r}_B = \mathbf{R}_2 - \mathbf{R}_1$, we finally find an expression for the pair distribution function as

$$g^{(2)}(\mathbf{R}_1,\mathbf{R}_2) = 1 - K(\mathbf{R}_1-\mathbf{R}_2)K(\mathbf{R}_2-\mathbf{R}_1). \qquad (14.32)$$

The result for $d = 3$ with Eq. 14.30 is well known [1, 2, 8].

14.3 Ternary Distribution Function

Next, we calculate the ternary distribution function,

$$g^{(3)}(\mathbf{R}_1,\mathbf{R}_2,\mathbf{R}_3) = \left\langle \frac{1}{N^3}\sum_{i,j,k=1}^{N}\delta(\mathbf{R}_1-\mathbf{r}_i)\delta(\mathbf{R}_2-\mathbf{r}_j)\delta(\mathbf{R}_n-\mathbf{r}_k)\right\rangle. \qquad (14.33)$$

Again, after introducing $\mathbf{r}_A = \mathbf{R}_1 - \mathbf{R}_2$, $\mathbf{r}_B = \mathbf{R}_2 - \mathbf{R}_3$ and $\mathbf{r}_C = \mathbf{R}_3 - \mathbf{R}_1$, and making a symmetrization,

$$\delta(\mathbf{R}_1 - \mathbf{r}_i)\,\delta(\mathbf{R}_2 - \mathbf{r}_j)\,\delta(\mathbf{R}_3 - \mathbf{r}_k) \longrightarrow \delta(\mathbf{r}_A + \mathbf{r}_j - \mathbf{r}_i)\,\delta(\mathbf{r}_B + \mathbf{r}_k - \mathbf{r}_j)\,\delta(\mathbf{r}_C + \mathbf{r}_i - \mathbf{r}_k),$$
(14.34)

we start with an expression for three-body distribution function as

$$f^{(3)}(\mathbf{r}_A, \mathbf{r}_B, \mathbf{r}_C) = \left\langle \frac{1}{N^3} \sum_{i,j,k=1}^{N} \delta(\mathbf{r}_A + \mathbf{r}_j - \mathbf{r}_i)\,\delta(\mathbf{r}_B + \mathbf{r}_k - \mathbf{r}_j)\,\delta(\mathbf{r}_C + \mathbf{r}_i - \mathbf{r}_k) \right\rangle.$$
(14.35)

Carrying out the Fourier transformation from the variables $\mathbf{r}_A$, $\mathbf{r}_B$ and $\mathbf{r}_C$ to $\mathbf{k}_1$, $\mathbf{k}_2$ and $\mathbf{k}_3$, the three-body structure factor is expressed by the density fluctuation operators as

$$S^{(3)}(\mathbf{k}_1, \mathbf{k}_2, \mathbf{k}_3)$$
$$= \int d\mathbf{r}_A \int d\mathbf{r}_B \int d\mathbf{r}_C \exp(-i\mathbf{k}_1 \cdot \mathbf{r}_A - i\mathbf{k}_2 \cdot \mathbf{r}_B - i\mathbf{k}_3 \cdot \mathbf{r}_C)\, f^{(3)}(\mathbf{r}_A, \mathbf{r}_B, \mathbf{r}_C)$$
$$= \frac{1}{N^3} \left\langle \sum_{i=1}^{N} \exp[-i(\mathbf{k}_1 - \mathbf{k}_3) \cdot \mathbf{r}_i] \sum_{j=1}^{N} \exp[-i(\mathbf{k}_2 - \mathbf{k}_1) \cdot \mathbf{r}_j] \sum_{k=1}^{N} \exp[-i(\mathbf{k}_3 - \mathbf{k}_2) \cdot \mathbf{r}_k] \right\rangle$$
$$= \frac{1}{N^3} \left\langle \rho_{\mathbf{k}_1 - \mathbf{k}_3} \rho_{\mathbf{k}_2 - \mathbf{k}_1} \rho_{\mathbf{k}_3 - \mathbf{k}_2} \right\rangle.$$
(14.36)

The product of the density fluctuation operators is then expressed in terms of the creation and annihilation operators as

$$\rho_{\mathbf{k}_1 - \mathbf{k}_3} \rho_{\mathbf{k}_2 - \mathbf{k}_1} \rho_{\mathbf{k}_3 - \mathbf{k}_2}$$
$$= -\sum_{\mathbf{p},\mathbf{q},\mathbf{l}} c_{\mathbf{p}+\mathbf{k}_1-\mathbf{k}_2}^{\dagger} c_{\mathbf{q}+\mathbf{k}_2-\mathbf{k}_3}^{\dagger} c_{\mathbf{l}+\mathbf{k}_3-\mathbf{k}_1}^{\dagger} c_{\mathbf{p}} c_{\mathbf{q}} c_{\mathbf{l}}$$
$$- \sum_{\mathbf{p},\mathbf{q}} \left(c_{\mathbf{p}+\mathbf{k}_1-\mathbf{k}_2}^{\dagger} c_{\mathbf{q}+\mathbf{k}_2-\mathbf{k}_1}^{\dagger} c_{\mathbf{p}} c_{\mathbf{q}} + c_{\mathbf{p}+\mathbf{k}_2-\mathbf{k}_3}^{\dagger} c_{\mathbf{q}+\mathbf{k}_3-\mathbf{k}_2}^{\dagger} c_{\mathbf{p}} c_{\mathbf{q}} + c_{\mathbf{p}+\mathbf{k}_3-\mathbf{k}_1}^{\dagger} c_{\mathbf{q}+\mathbf{k}_1-\mathbf{k}_3}^{\dagger} c_{\mathbf{p}} c_{\mathbf{q}} \right)$$
$$+ \sum_{\mathbf{p}} c_{\mathbf{p}}^{\dagger} c_{\mathbf{p}},$$
(14.37)

using the anticommutation relations (14.12)–(14.14).

It is noted for the ground state of fermions that the terms for $\mathbf{q} = \mathbf{l} + \mathbf{k}_3 - \mathbf{k}_1$ and $\mathbf{l} = \mathbf{p} + \mathbf{k}_1 - \mathbf{k}_2$, or for $\mathbf{q} = \mathbf{p} + \mathbf{k}_1 - \mathbf{k}_2$ and $\mathbf{p} = \mathbf{l} + \mathbf{k}_3 - \mathbf{k}_1$ survive in the first summation of Eq. 14.37 when all the wavenumber vectors $\mathbf{k}_1$, $\mathbf{k}_2$ and $\mathbf{k}_3$ are different from each other. Analogous rules apply for the second summation of Eq. 14.37 as well. Thus, in this case the structure factor is written as

$$S^{(3)}(\mathbf{k}_1, \mathbf{k}_2, \mathbf{k}_3) = \frac{1}{N^3} \sum_{\mathbf{p}} \left(n_{\mathbf{p}+\mathbf{k}_1} n_{\mathbf{p}+\mathbf{k}_2} n_{\mathbf{p}+\mathbf{k}_3} + n_{\mathbf{p}-\mathbf{k}_1} n_{\mathbf{p}-\mathbf{k}_2} n_{\mathbf{p}-\mathbf{k}_3} \right)$$

$$- \frac{1}{N^3} \sum_{\mathbf{p}} \left(n_{\mathbf{p}+\mathbf{k}_1} n_{\mathbf{p}+\mathbf{k}_2} + n_{\mathbf{p}+\mathbf{k}_2} n_{\mathbf{p}+\mathbf{k}_3} + n_{\mathbf{p}+\mathbf{k}_3} n_{\mathbf{p}+\mathbf{k}_1} \right) + \frac{1}{N^2},$$

$$(14.38)$$

using the occupation number density $n_{\mathbf{k}}$. Next, when two wavenumber vectors of $\mathbf{k}_1$, $\mathbf{k}_2$ and $\mathbf{k}_3$ are identical, the three-body structure factor is given by

$$S^{(3)}(\mathbf{k}_1, \mathbf{k}_2, \mathbf{k}_3) = \frac{1}{N} \left(1 - \frac{1}{N} \sum_{\mathbf{p}} n_{\mathbf{p}+\mathbf{k}_3} n_{\mathbf{p}+\mathbf{k}_1} \right) \delta_{\mathbf{k}_1, \mathbf{k}_2}$$

$$+ \frac{1}{N} \left(1 - \frac{1}{N} \sum_{\mathbf{p}} n_{\mathbf{p}+\mathbf{k}_1} n_{\mathbf{p}+\mathbf{k}_2} \right) \delta_{\mathbf{k}_2, \mathbf{k}_3}$$

$$+ \frac{1}{N} \left(1 - \frac{1}{N} \sum_{\mathbf{p}} n_{\mathbf{p}+\mathbf{k}_2} n_{\mathbf{p}+\mathbf{k}_3} \right) \delta_{\mathbf{k}_3, \mathbf{k}_1}. \qquad (14.39)$$

Finally, in the case of $\mathbf{k}_1 = \mathbf{k}_2 = \mathbf{k}_3$, the structure factor is expressed as

$$S^{(3)}(\mathbf{k}_1, \mathbf{k}_2, \mathbf{k}_3) = \delta_{\mathbf{k}_1, \mathbf{k}_2} \delta_{\mathbf{k}_1, \mathbf{k}_3}. \qquad (14.40)$$

By the inverse Fourier transformation with the aid of Eq. 14.24, the three-body distribution function is thus given by

$$f^{(3)}(\mathbf{r}_A, \mathbf{r}_B, \mathbf{r}_C) = \sum_{\mathbf{k}_1} \sum_{\mathbf{k}_2} \sum_{\mathbf{k}_3} \exp\left(i\mathbf{k}_1 \cdot \mathbf{r}_A + i\mathbf{k}_2 \cdot \mathbf{r}_B + i\mathbf{k}_3 \cdot \mathbf{r}_C \right) S^{(3)}(\mathbf{k}_1, \mathbf{k}_2, \mathbf{k}_3)$$

$$= G(\mathbf{r}_A, \mathbf{r}_B, \mathbf{r}_C) \delta(\mathbf{r}_A + \mathbf{r}_B + \mathbf{r}_C) + O(1/N) \qquad (14.41)$$

with

$$G(\mathbf{r}_A, \mathbf{r}_B, \mathbf{r}_C) = 1 - K(\mathbf{r}_A) K(-\mathbf{r}_A) - K(\mathbf{r}_B) K(-\mathbf{r}_B) - K(\mathbf{r}_C) K(-\mathbf{r}_C)$$

$$+ K(\mathbf{r}_A) K(\mathbf{r}_B) K(\mathbf{r}_C) + K(-\mathbf{r}_A) K(-\mathbf{r}_B) K(-\mathbf{r}_C). \qquad (14.42)$$

In Eq. 14.41, the singular terms associated with the self-correlation are included in the higher-order contributions with respect to $1/N$. Removing the irrelevant $\delta(\mathbf{r}_A + \mathbf{r}_B + \mathbf{r}_C)$ factor that is ascribed to the symmetrization of coordinates, we find the form of the ternary distribution function of ideal Fermi gas as

$$g^{(3)}(\mathbf{R}_1, \mathbf{R}_2, \mathbf{R}_3)$$

$$= 1 - K(\mathbf{R}_1 - \mathbf{R}_2) K(\mathbf{R}_2 - \mathbf{R}_1) - K(\mathbf{R}_2 - \mathbf{R}_3) K(\mathbf{R}_3 - \mathbf{R}_2)$$

$$- K(\mathbf{R}_3 - \mathbf{R}_1) K(\mathbf{R}_1 - \mathbf{R}_3) + K(\mathbf{R}_1 - \mathbf{R}_2) K(\mathbf{R}_2 - \mathbf{R}_3) K(\mathbf{R}_3 - \mathbf{R}_1)$$

$$+ K(\mathbf{R}_2 - \mathbf{R}_1) K(\mathbf{R}_3 - \mathbf{R}_2) K(\mathbf{R}_1 - \mathbf{R}_3). \qquad (14.43)$$

14.4 Multiparticle Distribution Functions

By inspection of Eqs. 14.32 and 14.43, it is conjectured that the n-particle distribution function of ideal Fermi gas has a determinant form as

$$g^{(n)}(\mathbf{R}_1, \mathbf{R}_2, \cdots, \mathbf{R}_n) = \det \mathbf{K}^{(n)}. \tag{14.44}$$

Here, $\mathbf{K}^{(n)}$ is an $n \times n$ matrix whose component $(i, j = 1, 2, \cdots, n)$ is expressed as

$$\mathbf{K}^{(n)} = (K_{ij}), \tag{14.45}$$

$$K_{ij} = K(\mathbf{R}_i - \mathbf{R}_j), \tag{14.46}$$

and

$$K_{ii} = \lim_{\mathbf{R}_i \to \mathbf{R}_j} K(\mathbf{R}_i - \mathbf{R}_j) = 1, \tag{14.47}$$

using the correlation kernel (Eq. 14.29). In the following, it is shown that this conjecture is actually the case in the limit of $N \to \infty$.

In order to prove the formula (Eq. 14.44), we consider the N-body wave function of ideal Fermi gas in the ground state as a Slater determinant [1, 2, 8],

$$|0\rangle = \frac{1}{\sqrt{N!}} \det \mathbf{A}. \tag{14.48}$$

The component of the matrix is given by

$$\mathbf{A} = (a_{ij}) \tag{14.49}$$

and

$$a_{ij} = \exp(-i\mathbf{k}_i \cdot \mathbf{r}_j), \tag{14.50}$$

where $i, j = 1, 2, \cdots, N$ and the wavenumber vectors $\mathbf{k}_i$ $(i = 1, 2, \cdots, N)$ fill the Fermi sphere with the radius of k_F. The adjoint wave function is then introduced as

$$\langle 0| = \frac{1}{\sqrt{N!}} \det \mathbf{A}^\dagger \tag{14.51}$$

with

$$\mathbf{A}^\dagger = \left(a_{ij}^\dagger\right) \tag{14.52}$$

and

$$a_{ij}^\dagger = \exp(i\mathbf{k}_j \cdot \mathbf{r}_i). \tag{14.53}$$

The normalization condition,

$$\int d\mathbf{r}_1 \cdots d\mathbf{r}_N \langle 0|0\rangle = 1, \tag{14.54}$$

is shown by noting the relation,

$$\det \mathbf{A}^{\mathrm{T}} = \det \mathbf{A}, \tag{14.55}$$

for the transposed matrix $\mathbf{A}^{\mathrm{T}}$ and the definition of the determinant,

$$\det \mathbf{A} = \sum_{P^{(N)}} \left[\mathrm{sgn} P^{(N)} \right] a_{1p_1} a_{2p_2} \cdots \times a_{Np_N}, \tag{14.56}$$

where

$$P^{(N)} = \begin{pmatrix} 1 & 2 & \cdots & N \\ p_1 & p_2 & \cdots & p_N \end{pmatrix} \tag{14.57}$$

means the permutations of order N and

$$\mathrm{sgn} P^{(N)} = \begin{cases} +1, & \text{for even permutation} \\ -1, & \text{for odd permutation} \end{cases} \tag{14.58}$$

is their signature.

Next, we note a relation,

$$\langle 0|0 \rangle = \frac{1}{N!} \det \mathbf{A}^{\dagger} \det \mathbf{A} = \frac{1}{N!} \det \left(\mathbf{A}^{\dagger} \mathbf{A} \right), \tag{14.59}$$

where the component of the product of matrices is given by

$$\left(\mathbf{A}^{\dagger} \mathbf{A} \right)_{ij} = \sum_{k=1}^{N} a_{ik}^{\dagger} a_{kj} = \sum_{k=1}^{N} \exp \left(i \mathbf{k}_k \cdot \mathbf{r}_i \right) \exp \left(-i \mathbf{k}_k \cdot \mathbf{r}_j \right) = \sum_{k=1}^{N} \exp \left[i \mathbf{k}_k \cdot (\mathbf{r}_i - \mathbf{r}_j) \right]. \tag{14.60}$$

Then, N-body distribution function of N-particle Fermi gas system can be calculated as

$$g^{(N)} (\mathbf{R}_1, \mathbf{R}_2, \cdots, \mathbf{R}_N)$$

$$= \left\langle \frac{1}{N^N} \sum_{i_1, i_2, \cdots, i_N = 1}^{N} \delta (\mathbf{R}_1 - \mathbf{r}_{i_1}) \delta (\mathbf{R}_2 - \mathbf{r}_{i_2}) \cdots \times \delta (\mathbf{R}_N - \mathbf{r}_{i_N}) \right\rangle$$

$$= \int d\mathbf{r}_1 \cdots d\mathbf{r}_N \frac{1}{N!} \sum_{P^{(N)}} \left[\mathrm{sgn} P^{(N)} \right] \sum_{k_1=1}^{N} \exp \left[i \mathbf{k}_{k_1} \cdot (\mathbf{r}_1 - \mathbf{r}_{p_1}) \right]$$

$$\times \sum_{k_2=1}^{N} \exp \left[i \mathbf{k}_{k_2} \cdot (\mathbf{r}_2 - \mathbf{r}_{p_2}) \right]$$

$$\cdots \times \sum_{k_N=1}^{N} \exp \left[i \mathbf{k}_{k_N} \cdot (\mathbf{r}_N - \mathbf{r}_{p_N}) \right] \frac{1}{N^N} \sum_{i_1, i_2, \cdots, i_N = 1}^{N} \delta (\mathbf{R}_1 - \mathbf{r}_{i_1}) \delta (\mathbf{R}_2 - \mathbf{r}_{i_2})$$

$$\cdots \times \delta (\mathbf{R}_N - \mathbf{r}_{i_N})$$

$$= \sum_{P^{(N)}} \left[\mathrm{sgn} P^{(N)} \right] K_{1p_1} K_{2p_2} \cdots \times K_{Np_N}$$

$$= \det \mathbf{K}^{(N)} \tag{14.61}$$

when all the coordinates $\mathbf{R}_i$ are different from each other.

Now, let us assume that Eq. 14.44 holds for the case of n-particle distribution function. The $(n-1)$-particle distribution function is then calculated as

$$g^{(n-1)}(\mathbf{R}_1, \mathbf{R}_2, \cdots, \mathbf{R}_{n-1}) = \frac{N}{N-(n-1)} \int d\mathbf{R}_n \, g^{(n)}(\mathbf{R}_1, \mathbf{R}_2, \cdots, \mathbf{R}_n)$$

$$= \frac{N}{N-(n-1)} \int d\mathbf{R}_n \sum_{P^{(n)}} \left[\mathrm{sgn} P^{(n)} \right] K_{1p_1} K_{2p_2} \cdots \times K_{np_n}$$

$$\tag{14.62}$$

with the permutations of order n,

$$P^{(n)} = \begin{pmatrix} 1 & 2 & \cdots & n \\ p_1 & p_2 & \cdots & p_n \end{pmatrix}, \tag{14.63}$$

where the denominator $N-(n-1)$ in Eq. 14.62 comes from $[N-(n-1)]!/(N-n)!$.

The right-hand side of Eq. 14.62 can be calculated by considering the following three cases: First, in the case of $p_n = n$, it is expressed as

$$\frac{N}{N-(n-1)} \int d\mathbf{R}_n \sum_{P^{(n-1)}} \left[\mathrm{sgn} P^{(n-1)} \right] K_{1p_1} K_{2p_2} \cdots \times K_{n-1,p_{n-1}} K_{nn}$$

$$= \frac{N}{N-(n-1)} \sum_{P^{(n-1)}} \left[\mathrm{sgn} P^{(n-1)} \right] K_{1p_1} K_{2p_2} \cdots \times K_{n-1,p_{n-1}}$$

$$= \frac{N}{N-(n-1)} \det \mathbf{K}^{(n-1)}. \tag{14.64}$$

Next, in the case of $p_n \neq n$, we focus on a part,

$$\begin{pmatrix} m & n \\ n & p_n \end{pmatrix}, \tag{14.65}$$

in the permutation $P^{(n)}$. The number of ways of choosing m is $n-1$, and we consider the case of $m = n-1$ without the loss of generality. When $p_n = n-1$ in the transposition (Eq. 14.65), the right-hand side of Eq. 14.62 gives

$$-\frac{N(n-1)}{N-(n-1)} \int d\mathbf{R}_n \sum_{P^{(n-2)}} \left[\mathrm{sgn} P^{(n-2)} \right] K_{1p_1} K_{2p_2} \cdots \times K_{n-2,p_{n-2}} K_{n-1,n} K_{n,n-1}$$

$$= -\frac{N(n-1)}{N-(n-1)} \sum_{P^{(n-2)}} \left[\mathrm{sgn} P^{(n-2)} \right] K_{1p_1} K_{2p_2} \cdots \times K_{n-2,p_{n-2}}$$

$$\times \int d\mathbf{R}_n \frac{1}{N^2} \sum_{k_1=1}^{N} \exp\left[i\mathbf{k}_{k_1} \cdot (\mathbf{R}_{n-1} - \mathbf{R}_n)\right] \sum_{k_2=1}^{N} \exp\left[i\mathbf{k}_{k_2} \cdot (\mathbf{R}_n - \mathbf{R}_{n-1})\right]$$

$$= -\frac{N(n-1)}{N-(n-1)} \sum_{P^{(n-2)}} \left[\mathrm{sgn}P^{(n-2)}\right] K_{1p_1}K_{2p_2} \cdots \times K_{n-2,p_{n-2}} \times \frac{1}{N}$$

$$= -\frac{n-1}{N-(n-1)} \sum_{P^{(n-2)}} \left[\mathrm{sgn}P^{(n-2)}\right] K_{1p_1}K_{2p_2} \cdots \times K_{n-2,p_{n-2}}$$

$$= -\frac{n-1}{N-(n-1)} \sum_{P_a^{(n-1)}} \left[\mathrm{sgn}P_a^{(n-1)}\right] K_{1p_1}K_{2p_2} \cdots \times K_{n-2,p_{n-2}}K_{n-1,n-1}. \quad (14.66)$$

In Eq. 14.66, $P_a^{(n-1)}$ means the permutations of order $n-1$ under the constraint of $p_n \to p_{n-1} = n-1$. On the other hand, when $p_n \neq n-1$ (and $p_n \neq n$) in Eq. 14.65, the right-hand side of Eq. 14.62 gives

$$\frac{N(n-1)}{N-(n-1)} \int d\mathbf{R}_n \sum_{P^{(n-2)}} \left[\mathrm{sgn}P^{(n-2)}\right] K_{1p_1}K_{2p_2} \cdots \times K_{n-2,p_{n-2}}K_{n-1,n}K_{n,p_n}$$

$$= \frac{N(n-1)}{N-(n-1)} \sum_{P^{(n-2)}} \left[\mathrm{sgn}P^{(n-2)}\right] K_{1p_1}K_{2p_2} \cdots \times K_{n-2,p_{n-2}}$$

$$\times \int d\mathbf{R}_n \frac{1}{N^2} \sum_{k_1=1}^{N} \exp\left[i\mathbf{k}_{k_1} \cdot (\mathbf{R}_{n-1} - \mathbf{R}_n)\right] \sum_{k_2=1}^{N} \exp\left[i\mathbf{k}_{k_2} \cdot (\mathbf{R}_n - \mathbf{R}_{p_n})\right]$$

$$= \frac{N(n-1)}{N-(n-1)} \sum_{P^{(n-2)}} \left[\mathrm{sgn}P^{(n-2)}\right] K_{1p_1}K_{2p_2} \cdots \times K_{n-2,p_{n-2}}$$

$$\times \frac{1}{N^2} \sum_{k_1=1}^{N} \exp\left[i\mathbf{k}_{k_1} \cdot (\mathbf{R}_{n-1} - \mathbf{R}_{p_n})\right]$$

$$= \frac{n-1}{N-(n-1)} \sum_{P^{(n-2)}} \left[\mathrm{sgn}P^{(n-2)}\right] K_{1p_1}K_{2p_2} \cdots \times K_{n-2,p_{n-2}}K_{n-1,p_n}$$

$$= -\frac{n-1}{N-(n-1)} \sum_{P_b^{(n-1)}} \left[\mathrm{sgn}P_b^{(n-1)}\right] K_{1p_1}K_{2p_2} \cdots \times K_{n-2,p_{n-2}}K_{n-1,p_{n-1}}. \quad (14.67)$$

In Eq. 14.67, $P_b^{(n-1)}$ means the permutations of order $n-1$ under the constraint of $p_n \to p_{n-1} \neq n-1$.

Summing up the contributions of Eqs. 14.64, 14.66 and 14.67, we obtain

$$g^{(n-1)}(\mathbf{R}_1, \mathbf{R}_2, \cdots, \mathbf{R}_{n-1}) = \left[\frac{N}{N-(n-1)} - \frac{n-1}{N-(n-1)}\right] \det \mathbf{K}^{(n-1)}$$

$$= \det \mathbf{K}^{(n-1)}. \quad (14.68)$$

We can thus show that Eq. 14.44 holds also for the $(n-1)$-particle distribution function. Considering also the case of N-particle distribution function, Eq. 14.61, we see that Eq. 14.44 holds for $2 \leq n \leq N$. The case of $n = 1$ is trivial.

14.5 Description by Density Matrices

In this section we consider the N-body density matrix [8] for the Fermi system expressed in terms of the product of the Slater determinants with the plane waves [1, 2, 8] as

$$
\rho^{(N)}\left(\mathbf{r}_1,\mathbf{r}_2,\cdots,\mathbf{r}_N\right)
$$

$$
= \frac{1}{N!} \sum_{P^{(N)}} \left[\mathrm{sgn}P^{(N)}\right] \exp\left(i\mathbf{k}_1 \cdot \mathbf{r}_{p_1} + i\mathbf{k}_2 \cdot \mathbf{r}_{p_2} + \cdots + i\mathbf{k}_N \cdot \mathbf{r}_{p_N}\right)
$$

$$
\times \sum_{P'^{(N)}} \left[\mathrm{sgn}P'^{(N)}\right] \exp\left(-i\mathbf{k}_1 \cdot \mathbf{r}_{p'_1} - i\mathbf{k}_2 \cdot \mathbf{r}_{p'_2} - \cdots - i\mathbf{k}_N \cdot \mathbf{r}_{p'_N}\right)
$$

$$
= \frac{1}{N!} \sum_{P^{(N)}} \left[\mathrm{sgn}P^{(N)}\right] \exp\left(i\mathbf{k}_{p_1} \cdot \mathbf{r}_1 + i\mathbf{k}_{p_2} \cdot \mathbf{r}_2 + \cdots + i\mathbf{k}_{p_N} \cdot \mathbf{r}_N\right)
$$

$$
\times \sum_{P'^{(N)}} \left[\mathrm{sgn}P'^{(N)}\right] \exp\left(-i\mathbf{k}_{p'_1} \cdot \mathbf{r}_1 - i\mathbf{k}_{p'_2} \cdot \mathbf{r}_2 - \cdots - i\mathbf{k}_{p'_N} \cdot \mathbf{r}_N\right), \quad (14.69)
$$

where $\mathbf{r}_i$ and $\mathbf{k}_i$ refer to the spatial coordinates and the wavenumber vectors, respectively. The density matrix given by Eq. 14.69 satisfies the normalization condition as

$$
\int d\mathbf{r}_1 \cdots d\mathbf{r}_N \rho^{(N)}\left(\mathbf{r}_1,\mathbf{r}_2,\cdots,\mathbf{r}_N\right) = \frac{1}{N!} \sum_{P^{(N)}} \left[\mathrm{sgn}P^{(N)}\right]^2 = 1. \quad (14.70)
$$

Here, noting the identities for $N \times N$ matrices such as Eqs. 14.55 and 14.59, we find

$$
\rho^{(N)}\left(\mathbf{r}_1,\mathbf{r}_2,\cdots,\mathbf{r}_N\right) = \frac{N^N}{N!} \sum_{P^{(N)}} \left[\mathrm{sgn}P^{(N)}\right] K_{1p_1} K_{2p_2} \cdots \times K_{Np_N}
$$

$$
= \frac{N^N}{N!} \det \mathbf{K}^{(N)}, \quad (14.71)
$$

where

$$
\mathbf{K}^{(N)} = (K_{ij}) \quad (14.72)
$$

is an $N \times N$ matrix whose components are

$$K_{ij} = \frac{1}{N} \sum_{k=1}^{N} \exp\left[i\mathbf{k}_k \cdot (\mathbf{r}_i - \mathbf{r}_j)\right] \tag{14.73}$$

for $1 \leq i,j \leq N$.

The n-body ($1 \leq n \leq N$) density matrix [8] is then given by

$$\rho^{(n)}(\mathbf{r}_1,\mathbf{r}_2,\cdots,\mathbf{r}_n) = \int d\mathbf{r}_{n+1}\cdots d\mathbf{r}_N \rho^{(N)}(\mathbf{r}_1,\mathbf{r}_2,\cdots,\mathbf{r}_N)$$

$$= \frac{(N-n)!}{N!} N^n \sum_{P^{(n)}} \left[\mathrm{sgn}P^{(n)}\right] K_{1p_1} K_{2p_2} \cdots \times K_{np_n}$$

$$= \frac{(N-n)!}{N!} N^n \det \mathbf{K}^{(n)} \tag{14.74}$$

with an $n \times n$ matrix $\mathbf{K}^{(n)}$, as in Sect. 14.4. It is noted that the relations above hold for finite N and that the prefactor in front of $\det \mathbf{K}^{(n)}$ in Eq. 14.74 becomes unity for finite n and $N \to \infty$.

14.6 Correlation Kernel

Here we derive an explicit expression for the correlation kernel, Eq. 14.29, in any dimension d, which represents Eq. 14.73 in the limit of infinite N. Since the occupation number density satisfies the normalization condition, Eq. 14.18, the Fermi wavenumber k_F is given by

$$\frac{1}{(2\pi)^d} V_d(k_F) = N, \tag{14.75}$$

where

$$V_d(k_F) = C_d k_F^d \tag{14.76}$$

with

$$C_d = \frac{\pi^{d/2}}{\Gamma\left(\frac{d}{2}+1\right)} \tag{14.77}$$

is the volume of the d-dimensional sphere with the radius of k_F; $\Gamma(s)$ is the gamma function [9, 10]. The Fermi wavenumber is thus expressed as

$$k_F = 2\sqrt{\pi}\left[\Gamma\left(\frac{d}{2}+1\right)N\right]^{1/d}, \tag{14.78}$$

where it is noted that we employ the convention of $V = 1$, so that N represents the number density.

By choosing the direction of $\mathbf{r}$ in parallel with the d-th component of the wavenumber vector $\mathbf{k}$, the correlation kernel is expressed as

$$
\begin{aligned}
K(\mathbf{r}) &= \frac{1}{N(2\pi)^d} \int_{-1}^{1} dt\, k_F \exp(ik_F rt) V_{d-1}\left(k_F \sqrt{1-t^2}\right) \\
&= \frac{C_{d-1} k_F^d}{N(2\pi)^d} \int_{-1}^{1} dt \exp(ik_F rt)\left(1-t^2\right)^{\frac{d-1}{2}}
\end{aligned}
\tag{14.79}
$$

with $r = |\mathbf{r}|$. Then, employing Poisson's formula [9, 10],

$$
\int_{-1}^{1} dt \exp(izt)\left(1-t^2\right)^{\nu-\frac{1}{2}} = \frac{\sqrt{\pi}\Gamma\left(\nu+\frac{1}{2}\right)}{(z/2)^\nu} J_\nu(z),
\tag{14.80}
$$

we find an explicit expression for the correlation kernel as

$$
K(\mathbf{r};\nu) = \Gamma(\nu+1)\left(\frac{2}{k_F r}\right)^\nu J_\nu(k_F r)
\tag{14.81}
$$

with $\nu = d/2$, where $J_\nu(z)$ is the Bessel function of the first kind of order ν [9, 10]. It is remarked that, though Eq. 14.81 has been derived for integral values of spatial dimension d, $K(\mathbf{r};\nu)$ may be regarded as a continuous function of the auxiliary order variable ν, which follows from its series representation [9, 10].

Let us here consider the case of $d = 1$ and $\nu = 1/2$. We then find

$$
\begin{aligned}
K(r) &= \Gamma\left(\frac{3}{2}\right)\left(\frac{2}{k_F r}\right)^{1/2} J_{1/2}(k_F r) \\
&= j_0(k_F r) \\
&= \frac{\sin k_F r}{k_F r},
\end{aligned}
\tag{14.82}
$$

where

$$
j_n(z) = \left(\frac{\pi}{2z}\right)^{1/2} J_{n+\frac{1}{2}}(z)
\tag{14.83}
$$

is the spherical Bessel function of the first kind of order n [9, 10]. Recalling Eq. 14.44, this correlation structure is essentially the same as that for the eigenvalues of random matrices in the Gaussian unitary ensemble [3–5]. Interestingly, it has been known that this type of correlation structure may hold also for the distribution of zeros in the Riemann zeta function [6].

14.7 Similarity to Random Matrices and the Riemann Zeros

In the preceding sections we have shown that the n-particle distribution function of ideal Fermi gas is expressed in terms of a simple determinant form (Eq. 14.44). A very analogous finding has long been known in the theory for random matrix which was initially introduced to describe the statistical distribution of nuclear energy levels [11]. Let us represent the eigenvalues of random unitary matrices $U(N)$ as $\exp(i\theta_j)$ with $1 \le j \le N$ and $\theta_j \in \mathbf{R}$. For unfolded eigenphases defined by

$$\phi_j = \theta_j \frac{N}{2\pi}, \tag{14.84}$$

Dyson[3–5] showed in the limit of $N \to \infty$ that the n-point correlation function is expressed by

$$R^{(n)}(\phi_1, \cdots, \phi_n) = \det\left[S(\phi_j - \phi_k)\right]_{j,k=1,\cdots,n} \tag{14.85}$$

with

$$S(x) = \frac{\sin \pi x}{\pi x} \tag{14.86}$$

and $1 \le n \le N$. The expression (Eq. 14.85) apparently has a mathematical structure essentially equivalent to Eq. 14.44 in the one-dimensional ($d = 1$) case, Eq. 14.82, of ideal Fermi gas system.

Interestingly, it has been known that this type of correlation structure holds also for the distribution of zeros in the Riemann zeta function. The Riemann zeta function [12] for complex variable s is defined by

$$\zeta(s) = \sum_{n=1}^{\infty} \frac{1}{n^s} = \prod_p \left(1 - p^{-s}\right)^{-1} \tag{14.87}$$

for Re $s > 1$, where n and p mean the natural numbers and the prime numbers, respectively. After the analytic continuation over the whole complex plane, the $\zeta(s)$ has non-trivial zeros in the critical strip, $0 < \text{Re } s < 1$, and the Riemann hypothesis states that all of them lie on the critical line Re $s = 1/2$; that is,

$$\zeta\left(\frac{1}{2} + it\right) = 0 \tag{14.88}$$

has non-trivial solutions only when $t = t_j \in \mathbf{R}$.

The mean density of the non-trivial zeros of $\zeta(s)$ increases logarithmically with height t up to the critical line. We then define unfolded zeros by

$$w_j = \frac{t_j}{2\pi} \log \frac{t_j}{2\pi}. \tag{14.89}$$

Montgomery [6] showed, assuming the Riemann hypothesis, that the pair correlation function of the unfolded zeros for $j \to \infty$ has a form,

$$R^{(2)}(w) = 1 - \left(\frac{\sin \pi w}{\pi w} \right)^2, \qquad (14.90)$$

under a restriction on the correlation range, which is identical to that in the random matrix theory, Eq. 14.85, for $n = 2$. He also conjectured that this form applies without the restriction on the correlation range. This conjecture has been supported through substantial numerical calculations [13]. Further, the analysis by Montgomery was generalized for all the n-point correlations by Rudnick and Sarnak [14]. In addition, Bogomolny and Keating [15, 16] showed that the n-point correlation function of the Riemann zeros is equivalent to the corresponding result in the random matrix theory for the Gaussian unitary ensemble in an appropriate asymptotic limit with the aid of Hardy-Littlewood prime-correlation conjecture [17].

Thus, it is currently believed that the structure of distribution of the Riemann zeros is mathematically analogous to that of the eigenvalues of random matrices. This conjecture has activated many studies to investigate the mathematical structure of the zeta function in the light of that of the random matrices. For example, the connections between the random matrix theory and the theory for L-functions, which can be regarded as a broader class of functions including the zeta function, have been extensively studied [7, 18]. Further, their relationships with quantum chaos have also been discussed [19]. In the present analysis, on the other hand, an intimate analogy between the distribution function of ideal Fermi gas system and that in the random matrix theory has been demonstrated, especially in the one-dimensional ($d = 1$) case. Accordingly, further investigations on the Fermi gas systems and their extensions would be expected to provide more insights into the random matrices and the zeta functions, and vice versa. As for the extensions of the studies on the one-dimensional ideal Fermi gas in the ground state, those to other dimensions as addressed above are remarked first. Other examples would include the extensions to finite temperatures or excited states, the introduction of particle interactions, and the consideration of finite N systems.

14.8   Concluding Remarks

In this article we have addressed a general expression for the multiparticle distribution functions of ideal Fermi gas in the ground state. The expression is of a determinant form for any spatial dimension, and has a mathematical structure similar to that for the correlation function of the eigenvalues of random unitary matrices, especially in the one-dimensional case. Noting intimate relationships between the random matrix theory and the Riemann hypothesis, the present observation may provide deeper insights into the underlying mathematical structure that connects these analogous subjects. Further studies on the correlation properties of fermion systems would thus be expected to play important roles, not only for their own, but also for the deeper understandings of the statistical properties of number-theoretical zeta-related functions and random matrices.

Thus, it has been revealed that a correlation structure similar to those observed in the eigenvalues of the random unitary matrices and in the Riemann zeros is embedded in the Fermi gas system as well. The case of $v = d/2 \to 1/2$ in the Fermi gas system gives a special correlation structure already discussed in the random matrix and zeta function theories. Therefore, the behaviors of the multiparticle correlations of the Fermi gas system with the correlation kernel $K(\mathbf{r}; v)$ at and around $v = 1/2$ may provide useful information about the random matrices and the zeta function by regarding $K(\mathbf{r}; v)$ as a continuous function of v. Another challenge, which would be more ambitious, is to look for a family of functions or matrices whose zero or eigenvalue distributions are described by the correlation functions given by Eqs. 14.44 and 14.81 for arbitrary values of v.

References

1. Wigner E, Seitz F (1933) Phys Rev 43:804
2. Pines D (1963) Elementary excitations in solids. Benjamin, New York
3. Dyson FJ (1962) J Math Phys 3:140
4. Dyson FJ (1962) J Math Phys 3:157
5. Dyson FJ (1962) J Math Phys 3:166
6. Montgomery HL (1973) Proc Symp Pure Math 24:181
7. Keating JP, Snaith NC (2003) J Phys A 36:2859
8. Mahan GD (1990) Many-Particle physics. Plenum, New York
9. Abramowitz M, Stegun IA (1965) Handbook of mathematical functions. Dover Publications, New York
10. Gradshteyn IS, Ryzhik IM (1994) Table of integrals, series, and products. Academic, San Diego
11. Wigner EP (1951) Ann Math 53:36
12. Titchmarsh EC, Heath-Brown DR (1986) The theory of the Riemann Zeta-function. Clarendon, Oxford
13. Odlyzko AM (2001) In: van Frankenhuysen M, Lapidus ML (ed) Dynamical, spectral, and arithmetic zeta functions. Am Math Soc, Contemporary Math Series, Providence, RI, vol 290. p. 139
14. Rudnick Z, Sarnak P (1996) Duke Math J 81:269
15. Bogomolny EB, Keating JP (1995) Nonlinearity 8:1115
16. Bogomolny EB, Keating JP (1996) Nonlinearity 9:911
17. Hardy GH, Littlewood JE (1923) Acta Math 44:1
18. Katz NM, Sarnak P (1999) Bull Am Math Soc 36:1
19. Berry MV, Keating JP (1999) SIAM Rev 41:236

Part V
Dynamics and Quantum Monte-Carlo Methodology

Chapter 15
Hierarchical Effective-Mode Approach for Extended Molecular Systems

Rocco Martinazzo, Keith H. Hughes, and Irene Burghardt

Abstract Photoinduced processes in extended molecular systems are often ultrafast and involve strong electron-vibration (vibronic) coupling effects which necessitate a non-perturbative treatment. In the approach presented here, high-dimensional vibrational subspaces are expressed in terms of effective modes, and hierarchical chains of such modes which sequentially resolve the dynamics as a function of time. This permits introducing systematic reduction procedures, both for discretized vibrational distributions and for continuous distributions characterized by spectral densities. In the latter case, a sequence of spectral densities is obtained from a Mori/Rubin-type continued fraction representation. The approach is suitable to describe nonadiabatic processes at conical intersections, excitation energy transfer in molecular aggregates, and related transport phenomena that can be described by generalized spin-boson models.

R. Martinazzo
Department of Physical Chemistry and Electrochemistry, University of Milan, Via Golgi 19, 20122 Milan, Italy
e-mail: rocco.martinazzo@unimi.it

K.H. Hughes
School of Chemistry, Bangor University, Bangor, Gwynedd LL57 2UW, UK
e-mail: keith.hughes@bangor.ac.uk

I. Burghardt (✉)
Département de Chimie, Ecole Normale Supérieure, 24 rue Lhomond, 75231 Paris, France

Present address: Institute of Physical and Theoretical Chemistry, Goethe University Frankfurt, Max-von-Laue-Str. 7, 60438 Frankfurt, Germany
e-mail: burghardt@theochem.uni-frankfurt.de

P.E. Hoggan et al. (eds.), *Advances in the Theory of Quantum Systems in Chemistry and Physics*, Progress in Theoretical Chemistry and Physics 22,
DOI 10.1007/978-94-007-2076-3_15, © Springer Science+Business Media B.V. 2012

15.1 Introduction

Photoinduced processes in extended molecular systems often necessitate a quantum dynamical treatment based upon the explicit representation of a number of relevant vibrational (phonon) degrees of freedom evolving on coupled potential energy surfaces. This applies to molecule-solvent complexes, biological systems like chromophore-protein complexes, natural and artificial light-harvesting systems, organic materials like semiconducting polymers, and various other molecular assemblies. Many of the relevant photoinduced processes are ultrafast, as a result of nonadiabatic interactions by which electronically excited states decay non-radiatively. These processes often involve conical intersection topologies [1–4], which play a landmark role in the photochemistry of polyatomic molecular systems.

In line with the quantum dynamical nature of the photoinduced processes, recent experimental observations have provided compelling evidence that quantum coherence plays an important role even in high-dimensional, extended molecular assemblies, and in solvent or protein environments. In particular, ultrafast energy transport, i.e., excitation energy transfer (EET) in biological light-harvesting systems [5] and semiconducting polymer materials [6] has been shown to conserve excitonic coherence, rather than involving a Förster type hopping [7, 8] between neighboring sites. Contrary to the conventional assumption that environment-induced decoherence sets in within several tens of femtoseconds [9], long-lived coherences have been observed on a time scale of a picosecond [5, 6]. Conjectures have been made regarding the possible role of correlated environmental fluctuations in generating these long-lived coherences [10, 11].

From a system-bath theory perspective, the systems of interest fall into a markedly non-Markovian regime [12, 13], since the coupling between the electronic and vibrational degrees of freedom is strong and the environment is neither static nor rapidly fluctuating (Markovian). Instead, photoexcitation of the subsystem, or chromophore, induces a nonequilibrium response of the environment that is interleaved with the subsystem evolution, thus generating a dynamical evolution in the high-dimensional system-plus-environment space. In view of this, it is not clear a priori how to systematically construct a reduced dynamical description including the effect of the vibrational (phonon) modes on the electronic subsystem.

Over the past decade, various theoretical approaches and simulation techniques have been developed to tackle photoinduced dynamics in high-dimensional molecular systems. Broadly, two types of approaches can be distinguished: First, approaches which rely on quantum dynamical calculations in conjunction with model Hamiltonians of vibronic coupling type [14–17], or generalized spin-boson models [12, 13, 18]. Here, the phonon modes are either treated explicitly or are (partially) integrated out by a reduced dynamics procedure. The relevant model Hamiltonians are generally parametrized using electronic structure calculations. The second type of approach relates to an explicit, on-the-fly treatment of the electronic structure *and* dynamics of the high-dimensional, supermolecular system [19–21]. Here, the dynamics is often approximated by classical trajectory ensembles or Gaussian wavepackets [22–24]. The main strength of the first approach is its accurate description of the underlying quantum dynamics, at the expense of approximate potential surfaces, while the strength of the second approach lies in the more accurate treatment of the electronic structure properties.

The approach which is described in the present paper focuses on an *explicit but reduced-dimensional representation* of the dynamical problem. Our starting point is a class of generalized spin-boson models including a (large) number of degrees of freedom which can be treated within a linear vibronic coupling (LVC) approximation [1]. Using this model, we seek to extract a set of effective environmental modes which are generated by suitable coordinate transformations [25, 26]. The modes in question are collective coordinates which are constructed in such a way that they capture (1) the effects of the system-environment interaction on short time scales, and beyond this, provide (2) a general procedure by which chains of effective environmental modes are generated which unravel the dynamics as a function of time. The effective modes in question can be interpreted as generalized Brownian oscillator modes [27, 28], and the effective-mode chains are related to Mori chains [29–31] known from statistical mechanics. If the environment is effectively infinite-dimensional, the irreversible nature of the dynamics is maintained by a Markovian closure of the truncated chain representation.

The effective-mode construction can thus be employed both for a discrete set of vibrational modes (e.g., in a polyatomic molecule) and for typical system-bath type situations where the spectrum of bath modes is dense. In the latter case, the environment and its coupling to the electronic subsystem are entirely characterized by a spectral density. Approximate spectral densities can be constructed from few effective modes, representing a simplified realization of the true environmental spectral density that is designed to give a faithful representation of the dynamics on short time scales [32, 33]. Thus, even a highly structured, multi-peaked spectral density can be reduced to an effective, simplified spectral density on ultrafast time scales. Importantly, the procedure converges, as has recently been shown in Ref. [34].

In the present contribution, the effective-mode decomposition of spectral densities is demonstrated for the case of oligomers of poly-phenylene-vinylene (PPV) type. Here, the relevant spectral density is constructed from a classical-statistical correlation function that is in turn obtained from molecular dynamics (MD) simulations. Together with the effective-mode decomposition, this provides a practicable procedure for characterizing and reducing spectral densities within the high-temperature limit.

Overall, the present approach provides an alternative to other hierarchical schemes which seek to unravel the effects of a non-Markovian environment, see, e.g., Refs. [35–37]. The distinguishing feature of the present development is that its starting point is a coordinate transformation which facilitates the subsequent dynamical treatment.

The remainder of this contribution is organized as follows. Section 15.2 introduces the effective mode transformation techniques of Refs. [25, 26, 38], based upon a discretized version of a general LVC model. Section 15.3 focuses upon the parallel continued fraction development for spectral densities [32, 34, 39]. Finally, Sect. 15.4 demonstrates the convergence, in the time domain, of the effective-mode decomposition, and Sect. 15.5 concludes.

15.2 Effective-Mode Transformations

In this section, a general model Hamiltonian is introduced which is of LVC type for a subset of bath coordinates (Sect. 15.2.1). This Hamiltonian is subsequently transformed to the effective-mode representation mentioned above (Sects. 15.2.2 and 15.2.3). The development is presented in a system-bath theory setting, since this facilitates the transition to Sect. 15.3 where we focus upon a description of the vibrational distribution in terms of bath spectral densities.

15.2.1 Generalized Spin-Boson Models

We consider a class of generalized spin-boson models, assuming that a system-bath partitioning is appropriate to describe the high-dimensional system of interest,

$$\hat{H} = \hat{H}_S + \hat{H}_{SB} + \hat{H}_B \tag{15.1}$$

where the system Hamiltonian $\hat{H}_S$ takes the general form,

$$\hat{H}_S = \hat{V}_\Delta + \sum_{i=1}^{N_S} \frac{1}{2}\hat{p}_{S,i}^2 + \frac{\omega_{S,i}^2}{2}\hat{x}_{S,i}^2$$

$$+ \sum_{\xi=1}^{N_\xi} \sum_{\xi'<\xi}^{N_\xi} \hat{v}_{S,\xi\xi'}(\hat{x}_{S,1},\dots,\hat{x}_{S,N_S})\left(|\xi\rangle\langle\xi'| + |\xi'\rangle\langle\xi|\right) \tag{15.2}$$

where $\hat{V}_\Delta$ is an offset and the $\{|\xi\rangle\}$ define a discretized representation for N_ξ states, e.g., of the electronic subspace, while the coordinates $\hat{x}_S = \{\hat{x}_{S,1},\dots,\hat{x}_{S,N_S}\}$ refer to a subset of vibrational modes which are associated with the system part. In the present discussion, mass-weighted coordinates are employed as in Refs. [32, 39]; see Refs. [25, 26, 40–42] for a complementary development using mass-and-frequency weighted coordinates.

The system-bath coupling is restricted to a form which is linear in the bath coordinates $\hat{x}_B = \{\hat{x}_{B,1},\dots,\hat{x}_{B,N_B}\}$ and involves the discretized system operators,

$$\hat{H}_{SB} = \sum_i^{N_B} \sum_{\xi\xi'}^{N_\xi} c_{i,\xi\xi'}\hat{x}_{B,i}\left(|\xi\rangle\langle\xi'| + |\xi'\rangle\langle\xi|\right) \tag{15.3}$$

Thus, no direct coordinate couplings between the system and bath subspaces appear at this point.

Finally, the bath Hamiltonian $\hat{H}_B$ represents the zeroth-order Hamiltonian for N_B environmental modes,

$$\hat{H}_B = \sum_{i=1}^{N_B} \left(\frac{1}{2}\hat{p}_{B,i}^2 + \frac{\omega_{B,i}^2}{2}\hat{x}_{B,i}^2 \right) \tag{15.4}$$

The Hamiltonian equations 15.1–15.4 is applicable to various processes character-istic of molecular systems, including the dynamics at conical intersections (CoIn's) [1–4] and excitation energy transfer (EET) processes [5,6,8]. Its simplest realization corresponds to a single system operator, in which case the classical spin-boson Hamiltonian [12, 18] is obtained, where the bath coordinates couple to the energy gap operator $\hat{\sigma}_z = |\alpha\rangle\langle\alpha| - |\beta\rangle\langle\beta|$ of a two-level system (TLS).

The model Hamiltonian described above is thus of LVC type for the bath part, while the potentials pertaining to the system coordinates can take an anharmonic, non-separable form, see Eq. 15.2. For example, in the context of CoIn's, the system coordinates could be described by a quadratic coupling model whereas a (large) number of bath coordinates are approximated by the LVC form, as illustrated in the example of Sect. 15.4.

15.2.2 Effective-Mode Construction

The LVC model employed for the bath subspace allows one to introduce coordinate transformations by which a set of effective, or collective modes are extracted that act as generalized reaction coordinates for the dynamics. As shown in Refs. [25,26], $N_{\text{eff}} = N_\xi (N_\xi + 1)/2$ such coordinates can be defined for an N_ξ-state system. Thus, three effective modes are introduced for an electronic two-level system, six effective modes for a three-level system etc., for an arbitrary number of phonon modes that couple to the subsystem according to the LVC model. The subset of effective modes entirely determine the short-time dynamics, if the initial excitation is localized in the system subspace [26]. In order to capture the dynamics on longer time scales, chains of such effective modes can be introduced [38, 43]. These transformations, which are summarized below, will be shown to yield a unique perspective on high-dimensional dynamics in extended systems.

From the interaction Hamiltonian equation (15.3), we note that the N_B bath modes produce cumulative effects by their coupling to the discretized subsystem. This suggests that the interaction Hamiltonian can be formally re-written in terms of a set of collective coordinates $\hat{X}'_{B,\xi\xi'} = 1/C'_{\xi\xi'} \sum_i^{N_B} c_{i,\xi\xi'}\hat{x}_{B,i}$, such that

$$\hat{H}_{SB} = \sum_{\xi\xi'} C'_{\xi\xi'}\hat{X}'_{B,\xi\xi'}(|\xi\rangle\langle\xi'| + |\xi'\rangle\langle\xi|) \tag{15.5}$$

This definition corresponds to introducing Brownian-oscillator modes, in keeping with the transformation introduced, e.g., in Ref. [28]. However, in the case where several subsystem operators are present, these modes are not generally orthogonal on the space defined by the original bath coordinates $\{\hat{x}_B\}$. One therefore has to introduce an orthogonalization procedure [25, 26] leading to a set of N_{eff} orthogonal modes $\{\hat{X}_{B,i}\}$. With these, the interaction Hamiltonian takes the final form

$$\hat{H}_{SB} = \sum_i^{N_{\text{eff}}} \sum_{\xi\xi'} C_{i,\xi\xi'} \hat{X}_{B,i} (|\xi\rangle\langle\xi'| + |\xi'\rangle\langle\xi|) \tag{15.6}$$

This expression is formally the same as the one of the original system-bath coupling, Eq. 15.3, except that the system-bath interaction is entirely absorbed by N_{eff} effective modes. Depending on the orthogonalization procedure, different couplings can result [44].

15.2.3 Residual Bath Subspace: Stars, Chains, and Truncated Chains

The introduction of a set of effective modes $\{\hat{X}_{B,i}\}$ is the first step in defining an overall orthogonal transformation which leaves the subsystem coordinates $\{\hat{x}_S\}$ unaffected while transforming the bath coordinates,

$$\hat{X}_B = \mathbf{T}\hat{x}_B \tag{15.7}$$

As a result, one obtains the bath Hamiltonian in the following form,

$$\hat{H}_B = \sum_{i=1}^{N_B} \left(\frac{1}{2}\hat{P}_{B,i}^2 + \frac{\Omega_{B,i}^2}{2}\hat{X}_{B,i}^2 \right) + \sum_{i,j=1,j>i}^{N_B} d_{ij}\hat{X}_{B,i}\hat{X}_{B,j} \tag{15.8}$$

where bilinear coupling terms now appear in the bath subspace. The new frequencies $\Omega_{B,i}$ and couplings d_{ij} result from the coordinate transformation introduced above, such that $\Omega_{B,i}^2 = \sum_{j=1}^{N_B} \omega_{B,j}^2 t_{ji}^2$ and $d_{ij} = \sum_{k=1}^{N_B} \omega_k^2 t_{ki} t_{kj}$, where t_{ji} are the elements of the transformation matrix $\mathbf{T}$.

The transformed interaction Hamiltonian $\hat{H}_{SB}$ of Eq. 15.6 and the bath Hamiltonian of Eq. 15.8 define the system-bath Hamiltonian in the new coordinates. The subsystem part, comprising the electronic subspace and possibly a subset of strongly coupled vibrational modes, has remained unchanged.

In the new coordinates, the bath Hamiltonian takes a hierarchical form: The effective modes $\{\hat{X}_{B,i}\}$ couple directly to the electronic subsystem, while the remaining (residual) $N_B - N_{\text{eff}}$ bath modes couple in turn to the effective modes. The new bath Hamiltonian $\hat{H}_B$ of Eq. 15.8 can thus be split as follows:

$$\hat{H}_B = \hat{H}_B^{\text{eff}} + \hat{H}_B^{\text{eff-res}} + \hat{H}_B^{\text{res}} \tag{15.9}$$

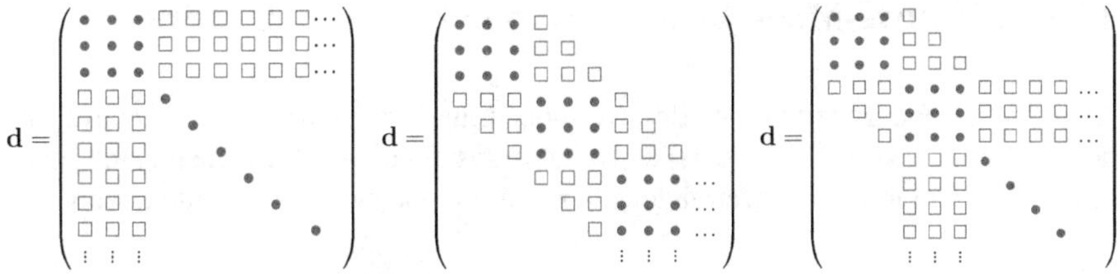

$$
\mathbf{d} = \begin{pmatrix} \bullet & \bullet & \bullet & \square & \square & \square & \square & \square & \cdots \\ \bullet & \bullet & \bullet & \square & \square & \square & \square & \square & \cdots \\ \bullet & \bullet & \bullet & \square & \square & \square & \square & \square & \cdots \\ \square & \square & \square & \bullet & & & & \\ \square & \square & \square & & \bullet & & & \\ \square & \square & \square & & & \bullet & & \\ \square & \square & \square & & & & \bullet & \\ \square & \square & \square & & & & & \bullet \\ \vdots & \vdots & \vdots & & & & & \end{pmatrix}
\qquad
\mathbf{d} = \begin{pmatrix} \bullet & \bullet & \bullet & \square & & & & \\ \bullet & \bullet & \bullet & \square & \square & & & \\ \bullet & \bullet & \bullet & \square & \square & & & \\ \square & \square & \square & \bullet & \bullet & \square & \square & \\ \square & \square & \square & \bullet & \bullet & \square & \square & \\ & \square & \square & \bullet & \bullet & \square & \square & \\ & & \square & \square & \bullet & \bullet & \cdots \\ & & & \square & \square & \bullet & \bullet & \cdots \\ & & & & \vdots & \vdots & \vdots & \end{pmatrix}
\qquad
\mathbf{d} = \begin{pmatrix} \bullet & \bullet & \bullet & \square & & & & \\ \bullet & \bullet & \bullet & \square & \square & & & \\ \bullet & \bullet & \bullet & \square & \square & & & \\ \square & \square & \square & \bullet & \bullet & \square & \square & \square & \cdots \\ & \square & \square & \bullet & \bullet & \square & \square & \square & \cdots \\ & \square & \square & \bullet & \bullet & \square & \square & \square & \cdots \\ & & \square & \square & \square & \bullet & & \\ & & & \square & \square & \square & \bullet & \\ & & & \square & \square & \square & & \bullet \\ & & & \vdots & \vdots & \vdots & & \end{pmatrix}
$$

Fig. 15.1 Schematic illustration of the coupling patterns within the residual-mode subspace, as described in Sect. 15.2.3. The number of effective modes is $N_{\mathrm{eff}} = 3$ in the present example. *Left*: Star type pattern, with a diagonal form of the $\{d_{ij}\}$ matrix in the residual $(N_B - N_{\mathrm{eff}})$-dimensional subspace; *middle*: chain type configuration with a band-diagonal form; *right*: truncated chain pattern which re-introduces the star pattern at a higher order of the chain

with the effective (eff) N_{eff}-mode bath portion

$$
\hat{H}_B^{\mathrm{eff}} = \sum_{i=1}^{N_{\mathrm{eff}}} \left(\frac{1}{2} \hat{P}_{B,i}^2 + \frac{\Omega_{B,i}^2}{2} \hat{X}_{B,i}^2 \right) + \sum_{i,j=1}^{N_{\mathrm{eff}}} d_{ij} \hat{X}_{B,i} \hat{X}_{B,j} \tag{15.10}
$$

the effective-residual (eff-res) mode interaction

$$
\hat{H}_B^{\mathrm{eff\text{-}res}} = \sum_{i=1}^{N_{\mathrm{eff}}} \sum_{j=N_{\mathrm{eff}}+1}^{N_B} d_{ij} \hat{X}_{B,i} \hat{X}_{B,j} \tag{15.11}
$$

and a definition analogous to Eq. 15.10 for the residual (res) Hamiltonian $\hat{H}_B^{\mathrm{res}}$ comprising the $(N_B - N_{\mathrm{eff}})$ remaining bath modes.

Since the orthogonal transformations leading to the form Eq. 15.9 of the transformed Hamiltonian are not unique, several construction schemes are possible [26, 32]. We have explored the following schemes in our recent work:

- Star-type configuration of the residual bath (see Fig. 15.1, left panel). Here, all residual bath modes are coupled to the N_{eff}-dimensional effective-mode subspace, while the bilinear coupling matrix $\{d_{ij}\}$ is diagonalized in the subspace of residual bath modes $\{N_{\mathrm{eff}} + 1, \ldots, N_B\}$
- Chain-type configuration of the residual bath (see Fig. 15.1, center panel). In this scheme, the bilinear coupling matrix is cast into a band-diagonal form, such that only the first layer of the residual bath is coupled to the effective-mode subspace. We have referred to this model as a hierarchical electron-phonon (HEP) model [38]
- Truncated chain-type configuration of the residual bath (see Fig. 15.1, right panel). This variant again employs the band-diagonal construction of the chain model, but is terminated at a given order M by a star construction as described above, now taken to represent a Markovian closure acting on the end of the chain [32]. This model mimics the Markovian closure which naturally terminates the chain as shown in the next section.

Figure 15.1 illustrates these three construction schemes for the residual bath. Various applications based upon these schemes can be found in Refs. [38, 41, 45].

15.3 Effective-Mode Decomposition of Spectral Densities

Against the background of the effective-mode transformation introduced above, we now address the system-bath perspective more systematically, by recasting the chain development in terms of spectral densities defining the system-bath interaction.

15.3.1 Mori/Rubin Type Continued Fractions

If the frequency distribution of the bath modes is dense, it is natural to characterize the influence of the bath on the subsystem in terms of a spectral density, or its discretized representation. In the case where the bath modes couple only to one of the subsystem operators, for instance

$$\hat{H}_{SB} = \sum_{i=1}^{N_B} c_{B,i}\hat{x}_{B,i}\hat{\sigma}_z \equiv D\hat{X}_{B,1}\hat{\sigma}_z \tag{15.12}$$

the definition of the spectral density corresponds to the form known for the spin-boson Hamiltonian [12, 13, 18],

$$J(\omega) = \sum_{i=1}^{N_B} \frac{c_{B,i}^2}{\omega_{B,i}}\delta(\omega - \omega_{B,i}) \tag{15.13}$$

This spectral density characterizes a bath that induces energy gap fluctuations in the TLS subsystem.

For the more general form of the system-bath coupling Eq. 15.3 where the bath modes couple to the $(\xi\xi')$ subsystem components, spectral densities are defined component-wise. Furthermore, if the bath modes couple simultaneously to several subsystem operators, we will refer to a *correlated bath*. The subsystem variables then do not experience independent fluctuations, and this is reflected in the definition of the spectral densities which involve cross-correlated contributions.

As a result of the transformation from the original Hamiltonian equations 15.1–15.4 to the effective-mode Hamiltonian equations 15.6–15.11, the spectral density has to be re-written in terms of the transformed quantities. As shown in Ref. [32], $J(\omega)$ then takes a continued fraction form which is close to the results obtained in Mori theory [29–31] or the Rubin model [12, 46].

15.3.1.1 Spectral Densities for Truncated Chains

In the simplest case of the spin-boson Hamiltonian (i.e., with a single subsystem operator such that $N_{eff}=1$), the spectral density takes the following continued-fraction form [32]

$$J^{(M)}(\omega) = -\lim_{\varepsilon\to 0^+} \text{Im}\, L^{(M)}(\omega + i\varepsilon) \tag{15.14}$$

where $L^{(M)}$ is a Heisenberg-domain propagator,

$$\hat{L}^{(M)}(z) = -z^2 - \cfrac{D^2}{\Omega_1^2 - z^2 - \cfrac{d_{1,2}^2}{\Omega_2^2 - z^2 - \cdots \cfrac{d_{M-2,M-1}^2}{\Omega_{M-1}^2 - z^2 - \cfrac{d_{M-1,M}^2}{\Omega_M^2 - z^2 - i\gamma z}}}} \qquad (15.15)$$

The order M corresponds to the number of modes which are included in the chain. The final chain member is then taken to undergo Markovian (Ohmic) dissipation with a friction coefficient γ.

According to Eqs. 15.14–15.15, Mth order truncated effective-mode chains translate to a series of approximate, coarse-grained spectral densities that are explored by the subsystem as a function of time [32, 33].

15.3.1.2   Residual Spectral Densities: Convergence

From the above, the question arises whether the Markovian termination of the chain becomes exact at some order M. In Ref. [34], this question has been posed in a complementary fashion, by analyzing the properties of the *residual* spectral densities J_M^{res}; these represent the spectral densities acting upon the end of the Mth-order chains, which are constituted by the remaining $N_B - M$ modes, for $N_B \to \infty$. If the residual spectral densities tend towards an Ohmic (Markovian) form for a given order M, the representation of Eqs. 15.14–15.15 will be exact, and the effective-mode series can be considered converged. If this limit can always be identified, the procedure provides a *universal* approach to decomposing non-Markovian environments.

Following Ref. [34], it can be shown that this is actually the case, i.e., convergence is guaranteed under very general conditions. The proof is based on the definition of the residual spectral densities in terms of their Cauchy transform [34],

$$J_M^{\text{res}}(\omega) = \lim_{\varepsilon \to 0^+} \text{Im} \, W_M(\omega + i\varepsilon) \qquad (15.16)$$

where the W_M obey the following one-term recurrence relation,

$$W_{M+1}(z) = \Omega_{M+1}^2 - z^2 - \frac{d_{M,M+1}^2}{W_M(z)} \qquad (15.17)$$

The limiting condition of this recurrence relation yields a (quasi-)Ohmic Rubin spectral density [34].

Effective-mode chain representations of the bath thus converge rigorously, in that the residual bath converges to an Ohmic spectral density. This result will be illustrated in the next section for a realistic spectral density.

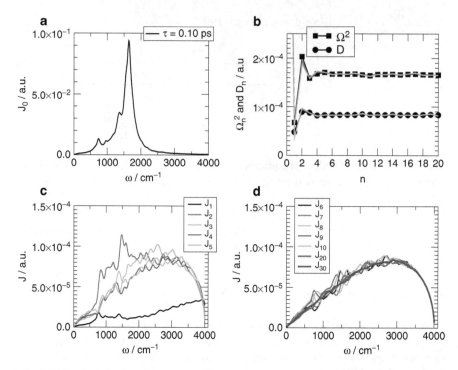

Fig. 15.2 Spectral density of S_1-S_0 energy gap fluctuations in a PPV oligomer (panel (**a**)) [47], and decomposition of the spectral density in terms of an effective-mode chain (panel (**b**)). Panels (**c**) and (**d**) illustrate the sequence of residual spectral densities J_M^{res} which converge towards a quasi-Ohmic form (with cutoff)

15.3.2 Spectral Density Decomposition: Poly-phenylene Vinylene

Figure 15.2 illustrates the effective-mode decomposition of a spectral density $J(\omega)$ for a spin-boson type system ($N_{eff} = 1$) which was obtained from classical molecular dynamics (MD) simulations [47]. The latter were combined with the construction of semi-empirical ground-state (S_0) and excited-state (S_1) potential energy surfaces for poly-phenylene-vinylene (PPV) oligomers, in line with Ref. [48]. The time-evolving energy gap $\Delta E(t) = E_{S_1}(\mathbf{r}(t)) - E_{S_0}(\mathbf{r}(t))$ was recorded along an excited-state (S_1) trajectory of 5 ps duration, yielding the real-valued correlation function [47]

$$C_{cl}(t) = \langle \Delta E(0) \Delta E(t) \rangle \tag{15.18}$$

In the high-temperature limit, the spectral density can be computed as follows from Eq. 15.18 [27],

$$J(\omega) = \frac{\omega}{2 k_B T} \tilde{C}_{cl}(\omega) \tag{15.19}$$

where $\tilde{C}_{cl}(\omega)$ is the Fourier transform of $C_{cl}(t)$. An exponential damping factor with a time constant $\tau = 0.1$ ps was introduced, yielding the spectral density of Fig. 15.2a. This spectral density is strongly peaked in the region around $1,500\,\mathrm{cm}^{-1}$, corresponding to high-frequency stretching and bond-length alternation modes. These modes are known to have a dominant effect in modulating the excited vs. ground state energy gap [38,48].

The remaining panels of Fig. 15.2 illustrate the effective-mode decomposition of the spectral density of Fig. 15.2a. Panel b shows that a series of high-frequency modes are sequentially extracted from the spectral density, with frequencies and couplings which remain almost constant with increasing orders of the effective-mode chain.

Panels c and d illustrate the sequence of residual spectral densities J_M^{res}, which clearly tend towards an Ohmic spectral density (with cut-off) as the order of the effective-mode decomposition increases. The cutoff frequency ω_R has to be larger than any frequency of interest (i.e., $J(\omega) \sim 0$ for $\omega > \omega_R$); here, we have chosen $\omega_R = 4,000\,\mathrm{cm}^{-1}$. The quasi-Ohmic spectral density towards which the successive J_M^{res}'s converge coincides with the spectral density generated by the Rubin chain model [12, 34, 39] with mode frequency $\omega_R/\sqrt{2}$. The convergence of the sequence towards the Rubin limit follows from the limiting behavior $\Omega_n^2/D_n \to 2$ that can be inferred from Fig. 15.2b.

In the case reported here, a chain with $M = 10\text{--}15$ modes should be able to accurately capture bath memory effects on the system dynamics, for times $t \leq \tau_P \propto M$, where τ_P is the Poincaré recurrence time of the chain. If the relevant system dynamics lasts longer, a Markovian closure acting on the last member of the chain (with $\gamma = \omega_R/2$) provides a reasonably good approximation valid for *all* times.

15.4 Effective-Mode Dynamics

In this section, the effective-mode approach is illustrated in the time domain. Following Ref. [39], we consider a tuning mode bath that is coupled to a 4-mode subsystem model of the S_2-S_1 conical intersection in pyrazine, described within the second-order model of Raab et al. [49]. In keeping with the general form of the Hamiltonian presented in Sect. 15.2.1, a linear vibronic coupling approximation is thus only made for the bath part. A continuous reference spectral density – similar to the one addressed in the previous section – is constructed by a Lorentzian convolution procedure from an $N_B = 20$ tuning mode distribution obtained in Ref. [50] as a weighted random ensemble. Following the continued-fraction construction described above, a sequence of approximate spectral densities $J^{(M)}$, $M = 1, 2, 3$, are then generated which in turn result in several approximate realizations of the dynamics. The M-th order spectral densities are re-discretized, here again for $N_B = 20$ bath modes.

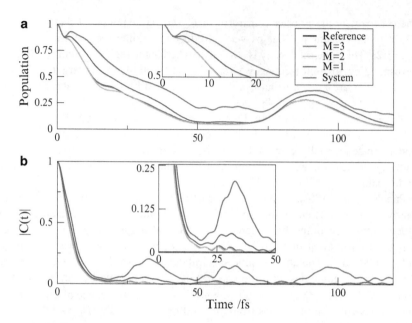

Fig. 15.3 (a) Time dependence of the S_2 state (diabatic) population for successive 4(system)+20(bath)-mode models based on the lowest-order bath spectral densities $J^{(M)}$, $M = 1, 2, 3$, as well as the reference spectral density of Ref. [39]. The 4-mode (system) dynamics is also shown for comparison (slowly decaying trace). (b) Time dependence of the wavepacket autocorrelation function, $|\langle \psi(0)|\psi(t)\rangle|$. At the level of the $M = 3$ approximation, the dynamics is indistinguishable from the reference dynamics (Adapted from Ref. [39])

Wavepacket calculations at $T = 0$ K were carried out for the combined 4-mode subsystem plus 20-mode bath, using the multiconfiguration time-dependent Hartree (MCTDH) method [51–53]. The explicit representation of all bath modes is not a necessity (and, in fact, the general method is designed so as to treat only the effective modes explicitly); however, an explicit wavepacket dynamics for all modes is convenient to demonstrate the convergence of the procedure for a zero-temperature system. In Refs. [32, 33], we have shown that explicit calculations for high-dimensional system-plus-bath wavefunctions are in excellent agreement with reduced density matrix calculations.

Initial conditions for the wavepacket calculations correspond to the Franck-Condon geometry, and the effective-mode expansion is defined with respect to this reference geometry. Figure 15.3 shows the time-dependent diabatic S_2 populations and autocorrelation functions $|C(t)| = |\langle \psi(0)|\psi(t)\rangle|$ generated from the successive spectral density approximants $J^{(M)}(\omega)$, $M = 1, \ldots, 3$. All orders agree over the shortest time scale (~ 5 fs), and the orders $M = 2, 3$ are found to be very close over the complete observation interval. The $M = 3$ result is virtually indistinguishable from the result obtained for the reference spectral density and can be considered converged.

As discussed in Ref. [39], time-domain convergence is often found to be more rapid than the frequency-domain convergence of the spectral densities as discussed in the preceding section. In the present example, the dynamics is well converged for $M = 3$ while the corresponding spectral density is still approximate as compared with the reference spectral density [39].

15.5 Conclusions

Photoinduced dynamics in extended molecular systems often fall into a short-time regime where inertial, coherent effects dominate and the many-particle dissipative dynamics has not yet set in. This generally precludes the use of standard system-bath approaches and necessitates an explicit dynamical treatment of the combined subsystem-plus-environment supermolecular system. The powerful QM/MM based simulation techniques that have been developed over recent years for the explicit simulation of photochemical processes in chromophore-solvent and chromophore-protein complexes, as well as extended systems like semiconducting polymers and various types of molecular aggregates have made great strides in this direction [19–21, 54]. Even so, the need for complementary reduced-dimensional models and dynamical interpretations persists. In the present contribution, we have presented such a complementary perspective.

Starting from an LVC model representing the system-environment interaction, the present approach identifies a set of $N_\xi (N_\xi + 1)/2$ effective modes which describe the collective environmental effects in an N_ξ-state system [25, 26]. Beyond the identification of the set of modes that predominate on the shortest time scale, further transformations are introduced by which chains of residual modes are created that successively unravel the dynamics [38, 43]. As shown above, this construction has proven useful, e.g., in the analysis of photoinduced dynamics in extended systems like semiconducting polymers [40, 55]. For this type of system, where distinct high-frequency vs. low-frequency phonon branches exist, one can further envisage alternative transformations by which effective modes are assigned to each phonon branch separately [56, 57].

Even the approximation where the residual bath is disregarded altogether can give good results in certain cases, e.g., for the short-time evolution at conical intersections [25, 45]. In general, convergence depends on the coupling strength between successive orders of the effective mode hierarchy.

Based upon the effective-mode construction, a systematic approximation procedure for the environment can be formulated in terms of a series of coarse-grained spectral densities [32, 33]. These spectral densities are generated from successive orders of a truncated chain model with Markovian closure. Analytical expressions can be given in terms of Mori type continued fractions. Assuming that an – a priori arbitrarily complicated – reference spectral density can be obtained independently, e.g., from experiments or classical simulations, one can thus (1) extract those features of the spectral density that determine the interaction with the subsystem on successive time scales, and (2) carry out reduced-dimensional simulations which are in exact agreement with the complete system-bath dynamics up to a certain time. Typically, quantum dynamical simulations will be carried out for the subsystem degrees of freedom augmented by the environment's effective modes, while the remaining modes are treated by a master equation [33]. Importantly, this procedure has been shown to converge rigorously, in that the residual spectral densities resulting from the chain development converge towards a (quasi-)Ohmic form [34].

In certain cases, diagonal interactions which give rise to energy gap fluctuations dominate. The tuning-mode models that were addressed in Sects. 15.3.2 and 15.4 are adapted to this case. Solute-solvent interactions can often be mapped upon such a model as well, such that the picture of a solvent coordinate [58–60] can be accommodated within the present class of models, even if the actual microscopic interactions cannot be described at the level of a harmonic oscillator bath.

We expect the effective-mode models described here to be versatile tools that can predict general trends, and that can be used in conjunction with microscopic information provided from other sources, i.e., spectral densities, energy gap correlation functions, and possibly cross-correlation functions. Further, model parametrizations could be provided by QM/MM type simulations, and the model-based dynamics could be employed to analyse the wealth of microscopic information provided by such simulations. Such complementary strategies would bridge the gap between system-bath theory approaches and explicit multi-dimensional simulations for ultrafast photochemical processes in various types of environments.

Acknowledgments Various contributions to the effective-mode developments described here, notably by Lorenz Cederbaum, Etienne Gindensperger, Eric Bittner, and Hiroyuki Tamura, are gratefully acknowledged.

References

1. Köppel H, Domcke W, Cederbaum LS (1984) Adv Chem Phys 57:59
2. Köppel H, Domcke W, Cederbaum LS (2004) In: Domcke W, Yarkony DR, Köppel H (eds) Conical intersections, vol 15. World Scientific, New Jersey, p 323
3. Yarkony DR (1996) Rev Mod Phys 68:985
4. Yarkony DR (1998) Acc Chem Res 31:511
5. Lee H, Cheng Y-C, Fleming GR (2007) Science 316:1462
6. Collini E, Scholes GD (2009) Science 323:369
7. Förster T (1948) Ann Phys 2:55
8. May V, Kühn O (2003) Charge and energy transfer dynamics in molecular systems. VCH-Wiley, Weinheim
9. Prezhdo OV, Rossky PJ (1998) Phys Rev Lett 81:5294
10. Brédas J-L, Silbey RJ (2009) Science 323:348
11. Chen X, Silbey RJ (2010) J Chem Phys 132:204503
12. Weiss U (1999) Quantum dissipative systems. World Scientific, Singapore
13. van Kampen NG (1992) Stochastic processes in physics and chemistry. North-Holland, Amsterdam
14. Kühl A, Domcke W (2002) J Chem Phys 116:263

15. Gelman D, Katz G, Kosloff R, Ratner MA (2005) J Chem Phys 123:134112
16. Worth GA, Meyer H-D, Köppel H, Cederbaum LS, Burghardt I (2008) Int Revs Phys Chem 27:569
17. Burghardt I, Giri K, Worth GA (2008) J Chem Phys 129:174104
18. Leggett AJ, et al (1987) Rev Mod Phys 59:1
19. Virshup AM, et al (2009) J Phys Chem B 113:3280
20. Boggio-Pasqua M, Robb MA, Groenhof G (2009) J Am Chem Soc 131:13580
21. Sinicropi A, et al (2007) Proc Natl Acad Sci U S A 105:17642
22. Toniolo A, Olsen S, Manohar L, Martínez TJ (2004) Faraday Discuss Chem Soc 127:149
23. Levine BG, Martínez TJ (2007) Annu Rev Phys Chem 58:613
24. Worth GA, Robb MA, Burghardt I (2004) Faraday Discuss 127:307
25. Cederbaum LS, Gindensperger E, Burghardt I (2005) Phys Rev Lett 94:113003
26. Gindensperger E, Burghardt I, Cederbaum LS (2006) J Chem Phys 124:144103
27. Mukamel S (1995) Principles of nonlinear optical spectroscopy. Oxford University Press, New York/Oxford
28. Garg A, Onuchic JN, Ambegaokar V (1985) J Chem Phys 83:4491
29. Mori H (1965) Prog Theor Phys 34:399
30. Dupuis M (1967) Prog Theor Phys 37:502
31. Grigolini P, Parravicini GP (1982) Phys Rev B 25:5180
32. Hughes KH, Christ CD, Burghardt I (2009) J Chem Phys 131:024109
33. Hughes KH, Christ CD, Burghardt I (2009) J Chem Phys 131:124108
34. Martinazzo R, Vacchini B, Hughes KH, Burghardt I (2011) J Chem Phys 134:011101
35. Tanimura Y, Kubo R (1989) J Phys Soc Jpn 58:101
36. Ishizaki A, Tanimura Y (2005) J Phys Soc Jpn 74:3131
37. Shi Q, Chen L, Nan G, Xu R-X, Yan Y (2009) J Chem Phys 130:164518
38. Tamura H, Bittner ER, Burghardt I (2007) J Chem Phys 127:034706
39. Martinazzo R, Hughes KH, Martelli F, Burghardt I (2010) Chem Phys 377:21
40. Tamura H, Bittner ER, Burghardt I (2007) J Chem Phys 126:021103
41. Tamura H, Ramon J, Bittner ER, Burghardt I (2008) J Phys Chem B 112:495
42. Gindensperger E, Köppel H, Cederbaum LS (2007) J Chem Phys 126:034106
43. Gindensperger E, Cederbaum LS (2007) J Chem Phys 127:124107
44. Burghardt I, Gindensperger E, Cederbaum LS (2006) Mol Phys 104:1081
45. Gindensperger E, Burghardt I, Cederbaum LS (2006) J Chem Phys 124:144104
46. Rubin RJ (1963) Phys Rev 131:964
47. Sterpone F, Martinazzo R, Burghardt I (2011) Coherent excitation transfer driven by torsional dynamics: a model Hamiltonian for PPV type systems. J Chem Phys (to be submitted)
48. Sterpone F, Rossky PJ (2008) J Phys Chem B 112:4983
49. Raab A, Worth GA, Meyer H-D, Cederbaum LS (1999) J Chem Phys 110:936
50. Krempl S, Winterstetter M, Plöhn H, Domcke W (1994) J Chem Phys 100:926
51. Meyer H-D, Manthe U, Cederbaum LS (1990) Chem Phys Lett 165:73
52. Manthe U, Meyer H-D, Cederbaum LS (1992) J Chem Phys 97:3199
53. Beck MH, Jäckle A, Worth GA, Meyer HD (2000) Phys Rep 324:1
54. Groenhof G, et al (2004) J Am Chem Soc 126:4228
55. Tamura H, Ramon J, Bittner ER, Burghardt I (2008) Phys Rev Lett 100:107402
56. Pereverzev A, Burghardt I, Bittner ER (2009) J Chem Phys 131:034104
57. Halperin B, Englman R (1974) Phys Rev B 9:2264
58. Warshel A, Weiss RM (1980) J Am Chem Soc 102:6218
59. Benjamin I, Barbara P, Gertner B, Hynes JT (1995) J Phys Chem 99:7557
60. Burghardt I, Hynes JT (2006) J Phys Chem A 110:11411

Chapter 16
Short-Time Dynamics Through Conical Intersections in Macrosystems: Quadratic Coupling Extension

Gábor J. Halász, Attila Papp, Etienne Gindensperger,
Horst Köppel, and Ágnes Vibók

Abstract We present an approach based on the quadratic vibronic coupling (QVC) Hamiltonian [New J Chem 17:7–29,1993] and the effective-mode formalism [Phys Rev Lett 94:113003, 2005] for the short-time dynamics through conical intersections in complex molecular systems. Within this scheme the nuclear degrees of freedom of the whole system are split as system modes and as environment modes. To describe the short-time dynamics in the macrosystem precisely, only three effective environmental modes together with the system's modes are needed.

G.J. Halász
Department of Information Technology, University of Debrecen, H-4010 Debrecen,
PO Box 12, Hungary

Theoretische Chemie, Physikalisch-Chemisches Institut, Universität Heidelberg,
D-69120, Germany
e-mail: halasz@inf.unideb.hu

A. Papp
Department of Theoretical Physics, University of Debrecen, H-40410 Debrecen,
PO Box 5, Hungary
e-mail: pappati@phys.unideb.hu

E. Gindensperger
Institut de Chimie, Laboratoire de Chimie Quantique, UMR 7177 CNRS/Université
Louis Pasteur, 4 rue Blaise Pascal, 67000 Strasbourg, France
e-mail: egindensperger@unistra.fr

H. Köppel
Theoretische Chemie, Physikalisch-Chemisches Institut, Universität Heidelberg,
D-69120, Germany
e-mail: horst.koeppel@pci.uni-heidelberg.de

Á. Vibók (✉)
Department of Theoretical Physics, University of Debrecen, H-40410 Debrecen,
PO Box 5, Hungary

Theoretische Chemie, Physikalisch-Chemisches Institut, Universität Heidelberg,
D-69120, Germany
e-mail: vibok@phys.unideb.hu

P.E. Hoggan et al. (eds.), *Advances in the Theory of Quantum Systems in Chemistry and Physics*, Progress in Theoretical Chemistry and Physics 22,
DOI 10.1007/978-94-007-2076-3_16, © Springer Science+Business Media B.V. 2012

Based on this decomposition, in the cumulant expansion of the autocorrelation function, the exact cumulants are recovered up to the second order. To demonstrate the capability of our method and for comparison with other results, the pyrazine molecule is chosen as a numerical example.

16.1 Introduction

Conical intersections (CIs) between electronic potential energy surfaces play a key mechanistic role in nonadiabatic molecular processes [1–4]. In this case the nuclear and electronic motions can couple and the energy exchange between the electrons and nuclei may become significant. In several important cases like dissociation, proton transfer, isomerization processes of polyatomic molecules or radiationless deactivation of the excited state systems [5, 6] the CIs can provide very efficient channels for ultrafast interstate crossing on the femtosecond time scale.

The nonadiabatic coupling terms (NACT) couple the different electronic states and may become the largest possible (as is well known from the Hellman-Feynman theorem) in the vicinity of the CIs [1, 3]. Therefore, approaching the CIs, the NACT become singular and provide the source for numerous phenomena that are considered as topological effects and lead to several interesting subjects, including the Longuet-Higgins or Berry phase [7,8], the open-path phase and the quantization feature of the NACTs and so forth. CIs can be evolved between different electronic states starting from triatomic systems to truly large polyatomic molecules. Several important books, review articles and publications have demonstrated the existence and relevance of such intersections in recent years [1–4,9–11].

The nonadiabatic dynamics through conical intersections are inherently quantum mechanical, involving strong mixing of several electronic states by nuclear displacements. Early attempts to treat the dynamics at CIs were based on the Landau-Zener-Stückelberg approach [12]. In parallel and later exact time-dependent quantum wave packet methods have been employed to explore the dynamics through conical intersections [13, 14]. Kuppermann et al. performed systematic calculations for H_3 to explore the geometric phase effect [15, 16]. Schinke and coworkers performed many calculations of the photodissociation dynamics of small molecules, which showed pretty good agreement with the experimental observations [17]. Although these approaches are quite accurate, they can only handle systems with very limited number of nuclear degrees of freedom (about 5–6 modes). A quite different approach to treat the multidimensional quantum dynamics through conical intersections is to apply the multi-configuration time-dependent Hartree (MCTDH) method [18–20]. This method is the only one at present which can treat the multi-mode quantum dynamics of polyatomic systems with controllable accuracy up to 20–30 modes.

In 2005 Cederbaum and collaborators published [21] an important study to describe the short-time dynamics through conical intersections in macrosystems. In this three-effective mode model all the modes of the macrosystem were decomposed into a system part and an environment part, and then, using an orthogonal transformation, a new scheme was proposed to decompose further the environment modes. As a result only three effective modes from the environment were obtained, which together with the system modes govern the short time dynamics in macrosystems [22, 23]. This method permits to undertake quantum dynamical calculations with reasonable accuracy on a short time scale for conical intersection situations, describing the effect of the environment with only three effective modes instead of treating all environmental modes explicitly. Later on they suggested a further step along this line by the construction of additional effective modes, which allow to describe accurately the intermediate-time dynamics [24–26]. An alternative approach to achieve another possible extension of this effective mode model has been developed and successfully applied by I. Burghardt and coworkers [27–31] too.

As the three effective mode model starts from the linear vibronic coupling Hamiltonian (LVC) [9] it may also have some relevance to generalize it and start from the quadratic vibronic coupling Hamiltonian (QVC) to obtain the appropriate quadratically extended (three)-effective mode equations. The motivation for this work has arisen that, in addition to the numerous applications of the LVC model, some other works in which the QVC model is used are also available [32, 35]. Our aim is to proceed along this direction. Following [21], we set up the QVC three-effective mode Hamiltonian and, using it for the pyrazine molecule we can calculate the autocorrelation function, the spectrum and the diabatic populations. The obtained results can be compared to those calculated by the LVC three-effective mode method.

The present article contains the following sections: In Sect. 16.2 the three-effective mode model is developed for the quadratically extended case and analyzed. In Sect. 16.3, the numerical results for the pyrazine molecule are presented and discussed. Conclusions are given in the final section.

16.2   Theory

In the first part of this section the effective mode formalism based upon the QVC Hamiltonian will be presented. Then, the impact of the outcome of this decomposition of the environment Hamiltonian on observable properties like the molecular spectrum will be discussed.

16.2.1   The Hamiltonian

Let us start with the Hamiltonian for an N-mode system described by the quadratic vibronic coupling model. For a two state conical intersection in the diabatic representation it has the form

$$H^{(QVC)} = \begin{pmatrix} E_1 & 0 \\ 0 & E_2 \end{pmatrix} + \sum_{k=1}^{N} H_k + \frac{1}{2} \sum_{k,l=1}^{N} H_{k,l} \qquad (16.1)$$

where

$$H_k = \frac{\omega_k}{2}\left(p_k^2 + x_k^2\right)\mathbf{1} + \begin{pmatrix} \kappa_k^{(1,1)} x_k & \kappa_k^{(1,2)} x_k \\ \kappa_k^{(2,1)} x_k & \kappa_k^{(2,2)} x_k \end{pmatrix} \tag{16.2}$$

and

$$H_{k,l} = \begin{pmatrix} \gamma_{kl}^{(1,1)} x_k x_l & \gamma_{kl}^{(1,2)} x_k x_l \\ \gamma_{kl}^{(2,1)} x_k x_l & \gamma_{kl}^{(2,2)} x_k x_l \end{pmatrix}. \tag{16.3}$$

Here x_k is the coordinate for the kth vibrational mode, p_k is the canonical momentum, and $\mathbf{1}$ is the 2×2 unit matrix. Mass- and frequency-weighted coordinates are used here, as well as atomic units ($\hbar = 1$). Each individual Hamiltonian H_k consists of three different parts: The first one is a harmonic 0th-order Hamiltonian with frequency ω_k, the second term represents the linear elements which couple the two electronic states, while the third contribution contains the quadratic and bilinear terms. The quantities $\kappa_k^{i,i}$, $\gamma_{k,l}^{i,i}$ and $\kappa_k^{i,j}$, $\gamma_{k,l}^{i,j}$ ($i \neq j$) are the intrastate and interstate coupling constants, respectively.

Next, we decompose the Hamiltonian equation (16.1) into a "system" Hamiltonian H_{System} and a "bath" Hamiltonian H_{Bath}

$$H = H_{System} + H_{Bath} \tag{16.4}$$

where

$$H_{System} = \begin{pmatrix} E_1 & 0 \\ 0 & E_2 \end{pmatrix} + H_S\left(y_1, y_2, \cdots, y_{N_S}\right) \tag{16.5}$$

and

$$H_{Bath} = \sum_{k=1}^{N_B} \frac{\omega_k}{2}\left(p_k^2 + x_k^2\right)\mathbf{1} + \sum_{k=1}^{N_B} \begin{pmatrix} \kappa_k^{(1,1)} x_k & \kappa_k^{(1,2)} x_k \\ \kappa_k^{(2,1)} x_k & \kappa_k^{(2,2)} x_k \end{pmatrix}$$
$$+ \frac{1}{2} \sum_{k,l=1}^{N_B} \begin{pmatrix} \gamma_{kl}^{(1,1)} x_k x_l & \gamma_{kl}^{(1,2)} x_k x_l \\ \gamma_{kl}^{(2,1)} x_k x_l & \gamma_{kl}^{(2,2)} x_k x_l \end{pmatrix} \tag{16.6}$$

The separation of the full Hamiltonian can be arbitrary. If the system is a single large molecule then one may collect the most relevant modes into H_{System}, and the remaining ones form H_{Bath}. However, if our system is a small molecule embedded in an environment, the partition is obvious.

16.2.2 Effective Modes for the Environment

In Refs. [22, 25] the effective mode approach developed for the case of the LVC Hamiltonian was presented and discussed in very detailed form. In what follows, we start from the Hamiltonian H_{Bath} (Eq. 16.6) obtained from the QVC approximation and use exactly the same type of orthogonal transformations as before. To obtain the QVC effective-mode Hamiltonian we repeat the same derivation.

Having the operator H_{Bath} we can split it further into two parts. To this end it is useful to introduce a unitary transformation of the bath modes which decomposes H_{Bath} into H_{eff} and V_{Bath} components

$$H_{Bath} = H_{eff} + V_{Bath}. \tag{16.7}$$

Within this separation the sum of the (H_{System} and H_{eff}) Hamiltonians,

$$H' = H_{System} + H_{eff} \tag{16.8}$$

governs the short-time dynamics of the system, while the remaining part V_{Bath} plays a role only at longer times. As in Refs. [22, 25] the H_{eff} operator is built up only from three (effective) modes, which couple the two electronic states. Now we can consider the elements $\sum_{k=1}^{N_B} \kappa_k^{(1,1)} x_k = \bar{\kappa}^{(1,1)} \tilde{X}_1$, $\sum_{k=1}^{N_B} \kappa_k^{(1,2)} x_k = \sum_{k=1}^{N_B} \kappa_k^{(2,1)} x_k = \bar{\kappa}^{(1,2)} \tilde{X}_2$, and $\sum_{k=1}^{N_B} \kappa_k^{(2,2)} x_k = \bar{\kappa}^{(2,2)} \tilde{X}_3$ appearing in H_{Bath} as effective modes. Here $\bar{\kappa}^{(1,1)} \equiv \sqrt{\sum_{k=1}^{N_B} \kappa_k^{(1,1)^2}}$, $\bar{\kappa}^{(1,2)} \equiv \sqrt{\sum_{k=1}^{N_B} \kappa_k^{(1,2)^2}}$, and $\bar{\kappa}^{(2,2)} \equiv \sqrt{\sum_{k=1}^{N_B} \kappa_k^{(2,2)^2}}$ are the effective coupling constants [22, 25]. These modes are, however, neither orthogonal to each other nor of physical relevance. Nevertheless these terms can be expressed as linear combination of three orthogonal modes. They may be constructed as follows [22, 25]

$$(X_1, X_2, X_3)^T = U_{3 \times 3} (\tilde{X}_1, \tilde{X}_2, \tilde{X}_3)^T = U_{3 \times 3} V_{3 \times N_B} (x_1, x_2, \cdots, x_{N_B})^T$$
$$= T_{3 \times N_B} (x_1, x_2, \cdots, x_{N_B})^T \tag{16.9}$$

Here the X_1, X_2 and X_3 vectors are normalized and orthogonal to each other and $U_{3 \times 3}$ is a matrix which orthogonalizes the modes $\tilde{X}_l$, $l = 1, 2, 3$. Combining this matrix $U_{3 \times 3}$ with the matrix $V_{3 \times N_B}$ one can obtain the $T_{3 \times N_B}$ transformation matrix ($U_{3 \times 3}^{-1} = V_{3 \times N_B} T_{3 \times N_B}^T$) between the initial environmental modes and corresponding orthonormalized ones. The $V_{3 \times N_B}$ transformation matrix gives the connection between the initial environmental modes and the intermediate normalized ones $(\tilde{X}_1, \tilde{X}_2, \tilde{X}_3)^T = V_{3 \times N_B} (x_1, x_2, \cdots, x_{N_B})^T$. Where

$$V_{3 \times N_B} = \begin{pmatrix} \kappa_1^{(1,1)} / \bar{\kappa}^{(1,1)} & \cdots & \kappa_{N_B}^{(1,1)} / \bar{\kappa}^{(1,1)} \\ \kappa_1^{(1,2)} / \bar{\kappa}^{(1,2)} & \cdots & \kappa_{N_B}^{(1,2)} / \bar{\kappa}^{(1,2)} \\ \kappa_1^{(2,2)} / \bar{\kappa}^{(2,2)} & \cdots & \kappa_{N_B}^{(2,2)} / \bar{\kappa}^{(2,2)} \end{pmatrix}. \tag{16.10}$$

By applying the transformation $T_{3 \times N_B}$ to the original modes of the environment, the terms $\sum_{k=1}^{N_B} \kappa_k^{(i,j)} x_k$ and $\frac{1}{2} \sum_{k,l=1}^{N_B} \gamma_{kl}^{(i,j)} x_k x_l$ of H_{Bath}(Eq. 16.6) can be expressed as

$$\sum_{k=1}^{N_B} \kappa_k^{(i,j)} x_k = \bar{\kappa}^{(i,j)} \sum_{k=1}^{3} K_k^{(i,j)} X_k \tag{16.11}$$

$$\frac{1}{2}\sum_{k,l=1}^{N_B}\gamma_{kl}^{(i,j)}x_kx_l = \sum_{k=1}^{N_B}\frac{d_{kk}^{(i,j)}}{2}X_k^2 + \sum_{k,l=1;k<l}^{N_B}d_{kl}^{(i,j)}X_kX_l \qquad (16.12)$$

with the coefficients $K_k^{(i,j)}$ and $d_{kl}^{(i,j)}$ given by

$$K_k^{(i,j)} \equiv \sum_{l=1}^{N_B}\frac{\kappa_l^{(i,j)}}{\bar{\kappa}^{(1,1)}}t_{kl} \quad \text{and} \quad d_{kl}^{(i,j)} \equiv \sum_{m,m'=1}^{N_B}\gamma_{mm'}^{(i,j)}t_{km}t_{lm'} \qquad (16.13)$$

where the t_{kl} are the elements of the matrix $T_{3\times N_B}$.

Now we are ready to derive the effective Hamiltonian envisaged in Eq. 16.7. For the sake of completeness we also give the form of the operator V_{Bath}. The results are

$$H_{eff}^{(i,j)} = \mathcal{E}_{eff}^{(i,j)} + H_A^{(i,j)} + H_B^{(i,j)} + H_C^{(i,j)} + H_D^{(i,j)} + H_E^{(i,j)}$$

$$V_{Bath}^{(i,j)} = \mathcal{E}_{Bath}^{(i,j)} + H_b^{(i,j)} + H_c^{(i,j)} + H_d^{(i,j)} + H_e^{(i,j)} \qquad (i,j=1,2) \ (16.14)$$

The explicit formulas for the different terms of the operators $H_{eff}^{(i,j)}$ and $V_{Bath}^{(i,j)}$ are listed below

$$\mathcal{E}_{eff}^{(i,i)} \equiv \Sigma_{k=1}^3\left(\frac{1}{2}\Omega_k + \frac{1}{4}d_{kk}^{(i,i)}\right) \qquad\qquad \mathcal{E}_{Bath}^{(i,i)} \equiv \Sigma_{k=4}^{N_B}\left(\frac{1}{2}\Omega_k + \frac{1}{4}d_{kk}^{(i,i)}\right)$$

$$\mathcal{E}_{eff}^{(1,2)} \equiv \mathcal{E}_1^{(2,1)} \equiv \Sigma_{k=1}^3\frac{1}{4}d_{kk}^{(1,2)} \qquad\qquad \mathcal{E}_{Bath}^{(1,2)} \equiv \mathcal{E}_{r1}^{(2,1)} \equiv \Sigma_{k=4}^{N_B}\frac{1}{4}d_{kk}^{(1,2)}$$

$$H_A^{(i,j)} \equiv \bar{\kappa}^{(i,j)}\Sigma_{k=1}^3 K_k^{(i,j)}X_k \qquad\qquad \left(H_a^{(i,j)} \equiv 0\right)$$

$$H_B^{(i,i)} \equiv \Sigma_{k=1}^3\frac{1}{2}\Omega_k\left(P_k^2 + X_k^2 - 1\right) \qquad\quad H_b^{(i,i)} \equiv \Sigma_{k=4}^{N_B}\frac{1}{2}\Omega_k\left(P_k^2 + X_k^2 - 1\right)$$

$$H_C^{(i,i)} \equiv \Sigma_{k,l=1;k<l}^3 d_{kl}\left(P_kP_l + X_kX_l\right) \qquad H_c^{(i,i)} \equiv \Sigma_{k=1;l=4;k<l}^{N_B} d_{kl}\left(P_kP_l + X_kX_l\right)$$

$$H_B^{(1,2)} \equiv H_B^{(2,1)} \equiv H_C^{(1,2)} \equiv H_C^{(2,1)} \equiv 0 \qquad H_b^{(1,2)} \equiv H_b^{(2,1)} \equiv H_c^{(1,2)} \equiv H_c^{(2,1)} \equiv 0$$

$$H_C^{(i,j)} \equiv 0 \qquad\qquad\qquad\qquad H_c^{(i,i)} \equiv \Sigma_{k=1;l=4}^3\Sigma_{l=4}^{N_B} d_{kl}\left(P_kP_l + X_kX_l\right)$$

$$H_D^{(i,j)} \equiv \Sigma_{k=1}^3\frac{1}{2}d_{kk}^{(i,j)}\left(X_k^2 - \frac{1}{2}\right) \qquad H_d^{(i,j)} \equiv \Sigma_{k=4}^{N_B}\frac{1}{2}d_{kk}^{(i,j)}\left(X_k^2 - \frac{1}{2}\right)$$

$$H_E^{(i,j)} \equiv \Sigma_{k,l=1;k<l}^3 d_{kl}^{(i,j)}X_kX_l \qquad\qquad H_e^{(i,j)} \equiv \Sigma_{k=1;l=4;k<l}^{N_B} d_{kl}^{(i,j)}X_kX_l$$

$$i,j=1,2$$

$$(16.15)$$

where $\bar{\kappa}^{(i,j)}$ $(i,j=1,2)$ are the effective coupling constants, and

$$d_{kl} \equiv \sum_{m=1}^{N_B}\omega_m t_{km}t_{lm} \quad \Omega_k = d_{kk} = \sum_{m=1}^{N_B}\omega_m t_{km}^2 \quad d_{kl}^{(i,j)} = \sum_{m,m'=1}^{N_B}\gamma_{mm'}^{(i,j)}t_{km}t_{lm'}. \quad (16.16)$$

As can be seen from Eq. 16.15 above, only three modes contribute to H_{eff}. The remaining $N_B - 3$ modes (N_B is the number of bath modes) of the environment appear in V_{Bath}. In the numerical calculations we will (completely) neglect V_{Bath}, because this part of the Hamiltonian does not couple directly to the electronic subsystem.

16.2.3   Cumulants and Short-Time Dynamics in the Quadratic Extension

Having the explicit form for the QVC Hamiltonian H_{eff} we are able to study the short-time dynamics through a conical intersection in a macrosystem. At this stage we make contact with the autocorrelation function [36] and with the spectra which can be experimentally measured. The autocorrelation function is

$$P(t) = < 0| \exp(-iHt)|0 > \qquad (16.17)$$

and represents the overlap between the initial wave function and the wave function at time t after excitation from the noninteracting electronic ground state to the coupled electronic states. The resulting spectral intensity distribution $P(E)$ is the Fourier transform of the autocorrelation function. As is known, the short-time dynamics is controlled by the first few cumulants [37]. These are related to the autocorrelation function as they are the coefficients of an expansion of $\ln[P(t)]$ at $t = 0$. Due to the above mentioned relation between the $P(t)$ and $P(E)$ functions, cumulants are related to observable features of the spectrum. Namely, the zeroth and first cumulants give the total intensity and the center of gravity of the spectrum, respectively. The second and third orders describe the width and major asymmetry of it.

In Refs. [22–25], it was proved that H' resulting from the decomposition of the LVC Hamiltonian as

$$H = H_{System} + H_{Bath} = H_{System} + H_{eff} + V_{Bath} = H' + V_{Bath} \qquad (16.18)$$

reproduces the cumulants of the total Hamiltonian H up to third order. It means, that using either the H or the "effective" H' operators in the dynamical calculations one obtains the same results for those properties of the spectra which are related to the first four cumulants.

Based on this reasoning we have performed a similar analysis for the case of our quadratic extension. As a result we could prove the equality of the moments only up to second order [38]. One has to emphasize, that in the former case the results obtained by using the LVC Hamiltonian H' were compared to those obtained from the LVC Hamiltonian H. Correspondingly, in the present case, we compare the cumulants to the QVC Hamiltonians H' and H. It should be recognized, that generally the QVC Hamiltonian H is more accurate than the LVC Hamiltonian, as it provide a better description of the actual electronic potential energy surfaces.

16.3   Numerical Results and Discussion

To test the performance of the QVC effective-mode scheme, one has to find a numerically solvable sample system. For this purpose we choose the pyrazine molecule as an example. In the decomposition of $H = H_{System} + H_{Bath}$, and $H_{Bath} = H_{eff} + V_{Bath}$

we put all the 24 modes of the molecule into the H_{Bath} operator. In this situation the number of system modes is zero $(N_S = 0)$, and only three effective modes describe the short-time dynamics in this system.

In our computational work, on which the following numerical analysis is based, the multiconfiguration time-dependent Hartree (MCTDH) method is used [18–20]. This method is, the only one at present, which can propagate the multidimensional wave packet and treat the multi-mode quantum dynamics of polyatomic systems with controllable accuracy up to 20–30 modes.

In the actual calculations, we will present autocorrelation functions, (photoelectron) spectra and diabatic populations. The autocorrelation is calculated according to Eq. 16.17. The photoelectron spectrum is given by a Fourier transform of the autocorrelation function obtained from a long propagation in real time. To compute the diabatic population we use the formula

$$P^{SF}(t) = < \psi^{SF}(t) | \psi^{SF}(t) >, \tag{16.19}$$

where $P^{SF}(t)$ corresponds to the population or probability of being on the ground $(SF = G)$ or excited $(SF = E)$ state diabatic surface. Here ψ^{SF} is the diabatic nuclear wave function for the ground $(SF = G)$ or excited $(SF = E)$ electronic state, respectively.

Four kinds of computations were performed and are compared to each other. (A) An exact calculation (based on the QVC Hamiltonian) taking into account all the 24 modes of pyrazine. This calculation has already been presented in [32]. (B) A calculation where the LVC Hamiltonian was used. In this case, due to symmetry reasons only six vibrational modes are relevant. Using different parameters, and including four of these six modes, similar work has also been done [33, 34]. (C) A calculation using the effective mode scheme based on the LVC Hamiltonian. Here only three effective modes were chosen to describe the short-time dynamics [26]. (D) Calculations with our method (effective mode formalism based on the QVC Hamiltonian) described in the present paper taking again into account three effective modes. These modes of course are different from those ones used in (C) as they come from different Hamiltonians.

Figure 16.1 displays the autocorrelation function up to 50 fs for the four calculations listed above, which we abbreviate as the 24-mode, 6-mode, 3(LVC)-mode, and 3(QVC)-mode cases, respectively. Comparing the curves one notices that three of them corresponding to the 24-mode, 6-mode, and 3(QVC)-mode, are in excellent agreement up to 20 fs. The 3(LVC)-mode curve differs from these after 10 fs. Then, around 20 fs these three curves deviate from each other and a recurrence occurs between (∼20–40 fs) for each of the four functions. This recurrence is the smallest for the 24-mode curve and the largest for the two 3(LVC, QVC)-mode one. The 6-mode curve goes in the middle within this interval. The two "3(LVC, QVC)-mode" recurrences have similar magnitude but they are shifted to each other by ∼5–6 fs. At longer times, after 45 fs the three curves (except for the 3(LVC)-mode) go again roughly together and only the 3(LVC)-mode differs from them. We

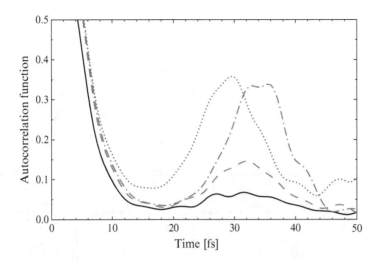

Fig. 16.1 The autocorrelation functions for the pyrazine molecule. *Solid line*: exact 24-mode result. *Dashed line*: result for the linear vibronic coupling model (6-mode). *Dotted line*: result for the three effective mode model. *Dashdotted line*: result for our quadratically extended three effective mode approach

conclude that although the shape of the 3(LVC)-mode and 3(QVC)-mode curves are very similar and they are only shifted to each other by ~5–6 fs, the first 20 fs the present approach is much better. Practically, it very accurately reproduces the result of the exact short-time dynamics of the 24-mode calculation.

In Fig. 16.2(A–D) we present the spectra as a function of energy for all four cases. In Fig. 16.2A we show the exact 24-mode result and for the sake of comparison the remaining Fig. 16.2(B–D) also display this curve. It was already mentioned in the theory section that the 3(QVC)-mode model reproduces the zeroth- to second-order cumulants of the molecule, connected to observable properties of the spectrum in such experiments. These are the total intensity, the position of the maximum and the with of the spectrum. Due to the absence of theoretical proof concerning the "conservation" of the third cumulant, which is related to the major asymmetry of the spectra, we can not discuss this feature with certainty. As expected the spectra obtained by the 6-mode model (see Fig. 16.2B) is most similar to the exact one. However the global shape of the exact spectrum is more or less reproduced by our 3(QVC)-mode formulas too. Some oscillations appear in particular within the (0–0.3 eV) interval, but the picture is definitely closer to the exact one than that obtained by the 3(LVC)-mode approach. Apart from these oscillations one can see that the position of the maximum, the width and even the main asymmetry of the spectrum are satisfactorily reproduced by our present method.

The last bundle of figures shows the diabatic populations. Each panel on Fig. 16.3 shows the diabatic lower and diabatic upper states for one particular

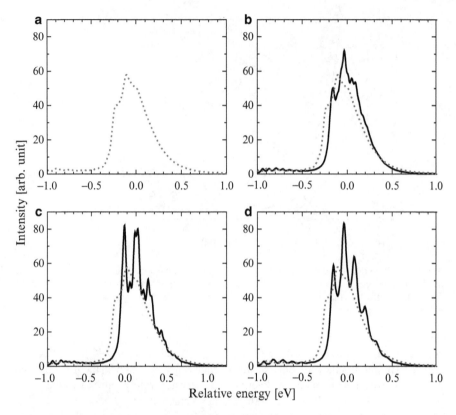

Fig. 16.2 The spectra for the pyrazine molecule. The damping parameter is 30 fs. All panels display the result for the exact 24-mode model (with *dotted lines* in the background). Panel **A**: result (*dotted line*) for the exact 24-mode model. Panel **B**: result (*solid line*) for the linear vibronic coupling model (6-mode). Panel **C**: result (*solid line*) for the three effective mode model. Panel **D**: result (*solid line*) for our quadratically extended three effective mode approach

case among the four investigated ones. It is now noticed that there is practically no difference between the curves up to 4 fs. At longer times, some effect can be seen. The oscillation "frequency" of the population functions for the three other situations are more or less the same, but definitely smaller than that produced by the 3(QVC)-mode. If we compare the exact result (panel A) with that of 3(QVC)-mode (panel D), we see an overall good agreement concerning the shape of the functions, but the values of them are shifted to each other quite significantly. Moreover, the functions in the 24-mode, and 6-mode pictures are smoother than those obtained by the two effective-mode models. For this latter case, there is additionally a highly structured shape of the diabatic population functions too.

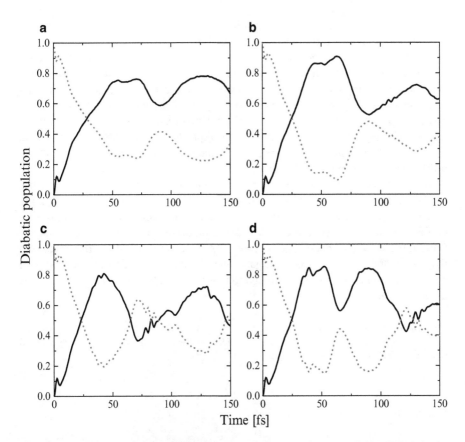

Fig. 16.3 The diabatic-state populations for the pyrazine as a function of time. Panel **A**: the population for the lower state (*solid line*) and for the upper state (*dotted line*) using the exact 24-mode model. Panel **B**: the population for the lower state (*solid line*) and for the upper state (*dotted line*) using the linear vibronic coupling model (6-mode). Panel **C**: the population for the lower state (*solid line*) and for the upper state (*dotted line*) using the three effective mode approach. Panel **D**: the population for the lower state (*solid line*) and for the upper state (*dotted line*) using our quadratically extended three effective mode approach

16.4   Conclusions

A formalism, based on the QVC Hamiltonian is developed and used to describe the short-time dynamics of large molecules or molecule-environment systems for a conical intersection situation. Our method might be considered as one possible extension of the earlier three effective-mode model based on the LVC Hamiltonian. In both schemes, the macrosystem is decomposed into system modes and an environment. It is demonstrated that the short-time dynamics of the full system can reasonably be described using the system modes, augmented by only three effective modes. These effective coordinates can be obtained from the environmental modes by applying a suitable orthogonal transformation.

It is known that the short-time dynamics is controlled by the first few cumulants. In a cumulant expansion of the autocorrelation function, we recover the exact cumulants up to the second order. It means that the zeroth-, first-, and second-order cumulants of H, Eq. 16.1, and those of H' (Eq. 16.8), are identical. This provides a guarantee that using the operator H' in the calculations, the intensity, the center of gravity and the width of the spectrum can be reproduced reasonably well. We note, however, that this kind of "conservation" of the cumulants are fulfilled up to third order when concerning the LVC three-effectiv mode decomposition.

As a sample system we used the pyrazine molecule. Fortunately, we could use the MCTDH method to compute the dynamics of all the 24 modes, thus providing numerically exact results for comparison. Four different kinds of computations were performed and compared to each other. Based on these results we conclude that our QVC three-effective mode scheme well reproduces the short-time dynamics and the overall shape of the spectra. In particular, the autocorrelation function is more accurate up to the first 20 fs, compared to that obtained by the LVC three-effective mode approach.

Acknowledgements Á.V. acknowledges the OTKA Grant No. 80095. The financial support by the COST Action CM0702 CUSPFEL, the Deutsche Forschungsgemeinschaft (H.K.), and Egide/DAAD Procope (20139VE) is greatly acknowledged.

References

1. Domcke W, Yarkony DR, Köppel H (eds) (2004) Conical intersections: electronic structure, dynamics and spectroscopy. World Scientific, Singapore
2. Yarkony DR (1996) Rev Mod Phys 68:985
3. Baer M (2006) Beyond Born Oppenheimer: electronic non-adiabatic coupling terms and conical intersections. Wiley, Hoboken
4. Worth GA, Cederbaum LS (2004) Annu Rev Phys Chem 55:127
5. Klessinger M, Michl J (1995) Excited states and photochemistry of organic molecules. VCH Publishers Inc, New York
6. Andruniow T, Ferre N, Olivucci M (2004) Proc Natl Acad Sci 101:17908
7. Herzberg G, Longuet-Higgins HC (1963) Discuss Faraday Soc 35:77
8. Longuet-Higgins HC (1975) Proc R Soc London Ser A 344:147
9. Köppel H, Domcke W, Cederbaum LS (1984) Adv Chem Phys 57:59
10. Baer M, Billing GD (2002) The role of degenerate states in chemistry. In: Advances in chemical physics, vol 124. Wiley-Interscience, New York
11. Matsika S (2007) Rev Comp Chem 23:83
12. Desouter-Lecomte M, Dehareng D, Leyh-Nihant B, Praet AJ, Lorquet MT, Lorquet JC (1985) J Chem Phys 89:214
13. Köppel H (1983) Chem Phys 77:359375
14. Kuppermann A (1996) In: Wyatt RE, Zhang JZ (eds) Dynamics of molecules and chemical reactions. Marcel Dekker, New York
15. Lepetit B, Kuppermann A (1990) Chem Phys Lett 166:581
16. Kuppermann A, Abrol R (2002) Adv Chem Phys 124:323

17. Schinke R (1993) Photodissociation dynamics. Cambridge University Press, Cambridge
18. Meyer HD, Manthe U, Cederbaum LS (1990) Chem Phys Lett 165:73
19. Beck MH, Jäckle A, Worth GA, Meyer H-D (2000) Phys Rep 324:1; Worth GA, et al (2000) The MCTD package, version 8.2; Meyer H-D 2002 The MCTH package, version 8.3 see http://www.pci.uni-heidelberg.de/cms/mctdh/
20. Meyer H-D, Gatti F, Worth GA (eds) (2009) Multidimensional quantum dynamics: MCTDH theory and applications. Wiley-VCH, Weinheim
21. Cederbaum LS, Gindensperger E, Burghardt I (2005) Phys Rev Lett 94:113003-0
22. Gindensperger E, Burghardt I, Cederbaum LS (2006) J Chem Phys 124:144103
23. Gindensperger E, Burghardt I, Cederbaum LS (2006) J Chem Phys 124:144104
24. Gindensperger E, Köppel H, Cederbaum LS (2007) J Chem Phys 126:034106
25. Gindensperger E, Cederbaum LS (2007) J Chem Phys 127:124107
26. Basler M, Gindensperger E, Meyer HD, Cederbaum LS (2008) Chem Phys 347:78
27. Tamura H, Bittner ER, Burghardt I (2007) J Chem Phys 126:021103
28. Tamura H, Bittner ER, Burghardt I (2007) J Chem Phys 127:021103
29. Hughes KH, Christ CD, Burghardt I (2009) J Chem Phys 131:124108
30. Hughes KH, Christ CD, Burghardt I (2009) J Chem Phys 131:024109
31. Burghardt I, Hughes KH, Tamura H, Gindensperger E, Koeppel H, Cederbaum LS (2011) Conical intersections coupled to an environment. In: Domcke W, Yarkony DR, Koeppel H (eds) Conical intersections: theory, computation, and experiment. World Scientific, New Jersey (in press)
32. Raab A, Worth GA, Meyer H-D, Cederbaum LS (1999) J Chem Phys 110:936
33. Worth GA, Meyer HD, Cederbaum LS (1996) J Chem Phys 105:4412
34. Worth GA, Meyer HD, Cederbaum LS (1998) J Chem Phys 109:3518
35. Mahapatra S, Worth GA, Meyer H-D, Cederbaum LS, Köppel H (2001) J Phys Chem A 105:5567
36. Levine RD (1969) Quantum mechanics of molecular rate processes. Oxford University Press, New York
37. Cramér H (1946) Mathematical methods of statistics. Princeton University Press, Princeton
38. Halász GJ, Csehi A, Gindensperger E, Köppel H, Vibók Á submitted to J Phys Chem A

Chapter 17
Theoretical Methods for Nonadiabatic Dynamics "on the fly" in Complex Systems and its Control by Laser Fields

Roland Mitrić, Jens Petersen, Ute Werner, and Vlasta Bonačić-Koutecký

Abstract We present a general theoretical approach for the simulation and control of ultrafast processes in complex molecular systems. It is based on the combination of quantum chemical nonadiabatic dynamics "on the fly" with the Wigner distribution approach for simulation and control of laser-induced ultrafast processes. Specifically, we have developed a procedure for the nonadiabatic dynamics in the framework of time-dependent density functional theory using localized basis sets, which is applicable to a large class of molecules and clusters. This has been combined with our general approach for the simulation of time-resolved photoelectron spectra that represents a powerful tool to identify the mechanism of nonadiabatic processes, which has been illustrated on the example of ultrafast photodynamics of furan. Furthermore, we present our "field-induced surface hopping" (FISH) method which allows to include laser fields directly into the nonadiabatic

R. Mitrić
Fachbereich Physik, Freie Universität Berlin, Arnimallee 14, D-14195 Berlin, Germany
e-mail: mitric@zedat.fu-berlin.de

J. Petersen
Fachbereich Physik, Freie Universität Berlin, Arnimallee 14, D-14195 Berlin, Germany

Institut für Chemie, Humboldt-Universität zu Berlin, Brook-Taylor-Straße 2,
D-12489 Berlin, Germany
e-mail: jpetersen@zedat.fu-berlin.de

U. Werner
Institut für Chemie, Humboldt-Universität zu Berlin, Brook-Taylor-Straße 2,
D-12489 Berlin
e-mail: ute.werner@chemie.hu-berlin.de Germany

V. Bonačić-Koutecký (✉)
Institut für Chemie, Humboldt-Universität zu Berlin, Brook-Taylor-Straße 2,
D-12489 Berlin, Germany

Interdisciplinary Center for Science and Technology, University of Split,
Meštrovićevo Šetalište bb., HR-21000 Split, Croatia
e-mail: vbk@chemie.hu-berlin.de

P.E. Hoggan et al. (eds.), *Advances in the Theory of Quantum Systems in Chemistry and Physics*, Progress in Theoretical Chemistry and Physics 22,
DOI 10.1007/978-94-007-2076-3_17, © Springer Science+Business Media B.V. 2012

molecular dynamics simulations and thus to realistically model their influence on ultrafast processes. On the example of optimal dynamic discrimination of two almost identical flavin molecules we demonstrate that experimentally optimized laser fields can be directly used in the framework of the FISH method to reveal the dynamical processes behind the optimal control.

17.1 Introduction

The challenge for the theory in the field of ultrafast dynamics is two-fold: The appropriate description of nonadiabatic processes due to the breakdown of the Born-Oppenheimer approximation in the vicinity of conical intersections or avoided crossings, and the simulation of time-resolved spectroscopic signals allowing for the interpretation of experimental observables. The full quantum mechanical treatment of nuclear dynamics is severely limited to small systems or models with reduced dimensionality since it requires the precalculation of global potential energy surfaces (PES). In contrast, this is not necessary for more approximate methods based on the propagation of classical trajectories which can be carried out "on the fly" and thus offer a convenient alternative. The basic idea of the molecular dynamics (MD) "on the fly" [1] is to compute the forces acting on the nuclei from the electronic structure calculations only when they are needed during the propagation. This is in particular advantageous for systems which do not contain "chromophore-type" subunits and thus no separation in active and passive degrees of freedom is possible.

The conceptual framework for the – semiclassical simulation of ultrafast spectro-scopic observables is provided by the Wigner representation of quantum mechanics [2, 3]. Specifically, for the ultrafast pump-probe spectroscopy using classical trajectories, methods based on the semiclassical limit of the Liouville-von Neu-mann equation for the time evolution of the vibronic density matrix have been developed [4–8]. Our approach [4, 6–8] is related to the Liouville space theory of nonlinear spectroscopy developed by Mukamel et al. [9]. It is characterized by the ability to approximately describe quantum phenomena such as optical transitions by averaging over the ensemble of classical trajectories. Moreover, quantum correc-tions for the nuclear dynamics can be introduced in a systematic manner, e.g. in the framework of the "entangled trajectory method" [10,11]. Alternatively, these effects can be also accounted for in the framework of the multiple spawning method [12]. In general, trajectory-based methods require drastically less computational effort than full quantum mechanical calculations and provide physical insight in ultrafast processes. Additionally, they can be combined directly with quantum chemistry methods for the electronic structure calculations.

In this context, one of the most efficient approaches is based on mixed quantum-classical dynamics in which the nonadiabatic effects are simulated using Tully's surface hopping (TSH) method [13, 14]. It is applicable to a large variety of systems ranging from isolated molecules and clusters to complex nanostructures interacting with different environments. In the TSH method classical trajectories are propagated in different electronic states and exhibit stochastic transitions between the states according to quantum mechanical hopping probabilities. The necessary ingredients for TSH simulations are forces in ground and excited electronic states as well as nonadiabatic couplings. These can be calculated using the whole spectrum of methods for the electronic structure, such as ab initio "frozen ionic bond" approximation [4], ab-initio configuration interaction (CI) [15], restricted open-shell Kohn-Sham density functional theory (DFT) [16], linear response time-dependent density functional theory (TDDFT) [17–23] as well as semiempirical methods [24–27]. In addition, recently the applications have been extended to the mixed quantum mechanical-molecular mechanical (QM/MM) methods allowing to treat complex systems such as photoactive proteins [28–32] or chromophores interacting with the environment [33].

A further important aspect in the field of ultrafast science is the introduction of electric laser fields into molecular dynamics. This opens the perspective for controlling molecular processes by shaped laser pulses and allows for new applications in which the light is used as photonic catalyst in chemical reactions [34,35]. The idea to control the selectivity of product formation in a chemical reaction using ultrashort pulses is based on exploitation of the coherence properties of laser radiation due to quantum mechanical interference effects. For this, either the proper choice of the pulse phase or of time duration and delay between the pump and the probe (or dump) step can be employed. Pioneering conceptual work [36–40] was followed by studies using variationally optimized electric fields, [41, 42] allowing to address further application aspects [42–47]. Technological progress due to fs-pulse shapers allowed manipulation of ultrashort laser pulses [48–52]. Finally, closed-loop learning control (CLL) was introduced by Judson and Rabitz [53]. The first experimental realizations of this approach [54, 55] opened the possibility to apply optimal control to more complex systems.

In order to establish the connection of the experimentally optimized pulse shapes with the underlying dynamical processes as well as between theoretically and experimentally optimized pulses, developments of theoretical methods are needed which allow for the design of interpretable laser pulses for complex systems. To avoid the obstacle of precalculating multidimensional PES, ab-initio adiabatic and in particular nonadiabatic MD "on the fly" is particularly suitable provided that an accurate description of the electronic structure is feasible [56]. In addition, this approach offers the advantages that the MD "on the fly" can be applied to relatively complex systems and can be also directly connected with different procedures for optimal control [56–58]. Moreover, as recently proposed by us, it is particularly convenient to introduce the field directly in the nonadiabatic dynamics which can be then optimized as desired [59].

In this contribution we present the development of theoretical methods for the simulation of nonadiabatic dynamics and its manipulation by laser fields in complex systems accounting for all degrees of freedom. Therefore, we will first describe nonadiabatic dynamics "on the fly" in the frame of TDDFT in Sect. 17.2 and then the procedure for the simulation of time-resolved photoelectron spectra (TRPES) based on the nonadiabatic dynamics in Sect. 17.3. The application of our theoretical approach will be illustrated on the ultrafast photodynamics of furan and compared with the experimental TRPES in Sect. 17.4. Furan is a fundamental five membered aromatic heterocycle serving as a structural unit in various biological substances. Then, in Sect. 17.5 we introduce the field-induced surface hopping method (FISH) which is based on the combination of quantum electronic state population dynamics with classical nuclear dynamics carried out "on the fly". FISH opens the possibility of broad applications for simulation of spectroscopic observables as well as to control dynamics employing shaped laser fields. The scope of the FISH method will be illustrated in Sect. 17.6 on the optimal dynamic discrimination (ODD) [60, 61] of the two molecules flavin mononucleotide (FMN) and riboflavin (RBF), which exhibit almost identical spectroscopical features [62,63]. The selective identification of target molecules in the presence of a structurally and spectroscopically similar background by optimally shaped laser fields opens prospects for new applications in multiple areas of science and engineering. Finally, conclusions and outlook will be given in Sect. 17.7.

17.2   Nonadiabatic Dynamics "on the Fly" in the Framework of Time-Dependent Density Functional Theory (TDDFT)

The TDDFT represents an efficient generally applicable method for the treatment of the optical properties in complex systems whose performance and accuracy have been steadily improved [64, 65]. Due to this fact, a variety of approaches for performing TDDFT-based nonadiabatic dynamics simulations "on the fly" have been developed and successfully applied in recent years [16–18, 20–23]. TDDFT is still one of the most practical means to address a large class of problems if for the chosen system no description of long-range charge transfer transitions, dispersion interaction and multireference character is needed. Moreover, recent developments of new functionals promise to substantially improve the description of long-range charge transfer transitions [66–70]. In connection with nonadiabatic processes, the ability of linear response TDDFT to describe conical intersections between excited states and the ground state has been critically examined in the literature [22]. The conclusion has been made that while the topology of the S_1-S_0 crossing region may be not exact, this does not substantially influence the relaxation pathways and photochemistry of the studied examples. Successful applications of TDDFT nonadiabatic dynamics steadily grow and have already significantly contributed towards understanding of the mechanisms of photochemical processes in complex systems [17–19,22,71] and have also been verified by comparison with experimental data [19,72].

In this section we briefly outline our formulation of the nonadiabatic dynamics in the framework of TDDFT using localized Gaussian basis sets combined with Tully's surface hopping (TSH) method [13]. Within the TSH procedure, nonadiabatic processes are simulated by propagating ensembles of classical trajectories parallel to the solution of the time-dependent Schrödinger equation which determines the quantum-mechanical electronic state populations. For this purpose, along each classical trajectory an electronic wavefunction $|\Psi(\mathbf{r};\mathbf{R}(t))\rangle$ is defined in terms of the adiabatic electronic state wavefunctions according to:

$$|\Psi(\mathbf{r};\mathbf{R}(t))\rangle = \sum_K C_K(t)|\Psi_K(\mathbf{r};\mathbf{R}(t))\rangle, \qquad (17.1)$$

where $|\Psi_K(\mathbf{r};\mathbf{R}(t))\rangle$ represents the wavefunction for the electronic state K while the $C_K(t)$ are the time-dependent expansion coefficients. The time evolution of the expansion coefficients for a given trajectory can be obtained by numerical solution of the time-dependent Schrödinger equation:

$$i\hbar \dot{C}_K(t) = E_K(\mathbf{R}(t))C_K(t) - i\hbar \sum_I C_I(t)\left\langle \Psi_K(\mathbf{r};\mathbf{R}(t))\left|\frac{\partial \Psi_I(\mathbf{r};\mathbf{R}(t))}{\partial t}\right.\right\rangle \qquad (17.2)$$

where the bracket in the second term corresponds to the nonadiabatic coupling D_{KI} between the states I and K, which can be approximately calculated using the finite difference approximation for the time derivative [73]:

$$D_{KI}\left(\mathbf{R}\left(t+\frac{\Delta t}{2}\right)\right) \approx \frac{1}{2\Delta t}\left(\langle \Psi_K(\mathbf{r};\mathbf{R}(t))|\Psi_I(\mathbf{r};\mathbf{R}(t+\Delta t))\rangle\right.$$
$$\left. - \langle \Psi_K(\mathbf{r};\mathbf{R}(t+\Delta t))|\Psi_I(\mathbf{r};\mathbf{R}(t))\rangle\right) \qquad (17.3)$$

where Δt is the timestep used for the integration of the classical Newton's equations of motion.

The numerical solution of the Eq. 17.2, obtained e.g. using the fourth order Runge-Kutta procedure, provides the time-dependent electronic state coefficients $C_K(t)$ which can be used to define the hopping probabilities that are needed for the electronic state switching procedure in the frame of the TSH approach. The hopping probabilities $P_{I\rightarrow K}$ for switching from state I to state K can be either calculated after each nuclear dynamics time step Δt or, alternatively, after each of the much smaller time steps $\Delta \tau$ used for the integration of the electronic Schrödinger equation (17.2), as recently introduced by us [18].

In the latter case, the hopping probability is defined as:

$$P_{I\rightarrow K}(\tau) = -2\frac{\Delta \tau [Re(C_K^*(\tau)C_I(\tau)D_{KI}(\tau))]}{C_I(\tau)C_I^*(\tau)}. \qquad (17.4)$$

An alternative procedure for calculating the hopping probabilities can be also based only on the occupations of the electronic states, represented by diagonal density matrix elements $\rho_{II} = C_I^*(t)C_I(t)$ and $\rho_{KK} = C_K^*(t)C_K(t)$ [74,75].
The probability for hopping from state I to state K can then be defined as:

$$P_{I\rightarrow K} = \Theta(-\dot{\rho}_{II})\Theta(\dot{\rho}_{KK})\frac{-\dot{\rho}_{II}}{\rho_{II}}\frac{\dot{\rho}_{KK}}{\sum_L \Theta(\dot{\rho}_{LL})\dot{\rho}_{LL}}\Delta t \qquad (17.5)$$

This probability is nonzero only if the population of the Ith state is decreasing and the population of the Kth state is increasing, which is represented by the Θ function. The summation in the denominator is performed over all states L whose population is also growing. It should be pointed out that this equation requires only the calculation of the hopping probabilities at each nuclear time step due to the fact that populations generally vary more slowly than the coherences $C_K^*(\tau)C_I(\tau)$ which are employed in Eq. 17.4. This is particularly useful in the context of field-driven multistate dynamics during which the laser field is varying very fast as it will be shown in Sect. 17.6.

The necessary ingredients for performing TSH simulations are the forces (energy gradients) in the ground and excited electronic states as well as the nonadiabatic couplings $D_{KI}(\mathbf{R}(t + \frac{\Delta}{2}))$. While the calculation of excited state forces in the framework of TDDFT is already a standard procedure available in many commonly used quantum chemical program packages, the procedure for the calculation of nonadiabatic couplings in the framework of linear response TDDFT has been developed only recently using plane wave basis sets by Röthlisberger et al. [21–23], as well as using localized Gaussian basis sets by us [17, 18]. In the following after introducing the representation of the electronic wave function within the Kohn-Sham linear response method, we briefly outline our approach for the calculation of the nonadiabatic couplings using localized Gaussian basis sets.

In order to calculate nonadiabatic couplings in the framework of the TDDFT method a representation of the wavefunction based on Kohn-Sham (KS) orbitals is required. Since in the linear response TDDFT method the time-dependent electron density contains only contributions of single excitations from the manifold of occupied to virtual KS orbitals, a natural ansatz for the excited state electronic wavefunction is the configuration interaction singles (CIS)-like expansion:

$$|\Psi_K(\mathbf{r}; \mathbf{R}(t))\rangle = \sum_{i,a} c_{i,a}^K \left| \Phi_{i,a}^{CSF}(\mathbf{r}; \mathbf{R}(t)) \right\rangle. \tag{17.6}$$

where $\left| \Phi_{i,a}^{CSF}(\mathbf{r}; \mathbf{R}(t)) \right\rangle$ represents a singlet spin adapted configuration state function (CSF) defined as:

$$\left| \Phi_{i,a}^{CSF}(\mathbf{r}; \mathbf{R}(t)) \right\rangle = \frac{1}{\sqrt{2}} \left(\left| \Phi_{i\alpha}^{a\beta}(\mathbf{r}; \mathbf{R}(t)) \right\rangle + \left| \Phi_{i\beta}^{a\alpha}(\mathbf{r}; \mathbf{R}(t)) \right\rangle \right), \tag{17.7}$$

where $\left| \Phi_{i\alpha}^{a\beta}(\mathbf{r}; \mathbf{R}(t)) \right\rangle$ and $\left| \Phi_{i\beta}^{a\alpha}(\mathbf{r}; \mathbf{R}(t)) \right\rangle$ are Slater determinants in which one electron has been promoted from the occupied orbital ϕ_i to the virtual orbital ϕ_a with spin α or β, respectively. This ansatz can be used to calculate the nonadiabatic coupling as described below, but more generally, it can provide the expectation values of any observable of interest, e.g. transition dipole moments between excited states as we have shown in Ref. [19]. In the context of nonadiabatic dynamics the accuracy of this representation of the wavefunction has been previously demonstrated in our work on pyrazine [17, 19] and benzylideneaniline [18]. The expansion coefficients

$c_{i,a}^K$ in Eq. 17.6 are determined by requiring that the wavefunction in Eq. 17.6 gives rise to the same density response as the one obtained by the linear response TDDFT procedure. Their precise connection to the TDDFT eigenvectors has been shown in Ref. [18].

The electronic structure of isolated molecular systems is most naturally described by using Gaussian type atomic orbitals (AO's) as basis functions in contrast to plane waves, which represent the natural choice in extended periodic systems. Here we present the approach for the calculation of the nonadiabatic couplings using KS orbitals expanded in terms of localized Gaussian atomic basis sets. This formulation is particularly convenient since it can be coupled with commonly used quantum chemical DFT codes.

In order to calculate the nonadiabatic couplings according to the discrete approximation given by Eq. 17.3 the overlap between two CI wavefunctions at times t and $t + \Delta t$ along the nuclear trajectory $\mathbf{R}(t)$ is needed:

$$\langle \Psi_K(\mathbf{r}; \mathbf{R}(t)) | \Psi_I(\mathbf{r}; \mathbf{R}(t + \Delta t)) \rangle$$
$$= \sum_{ia} \sum_{i'a'} c_{i,a}^{*K} c_{i',a'}^I \left\langle \Phi_{i,a}^{CSF}(\mathbf{r}; \mathbf{R}(t)) \middle| \Phi_{i',a'}^{CSF}(\mathbf{r}; \mathbf{R}(t + \Delta t)) \right\rangle \qquad (17.8)$$

The overlap of the CSF's in Eq. 17.8 can be reduced to the overlap of singly excited Slater determinants using Eq. 17.7, which can be further decomposed to the overlap of spatial KS orbitals $\phi_i(t)$ and $\phi_{i'}'(t + \Delta t)$ as described in Refs. [17, 18]. The spatial KS orbitals can be expressed in terms of atomic basis functions $b_k(R(t))$ according to:

$$\phi_i(t) = \sum_{k=1}^n c_{ik}(t) \, b_k(\mathbf{R}(t)) \qquad (17.9)$$

with the molecular orbital (MO) coefficients $c_{ik}(t)$. This leads to the final expression for the overlap integral of two spatial KS orbitals at times t and $t + \Delta t$:

$$\left\langle \phi_i(t) \middle| \phi_{i'}'(t + \Delta t) \right\rangle = \sum_{k=1}^n \sum_{m=1}^n c_{ik}(t) c_{jm}'(t + \Delta t) \left\langle b_k(\mathbf{R}(t)) \middle| b_m'(\mathbf{R}(t + \Delta t)) \right\rangle.$$
$$(17.10)$$

It should be noticed that since the two sets of basis functions $b_k(\mathbf{R}(t))$ and $b_m'(\mathbf{R}(t + \Delta t))$ are centered at different positions $\mathbf{R}(t)$ and $\mathbf{R}(t + \Delta t)$ they do not form an orthonormal basis set. Therefore, in order to calculate nonadiabatic couplings along each classical trajectory the overlap integrals between moving basis functions are calculated at successive nuclear time steps and the KS MO coefficients and linear response eigenvectors are utilized to transform the overlap integrals. In order to eliminate possible random phase variations of the nonadiabatic coupling, the phases of the CI-like wavefunction coefficients (cf. Eq. 17.6) and of the Kohn-Sham orbital coefficients (cf. Eq. 17.9) are aligned in each nuclear timestep to the phases of the previous step.

17.3 Simulation of Time-Resolved Photoelectron Spectra (TRPES)

The time-resolved photoelectron spectroscopy represents a powerful approach for interrogation of nonadiabatic processes. The basic principle of this technique involves the creation of a coherent superposition of the ground and excited electronic states of the studied system by an ultrashort laser pulse. This gives rise to a wavepacket in the excited electronic states whose time evolution is subsequently probed through the photoionization by a time-delayed ultrashort probe pulse. The kinetic energy and angular distribution of the released photoelectrons reflect therefore the character of the electronic state which has been ionized. Since during the excited state dynamics this character can change, e.g. due to the passage through a conical intersection, the above-mentioned observables offer a sensitive probe for the nonadiabatic transitions [76].

Our method for the simulation of TRPES pump-probe signals in the frame of the Wigner distribution approach [6,56] is based on the propagation of an ensemble of classical trajectories "on the fly". For weak electric fields of Gaussian form, a perturbation theory expression for the final quantum state populations leads to an analytical formula for the pump-probe signal. This approach provides a general tool for simulation of ultrafast processes and femtosecond signals in complex systems, involving both adiabatic and nonadiabatic dynamics [56]. However, for the simulation of TRPES a modification has to be introduced which takes into account that a part of the probe-pulse energy E_{pr} changes into the photoelectron kinetic energy (PKE). Furthermore, the vibrational states of the ionized system can be also taken into account as discussed in Ref. [77]. The photoionization process produces photoelectrons with kinetic energies E ranging from zero up to the maximal value of $PKE_{max} = E_{pr} - IP(t_D)$, where IP is the ionization potential. The intensity of the photoelectrons at a particular PKE is proportional to the electronic transition dipole moments μ_{ik} between the bound state i and the ionized continuum state k as well as to the Franck-Condon (FC) factors $F_{ik,v}$ between the neutral and the individual cationic vibrational states. The TRPES signal at the time delay t_D in the frame of the Wigner distribution approach assumes then the following analytic form:

$$S_{TRPES}(t_D, E) \sim \int\int d\mathbf{q}_0 d\mathbf{p}_0 \int_0^\infty d\tau_1 \exp\left\{-\frac{(\tau_1 - t_D)^2}{\sigma_{pu}^2 + \sigma_{pr}^2}\right\} \sum_{i,k} |\mu_{ik}(\mathbf{q}(\tau_1; \mathbf{q}_0, \mathbf{p}_0), E)|^2$$

$$\times \int_0^\infty dE_{k,v} F_{ik,v} \exp\left\{\frac{-\sigma_{pr}^2}{\hbar^2}\left[E_{pr} - V_{ik}(q(\tau_1; \mathbf{q}_0, \mathbf{p}_0)) - E_{k,v} - E\right]^2\right\}$$

$$\times \exp\left\{\frac{-\sigma_{pu}^2}{\hbar^2}\left[E_{pu} - V_{i0}(\mathbf{q}_0)\right]^2\right\} P_{00}(\mathbf{q}_0, \mathbf{p}_0) \qquad (17.11)$$

Since in our classical simulation the FC factors are not available, we have assigned them a constant value for the whole PKE interval $[0, PKE_{max}]$ [56, 78]. Therefore the integration over the vibrational levels $E_{k,v}$ of the ionized system can be performed analytically. This approximate treatment is verified by comparison with experimental TRPES signals [79]. In the above expression σ_{pu} (σ_{pr}) and $E_{pu} = \hbar\omega_{pu}$ ($E_{pr} = \hbar\omega_{pr}$) are the pulse durations and excitation energies for the pump and probe step with time delay t_D. $V_{ki}(\mathbf{q}_1(\tau_1; \mathbf{q}_0, \mathbf{p}_0))$ labels the time-dependent energy gap between the electronic state i in which the dynamics takes place and the ionized electronic state k that is used for probing. Both are obtained from the ab initio MD "on the fly" [56]. The initial coordinates and momenta $\mathbf{q}_0$ and $\mathbf{p}_0$ needed for the dynamics simulation can be sampled from a canonical Wigner distribution for all normal modes at the given temperature according to:

$$P_{00}(\mathbf{q}_0, \mathbf{p}_0) = \prod_{i=1}^{N} \frac{\alpha_i}{\pi\hbar} \exp\left[-\frac{\alpha_i}{\hbar\omega_i}(p_{i0}^2 + \omega_i^2 q_{i0}^2)\right], \qquad (17.12)$$

where ω_i represents the frequency of the i'th normal mode and $\alpha_i = \tanh(\hbar\omega_i/2k_bT)$ [56]. $V_{i0}(\mathbf{q}_0)$ are the excitation energies of the initial ensemble. The signal is calculated by averaging over the whole initial distribution $P_{00}(\mathbf{q}_0, \mathbf{p}_0)$ given by the ensemble of trajectories. Notice, that expression (17.11) is valid under the assumption of weak electric fields due to the perturbation theory treatment [6, 56].

The simulation of the TRPES thus involves three steps: (1) The ensemble of initial conditions is generated by sampling the Wigner distribution function corresponding to the canonical ensemble at the given temperature. (2) The ensemble of trajectories is propagated using nonadiabatic MD "on the fly". (3) The TRPES is calculated by averaging over the ensemble of trajectories employing the analytical expression (17.11).

17.4 Application of the Nonadiabatic Dynamics "on the fly" for the Simulation of Ultrafast Observables of Furan: Comparison with Experiment

The simulation of ultrafast observables such as TRPES allows to make direct comparison with experimental data and thus to reveal the dynamical processes involved in the excited state relaxation and their time scales. Moreover, the new methods for simulation of ultrafast processes challenge also the development of novel experimental techniques with increasing resolution.

We wish to show that ultrafast time-resolved photoelectron imaging (TR-PEI) together with nonadiabatic ab initio dynamics "on the fly" accounting for all degrees of freedom allows to elucidate precisely the photophysics and photochemistry of furan. Our theoretical simulation of photoionization is based on the methods described in Sects. 17.2 and 17.3. The theoretical analysis is focused on the time-dependent photoelectron signal intensity and PKE distribution. The complementary

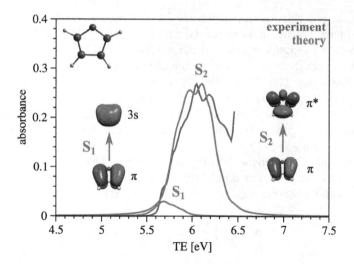

Fig. 17.1 Comparison of the theoretical thermally broadened absorption spectrum of furan (*red*) for the first $S_1[^1A_2(\pi3s)] \leftarrow S_0(^1A_1)$ and second $S_2[^1B_2(\pi\pi^*)] \leftarrow S_0(^1A_1)$ excited state obtained from 240 structures sampled from the thermal ensemble at T = 300 K with the measured absorption spectrum at room temperature (*blue*). The discrete absorption lines for each member of the ensemble were convoluted with a Lorentzian function with a width of 0.1 eV and added together. The equilibrium structure of furan in the neutral ground state as well as the dominant excitations of the transitions to the S_1 ($\pi3s$) and S_2 ($\pi\pi^*$) states are also shown

experimental data have been obtained by TR-PEI with an unprecedented time resolution of 22 fs [79] using sub-20 fs pulses at 260 and 200 nm generated by the multi-colour filamentation method [80, 81]. The combination of the experimental findings with the theoretical simulations reveals ultrafast deactivation of excited furan through internal conversion from S_2 over S_1 to the ground state [79].

The simulations have been performed in a manifold consisting of the ground and the three lowest excited states. The energies, gradients as well as nonadiabatic couplings needed to carry out the nonadiabatic dynamics have been calculated "on the fly" using the hybrid PBE0 functional [82] combined with the 6-311G**++ basis set [83] containing also diffuse functions. This level of theory for electronic structure describes accurately the stationary absorption properties and is suitable for performing the dynamics simulations as discussed in Ref. [79]. Notice that recently the accuracy of the TDDFT method for the description of nonadiabatic dynamics in heterocyclic organic molecules has been validated against the highly correlated multireference ab initio methods on the example of the pyrrole molecule [84]. For the further computational details cf. Ref. [79] Based on the nonadiabatic MD trajectories, the TRPES signal was calculated according to Eq. 17.11, assuming a constant value for the transition dipole moments μ_{ik} in the whole energy range.

The experimental photoabsorption spectrum of furan vapour at room temperature as well as our TDDFT absorption spectrum simulated also at room temperature are shown in Fig. 17.1. The good agreement between experiment and theory allows

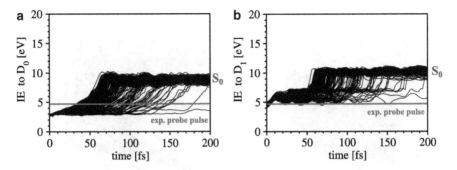

Fig. 17.2 Time-dependent ionization energy IE calculated along 240 nonadiabatic trajectories for (**a**) the cationic ground state D_0 and (**b**) the cationic first excited state D_1. The *red line* at 4.7 eV indicates the experimental probe pulse energy. Reprinted with permission from Ref. [79]. Copyright 2010, American Institute of Physics

for straightforward assignment of the transitions. The strong absorption feature between 5.8 and 6.2 eV is caused by the $S_2[^1B_2(\pi\pi^*)] \leftarrow S_0(^1A_1)$ transition. The weak feature in the low energy part (5.6–6.0 eV) of the absorption spectrum is due to the $S_1[^1A_2(\pi 3s)] \leftarrow S_0(^1A_1)$ transition involving the Rydberg 3s state, which is in agreement with previous work [85, 86]. Since both the oscillator strength of this state and the overlap with the pump pulse spectrum peaked at 6.2 eV are very small we expect that the dominant excitation occurs to the S_2 state which overlaps well with the pump pulse spectrum. This state is thus used as the starting point for the nonadiabatic dynamics simulations discussed below. The probe pulse (260 nm, 4.7 eV) has no overlap with the UV absorption spectrum of furan. Since the sum of the pump (6.2 eV) and probe (4.7 eV) photon energies is 10.9 eV, it is energetically possible to ionize furan at equilibrium geometry to two cation states, D_0 (ionization energy, IE = 8.9 eV [87]) and D_1 (IE = 10.3 eV [88]) by (1+1') resonance-enhanced multiphoton ionization.

The nature of the ionization process can be determined from the calculated time-dependent ionization energies between the current excited state in which the dynamics takes place, and the cationic D_0 and D_1 states as shown in Fig. 17.2. It can be seen that at very short time <10 fs both cationic states are accessible by the experimental probe pulse, but after t >10 fs the only energetically possible transition occurs to the D_0 state. After the ensemble of trajectories returns to the ground state S_0 no cationic states are accessible anymore, therefore, no photoionization occurs. This is clearly evidenced by the simulated TRPES shown in Fig. 17.3a which reflects ultrafast deactivation of furan by the decreasing intensity of the signals in the time regime between 10 and 100 fs. The calculated photoelectron intensities at selected PKEs presented in Fig. 17.3c show an increase of the signals for decreasing PKEs, exhibiting maxima at short time delays which are shifting to longer time delays for lower PKE values. This is in agreement with the experimental findings presented in Fig. 17.3b and d, which were obtained by photoelectron imaging spectroscopy with a time resolution of 22 fs [79]. It should be noticed that for very short time

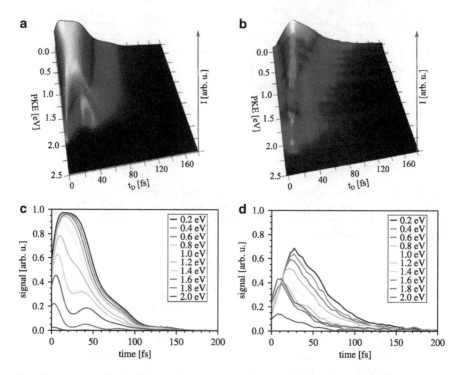

Fig. 17.3 (**a**) Simulated TRPES of furan. (**b**) Experimental TRPES obtained from pump-probe photoelectron imaging spectroscopy of furan. (**c**) Time evolution of theoretically obtained photoelectron intensities at selected PKE values. (**d**) Time-evolution of experimental photoelectron intensities at selected PKEs

delays our simulation provides only qualitative results for TRPES, since we excite S_2 instantaneously and do not consider the overlap between pump and probe pulses.

The agreement between simulated and measured TRPES in the low energy regime as evidenced by Fig. 17.3 allows for complete assignment of the underlying ultrafast processes to the measured features. For most trajectories, the transition from S_2 to S_1 occurs at very short times, resulting in a lifetime of the adiabatic S_2 state of 9.2 fs, as can be seen from the time-dependent adiabatic populations shown in Fig. 17.4a. The S_2 population is almost completely transferred into the S_1 state after 20 fs. The lifetime of the adiabatic S_1 state is ~60 fs, and return to the ground state is completed after 140 fs (cf. Fig. 17.4a) which is also reflected in the decrease of TRPES signal intensities. Despite the very fast S_2–S_1 transition, the $\pi - \pi^*$ diabatic character remains largely preserved, such that transition to the ground state occurs mostly directly from the $\pi - \pi^*$ state. This is illustrated in Fig. 17.4b which shows diabatic state populations obtained by decomposing the adiabatic populations in terms of the diabatic characters of the involved states. As can be seen, the diabatic $\pi - 3s$ Rydberg state is only weakly populated during the simulation. Although the theoretically obtained lifetime of the S_1 state of ~60 fs is

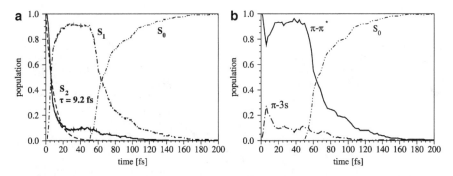

Fig. 17.4 Time-dependent (**a**) calculated adiabatic and (**b**) approximate diabatic populations of the ground and two excited states of furan. The characters of the adiabatic states (S_0, S_1 and S_2) as well as the diabatic states (S_0, $\pi - \pi^*$ and π-3s) are given. The lifetime $\tau = 9.2$ fs of the adiabatic S_2 state was determined by exponential fit (*dashed line*) (Reprinted with permission from Ref. [79], Copyright 2010, American Institute of Physics)

longer than the experimental value of 29 fs [79], the theory correctly predicts the trends for timescales of internal conversions: short for S_2–S_1 ($\sim$9 fs) and longer for S_1–S_0 ($\sim$60 fs). Analysis of the nonadiabatic MD trajectories reveals that the geometric relaxation in excited states takes place within the C–O–C subunit [79] and the main channel after returning to the ground state leads to formation of hot furan (I), as also illustrated in Fig. 17.5. Two other possible minor channels involving breaking of the C–O bond and leading to formation of 2,3-butadienal (II) and cyclopropen-3-carbaldehyde (III) are reached with very low probability (cf. Fig. 17.5) at later times after the transition to the ground state. Thus, bond breaking occurs sequentially in the ground state.

In summary, time-resolved photoelectron imaging spectroscopy with the very high time-resolution of 22 fs using two-colour deep UV pulses and ab initio nonadiabatic dynamics simulations have for the first time revealed the ultrafast deactivation processes from S_2 to S_0 state in furan. Joint theoretical and experimental results represent a general approach for investigation of ultrafast photochemical reactions, allowing to identify the fingerprints of the character of electronic states with an unprecedented precision.

17.5 Field-Induced Surface-Hopping Method (FISH) for Simulation and Control of Ultrafast Photodynamics

The simulation of laser-induced dynamical phenomena provides a basis for a deeper understanding of molecular processes under the influence of light. This is particularly interesting in the context of optimal control by shaped laser pulses.

In order to address complex systems, semiclassical methods in which the nuclear degrees of freedom can be treated efficiently have been developed in the frame of the Wigner distribution approach [56–58]. However, since the interaction with the laser

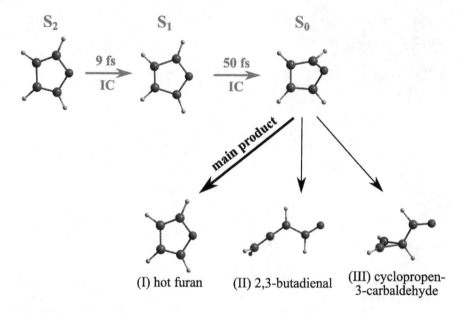

Fig. 17.5 Schematic representation of the photodynamics of furan obtained from nonadiabatic dynamics. The S_2/S_1 and S_1/S_0 internal conversions and the corresponding time scales are shown in *red* while the products in the ground state are indicated by *black arrows* (main product *bold arrow*, other two products *thin arrows*). The minor channels have energies of 1.2 (II) and 2.1 eV (III) above the main channel (I)

field has been described using perturbation theory these methods are limited only to processes in relatively weak fields. For this reason, new theoretical approaches for the simulation of dynamics driven by moderately strong laser fields (below the multielectron ionization limit) are particularly desirable. Such fields open a rich variety of pathways for the control of ultrafast dynamics in complex systems.

Therefore, we present here our semiclassical "Field-Induced Surface Hopping" (FISH) method [59] for the simulation and control of the laser-driven coupled electron-nuclear dynamics in complex molecular systems including all degrees of freedom. It is based on the combination of quantum electronic state population dynamics with classical nuclear dynamics carried out "on the fly". The idea of the method is to propagate independent trajectories in the manifold of adiabatic electronic states and allow them to switch between the states under the influence of the laser field. The switching probabilities are calculated fully quantum mechanically. The application of our FISH method will be illustrated in Sect. 17.6 on the example of optimal dynamic discrimination (ODD) of two almost identical flavin molecules.

The starting point for the description of laser-driven multistate dynamics is the semiclassical limit of the Liouville-von Neumann (LvN) equation for the quantum mechanical density operator $\hat{\rho}$,

$$i\hbar\dot{\hat{\rho}} = [\hat{H}_0 - \mu \cdot \mathbf{E}(t), \hat{\rho}].$$ (17.13)

$\hat{H}_0$ represents the field-free nuclear Hamiltonian for a molecular system with several electronic states in the Born-Oppenheimer approximation, and the interaction with the laser field $\mathbf{E}(t)$ is described using the dipole approximation. The semiclassical limit can be straightforwardly derived in the framework of the Wigner phase space representation [2, 3] of quantum mechanics. The equations of motion for the phase space representation of the density matrix elements involving an arbitrary number of electronic states then read:

$$\dot{\rho}_{ii} = \{H_i, \rho_{ii}\} - \frac{2}{\hbar} \sum_j \text{Im}\left(\mu_{ij} \cdot \mathbf{E}(t)\rho_{ji}\right) \tag{17.14}$$

$$\dot{\rho}_{ij} = -i\omega_{ij}\rho_{ij} + \frac{i}{\hbar}\mu_{ij} \cdot \mathbf{E}(t)\left(\rho_{jj} - \rho_{ii}\right) + \frac{i}{\hbar} \sum_{k \neq i,j} \left(\mu_{ik} \cdot \mathbf{E}(t)\rho_{kj} - \mu_{kj} \cdot \mathbf{E}(t)\rho_{ik}\right)$$

$$\tag{17.15}$$

where the diagonal elements ρ_{ii} determine the quantum mechanical state populations and the off-diagonal elements ρ_{ij} describe the coherence. The curly braces denote the Poisson brackets, H_i are the Hamiltonian functions for the respective electronic state i. The quantity ω_{ij} is the energy gap and μ_{ij} the transition dipole moment between the electronic states i and j.

In order to connect Eqs. 17.14–17.15 with classical molecular dynamics "on the fly" the diagonal density matrix elements $\rho_{ii}(\mathbf{q}, \mathbf{p}, t)$ which are functions of the coordinates $\mathbf{q}$ and momenta $\mathbf{p}$ can be represented by independent trajectories propagated in the ground and excited electronic states, respectively. Thus, employing a number of N_{traj} trajectories, $\rho_{ii}(\mathbf{q}, \mathbf{p}, t)$ can be represented by a swarm of time-dependent δ functions

$$\rho_{ii}(\mathbf{q}, \mathbf{p}, t) = \frac{1}{N_{traj}} \sum_k \theta_i^k(t)\, \delta(\mathbf{q} - \mathbf{q}_k^i(t; \mathbf{q}_0, \mathbf{p}_0))\delta(\mathbf{p} - \mathbf{p}_k^i(t; \mathbf{q}_0, \mathbf{p}_0)) \tag{17.16}$$

where $(\mathbf{q}_k^i, \mathbf{p}_k^i)$ signifies a trajectory propagated in the electronic state i and the parameter $\theta_i^k(t)$ is one if the trajectory k resides in the state i and zero otherwise [56]. The population transfer between the electronic states is achieved by a process in which the trajectories are allowed to switch between the states. This procedure is related to Tully's surface hopping method [13] for field-free nonadiabatic transitions in molecular systems. However, in our case the coupling between the states is induced by the applied laser field. The probabilities for switching the electronic state can be calculated according to Eq. 17.5 given in Sect. 17.2.

The simulation of the laser-induced dynamics in the framework of our FISH method using the above derived approach is performed in the following three steps:

1. We generate initial conditions for an ensemble of trajectories by sampling e.g. the canonical Wigner distribution function (cf. Eq. 17.12) or a long classical trajectory in the electronic ground state.

2. Along each trajectory $\mathbf{R}(t)$ which is propagated in the framework of MD "on the fly", we calculate the density matrix elements ρ_{ij} by numerical integration. If the initial electronic state is a pure state as it is in our case, the set of equations 17.14–17.15 is equivalent to the time-dependent Schrödinger equation in the representation of adiabatic electronic states:

$$i\hbar \dot{c}_i(t) = E_i(\mathbf{R}(t))c_i(t) - \sum_j \mu_{ij}(\mathbf{R}(t)) \cdot \mathbf{E}(t)c_j(t) \qquad (17.17)$$

where $c_i(t)$ are the expansion coefficients of the electronic wavefunction from which the density matrix elements can be calculated as $\rho_{ij} = c_i^* c_j$.

If the intrinsic nonadiabatic coupling D_{ij} of the electronic states (cf. Eq. 17.3) also has to be taken into account, the Eq. 17.17 can be generalized to

$$i\hbar \dot{c}_i(t) = E_i(\mathbf{R}(t))c_i(t) - \sum_j [\mu_{ij}(\mathbf{R}(t)) \cdot \mathbf{E}(t) + i\hbar D_{ij}(\mathbf{R}(t))] c_j(t). \qquad (17.18)$$

In this way, after the duration of the applied field is over, field free multistate nonadiabatic dynamics can be further carried out. The Eqs. 17.17 or 17.18 are solved numerically using e.g. the fourth order Runge-Kutta procedure.

The nuclear trajectories $\mathbf{R}(t)$ are obtained by solution of Newton's equations of motion where the necessary forces are obtained from the energy gradients in the actual electronic state in which the trajectory is propagated.

In contrast to field-free nonadiabatic dynamics, in the presence of electric fields the energy of a molecular system is not conserved due to the interaction with the field. Therefore, when exposed to a long intense laser pulse, molecules can accumulate energy and eventually get heated, which for isolated molecules can finally lead to fragmentation. However, if the molecule is interacting with an environment such as solution, the excess thermal energy can be dissipated. For approximate inclusion of these effects dissipative Langevin dynamics instead of Newtonian dynamics can be employed. The solution of the Newton or Langevin equations of motion provides continuous nuclear trajectories which reside in different electronic states according to the quantum mechanical occupation probabilities given by ρ_{ii}.

3. In order to determine in which electronic state the trajectory is propagated we calculate the hopping probabilities under the influence of the field and decide if the trajectory is allowed to change the electronic state by using a random number generator. For a general number of states the hopping probability can be calculated according to Eq. 17.5.

Notice that while the trajectories jump between the electronic states at a given time, all density matrix elements are propagated continuously over the entire time according to Eqs. 17.14–17.15, or alternatively either Eq. 17.17 or 17.18. Although the individual trajectory is allowed to jump, the fraction of trajectories in a given state, which represents ρ_{ii} as an ensemble average, is also a continuous function of time. The phase of the electronic wavefunction is preserved and our procedure gives rise to the full quantum mechanical coherent state population, therefore being able to mimic laser-induced processes such as coherent Rabi oscillations between the electronic states [59].

Our semiclassical FISH method is a valuable tool for the simulation of ultrafast laser-driven coupled electron-nuclear dynamics involving several electronically excited states in complex molecular systems. It can be applied to simulate spectroscopic observables [77] as well as to control the dynamics employing shaped laser fields and thus to steer molecular processes. Since the laser field enters the equations for population dynamics directly, the combination with the optimal control theory is straightforward. The electric field can be iteratively optimized using e.g. evolutionary algorithms [58, 89] as it has been illustrated in Ref. [59]. For the propagation of classical trajectories the whole spectrum of methods ranging from empirical force fields, semiempirical to ab initio quantum chemical methods can be employed. Moreover, in addition to isolated systems in the gas phase, molecular systems interacting with different environments such as solvent, bioenvironment, surfaces or metallic nanostructures can be also treated. The FISH method allows not only to obtain optimized pulses but also to analyze their shapes on the basis of molecular dynamics "on the fly". In this way the comparison between theoretically optimized laser fields with those obtained from experiments, e.g. using the CLL procedure, allows to assign the underlying processes to the specific forms of the pulses. By this means the inversion problem can be addressed and important parts of the PES could be constructed. Altogether, the FISH method opens new avenues to perform the optimization of laser pulses for different exciting applications as it will be illustrated in Sect. 17.6.

17.6 Application of the FISH Method for the Optimal Dynamic Discrimination

We wish to reveal the mechanism for the optimal dynamic discrimination between the very similar biochromophores riboflavin (RBF) and flavin mononucleotide (FMN) using optimally shaped laser fields. Our FISH simulations utilize experimentally optimized laser fields and show that the fluorescence depletion ratio between two molecules can be manipulated with such fields, eventually achieving discrimination between them. Moreover, these results validate for the first time the experimental optimal control technique applied on complex systems [63].

The general concept of the optimal dynamic discrimination (ODD) has been recently proposed by Rabitz and Wolf et al. [60, 61]. The idea of the ODD relies on a theoretical analysis which has shown that quantum systems differing even infinitesimally may be distinguished by means of their dynamics when a suitably shaped ultrafast control field is applied. In the case of the two similar flavins (differing only by replacement of H by $PO(OH)_2$ in the side chain) the controlled depletion of the fluorescence signal has been used as a discriminating observable [62]. The schematic representation of the discrimination process is presented in

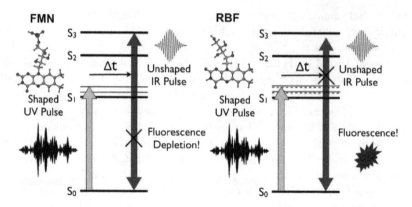

Fig. 17.6 Schematic illustration of the discrimination of FMN and RBF by fluorescence depletion. Excitation with a shaped UV laser pulse leads to transition from S_0 to S_1 state, as indicated by the *light grey arrow*. After a time-delay Δt during which dynamical processes take place, an unshaped IR pulse is applied. In the case of FMN (*left* part of the figure), this leads to transitions to higher excited states where irreversible processes such as ionization can occur (*dark arrow*), consequently the fluorescence gets depleted (*crossed dark arrow*). For RBF (*right* part of the figure), excitation to higher states is less favorable (*crossed dark arrow*), and fluorescence will remain stronger than in FMN (*dark arrow*). With differently shaped UV pulses, also the reverse situation is possible

Fig. 17.6. In general, a shaped ultraviolet (UV) pulse excites both molecules to the S_1 state and induces ultrafast dynamics which can follow slightly different pathways in both molecules. After a specified time delay Δt a second unshaped infrared (IR) pulse excites the molecule further to higher excited states and can induce dissipative processes such as ionization which lead to irreversible depopulation of the S_1 state, and thus to depletion of the fluorescence signal in one of the species (cf. left part of Fig. 17.6) and not in the other one (cf. right part of Fig. 17.6). Since for both molecules depletion can be minimized and maximized independently, the total fluorescence yield can be used to quantitatively determine the amounts of both species [62]. Although in this study only flavins have been considered, the results should be broadly applicable to control systems whose static spectra show essentially indistinguishable features. In particular, this should allow in the future for the selective identification of target molecules in the presence of structurally and spectroscopically similar background. This is an important issue in multiple areas of science and engineering. Our FISH method offers a unique opportunity not only to perform multistate dynamics "on the fly" and to optimize the laser pulses but also to apply directly the experimentally optimized pulses and thus to reveal the processes which enable discrimination of similar chromophores.

Our simulation of ODD between RBF and FMN is based on FISH dynamics "on the fly" in the ground and the nine lowest excited singlet states (S_0–S_9) under the influence of the experimentally optimized laser fields. We describe the electronic structure using the semiempirical PM3 CI method [90] and calculate the nonadiabatic couplings and transition dipole moments between all electronic states using the method of Thiel et al. [91, 92] The nuclear dynamics is performed employing the Langevin equation in order to approximately account for dissipative effects of the water environment present in the experiment [93]. Along the nuclear trajectories, the time-dependent Schrödinger equation (17.18) is integrated and the hopping probabilities are obtained from the electronic state populations according to Eq. 17.5. The shaped UV laser fields with a central wavelength of 400 nm employed in the simulation are reconstructed from the experimental spectral amplitudes A_n, phases ϕ_n and frequencies ω_n [62] according to $E(t) = \sum_n A_n \exp\left[i\left(\omega_n t + \phi_n\right)\right]$. The pulses obtained in this way have a duration of $\sim$5 ps and a maximum amplitude of $\sim$ $6 \cdot 10^{11}$ W cm^{-2}. The unshaped IR probe pulse with a wavelength of 800 nm has a maximum amplitude of $\sim 3 \cdot 10^{12}$ W cm^{-2} and a Gaussian envelope with a width of 100 fs (cf. Fig. 17.8).

The irreversible processes such as ionization, which lead to fluorescence depletion, are approximately introduced in Eq. 17.18 by adding an imaginary component $i\Gamma$ to the energy of the highest excited state S_9 which lies close to the experimentally determined ionization limit in water [94]. In this way irreversible population decay from the S_9 state is introduced. Subsequently, the time-dependent coefficients along the trajectories are recalculated and the hopping from the S_9 state to the ionized state is accounted for. Thus, an ionized population P_{ion} is obtained by averaging over all trajectories, which can be used as a measure for the decrease of the excited state population and thus for fluorescence depletion. In the experiment, the latter is quantified by the fluorescence intensities after application of the UV pulse alone, $F(UV)$, and application of both the UV and IR pulses, $F(UV + IR)$, according to $D_{exp} = [F(UV) - F(UV + IR)]/F(UV)$. In order to calculate the equivalent depletion signal from our ionized populations P_{ion}, we determine the fluorescence depletion D as

$$D = \frac{P_{ion}(UV + IR) - P_{ion}(UV)}{1 - P_{ion}(UV)}$$

For further computational details see Ref. [59, 74].

The RBF and FMN molecules represent particularly challenging systems for the optical discrimination since they have nearly identical stationary absorption and fluorescence spectra. The electronic spectroscopy of flavins is primarily associated with their common chromophore $\pi - \pi^*$ type transitions at 400 nm localized on the isoalloxazine ring and is influenced only very slightly by the terminal chemical moieties (H versus PO(OH)$_2$) on the side chains. The optimal UV pulses allowing for discrimination have been obtained experimentally by closed-loop optimization of the fluorescence depletion ratio of FMN over RBF and vice versa. Specifically, the maximization of the FMN over RBF ratio yielded a shaped UV/IR pulse pair (termed pulse 1 in the following) that leads to distinguishable fluorescence depletions values of 12.6% for RBF and 16.4% for FMN (cf. Fig. 17.7a). Oppositely, minimization of the FMN over RBF ratio has yielded a second pulse pair (pulse 2) that achieves approximately the same level of discrimination but reverses the ordering of the depletion signals (cf. Fig. 17.7b). In contrast, the excitation with an unshaped UV component leads to indistinguishable fluorescence depletion signals of 26% for RBF and FMN as also shown in Fig. 17.7. In order to

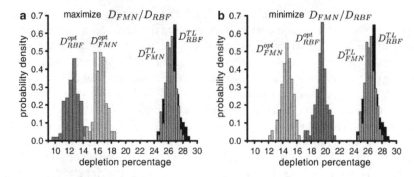

Fig. 17.7 Absolute experimental RBF and FMN depletion signals for optimized UV pulse shapes for maximizing (**a**) and minimizing (**b**) the ratio D(FMN)/D(RBF). Reprinted with permission from Roth et al. [62]. Copyright 2009 by the American Physical Society. The time delay for the IR pulse is 500 fs. Absolute depletions induced by the transform-limited pulse for both RBF (*black*) and FMN (*grey*) are statistically equivalent at 26%. Optimal pulses pull apart the RBF and FMN distributions to achieve discrimination between the two molecules

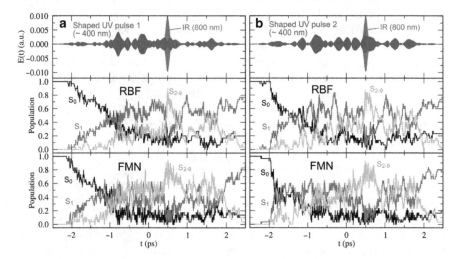

Fig. 17.8 (**a**) *Upper panel*: Temporal structure of the shaped pulse 1 for maximization of D(FMN)/D(RBF) (*blue*) and of the unshaped IR probe pulse (*red*). *Middle panel*: Time-dependent populations of the electronic states S_0 (*black*), S_1 (*red*), and S_2–S_9 (*orange*) in RBF driven by the pulses shown in the *upper panel*. *Lower panel*: Time-dependent populations of the electronic states S_0 (*black*), S_1 (*red*), and S_2–S_9 (*orange*) in FMN driven by the pulses shown in the *upper panel*. (**b**) The same as (**a**), but for pulse 2

discover the processes responsible for ODD of RBF and FMN, these experimentally optimized pulses for maximization and minimization of the depletion ratios [62] (pulses 1 and 2, respectively) have been used in our FISH simulations. The population dynamics induced by both pulses in RBF and FMN is shown in Fig. 17.8 together with the temporal pulse structure. Both pulses 1 and 2 lead to a smaller population of the higher excited states (S_2–$_9$) in RBF compared to FMN before the

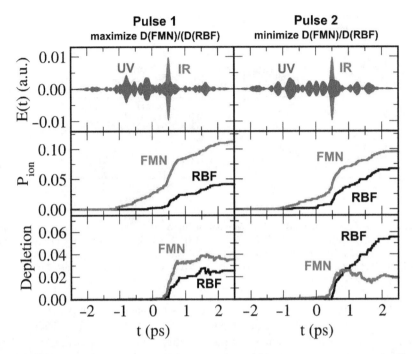

Fig. 17.9 *Upper panel*: Temporal structure of pulse 1 for maximization (*left*) and of pulse 2 for minimization (*right*) of the depletion ratio D(FMN)/D(RBF). *Middle panel*: Ionized populations P_{ion} of RBF (*black*) and FMN (*red*) due to pulse 1 (*left*) and pulse 2 (*right*). *Lower panel*: Fluorescence depletion D of RBF (*black*) and FMN (*red*) due to pulse 1 (*left*) and pulse 2 (*right*) (Reprinted from Ref. [63]. Copyright 2010 by the American Physical Society)

IR component has been applied. After the IR pulse, the S_{2-9} population raises for both pulses and both molecules. Although transient differences in the excited state populations are present during the pulses, the population returns in all cases from the higher excited states to the S_1 state after the pulses have ceased, if no irreversible processes such as ionization from these states are taken into account. Therefore, in order to describe fluorescence depletion the irreversible population decay from the higher excited states is modeled by adding an imaginary component to the energy of highest excited state S_9 as described above. The value of the imaginary component has been calibrated such that with an unshaped UV pulse both molecules exhibit identical depletion ratios. The ionized state populations P_{ion} obtained in this way are shown in Fig. 17.9. It can be seen that the IR subpulse is mainly responsible for the ionization, which in the case of RBF sets in at about +0.5 ps. Although for FMN, there is some ionization at earlier times, the main part of the ionized population is also generated at about +0.5 ps. The ionization yield of RBF is lower for pulse 1 than for pulse 2, whereas the reversed effect is found for FMN, proving that the shaped laser fields can selectively and independently modulate the ionization efficiency. The fluorescence depletion D (cf. lower part of Fig. 17.9) relies upon the relative decrease of the excited state population (S_1–S_9) due to both the UV and

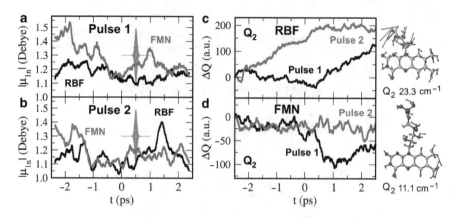

Fig. 17.10 *Left*: Average transition dipole moments for $S_1 \rightarrow S_2$–S_9 transitions for the dynamics driven by pulse 1 (**a**) and 2 (**b**) for FMN (*red*) and RBF (*black*). The average is performed over the states S_2–S_9 and over the ensemble of trajectories. *Right*: Selected averaged ground state normal mode displacements for RBF (**c**) and FMN (**d**) induced by pulse 1 (*black*) and pulse 2 (*red*)

IR pulses compared to the UV pulse alone. It is initiated for both pulses and both molecules by the IR subpulse at $+0.5$ ps. For pulse 1, D is systematically larger for FMN than for RBF, whereas for pulse 2 after 1 ps it becomes larger for RBF than for FMN. The final depletion ratios D(FMN)/D(RBF) after the pulses have ceased are calculated to be 1.4 for pulse 1 and 0.4 for pulse 2. These values are in good agreement with the experimental ODD values of 1.3 for pulse 1 and 0.7 for pulse 2 [62], thus confirming the experimental optical discrimination between FMN and RBF.

Our simulations offer a unique opportunity not only to reproduce the experimental findings but to gain an insight into the mechanism of dynamical processes responsible for discrimination. The fluorescence depletion is directly related to the ionization yield, which depends on the efficiency of populating excited states above S_1. Therefore the averaged transition dipole moments between S_1 and the higher excited states along the trajectories driven by the optimal laser fields have been calculated. It can be seen from Fig. 17.10a that pulse 1 induces dynamical pathways which exhibit systematically larger transition dipole moments for FMN than for RBF, indicating the stronger ionization and accordingly stronger depletion of fluorescence in FMN. In contrast, for pulse 2 (cf. Fig. 17.10b) at times after $+0.5$ ps this behavior is reversed leading to higher transition dipole moments for RBF, and thus in this case the fluorescence depletion should become stronger in RBF. These findings are consistent with the ionization yields presented in Fig. 17.9.

In order to establish the connection between higher transition dipole moments and the structural changes during the dynamics the averaged time-dependent normal mode displacements along the trajectories have been analyzed. In general, the conformational differences induced by the discriminating pulses are localized mainly in the polar side chains of both molecules. In Fig. 17.10c and d one prototype

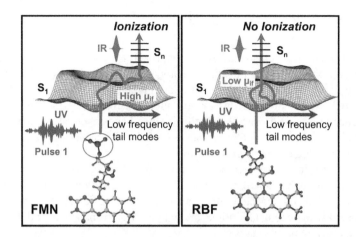

Fig. 17.11 Schematic illustration of the optimal dynamic discrimination by shaped laser fields on the example of shaped pulse 1 maximizing the FMN/RBF fluorescence depletion ratio (Reprinted from Ref. [63]. Copyright 2010 by the American Physical Society)

low-frequency normal mode of each molecule exhibiting large displacements induced by pulses 1 and 2 is shown. The pulse 1 invokes smaller deviations for the normal coordinate Q_2 in RBF and larger deviations after 0 ps for Q_2 in FMN compared to pulse 2. Thus, the excitation of low-frequency normal modes leads to conformations which have systematically higher or lower transition dipole moments to higher excited states leading to ionization, depending on which of two discriminating pulses is acting. Since RBF and FMN only differ in the side chain, differences in the dynamical behavior are expected to occur here due to the interplay between the effect of the heavy phosphorus atom in FMN and the differences of the vibrational density of states in both molecules.

In summary, the discrimination mechanism can be depicted as shown in Fig. 17.11: UV excitation of the molecule induces dynamical processes in excited states which mainly affect low-frequency tail vibrational modes. The discriminating pulse efficiently drives one of the molecules to regions of the PES where the transition dipole moments to higher excited states are large, such that the ionization and thus the fluorescence depletion are enhanced (pulse 1 for FMN). The same pulse acting on the other molecule (RBF, cf. right part of Fig. 17.11) suppresses the ionization (depletion of fluorescence) by keeping it in regions of the PES with lower transition dipole moments. Thus, in general, the shaped pulses can take advantage of minute differences in vibrational dynamics and exploit them to manipulate observables such as transition dipoles allowing for the selective molecular discrimination. This mechanism represents a general feature that can be exploited for the discrimination between similar molecules and offers a promising tool for using optimally shaped laser pulses in bioanalytical applications, thus increasing the selectivity beyond the current capability.

17.7 Conclusions and Outlook

We have presented a general theoretical approach for the simulation and control of ultrafast processes in complex molecular systems. Our methodological developments are based on the combination of quantum chemical nonadiabatic dynamics "on the fly" with the Wigner distribution approach for simulation and control of laser-induced ultrafast processes. Specifically, we have developed an approach for the nonadiabatic dynamics in the framework of TDDFT using localized basis sets, which is applicable to a large class of molecules and clusters.

Furthermore, the FISH method is introduced, allowing to include laser fields directly into the nonadiabatic molecular dynamics simulations and thus to realistically model their influence on ultrafast processes. In particular, this approach can be combined with genetic algorithms allowing to design shaped laser pulses which can drive a variety of processes.

The applications of our approaches have been illustrated on selected examples which serve to demonstrate their scope as well as the ability to accurately simulate experimental ultrafast observables and to assign them to underlying dynamical processes. In particular, a general approach for the simulation of TRPES has been developed, representing a powerful tool to identify nonadiabatic processes. Moreover, we have demonstrated for the first time that in the framework of the FISH method experimentally optimized laser fields can be directly used to reveal dynamical processes behind the optimal control. In addition, the FISH method combined with the optimal control theory allows to predict forms of laser fields capable to steer molecular dynamics in complex systems such as large molecules and nanosystems in different environments. Altogether, our approaches based on the classical molecular dynamics accounting for electronic transitions induced by both nonadiabatic effects as well as by light open new avenues for studying femtochemistry of attractive molecular and nano-systems which were not accessible earlier due to their complexity.

Acknowledgements We wish to acknowledge the contribution of our experimental partner Prof. T. Suzuki. We extend our thanks to Prof. J.-P. Wolf and Prof. H. Rabitz for stimulating cooperation and for providing us with experimental results on the discrimination of flavin molecules. Prof. W. Thiel we thank for providing us with the nonadiabatic MNDO code. Finally, we would also like to acknowledge the financial support from the Deutsche Forschungsgemeinschaft in the frame of SPP 1391, FOR 1282, the Emmy Noether Programme, MI-1236 (R.M.), as well as the Fonds der Chemischen Industrie (J.P.).

References

1. Car R, Parrinello M (1985) Phys Rev Lett 55:2471
2. Wigner E (1932) Phys Rev 40:749
3. Hillery M, O'Connel RF, Scully MO, Wigner EP (1984) Phys Rep 106:121
4. Hartmann M, Pittner J, Bonačić-Koutecký V (2001) J Chem Phys 114:2123

5. Li Z, Fang J-Y, Martens CC (1996) J Chem Phys 104:6919
6. Hartmann M, Pittner J, Bonačić-Koutecký V, Heidenreich A, Jortner J (1998) J Chem Phys 108:3096
7. Hartmann M, Pittner J, Bonačić-Koutecký V, Heidenreich A, Jortner J (1998) J Phys Chem A 102:4069
8. Hartmann M, Pittner J, Bonačić-Koutecký V (2001) J Chem Phys 114:2106
9. Mukamel S (1995) Principles of nonlinear optical spectroscopy. Oxford University Press, New York, NY
10. Donoso A, Martens CC (2001) Phys Rev Lett 87:223202
11. Donoso A, Zheng Y, Martens C (2003) J Chem Phys 119:5010
12. Ben-Nun M, Quenneville J, Martínez TJ (2000) J Phys Chem A 104:5161
13. Tully JC (1990) J Chem Phys 93:1061
14. Tully JC (1998) Faraday Discuss 110:407
15. Mitrić R, Bonačić-Koutecký V, Pittner J, Lischka H (2006) J Chem Phys 125:024303
16. Doltsinis NL, Marx D (2002) Phys Rev Lett 88:166402
17. Werner U, Mitrić R, Suzuki T, Bonačić-Koutecký V (2008) Chem Phys 349:319
18. Mitrić R, Werner U, Bonačić-Koutecký V (2008) J Chem Phys 129:164118
19. Werner U, Mitrić R, Bonačić-Koutecký V (2010) J Chem Phys 132:174301
20. Craig CF, Duncan WR, Prezhdo OV (2005) Phys Rev Lett 95:163001
21. Tapavicza E, Tavernelli I, Rothlisberger U (2007) Phys Rev Lett 98(2):023001
22. Tapavicza E, Tavernelli I, Rothlisberger U, Filippi C, Casida ME (2008) J Chem Phys 129:124108
23. Tavernelli I, Tapavicza E, Rothlisberger U (2009) J Chem Phys 130:124107
24. Granucci G, Persico M, Toniolo A (2001) J Chem Phys 114:10608
25. Fabiano E, Thiel W (2008) J Phys Chem A 112:6859
26. Fabiano E, Keal TW, Thiel W (2008) Chem Phys 349:334
27. Lan Z, Fabiano E, Thiel W (2009) J Phys Chem B 113:3548
28. Toniolo A, Olsen S, Manohar L, Martinez TJ (2004) Faraday Discuss 127:149
29. Virshup AM, Punwong C, Pogorelov TV, Lindquist BA, Ko C, Martinez TJ (2009) J Phys Chem B 113:3280
30. Schäfer LV, Groenhof G, Boggio-Pasqua M, Robb MA, Grubmüller H (2008) PLoS Comp Biol 4:e1000034
31. Groenhof G, Schäfer LV, Boggio-Pasqua M, Grubmüller H, Robb MA (2008) J Am Chem Soc 130:3250
32. Boggio-Pasqua M, Robb MA, Groenhof G (2009) J Am Chem Soc 131:13580
33. Ciminelli C, Granucci G, Persico M (2008) Chem Phys 349:325
34. Brixner T, Gerber G (2003) ChemPhysChem 4:418
35. Dantus M, Lozovoy VV (2004) Chem Rev 104:1813
36. Tannor DJ, Rice SA (1985) J Chem Phys 83:5013
37. Tannor DJ, Rice SA (1988) Adv Chem Phys 70:441
38. Brumer P, Shapiro M (1986) Faraday Discuss Chem Soc 82:177
39. Shapiro M, Brumer P (1986) J Chem Phys 84:4103
40. Shapiro M, Brumer P (1994) Int Rev Phys Chem 13:187
41. Tannor DJ, Rice SA (1986) J Chem Phys 85:5805
42. Peirce AP, Dahleh MA, Rabitz H (1988) Phys Rev A 37:4950
43. Shi S, Rabitz H (1989) Chem Phys 139:185
44. Kosloff R, Rice SA, Gaspard P, Tersigni S, Tannor DJ (1989) Chem Phys 139:201
45. Tersigni SH, Gaspard P, Rice SA (1990) J Chem Phys 93:1670
46. Rabitz H, Shi S (1991) Adv Mol Vibr Collision Dyn 1A:187
47. Warren WS, Rabitz H, Dahleh M (1993) Science 259:1581
48. Baumert T, Brixner T, Seyfried V, Strehle M, Gerber G (1997) Appl Phys B 65:779
49. Yelin D, Meshulach D, Silberberg Y (1997) Opt Lett 22:1793
50. Efimov A, Moores MD, Beach NM, Krause JL, Reitze DH (1998) Opt Lett 23:1915

51. Zeek E, Maginnis K, Backus S, Russek U, Murnane MM, Mourou G, Kapteyn HC, Vdovin G (1999) Opt Lett 24:493
52. Zeek E, Bartels R, Murnane MM, Kapteyn HC, Backus S, Vdovin G (2000) Opt Lett 25:587
53. Judson RS, Rabitz H (1992) Phys Rev Lett 62:1500
54. Assion A, Baumert T, Bergt M, Brixner T, Kiefer B, Seyfried V, Strehle M, Gerber G (1998) Science 282:919
55. Wöste L, Kühn O (eds) (2007) Analysis and control of ultrafast photoinduced reactions. In: Springer series in chemical physics, vol 87. Springer, Berlin
56. Bonačić-Koutecký V, Mitrić R (2005) Chem Rev 105:11
57. Mitrić R, Hartmann M, Pittner J, Bonačić-Koutecký V (2002) J Phys Chem A 106:10477
58. Mitrić R, Bonačić-Koutecký V (2007) Phys Rev A 76:031405(R)
59. Mitrić R, Petersen J, Bonačić-Koutecký V (2009) Phys Rev A 79:053416
60. Li B, Turinici G, Ramakrishna V, Rabitz H (2002) J Phys Chem B 106:8125
61. Li B, Rabitz H, Wolf J-P (2005) J Chem Phys 122:154103
62. Roth M, Guyon L, Roslund J, Boutou V, Courvoisier F, Wolf J-P, Rabitz H (2009) Phys Rev Lett 102:253001
63. Petersen J, Mitrić R, Bonačić-Koutecký V, Wolf J-P, Roslund J, Rabitz H (2010) Phys Rev Lett 105:073003
64. Dreuw A, Head-Gordon M (2005) Chem Rev 105:4009
65. Kümmel S, Kronik L (2008) Rev Mod Phys 80:3
66. Leininger T, Stoll H, Werner H-J, Savin A (1997) Chem Phys Lett 275:151
67. Iikura H, Tsuneda T, Yanai T, Hirao KJ (2001) Chem Phys 115:3540
68. Yanai T, Tew DP, Handy NC (2004) Chem Phys Lett 393:51
69. Baer R, Neuhauser D (2005) Phys Rev Lett 94:043002
70. Stein T, Kronik L, Baer R (2009) J Am Chem Soc 131:2818
71. Prezhdo OV (2009) Acc Chem Res 42:2005
72. Suzuki Y-I, Fuji T, Horio T, Suzuki T (2010) J Chem Phys 132:174302
73. Hammes-Schiffer S, Tully JC (1994) J Chem Phys 101(6):4657
74. Mitrić R, Petersen J, Bonačić-Koutecký V (2011) In: Conical intersections – theory, computation and experiment. Advanced series in physical chemistry, vol 17. World Scientific, Singapore
75. Lisinetskaya PG, Mitrić R (2011) Phys Rev A 83:033408
76. Horio T, Fuji T, Suzuki Y-I, Suzuki T (2009) J Am Chem Soc 113:10392
77. Mitrić R, Petersen J, Wohlgemuth M, Werner U, Bonačić-Koutecký V, Wöste L, Jortner J (2010) J Phys Chem A 115:3755
78. Hudock HR, Levine BG, Thompson AL, Satzger H, Tonwnsend D, Gador N, Ullrich S, Stolow A, Martinez TJ (2007) J Phys Chem A 111:8500
79. Fuji T, Suzuki Y-I, Horio T, Suzuki T, Mitrić R, Werner U, Bonačić-Koutecký V (2010) J Chem Phys 133:234303
80. Fuji T, Horio T, Suzuki T (2007) Opt Lett 32(17):2481
81. Fuji T, Suzuki T, Serebryannikov EE, Zheltikov A (2009) Phys Rev A 80:063822
82. Adamo C, Barone V (1999) J Chem Phys 110:6158
83. Krishnan R, Binkley JS, Seeger R, Pople JA (1980) J Chem Phys 72:650
84. Barbatti M, Pittner J, Pederzoli M, Werner U, Mitrić R, Bonačić-Koutecký V (2010) Chem Phys 375:26
85. Gromov EV, Tromifov AB, Vitkovskaya NM, Schirmer J, Köppel H (2003) J Chem Phys 119:737
86. Gromov EV, Tromifov AB, Vitkovskaya NM, Köppel H, Schirmer J, Meyer HD, Cederbaum LS (2004) J Chem Phys 121:4585
87. Yang J, Li J, Mo Y (2006) J Chem Phys 125:174313
88. Derrick P, Asbrink L, Edqvist O, Lindholm E (1971) Spectrochim Acta A 27:2525
89. Goldberg DE (1989) Genetic algorithms in search, optimization and machine learning. Addison-Wesley, Reading

90. Stewart JJP (1989) J Comput Chem 10:209
91. Koslowski A, Beck ME, Thiel W (2003) J Comput Chem 24:714
92. Patchkovskii S, Koslowski A, Thiel W (2005) Theor Chem Acc 114:84
93. van Gunsteren WF, Berendsen HJC (1982) Mol Phys 45:637
94. Getoff N, Solar S, McCormick DB (1978) Science 201:616

Chapter 18
A Survey on Reptation Quantum Monte Carlo

Wai Kong Yuen and Stuart M. Rothstein

Abstract We review the conceptual and mathematical foundations of reptation quantum Monte Carlo and its variants, placing them in the context of other path integral-based methods and the commonly-used diffusion Monte Carlo method. We describe quantum Monte Carlo sampling from the pure distribution, and strategies to improve the efficiency of this sampling. This is followed by a compilation of applications to electronic structure problems and to those in condensed matter physics. We conclude by reflecting on potential improvements of quantum Monte Carlo algorithms and how they will evolve with developments in high performance computing.

18.1 Introduction

Quantum Monte Carlo methods are among the most powerful tools for the treatment of the electron gas, atoms and molecules, including weakly-interacting systems, bio-molecules and solids, in addition to bose systems such as doped clusters and quantum gases (recent overview [1]). The most widely used quantum Monte Carlo algorithm is diffusion Monte Carlo (DMC; [2, 3]), used in over 800 articles published since 1987. Applications are to ground states energies for systems as large as transition metal systems and free base porphyrin [4–8], but due to the following inherent limitation, applications to other important physical properties are somewhat sparse.

W.K. Yuen
Department of Mathematics, Brock University, St. Catharines, Ontario, Canada L2S 3A1
e-mail: wyuen@brocku.ca

S.M. Rothstein (✉)
Departments of Chemistry and Physics, Brock University, St. Catharines, Ontario,
Canada L2S 3A1
e-mail: srothste@brocku.ca

P.E. Hoggan et al. (eds.), *Advances in the Theory of Quantum Systems in Chemistry and Physics*, Progress in Theoretical Chemistry and Physics 22,
DOI 10.1007/978-94-007-2076-3_18, © Springer Science+Business Media B.V. 2012

Diffusion Monte Carlo samples from the mixed distribution, $\Psi\Phi$, where Ψ is an inputted wave function and Φ is the unknown, exact wave function [9]. Sampling from this distribution allows for an accurate evaluation of the energy, albeit one biased by the incorrect exchange nodes introduced to the sampling by Ψ, which are kept fixed. This is the so-called "fixed-node" approximation [10]. However, in principle, expectation values for properties represented by operators that do not commute with the Hamiltonian are not accurate. Experimentally important properties of molecules, such as electrical and magnetic response properties, are represented by non-commutative operators. To accurately estimate these properties it is essential to use a method to sample from the pure electron distribution, Φ^2. Various "pure" methods have been proposed to indirectly ameliorate the Ψ-contamination bias in the density: descendent counting and averaging quantities with statistical weights $\propto \Psi/\Phi$, but with loss of precision and population-control bias: Refs. [11, 12] and [13, 14], respectively.

Reptation quantum Monte Carlo (RQMC) [15, 16] allows pure sampling to be done directly, albeit in common with DMC, with a bias introduced by the time-step (large, but controllable in DMC; e.g. [17]) and the fixed-node approach (small, but not controllable; e.g. [18]). Property estimation in this manner is free from population-control bias that plagues calculation of properties in diffusion Monte Carlo (e.g. [19]). Inverse Laplace transforms of the imaginary time correlation functions allow simulation of dynamic structure factors and other properties of physical interest.

In RQMC a configuration (set of m-electron coordinates) of the system is propagated in imaginary time for $L+1$ iterations by using a drift and diffusion process, thereby forming a "reptile", X, containing $L+1$ electron configurations. (Here, and below, we are assuming that the electronic Schroedinger equation is being simulated in imaginary time.) Once we have the reptile, new configurations are added to X by further propagation in imaginary time. But in order to keep the length of the reptile constant by removing, say, l iterations from the original reptile's head (tail), we will add l new ones to its tail (head). The new reptile is denoted as Y. If we assume that microscopic reversibility holds during the configurations' time evolution, the quantity that decides if the propagation $X \to Y$ is accepted or refused is given by a Metropolis decision.

These Metropolis-accepted or -rejected moves are taken in the distant past or far future in imaginary time, and ensure that the Markov chain converges to its target distribution. The algorithm's failure to meet the assumed criterion of microscopic reversibility causes the time-step bias to accumulate at the middle of the reptile, where the desired pure distribution is being sampled. Therefore, RQMC variants were introduced to avoid these difficulties [20–24].

This review article is divided into two major sections, the first of which details the theoretical basis of RQMC (Sect. 18.2). Initially we describe quantum Monte Carlo sampling from the pure distribution Φ^2 and mixed distribution $\Phi\Psi$, showing that the RQMC approach to sample from the pure distribution rests on Metropolis-Hastings (MH; [25, 26]) sampling, as does the variational path integral (VPI; [27]) method. As already mentioned, RQMC proposes reptation-type "moves" while those in VPI are based on other strategies. This leads us to sections that provide an intuitive justification for RQMC sampling and its variants: "bounced" versions [28, 29] and some that have been proposed by us [20–24].

The other major section is devoted to applications of RQMC variants and the original RQMC algorithm in chemistry and physics (Sect. 18.3). Therein we summarize calculations on electronic structure and on various properties of importance in condensed matter physics. The article concludes with our reflections on future developments (Sect. 18.4).

18.2 Reptation Quantum Monte Carlo: Theory

18.2.1 Sampling from the Pure and Mixed Distributions

Reptation quantum Monte Carlo (RQMC; [15]) and variational path integral (VPI; [27]), also known as path integral ground state (PIGS; [30]) methods are conceptually similar alternatives to more traditional methods such as variational Monte Carlo (VMC; [31]) and diffusion Monte Carlo (DMC; [2]). The fundamental idea behind both methods is to generate samples of paths from a probability distribution Π that describes the imaginary time evolution of the electronic configurations. A proper choice of Π allows one to sample from both the pure distribution and mixed distribution at different locations of the path. While VPI typically depends on more generic sampling techniques, RQMC uses a more intuitive "reptation" process based on drift and diffusion moves, with the paths being referred to as "reptiles". Formally, let $x \in R^{3m}$ denote the three dimensional positions of the m electrons in a given quantum system at an instant of the imaginary time. These configurations are linked together to form a reptile $X = x_0 x_1 ... x_L \in R^{3m(L+1)}$ of length L (with $L+1$ configurations), based on the drift and diffusion moves described by the time-discretized Langevin diffusion equation

$$x' = x + \tau \frac{\nabla \Psi(x)}{\Psi(x)} + \sqrt{\tau} \chi \tag{18.1}$$

where Ψ is the inputted importance sampling wave function, τ is the size of the time-step in the discretization, and $\chi \in R^{3m}$ is a $3m$-dimensional standard normal variate. While both VMC and DMC linked the configurations together through (18.1) only, RQMC and VPI also include the accumulated branching factor $e^{-S(X)}$ in the link, where

$$S(X) = \tau \left(\frac{1}{2} E_{loc}(x_0) + E_{loc}(x_1) + ... + E_{loc}(x_{L-1}) + \frac{1}{2} E_{loc}(x_L) \right) \tag{18.2}$$

is the sum of local energies

$$E_{loc}(x_i) = \frac{\hat{H} \Psi(x_i)}{\Psi(x_i)} \tag{18.3}$$

accumulated over the reptile X. Mathematically, samples of reptiles are generated from the probability distribution

$$\Pi(X) \propto \Psi^2(x_0) G(x_0 \to x_1) ... G(x_{L-1} \to x_L) \exp(-S(X)) \qquad (18.4)$$

where

$$G(x \to x') = (2\pi\tau)^{-3m/2} \exp\left[-\left(x' - x - \tau \frac{\nabla\Psi(x)}{\Psi(x)} \right)^2 / (2\tau) \right] \qquad (18.5)$$

is the Green's function of the drift and diffusion moves [2]. The inclusion of the branching factor yields the pure distribution Φ^2 at the middle of the reptile ($x_{L/2}$) and the mixed distribution $\Psi\Phi$ at both its head and tail (x_0 and x_L), providing the estimations of energy (sampling at the head and tail) and various physical properties represented by operators that commute with the position operator (sampling at the middle).

18.2.2 Metropolis-Hastings Sampling from Π

Due to the high dimension and the complexity of Π, it is important to find an implementable and efficient sampling method to obtain reliable estimations. It is standard to use a Markov chain Monte Carlo (MCMC) method, in which a Markov chain is constructed on the space of reptiles ($R^{3m(L+1)}$) with stationary distribution Π. Under some regularity conditions, the chain converges to Π and consequently, the reptiles simulated from the chain can be treated as an approximate sample from Π. Samples of the pure and mixed distributions are then "extracted" from the middle and head/tail configurations of the reptiles to compute Monte Carlo estimates of various physical properties.

Both RQMC and VPI are based on the Metropolis-Hastings (MH; [25, 26]) algorithm, one of the most popular MCMC algorithms. Given the current state of the algorithm at $X \in R^{3m(L+1)}$, a trial move to $Y \in R^{3m(L+1)}$ is proposed with a predetermined density $W(X \to Y)$. The move is accepted with probability

$$A(X \to Y) = \min\left\{ 1, \frac{\Pi(Y)W(Y \to X)}{\Pi(X)W(X \to Y)} \right\}, \qquad (18.6)$$

otherwise the algorithm remains at state X. Such a choice of A guarantees the detailed balance, which implies that Π is the stationary distribution of the underlying Markov chain. Based on the theory of general state space Markov chain, W can be arbitrary but must be chosen so that the chain is irreducible (i.e. able to explore the whole reptile space $R^{3m(L+1)}$) and aperiodic (which is always satisfied since our proposed move has a density) to ensure convergence to Π (see e.g. [32]).

While theoretical convergence is easy to establish, in practice, it is not clear how many moves are required to obtain a good enough approximate distribution of Π. For example, a random walk type uniform proposal density W is already enough to guarantee convergence to Π. However, due to the high correlations among the configurations, the algorithm is likely too sensitive to the random walk size and not moving efficiently in the whole reptile space $R^{3m(L+1)}$, even after a very large number of iterations. Even worse, sometimes the algorithm may seem to have converged but the reality is that it is only exploring regions near a few local maximums of $\Pi(X)$. These factors contribute to unreliable estimates, particular in high dimensional problems. Theoretically, it is well-known that to achieve optimal efficiency, the covariance matrix of W should match that of Π, which is typically unknown [33]. Therefore, it is far from trivial to propose good moves.

A simple generalization of MH is to consider a *mixture* of lower dimensional trial moves $W_1, ..., W_k$ in each iteration instead of one move per iteration. Here, W_i are chosen so that only a small number of configurations in X are being moved and/or follow some specific strategies (e.g., multilevel sampling with bisection methods [27]) to improve the chance of proposing a good move. This is essentially the main ingredient of VPI. While their strategies perform reasonably well in some problems, the choice of W_i still requires much experimentation.

18.2.3   The Original Reptation Quantum Monte Carlo Algorithm

Baroni and Moroni [15] proposed RQMC that takes advantage of the evolution of the configurations in the imaginary time. An intuitive justification of this approach can be based on the fact that the density $\Pi(X)$ is the product of two factors: $\Pi_s(X) \equiv \Psi^2(x_0)G(x_0 \to x_1)...G(x_{L-1} \to x_L)$ and $\exp(-S(X))$. This leads to two important observations. First, the factor $\Pi_s(X)$ is completely determined by the time-discretized Langevin diffusion equation in (18.1), so that each configuration x_i has the stationary density $\Psi^2(x_i)$. Therefore, it is straightforward to sample from Π_s using simple drift and diffusion moves as in VMC and DMC. Second, although the additional factor of $\exp(-S(X))$ makes it impossible to sample directly from Π, we can make the following intuitive interpretation: a sample reptile X from the density $\Pi(X)$ can be treated as a sample from $\Pi_s(X)$ that has a large value of $\exp(-S(X))$.

The first observation suggests that a good proposed move in MH from reptiles X to Y should somehow linked them together using simple drift and diffusion moves. The most natural way to do this is to add $l(\leq L)$ consecutive configurations generated by these moves. Such an addition also means that we must first remove some configurations from X to keep Y in the reptile space $R^{3m(L+1)}$. In addition, such removal should cause minimal disturbance to the links among x_i's, described by (18.1). The simplest way to accomplish this is to remove only from the "head" or "tail" of X. The second observation suggests that the number of configurations removed must be controlled so that Y is still "close" enough to X to maintain a large enough value of $\exp(-S(Y))$, or the proposed move is likely to be rejected.

Taking into account of all these suggestions and the detailed balance, we describe the proposed move from $X = x_0 x_1 ... x_L$ to $Y = y_0 y_1 ... y_L$ in the basic version of RQMC as follows. With probability 1/2, l configurations from the head of X is removed (the head chop), and l new configurations are added to tail of X based on (18.1) to form a new reptile Y. With probability 1/2, l configurations from the tail of X is removed (the tail chop), and l new configurations are added to head of X to form Y. Formally, the proposal density is given by

$$W(X \to Y) = \frac{1}{2} W^h(X \to Y) + \frac{1}{2} W^t(X \to Y) \tag{18.7}$$

where W^h is the head chop proposal density, given by

$$W^h(X \to Y) = G_0(y_{L-l} \to y_{L-l+1}) G_0(y_{L-l+1} \to y_{L-l+2}) ... G_0(y_{L-1} \to y_L) \tag{18.8}$$

whenever $y_0 = x_l$, $y_1 = x_{l+1}, ..., y_{L-l} = x_L$ and 0 otherwise, and W^t is the tail chop proposal density, given by

$$W^t(X \to Y) = G(y_l \to y_{l-1}) G(y_{l-1} \to y_{l-2}) ... G(y_1 \to y_0) \tag{18.9}$$

whenever $y_l = x_0$, $y_{l+1} = x_1, ..., y_L = x_{L-l}$ and 0 otherwise. Furthermore, if we assume that the propagator G satisfies the micro-reversibility condition, i.e.

$$\Psi^2(x) G(x \to x') = \Psi^2(x') G(x' \to x), \tag{18.10}$$

the acceptance probability in (18.6) simplifies to

$$A(X \to Y) = \min \left\{ 1, \frac{e^{-S(Y)}}{e^{-S(X)}} \right\}. \tag{18.11}$$

We denote this algorithm by RQMC-MH. The description of this version is slightly different from the original RQMC in [15], but they are essentially equivalent [20, 23].

In the basic version of RQMC-MH, the chop size l is fixed. If l is too small, most of the moves are accepted but it will take a large number of iterations for the updated configurations to be "translated" to the middle of the reptiles, where the pure distribution is located. This is analogous to "shaking" the reptile and only making changes on the head and tail without actually update much of the middle configuration, which is most important. In this situation, we are only exploring a few regions near local maximums of Π. On the other hand, if l is too large, very few moves are accepted, although only a few accepted moves are enough to update the whole reptile. In this case, the chain is able to reach many regions in the reptile space, but it is not exploring within each region. Therefore, the optimal chop size should be somewhere between the two extremes and it is usually chosen by minimizing the correlation times of interested quantities in trial runs. It is also possible to randomized the choice of l at each iteration, as proposed in the original RQMC, to make sure that the algorithm will propose both small and large moves. Another alternative is to use "bounced" moves, in which the algorithm continues in the same direction if the previous move in that direction was accepted [28, 29]. These techniques improve the speed of the chain and hence the sampling efficiency of the algorithm.

18.2.4   Variants of Reptation Quantum Monte Carlo

While the choice of a good proposal density is essential to improve the sampling efficiency of the algorithm, another approach to improve RQMC is to study the "quality" of the target distribution Π. In fact, to obtain good estimates of some properties, it is important to sample reptiles from Π for large value of L. One problem of RQMC is that the propagator G only satisfies the micro-reversibility (18.10) as $\tau \to 0$. In practice, to obtain good estimates, sampling of Π must be done for very large L. Therefore, even a small violation of (18.10) at each time-step accumulates to a significant error in the middle configuration of the reptile – the so-called time-step bias.

Yuen et al. [20] proposed the removal of assumption (18.10) to improve the estimations. Consequently, the acceptance probability depends also on the trial wave function Ψ and the propagator G explicitly, and no longer takes the simple form (18.11). Since then, a few other variants of RQMC have been proposed to address the time-step bias. These variants are all based on various kinds of adjustments made to the target density Π.

These adjustments can be explained by considering the following two densities:

$$\hat{\Pi}(X) \propto G(x_1 \to x_0)...G(x_{L/2} \to x_{L/2-1})\Psi^2(x_{L/2})G(x_{L/2} \to x_{L/2+1})$$
$$...G(x_{L-1} \to x_L)\exp(-S(X)), \tag{18.12}$$

and

$$\tilde{\Pi}(X) \propto \Psi^2(x_L)G(x_1 \to x_0)...G(x_L \to x_{L-1})\exp(-S(X)), \tag{18.13}$$

Obviously, under assumption (18.10), we have $\Pi = \hat{\Pi} = \tilde{\Pi}$. Therefore, the MH algorithm for all three densities are identical with acceptance probability given by (18.11). However, once the assumption is dropped, some terms in the calculation of acceptance probabilities in (18.6) cannot be canceled. The placing of Ψ^2 at strategic locations in the reptile means that the Metropolis decision needs to be adjusted accordingly. All the variants proposed, which we outline below, take advantage of specific properties of the chosen target densities.

18.2.4.1 Middle Adjusted Reptation Quantum Monte Carlo (RQMC-MI)

Since our main interest is to sample from the pure distribution, it is natural to locate the factor Ψ^2 at the middle of the reptile by choosing $\hat{\Pi}$ in (18.12) as our target density. To understand how this adjustment improves the quality of Π, first observe that without the factor $e^{-S(X)}$, all the configurations in the reptile have the same stationary density $\Psi^2(x)$ under assumption (18.10). Therefore, this adjustment ensures that the density of the pure distribution at the middle has the "correct" factor $\Psi^2(x_{L/2})$. As a comparison, the use of the original Π means that this factor is now approximated by $\Psi^2(x_0)G(x_0 \to x_1),...,G(x_{L/2-1} \to x_{L/2})$, incurring unnecessary time-step bias error at the middle. In RQMC-MI, the acceptance probability (18.6) simplifies to

$$A_{MI}^h(X \to Y) = \min\left(1, \frac{\Psi^2(x_{L/2+l})G(x_{L/2+1} \to x_{L/2})...G(x_{L/2+l} \to x_{L/2+l-1})e^{-S(Y)}}{\Psi^2(x_{L/2})G(x_{L/2} \to x_{L/2+1})...G(x_{L/2+l-1} \to x_{L/2+l})e^{-S(X)}}\right)$$

(18.14)

when a head chop occurs; and to

$$A_{MI}^t(X \to Y) = \min\left(1, \frac{\Psi^2(x_{L/2-l})G(x_{L/2-1} \to x_{L/2})...G(x_{L/2-l} \to x_{L/2-l+1})e^{-S(Y)}}{\Psi^2(x_{L/2})G(x_{L/2} \to x_{L/2-1})...G(x_{L/2-l+1} \to x_{L/2-l})e^{-S(X)}}\right)$$

(18.15)

when a tail chop occurs. Note that in the Metropolis decision, the functional values of Ψ^2 at the middles of both X and Y, and the Green's functional values near the middle are included.

18.2.4.2 Head Adjusted Reptation Quantum Monte Carlo (RQMC-HE)

Similarly, if our main interest is to sample from the mixed distribution at the head, we choose the original Π as our target density [24]. In RQMC-HE, the head chop and tail chop acceptance probabilities are given by

$$A_{HE}^h(X \to Y) = \min\left(1, \frac{\Psi^2(x_l)G(x_1 \to x_0)...G(x_l \to x_{l-1})e^{-S(Y)}}{\Psi^2(x_0)G(x_0 \to x_1)...G(x_{l-1} \to x_l)e^{-S(X)}}\right)$$

(18.16)

and

$$A_{HE}^t(X \to Y) = \min\left(1, \frac{\Psi^2(y_0)G(y_0 \to y_1)...G(y_{l-1} \to y_l)e^{-S(Y)}}{\Psi^2(y_l)G(y_1 \to y_0)...G(y_l \to y_{l-1})e^{-S(X)}}\right)$$

(18.17)

respectively. This adjustment includes the values of the wave function and the Green's function near the head of both X and Y.

18.2.4.3   Head-Tail Adjusted Reptation Quantum Monte Carlo (RQMC-HT)

In both RQMC-MI and RQMC-HE, the adjustments are reflected in the functional values near the locations of corresponding designated configurations, the middle and head locations, respectively. A closer examination reveals that for the tail chop in RQMC-HE, we can make another interpretation: the Metropolis decision $(A_{HE}^t(X \rightarrow Y))$ essentially depends only on the added configurations (except the original head of X). This motivates the consideration of a slightly different kind of adjustment that improves the quality of the added configurations, which can be accomplished by modifying RQMC-HE as follows. When a head chop occurs, we use the target density $\tilde{\Pi}$ in calculating the acceptance probability, which simplifies to

$$A_{HT}^h(X \rightarrow Y) = \min\left(1, \frac{\Psi^2(y_L)G(y_L \rightarrow y_{L-1})...G(y_{L-l+1} \rightarrow y_{L-l})e^{-S(Y)}}{\Psi^2(y_{L-l})G(y_{L-1} \rightarrow y_L)...G(y_{L-l} \rightarrow y_{L-l+1})e^{-S(X)}}\right).$$

(18.18)

When a tail chop occurs, Π is used as in RQMC-HE, which gives $A_{HT}^t = A_{HE}^t$. Not only does this algorithm (RQMC-HT; [22]) improve the quality of all the added configurations, it is also symmetric about the middle of the reptile, a property that is lacking in RQMC-HE. This symmetry also indirectly affects the quality of the pure distribution at the middle of the reptile in a positive way. Intuitively, since all the configurations are linked together by drift and diffusion moves, a reptile which has good properties at both ends should automatically have good properties at the middle.

18.3   Applications of Reptation Quantum Monte Carlo

18.3.1   Variants of Reptation Quantum Monte Carlo

Given that the pure density is sampled at the middle of the reptile, RQMC-MI (Sect. 18.2.4.1) was the first variant to be developed and tested [20]. (In that work RQMC-MI was denoted as RQMC-NC.) To provide a proof in principle, the application was to ground-state hydrogen atom, where moments of the electron density were calculated for variational densities of crude and good quality. Values for $\langle r \rangle, \langle r^2 \rangle, \langle r^3 \rangle$, and $\langle 1/r \rangle$ were found to agree within statistical error to the analytical determinations for the exact density. The time-step bias for RQMC-MI was under better control than for RQMC-MH, the approach equivalent to that of Baroni and Moroni's original RQMC algorithm.

RQMC-MI and RQMC-MH were employed along with a variety of quantum Monte Carlo approaches to sample the pure density for ground-state dihydrogen [21]. RQMC-MI exhibited the smallest integrated absolute error in the simulated electron-nuclear and electron-electron distributions, closely followed by RQMC-MH. The closest competitor was fixed-node diffusion Monte Carlo with detailed balance and descendent counting [11, 12].

Shortly thereafter RQMC-HT (Sect. 18.2.4.3) was developed and tested against RQMC-MI and RQMC-MH for the energy and several interelectronic properties of LiH: $\langle r_{12}^m \rangle$ ($m = -1, 1, 2$); $\langle u_c^m \rangle$ ($m = -1, 1, 2$); and $\langle u_z^m \rangle$ ($m = 1, 2$) [22]. Here RQMC-HT estimated the ground-state energy with less time-step bias, and thus more efficiently, than did RQMC-MI and RQMC-MH. The accuracy and precision of the interelectronic properties were not adversely affected by implementing RQMC-HT.

The most-recently developed variant was RQMC-HE (Sect. 18.2.4.2). For the ground state of water molecule, its performance was tested against all the others: RQMC-HT, RQMC-MI, and RQMC-MH [24]. Similar in spirit to Ref. [20], trial densities ranging in quality from crude (single-ζ and double-ζ, SZ and DZ) to high-quality (double-ζ and triple-ζ with polarization; DZP and TZP) were employed. The energy and the following one-electron properties were calculated and compared with experiment and CBS/FCI results: dipole moment, components of the quadrupole and octopole moment tensors, diamagnetic shielding at the nuclei, spherically-averaged diamagnetic susceptibility, and electric field gradient tensors.

As was the case for LiH, RQMC-HT outperformed the others, but to a much lesser extent as the quality of the importance sampling wave function improved. These results suggest that the simplest algorithm, RQMC-MH, is the optimal choice when given importance sampling functions of the highest quality.

18.3.2   Original Reptation Quantum Monte Carlo Algorithm

18.3.2.1   Electronic Structure Calculations

RQMC (Sect. 18.2.3) was utilized in a study of transition metal oxides (ScO, TiO, VO, CrO, and MnO): their energetics and dipole moments [34, 35]. Despite excellent agreement of the energetics with experiment, the dipole moments of these molecules significantly differed from experiment. After determining the errors associated with the pseudopotential approximation and the breakdown of the Hellmann-Feynman theorem to be small, the authors focused on the fixed-node error and the localization approximation employed in density functional theory. A multi-determinantal guiding function (better nodes) for TiO leads to an improved dipole moment, consistent with CCSD(T), but still somewhat larger than the value reported by experiment.

RQMC and diffusion Monte Carlo were employed to calculate the potential energy curve of helium dimer from a single-determinantal wave function with a large basis set (19s9p8d/8s7p6d) and a three-body-Jastrow factor [36]. This van der Waals system presents well-known challenges both to experimentalists and theorists. They found very accurate results from both quantum Monte Carlo approaches, but RQMC generally gave more accurate interaction energies and smaller error bars.

RQMC was employed in calculations of benzene-dyhydrogen complex, another challenging van der Waals system [37]. Here the objective was to analyze the electron density and reduced density gradient calculated from density functional theory to quantify the role of exchange in the exchange-correlation approximation for describing non-bonded interactions. A single determinant of Kohn-Sham orbitals multiplied by a two-body Jastrow function was used for the importance sampling in RQMC, in addition to pseudo-potentials to describe the core electrons. The densities generated from a variety of functionals were negligibly different from that of RQMC, except for that generated using the local density approximation. Similarly the reduced density gradients quantitatively agreed with RQMC. The authors found that enhancing the exchange energy density where there is a large reduced density gradient (i.e.; non-bonded regions) is crucial for an accurate description of weak interactions.

18.3.2.2   Condensed Matter Physics

RQMC plays a central role in the so-called "coupled electron-ion Monte Carlo" [CEIMC] approach to systems of many electrons and ions within the Born-Oppenheimer approximation, applied to high pressure hydrogen. Here one samples the ionic degrees of freedom by a Metropolis decision at inverse temperature $\beta = (k_B T)^{-1}$, where the difference between the Born-Oppenheimer energies of proton state S and trial state S' is computed by RQMC [28, 38–40]. The electronic degrees of freedom are sampled from the RQMC probability distributions for the electron configurations at the current and proposed nuclear positions: $(\Pi(s; S) + \Pi(s; S'))$. To improve the efficiency of RQMC sampling, by increasing the number of accepted moves, a "bounce" algorithm was proposed whereby the growth direction of the Markov chain is maintained until reversal upon a rejected move [28]. In an application to high pressure hydrogen, simulation results were obtained for the equation of state over a wide range of the phase diagram, and the energy parameterized as a function of temperature and pressure [41].

One of the first applications of RQMC was to the rotational dynamics of carbonyl sulfide (OCS) molecules solvated in helium clusters, for cluster sizes $(N = 3, 10)$ [42]. This and related work, described shortly, rest on the absorption spectrum given by the Fourier transform of the reptilian imaginary time electric dipole correlation function. Similarly, the optical activity is extracted from the autocorrelation of the molecular orientation vector. This work by Moroni and co-workers and/or Boroini and co-workers was closely followed by several other investigations of rotational dynamics in doped clusters, summarized as follows:

- One was an investigation of CO solvated in small helium clusters, for cluster sizes (N) up to 30 [43]. Here binding energies were calculated and related to the change in helium density with N; in addition to positions and spectral weights of rotational and infrared lines as a function of N. The propensity of He atoms to cluster on a side of the molecule decreased with N until the first solvation shell was completed, and then increased while the second shell was being built. This longitudinal asymmetry is a necessary condition for the doubling lines in the rotational spectrum.

- Similarly, RQMC calculations were performed on the pure rotational spectra of CO_2 solvated in helium clusters: for cluster sizes up to 17 [44]. Analysis of the excitation energies yielded rotational constants and distortion constants, and the agreement with experiment of the former was almost perfect. The latter quantities were generally smaller than experimentally observed, but the general trend with cluster size and its very large value relative to carbonyl sulfide (OCS) and nitrous oxide (N_2O) – doped He clusters (see below) was reproduced. Comparing density profiles for CO_2-He_N clusters, for sizes $N = 5$ and 6, with their analogs for OCS and N_2O doping was interpreted by the so-called donut model: the first five atoms fill an equatorial ring around the dopant and the sixth orients towards its poles, leaving the density in the ring essentially unchanged. This interpretation was also supported by trends in the incremental binding energy per added helium atom.

- A similar study of rotational constants and incremental binding energies using RQMC was performed on nitrous oxide (N_2O) – doped He clusters for cluster sizes (N) 3–20, 25, and 30 [45]. As the RQMC calculations are at 0 K, PIMC for Boltzmann statistics was used to introduce finite temperature to assess the role of exchange effects, helping to understand the rotational dynamics for small cluster sizes. The calculations were based on a multilevel bisection algorithm [27] that incorporated all degrees of freedom of the complex. Evidence of exchange effects was provided for a decoupling between N_2O motion, with some He atoms attached to the impurity, and the rest of the helium, as well as the related turn-around of the effective rotational constant at $N = 8$.

- Another publication considered the rotational spectrum of CO solvated in $para-(H_2)_N$ clusters, for cluster sizes (N) 2–17 [46]. Here $R(0)$ transitions and their spectral weights were assigned up to $N = 9$ for b-type series (free molecule rotations) and $N = 14$ for a-type series (end-over-end rotations). As was the case for CO-He_N, there was a decreasing tendency of the hydrogen molecules to dynamically cluster on one side of the CO molecular axis as completion of the first solvation shell was approached. Theory and experiment agreed well, except that theory tended to overestimate the b-type energies.

- This work was followed-up by a publication on simulation of quantum melting in hydrogen clusters [47]. Here a multipole dynamic correlation criterion was introduced and calculated within RQMC to discriminate between melting and freezing behavior of quantum clusters. The focus was on small clusters of $para$-hydrogen molecules for cluster sizes near $N = 13$. Despite their similar geometric structures, $para-(H_2)_{13}$ behaves like a superfluid, while $CO@para-(H_2)_{12}$ has a rigid, crystalline behavior.

- Superfluity was also observed in the rotational spectrum of cyanoacetylene (HCCCN) solvated in helium clusters for size ranges $N = 1$–18 and 25–31 [48].

The rotational energies were computed by RQMC allowing an assignment of R(0) transitions from $N > 6$ and R(1) for all $N > 1$. Oscillatory behavior of the rotational transition frequencies for $N > 6$ suggested the onset of superfluity. In addition, PIMC results suggest that after completion of the first solvation shell at $N = 9$, exchanges between helium atoms in the first and second shell become dominant, as proposed in previous literature [46].

• Returning now to CO doped clusters, the RQMC method was used to calculate the ground-state energy, structural properties, and imaginary time correlation functions of $CO@He_N$, from which an inverse Laplace transform extracts the rotational energies [49]. The vibrational shift, which describes difference between rotational ground-state energies of the cluster where the helium atoms are interacting with the molecule in its vibrational ground-state and first excited-state, was calculated from perturbation theory, as well as radial density profiles for various cluster sizes, $N \leq 100$. Rotational energies for a-type and b-type transitions (see above) were reported for $^{12}C^{16}O$, $^{13}C^{16}O$, $^{12}C^{18}O$, and $^{13}C^{18}O$ for $N \leq 20$ and for $N \leq 11$, respectively. The a-type R(0) transitions for $^{12}C^{16}O$ for $N > 15$ and their changes upon isotopic substitution to $^{13}C^{18}O$ for $N \leq 20$ were compared with experiment. General agreement was good, especially for small cluster sizes.

• Several of the above-described publications extracted rotational spectra from inverse Laplace transforms of imaginary-time autocorrelation functions, quantities readily calculated with RQMC. The utility of defining a larger set of correlation functions, so-called "symmetry-adapted imaginary-time autocorrelation functions" was explored in a recent paper [50]. Computational efficiency in the calculation of weak spectral features was demonstrated by a study of He-CO binary complex. Some preliminary results of an analysis of a recently observed satellite band in the IR spectrum of CO_2 doped He clusters were presented.

RQMC was used to generate distribution functions for the two-dimensional electron gas [51]. Considered were the spin-summed and spin-resolved pair distribution functions, in addition to the spin-resolved potential energy for a range of electron densities and polarization.

RQMC is an important tool in the study of quantum gases. Calculations were performed on one-dimensional dipolar quantum gases. This is a system of N atoms, with linear density n, and permanent dipoles considered to be arranged parallel to each other, resulting in purely repulsive interactions. RQMC used in combination with bosonization techniques provided a unifying theory for the crossover physics for this system, from the low-density, so-called "Tonks-Girardeau" regime to the high-density classical, quasiordered state [52]. Analysis of the static structure factor being consistent with Luttinger-liquid theory provided firm evidence for Luttinger-liquid behavior in the entire cross-over region. Simulations were done for up to $N = 200$ bosons placed in a square box with periodic boundary conditions. The thermodynamic energy per particle was presented as an analytical function of nr_0, where r_0 is the effective Bohr radius of the gas molecules [53]. Furthermore, by using the equation of state for the homogeneous dipolar gas, the same authors calculated excitation modes when the gas is confined in a harmonic trap [54]. The most recent paper in this series reports the low-energy spectrum from dynamic density-density correlation functions obtained from RQMC [55]. This simulation data together with arguments based on the uncertainty-principle proves the absence of long-range order in this strongly correlated system, again confirming that the dipolar gas is in a Luttinger-liquid state.

Quantum dimers provide models for anti-ferromagnets. Using a combination of continuous-time lattice diffusion Monte Carlo [56] and RQMC, the square lattice quantum dimer model for lattice sizes up to 48×48 sites was investigated to estimate the location of the columnar to plaquette phase transition [57]. The former phase has no parallel dimers, while the latter one has sets of plaquettes with parallel dimers continuously changing orientation. The author found significant finite-size corrections to scaling for the plaquette phase and liquid phase.

This approach was generalized and improved by adaptation of directed updates [28], to reduce the correlation time in path sampling, and a worm algorithm to sample expectation values of off-diagonal observables [29]. In addition, a strategy was introduced to improve upon the fixed-node approximation; see Sect. 18.4. Applications included the one-dimensional Heisenberg model and the fermionic Hubbard model.

18.4   Future Directions

Each of the above-described pure sampling algorithms rests on the fixed-node approximation [10], where the simulated density is biased by the incorrect exchange nodes of the importance sampling function, Ψ. It is beyond the scope of this review to summarize the several attempts to understand the fermion nodes, aimed towards algorithms which go beyond the fixed-node approximation: [58–67]. To date none of these efforts have lead to a host of applications to realistic systems. Therefore, the exchange-node problem remains an active area of path integral Monte Carlo research; e.g. [29], to our knowledge the most recent one. In another vein, a quantum Monte Carlo method that is based on random walks in Slater determinant space is a recent development [68,69], albeit without an upper bound to the energy and having population control bias, similar to that of diffusion Monte Carlo.

Quantum Monte Carlo algorithms run in parallel on modern computers and are naturally scalable to a large number of processors. There is no question that calculations will be performed on realistic systems as these algorithms are improved in parallel with ongoing advances in high performance computing technology, such as the emergence of petascale computers. There already exists a quantum Monte Carlo package (QMCPACK; [70]) designed to take advantage of multi-core processors and Graphics Processing Units. This and related developments are described in [71].

Acknowledgements This work was supported, in part, by grants from the Natural Sciences and Engineering Research Council of Canada (NSERC).

References

1. Ceperley DM (2010) Theoretical and computational methods in mineral physics: geophysical applications. Rev Mineral Geochem 71:129
2. Anderson JB (1975) J Chem Phys 63:1499
3. Ceperley DM, Alder BJ (1980) Phys Rev Lett 45:566
4. Hammond BL, Lester WA Jr, Reynolds PJ (1994) Monte Carlo methods and ab initio quantum chemistry. World Scientific, Singapore
5. Anderson JB, Rothstein SM (eds) (2007) Advances in quantum Monte Carlo. American Chemical Society, Washington, DC
6. Anderson JB (2007) Quantum Monte Carlo: origins, development, applications. Oxford University Press, Oxford, New York
7. Huang P, Carter EA (2008) Annu Rev Phys Chem 59:261
8. Lester WA Jr, Mitas L, Hammond B (2009) Chem Phys Lett 478:1
9. Reynolds PJ, Ceperley DM, Alder BJ, Lester WA Jr (1982) J Chem Phys 77:5593
10. Anderson JB (1976) J Chem Phys 65:4121
11. Reynolds PJ, Barnett RN, Hammond BL, Lester WA Jr (1986) J Stat Phys 43:1017
12. Barnett RN, Reynolds PJ, Lester WA Jr (1992) J Chem Phys 96:2141
13. Langfelder P, Rothstein SM, Vrbik J (1997) J Chem Phys 107:8525
14. Hornik M, Rothstein SM (2002) In: Lester WA Jr, Rothstein SM, Tanaka S (eds) Recent advances in quantum Monte Carlo methods; Part II. World Scientific, Singapore, pp 71–94
15. Baroni S, Moroni S (1999) Phys Rev Lett 82:4745
16. Baroni S, Moroni S (1999) In: Nightingale MP, Umrigar CJ (eds) Quantum Monte Carlo methods in physics and chemistry. Kluwer, Dordrecht
17. Umrigar CJ, Nightingale MP, Runge KJ (1993) J Chem Phys 99:2865
18. Grossman JC (2002) J Chem Phys 117:1434
19. Bosá I, Rothstein SM (2004) J Chem Phys 121:4486
20. Yuen WK, Farrar T, Rothstein SM (2007) J Phys A 40:F639
21. Coles B, Vrbik P, Giacometti RD, Rothstein SM (2008) J Phys Chem A 112:2012
22. Yuen WK, Oblinsky DG, Giacometti RD, Rothstein SM (2009) Intern J Quantum Chem 109:3229
23. Yuen WK, Oblinsky DG, Rothstein SM (2009) (unpublished)
24. Oblinsky DG, Yuen WK, Rothstein SM (2010) J Molec Struct Theochem 961:29
25. Metropolis N, Rosenbluth AW, Teller AH, Teller E (1953) J Chem Phys 21:1087
26. Hastings WK (1970) Biometrica 57:97
27. Ceperley DM (1995) Rev Mod Phys 67:279
28. Pierleoni C, Ceperley DM (2005) ChemPhysChem 6:1872
29. Carleo G, Becca F, Moroni S, Baroni S (2010) Phys Rev E 82:046710
30. Cuervo JE, Roy P-N, Boninsegni M (2005) J Chem Phys 122:114504
31. Ceperely DM, Chester GV, Kalos MH (1977) Phys Rev B 16:3081
32. Tierney L (1994) Ann Stat 22:1701
33. Roberts GO, Rosenthal JS (2001) Stat Sci 16:351
34. Wagner LK, Mitas L (2007) J Chem Phys 126:034105
35. Wagner LK (2007) J Phys Condens Matter 19:343201
36. Wu X, Hu X, Dai Y, Du C, Chu S, Hu L, Deng J, Feng Y (2010) J Chem Phys 132:204304
37. Kanai Y, Grossman JC (2009) Phys Rev A 80:032504
38. Dewing M, Ceperley DM (2002) In: Lester WA Jr, Rothstein SM, Tanaka S (eds) Recent advances in quantum Monte Carlo II. World Scientific, River Edge
39. Ceperley DM, Dewing M, Pierleoni C (2002) Lect Notes Phys 605:473
40. Pierleoni C, Ceperley DM (2006) Lect Notes Phys 703:641
41. Morales MA, Pierleoni C, Ceperley DM (2010) Phys Rev E 81:021202
42. Moroni S, Sarsa A, Fantoni S, Schmidt KE, Baroni S (2003) Phys Rev Lett 90:143401
43. Cazzato P, Paolini S, Moroni S, Baroni S (2004) J Chem Phys 120:9071

44. Tang J, McKellar ARW, Mezzacapo F, Moroni S (2004) Phys Rev Lett 92:145503
45. Moroni S, Blinov N, Roy P-N (2004) J Chem Phys 121:3577
46. Moroni S, Botti M, DePalo S, McKellar ARW (2005) J Chem Phys 122:094314
47. Baroni S, Moroni S (2005) ChemPhysChem 6:1884
48. Topic W, Jäger W, Blinov N, Roy P-N, Botti M, Moroni S (2006) J Chem Phys 125:144310
49. Škrbić T, Moroni S, Baroni S (2007) J Phys Chem A 111:7640
50. Škrbić T, Moroni S, Baroni S (2007) J Phys Chem A 111:12749
51. Gori-Giorgi P, Moroni S, Bachelet GB (2004) Phys Rev B 70:115102
52. Citro R, Orignac E, De Palo S, Chiofalo ML (2007) Phys Rev A 75:051602
53. Citro R, De Palo S, Orignac E, Pedri P, Chiofalo ML (2008) New J Phys 10:045011
54. Pedri P, De Palo S, Orignac E, Citro R, Chiofalo ML (2008) Phys Rev A 77:015601
55. De Palo S, Orignac E, Citro R, Chiofalo ML (2008) Phys Rev B 77:212101
56. Syljuåsen OF (2005) Phys Rev B 71:020401 (R)
57. Syljuåsen OF (2006) Phys Rev B 73:245105
58. Anderson JB (1987) Phys Rev A 35:3550
59. Ceperley DM (1991) J Stat Phys 63:1237
60. Glauser WA, Brown WR, Lester WA Jr, Bressanini D, Hammond BL, Koszykowski ML (1992)
    J Chem Phys 97:9200
61. Bressanini D, Ceperely DM, Reynolds PJ (2002) In: Lester WA Jr, Rothstein SM, Tanaka S
    (eds) Recent advances in quantum Monte Carlo II. World Scientific, Edge, pp 3–11
62. Foulkes WMC, Hood RQ, Needs RJ (1999) Phys Rev B 60:4558
63. Mitas L (2006) Phys Rev Lett 96:240402
64. Bajdick M, Mitas L, Wagner LK, Schmidt KE (2008) Phys Rev B 77:115112
65. Kruger F, Zaanen J (2008) Phys Rev B 78:035104
66. Bressanini D, Morosi G (2008) J Chem Phys 129:054103
67. Bouabça T, Braïda B, Caffarel M (2010) J Chem Phys 133:044111
68. Booth GH, Thom AJW, Alavi A (2009) J Chem Phys 131:054106
69. Booth GH, Alavi A (2010) J Chem Phys 132:174104
70. Kim J, Esler K, McMinis J, Clark B, Gergely J, Chiesa S, Delaney K, Vincent J, Ceperley DM.
    QMCPACK simulation suite. http://qmcpack.cmscc.org
71. Kim J, Esler KP, McMinis J, Ceperley DM (2010) SciDAC Proceedings

Chapter 19
Quantum Monte Carlo Calculations of Electronic Excitation Energies: The Case of the Singlet $n \to \pi^*$ (CO) Transition in Acrolein

Julien Toulouse, Michel Caffarel, Peter Reinhardt, Philip E. Hoggan, and C.J. Umrigar

Abstract We report state-of-the-art quantum Monte Carlo calculations of the singlet $n \to \pi^*$ (CO) vertical excitation energy in the acrolein molecule, extending the recent study of Bouabça et al. [J Chem Phys 130:114107, 2009]. We investigate the effect of using a Slater basis set instead of a Gaussian basis set, and of using state-average versus state-specific complete-active-space (CAS) wave functions, with or without reoptimization of the coefficients of the configuration state functions (CSFs) and of the orbitals in variational Monte Carlo (VMC). It is found that, with the Slater basis set used here, both state-average and state-specific CAS(6,5) wave functions give an accurate excitation energy in diffusion Monte Carlo (DMC), with or without reoptimization of the CSF and orbital coefficients in the presence of the Jastrow factor. In contrast, the CAS(2,2) wave functions require reoptimization of the CSF and orbital coefficients to give a good DMC excitation energy. Our best estimates of the vertical excitation energy are between 3.86 and 3.89 eV.

J. Toulouse (✉) • P. Reinhardt
Laboratoire de Chimie Théorique, Université Pierre et Marie Curie and CNRS, Paris, France
e-mail: julien.toulouse@upmc.fr; peter.reinhardt@upmc.fr

M. Caffarel
Laboratoire de Chimie et Physique Quantiques, IRSAMC, CNRS and Université de Toulouse, Toulouse, France
e-mail: caffarel@irsamc.ups-tlse.fr

P.E. Hoggan
LASMEA, CNRS and Université Blaise Pascal, Aubière, France
e-mail: pehoggan@yahoo.com

C.J. Umrigar
Laboratory of Atomic and Solid State Physics, Cornell University, Ithaca, New York, USA
e-mail: cyrusumrigar@cornell.edu

P.E. Hoggan et al. (eds.), *Advances in the Theory of Quantum Systems in Chemistry and Physics*, Progress in Theoretical Chemistry and Physics 22,
DOI 10.1007/978-94-007-2076-3_19, © Springer Science+Business Media B.V. 2012

19.1   Introduction

Quantum Monte Carlo (QMC) methods (see, e.g., Refs. [1–3]) constitute an alternative to standard quantum chemistry approaches for accurate calculations of the electronic structure of atoms, molecules and solids. The two most commonly used variants, variational Monte Carlo (VMC) and diffusion Monte Carlo (DMC), use a flexible trial wave function, generally consisting for atoms and molecules of a Jastrow factor multiplied by a short expansion in configuration state functions (CSFs), each consisting of a linear combination of Slater determinants. Although VMC and DMC have mostly been used for computing ground-state energies, excitation energies have been calculated as well (see, e.g., Refs. [4–15]).

The simplest QMC calculations of excited states have been performed without reoptimizing the determinantal part of the wave function in the presence of the Jastrow factor. It has recently become possible to optimize in VMC both the Jastrow and determinantal parameters for excited states, either in a state-specific or a state-average approach [6, 7, 9, 10, 12, 14, 15]. Although this leads to very reliable excitation energies, reoptimization of the orbitals in VMC can be too costly for large systems.

In this context, Bouabça et al. [13] studied how to obtain a reliable excitation energy in QMC for the singlet $n \to \pi^*$ (CO) vertical transition in the acrolein molecule without reoptimization of the determinantal part of the wave function. The acrolein molecule is the simplest member of the unsaturated aldehyde family whose photochemistry is of great interest. They showed that a good DMC excitation energy can be obtained by using *non-reoptimized* complete-active-space (CAS) wave functions if two conditions are fulfilled: (a) The wave functions come from a *state-average* multiconfiguration self-consistent-field (MCSCF) calculation (using the same molecular orbitals for the two states is indeed expected to improve the compensation of errors due to the fixed-node approximation in the excitation energy), and (b) a sufficiently large active space including all chemically relevant molecular orbitals for the excitation process is used. In comparison, all the small CAS wave functions and the large *state-specific* CAS wave functions (coming from two separate MCSCF calculations) were found to lead to quite unreliable DMC excitation energies, with a strong dependence on the size of the basis set. These results were obtained using standard all-electron QMC calculations with Gaussian basis sets, with orbitals appropriately modified near the nuclei to enforce the electron-nucleus cusp condition, in the same spirit as in Ref. [16].

In this work, we extend the study of Bouabça et al. by testing the use of a Slater basis set and the effect of reoptimization of the determinantal part of the wave function in VMC. The use of Slater basis functions is motivated by the observation that they are capable of correctly reproducing the electron-nucleus cusp condition as well as having the correct exponential decay at large distances. In contrast, Gaussian basis functions have no cusp at the nucleus and a too rapid decay at large distances. As regards the effect of reoptimization, conclusions about the validity of using non-reoptimized CAS wave functions are drawn. The paper is organized as follows. In Sect. 19.2, we explain the methodology used. In Sect. 19.3, we present and discuss our results. Finally, Sect. 19.4 summarizes our conclusions.

19.2   Methodology

We are concerned with the vertical electronic transition in the acrolein (or propenal) molecule, $CH_2 = CH–CHO$ (symmetry group C_s), from the spin-singlet ground state (symmetry A') to the first spin-singlet excited state (A''). This transition is identified as the excitation of an electron from the lone pair (n) of the oxygen to the antibonding π^* orbital of the CO moiety. We use the *s-trans* experimental geometry of Ref. [17], obtained by microwave spectroscopy in the gas phase (Fig. 19.1).

We use Jastrow-Slater wave functions parametrized as [18, 19]

$$|\Psi(\mathbf{p})\rangle = \hat{J}(\alpha)e^{\hat{\kappa}(\kappa)} \sum_{I=1}^{N_{CSF}} c_I|C_I\rangle, \qquad (19.1)$$

where $\hat{J}(\alpha)$ is a Jastrow factor operator, $e^{\hat{\kappa}(\kappa)}$ is the orbital rotation operator and $|C_I\rangle$ are CSFs. Each CSF is a symmetry-adapted linear combination of Slater determinants of single-particle orbitals which are expanded in Slater basis functions. The parameters $\mathbf{p} = (\alpha, \mathbf{c}, \kappa)$ that are optimized are the Jastrow parameters α, the CSF coefficients $\mathbf{c}$ and the orbital rotation parameters κ. The exponents of the basis functions are kept fixed in this work. We use a Jastrow factor consisting of the exponential of the sum of electron-nucleus, electron-electron, and electron-electron-nucleus terms, written as systematic polynomial and Padé expansions [20] (see also Refs. [21, 22]).

For each state, we start by generating standard restricted Hartree-Fock (RHF), and state-average and state-specific MCSCF wave functions with a complete active space generated by distributing N valence electrons in M valence orbitals [CAS(N,M)], using the quantum chemistry program GAMESS [23]. As in Ref. [13], we consider a minimal CAS(2,2) active space containing the two molecular orbitals n (A') and π^*_{CO} (A'') involved in the excitation, and a larger CAS(6,5) active space containing the five molecular orbitals that are expected to be chemically

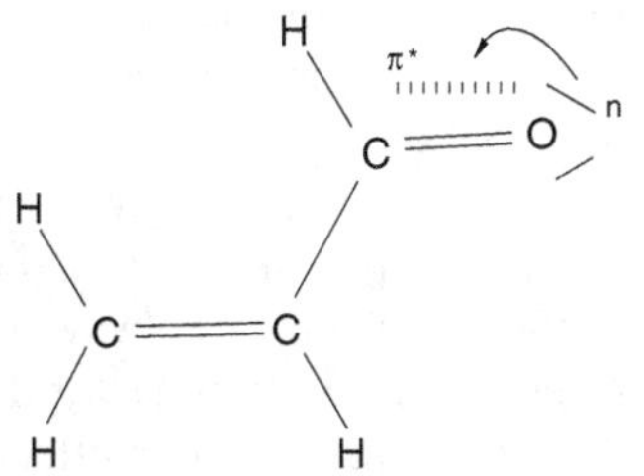

Fig. 19.1 Schematic representation of the singlet $n \rightarrow \pi^*$ excitation in the CO moiety of the acrolein molecule

Table 19.1 Ground-state energy E_0, first excited-state energy E_1, and vertical excitation energy $E_1 - E_0$ for the singlet $n \to \pi^*$ transition in the acrolein molecule at the experimental geometry calculated in DMC with different time steps τ using the VB1 Slater basis set and a state-specific Jastrow-Slater CAS(6,5) wave function with Jastrow, CSF and orbital parameters optimized by energy minimization in VMC

τ (hartree^{-1})	E_0 (hartree)	E_1 (hartree)	$E_1 - E_0$ (eV)
0.01	−191.8734(4)	−191.7312(4)	3.87(2)
0.005	−191.8753(4)	−191.7319(4)	3.90(2)
0.0025	−191.8762(4)	−191.7330(4)	3.90(2)
0.001	−191.8769(3)	−191.7350(3)	3.86(1)

relevant: π_{CO} (A″), n (A′), π_{CC} (A″), π_{CO}^* (A″), π_{CC}^* (A″). Note that, since the two states have different symmetries, the purpose behind using the state-average procedure is not the usual one of avoiding a variational collapse of the excited state onto the ground state, but rather to possibly improve the compensation of errors in the excitation energy by using the same molecular orbitals for the two states. We use the triple-zeta quality VB1 Slater basis of Ema et al. [24]. For C and O, this basis contains two $1s$, three $2s$, three $2p$ and one $3d$ sets of functions; for H, it contains three $1s$ and one $2p$ sets of functions. Each Slater function is actually approximated by a fit to ten Gaussian functions [25–27] in GAMESS. These wave functions are then multiplied by the Jastrow factor, imposing the electron-electron cusp condition, and QMC calculations are performed with the program CHAMP [28] using the true Slater basis set rather than its Gaussian expansion. The wave function parameters are optimized with the linear energy minimization method in VMC [18, 19, 29], using an accelerated Metropolis algorithm [30, 31]. Two levels of optimization are tested: optimization of only the Jastrow factor while keeping the CSF and orbital parameters at their RHF or MCSCF values, and simultaneous optimization of the Jastrow, CSF and orbital parameters. For all wave functions, even the state-average ones, we always optimize a separate Jastrow factor for each state, rather than a common Jastrow factor for the two states. Although the electron-nucleus cusp condition is not enforced during the optimization in our current implementation, the orbitals obtained from Slater basis functions usually nearly satisfy the cusp condition. Once the trial wave functions have been optimized, we perform DMC calculations within the short-time and fixed-node (FN) approximations (see, e.g., Refs. [32–36]). We use an efficient DMC algorithm with very small time-step errors [37]. For a given trial wave function, the evolution of the ground- and excited-state total DMC energies and of the corresponding excitation energy when the imaginary time step τ is decreased from 0.01 to 0.001 hartree^{-1} is shown in Table 19.1. While the time-step bias is clearly seen for the total energies, it largely cancels out for the excitation energy for all the time steps tested here and cannot be resolved within the statistical uncertainty. In the following, we always use an imaginary time step of $\tau = 0.001$ hartree^{-1}. Note that the Jastrow factor does not change the nodes of the wave function, and therefore it has no direct effect on the fixed-node DMC total energy (aside from of course the time-step bias and the population-control bias). Improving the trial wave function by optimization of the Jastrow factor is nevertheless important for DMC calculations in order to reduce the fluctuations and to make the time-step error very small and the population-control bias negligible. Of course, when the Jastrow factor is optimized together with the CSF and/or orbital parameters, then it has an indirect effect through those parameters on the nodes of the wave function.

19.3   Results and Discussion

Table 19.2 reports the ground-state energy E_0, the first excited-state energy E_1, and the excitation energy $E_1 - E_0$ calculated by different methods. Since the excited state is a spin-singlet open-shell state, it cannot be described by a restricted single-determinant wave function; however, we report single-determinant results for the ground state for comparison of total energies. We take our best estimates of the vertical excitation energy to be those obtained with the CAS(6,5) wave functions in DMC. They range from 3.86 to 3.89 eV, depending whether a state-average or state-specific approach is used and whether the determinantal part of the wave function is reoptimized in QMC. Previously reported calculations include (a) time-dependent density-functional theory (TDDFT): 3.66 eV [38] and 3.78 eV [39]; (b) complete-active-space second-order perturbation theory (CASPT2): 3.63 eV [38], 3.69 eV [40], and 3.77 eV [41]; (c) multireference configuration interaction: 3.85 eV [42]; (d) different variants of coupled cluster: 3.83 eV [43], 3.93 eV [39], 3.75 eV [39]. The most recent experimental estimate is 3.69 eV, which corresponds to the maximum in the UV absorption band in gas phase and which is in agreement with previous experimental data [44–47]. Beside different treatment of electron correlation, the discrepancies between these values may be due to the high sensitivity of the excitation energy to the $C=C$ and $C=O$ bond lengths [39]. Moreover, the comparison with experiment relies on the approximation that the vertical excitation energy corresponds to the maximum of the broad UV absorption band. In view of all these data, a safe estimate range for the exact vertical excitation energy is from about 3.60 to 3.90 eV.

Even without reoptimization of the CSF and orbital coefficients, our state-specific Jastrow-Slater CAS(6,5) wave functions give a DMC excitation energy, 3.88(2) eV, as accurate as the one obtained with the fully optimized wave functions, even though the total energies E_0 and E_1 are about 20 mhartree higher. Also, our non-reoptimized state-average Jastrow-Slater CAS(6,5) wave functions give an essentially identical DMC excitation energy of 3.89(2) eV. This agrees well with the DMC result of Bouabça et al. [13], 3.86(7) eV, obtained with non-reoptimized state-average Jastrow-Slater CAS(6,5) wave functions with a Gaussian basis set.

Thus it appears possible to obtain an accurate excitation energy using non-reoptimized state-specific CAS(6,5) wave functions in DMC. This is different from what was observed in Ref. [13] where state-specific CAS(6,5) wave functions were found to give unreliable excitation energies. The difference is that we use

Table 19.2 Ground-state energy E_0, first excited-state energy E_1, and vertical excitation energy $E_1 - E_0$ for the singlet $n \to \pi^*$ transition in the acrolein molecule at the experimental geometry calculated by different methods using the VB1 Slater basis set

	E_0 (hartree)	E_1 (hartree)	$E_1 - E_0$(eV)
RHF	−190.83430261		
MCSCF CAS(2,2) SA	−190.82258836	−190.68568203	3.73
MCSCF CAS(2,2) SS	−190.83891553	−190.71709289	3.31
MCSCF CAS(6,5) SA	−190.88736483	−190.74691372	3.82
MCSCF CAS(6,5) SS	−190.89520291	−190.75181511	3.90
VMC JSD [J]	−191.7107(5)		
VMC JSD [J+o]	−191.7636(5)		
VMC JCAS(2,2) SA [J]	−191.7121(5)	−191.5619(5)	4.09(2)
VMC JCAS(2,2) SS [J]	−191.7099(5)	−191.5652(5)	3.94(2)
VMC JCAS(2,2) SS [J+c+o]	−191.7643(5)	−191.6247(5)	3.80(2)
VMC JCAS(6,5) SA [J]	−191.7182(5)	−191.5747(5)	3.90(2)
VMC JCAS(6,5) SS [J]	−191.7221(5)	−191.5776(5)	3.93(2)
VMC JCAS(6,5) SS [J+c+o]	−191.7795(5)	−191.6342(5)	3.95(2)
DMC JSD [J]	−191.8613(4)		
DMC JSD [J+o]	−191.8698(3)		
DMC JCAS(2,2) SA [J]	−191.8608(5)	−191.7133(5)	4.01(2)
DMC JCAS(2,2) SS [J]	−191.8606(4)	−191.7113(4)	4.06(2)
DMC JCAS(2,2) SS [J+c+o]	−191.8700(3)	−191.7293(3)	3.83(1)
DMC JCAS(6,5) SA [J]	−191.8568(5)	−191.7138(5)	3.89(2)
DMC JCAS(6,5) SS [J]	−191.8585(4)	−191.7160(4)	3.88(2)
DMC JCAS(6,5) SS [J+c+o]	−191.8769(3)	−191.7350(3)	3.86(1)
DMC JCAS(6,5) SA [J][a]	−191.8504(20)	−191.7086(23)	3.86(7)
Experimental estimate[b]			3.69

The QMC calculations are done with Jastrow-Slater wave functions using a single determinant (JSD), or a state-average (SA) or state-specific (SS) complete-active-space multideterminant expansion (JCAS). The lists of parameters optimized by energy minimization in VMC are indicated within square brackets: Jastrow (J), CSF coefficients (c), and orbitals (o). For comparison, the DMC results of Ref. [13] obtained with state-average CAS(6,5) wave functions and a Gaussian basis set are also shown
[a] QMC calculations with a Gaussian basis, Ref. [13]
[b] Maximum in the UV absorption band in gas phase, Ref. [39]

here a Slater basis set rather than the Gaussian basis set employed in Ref. [13]. Even though the Gaussian basis contains more basis functions than the VB1 Slater basis, it gives a higher DMC energy for both states and tends to favor one state over the other in state-specific calculations. This example shows the importance of using a well-balanced basis set in state-specific calculations, even in DMC.

We comment now on the results obtained with the CAS(2,2) wave functions. The state-specific MCSCF CAS(2,2) excitation energy, 3.31 eV, is a strong under-estimate. The corresponding VMC and DMC state-specific calculations without reoptimization of the CSF and orbital coefficients, give slightly overestimated excitation energies, 3.94(2) and 4.06(2) eV, respectively. Whereas the state-average MCSCF CAS(2,2) calculation gives a much better excitation energy, 3.73 eV, compared to the state-specific MCSCF calculation, the non-reoptimized state-average CAS(2,2) wave functions do not seem to improve the excitation energies in VMC and DMC. In fact, they give a worse VMC excitation energy of 4.09(2) eV, and a DMC excitation energy of 4.01(2) eV which is not significantly better than with the non-reoptimized state-specific wave functions.

The excitation energies obtained from the CAS(6,5) wave functions depend very little on whether (a) they are calculated in MCSCF, VMC or DMC, (b) the state-average or the state-specific approach is employed, and (c) the CSF and orbital coefficients are reoptimized or not in the presence of the Jastrow factor. In contrast, the excitation energies obtained from CAS(2,2) wave functions do depend on all of the above and, in particular the reoptimization of the CSF and orbital coefficients in the presence of the Jastrow factor significantly improves the VMC and DMC excitation energies, to 3.80(2) and 3.83(1) eV, respectively. The importance of reoptimizing in VMC the CAS(2,2) expansions but not the CAS(6,5) expansions suggests that the Jastrow factor includes important correlation effects that are present in CAS(6,5) but not in CAS(2,2).

Finally, we note that without reoptimization of the determinantal part of the wave functions, the ground-state VMC and DMC energies can actually increase when going from a single-determinant wave function to a CAS(2,2) or CAS(6,5) wave function. This behavior has been observed in other systems as well, e.g. in C_2 and Si_2 [29]. Of course, if the CSF and orbital coefficients are reoptimized in VMC, then the VMC total energies must decrease monotonically upon increasing the number of CSFs. In practice, it is found that the DMC total energies also decrease monotonically although there is in principle no guarantee that optimization in VMC necessarily improves the nodes of the wave function.

19.4   Conclusion

In this work, we have extended the study of Bouabça et al. [13] on how to obtain a reliable excitation energy in QMC for the singlet $n \rightarrow \pi^*$ (CO) vertical transition in the acrolein molecule. We have tested the use of a Slater basis set and the effect of reoptimization of the determinantal part of the wave function in VMC and of the corresponding changes in the nodal structure in fixed-node DMC. Putting together the conclusions of the study of Bouabça et al. and the present one, we can summarize the findings on acrolein as follows:

(a) It is possible to obtain an accurate DMC excitation energy with non-reoptimized CAS wave functions, provided that a sufficiently large chemically relevant active space is used. In the case of too small an active space, reoptimization of the CSF and orbital coefficients in the presence of the Jastrow factor appears to be necessary in order to get a good DMC excitation energy.

(b) When using Gaussian basis sets of low or intermediate quality, reliable DMC excitation energies could be obtained only by using state-average wave functions (i.e., with the same molecular orbitals for the two states). In contrast, when using a good quality Slater basis set such as the VB1 basis, state-specific wave functions were found to also give reliable DMC excitation energies. Thus, this provides some support for using Slater, rather than Gaussian, basis sets in all-electron QMC calculations. Note that other authors also advocate the use of Slater basis sets in all-electron QMC calculations (see, e.g., Refs. [48–50]).

It remains to check whether these conclusions are generally true for other systems. It would be indeed desirable for calculations on large molecular systems if accurate DMC excitation energies could be obtained with state-specific or state-average CAS expansions without the need of an expensive reoptimization of the determinantal part of the wave functions in QMC.

Acknowledgements Most QMC calculations have been done on the IBM Blue Gene of Forschungszentrum Jülich (Germany) within the DEISA project STOP-Qalm. CJU acknowledges support from NSF grant number CHE-1004603.

References

1. Hammond BL, Lester JWA, Reynolds PJ (1994) Monte Carlo methods in ab initio quantum chemistry. World Scientific, Singapore
2. Nightingale MP, Umrigar CJ (eds) (1999) Quantum Monte Carlo methods in physics and chemistry. NATO ASI ser. C 525. Kluwer, Dordrecht
3. Foulkes WMC, Mitas L, Needs RJ, Rajagopal G (2001) Rev Mod Phys 73:33
4. Grossman JC, Rohlfing M, Mitas L, Louie SG, Cohen ML (2001) Phys Rev Lett 86:472
5. Aspuru-Guzik A, Akramine OE, Grossman JC, Lester WA (2004) J Chem Phys 120:3049
6. Schautz F, Filippi C (2004) J Chem Phys 120:10931
7. Schautz F, Buda F, Filippi C (2004) J Chem Phys 121:5836
8. Drummond ND, Williamson AJ, Needs RJ, Galli G (2005) Phys Rev Lett 95:096801
9. Scemama A, Filippi C (2006) Phys Rev B 73:241101
10. Cordova F, Doriol LJ, Ipatov A, Casida ME, Filippi C, Vela A (2007) J Chem Phys 127:164111
11. Tiago ML, Kent PRC, Hood RQ, Reboredo FA (2008) J Chem Phys 129:084311
12. Tapavicza E, Tavernelli I, Rothlisberger U, Filippi C, Casida ME (2008) J Chem Phys 129:124108
13. Bouabça T, Ben Amor N, Maynau D, Caffarel M (2009) J Chem Phys 130:114107
14. Filippi C, Zaccheddu M, Buda F (2009) J Chem Theory Comput 5:2074
15. Zimmerman PM, Toulouse J, Zhang Z, Musgrave CB, Umrigar CJ (2009) J Chem Phys 131:124103
16. Manten S, Lüchow A (2001) J Chem Phys 115:5362
17. Blom CE, Grassi G, Bauder A (1984) J Am Chem Soc 106:7427
18. Toulouse J, Umrigar CJ (2007) J Chem Phys 126:084102
19. Toulouse J, Umrigar CJ (2008) J Chem Phys 128:174101
20. Umrigar CJ. Unpublished
21. Filippi C, Umrigar CJ (1996) J Chem Phys 105:213
22. Güçlü AD, Jeon GS, Umrigar CJ, Jain JK (2005) Phys Rev B 72:205327

23. Schmidt MW, Baldridge KK, Boatz JA, Elbert ST, Gordon MS, Jensen JH, Koseki S, Matsunaga N, Nguyen KA, Su SJ, Windus TL, Dupuis M, Montgomery JA (1993) J Comput Chem 14:1347
24. Ema I, García de la Vega JM, Ramírez G, López R, Fernández Rico J, Meissner H, Paldus J (2003) J Comput Chem 24:859
25. Hehre WJ, Stewart RF, Pople JA (1969) J Chem Phys 51:2657
26. Stewart RF (1970) J Chem Phys 52:431
27. Kollias A, Reinhardt P, Assaraf R. Unpublished
28. Umrigar CJ, Filippi C, Toulouse J. CHAMP, a quantum Monte Carlo program http://pages.physics.cornell.edu/~cyrus/champ.html. Accessed 8 Aug 2011
29. Umrigar CJ, Toulouse J, Filippi C, Sorella S, Hennig RG (2007) Phys Rev Lett 98:110201
30. Umrigar CJ (1993) Phys Rev Lett 71:408
31. Umrigar CJ (1999) In: Nightingale MP, Umrigar CJ (eds) Quantum Monte Carlo methods in physics and chemistry. NATO ASI Ser. C 525. Kluwer, Dordrecht, p 129
32. Grimm R, Storer RG (1971) J Comput Phys 7:134
33. Anderson JB (1975) J Chem Phys 63:1499
34. Anderson JB (1976) J Chem Phys 65:4121
35. Reynolds PJ, Ceperley DM, Alder BJ, Lester WA (1982) J Chem Phys 77:5593
36. Moskowitz JW, Schmidt KE, Lee MA, Kalos MH (1982) J Chem Phys 77:349
37. Umrigar CJ, Nightingale MP, Runge KJ (1993) J Chem Phys 99:2865
38. Aquilante F, Barone V, Roos BO (2003) J Chem Phys 119:12323
39. Aidas K, Møgelhøj A, Nilsson EJK, Johnson MS, Mikkelsen KV, Christiansen O, Söderhjelm P, Kongsted J (2008) J Chem Phys 128:194503
40. Martín ME, Losa AM, Fdez.-Galván I, Aguilar MA (2004) J Chem Phys 121:3710
41. Losa AM, Fdez.-Galván I, Aguilar MA, Martín ME (2007) J Phys Chem B 111:9864
42. do Monte SA, Müller T, Dallos M, Lischka H, Diedenhofen M, Klamt A (2004) Theor Chem Acc 111:78
43. Saha B, Ehara M, Nakatsuji H (2006) J Chem Phys 125:014316
44. Blacet FE, Young WG, Roof JG (1937) J Am Chem Soc 59:608
45. Inuzuka K (1960) Bull Chem Soc Jpn 33:678
46. Hollas JM (1963) Spectrochim Acta 19:1425
47. Moskvin AF, Yablonskii OP, Bondar LF (1966) Theor Exp Chem 2:636
48. Galek PTA, Handy NC, Cohen AJ, Chan GKL (2005) Chem Phys Lett 404:156
49. Galek PTA, Handy NC, Lester WA Jr (2006) Mol Phys 104:3069
50. Nemec N, Towler MD, Needs RJ (2010) J Chem Phys 132:034111

Part VI
Structure and Reactivity

Chapter 20
Analysis of the Charge Transfer Mechanism in Ion-Molecule Collisions

E. Rozsályi, E. Bene, G.J. Halász, Á. Vibók, and M.C. Bacchus-Montabonel

Abstract The collision of C^{2+} ions on a series of molecular targets, OH, CO and HF is investigated in relation with indirect processes in the action of radiations with the biological medium. The charge transfer cross sections are determined with regard to the orientation of the projectile towards the molecular target, and consideration of the vibration of the diatomics during the collision process. Correlations may be pointed out between the non-adiabatic interactions and the charge transfer cross sections and general rules for the corresponding mechanism are proposed.

E. Rozsályi • Á. Vibók
Department of Theoretical Physics, University of Debrecen,
P.O.Box 5, H-4010 Debrecen, Hungary
e-mail: remeset@gmail.com; vibok@phys.unideb.hu

E. Bene
Institute of Nuclear Research, Hungarian Academy of Sciences,
P.O.Box 51, H-4001 Debrecen, Hungary
e-mail: bene@namafia.atomki.hu

G.J. Halász
Department of Information Technology, University of Debrecen,
P.O.Box 12, H-4010 Debrecen, Hungary
e-mail: halasz@inf.unideb.hu

M.C. Bacchus-Montabonel (✉)
Laboratoire de Spectrométrie Ionique et Moléculaire, Université de Lyon I, CNRS UMR 5579,
43 Boulevard du 11 Novembre 1918, F69622 Villeurbanne Cedex, France
e-mail: bacchus@lasim.univ-lyon1.fr

P.E. Hoggan et al. (eds.), *Advances in the Theory of Quantum Systems in Chemistry and Physics*, Progress in Theoretical Chemistry and Physics 22,
DOI 10.1007/978-94-007-2076-3_20, © Springer Science+Business Media B.V. 2012

20.1   Introduction

Experimental and theoretical investigations on collisions between ions and molecular [1–7] or even biomolecular targets [8–13] have been developed recently in relation with possible direct or indirect processes occurring in the irradiation of the biological medium. Effectively, important damage induced by interaction of ionizing radiations with biological tissues is due to the secondary particles, low-energy electrons, radicals or singly and multiply charged ions, generated along the track after interaction of the radiation with the biological medium [14]. The analysis of such mechanisms at the molecular level is of preponderant importance in order to provide detailed information on the different processes occurring during the collision, as charge transfer between the projectile ion and the molecule or fragmentation dynamics after removal of electrons from the target.

In that sense, we have investigated a series of ion-diatomic collision systems in order to analyze the charge transfer mechanism and, if possible, establish a number of general rules for these reactions. Effectively, in indirect processes where ions are not interacting directly with biomolecules, but with the environment, generally the water solvent, very reactive species may be produced, as for example the OH radical by action of ions with the water molecules. Such species may induce severe damage to the biological environment and indirect processes have been shown to be determinant for the physiological point of view [15].

In order to have a better understanding of the mechanism, we have performed a comparative analysis for the collision of the C^{2+} projectile with different targets, differing one to another by one atom. We have studied first of all the $C^{2+} + OH$ collision system, and compared its mechanism to the $C^{2+} + CO$ and $C^{2+} + HF$ reactions [5–7]. Our attention has been focused on two main points: the anisotropy of the charge transfer with regard to the orientation of the projectile towards the target and the influence of the vibration of the diatomic molecule during the collision. All along this work, a detailed analysis of the mechanism of the charge transfer in relation with the non-adiabatic interactions between the different molecular states involved in the process has been performed [16]. The molecular calculations were carried out using *ab initio* quantum chemistry methods followed by a semi-classical collision treatment in the keV laboratory energy range.

20.2   Theoretical Treatment

20.2.1   Molecular Calculations

The geometry is described using the internal Jacobi coordinates $\{R,r,\theta\}$ with the origin at the centre of mass of the target molecule, as defined in Fig. 20.1.

The diatomic molecule corresponds respectively to AB = OH, OC and FH for the collision of the C^{2+} ion on OH, CO and HF diatomic targets, such as, in the

Fig. 20.1 Internal Jacobi
coordinates

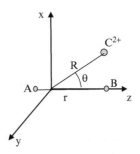

linear approach, the collision of the C^{2+} ion towards the more electronegative
atom, oxygen or fluorine, would correspond to an angle $\theta = 180°$. The molecular calculations were performed by means of the MOLPRO suite of *ab initio*
programs [17]. The molecular orbitals have been optimized at the state-average
Complete Active Space Self Consistent Field (CASSCF) level with Multi Reference
Configuration Interaction (MRCI) calculations using the correlation-consistent basis
sets of Dunning [18]. The active space includes all valence electrons distributed
among $n = 2, 3$ orbitals for carbon, oxygen and fluorine and the 1s orbital for
hydrogen. The 1s orbitals of carbon, oxygen and fluorine were frozen in the
calculation. The equilibrium geometry r_{eq} of the ground state of each diatomic
molecule has been optimized at the MRCI level of theory and lead to vertical
ionization potentials in good agreement with experimental and previous theoretical
values [5–7]. We have performed calculations for different geometries r around
the equilibrium distance in order to take account of the vibration effect during the
collision process. The anisotropy of the charge transfer has been investigated by
performing a series of calculations for different orientations of the projectile towards
the molecular target corresponding to specific values of the angle θ, about every 20°,
from $\theta = 0°$ to $\theta = 180°$. In that case, the diatomic targets have been considered at
equilibrium. The molecular calculations have been performed in the C_{2v} symmetry
group in linear geometries, and using the C_s symmetry group for non-linear ones
taking the plane of the molecular system as plane of symmetry. Spin-orbit coupling
being negligible in the energy range of interest, we assume electron spin to be
conserved in the collision process.

The charge transfer process is driven mainly by non-adiabatic interactions in the
vicinity of the avoided crossings [19] and radial coupling matrix elements between
all pairs of states of the same symmetry have been calculated by means of the finite
difference technique:

$$g_{KL}(R) = \langle \psi_K | \partial/\partial R | \psi_L \rangle = \lim_{\Delta \to 0} \frac{1}{\Delta} \langle \psi_K(R) | \psi_L(R+\Delta) \rangle,$$

with the parameter $\Delta = 0.0012$ a.u. previously tested [20].The rotational coupling
matrix elements $\langle \psi_K(R) | iL_y | \psi_L(R) \rangle$ between states of angular moment $\Delta\Lambda = \pm 1$
have been calculated directly from the quadrupole moment tensor with the centre of
mass of the system at origin of electronic coordinates.

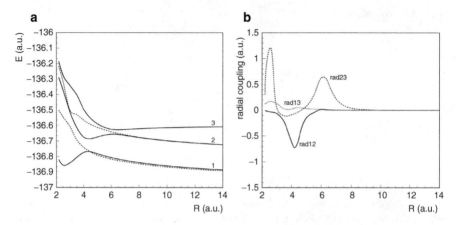

Fig. 20.2 (**a**) Potential energy curves for the $^1\Sigma^+$ (*full line*) and $^1\Pi$ (*broken line*) states of the C^{2+} + HF molecular system at equilibrium $r_{HF} = 1.738368$ a.u., $\theta = 160°$. 1, $C^+(1s^2 2s^2 2p)^2P$ + HF$^+$ ($^2\Pi$); 2, $C^+(1s^2 2s^2 2p)^2P$ + HF$^+(^2\Sigma^+)$; 3, $C^{2+}(1s^2 2s^2)^1S$ + HF($^1\Sigma^+$) entry channel. (**b**) Radial coupling matrix elements between $^1\Sigma^+$ states of the C^{2+} + HF molecular system at equilibrium, $\theta = 160°$. Same labels as in Fig. 20.2a

The molecular calculations have been performed for the different collision systems taking account of all the exit channels which may be correlated with the ground state entry channel, respectively $C^{2+}(1s^2 2s^2)^1S$ + OH($^2\Pi$), $C^{2+}(1s^2 2s^2)^1S$ + CO($^1\Sigma^+$), and $C^{2+}(1s^2 2s^2)^1S$ + HF($^1\Sigma^+$). As an example, we detail the different molecular states involved in the $C^{2+}(1s^2 2s^2)^1S$ + HF($^1\Sigma^+$) charge transfer system. Taking account of the $^1\Sigma^+$ symmetry of the $C^{2+}(1s^2 2s^2)^1S$ + HF($^1\Sigma^+$) entry channel, three $^1\Sigma^+$ states and two $^1\Pi$ states have to be considered in this process with regard to the different excited states of HF$^+$ and spin considerations:

$$
\begin{array}{ll}
C^{2+}(1s^2 2s^2)^1S + HF(^1\Sigma^+) & ^1\Sigma^+ \\
C^+(1s^2 2s^2 2p)^2P + HF^+(^2\Sigma^+) & ^1\Sigma^+, {}^1\Pi \\
C^+(1s^2 2s^2 2p)^2P + HF^+(^2\Pi) & ^1\Sigma^+, {}^1\Pi
\end{array}
$$

The potential energy curves for the equilibrium distance, associated radial and rotational coupling matrix elements between $^1\Sigma^+$ and $^1\Pi$ states have been calculated in the [2.0–14.0] a.u. internuclear distance range. The main features are presented in Fig. 20.2a, b for an orientation angle $\theta = 160°$ and the equilibrium r_{HF} distance. Two avoided crossings are clearly visualized on the $^1\Sigma^+$ potential energy curves, one between the entry channel and the $2^1\Sigma^+\{C^+(1s^2 2s^2 2p)^2P + HF^+(^2\Sigma^+)\}$ exit channel around 6. a.u. and another one, at shorter distance range around R = 4. a.u., between the $2^1\Sigma^+$ and $1^1\Sigma^+\{C^+(1s^2 2s^2 2p)^2P + HF^+(^2\Pi)\}$ exit channels. Peaks may be observed for the radial coupling matrix elements in correspondence to these avoided crossings, as shown on Fig. 20.2b. The radial coupling rad23 exhibits besides a sharp peak at short range corresponding to an interaction in the repulsive part of the potential energy curve. Such an interaction would certainly be considered as quasi-diabatic in the collision treatment [21, 22]. The radial coupling matrix elements between $^1\Pi$ states have not been presented as no significant avoided crossing may be observed in the distance range of interest. Of course, the potential energy curves and couplings are depending on the geometry of the collision system, in particular on the orientation angle θ and the r_{HF} distance. This may be relied directly to the anisotropy of the process and the vibration of the diatomic target and will be detailed in further chapters.

20.2.2   Collision Dynamics

The collision dynamics has been performed by means of the EIKONX code [23] in the keV laboratory energy range where semi-classical approaches may be used with a good accuracy. In such charge transfer processes, electronic transitions can be assumed to occur so fast that vibration and rotation motions may be neglected during the collision. We can thus use the sudden approximation hypothesis and determine the partial and total cross sections considering the internuclear distance of the molecular target fixed in a given geometry. This approach is, of course, relatively crude but it is widely used in the field of ion-molecule collisions and has shown its efficiency in the keV energy range we are dealing with [24]. Such a treatment has been performed for different orientations θ and different r distances, for the collision of C^{2+} on the diatomic targets CO and HF taking account of all the transitions driven by radial and rotational couplings. The $C^{2+} + OH$ collision system being extremely complex, the transitions driven by rotational coupling matrix elements have been analyzed with regard to the collision energy [6]. Translation effects have not been included in this study. However, the origin of coordinates has been chosen in order to expect accurate enough values of total cross sections in the $[1-75]$ keV collision energy range we are dealing with [25].

The partial and total cross sections between the different states involved in the charge transfer process are presented in Fig. 20.3 for the $C^{2+}(1s^22s^2)^1S + HF(^1\Sigma^+)$ system at equilibrium and corresponding to the orientation $\theta = 160°$, detailed in previous paragraph. At low collision energies, the process is driven mainly by the radial coupling rad23 between the entry channel and the $2^2\Sigma^+\{C^+(1s^22s^22p)^2P + HF^+(^2\Sigma^+)\}$ level. It is characterized by a low-energy bump on the corresponding partial cross section sec32 which may be observed also on the total charge transfer cross section. At higher collision energies, this non-adiabatic interaction becomes less efficient, and the process is more likely driven by the radial coupling rad12 with increase of the corresponding sec31 partial cross section on the $1^1\Sigma^+\{C^+(1s^22s^22p)^2P + HF^+(^2\Pi)\}$ charge transfer channel. Besides, an important rotational effect may be pointed out at higher collision energies, with a significant increase of the secpi32 partial cross section. Such remarks may be extended to the different geometries of the charge transfer system, considering the variation of the cross sections taking account of the evolution of the different radial coupling matrix elements. This could lead to interesting correlations

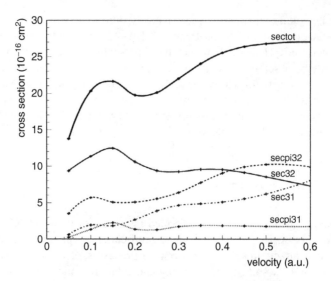

Fig. 20.3 Total and partial charge transfer cross sections for the $C^{2+} + HF$ system at equilibrium $r_{HF} = 1.738368$ a.u., $\theta = 160°$; *sectot*, total cross section; *sec32*, partial cross section on $^1\Sigma^+ \{C^+(1s^2 2s^2 2p)^2P + HF^+(^2\Sigma^+)\}$; *secpi32*, partial cross section on $^1\Pi\{C^+(1s^2 2s^2 2p)^2P + HF^+(^2\Sigma^+)\}$; *sec31*, partial cross section on $^1\Sigma^+\{C^+(1s^2 2s^2 2p)^2P + HF^+(^2\Pi)\}$; *secpi31*, partial cross section on $^1\Pi\{C^+(1s^2 2s^2 2p)^2P + HF^+(^2\Pi)\}$

between non-adiabatic interactions at the vicinity of avoided crossings and the observable cross sections. The analysis is deepened by comparison of similar collision systems as general behaviours could be established; we have considered the collision of C^{2+} on HF, OH and CO targets in order to extract more general rules regarding the anisotropy of the process and the vibration effect.

20.3 Vibration Effect

A complete treatment has been performed for the different collision systems for a series of values of the vibration coordinate r around the equilibrium distance in the linear approach. The main features can be exhibited by discussing the results on the $C^{2+} + OH$ and $C^{2+} + HF$ collision systems.

As shown on Fig. 20.4, the $C^{2+} + OH$ charge transfer process involves a very great number of molecular states, very close in energy, and leads to intricate molecular calculations. However, the interaction between the entry channel $^2B_1\{C^{2+}(^2P) + OH(^2\Pi)\}$ and the $^2B_1\{C^+(^2P) + OH^+(^1\Pi)\}$ exit channel is strong and can be considered to mainly drive the process. A very straightforward correlation can be established between the charge transfer cross sections and the corresponding g_{78} radial coupling matrix element. Effectively, as displayed in Fig. 20.5a, b, a regular increase is observed for the total cross sections in correspondence with the

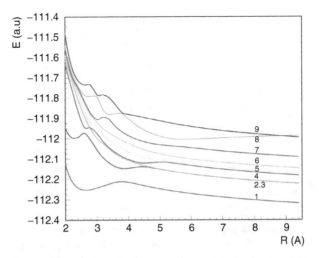

Fig. 20.4 Potential energy curves for the $^2B_1$ states of the $C^{2+} + OH$ molecular system at equilibrium $r_{OH} = 1.834967$ a.u., in the linear geometry. 1, $C^+(^2P) + OH^+(^3\Sigma^-)$; 2,3, $C^+(^2P) + OH^+(^1\Delta)$; 4, $C^+(^2P) + OH^+(^1\Sigma^+)$; 5, $C^+(^2P) + OH^+(^3\Pi)$; 6, $C^+(^4P) + OH^+(^3\Sigma^-)$; 7, $C^+(^2P) + OH^+(^1\Pi)$; 8, $C^{2+}(^1S) + OH(^2\Pi)$ entry channel; 9, $C^+(^4P) + OH^+(^3\Pi)$

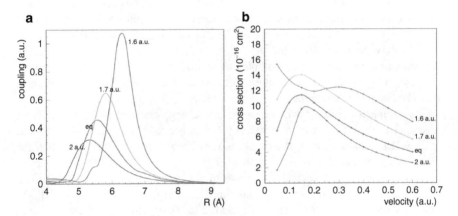

Fig. 20.5 (a) Variation of the radial coupling matrix element g_{78} for different geometries of the OH radical around the equilibrium distance $r_{OH} = 1.834967$ a.u. in the linear geometry. (b) Corresponding charge transfer cross sections for the $C^{2+} + OH$ system

regular increase of the g_{78} radial coupling around the equilibrium distance, between $r_{OH} = 2.0$ a.u. and $r_{OH} = 1.7$ a.u.. On the contrary, for very constraint geometries, corresponding to a very short r_{OH} distance, a completely different behaviour is observed for the charge transfer cross section. The collision of C^{2+} ions with OH appears to present a two-step mechanism with first a relaxation of the system before the effective charge transfer process could occur, leading to a more regular variation

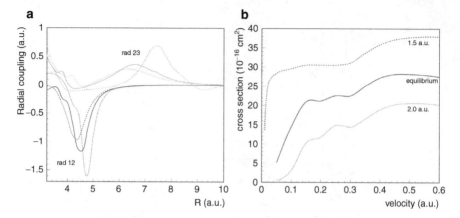

Fig. 20.6 (**a**) Radial coupling matrix elements between $^1\Sigma^+$ states of the $C^{2+} + HF$ system in the linear approach for different values of the vibration coordinate r_{HF}. Same labels as in Fig. 20.2a. *Upper curves*, rad23; *lower curves*, rad12. *Dotted line*, $r_{HF} = 2.0$ a.u.; *full line*, $r_{HF} = 1.73836832$ a.u. (equilibrium); *dashed line*, $r_{HF} = 1.5$ a.u. (**b**) Corresponding charge transfer cross sections for the $C^{2+} + HF$ system

of the cross sections at higher energies. Such behaviour can be correlated to the very sharp g_{78} radial coupling showing a strong increase of the non-adiabatic interaction.

Similar correlations between non-adiabatic effect and charge transfer cross sections may be established for the collision of C^{2+} ions on HF. However, in that case, the analysis is more complex as several non-adiabatic interactions may interfere and the concomitant evolution of several radial coupling matrix elements has to be taken into account.

The radial coupling matrix elements and charge transfer cross sections for different r_{HF} values are presented respectively in Fig. 20.6 a, b. As in the previous $C^{2+} + OH$ collision system, the cross sections show a regular increase when the vibration coordinate r_{HF} is reduced from 2.0 to 1.5 a.u. around the equilibrium distance, in agreement with the increase of the radial coupling matrix element rad23 between the entry channel and the $2^1\Sigma^+$ level. On the other hand, the radial coupling rad12 between $1^1\Sigma^+$ and $2^1\Sigma^+$ exit channels decreases with the r_{HF} vibration coordinate corresponding to a smoother interaction at shorter r_{HF} distances. The non-adiabatic interaction between the entry channel and the $2^1\Sigma^+\{C^+(1s^22s^22p)^2P + HF^+(^2\Sigma^+)\}$ level is thus clearly the driving step in the charge transfer process. It may be relied to the low-energy bump observed on the sec32 partial cross section (Fig. 20.3) which increases for shorter r_{HF} values [7]. The non-adiabatic interaction with the lower $1^1\Sigma^+\{C^+(1s^22s^22p)^2P + HF^+(^2\Pi)\}$ charge transfer channel may be more likely relied to the smooth high energy bump observed in charge transfer cross sections. At variance with the previous case, no specific behaviour is exhibited at very constrained HF geometry and a first relaxation process is not expected in the collision of C^{2+} with the HF target.

20.4 Anisotropic Effect

The orientation of the projectile towards the molecular target has also been studied in detail for the series of charge transfer systems. It may be developed on the $C^{2+} + CO$ and $C^{2+} + HF$ collisions. The calculations have been performed for specific values of the θ angle, from linear to perpendicular orientations. As exhibited previously for the vibration effect, a significant correlation can be established between non-adiabatic interactions and charge transfer cross sections. For the $C^{2+} + CO$ collision system, seven states have to be taken into account with regard to the different excited levels of CO^+ and consideration of radial and rotational couplings. They are displayed in Fig. 20.7 in the linear geometry.

Radial couplings and charge transfer cross sections are displayed in Fig. 20.8a, b. The most important interaction corresponds to the g_{34} radial coupling between the entry channel and the $3^1\Sigma^+\{C^+(1s^22s^22p)^2P + CO^+(B^2\Sigma^+)\}$ exit channel. A significant evolution is exhibited for the different orientations and correlations may be evidenced between this radial coupling and the charge transfer cross sections. Effectively, the charge transfer appears clearly more favourable in the linear direction towards the oxygen atom ($\theta = 180°$) where the radial coupling g_{34} is highest. On the contrary, the perpendicular orientation is markedly unfavoured with cross sections about six times lower, in agreement with a very low radial coupling matrix element g_{34}. The correlation however is not symmetrical, the decrease of cross sections from $\theta = 180°$ to $90°$ is regular, but the process appears globally less favourable on the carbon side.

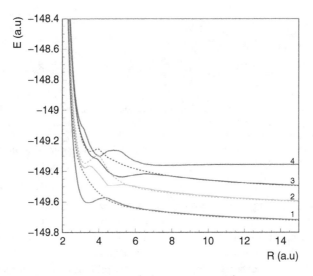

Fig. 20.7 Potential energy curves for the $^1\Sigma^+$ (full line) and $^1\Pi$ (broken line) states of the $C^{2+} + CO$ molecular system at equilibrium $r_{CO} = 2.140535$ a.u., $\theta = 180°$. 1, $C^+(1s^22s^22p)^2P + CO^+(A^2\Sigma^+)$; 2, $C^+(1s^22s^22p)^2P + CO^+(A^2\Pi)$; 3, $C^+(1s^22s^22p)^2P + CO^+(B^2\Sigma^+)$; 4, $C^{2+}(1s^22s^2)^1S + CO(^1\Sigma^+)$ entry channel

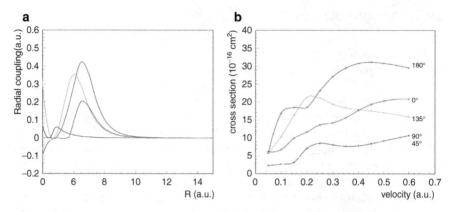

Fig. 20.8 (a) Variation of the g_{34} radial coupling matrix element between $^1\Sigma^+$ levels for different orientations of the C^{2+} projectile towards the CO molecule at equilibrium $r_{CO} = 2.140535$ a.u., *red* $\theta = 180°$; *green* $\theta = 135°$; *blue* $\theta = 90°$; *yellow* $\theta = 45°$; *magenta* $\theta = 0°$. (b) Corresponding charge transfer cross sections for the $C^{2+} + CO$ system

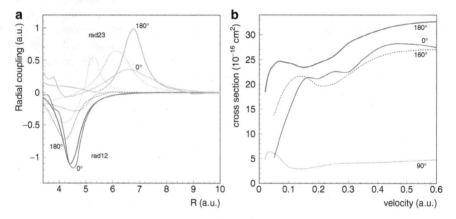

Fig. 20.9 (a) Radial coupling matrix elements between $^1\Sigma^+$ states of the $C^{2+} + HF$ system at equilibrium, $r_{HF} = 1.738368$ a.u., for different orientations. Same labels as in Fig. 20.2a. *Upper curves,* rad23; *lower curves,* rad12. *Dotted line,* $\theta = 90°$; *dashed line,* $\theta = 160°$; *thin full line,* $\theta = 0°$; *full line,* $\theta = 180°$. (b) Corresponding charge transfer cross-sections for the $C^{2+} + HF$ system at equilibrium, for different orientations θ from $0°$ to $180°$

The radial couplings and cross sections presented in Fig. 20.9a, b clearly drive the same conclusions for the $C^{2+} + HF$ collision system. The collision towards the fluorine atom in the linear approach ($\theta = 180°$) is particularly efficient. Globally speaking, the charge transfer process is favoured in the linear geometry and clearly less efficient in the perpendicular one. As a general rule, we can assess that in collisions on hetero-nuclear molecular targets, the charge transfer is favoured in a collinear approach towards the most electronegative atom, fluorine or oxygen preferentially than hydrogen for HF and OH, or carbon in the case of the CO molecular

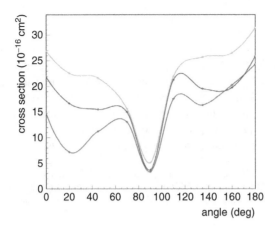

Fig. 20.10 Total charge transfer cross-sections for the C^{2+} + HF system at equilibrium $r_{HF} = 1.738368$ a.u., for different orientations θ from $0°$ to $180°$. *green* v $= 0.45$ a.u. ($E_{lab} = 60.75$ keV); *blue* v $= 0.25$ a.u. ($E_{lab} = 18.75$ keV); *red* v $= 0.1$ a.u. ($E_{lab} = 3$ keV)

target. As previously stated, correlations between charge transfer efficiency and non-adiabatic interactions may also be established for this system. Two avoided crossings are involved presently for the C^{2+} + HF collision system, the avoided crossing between the entry channel and the $2^1\Sigma^+\{C^+(1s^22s^22p)^2P + HF^+(^2\Sigma^+)\}$ level corresponding to the radial coupling rad23, and the avoided crossing between the $2^1\Sigma^+$ and $1^1\Sigma^+\{C^+(1s^22s^22p)^2P + HF^+(^2\Pi)\}$ exit channels characterised by the radial coupling rad12. Both of them are maximum for the linear geometry towards fluorine ($\theta = 180°$) and decrease significantly for perpendicular orientation. The radial coupling rad23 remains anyway not negligible at $\theta = 90°$. It is globally smoother for collisions on the hydrogen side, as observed for the CO target on the carbon side. The radial coupling rad12 is less sensitive to this preferred orientation. It is almost symmetric for both sides of the collision and significantly sharper in both collinear orientations.

The present analysis may be supported by looking at the evolution of the charge transfer cross section with regard to the θ angle, for given collision velocities displayed in Fig. 20.10. The conclusions depend a bit on the collision energy, but the perpendicular orientation is always shown to be strongly unfavourable whatever the collision velocity and the side of fluorine, the most electronegative atom, leads clearly to a more efficient charge transfer process.

20.5   Concluding Remarks

This paper presents a theoretical treatment of charge transfer processes induced by collision of the C^{2+} projectile ions on a series of diatomic molecules, OH, CO and HF. An interesting insight into the mechanism of the charge transfer process has been exhibited, with regard to the behaviour of the collision system in different orientations, as well as for different values of the vibration coordinate. The collision process is highly anisotropic: the charge transfer is favoured in the linear approach with collision of the C^{2+} ion towards the more electronegative atom, and, on the contrary, very significantly non-favoured in the perpendicular approach. Correlations may be driven between the partial cross section values and the non-adiabatic interactions shown by the molecular system, in particular when the charge transfer process is clearly driven by one non-adiabatic interaction. Such an approach could be extended to more complicated targets.

Acknowledgments This work was granted access to the HPC resources of [CCRT/CINES/IDRIS under the allocation 2010- [i2010081566] made by GENCI [Grand Equipement National de Calcul Intensif]. The support of the COST Action CM0702 CUSPFEL is greatly acknowledged.

References

1. Sobocinski P, Rangama J, Laurent G, Adoui L, Cassimi A, Chesnel J-Y, Dubois A, Hennecart D, Husson X, Frémont F (2002) J Phys B 35:1353
2. Laurent G, Fernandez J, Legendre S, Tarisien M, Adoui L, Cassimi A, Flechard X, Frémont F, Gervais B, Giglio E, Grandin JP, Martin F (2006) Phys Rev Lett 96:173201
3. Frémont F, Martina D, Kamalou O, Sobocinski P, Chesnel J-Y, McNab IR, Bennett FR (2005) Phys Rev A 71:042706
4. Bacchus-Montabonel MC (1999) Phys Rev A 59:3569
5. Bene E, Vibók Á, Halász GJ, Bacchus-Montabonel MC (2008) Chem Phys Lett 455:159
6. Bene E, Martínez P, Halász GJ, Vibók Á, Bacchus-Montabonel MC (2009) Phys Rev A 80:012711
7. Rozsályi E, Bene E, Halász GJ, Vibók Á, Bacchus-Montabonel MC (2010) Phys Rev A 81:062711
8. Cabrera-Trujillo R, Deumens E, Ohrn Y, Quinet O, Sabin JR, Stolterfoht N (2007) Phys Rev A 75:052702
9. Alvarado F, Bari S, Hoekstra R, Schlatölter T (2007) J Chem Phys 127:034301
10. de Vries J, Hoekstra R, Morgenstern R, Schlathölter T (2003) Phys Rev Lett 91:053401
11. Bacchus-Montabonel MC, Łabuda M, Tergiman YS, Sienkiewicz JE (2005) Phys Rev A 72:052706
12. Bacchus-Montabonel MC, Tergiman YS (2006) Phys Rev A 74:054702
13. Bacchus-Montabonel MC, Tergiman YS, Talbi D (2009) PhysRev A 79:012710
14. Michael BD, O'Neill PD (2000) Science 287:1603
15. Spotheim-Maurizot M, Bergusova M, Charlier M (2003) Actual Chim 1112:97
16. Bacchus-Montabonel MC, Vaeck N, Lasorne B, Desouter-Lecomte M (2003) Chem Phys Lett 374:307
17. Werner HJ, Knowles P MOLPRO (version 2009.1) package of ab-initio programs
18. Woon DE, Dunning TH Jr (1993) J Chem Phys 98:1358
19. Vaeck N, Bacchus-Montabonel MC, Baloïtcha E, Desouter-Lecomte M (2001) Phys Rev A 63:042704
20. Bacchus-Montabonel MC, Courbin C, McCarroll R (1991) J Phys B 24:4409
21. Honvault P, Gargaud M, Bacchus-Montabonel MC, McCarroll R (1995) Astron Astrophys 302:931

22. Chenel A, Mangaud E, Justum Y, Talbi D, Bacchus-Montabonel MC, Desouter-Lecomte M (2010) J Phys B 43:245701
23. Allan RJ, Courbin C, Salas P, Wahnon P (1990) J Phys B 23:L461
24. Stancil PC, Zygelman B, Kirby K (1998) In: Aumayr F, Winter HP (eds) Photonic, electronic, and atomic collisions. World Scientific, Singapore, p 537
25. Bates DR, McCarroll R (1958) Proc R Soc A 245:175

Chapter 21
Recombination by Electron Capture in the Interstellar Medium

M.C. Bacchus-Montabonel and D. Talbi

Abstract Rate constants for charge transfer processes in the interstellar medium are calculated using *ab-initio* molecular calculations. Two important reactions are presented: the recombination of Si^{2+} and Si^{3+} ions with atomic hydrogen and helium which is critical in determining the fractional abundances of silicon ions, and the $C^{+} + S \rightarrow C + S^{+}$ reaction, fundamental in both carbon and sulphur chemistry.

21.1 Introduction

A quantitative analysis of the emission-line spectra of ionized astronomical objects requires reliable data on the microscopic ionization and recombination processes involved. Recombination may occur either by radiative capture or by charge transfer from neutral species. The charge transfer recombination process with atomic hydrogen or helium is particularly important in astrophysical plasmas for many multiply-charged ions, whose emission lines are used to provide direct information of the ionization structure of astronomical objects [1–8]. For some doubly and triply charged ions, electron capture can lead directly to the formation of ground states and thus may induce rapid ionization via the inverse charge transfer process. This is the case for the $Si^{2+}(^{1}S) + H(^{2}S)$ and $Si^{3+}(^{2}S) + He(^{1}S)$ reactions which are critical in determining the fractional abundances of silicon ions [9–11]. The charge transfer $C^{+} + S$ reaction plays also a determinant role in the formation of sulphur bearing

M.C. Bacchus-Montabonel (✉)
Laboratoire de Spectrométrie Ionique et Moléculaire, Université de Lyon I, CNRS UMR5579,
43 Bd. du 11 Novembre 1918, 69622 Villeurbanne Cedex, France
e-mail: bacchus@lasim.univ-lyon1.fr

D. Talbi
Laboratoire Univers et Particules de Montpellier UMR 5299
de Montpellier II et CNRS, Place Eugène Bataillon, 34095 Montpellier cedex 05, France
e-mail: dahbia.talbi@univ-montp2.fr

P.E. Hoggan et al. (eds.), *Advances in the Theory of Quantum Systems in Chemistry and Physics*, Progress in Theoretical Chemistry and Physics 22,
DOI 10.1007/978-94-007-2076-3_21, © Springer Science+Business Media B.V. 2012

molecules [12], for instance in the abundance of H_2CS in the dense interstellar clouds. It is crucial in the chemistry of the photon dominated regions (PDR's) of the interstellar medium [13, 14] and allows the enhancement of the ionic carbon chemistry at the origin of the formation of the complex carbon molecules observed in the PDR's.

These examples present a significant overview of charge transfer in space chemistry. We thus report a complete ab-initio treatment of the $Si^{2+}(^1S) + H(^2S)$ and $Si^{3+}(^2S) + He(^1S)$ reactions as well as new calculations of the $C^+(2s^22p)^2P + S(3s^23p^4)^3P$ process and its reverse $C + S^+$ reaction in order to provide an interpretation of the mechanisms involved and determine the corresponding rate coefficients at different temperatures which are crucial data in the modellisation of the interstellar medium.

21.2 Theoretical Treatment

21.2.1 Molecular Calculations

The potential energy curves have been calculated by means of *ab-initio* methods. For the $Si^{2+} + H$ and $Si^{3+} + He$ collision systems, MCSCF/CI calculations have been performed using the CIPSI algorithm [15]. A non-local pseudopotential was used to represent the core electrons of the silicon atom [16] and 9s7p2d basis of Gaussian functions have been optimized on $Si^{2+}(3s^2)^1S$, $Si^{2+}(3s3p)^2P$ and $Si^{3+}(3s)^2S$ from the basis sets of McLean and Chandler [17]. Previously used 4s1p and 5s3p basis have been used for helium and hydrogen atoms [18,19]. This basis of atomic functions may be compared to the larger coupled-cluster polarized valence triple zeta and augmented quadruple zeta basis sets of Dunning [20] with errors of the order 10^{-4} a.u. on Hartree-Fock energies. Special care was taken to construct sets of determinants providing the same level of accuracy over the whole distance range. For the CS^+ molecular system, potential energy curves have been carried out using the MOLPRO suite of *ab-initio* programs [21] at the state average CASSCF-MRCI level of theory. The active space includes the $n = 2$ orbitals for carbon and $n = 3$ orbitals for sulphur. The ECP10sdf relativistic pseudo-potential has been used to describe the 10 core-electrons of sulphur [22] with the correlation-consistent aug-cc-pVQZ basis sets of Dunning [20] for all atoms. The spin-orbit effects may be neglected in the collision energy range of interest so doublet and quartet manifolds can be considered separately.

The charge transfer process is driven mainly by non-adiabatic interactions in the vicinity of avoided crossings [23,24]. The corresponding radial coupling matrix elements between all pairs of states of the same symmetry were calculated by means of the finite difference technique:

$$g_{KL}(R) = \langle \psi_K \,|\partial/\partial R|\, \psi_L \rangle = \lim_{\Delta \to 0} \frac{1}{\Delta} \langle \psi_K(R)|\, \psi_L(R+\Delta) \rangle,$$

with the parameter $\Delta = 0.0012$ a.u. previously tested [25]. For reasons of numerical accuracy, we performed a three-point numerical differentiation using calculations at $R + \Delta$ and $R–\Delta$ for a very large number of interatomic distances in the avoided-crossing region.

The rotational coupling matrix elements $\langle \psi_K | iL_y | \psi_L \rangle$ between $\Sigma - \Pi$ molecular states were determined directly from the quadrupole moment tensor which allows the consideration of translation effects in the collision dynamics [26]. In the approximation of the common translation factor [27], the radial and rotational coupling matrix elements between states ψ_K and ψ_L may indeed be transformed respectively into:

$$\langle \psi_K | \partial / \partial R - (\varepsilon_K - \varepsilon_L) z^2 / 2R | \psi_L \rangle,$$

$$\langle \psi_K | iL_y + (\varepsilon_K - \varepsilon_L) zx | \psi_L \rangle,$$

where ε_K and ε_L are the electronic energies of states ψ_K and ψ_L and z^2 and zx are the component of the quadrupole moment tensor.

21.2.2 Collision Dynamics

The collision dynamics was treated in the eV energy range by a semi-classical approach using the EIKONXS program [28] in the case of the $Si^{2+} + H$ and $C^+ + S$ reactions. Both radial and rotational coupling matrix elements were taken into account, as well as translation effects, although they are expected to be low at these energies. For the $Si^{3+} + He$ collision system, a quantum mechanical approach was preferred. Allowance for translation effects was made by introducing appropriate reaction coordinates [6, 29] which induce a modification of the radial and rotational matrix elements similar in form to those resulting of the application of the common translation factor method [27]. The rate constants k(T) were calculated by averaging the cross sections $\sigma(E)$ over a Maxwellian velocity distribution at temperature T.

21.3 The $Si^{2+}+H$ and $Si^{3+}+He$ Collision Systems

Both systems, $Si^{2+} + H$ and $Si^{3+} + He$, present the characteristic to lead to the ground state $Si^{(q-1)+}$ ion by recombination from the Si^{q+} ion with, respectively, the hydrogen or helium atom and thus induce a rapid reverse ionization process. The charge exchange recombination of Si^{2+} ions with atomic hydrogen

$$Si^{2+}(3s^2)^1S + H(1s)^2S \rightarrow Si^+(3s^23p)^2P + H^+$$

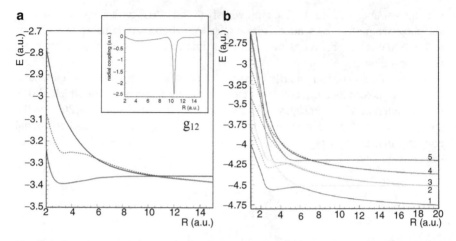

Fig. 21.1 (**a**) Adiabatic potential energy curves for the $^2\Sigma^+$ (*full lines*) and $^2\Pi$ (*dashed lines*) states of the collision system $Si^{2+} + H$. (*1*) $^2\Sigma^+,^2\Pi\{Si^+(3s^23p)^2P + H^+\}$. (*2*) $^2\Sigma^+$ $\{Si^{2+}(3s^2)^1S + H(1s)\}$, entry channel. (**b**) Adiabatic potential energy curves for the $^2\Sigma^+$ (*full lines*) and $^2\Pi$ (*dashed lines*) states of the collision system $Si^{3+} + He$. (*1*) $^2\Sigma^+\{Si^{2+}$ $(3s^2)^1S + He^+(1s)$. (*2*) $^2\Sigma^+\{Si^{3+}(3s)$ $^2S + He\}$, ground entry channel. (*3*) $^2\Sigma^+,^2\Pi$ $\{Si^{2+}$ $(3s3p)^3P + He^+\}$. (*4*) $^2\Sigma^+,^2\Pi\{Si^{2+}(3s3p)^1P + He^+\}$. (*5*) $^2\Sigma^+,^2\Pi\{Si^{3+}(3p)^2P + He\}$, metastable entry channel

is a relatively simple collision system where only three molecular states are involved, the $^2\Sigma^+$ entry channel and the $^2\Sigma^+$ and $^2\Pi$ states correlated to the one-electron capture channel $\{Si^+(3s^23p)^2P + H^+\}$. Such potentials are displayed in Fig. 21.1a and present a sharp avoided crossing around $R = 10.5$ a.u. corresponding to a peaked radial coupling matrix element, 2.47 a.u. high.

The $Si^{3+}(3s)^2S + He$ collision system is also a simple molecular system

$$Si^{3+}(3s)^2S + He(1s^2)^1S \rightarrow Si^{2+}(3s^2)^1S + He^+(1s)^2S,$$

but, for a complete treatment of the process, we have to take into account simultaneously the charge transfer from the metastable $Si^{3+}(3p)$ ion as molecular states are close in energy and can interact. We have thus to consider also the reaction

$$Si^{3+}(3p)^2P + He(1s^2)^1S \rightarrow Si^{2+}(3s3p)^{1,3}P + He^+(1s)^2S$$

which involves $^2\Sigma^+$ and $^2\Pi$ levels. The potential energy curves are presented in Fig. 21.1b. The ground state $Si^{3+}(3s)^2S + He$ entry channel leads to a simple electron capture process. The potential energy curves present a pronounced avoided crossing around $R = 6.0$ a.u. with the $\{Si^{2+}(3s^2)^1S + He^+(1s)^2S\}$ exit channel. A very sharp avoided crossing may also be observed around $R = 7.0$ a.u. between the metastable entry channels $^2\Sigma^+,^2\Pi$ $\{Si^{3+}(3p)^2P + He(1s^2)^1S\}$ and the $\{Si^{2+}(3s3p)^1P + He^+(1s)^2S\}$ in $^2\Sigma^+$ and $^2\Pi$ symmetries and a smoother one, around $R = 5.0$ a.u., between the $\{Si^{2+}(3s3p)^1P + He^+(1s)^2S\}$ and

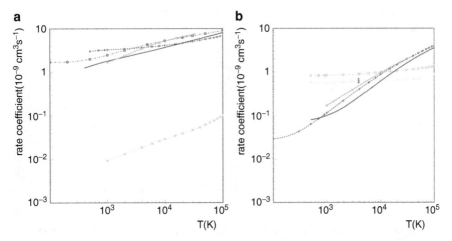

Fig. 21.2 (**a**) Rate coefficients $(10^{-9}\,\mathrm{cm^3\,s^{-1}})$ for the charge transfer recombination processes $\mathrm{Si^{2+}} + \mathrm{H}$. From $\mathrm{Si^{2+}}(3s^2)$, (*blue*) this work; (*red*) [30]; (*magenta*) [31]; (*green*) [32]. From $\mathrm{Si^{2+}}(3s3p)$, (*light blue*) [30]. (**b**) Rate coefficients $(10^{-9}\,\mathrm{cm^3\,s^{-1}})$ for the charge transfer recombination processes $\mathrm{Si^{3+}} + \mathrm{He}$. From $\mathrm{Si^{3+}}(3s)$, (*blue*) this work; (*red*) [35]; (*green*) [34]; (*magenta*) [33]. From $\mathrm{Si^{3+}}(3p)$, (*yellow*) this work; (*light blue*) [35]

$\{\mathrm{Si^{2+}}(3s3p)^3P + \mathrm{He^+}(1s)^2S\}^2\Sigma^+$ and $^2\Pi$ exit channels. At shorter internuclear distance, around $R = 3.2$ a.u., an important avoided crossing between the $^2\Sigma^+\{\mathrm{Si^{2+}}(3s3p)^3P + \mathrm{He^+}(1s)^2S\}$ level and the ground state entry channel may also be pointed out.

The collision dynamics has been performed for both systems. The rate coefficients for the $\mathrm{Si^{2+}}(3s^2)^1S + \mathrm{H}$ system are presented in Fig. 21.2a and compared to the SCVB *ab-initio* calculation of Clarke et al. [30] with radial coupling only, as well as with the close-coupling approach of Gargaud et al. [31] using model potentials, and the Landau-Zener analysis of Bates and Moiseiwitsch [32]. The different results are in globally good agreement. As already noticed by Clarke et al. [30], their quantal close-coupling approach differs slightly from the results of Gargaud et al. [31] and Bates and Moiseiwitsch [32]. On the contrary, they are in good agreement at high temperatures with the present ones using a semi-classical method, exhibiting, of course, some discrepancies at lower temperatures related to trajectory effects which are not considered in our theoretical approach. The Fig. 21.2a displays also the results for the electron capture from the $\mathrm{Si^{2+}}(3s3p)^3P^\circ$ metastable ion

$$\mathrm{Si^{2+}}(3s3p)^3P^\circ + \mathrm{H}(1s)^2S \rightarrow \mathrm{Si^+}(3s3p^2)^2D + \mathrm{H^+}$$
$$\rightarrow \mathrm{Si^+}(3s^23p)^2P^\circ + \mathrm{H^+}$$

determined by Clarke et al. [30] to be two orders of magnitude lower than the capture by the ground state ion. Such process appears non determinant and so has not been taken into account in our calculation.

Table 21.1 Rate coefficients for charge transfer and ionization processes in the $Si^{3+} + He$ collision (in $10^{-9} \, cm^3 \, s^{-1}$)

T(K)	k_{CT}	k_{CT} [34]	k_{ion}	k_{ion} [34]
500	0.08		-	
1,000	0.10	0.17	-	
2,000	0.15		-	
3,000	0.21	0.39	-	
5,000	0.33		-	
10,000	0.62	0.96	0.00002	0.00003
20,000	1.14		0.0065	
30,000	1.58	2.00	0.05	0.07
40,000	1.97		0.15	
50,000	2.31		0.29	
100,000	3.60		1.28	1.21

For the $Si^{3+} + He$ system, the coupling equations were solved simultaneously for all the levels involved in the charge transfer process from both the ground state and the excited entry channels. The rate coefficients are displayed in Fig. 21.2b and compared to the ion-trap experiment of Fang and Kwong [33]. For the capture process from the ground state $Si^{3+}(3s)$, a global agreement is observed between the Landau-Zener calculations [34], the present *ab-initio* treatment and the *ab-initio* calculations of Stancil et al. [35] with almost the same variation of rate constants with temperature. Nevertheless, all theoretical results provide rate coefficients lower than the experimental point of Fang and Kwong [33]. On the contrary, the rate coefficients calculated for the capture from the metastable $Si^{3+}(3p)$ ion are of the same order of magnitude than the experimental point. Some uncertainty on the temperature of the trap have to be considered, however, we could suggest certainly the presence of excited $Si^{3+}(3p)$ in the experiment.

At typical astrophysical temperatures, only the ground state $Si^{3+}(3s)$ is significantly populated and the charge transfer process leads to the ground $Si^{2+}(3s^2)$ level. The rate constant for the reverse ionization process k_{ion} may be determined easily by means of the microreversibility relation from the corresponding charge transfer rate constant k_{CT}:

$$k_{ion} = g \exp\left(-\frac{\Delta E}{kT}\right) k_{CT},$$

where g is the ratio of the statistical weights of initial and final states ($g = 1$), and ΔE is the energy gain of the charge transfer reaction. The ionization rate coefficients are presented in Table 21.1. They reach significant values for temperatures above 3×10^4 K, they are rapidly negligible for lower temperatures with regard to the exponential factor. They are in good agreement with the previous calculation of Butler and Dalgarno [34].

21.4 The C$^+$+S Collision System

The C$^+$ + S charge transfer is a determinant reaction for both carbon and sulphur chemistry. The rate constant generally considered for this process is $1.5 \times 10^{-9} \, \text{cm}^3 \, \text{s}^{-1}$ [36] between 10 and 41,000 K, but it remains uncertain for such a large temperature domain and detailed calculations have to be performed. At low temperatures where the process takes place, the different species may be in their ground state. With regard to the correlation diagram, only two molecular states $\{C^+(2s^2 2p)^2P + S(3s^2 3p^4)^3P\}$ and $\{C(2s^2 2p^2)^3P + S^+(3s^2 3p^3)^4S\}$ would thus have to be considered in the charge transfer reaction.

Correlation diagram

Configuration	Molecular states	Asymptotic energy (eV) [37]
$C(2s^2 2p^2)^1S + S^+(3s^2 3p^3)^4S$	$^4\Sigma$	2.68
$C^+(2s^2 2p)^2P + S(3s^2 3p^4)^1D$	$^2\Sigma, ^2\Pi, ^2\Delta, ^2F$	2.04
$C(2s^2 2p^2)^3P + S^+(3s^2 3p^3)^2D$	$^{2,4}\Sigma, ^{2,4}\Pi, ^{2,4}\Delta, ^{2,4}F$	1.86
$C(2s^2 2p^2)^1D + S^+(3s^2 3p^3)^4S$	$^4\Sigma, ^4\Pi, ^4\Delta$	1.26
$C^+(2s^2 2p)^2P + S(3s^2 3p^4)^3P$	$^{2,4}\Sigma, ^{2,4}\Pi, ^{2,4}\Delta$	0.92
$C(2s^2 2p^2)^3P + S^+(3s^2 3p^3)^4S$	$^{2,4,6}\Sigma, ^{2,4,6}\Pi$	0.0

Such two-channel process is presented in Fig. 21.3a for the doublet states. However, a strong interaction with the higher $\{C(2s^2 2p^2)^1D + S^+(3s^2 3p^3)^4S\}$ is pointed out for the quartet manifold as shown on Fig. 21.3b and three levels have to be taken into account for this spin multiplicity.

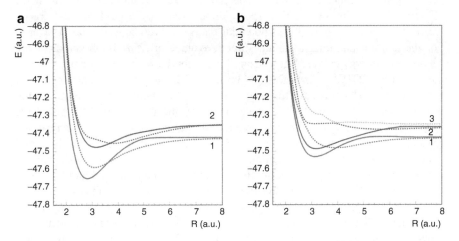

Fig. 21.3 (**a**) Adiabatic potential energy curves for the Σ (*full lines*) and Π (*dashed lines*) states of the doublet manifold of the CS$^+$ molecular system. (*1*) $\{C(2s^2 2p^2)^3P + S^+(3s^2 3p^3)^4S\}$. (*2*) $\{C^+(2s^2 2p)^2P + S(3s^2 3p^4)^3P\}$ entry channel. (**b**) Adiabatic potential energy curves for the Σ (*full lines*) and Π (*dashed lines*) states of the quartet manifold of the CS$^+$ molecular system. (*1*) and (*2*), same labels as in Fig. 21.3a. (*3*) $\{C(2s^2 2p^2)^1D + S^+(3s^2 3p^3)^4S\}$

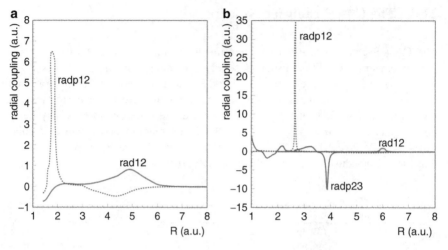

Fig. 21.4 (**a**) Radial coupling matrix elements between Σ (rad12, *red line*) and Π (radp12, *blue line*) states of the doublet manifold of the CS$^+$ molecular system. (*1*) {C(2s^{2}2p^2)^{3}P + S$^+$(3s^{2}3p^3)^{4}S}. (*2*) {C$^+$(2s^{2}2p)^{2}P + S(3s^{2}3p^4)^{3}P} entry channel. (**b**) Radial coupling matrix elements between Σ (rad12, *red line*) and Π (radp12, radp23, *blue lines*) states of the quartet manifold of the CS$^+$ molecular system. (*1*) and (*2*), same labels as in Fig. 21.4a. (*3*) {C(2s^{2}2p^2)^{1}D + S$^+$(3s^{2}3p^3)^{4}S}

The $^2\Sigma$ and $^2\Pi$ potentials present a smooth avoided crossing around R = 5 a.u., in agreement with the previous calculations of Larsson [38] and Honjou [39]. The corresponding radial coupling matrix elements are drawn in Fig. 21.4a. They show smooth peaks around R = 5 a.u., respectively, 0.823 a.u. and 0.459 a.u. high for $^2\Sigma$ and $^2\Pi$ states as well as a sharp radial coupling, 6.475 a.u. high, at R = 1.8 a.u. in the repulsive part of the potential energy curves between the $^2\Pi$ states. For the quartet manifold, a similar smooth avoided crossing is observed for the $^4\Sigma$ potential energy curves. But a strong interaction between the $^4\Pi$ entry channel and the upper $^4\Pi$\{C(2s^{2}2p^2)^{1}D + S$^+$(3s^{2}3p^3)^{4}S} level is exhibited around R = 4 a.u. and three $^4\Pi$ states have to be considered in the calculation. Such interaction is not observed between the $^4\Sigma$ levels and only the two lowest $^4\Sigma$ levels have been taken into account. The corresponding radial coupling matrix elements are presented in Fig. 21.4b. A smooth peak, 0.915 a.u. high is observed for the radial coupling between the $^4\Sigma$ states, relatively similar to the interaction between $^2\Sigma$ levels. However, the radial coupling between the $^4\Pi$ entry channel and the upper $^4\Pi$\{C(2s^{2}2p^2)^{1}D + S$^+$(3s^{2}3p^3)^{4}S} level reaches up to 10.093 a.u. in absolute value and may be determinant in the collision treatment. An extremely sharp radial coupling matrix element between the two lowest $^4\Pi$ levels is also exhibited at short range. It could certainly be considered as quasi-diabatic in the collision dynamics. The Δ states correlated by means of rotational coupling have not been considered in the calculation and the sextuplet states cannot be involved in the process, since there are no states of equivalent spin correlating to any higher asymptotic limits.

Fig. 21.5 Partial and total
cross sections for the CS$^+$
molecular system: doublet
manifold (*red, dashed line*);
quartet manifold (*red, dotted
line*); total cross section (*blue,
solid line*)

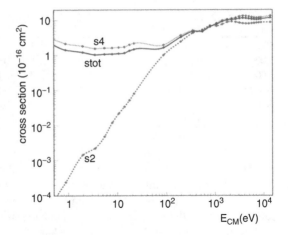

The collision dynamics has been performed for the direct reaction $C^+(2s^22p)^2P + S(3s^23p^4)^3P \rightarrow C(2s^22p^2)^3P + S^+(3s^23p^3)^4S$ for a wide range of collision velocities, in particular at low velocities where trajectory effects should be considered and results have to be considered as qualitative. As expected, the sharp peaks presented by the radial coupling matrix elements radp12 at short range appear as quasi-diabatic in the dynamical treatment. This is the case, of course for the $^4\Pi$ states, where radp12 is extremely sharp, but also for the corresponding coupling between $^2\Pi$ states. As spin-orbit effects may be neglected in the collision energy range of interest, calculations have been performed separately for doublet and quartet manifolds. With consideration of statistical weights between Σ and Π states, the cross sections for doublet and quartet manifolds is expressed from the cross sections σ^Σ and σ^Π for Σ and Π states respectively:

$$^{2,4}\sigma = 1/3\sigma^\Sigma + 2/3\sigma^\Pi.$$

The total cross section is then:

$$\sigma_{tot} = 1/3\,^2\sigma + 2/3\,^4\sigma$$

with regard to the statistical weights between doublet and quartet manifolds. They are presented in Fig. 21.5. The quartet states provide the main contribution to the total cross section at low collision energies and the consideration of the upper $^4\Pi\{C(2s^22p^2)^1D + S^+(3s^23p^3)^4S\}$ level is necessary for an accurate description of the system.

The rate constants for the direct reaction $C^+(2s^22p)^2P + S(3s^23p^4)^3P \rightarrow C(2s^22p^2)^3P + S^+(3s^23p^3)^4S$ are presented in Table 21.2 together with the rate coefficients for the reverse process deduced, as in previous paragraph, from the symmetry properties of the S-matrix. In that case, the degeneracy is g = 3 with regard to the multiplicity of initial and final states and the energy gain is $\Delta E = 0.92\,eV$.

Table 21.2 Rate coefficients for the $C^+ + S$ and reverse reaction (in 10^{-9} cm^3 s^{-1})

T(K)	$C^+(^2P) + S(^3P) \rightarrow C(^3P) + S^+(^4S)$	$C(^3P) + S^+(^4S) \rightarrow C^+(^2P) + S(^3P)$
500	0.018	–
1,000	0.038	0.0000026
5,000	0.072	0.026
10,000	0.073	0.075
50,000	0.13	0.31
100,000	0.20	0.55

The rate constants for the direct reaction are small, about 7.2×10^{-11} cm^{3}s^{-1} at 5,000 K. Such a value is significantly lower than the suggested one 1.5×10^{-9} cm^3 s^{-1} given in the UMIST data base [36] for the 10–41,000 K temperature range. However, the variation of the calculated rate coefficients is relatively weak in a wide temperature domain and a value of about 1×10^{-10} cm^3 s^{-1} may be assumed in the 5,000–50,000 K temperature range with a reasonable accuracy. This result is in global accordance with the constant value considered in astrophysical models; the usual value seems anyway to be overestimated by about a power of 10. The total rate constant for the reverse process $C(^3P) + S^+(^4S)$ reaches the value 2.6×10^{-11} cm^3 s^{-1} at 5,000 K but, as previously noticed, it becomes rapidly negligible for lower temperatures with the exponential factor.

21.5 Conclusion

This study provides reasonably accurate rate constants for charge transfer processes important to model the interstellar medium. The $Si^{2+} + H$ and $Si^{3+} + He$ reactions are rather efficient charge transfer processes with rate constants of the order of 10^{-9} cm^3 s^{-1}. On the contrary, the $C^+ + S \rightarrow C + S^+$ charge transfer and its reverse reaction appear to be less efficient, with a rate constant an order of magnitude lower than the one used in the astrochemical model. It might be wise to test the effect of a lower rate coefficient in the chemistry of carbon and sulphur in the interstellar medium. It is important to outline the importance of the $^4\Pi\{C(2s^22p^2)^1D + S^+(3s^23p^3)^4S\}$ level in the mechanism. This state is determinant for the efficiency of the reaction and has to be considered in order to have an accurate description of the collision system.

Acknowledgements This work was granted access to the HPC resources of [CCRT/CINES/IDRIS] under the allocation i2010081566 made by GENCI [Grand Equipement National de Calcul Intensif]. The support of the COST Action CM0702 CUSPFEL is gratefully acknowledged.

References

1. M.A. Hayes, H. Nussbaumer, *Astrophys. J.* **161**, 287 (1986).
2. G.D. Sandlin, J.D.F. Bartoe, G.F. Baureckner, R. Tousey, M.E. Van Hoosier, *Astrophys. J. Suppl.* **61**, 801 (1986).
3. H. Nussbaumer, *Astron. Astrophys.* **155**, 205 (1986).
4. P. Honvault, M.C. Bacchus-Montabonel, R. McCarroll, *J. Phys. B.*. **27**, 3115 (1994).
5. P. Honvault, M. Gargaud, M.C. Bacchus-Montabonel, R. McCarroll, *Astron. Astrophys.* **302**, 931 (1995).
6. M. Gargaud, M.C. Bacchus-Montabonel, R. McCarroll, J. Chem. Phys. **99**, 4495 (1993).
7. M.C. Bacchus-Montabonel, P. Ceyzeriat, *Phys. Rev.* A**58**, 1162 (1998).
8. N. Vaeck, M.C. Bacchus-Montabonel, E. Baloïtcha, M. Desouter-Lecomte, Phys. Rev. A **63**, 042704 (2001).
9. S.L. Baliunas and S.E. Butler, *Astrophys. J.* **235**, L45 (1980).
10. M.C. Bacchus-Montabonel, *Theor. Chem. Acc.* **104**, 296 (2000); *Chem. Phys.* **237**, 245 (1998).
11. P. Honvault, M.C. Bacchus-Montabonel, M. Gargaud, R. McCarroll, *Chem. Phys.* **238**, 401 (1998).
12. M.C. Bacchus-Montabonel and D. Talbi, *Chem. Phys. Lett.* **467**, 28 (2008).
13. J. Le Bourlot, G. Pineau des Forêts, E. Roueff, D.R. Flower, *Astron. Astrophys.* **267**, 233 (1993).
14. D. Teyssier, D. Fosse, M. Gerin, J. Pety, A. Abergel, E. Roueff, *Astron. Astrophys.* **417**, 135 (2004).
15. B. Huron, J.P. Malrieu, P. Rancurel, *J. Chem. Phys.* **58**, 5745 (1973).
16. M. Pélissier, N. Komiha, J.P. Daudey, *J. Comput. Chem.* **9**, 298 (1988).
17. A.D. McLean, G.S. Chandler, *J. Chem. Phys.* **72**, 5639 (1980).
18. M.C. Bacchus-Montabonel, *Phys. Rev.* A**46**, 217 (1992).
19. M.C. Bacchus-Montabonel and F. Fraija, *Phys. Rev.* A**49**, 5108 (1994).
20. D.E. Woon, T.H. Dunning Jr. *J. Chem. Phys.* **98**, 1358 (1993).
21. H.J. Werner, P.J. Knowles, MOLPRO (version 2009.1) package of ab-initio programs.
22. A. Nicklass, M. Dolg, H. Stoll, H. Preuss, J. Chem. Phys. 102, 8942 (1995).
23. M.C. Bacchus-Montabonel, N. Vaeck, M. Desouter-Lecomte, Chem. Phys. Lett. **374**, 307 (2003).
24. M.C. Bacchus-Montabonel, Y.S. Tergiman, Phys. Rev. A **74**, 054702 (2006).
25. M.C. Bacchus-Montabonel, C. Courbin, R. McCarroll, J. Phys. B **24**, 4409 (1991).
26. F. Fraija, A.R. Allouche, M.C. Bacchus-Montabonel, Phys. Rev. A **49**, 272 (1994).
27. L.F. Errea, L. Mendez, A. Riera, *J. Phys. B.*. **15**, 101 (1982).
28. R.J. Allan, C. Courbin, P. Salas, P. Wahnon, *J. Phys. B***23**, L461 (1990).
29. M. Gargaud, R. McCarroll, P. Valiron, *J. Phys. B* **20**, 1555 (1987).
30. N.J. Clarke, P.C. Stancil, B. Zygelman, D.L. Cooper, *J. Phys. B***31**, 533 (1998).
31. M. Gargaud, R. McCarroll, P. Valiron, *Astron. Astrophys.* **106**, 197 (1982).
32. D.R. Bates, B.L. Moiseiwitsch, *Proc. Phys. Soc.* A**67**, 805 (1954).
33. Z. Fang, V.H.S. Kwong, *Astrophys. J.* **483**, 527 (1997).
34. S.E. Butler, A. Dalgarno, *Astrophys. J.* **241**, 838 (1980).
35. P.C. Stancil, N.J. Clarke, B. Zygelman, D.L. Cooper, *J. Phys. B***32**, 1523 (1999).
36. The UMIST database for Astrochemistry. http://www.udfa.net.
37. NIST Atomic Spectra Database Levels. http://www.nist.gov/pml/data/asd.cfm
38. M. Larsson, *Chem. Phys. Lett.* **117**, 331 (1985).
39. N. Honjou, *Chem. Phys.* **344**, 128 (2008).

Chapter 22
Systematic Exploration of Chemical Structures and Reaction Pathways on the Quantum Chemical Potential Energy Surface by Means of the Anharmonic Downward Distortion Following Method

Koichi Ohno and Yuto Osada[†]

Abstract Anharmonic downward distortion (ADD) of potential energy surfaces has been used for automated global reaction route mapping of a given chemical formula of BCNOS. It is demonstrated that the ADD following method gives not only the larger numbers (122) of equilibrium structures (EQ) than those (103) of the earlier method by a stochastic approach but also the entire reaction pathways via 430 transition structures (TS) connecting the discovered EQ as well as 155 dissociation channels, 60 via TS and 95 without TS. Interesting propensities were found for chemical preference of isomeric structures and their dissociated fragments as well as characteristic reaction pathways, such as a fragment rotation mechanism.

22.1 Introduction

It has been a primitive but difficult problem to elucidate entire reaction channels for a given chemical composition of a chemical formula This problem includes several fundamental questions, what kinds of chemical species (isomers) are producible from a given chemical formula, how the isomers can be converted one another, and how they are decomposed into smaller species or conversely how they are made of smaller species. These questions are of great significance to discover unknown reaction channels and chemical species.

K. Ohno (✉)
Toyota Physical and Chemical Research Institute, Nagakute, Aichi 480-1192, Japan
e-mail: ohnok@m.tohoku.ac.jp

[†] Graduate School of Science, Tohoku University, Sendai 980-8578, Japan

P.E. Hoggan et al. (eds.), *Advances in the Theory of Quantum Systems in Chemistry and Physics*, Progress in Theoretical Chemistry and Physics 22,
DOI 10.1007/978-94-007-2076-3_22, © Springer Science+Business Media B.V. 2012

The above fundamental questions can be solved in principle theoretically from mathematical properties of the potential energy surface (PES) [1,2].

1. An individual equilibrium structure (EQ) on PES corresponds to a chemical species.
2. A first-order saddle point on PES, a maximum along only one direction and a minimum for all other perpendicular directions, is called a transition structure (TS), which connects the reactant with the product via minimum energy paths or intrinsic reaction coordinates (IRC) [3].
3. A valley leading to fragment species is denoted as a dissociation channel (DC).

The above questions for diatomic systems are trivial. In the case of three-atom systems, there are several isomers in general, but all isomers as well as all reaction channels can be studied easily. However, for four-atom systems such as H_2CO a full theoretical search of possible chemical species and reaction channels had long been eluded. In 1996 Bondensgård and Jensen first reported a global map of all isomers and reaction channels for H_2CO based on quantum chemical PES at the level of HF/STO-3G [4]. The global reaction route map for H_2CO was also reported by Quapp and coworkers in 1998 [5]. Because of considerably heavy computational demands for the global reaction route mapping (GRRM), a full search of all transition structures of systems with more than four atoms was seemed to be impossible [1].

The major obstacle for performing GRRM was the time-consuming quantum chemical sampling processes of PES, which requires 3×10^{10} years of computation time even for a five-atom system $(N = 5)$ with very rough samplings of 100 grid points in each directions of $3N$-6 $= 9$ variables, if the samplings are taken at conventional regular grids [6]. Similarly Mote Carlo samplings cannot avoid the difficulties. Such sampling methods inevitably include huge numbers of useless points far from EQ and TS on the PES.

The most efficient way of quantum chemical samplings on PES can be made, if samplings are confined around reaction pathways. The numbers of EQ and TS are finite, and their connections are also in the limited area along the reaction coordinates with essentially one dimensional nature which can be described by small numbers of sampling points. Downhill walks from TS toward EQ or DC along reaction pathways on PES can easily be made by conventional methods, such as the steepest decent method [1]. On the other hand for uphill walks from EQ toward TS or DC along reaction pathways on PES without any intuition, no algorithm has been reported before the anharmonic downward distortion (ADD) following [7].

The common feature of reaction channels from an EQ point can be summarized as ADD, as indicated by arrows in Fig. 22.1. On going toward DC, the potential energy curve becomes flattened over the long distance. The presence of another EQ leads to TS. Such propensities due to the existence of another EQ or DC affect the local properties of potentials around an EQ, which necessarily appear as ADD. It follows that ADD around an EQ point can be considered as a "compass" of the chemical reaction [7–9].

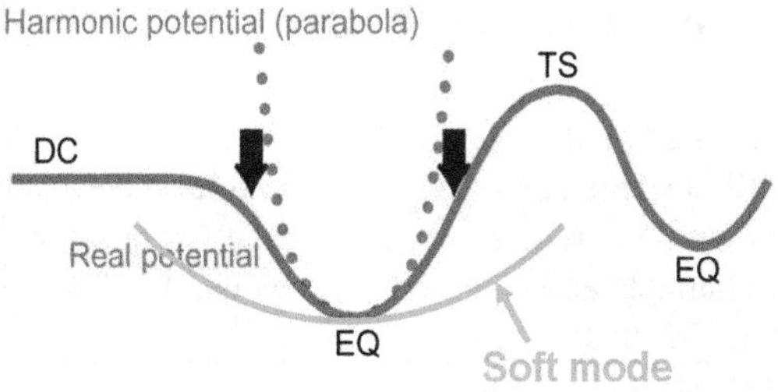

Fig. 22.1 Common features of potential energy curves for chemical reactions on going from an equilibrium structure (*EQ*) toward a dissociation channel (*DC*) or another equilibrium structure (*EQ*) via a transition structure (*TS*). Anharmonic downward distortion (*denoted by thick solid arrow*) of the real potential from the harmonic potential (*shown by dotted line*) indicates the direction of the chemical reactions toward DC or TS. A curve for a soft mode is also shown for a multi-dimensional case

Here, one should note that ADD means neither a gentle curvature nor a soft potential. Soft vibrations such as internal rotation and bending motion hardly break chemical bonds. The "anharmonic" downward distortion is the essential characteristics indicating directions of chemical reactions. If one follows the lowest energy part or the softest part on the PES, one cannot follow ADD to be diverted considerably from the right reaction channels. To avoid the softest part, scaled normal coordinates have been used in the scaled hypersphere search (SHS) technique [7–9].

By noting ADD, a novel method for finding reaction pathways around EQ has been established as a general uphill walking method for GRRM [7–9]. The fundamental questions for each chemical formula listed above can now be solved by the GRRM method. The GRRM method using the SHS technique based on the ADD following [7, 8] has been successfully applied to small molecules [7–23], clusters [24–28], and large molecules [29–32].

As long as locating one EQ, the conventional geometry optimization technique in quantum chemical packages can easily be used to obtain an EQ starting from any geometry. The structure of so searched EQ, however, crucially depends on the choice of the initial geometry. Thus, for locating all possible EQ based on the geometry optimization technique, a tremendous number of initial structures should be introduced in a stochastic way [33–35].

In the present paper, the GRRM method was applied to a five-atom system of B, C, N, O, and S (BCNOS), for which an automated stochastic search procedure had been made for locating all possible minima [35]. It is interesting to check the performance of the GRRM method for finding all possible structures and reaction pathways for the benchmark system of (BCNOS). It seems also worthy to investigate (a) structural propensities of (BCNOS), (b) fragment distributions in the dissociated products of (BCNOS), and (c) their reaction channels for syntheses.

22.2 Global Reaction Route Mapping

The GRRM method based on the ADD following can be used to elucidate for a given chemical formula all possible chemical species and reaction channels among them. The GRRM method does not require any information or intuition beforehand. Thus, unknown reaction channels and chemical species can be discovered automatically by the use of the GRRM method.

22.2.1 The GRRM Procedures

Since details of the GRRM procedures were reported previously [6–9], only outlines of the GRRM method are described here.

1. At first, normal coordinates are determined at an EQ point.
2. Around the EQ, PES is expanded in terms of scaled normal coordinates defined by $q_i = \lambda_i^{1/2} Q_i$, where Q_i is a normal coordinate with a respective eigenvalue λ_i.
3. Reaction path points are determined as energy minima on a scaled hypersphere with a center at the EQ.
4. Using several sizes of hypersphere, one can obtain series of points for the reaction pathways around the EQ.
5. Indication of a TS region can be recognized as change of signs of first order derivatives along the reaction path. Location of each TS can be determined precisely by a conventional technique [36, 37].
6. Asymptotic behavior separating a fragment from the remaining part indicates a DC. Searched DC in the uphill walking is denoted as upward DC (UDC), which is found without TS. On the other hand for the downhill walking, searched DC is denoted as downward DC (DDC), which is found via TS.
7. After arriving at TS, a conventional downhill technique [38] is used to reach an EQ or DC. During this procedure, the IRC determined from each TS toward both sides.
8. Structures of newly found TS, and EQ are compared with those of already found ones. New ones are numbered successively as TSn and EQn, where the initial EQ is denoted as EQ0. This numbering is temporary during the search. In the finally obtained global reaction route map, renumbering of EQ is made so that energies increase with the numbers starting from 0.
9. After those procedures (1)–(8) starting from EQ0 are finished, normal coordinate calculations corresponding to the process (1) are performed for EQ1. Then, next processes from (2) – (8) are repeated around EQ1 to discover successively TS, EQ, or DC (UDC, DDC). These cyclic procedures should be repeated for every new EQ, until no unprocessed EQ remains. All reaction channels via TS are confirmed as IRC during the above cyclic procedures.

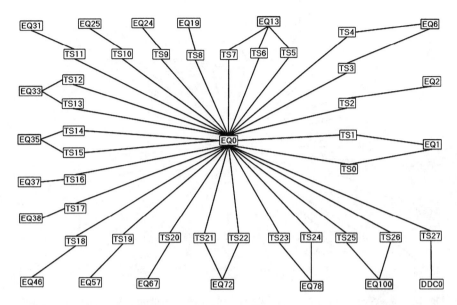

Fig. 22.2 Searched connections around the most stable structure (EQ0). Equilibrium structures (EQn) are numbered from the lowest $n = 0$ so that their energies may increase with the number n. Transition structures (TSn) are numbered according to the increasing numbering of the connected EQ. *DDC* is a downward dissociation channel via TS

10. Entire reaction path connections are discovered one after another in a systematic way by the above procedures to yield a global reaction route map for a given chemical composition automatically.

22.2.2   Automated Search for (BCNOS) by the GRRM Method

Quantum chemical calculations of PES were carried out at DFT B3LYP/6-31G* by Gaussian03 [39], for comparison with the earlier study by the kick method [35]. Although searched results are sometimes depend on the choice of the level of calculations, the employed level in this study is moderately reasonable in view of our recent experiences in GRRM calculations [6–32]. Zero-point energy (ZPE) correction can be made, but in the following it is disregarded, since ZPE corrections are not necessarily important to consider the performance of the GRRM.

The GRRM procedures for the lowest singlet electronic states of (BCNOS) automatically yielded 122 EQ and 430 TS. Searched EQs (EQn) are numbered from the most stable EQ0, the global minimum with the lowest energy. Then, TSs are numbered around each EQn, starting from EQ0, EQ1, EQ2, and so on. Figure 22.2 shows searched connections around EQ0, and Fig. 22.3 shows searched connections

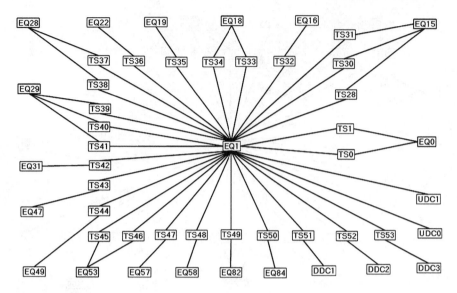

Fig. 22.3 Searched connections around the secondly stable structure (EQ1). *UDC* is an upward DC without TS. For other notations, see captions in Fig. 22.2

Table 22.1 List of linear structures

EQn	Atoms	Relative Energy/kJ mol^{-1}
EQ0	SCNBO	0
EQ1	SBNCO	14.4
EQ4	SNCBO	178.9
EQ9	SNBCO	312.7
EQ10	SBCNO	324.1
EQ12	SOBCN	369.0
EQ14	SOBNC	395.4
EQ17	SCBNO	455.8
EQ24	SCOBN	525.6
EQ32	SNBOC	580.6
EQ66	SCBON	718.4
EQ71	SOCBN	732.5

around EQ1. As can be seen in these Figures, each TSn is numbered systematically according to increasing numbers of connected EQ.

Our data handling program automatically classifies searched EQs into different symmetry species; among 122 searched EQs, 12 for C∞v (linear), 80 for Cs (planar), and 30 for C1 (nonplanar). The linear structures are listed in Table 22.1, where relative energies are shown with respect to the energy of EQ0. Searched connections among the lowest ten EQs (EQ0–EQ9) are shown in Fig. 22.4. The global minimum (EQ0) is a linear structure of SCNBO, which was confirmed to be also the global minimum at the level of cc-pVTZ/CCSD. The next one (EQ1) is a linear SBNCO,

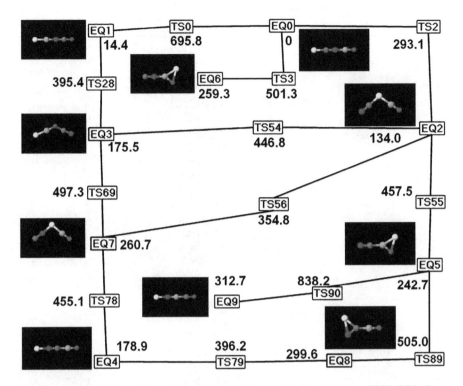

Fig. 22.4 Searched connections among the lowest ten equilibrium structures (EQ0–EQ9) for a five-atom system of B, C, N, O, and S. Illustrations produced by JMOL are given for EQ0–EQ9. In this map, paths connecting to the same EQ are omitted, and only the lower energy path is retained, if there are different connections for a pair of EQs. As for the numbering of EQ and TS, see text. Relative energies with respect to EQ0 are labeled in kJ/mol

then NCSBO with a bent at S (EQ2), NCOBS with a bent at O (EQ3), and a linear SNCBO (EQ4), which all agree with those reported by the kick method [35].

The total searched EQ number of 122 in our results apparently exceeds the reported number of 103 by the kick method [35]. Unfortunately, we cannot compare our results with the earlier one in detail, since precise structures were not reported in the earlier work. Nevertheless, the nearly 18% larger number of searched EQs in this work indicates the satisfactory performance of the GRRM method. Although the kick method seems to be incomprehensive, it may also be useful to find many possible structures automatically.

The GRRM procedures also gave many "dissociated" structures denoted as DDC or UDC. Table 22.2 lists the dissociation channels, in which dissociated fragments are shown as X + Y. For dissociation into fragments of a single-atom and four atoms (1 + 4), constituent four atoms are shown in parentheses, where the order should be disregarded. For dissociation into fragments of 2 + 3, three atoms are classified into ring structures or chain structures with a significant ordering. The numbers of classified types are shown for both DDC and UDC. The totals are 60 for DDC and 95 for UDC, respectively.

Table 22.2 List of
dissociation channels (DC)

Dissociated fragments	DDC	UDC
B + (CNOS)	0	21
C + (BNOS)	0	11
N + (BCOS)	0	9
O + (BCNS)	0	21
S + (BCNO)	0	15
BC + NOS	0	0
BC + NSO	0	0
BC + ONS	0	1
BN + ring − COS	0	1
BN + COS	0	1
BN + CSO	1	1
BN + SCO	2	0
BO + CNS	0	1
BO + CSN	0	0
BO + NCS	0	1
BS + CNO	0	1
BS + CON	0	0
BS + NCO	1	0
CN + BOS	0	2
CN + BSO	0	2
CN + OBS	0	1
CO + ring − BNS	2	0
CO + BNS	5	0
CO + BSN	1	0
CO + NBS	12	0
CS + ring − BNO	4	0
CS + BNO	2	0
CS + BON	3	0
CS + NBO	4	0
NO + ring − BCS	2	0
NO + BCS	18	0
NO + BSC	0	0
NO + CBS	3	0
NS + BCO	1	1
NS + BOC	0	0
NS + OBC	1	0
OS + BCN	0	2
OS + BNC	3	2
OS + CBN	0	1
Total	60	95

Downward DC via TS is denoted as *DDC*,
upward DC without TS is denoted as *UDC*

The total number of reaction channels was found to be 955; 800 for EQ-TS,
95 for EQ-UDC, and 60 for TS-DDC. The average number of reaction channels
around one EQ was found to be 7.3. Although numbers of reaction channels are

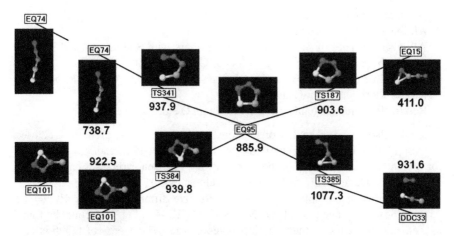

Fig. 22.5 Reaction pathways around a pentagonal ring structure of (O)-N-S-B-C-(O) (EQ95). Relative energies with respect to the most stable structure (EQ0) are labelled in kJ/mol. A dissociation channel (DDC33) into NO and SCB indicates a direct synthetic route to the ring isomer (EQ95) from two simple molecules

larger for stable isomers as can be seen in Figs. 22.1 and 22.2, in some isomers with the higher energies reaction channels become a few. Around a ring isomer of EQ95, only four reaction channels were found, as can be seen in Fig. 22.5. It is of note that EQ95 (O)-N-S-B-C-(O) is the only one pentagonal structure for this system. In this notation (O)-N-S-B-C-(O) for a ring structure, parenthesized symbols indicate the same atom.

22.3 Discussion

22.3.1 Chemical Structures of Searched Isomers

The number of topologically independent structures for five different atoms was suggested to be 577 altogether in the earlier work [35]. Among them, only 103 were really found to be stable structures by the kick method [35]. The present work gave 122 equilibrium structures, which is also much fewer than 577. It follows that really effective structures are much less than those imagined.

The number of possible permutations for five-atom linear chain structures with five different atoms is $5!/2 = 60$. The present results gave 47 chain structures, among which purely linear ones with a symmetry of C∞v were found to be only 12 (Table 22.1). The most stable one is SCNBO, and the next is SBNCO. Both of these include S and O atoms at the ends as well as N atom in the middle. It is interesting to note the chemical preference in the linear isomers listed in Table 22.1, as summarized as follows.

1. All linear isomers contain S atom at one end.
2. All linear isomers does not contain B atom at an end.
3. All possible types of 3! = 6 appear for isomers with S at an end and B in the middle, but much less for those with N or C in the middle.
4. All possible types of 3! = 6 appear for isomers with S and O at ends, but much less for those with N or C at an end.
5. Relatively more stable linear isomers contain O atom at an end.

It should also be noted that S or O atom inside tends to bend chain structures, because of its nonlinear valency with two single bonds. This tendency is stronger for S than O, as can be seen in the above feature (1). Bent structures can be found in Fig. 22.4; EQ2 and EQ7 with S in the middle, and EQ3 with O in the middle.

The number of possible ring structures for five different atoms is $4!/2 = 12$. Among these, only one case of (O)-N-S-B-C-(O) EQ95 in Fig. 22.5 could be found as a chemically stable isomer with a pentagonal ring. The energy of EQ95 is very high at 885.9 kJ/mol.

Although the number of isomers composed of a single triangle with a chain of three atoms like EQ8 O-B-(C)-N-S-(C) in Fig. 22.4 is simply expected to be 60 as suggested in earlier work, 31 EQ structures were really obtained in the present study. Among them, 21 are planar (Cs) and the remaining 10 are nonplanar (C1). Their energies widely range from 178.9 kJ/mol (EQ8, Cs) to 1108. 2 kJ/mol (C-B-(N)-S-O-(N), EQ116, Cs).

The number of isomers composed of a single tetragonal ring with an appendix of a single atom like B-(C)-O-N-S-(C) EQ101 in Fig. 22.5 has also been expected to be 60, but really searched structures were only seven; EQ82, EQ94, EQ100, EQ101, EQ108, and EQ120, whose structure can be expressed as B-(N)-O-S-C-(N), B-(C)-S-O-N-(C), B-(N)-O-S-C-(N), B-(C)-O-N-S-(C), B-(N)-S-C-O-(N), C-(B)-N-O-S-(B), and N-(O)-C-S-B-(O), respectively. It is of note that in five cases among seven B is located at an appendix. Tetragonal rings seem to be rather unstable, since the energies are in the higher region between 800 and 1,200 kJ/mol in comparison with the lower energy isomers (0–300 kJ/mol) shown in Fig. 22.4 and the linear isomers (0–730 kJ/mol) listed in Table 22.1.

22.3.2 Dissociation Channels Producing Fragments

Interesting chemical preference can also be noted for dissociation channels listed in Table 22.2, which includes all of $1 + 4$ and $2 + 3$ fragmentation channels. DDC denotes downward dissociation via TS, whereas UDC denotes upward dissociation from EQ.

As for $1 + 4$ fragmentation, many channels were found for UDC, though no channels were found for DDC. The number of UDC for ejection of B (21) or O (21) or S (15) is larger than the number of simple permutations of four-different-atom chains of $4!/2 = 12$. Some structures of four atoms were found to be cyclic.

There were spatially different dissociation channels even for the same combination of fragments. For example, fragmentation into $S + OBCN$ shows two independent channels with dissociation energies of ca. 490 kJ/mol, one is a channel releasing S atom from the O end of the OBCN moiety, and the other is a channel releasing S atom from the N end of the OBCN. Less numbers of UDC for ejection of C (11) and N (9) may be due to relatively firm chemical bonds by C and N atoms.

Concerning $2 + 3$ fragmentation, both DDC and UDC were found. These channels are interesting, because opposite directions forming five-atom structures can be considered as synthetic routes of the five-atom isomers without byproducts from a combination of a diatomic molecule and a triatomic molecule.

All possible ten types of diatomics, BC, BN, BO, BS, CN, CO, CS, NO, NS, and SO, appear in the $2 + 3$ channels, but their distribution for DDC and UDC are very much different. There are no channels yielding BC, BO, or CN for DDC, whereas there are no channels producing CO, CS, or NO for UDC. In other words, BC, BO, and CN are directly released without activation energies, and CO, CS, and NO are released only via TS. The other four types of BN, BS, NS, and SO can be released via both ways with or without TS.

Correspondingly, all possible ten combinations of three atoms were found in the $2 + 3$ channels, but five isomers, NOS, NSO, CSN, BSC, and BOC could not be obtained in either DDC or UDC.

22.3.3 Reaction Pathways and Synthetic Routes

Global reaction route mapping gives us the entire reaction channels around all EQs, which shows us connections among isomers as well as dissociation channels, as can be seen in Figs. 22.2 and 22.3.

As for very stable two linear isomers SCNBO (EQ0) and SBNCO (EQ1), there are direct connections via TS0 anTS1. Since these connections correspond to exchange processes of remote S and O atoms attached to both ends of the BNC fragment, these two reaction routes require high energy barriers of 695.8 kJ/mol (TS0) and 723.2 kJ/mol (TS1), respectively. If one looks at the global reaction route map carefully, one may find out more preferable routes with lower energy barriers. Such routes between EQ0 and EQ1 can be found in Fig. 22.4; (Route 1) EQ0-TS2-EQ2-TS54-EQ3-TS28-EQ1 has the highest barrier of 446.8 kJ/mol at TS54, and (Route 2) EQ0-TS2-EQ2-TS58-EQ7-TS69-EQ3-TS28-EQ1 has the highest barrier of 497.3 kJ/mol at TS69. These two indirect routes have much lower barriers than the direct routes between EQ0 and EQ1 via TS0 (695.8 kJ/mol) or TS1 (723.2 kJ/mol). Another indirect route via TS89 (505.0 kJ/mol) lower than the direct one may also be found in Fig. 22.4, although it is a very long series of reaction pathways.

It is interesting to note the mechanism of these more preferable indirect reaction routes. Output data of GRRM procedures contain all reaction pathways in detail. Although we cannot describe them here, interesting characteristics can be summarized as follows.

Chain of five atoms ABCDE can be divided into AB and CDE. Initially, B and C are connected. When the bonding between B and C is weakened, one of the two fragments can migrate around the other moiety. If AB migrates around CDE keeping a weak bonding at B, then an intermediate with a T shape can be formed with a connection between B and D. Further migration of AB toward the other end of CDE with keeping a weak bonding at B yields a bonding between B and E to result in a formation of ABEDC. This consecutive procedure is a migration of AB around CDE with B as an adhesive site. Alternatively, one may consider this procedure as a rotation of CDE keeping a weak bonding with B in AB. This fragment rotation mechanism can be schematically shown as follows.

$$
\begin{array}{c}
\text{C} \\
\text{AB} \cdot \text{CDE} \rightarrow \text{AB} \cdot \text{D} \rightarrow \text{AB} \cdot \text{EDC} \\
\text{E}
\end{array}
$$

The above Route 1 can be explained as follows.

$$
\begin{aligned}
\text{SCNBO (EQ0)} &\rightarrow \text{OB} \bullet \text{NCS} \rightarrow \text{OB} \bullet \text{SCN} \rightarrow \text{OBSCN (EQ2)} \\
&\rightarrow \text{OBS} \bullet \text{CN} \rightarrow \text{SBO} \bullet \text{CN} \rightarrow \text{SBOCN (EQ3)} \\
&\rightarrow \text{SB} \bullet \text{OCN} \rightarrow \text{SB} \bullet \text{NCO} \rightarrow \text{SBNCO (EQ1)}
\end{aligned}
$$

In the case of Route 2, CN moiety rotates between EQ2 and EQ7, and double rotation occurs between EQ7 and EQ3.

$$
\begin{aligned}
\text{SCNBO (EQ0)} &\rightarrow \text{OB} \bullet \text{NCS} \rightarrow \text{OB} \bullet \text{SCN} \rightarrow \text{OBSCN (EQ2)} \\
&\rightarrow \text{OBS} \bullet \text{CN} \rightarrow \text{OBS} \bullet \text{NC} \rightarrow \text{OBSNC (EQ7)} \\
&\rightarrow \text{OBS} \bullet \text{NC} \rightarrow \text{SBO} \bullet \text{CN} \rightarrow \text{SBOCN (EQ3)} \\
&\rightarrow \text{SB} \bullet \text{OCN} \rightarrow \text{SB} \bullet \text{NCO} \rightarrow \text{SBNCO (EQ1)}
\end{aligned}
$$

Figure 22.5 shows four types of synthetic routes for the pentagonal ring structure of (O)-N-S-B-C-(O) EQ95.

1. Ring-closing reaction from a chain: EQ74 → TS341 → EQ95
2. Ring expansion from a tetragonal ring: EQ101 → TS384 → EQ95
3. Ring expansion from a trigonal ring: EQ15 → TS187 → EQ95
4. Ring-closing reaction of 2 + 3: DDC33 → TS385 → EQ95

The route (1) starts from slightly bent chain of SBCON, and activated bending motion may make a new bond between S and N atoms to lead to the pentagonal form via TS341 with a barrier height of ca. 200 kJ/mol. The route (2) includes a bond transfer of CS into BS to lead to a ring expansion from four to five, which has the lowest barrier height of ca. 17 kJ/mol. The route (3) is a ring expansion from three to five; the BN bond in the trigonal ring opens to produce a new NO bond yields the pentagonal ring, which has a high barrier height of ca. 493 kJ/mol, since the starting structure of EQ15 is rather more stable than other starting compounds. The route (4) is a purely synthetic route from small species of NO and SCB; This 2 + 3 cyclization reaction is not so simple as it seems. In the initial step, O in NO adheres on to the middle part of SCB moiety to produce TS385 with new bonds of CO and SB as NO-(C)-S-B-(C). Then, CS bond opens to make a new bond between N and S to yield a pentagonal ring of (O)-N-S-B-C-(O) that is EQ 95. This 2 + 3 cyclization reaction needs an activation energy of ca. 146 kJ/mol.

Synthetic routes of a five-atom system composed of B, C, N, O, and S, can be designed in many ways by using dissociation channels listed in Table 22.2. By using UDC in the opposite way, this system can be produced via a 1 + 4 addition without activation energies. When a diatomic molecule and respective three atomic parts are available, a 2 + 3 addition reaction can be used to produce an isomer of BCNOS.

Since GRRM procedures yield spatial pathways for all reaction channels, one may design efficient reaction processes; in a 1 + 4 addition reaction, the best position on the target where the atom should approach can be elucidated, and in a 2 + 3 addition reaction, the best contact points and orientations of two reactants can be anticipated. If one can control orientation of molecules as well as directions of projectiles, spatial reaction pathways should be studied in detail, for which the GRRM method will provide valuable data.

22.4 Conclusions

Automated exploration of reaction channels on potential energy surfaces has become possible by the SHS method, in which ADD of the potential indicates the direction of chemical reactions. ADD can thus be used as a 'compass' to discover unknown chemistry played by atoms. GRRM procedures provide possible EQ and connections between them as well as dissociation channels into smaller species.

The present application of the GRRM method to a five atom system of BCNOS has given 120 EQs which are larger than the earlier record of 103 structures by a stochastic approach. Furthermore, the GRRM method gave 430 reaction channels going through TS as well as 60 DDC via TS and 95 UDC without TS. Searched results were found to be very much different from the expected numbers of classified types in the earlier work, and many interesting propensities of chemical preference were discovered.

Acknowledgements K.O. acknowledges the Grants-in-Aid for Scientific Research (No. 21350007 and 21655002) from the Ministry of Education, Science, Sports, and Culture. The authors thank to Dr Satoshi Maeda, for producing fundamental parts of the GRRM program.

References

1. Jensen F (1999) Introduction to computational chemistry. Wiley, Chichester
2. Schlegel HB (2003) J Comput Chem 24:1514
3. Fukui K (1981) Acc Chem Res 14:363
4. Bondensgård K, Jensen F (1996) J Chem Phys 104:8025
5. Quapp W, Hirsch M, Imig O, Heidrich D (1998) J Comput Chem 19:1087
6. Ohno K, Maeda S (2008) Phys Scripta 78:058122
7. Ohno K, Maeda S (2004) Chem Phys Lett 384:277
8. Maeda S, Ohno K (2005) J Phys Chem A 109:5742
9. Ohno K, Maeda S (2006) J Phys Chem A 110:8933
10. Maeda S, Ohno K (2004) Chem Lett 33:1372
11. Maeda S, Ohno K (2004) Chem Phys Lett 398:240
12. Maeda S, Ohno K (2005) Chem Phys Lett 404:95
13. Yang X, Maeda S, Ohno K (2005) J Phys Chem A 109:7319
14. Yang X, Maeda S, Ohno K (2006) Chem Phys Lett 418:208
15. Maeda S, Ohno K (2006) Astrophys J 640:823
16. Ohno K, Maeda S (2006) Chem Lett 35:492
17. Yang X, Maeda S, Ohno K (2007) J Phys Chem A 111:5099
18. Luo Y, Ohno K (2007) Organometallics 26:3597
19. Watanabe Y, Maeda S, Ohno K (2007) Chem Phys Lett 447:21
20. Maeda S, Ohno K (2008) Chem Phys Lett 460:55
21. Luo Y, Maeda S, Ohno K (2009) Chem Phys Lett 469:57
22. Maeda S, Ohno K, Morokuma K (2009) J Phys Chem A 113:1704
23. Moteki M, Maeda S, Ohno K (2009) Organometallics 28:2218
24. Maeda S, Ohno K (2006) J Chem Phys 124:174306
25. Maeda S, Ohno K (2007) J Phys Chem A 111:4527
26. Luo Y, Maeda S, Ohno K (2007) J Phys Chem A 111:10732
27. Maeda S, Ohno K (2008) J Phys Chem A 112:2962
28. Luo Y, Maeda S, Ohno K (2009) J Comput Chem 30:952
29. Maeda S, Ohno K (2007) J Phys Chem A 111:13168
30. Luo Y, Maeda S, Ohno K (2008) Tetrahedron Lett 49:6841
31. Maeda S, Ohno K (2008) J Am Chem Soc 130:17228
32. Maeda S, Ohno K, Morokuma K (2009) J Chem Theory Comput 5:2734
33. Lloyd LD, Johnston L (1998) Chem Phys 236:107
34. Saunders M (2004) J Comput Chem 25:621
35. Bera PP, Sattelmeyer KW, Sannders M, Schefer HF III, Schleyer PV (2006) J Phys Chem A 110:4287
36. Banerjee A, Adams N, Simons J, Shepard R (1985) J Phys Chem 89:52
37. Csaszar P, Pulay P (1984) J Mol Struct 114:31
38. Page M, McIver JW Jr (1988) J Chem Phys 88:922
39. Frisch MJ et al (2004) GAUSSIAN 03, Revision C.02. Gaussian, Inc., Wallingford

Chapter 23
Neutral Hydrolysis of Methyl Formate from *Ab initio* Potentials and Molecular Dynamics Simulation

S. Tolosa Arroyo, A. Hidalgo Garcia, and J.A. Sansón Martín

Abstract A study of chemical reactions in solution by means of molecular dynamics simulation and with solute-solvent interaction potentials derived from *ab initio* quantum calculations is realized in this work. We apply the procedure to the case of the neutral hydrolysis of methyl formate, $HCOOCH_3 + 3\ H_2O \rightarrow HCOOH + CH_3OH + 2\ H_2O$ in aqueous solution, via concerted and water-assisted mechanisms. We used the solvent fluctuation as reaction coordinate, and the free-energy curves for the calculation of the activation energies. The result for this hydrolysis reaction in aqueous solution, assisted by three water molecules, is in agreement with the available experimental information. In particular our study gives values of $\Delta G^{\neq} = 22.40\,kcal/mol$, close to the activation barrier experimental of $25.9\,kcal/mol$, and improving significantly the value found in another similar study using the PCM model.

23.1 Introduction

The hydrolysis of carboxylic esters (RCOOR') is one of the most intensively studied classes of chemical reaction due to its interest in chemistry, biology, and industrial processes [1]. Methyl formate ($HCOOCH_3$) is the simplest of these esters, and has found major applications. Due to its small size, this ester has been employed as a test case in both experimental and theoretical investigations. The reaction in a neutral aqueous medium [2], studied in the present work, leads to the formation of formic acid and methanol via different pathways. Since most hydrogen transfers occur in aqueous solution, one must consider the role of water molecules in this

S.T. Arroyo (✉) • A.H. Garcia • J.A.S. Martín
Departamento de Química Física, Universidad de Extremadura, 06071 Badajoz, Spain
e-mail: santi@unex.es; antonio@unex.es; jorge@unex.es

P.E. Hoggan et al. (eds.), *Advances in the Theory of Quantum Systems in Chemistry and Physics*, Progress in Theoretical Chemistry and Physics 22,
DOI 10.1007/978-94-007-2076-3_23, © Springer Science+Business Media B.V. 2012

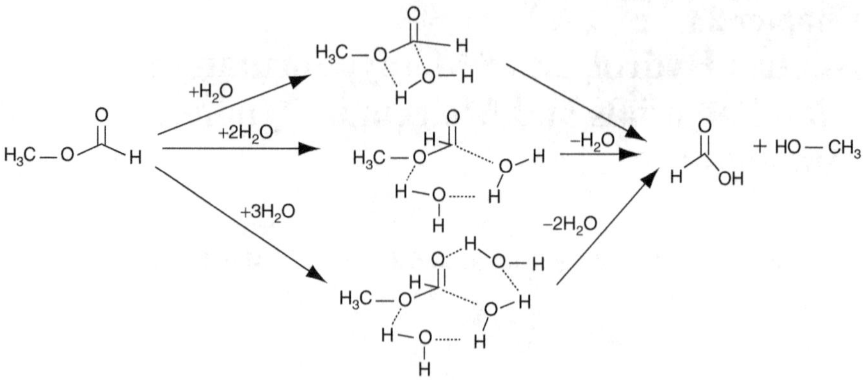

Scheme 23.1 Concerted mechanisms for the hydrolysis of methyl formate

transfer, since water can act not only as a solvent but also as a catalyst by donating both hydrogen H^+ and hydroxyl OH^- ions in an assisted mechanism.

Although the mechanism of gas phase neutral hydrolysis of this compound is known, much uncertainty on details of the reaction in the condensed phase remains. There are some mechanisms by which this reaction can proceed in an aqueous medium. In this work we study a concerted mechanism in which there is a nucleophilic attack of the hydroxyl OH^- ion from a water molecule on the ester's carbonyl carbon, while the hydrogen H^+ ion attacks the methoxy oxygen, electrophilically, via a transition state TS before forming the methanol and formic acid products. This concerted mechanism can be assisted by some water molecules.

When the reaction is assisted by two or three water molecules, the water molecules involved in the transition states give more flexible structures of six-membered rings. This structure make the mechanism assisted by three water molecules the most favorable one. The participation of more than three molecules of water in the reaction mechanism is not usually considered since the extra molecules are not directly involved in the reaction.

Kallies and Mitzner [2] studied this neutral hydrolysis reaction in gas and solution phases, but using the PCM model [3] to estimate the energies of the different structures present in the mechanisms. These authors analyze the lowering of the activation barrier when one or several water molecules are involved in the reaction mechanism, finding values that are higher than experiment. Their study was performed at a BLYP/6-31G* calculation level using the SCI-PCM model, and showed that the activation barrier can be lowered by some kcal/mol as the number of molecules increases from one to three in the two mechanisms they considered (Scheme 23.1).

Although there have been various studies of this reaction in an alkaline medium, the absence of work on the neutral hydrolysis using a discrete model to describe the solvent makes its study of particular interest. In the present work, we show how the calculation method and the model used for the solvent can affect the value of this process activation barrier in a neutral medium, and describe the reaction by the concerted water-assisted mechanisms. One must bear in mind that the thermodynamic study of hydrolysis reactions can lead to different results depending on the type of calculation used.

Continuing in the line of our studies of chemical reactivity in solution using the MD/ESIE method [4], in the present work we apply our methodology to methyl formate hydrolysis in aqueous solution. We show how the results are improved when the solvent is not described as a continuum but as a discrete system formed by numerous water molecules interacting with the solutes (reactant and transition state). The method differs from that used by other workers [5] in that it uses potentials with interaction parameters obtained from *ab initio* calculations, and free-energy curves are used to calculate the activation energy.

The aim of the present study was to show that the form of free-energy curves obtained via the MD/ESIE with *ab initio* potentials and a discrete solvent model is a reasonable option within the present limitations of MD simulations. The specific objectives were: (a) to apply the methodology based on free-energy curves to this hydrolysis reaction to provide values of its activation free energies in solution; (b) to compare our MD results using free-energy curves with results in the literature using other methods to study reactions in solution; and (c) to show the need to use large basis sets and *ab initio* solute-solvent potentials to construct free-energy curves.

23.2 Formalism and Calculation Details

About a thousand values of the SCF and MP2 solute-solvent interaction energy U_{sw} were calculated with the 6-311++G** basis [6]. In order to try to appropriately describe the attractive, repulsive, and long-range interactions, the grid of points was generated by placing the water molecule at different positions r_{ij} relative to the solute. The 6-311++G** basis set used contains polarization and diffuse functions for all of the atoms in order to improve the description of the outermost orbitals, and hence of the reactant and transition state energies and geometries.

The solute-solvent interaction is described by a Lennard-Jones (12-6-1) potential function. The net charges on each solute atom q_i^s were obtained with the ESIE procedure [7], fitting the values of the Coulomb component of the interaction energy, using the variational scheme of Morokuma and co-workers [8].

The Lennard-Jones parameters were obtained in a similar way to q_i^s, the energies used in the fits were those that describe the exchange (EX) and polarization (PL) components of the interaction energy at the SCF level, and the dispersion (DIS) component related to the MP2 correlation energy [7].

To construct curves of the free energy G, we used as reaction coordinate the solvent fluctuation, i.e. the difference in the interaction energy of a given set of solvent molecules in the presence of the reactant and transition state structures [9], for which one only needs the potential function that suitably describes this interaction.

Thus, to obtain the free-energy curve G_R associated with the reactant simulation one can use the differences in the solute-water interaction energies, (U_{sw}), between the diabatic states of the solute in its reactant (R) and transition state (TS) structures for a broad set of configurations of solvent molecules around the solute in a molecular dynamics simulation of the reactant solvation:

$$\Delta E_R = U_R(W, S_R) - U_R(W, S_{TS}), \tag{23.1}$$

where $U_R(W, S_R)$ denotes the solute-solvent interaction $U_{SW,R}$, at the reactant structure S_R and the solvent configuration W.

Transition state solvation is simulated likewise and the ΔE_{TS} values with respect to the other system result, considering the same direction in all the cases, i.e. the difference between the reactant, R, and transition state, TS, $\Delta E_{TS} = U_{TS}(W, S_R) - U_{TS}(W, S_{TS})$.

The difference ΔE_s fluctuates during the s simulation, and its values are collected as a histogram of the number of times that a particular value Δe of the macroscopic variable ΔE_S appears in the simulation. The probability $P_s(\Delta e)$ of finding the system in a given configuration can be expressed in terms of the delta function δ described in previous works [4, 10] and of the number of equally spaced steps N_s in the simulation:

$$P_S(\Delta e) = \frac{\sum\limits_{i=1}^{N} \delta(\Delta E_S(t_i) - \Delta e)}{N_S}. \tag{23.2}$$

This allows us to compute the free energy $G_S(\Delta e)$:

$$G_S(\Delta e) = -k_B T \ln P_S(\Delta e). \tag{23.3}$$

The values of free energy obtained presents some dispersion, so it is advisable to make a search for the polynomial function that best fits these free energies G_s, and the result is plotted. In all cases, the separation between the two minima, when the G_R curve intersects the G_{TS} curve at its lowest point $\Delta e_{TS}^{eq} = \Delta e_R^{TS,eq}$, is calculated as

$$\Delta G^{\#} = G_{TS}^{eq} - G_R^{eq} = a(\Delta e_{TS}^{eq} - \Delta e_R^{eq}) + b(\Delta e_{TS}^{eq} - \Delta e_R^{eq})^2 + c(\Delta e_{TS}^{eq} - \Delta e_R^{eq})^3 + \cdots \tag{23.4}$$

with Δe_R^{eq} and Δe_{TS}^{eq} being the most probable values of ΔE in the free-energy curves G_R and G_{TS}, respectively, and a, b, and c are the coefficients of the polynomial fit to the curve G_R [4].

Finally, some details of the simulations merit mention. Molecular dynamics simulations of an NVT ensemble of a solute molecule in an aqueous environment represented by about 200 water molecules were carried out at 298 K using the AMBER program [11]. The time considered for the simulations was 2,000 ps with time steps of 0.1 fs. The first 1,000 ps were used to ensure that equilibrium was reached completely, and the last 1,000 ps were to store the configurations of the water molecules required for the determination of the thermodynamic properties studied in this work. Those water molecules initially located at distances less than 1.6 Å from any solute atom were eliminated from the simulations. Long-range electrostatic interactions were treated by the Ewald method [12], and the solutes were kept rigid using the shake algorithm [13]. A cut-off of 7 Å was applied to the water-water interactions to simplify the calculations, and periodic boundary conditions were imposed to describe the liquid state. The grid of points used to fit the interaction potential to the Lennard-Jones 12-6-1 function was obtained with SCF and MP2 energies using the Gaussian/98 package [14], and the decomposition of the interaction energies was performed with the Gamess program [15]. The calculations were made on a QS16-2500C-X64Q QuantumCube multiprocessor computer at Extremadura University.

23.3   Results and Discussion

23.3.1   *Reactant and Transition State Structures*

Ground state geometries of the reactant and transition states in solution (see Fig. 23.1) were optimized at the MP2 level with the 6-311++G** basis [6] starting from standard geometries and using the PCM model. The transition state structure was confirmed by observing that there was one negative eigenvalue, of $1,256 \, cm^{-1}$ for the transition state, in the Hessian matrix corresponding to movement along the reaction path (see direction of the arrows on transition state structure in Fig. 23.1), in good agreement with the wavenumber of the imaginary vibration obtained by other workers [2].

In the concerted mechanism assisted by three water molecules, the reaction takes place when the oxygen of a water molecule attacks nucleophilically the carbonyl carbon, and there is a simultaneous proton transfer from a second water to the oxygen of the methoxy group, forming a six-membered ring when a hydrogen is transferred from the first to the second water molecule (Fig. 23.1). The third water molecule forms hydrogen bonds with the ester and with the first water molecule, with a second six-member ring appearing in this transition structure. In this process the O2-C1 distance undergoes a significant increase from 1.34 Å in the reactant to 1.62 Å in the transition state that facilitates their subsequent rupture to yield the products, and the H2w-O2w bond involved in the link with the ester increases its length relative to the value of 0.98 Å present in isolated water. The C1-O2-C2 and O1-C1-O2 angles decrease relative to the reactant structure when the transition state is formed, and the O1-C1-O2-C2 dihedral angle increases to 25.06° in the transition structure, moving away from the planarity observed in the reactant. These results are in agreement with those of Kallies and Mitzner for this transition structure [2]. In the next section we used these geometries to make the Molecular Dynamics Simulations in order to obtain the activation energy.

Fig. 23.1 Reactant and
transition state geometries
in solution

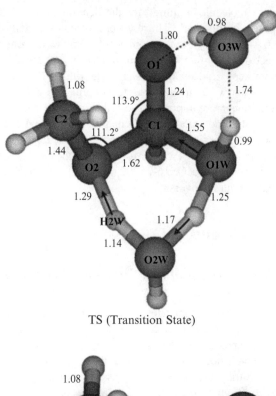

TS (Transition State)

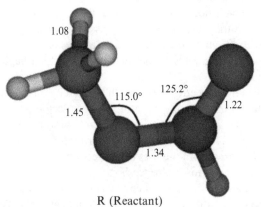

R (Reactant)

23.3.2 Energies

Table 23.1 lists the results for the activation free energies obtained for the hydrolysis
of methyl formate in solution, using different methods and the mechanisms described in this work. Our MD results were obtained from the free-energy curves G_R
and G_{TS} shown in Fig. 23.2, which were constructed from reactant and transition
state simulations using *ab initio* potentials.

Table 23.1 Activation free energies (kcal/mol) for methyl formate hydrolysis

Method	Concerted
Solution/SCI-PCM[a]	48.05
Solution/PCM[b]	55.81
Solution/MD[c]	22.40
Experimental[d]	25.90

[a] Values obtained by Kallies and Mitzner, Ref. [2]
[b] Value obtained in this work using the SCFVAC and radii=Pauling parameters in the PCM model
[c] Values obtained in this work using free-energy curves from *ab initio* potentials and MD simulations
[d] Experimental value obtained by Guthrie, Ref. [16]

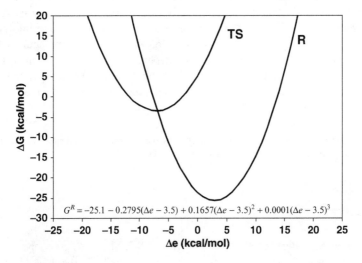

$$G^R = -25.1 - 0.2795(\Delta e - 3.5) + 0.1657(\Delta e - 3.5)^2 + 0.0001(\Delta e - 3.5)^3$$

Fig. 23.2 R, TS free-energy curves

In considering the activation energy, different situations need to be taken into account. The activation barrier ΔG may be considerably lowered as a result of the choice of solvent model used in the calculations and the basis set employed. Thus, the energy barrier in the solution is increased from 48.05 to 55.81 kcal/mol by considering the PCM solvation method instead of the SCI-PCM method used by Kallies and Mitzner [2], using the same level of calculation. In Table 23.1, we can also observe an important reduction in the energy barrier by considering the solvent as discrete instead of continuum, so the energy barrier is lowered to a value of 22.40 kcal/mol.

The positions of the minima of G_R and G_{TS}, together with the polynomial functions for all the curves given in Fig. 23.2, yielded values of 22.40 kcal/mol for the activation energy via the concerted mechanisms, using the Eq. 23.4. Our MD result would seem to agree better with experimental studies of esters in pure water which for this system give a value of 25.9 kcal/mol [16]. This good result is obtained by using a complete basis set containing polarization and diffuses functions, including the dispersion component of the interaction energy in the potential function, i.e. MP2/6-311++G** level.

Finally, the equation $k = \frac{k_B T}{h} e^{-\frac{\Delta G}{RT}}$ in transition state theory gives values of the rate constant using these activation barriers of $2.310^{-4} s^{-1}$ for the concerted mechanisms, which is in better agreement with the value found for the hydrolysis of this ester in a neutral medium ($k = 4.6 \cdot 10^{-9} s^{-1}$ [16]) that with the values of $5.5 \cdot 10^{-24} s^{-1}$ from the Kallies and Mitzner's studies [2].

In sum, the free-energy curves of the species involved in the reaction provide a good description of the thermodynamics of methyl formate hydrolysis in an aqueous medium. These curves, which are constructed on the basis of the solute-solvent interaction energies with the solvent fluctuation being chosen as reaction coordinate, respond acceptably to the activation barrier of this process. Also, to obtain reasonable results for this reaction in solution one must take into account a mechanism that includes the assistance of various water molecules. It was also observed that the activation barrier depends appreciably on other factors, in particular, the basis set used to describe the systems, the components of the interaction energy used in the fits of the potential functions, and the procedure employed to construct and move the curves.

Acknowledgments This research was sponsored by the Consejería de Infraestructuras y Desarrollo Tecnológico de la Junta de Extremadura (Project GR10036).

References

1. (a) Bender ML (1969) Chem Rev 60:53; (b) Johnson SL (1967) Adv Phys Org Chem 5:237; (c) Jencks WP (1972) Chem Rev 72:705; (d) Bamford CH, Tipper CFH (eds) (1972) Ester Formation and Hydrolysis. Elsevier, Amsterdam; (e) Ingold CK (1969) Structure and mechanism in Organic Chemistry 2nd ed. Cornell University Press, Ithaca, New York; (f) Jones RAY (1979) Physical and Mechanistic Organic Chemistry. Cambridge University Press, Cambridge; (g) McMurry JTH (1988) Organic Chemistry, 2nd ed. Cole Publishing, California; (h) Lowry TH, Richardson KS (1987) Mechanism and Theory in Organic Chemistry, 3rd ed. Herper and Row, New York; (i) Williams A (1987) In: Page MI, Williams A (eds) Enzyme Mechanisms Burlington. London; (j) Satchell DPN, Satchell RS (1992) In: Patai S (ed) The Chemistry of Acid Derivates. Wiley, Chistester; (k) Brown RS, Bennet AJ, Slebocka-Tilk H (1992) Acc Chem Res 25:481
2. Kallies B, Mitzner R (1998) J Mol Model 4:183
3. (a) Tomasi J (1994) Chem Rev 94:2007; (b) Tomasi J (1994) In: Cramer CJ, Truhlar DJ (eds) Structure and Reactivity in Aqueous Solution. American Chemical Society, Washington, pp 10; (c) Miertus S, Scrocco E, Tomasi J (1981) J Chem Phys 55:117
4. (a) Tolosa S, Corchado Martín-Romo, JC, Hidalgo A, Sansón JA (2007) J Phys Chem A 111: 13515; (b) Tolosa S, Sansón JA, Hidalgo A (2007) J Chem Phys A 111:339; (c) Tolosa S, Hidalgo A, Sansón JA (2008) Chem Phys 353:73; (d) Tolosa S, Hidalgo A, Sansón JA (2010) Theor Chem Acc 127:671

5. (a) Ando K, Hynes JT (1997) J Phys Chem B 101:10464; (b) Ando K, Hynes JT (1995) J Mol Liquids 64:25
6. (a) Ditchfield R, Hehre WJ, Pople JA (1971) J Chem Phys 54:724; (b) Hehre WJ, Ditchfield R, Pople JA (1972) J Chem Phys 56:2257
7. (a) Tolosa S, Sansón JA, Hidalgo A (2002) Chem Phys Lett 357:279; (b) Tolosa S, Sansón JA, Hidalgo A (2005) Chem Phys 315:76
8. (a) Kitaura K, Morokuma K (1976) Int J Quantum Chem 10:325; (b) Umeyama H, Morokuma K (1977) J Am Chem Soc 99:1316
9. Carter EA, Hynes JT (1989) J Phys Chem 93:2184
10. ((a) King G, Warshel A (1990) J Chem Phys 93:8682; (b) Kuharski RA, Bader JS, Chandler D, Sprik M, Klein ML, Impey RW (1998) J Chem Phys 89:3248
11. Case DA, Darden TA, Cheatham ITE, Simmerling CL, Wang J, Duke RE, Luo R, Mert KM, Wang B, Pearlman DA, Crowley M, Brozell S, Tsui V, Gohlke H, Mongan J, Hornak V, Cui G, Beroza P, Schafmeister C, Caldwell JW, Schevitz RW, Kollman PA (2004) AMBER 8. University of California, San Francisco, CA
12. Ewald P (1921) Ann Phys 64:253
13. Ryckaert P, Ciccotti G, Berendsen JJC (1977) J Comput Phys 23:237
14. Frisch MJ, Trucks GW, Schlegel HB,Scuseria GE, Robb MA, Cheeseman JR, Zakrzewski VG, Montgomery JA Jr, Stratmann RE, Burant JC, Dapprich S, Millam JM, Daniels AD, Kudin KN, Strian MC, Farkas O, Tomas J, Barone V, Cossi M, Cammi R, Mennucci B, Pomelli C, Adamo C, Clifford S, Ochterski J, Petersson GA, Ayala PY, Cui PY, Morokuma K, Malick DK, Rabuck AD, Raghavachari K, Foresman JB, Cioslowski J, Ortiz JV, Baboul AG, Stefanov BB, Liu G, Liashenko A, Piskorz P, Komaromi I, Gomperts R, Martin RL, Fox DJ, Keith T, Al-Laham MA, Peng CY, Nanayakkara A, Gonzalez C, Challacombe M, Gill PMW, Johnson B, Chen W, Wong MW, Andres JL, Gonzalez C, Head-Gordon M, Replogle ES, Pople JA (2002) Gaussian 98, revision A.11.3. Gaussian Inc., Pittsburgh, PA
15. Dupuis M, Spangler D, Wendoloski J (1980) GAMESS Program QG01, National Resource for Computations in Chemistry Software Catalog. University of California, Berkeley, CA
16. ((a) Guthrie JP (1973) J Am Chem Soc 95: 6999; (b) Guthrie JP (1978) J Am Chem Soc 100:5892

Chapter 24
Radial Coupling and Adiabatic Correction for the LiRb Molecule

I. Jendoubi, H. Berriche, H. Ben Ouada, and F.X. Gadea

Abstract The radial couplings between the adiabatic states dissociating into Rb(5s, 5p, 4d, 6s, 6p, 5d, 7s, 6d)+Li(2s, 2p), $Li^+ + Rb^-$ and $Li^- + Rb^+$ determined from accurate diabatic and adiabatic previous data for the LiRb molecule. The accuracy of adiabatic and diabatic results is shown by a comparison with previous *ab initio* calculations and experimental results. To evaluate the radial couplings we have used two methods which are numerical differentiation of the rotation matrix connecting the diabatic and adiabatic representations and the Hellmann-Feynman expression. The first and second derivatives present many peaks, associated to neutral-neutral and ionic-neutral crossings in the diabatic representation. These peaks can be interpreted from the diabatic potential energy curves. The radial coupling is then used to determine the adiabatic correction for several electronic states of LiRb molecule. This correction is about $100 \, cm^{-1}$ for some electronic states around particular distances related to avoided crossings and peaks of the second derivative. It is added to the Born-Oppenheimer potential energy curves to estimate the change in spectroscopic constants, which

I. Jendoubi • H.B. Ouada
Laboratoire de Physique et Chimie des Interfaces, Département de Physique, Faculté des Sciences de Monastir, Université de Monastir, Avenue de l'Environnement, 5019 Monastir, Tunisia
e-mail: jendoubi_ibtissem@yahoo.fr; hafedh.bouada@fsm.rnu.tn

H. Berriche (✉)
Laboratoire de Physique et Chimie des Interfaces, Département de Physique, Faculté des Sciences de Monastir, Université de Monastir, Avenue de l'Environnement, 5019 Monastir, Tunisia

Physics Department, College of Science, King Khalid University, P. O. B. 9004, Abha, Saudi Arabia
e-mail: hamidberriche@yahoo.fr; hamid.berriche@fsm.rnu.tn

F.X. Gadea
Laboratoire de Chimie et Physique Quantique, UMR5626 du CNRS Université Paul Sabatier, 118 Route de Narbonne, 31062 Toulouse Cedex 4, France
e-mail: gadea@irsamc.ups-tlse.fr

P.E. Hoggan et al. (eds.), *Advances in the Theory of Quantum Systems in Chemistry and Physics*, Progress in Theoretical Chemistry and Physics 22, DOI 10.1007/978-94-007-2076-3_24, © Springer Science+Business Media B.V. 2012

is significant mainly for the higher excited states. The vibrational levels are evaluated using corrected and uncorrected potential energies to determine the vibronic shift for the $^1\Sigma^+$ and $^3\Sigma^+$ states. This shift, which is the difference between the adiabatic levels and the corrected ones, has been determined for 20 singlet and triplet Σ^+ states. A shift of order $10\,\text{cm}^{-1}$ for some vibrational levels is observed, which shows the breakdown of the Born-Oppenheimer approximation.

24.1   Introduction

The current development of laser cooling and trapping of radioactive atoms has opened a new research domain in molecular dynamics and molecular spectroscopy. Great theoretical and experimental effort is currently motivated by possible applications e.g.; Bose-Einstein condensation, manipulation and controlling of ultracold molecules by photoassociation [1–5]. Alkali dimers thereby synthesized requires good knowledge of their electronic structure. Conversely, real time photodissociation experiments were done for alkali dimers that can be used to locate the crossings and coupling between the electronic states. This inspires much theoretical work [6–9] including quantum dynamics investigation [10].

The LiRb molecule has been extensively studied by experimental and theoretical groups [11–22]. We cite for example the work of Igel-Mann et al. [11] devoted to homonuclear and heteronuclear alkali dimers XY (X, Y = Li towards Cs). They have determined the spectroscopic constants for the ground state of almost all these systems. On the other hand, Urban et al. [12] have calculated the dipole moment and dipole polarizabilities of a series of alkali metal atoms including Li, Na, K and Rb. Recently, Korek et al. [13] have calculated the LiRb molecule *ab initio*, where they determined the potential energy curves for 28 electronic states. They have derived the spectroscopic constants (T_e, R_e, ω_e, B_e) of 7, 5 and 2 states of, respectively, $^{1,3}\Sigma^+$, $^{1,3}\Pi$, $^{1,3}\Delta$ symmetries. More recently, the same group of Korek et al. [14] have calculated the potential energy for the 58 lowest electronic states including the spin orbit effect within the range of 3.0–34.0 a.u. of the internuclear distance R. A detailed comparison between their results and ours was presented previously [15, 16]. In fact, the accurate adiabatic and diabatic states results appear as extended abstract [16] for the $^1\Sigma^+$ symmetry and in a full paper [15]. The spectroscopic constants (R_e, D_e, T_e, ω_e and B_e) were derived and compared to available theoretical studies. Good agreement was found for the ground and first excited $^{1,3}\Sigma^+$ states with previous work. This paper focuses on the use of both adiabtic and diabatic results obtained for LiRb to evaluate non adiabatic effects such as the radial couplings, the adiabatic correction and vibronic shift. In this context, we propose a dynamcsl study on LiRb molecule. The first and second derivatives neglected in the Born-Oppenheimer approximation, which are the source of many physical phenomena [17–29] such as predissociation [27], are determined for the first time for LiRb. Several authors have studied dynamics for the HeH$^+$, LiH, NaH, KH, RbH and CsH molecules [30–37]. They have used numerical methods or analytical expressions.

Therefore, this work follows our previous study on LiRb molecule where adiabatic and diabatic results were presented. Here, the aim is to use the diabatization procedure for extensive evaluation of nonadiabatic interaction using the rotation matrix connecting the adiabatic and diabatic representations. The radial coupling is computed between neighbouring adiabatic $^{1,3}\Sigma^+$ states and interpreted from the diabatic curves features. On the other hand, the radial coupling is used to determine the adiabatic correction and estimate the vibronic shift. The paper is organized as follows. In Sect. 24.2, a summary of the *ab initio* calculation and the numerical method based on the Hellman-Feynman expression is presented. The results and discussion are presented in Sect. 24.3, which is divided into five parts: diabatic and adiabatic results, radial coupling (first and second derivatives), adiabatic correction, spectroscopic constants, vibrational energy spacing and shift. Concluding remarks form Sect. 24.4.

24.2 Method of Calculation

The Born-Oppenheimer approximation plays a central role in molecular calculations; however there are a wide variety of problems where it breaks down. This is the case when several electronic states dynamically interact, as in conical intersections or at avoided crossings. In most cases, two equivalent representations can be adopted for the theoretical treatment of dynamics studies. These representations are adiabatic, where the radial coupling causes the non-adiabatic transitions, and the diabatic representation where the same role is imputed to the electronic coupling. The diabatization method, the results of which are used here, is based on effective Hamiltonian theory combined with an effective overlap matrix. Effective Hamiltonian theory is used in an unusual way [38–40] where the diabatic states lie in the target space and are linear combinations of the adiabatic states, allowing for variational properties of the effective Hamiltonian operator [38,41,42]. The diabatisation method was tested first for the CsH molecule [36,37] and applied later for the LiH, NaH, KH and RbH systems [32–36]. Recently, this same procedure was applied with care for alkali mixed dimers in our group for LiCs, LiRb, and NaCs [16,43,44].

24.2.1 First Derivative

As in ref. [33], we have evaluated the radial coupling according to two methods by making the hypothesis that the residual coupling in the diabatic basis is zero. Denoting the adiabatic states by $|\psi_i\rangle$ (with

$$|\psi_i\rangle = \sum_k C_{ki}|\phi_k\rangle$$

associated energy E_i), the diabatic states by $|\phi_k\rangle$ and the unitary matrix connecting the two sets by $C(C_{ki} = \langle \phi_k | \psi_i \rangle)$

(i) *Numerical differentiation of the rotation matrix*

$$\left\langle \psi_i \left| \frac{\partial}{\partial R} \right| \psi_j \right\rangle = \sum_\alpha C_{i\alpha} \frac{\partial C_{\alpha j}}{\partial R}$$

Or in matrix notation

$$= \left\langle C^+ \frac{\partial C}{\partial R} \right\rangle_{ij}$$

(ii) *Hellmann-Feynman expression*

$$\left\langle \psi_i \left| \frac{\partial}{\partial R} \right| \psi_j \right\rangle = (E_j - E_i)^{-1} \left\langle \psi_i \left| \frac{\partial H}{\partial R} \right| \psi_j \right\rangle$$

$$= (E_j - E_i)^{-1} \left\langle C^+ \frac{\partial H}{\partial R} C \right\rangle_{ij}$$

Where the electronic Hamiltonian is assumed known in and restricted to the diabatic basis,

$$H_{kl} = \langle \phi_k | H_{el} | \phi_l \rangle$$

$$\left\langle \frac{\partial H}{\partial R} \right\rangle_{kl} = \left\langle \phi_k \left| \frac{\partial H_{el}}{\partial R} \right| \phi_l \right\rangle = \frac{\partial H_{kl}}{\partial R}$$

As expected, both methods here give identical results. In both cases the matrix elements H_{ij} were interpolated by cubic splines in the first step and we determined the C matrix by diagonalization at all distances required by the three-point numerical differentiation. Interpolation of the rotation matrix C leads to numerical instability due to the loss of unitarity and should be avoided. It was demonstrated that the evaluation of the radial coupling is more stable and preferred by the Hellman-Feynman method.

24.2.2 Second Derivative

The second term not considered in the Born-Oppenheimer approximation is the second derivative. It is often neglected in the calculation of nonradiative lifetimes. The knowledge of the diabatic and adiabatic representations, as well as the rotation matrix, readily gives estimates of the second derivative matrix elements. It was shown for LiH [28, 29] that this second term is not negligible compared to the first derivative and has a significant effect on the nonradiative lifetimes. Deriving the rotational matrix C and use of closure give:

$$
\begin{aligned}
\frac{dC}{dR} &= \frac{d}{dR}\left\langle \psi \left| \frac{d}{dR} \right| \psi \right\rangle \\
&= \left\langle \frac{d}{dR}\psi \left| \frac{d}{dR}\psi \right.\right\rangle + \left\langle \psi \left| \frac{d^2}{dR^2} \right| \psi \right\rangle \\
&= \left\langle \frac{d}{dR}\psi \middle| \psi \right\rangle \left\langle \psi \left| \frac{d}{dR}\psi \right.\right\rangle + \left\langle \psi \left| \frac{d^2}{dR^2} \right| \psi \right\rangle \\
&= C^+ C + \left\langle \psi \left| \frac{d^2}{dR^2} \right| \psi \right\rangle
\end{aligned}
$$

Since C is an anti-hermitian matrix $(C^+ = -C)$ the second derivative can be easily expressed from the first and second derivatives:

$$
\left\langle \psi \left| \frac{d^2 C}{dR^2} \right| \psi \right\rangle = C^2 + \frac{dC}{dR}
$$

24.3 Results and Discussion

24.3.1 Diabatic and Adiabatic Potentials

The accurate adiabatic and diabatic results beyond the Born-Oppenheimer approximation are partially in [16] for the $^1\Sigma^+$ symmetry and later for all symmetries [15]. Their features, physical and chemical interests have been demonstrated and analysed in details. For a better understanding of the potential curve features and interactions, the dipole moment function was evaluated and related to the avoided crossings in the adiabatic representation and to the real crossings in the diabatic one. They are reproduced here in Figs. 24.1–24.4 for, respectively, the $^1\Sigma^+$ adiabatic, $^1\Sigma^+$ diabatic, $^3\Sigma^+$ adiabatic and $^3\Sigma^+$ diabatic states. The adiabatic potential energy curves and the resulting spectroscopic constants (R_e, D_e, T_e, ω_e and B_e) were compared previously to the available theoretical data and has shown excellent agreement. In this study, these spectroscopic properties will be recalculated by introducing the adiabatic correction and a new comparison will be made with available studies. The diabatic results, obtained for LiRb for the first time, are based on an effective Hamiltonian theory and overlap matrix. The determination of the diabatic states is founded on the condition that the wavefunction derivative is zero. It is difficult to satisfy this condition, except approximately, which corresponds to so-called

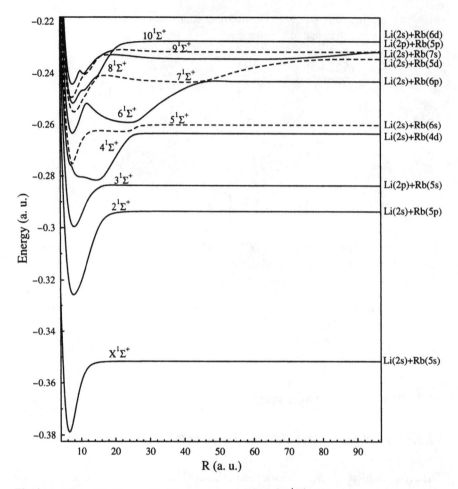

Fig. 24.1 LiRb adiabatic potential energy curves for the 1–10 $^1\Sigma^+$ states

quasi-diabatic states. This diabatization procedure has proved in the past to produce vanishingly small residual nonadiabatic couplings for the CsH [38, 42] molecule. This is a severe test since other usual methods lead to a residual coupling of the order of 10^{-1} a.u. [38]. This method was applied successfully later for the LiH, NaH, KH and RbH systems. Recently, the same diabatisation approach was applied for the first time for a mixed alkali diatomic molecule LiCs, by Mabrouk et al. [43]. The diabatic $^1\Sigma^+$ curves are plotted in Fig. 24.2. Note that the ionic curve noted D_1, which is associated to the Li^-Rb^+ ionic state, crosses all the neutral ones at different distances. These crossings occur with the electronic states named D_{2-10} dissociating into $Li(2s) + Rb(5s)$, $Li(2s) + Rb(5p)$, $Li(2p) + Rb(5s)$, $Li(2s) + Rb(4d)$, $Li(2s) + Rb(6s)$, $Li(2s) + Rb(6p)$, $Li(2s) + Rb(5d)$ and $Li(2s) + Rb(7s)$ states at internuclear distance around 10.66, 16.17, 16.70, 24.16, 26.48, 44.77, 72.34 and 91.36 a.u. Inspection of the coupling magnitude corroborates the conclusion that these crossings

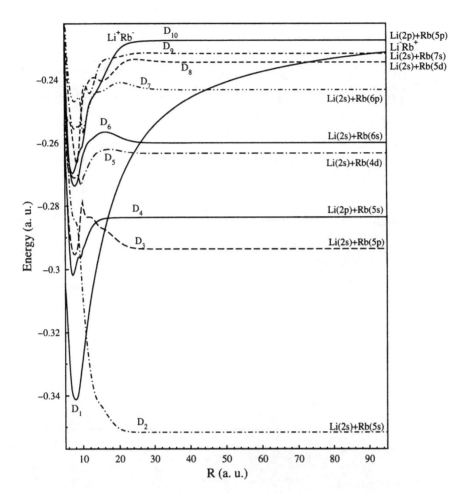

Fig. 24.2 LiRb diabatic potential energy curves for the $^1\Sigma^+$ states

are less and less avoided as the ionic curve asymptotically encounters higher-energy Rydbergs at increasing internuclear distances. In fact, the real crossings in the diabatic representation are transformed into avoided crossings in the adiabatic one. Figure 24.4 presents the first ten diabatic potential energy curves for the $^3\Sigma^+$ symmetry related to the adiabatic ones. They are also determined here for the first time. Similar crossings are detected for the triplet states. For example, the curve D_1 crosses the $D_3, D_4, D_5, D_6, D_7, D_8, D_9$ and D_{10} diabatic states at 13.3, 13.7, 18.5, 19.4, 22.5, 24.9, 26.1 and 28.3 a.u, respectively. The crossings for the triplet states occur at shorter distance than those observed for singlets.

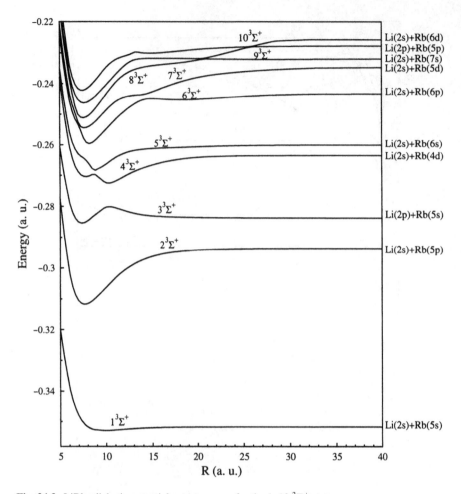

Fig. 24.3 LiRb adiabatic potential energy curves for the 1–10 $^3\Sigma^+$ states

24.3.2 First and Second Derivative

The procedure used to evaluate the radial coupling between the $^{1,3}\Sigma^+$ states for the LiRb molecule is based on the accurate adiabatic and diabatic data as explained in the previous section. In Fig. 24.5, we present the first derivative radial coupling ($\langle \psi_i | \frac{\partial}{\partial R} | \psi_j \rangle, |i-j| = 1$) between neighbour adiabatic $^1\Sigma^+$ states. The other radial couplings with $|i-j| \neq 1$ are much smaller and they are not reported. The first derivative presents many peaks, related to ionic-neutral and neutral-neutral couplings. This term, neglected in the Born-Oppenheimer approximation, will help us to study some new things and to find a correlation between the peaks of the radial coupling and the diabatic and adiabatic curves. Note that the 1–2 coupling is greater at short range, it presents a maximum at 8 a.u. and approaches zero at long range.

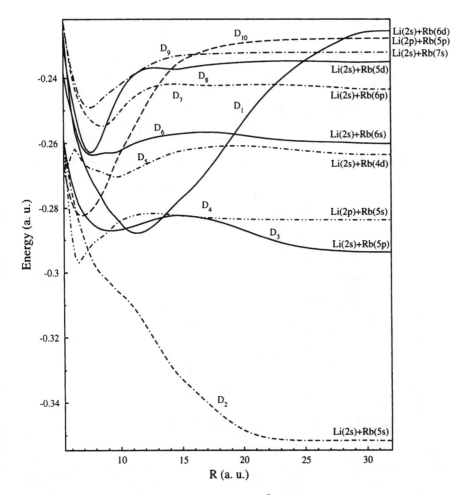

Fig. 24.4 LiRb diabatic potential energy curves for the $^3\Sigma^+$ states

This maximum reflects the avoided crossing between the related adiabatic states and a real crossing between the ionic diabatic curve named D_1 and the neutral one named D_2 at closed distance to that of the maximum. The 2–3 radial coupling has a unique form with a minimum and two maxima. They are associated with two neutral-neutral crossings between the state D_3 dissociating into $Li(2s) + Rb(5p)$ and the states D_2 and D_4 dissociating, respectively, into $Li(2s) + Rb(5s)$ and $Li(2p) + Rb(5s)$ and to another crossing between the ionic state D_1 with D_3. The 3–4 coupling has an intense peak at 9.5 a.u. and another small one at 16 a.u. The former, at short range, corresponds to the crossing between two neutral states D_2 and D_4 and or D_3 and D_4 in the diabatic representation, while the small one at long range is connected to the neutral-ionic crossing between the curves associated to $Li(2s) + Rb(5s)$ and $Li^- + Rb^+$. The short and intermediate distance intense peaks for the 4–5, 5–6,

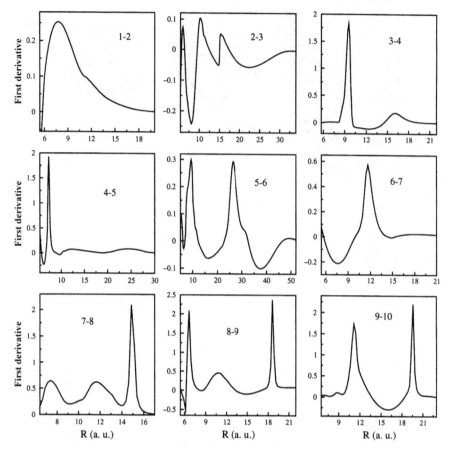

Fig. 24.5 First derivative between $(\langle \psi_i | \frac{\partial}{\partial R} | \psi_j \rangle \, |i - j| = 1)$ $^1\Sigma^+$ states of LiRb

6–7, 7–8, 8–9 and 9–10 couplings are due to the neutral-neutral and neutral-ionic crossings between the potential energy curves in the diabatic representation. The second ionic state is Li^+Rb^-. These peaks can also be explained by the avoided crossings between neighboring adiabatic states. Other peaks can appear at longer range, due to crossings between the first ionic $D_1(Li^-Rb^+)$ and higher neutral states.

The calculation of the second derivative, here, has no serious complications. We used the formalism described, based on the knowledge of the first derivative and the rotation matrix. As expected, this quantity is as large as the first derivative. This term multiplied by $\frac{1}{2\mu}$ corresponds to an energy correction omitted in the Born-Oppenheimer approximation, which contributes to the so-called adiabatic correction. Other lesser corrections [42], resulting from the radial dependence of the derivative of the electronic energy in the diabatic representation can be evaluated using the Virial theorem. The second derivative is not negligible and it will consequently contribute to the nonradiative lifetime. We display in

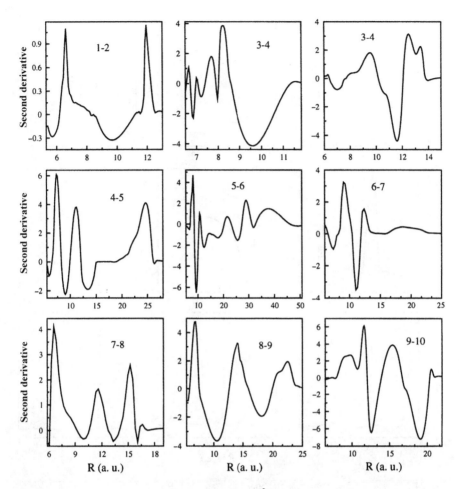

Fig. 24.6 Second derivative between neighbour $(\langle \psi_i | \frac{\partial^2}{\partial R^2} | \psi_j \rangle,\ |i - j| = 1)$ $^1\Sigma^+$ states of LiRb

Fig. 24.6 the second derivative radial coupling between neighboring $^1\Sigma^+$ adiabatic states:

$$\left(\left\langle \psi_i \left| \frac{\partial^2}{\partial R^2} \right| \psi_j \right\rangle, |i - j| = 1 \right)$$

It presents many peaks which can be also related to neutral-neutral and ionic-neutral crossings in the diabatic representation. Furthermore, such coupling disappears at internuclear distances corresponding to avoided crossings in the adiabatic representation.

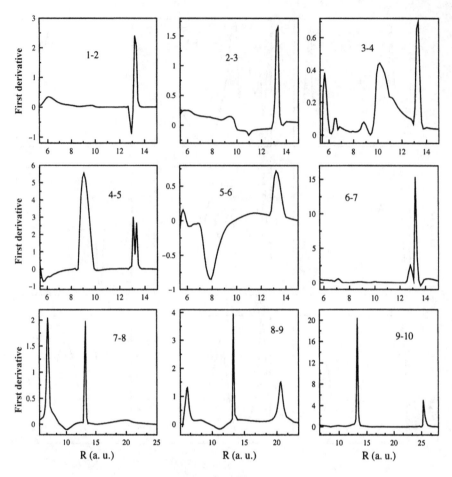

Fig. 24.7 First derivative between neighbour $^3\Sigma^+$ states of LiRb

The radial couplings $(\langle\psi_i|\frac{\partial}{\partial R}|\psi_j\rangle,\ \langle\psi_i|\frac{\partial^2}{\partial R^2}|\psi_j\rangle,\ |i-j|=1)$ between the $^3\Sigma^+$ adiabatic neighboring states are reported in Figs. 24.7 and 24.8. The i-j first and second derivative radial couplings for the triplet states are very similar in shape to they previous ones; however their positions are related to avoided and real crossing between related adiabatic and diabatic states. Since the radial coupling is known to be very sensitive to the details of the adiabatic wave functions, this has established a correlation between the peaks of the radial coupling and the diabatic curves. These peaks are related to the avoided crossings between adiabatic curves and real crossings between diabatic curves. In fact, the diabatic representation has helped us to understand the origin of the short range peaks (not readily explained by adiabatic curves).

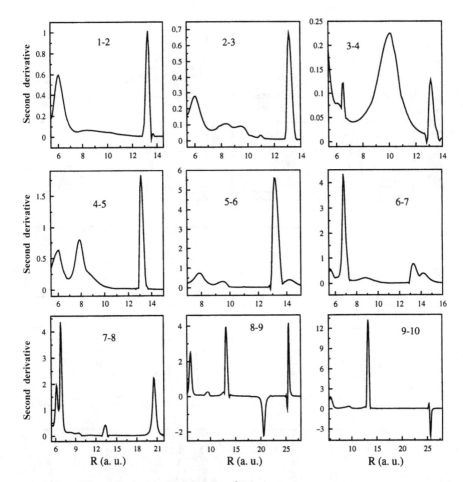

Fig. 24.8 Second derivative between neighbour $^3\Sigma^+$ states of LiRb

24.3.3 Adiabatic Correction

The Born-Oppenheimer approximation is generally a good one and is used in most molecular problems. However, it breaks down in many cases such as in avoided crossing regions, where transitions between potential energy surfaces can occur. The neglected interactions in this approximation are responsible for many physical processes of interest e.g.; predissociation, collisions or radiationless transitions. It is still possible to use this approximation if we take into account the neglected terms. The first theoretical calculation of the adiabatic correction was performed for the H_2^+ molecule [45] in 1941. For the heteronuclear molecules, the first theoretical calculation of the adiabatic correction for the HeH$^+$ molecule [30, 31] with two electrons, then for more complex systems such as LiH, KH, RbH and CsH

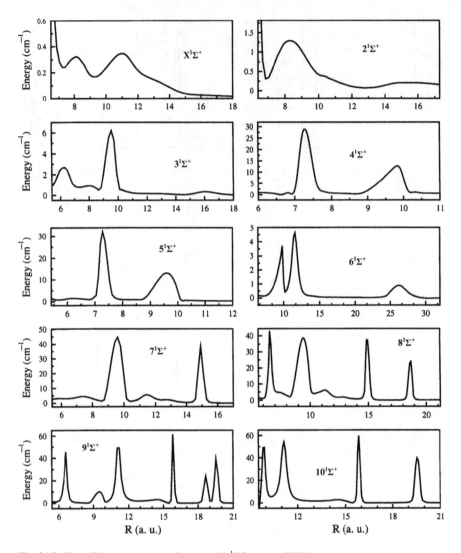

Fig. 24.9 The adiabatic correction for the 1–10 $^1\Sigma^+$ states of LiRb

molecules [32–37]. Most of these theoretical calculations are based on analytical and numerical derivation of the resulting *ab initio* electronic wavefunctions. However, in our study the adiabatic correction is evaluated in a simple way using the produced data in both adiabatic and diabatic representations.

We present in Fig. 24.9 the adiabatic correction for the $^1\Sigma^+$ states. This correction for the ground state, X $^1\Sigma^+$, is very small, less than 1 cm^{-1}. However, it presents two peaks and then vanishes at large intenuclear distance as the electronic wavefunction no longer depends on internuclear distance. This very small adiabatic correction can

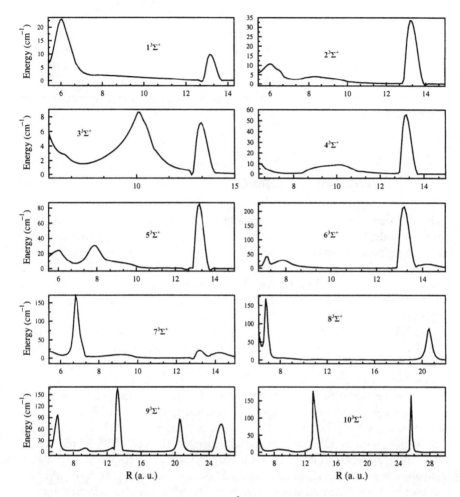

Fig. 24.10 The adiabatic correction for the 1–10 $^3\Sigma^+$ states of LiRb

be explained by the large avoided crossings between the ground and the first excited states. The adiabatic correction for the $2\,^1\Sigma^+$ state is also small as it does not exceed $1.5\,\mathrm{cm}^{-1}$ located around its equilibrium distance. The $3\,^1\Sigma^+$ state has two peaks, one at short range and the other one at longer internuclear distance. For the intense peak, the correction is around $6\,\mathrm{cm}^{-1}$ located at 9.5 a.u. For higher excited states, the correction becomes much more larger than the previous states, the adiabatic correction is about $30\,\mathrm{cm}^{-1}$ for the 4 and $5\,^1\Sigma^+$ states, $50\,\mathrm{cm}^{-1}$ for 7 and $8\,^1\Sigma^+$ states, and $60\,\mathrm{cm}^{-1}$ for the 9 and $10\,^1\Sigma^+$ states.

Figure 24.10 presents the adiabatic correction for the $^3\Sigma^+$ states. Note that this correction is greater than that found for the singlet states. It is of order $10\,\mathrm{cm}^{-1}$ for the 1–$5\,^3\Sigma^+$ states. However, it exceeds 100 of cm^{-1} for the 6–$10\,^3\Sigma^+$ states. The largest corrections are located at short and intermediate distances close to the equilibrium distances, which affect all spectroscopic constants and the vibrational energy levels. It is important to mention that the peaks positions of the adiabatic correction, which shows the region where this correction is large, are shifted to longer distances for higher excited states as the avoided crossings occur at larger internuclear distance. In addition, this correction is larger for such excited states as the crossings are much less avoided and potential energy curves are very close.

24.3.4 Spectroscopic Constants

Tables 24.1 and 24.2 give our spectroscopic constants (R_e = equilibrium distance, D_e = well depth, T_e = vertical transition energy, ω_e = vibration frequency and B_e = rotational constant) for the 20 singlet and triplet states. In addition, a comparison with our previous calculation without adiabatic correction and the available theoretical works is presented. Our results, especially for the ground state ($X^1\Sigma^+$), are compared with those of Korek et al. [13], Igel-Mann et al. [11] and Urban and Sadlej [12]. Their equilibrium distances R_e for this state are, respectively, 3.430, 3.450 and 3.497 Å. These values are in good agreement with our equilibrium distance R_e = 3.428 Å which differs from our previous value, [15] by only 0.001 Å. This is due to the small adiabatic correction found for the ground state. Our equilibrium distance for the $2^1\Sigma^+$, $1^3\Sigma^+$, $2^3\Sigma^+$ and $3^3\Sigma^+$ excited states are also in good agreement with the work of Korek et al. [13]. The same good agreement is observed for well depth and for ω_e and B_e spectroscopic constants. For example, our well depths for the $2^1\Sigma^+$, $3^1\Sigma^+$, $1^3\Sigma^+$ and $2^3\Sigma^+$ states are 7,053, 3,494, 276 and $3,969\,cm^{-1}$, respectively. They are very close to the values of 7,057, 3,516, 280 and $3,977\,cm^{-1}$ extracted from Korek et al. [46] on their web site. We also note good agreement for the other singlet and triplet Σ^+ states, especially, for the equilibrium distance, well depth and vertical transition energy, which are also extracted from the potential energy data [46].

Our spectroscopic constants are re-evaluated including the adiabatic correction. The shift in equilibrium distance for the ground state is about 10^{-3} Å which is of same order for many electronic states of small adiabatic correction. However, the shift for several excited $^{1,3}\Sigma^+$ states becomes larger, as does the adiabatic correction. For example, the change in the equilibrium distance for the $5^1\Sigma^+$ state, is equal to 0.069 Å for the first minimum and to $0.023 10^{-2}$ Å for the second one. A change is also observed for the well depth which is shifted by few to tens of cm^{-1} for the singlet and triplet Σ^+ states. The change in the D_e is $11\,cm^{-1}$ for the ground state and several tens of cm^{-1} for the higher excited states, which exhibit minima at large internuclear distance. For example, D_e for the $7^1\Sigma^+$ and $10^1\Sigma^+$ states are modified by 88 and $85\,cm^{-1}$. The same is observed for the vertical transition energy shifted by few to tens of cm^{-1}. The vibration frequency and the rotational constant are also changed due to the adiabatic correction. Similar changes in the spectroscopic

Table 24.1 Spectroscopic constants of the ground and excited $^1\Sigma^+$ states of the LiRb molecule

State	$R_e(\text{Å})$	$D_e(\text{cm}^{-1})$	$T_e(\text{cm}^{-1})$	$\omega_e(\text{cm}^{-1})$	$B_e(\text{cm}^{-1})$	Ref.
X $^1\Sigma^+$	3.427	5,957		191.30	0.226	This work
	3.428	5,968		196.02	0.223	[15]
	3.43	5,959[a]		194.0	0.220	[13]
	3.45			195.0		[11]
	3.497					[12]
(2) $^1\Sigma^+$	4.132	7,044	11,650	119.66	0.153	This work
	4.137	7,053	11,654	118.78	0.153	[15]
	4.138	7,057[a]	11,639	119.6	0.152	[13]
(3) $^1\Sigma^+$	4.243	3,485	17,377	111.67	0.145	This work
	4.243	3,494	17,382	114.24	0.145	[15]
	4.257	3,516[a]	17,348	113.00	0.144	[13]
(4) $^1\Sigma^+$	7.672	3,998	21,324	40.83	0.044	This work
	7.671	3,998	21,326	41.27	0.044	[15]
	7.629[a]	3,995[a]	21,318[a]	41.68[a]	0.044[a]	[46]
(5) $^1\Sigma^+$						
First min	4.031	3,316	22,743	206.78	0.162	This work
	3.962	3,360	22,702	211.81	0.167	[15]
	3.952[a]	3,372[a]	22,688[a]	212.27[a]	0.167[a]	[46]
Second min	11.88	595	25,463	15.17	0.018	This work
	11.650	598	25,473	13.10	0.019	[15]
	11.687[a]	595[a]	25,465[a]	14.70[a]	0.019[a]	[46]
(6) $^1\Sigma^+$						
First min	4.021	4,433	25,323	134.22	0.162	This work
	4.021	4,438	25,330	132.28	0.162	[15]
	4.026[a]	4,434[a]	25,325[a]	132.34[a]	0.162[a]	[46]
Second min	12.645	3,501	26,255	20.98	0.016	This work
	12.666	3,502	26,266	22.38	0.016	[15]
	12.624[a]	3,498[a]	26,260[a]	21.66[a]	0.016[a]	[46]
(7) $^1\Sigma^+$						
First min	4.328	4,517	27,150	98.81	0.140	This work
	4.301	4,535	27,143	98.04	0.141	[15]
	4.306[a]	4,509[a]	27,159[a]	103.52[a]	0.141[a]	[46]
Second min	21.65	1,891	29,794	11.28	0.002	This work
	21.470	1,979	29,700	11.10	0.005	[15]
	22.269[a]	1,910[a]	29,758[a]	8.70[a]	0.002[a]	[46]
(8) $^1\Sigma^+$	4.118	3,845	27,867	133.06	0.153	This work
	4.116	3,803	27,874	137.00	0.154	[15]
(9) $^1\Sigma^+$	3.947	3,966	28,356	142.67	0.168	This work
	3.947	907	28,362	141.10	0.168	[15]
(10) $^1\Sigma^+$						
First min	4.021	3,669	29,500	133.54	0.162	This work
	4.021	3,668	29,512	134.67	0.162	[15]
Second min	5.635	2,669	30,500	115.04	0.082	This work
	5.772	2,754	30,426	119.09	0.078	[15]

[a]These values are extracted from their potential energy curves available in their web site [46]

Table 24.2 Spectroscopic constants of the 1–10 $^3\Sigma^+$ states of the LiRb molecule

State	$R_e(\text{Å})$	$D_e(\text{cm}^{-1})$	$T_e(\text{cm}^{-1})$	$\omega_e(\text{cm}^{-1})$	$B_e(\text{cm}^{-1})$	Ref
(1) $^3\Sigma^+$	5.214	273	5,696	42.12	0.095	This work
	5.126	276	5,693	40.13	0.098	[15]
	5.216	280[a]	5,678	41.20	0.100	[13]
(2) $^3\Sigma^+$	4.058	3,963	14,742	129.66	0.153	This work
	4.058	3,969	14,737	128.63	0.159	[15]
	4.064	3,977[a]	14,719	129.90	0.157	[13]
(3) $^3\Sigma^+$	3.904	364	20,516	136.60	0.171	This work
	3.904	362	20,513	136.61	0.171	[15]
	3.915[a]	365[a]	20,499[a]	136.46[a]	0.171[a]	[46]
(4)$^3\Sigma^+$						
First min	4.137	1,526	23,798	117.83	0.153	This work
	4.137	1,527	23,797	117.85	0.153	[15]
	4.142[a]	1,529[a]	23,784[a]	117.50[a]	0.153[a]	[46]
Second min	5.444	1,968	23,356	107.55	0.089	This work
	5.433	1,983	23,340	109.58	0.088	[15]
	5.433[a]	1,997[a]	23,317[a]	109.44[a]	0.088[a]	[46]
(5) $^3\Sigma^+$	4.681	1,960	24,108	216.07	0.120	This work
	4.740	1,964	24,106	215.11	0.116	[15]
	4.671[a]	1,959[a]	24,101[a]	215.93[a]	0.120[a]	[46]
(6) $^3\Sigma^+$						
First min	4.365	3,569	26,200	176.23	0.137	This work
	4.343	3,598	26,170	175.31	0.139	[15]
	4.344[a]	3,596[a]	26,162[a]	175.65[a]	0.139[a]	[46]
Second min	9.560	444	29,325	19.63	0.028	This work
	9.560	444	29,324	19.35	0.028	[15]
	9.560[a]	433[a]	29,325[a]	19.36[a]	0.028[a]	[46]
(7) $^3\Sigma^+$	4.058	4,365	27,313	131.33	0.159	This work
	4.063	4,372	27,306	130.50	0.159	[15]
	4.063[a]	4,355[a]	27,314[a]	128.02[a]	0.159[a]	[46]
(8) $^3\Sigma^+$	3.994	4,233	28,037	152.56	0.164	This work
	3.989	4,243	28,026	147.88	0.165	[15]
(9) $^3\Sigma^+$						
First min	4.037	4,089	29,092	133.15	0.160	This work
	4.037	4,093	29,087	132.91	0.160	[15]
Second min	10.070	4,089	32,277	36.38	0.025	This work
	10.555	4,093	32,272	36.87	0.023	[15]
(10) $^3\Sigma^+$						
First min	3.936	3,621	29,991	147.77	0.169	This work
	3.936	628	29,984	146.90	0.169	[15]
Second min	7.809	960	32,652	32.60	0.044	This work
	7.798	961	32,650	34.41	0.045	[15]

[a]These values are extracted from their potential energy curves available in their web site

constants for the $^3\Sigma^+$ symmetry are observed. This shows that the spectroscopic constants are affected by the adiabatic correction, especially for excited states, which is not surprising as the correction for these states is more significant.

24.3.5 Vibrational Energy Levels and Shift

The vibrational shift is defined as the difference between the corrected and the adiabatic levels. The vibrational energy levels have been calculated using the adiabatic and the corrected potential energy curves for $1-10^1\Sigma^+$ and $1-10^3\Sigma^+$ electronic states of the LiRb molecule. These vibrational energy levels have been numerically obtained using the Numerov method in which the wavefunctions have been propagated for 30,000 points by cubic interpolation and by taking Rmax $=$ 100 a.u and $\mu = 6.4196$ amu. In Table 24.3 we report the vibrational energy level spacing $(E_v - E_{v-1})$, respectively, for the $1^{1,3}\Sigma^+$, $2^{1,3}\Sigma^+$, $3^{1,3}\Sigma^+$ and $4^{1,3}\Sigma^+$ states. To our knowledge, the vibrational level spacing for LiRb are determined here for the first time. Note that we obtain a large number of vibrational levels for the higher excited states, due to their large potential wells. Precisely, for the $2^{1,3}\Sigma^+$ and $4^{1,3}\Sigma^+$ states we obtained, respectively, $n_v = 77, 63, 107$ and 49 vibrational energy levels. The large numbers should be related to the wide wells and to the long range R^{-4} attractive wing of the potential curves. In Figs. 24.11 and 24.12, we present the vibrational shift for the $^1\Sigma^+$ and $^3\Sigma^+$ states, respectively. There is a small shift for the energy levels associated to the ground state with a maximum at $v = 23$. The shift for the $^1\Sigma^+$ higher excited states is more larger i.e. several cm^{-1} for $2^1\Sigma^+$ and $4^1\Sigma^+$ states and of few tens of cm^{-1} for $3^1\Sigma^+$, $5^1\Sigma^+$, $7-10^1\Sigma^+$ states. The vibrational shift for the $6^1\Sigma^+$ state is insignificant, except for the vibrational level $v = 14$ which is displaced by more than 20 cm^{-1} due to the adiabatic correction. The vibrational shift for the $^3\Sigma^+$ states is much greater than the singlet states as it reaches several tens of cm^{-1}. Note that the displacement in some vibrational levels for the 4, 5, and $8^3\Sigma^+$ states is about 100 cm^{-1}.

24.4 Conclusion

In this paper, the radial coupling and the adiabatic correction for the $1-10^{1,3}\Sigma^+$ electronic states of the LiRb molecule are presented, here, for the first time. We used accurate adiabatic and diabatic *ab initio* results determined previously [15] for the $1-10^{1,3}\Sigma^+$ states. To our best knowledge, no diabatic potential energy curves have been published for the LiRb molecule. To determine the radial coupling matrix elements, which correspond to the first and the second derivative, a computationally efficient method was use, based on a numerical differentiation of the rotation matrix connecting the diabatic and the adiabatic representations. The radial coupling (First and Second derivative) is known to be very sensitive to the details of the adiabatic

Table 24.3 Vibrational level spacings E_v–E_{v-1} (in cm^{-1}) of the $^{1,3}\Sigma^+$ symmetry

υ	X $^1\Sigma^+$	2 $^1\Sigma^+$	3 $^1\Sigma^+$	4 $^1\Sigma^+$	1 $^3\Sigma^+$	2 $^3\Sigma^+$	3 $^3\Sigma^+$	4 $^3\Sigma^+$
0								
1	192.373	118.083	112.429	40.506	36.887	128.036	133.834	106.898
2	192.488	116.525	111.607	39.962	33.859	126.084	130.766	103.617
3	188.422	115.637	110.766	39.035	31.19	124.177	102.837	98.969
4	183.527	114.05	109.959	37.667	28.114	122.285	96.837	92.897
5	177.091	112.871	109.161	35.715	24.975	120.444	90.381	58.533
6	171.601	111.899	108.315	32.864	21.901	118.494	85.440	27.728
7	167.189	110.636	107.406	28.261	18.901	116.468	81.397	75.066
8	160.52	109.794	106.391	23.027	15.911	114.335	73.907	15.465
9	292.37	109.392	105.27	22.843	13.036	112.155	72.213	60.685
10	258.337	108.332	104.117	25.45	10.293	110.094	70.340	45.186
11	206.633	106.165	102.947	27.421	7.749	108.204	68.295	42.874
12	225.292	103.839	101.704	28.999	5.379	106.335	66.075	53.365
13	232.786	102.351	100.403	30.197	3.372	104.267	63.699	48.488
14	237.579	101.722	99.058	31.26	1.95	102.229	61.290	48.785
15	239.46	101.665	97.624	32.28	0.996	100.263	58.837	50.474
16	234.226	101.52	96.138	33.237	0.393	98.154	56.462	49.325
17	229.533	100.196	94.582	34.115	0.024	96.141	54.085	48.182
18	221.821	97.23	92.967	34.855		94.039	51.783	47.923
19	214.313	95.362	91.274	35.603		91.961	49.502	47.311
20	204.081	93.375	89.475	36.306		89.846	47.313	46.238
21	194.997	90.975	87.64	36.94		87.724	45.141	45.17
22	175.968	88.233	85.672	37.492		85.57	43.044	44.195
23	169.127	85.714	83.66	38.033		83.418	40.965	43.258
24	297.258	82.878	81.519	38.554		81.215	38.968	42.179
25	245.475	79.722	79.309	39.007		79.027	36.969	40.991
26	103.783	136.911	77.003	39.412		76.797	34.971	39.786
27	90.42	62.109	74.57	39.803		74.558	33.012	38.627
28	78.58	204.906	72.087	40.142		72.275	31.075	37.457
29	67.497	113.34	69.534	40.434		69.982	29.091	36.229
30	56.278	117.502	66.9	40.644		67.678	27.113	34.94
31	46.19	123.003	64.289	40.82		65.354	25.102	33.62
32	36.375	127.074	61.673	40.883		62.989	23.097	32.249
33	27.253	129.94	59.124	40.789		60.622	21.024	30.888
34	19.503	126.824	56.719	40.537		58.23	18.915	29.457
35	13.42	130.207	54.391	40.014		55.838	16.740	27.973
36	8.01	131.346	52.183	39.242		53.395	14.748	26.449
37	5.176	133.18	50.02	38.435		50.978	12.529	24.819
38	2.916	134.582	47.813	38.111		48.515	10.120	23.159
39	1.438	136.528	45.31	38.518		46.061	8.338	21.398
40	0.732	265.069	42.346	39.317		43.604	6.366	19.567
41		131.622	38.424	40.032		41.125	4.475	17.637
42		127.924		40.471		38.677	2.565	15.652
43		125.922		40.685		36.236	1.383	13.814
44		127.06		40.796		33.801		11.551

(continued)

Table 24.3 (continued)

υ	X $^1\Sigma^+$	2 $^1\Sigma^+$	3 $^1\Sigma^+$	4 $^1\Sigma^+$	1 $^3\Sigma^+$	2 $^3\Sigma^+$	3 $^3\Sigma^+$	4 $^3\Sigma^+$
45		124.407		40.924		31.407		9.374
46		120.871		41.102		29.034		7.809
47		118.351		41.329		26.653		5.549
48		115.54		41.568		24.359		3.893
49		113.474		41.773		22.082		2.265
50		109.591		41.932		19.817		0.878
51		104.981		42.038		17.592		
52		100.849		42.126		15.412		
53		96.808		42.21		13.382		
54		92.705		42.306		11.449		
55		86.884		42.383		9.264		
56		81.04		42.472		7.857		
57		75.32		42.539		6.173		
58		70.601		42.581		4.654		
59		64.679		42.613		3.314		
60		58.925		42.633		2.233		
61		53.416		42.645		1.485		
62		47.088		42.64		0.714		
63		41.709		42.641		0.312		
64		35.616		42.632				
65		29.981		42.613				
66		25.535		42.581				
67		20.605		42.518				
68		16.312		42.432				
69		12.482		42.313				
70		9.414		42.14				
71		7.884		41.942				
72		5.143		41.773				
73		3.532		41.716				
74		2.453		41.827				
75		1.523		42.026				
76		0.586		42.098				
77		0.293		41.874				
78				41.439				
79				41.25				
80				41.615				
81				42.008				
82				41.561				
83				40.248				
84				39.747				
85				40.826				
86				41.264				
87				39.76				
88				38.709				
89				39.923				

(continued)

Table 24.3 (continued)

υ	X $^1\Sigma^+$	2 $^1\Sigma^+$	3 $^1\Sigma^+$	4 $^1\Sigma^+$	1 $^3\Sigma^+$	2 $^3\Sigma^+$	3 $^3\Sigma^+$	4 $^3\Sigma^+$
90				39.926				
91				38.011				
92				38.168				
93				38.738				
94				36.776				
95				36.156				
96				34.99				
97				31.078				
98				30.991				
99				32.265				
100				34.609				
101				32.395				
102				23.642				
103				21.208				
104				14.491				
105				9.748				
106				4.354				
107				1.792				

wavefunctions, this correspondence confirms the results themselves renewed validitation of the diabatization procedure we used. We observed many peaks related to ionic-neutral crossings in the diabatic representation. In addition, we observe some intense peaks related to the neutral-neutral crossings and interactions between diabatic states. As well as its physical interest, the radial coupling is used to determine the adiabatic correction. This term has been evaluated for all 1–$10^{1,3}\Sigma^+$ states. It varies from some cm^{-1} to a few hundred cm^{-1} for higher excited such as the 7 $^3\Sigma^+$ where it reached 150 cm^{-1}. The large correction observed for many excited states re-emphasizees the breakdown of the Born-Oppenheimer approximation, especially around the avoided crossings.

The spectroscopic constants of the 1–$10^{1,3}\Sigma^+$ adiabatic states including adiabatic correction are determined and compared with the available theoretical work [11–13] and our previous calculation without. Using the corrected energy, the spectroscopic constants for these states are recalculated. Note that these spectroscopic constants (R_e, D_e, T_e, ω_e and B_e) are changed by the adiabatic correction. Very good agreement is observed for them in the $^{1,3}\Sigma^+$ states with the theoretical work of Korek et al. [13]. The adiabatic correction is also used to calculate the shift in the vibrational energy levels, which is the difference between the corrected and the adiabatic levels for all studied states of $^{1,3}\Sigma^+$ symmetries. The spacing between the energy levels is determined for all studied states and presented here for the first time. The shift, difference between corrected and adiabatic energy levels, has been evaluated. A 50 cm^{-1} shift is observed for some vibrational levels showing again the breakdown of the Born-Oppenheimer approximation. In the future, attempts to gain new insight from these data, may establish a correlation between the

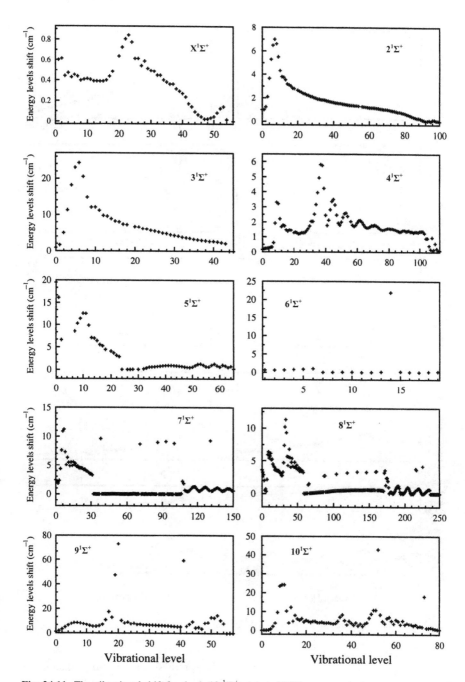

Fig. 24.11 The vibrational shift for the 1–10 $^1\Sigma^+$ states of LiRb

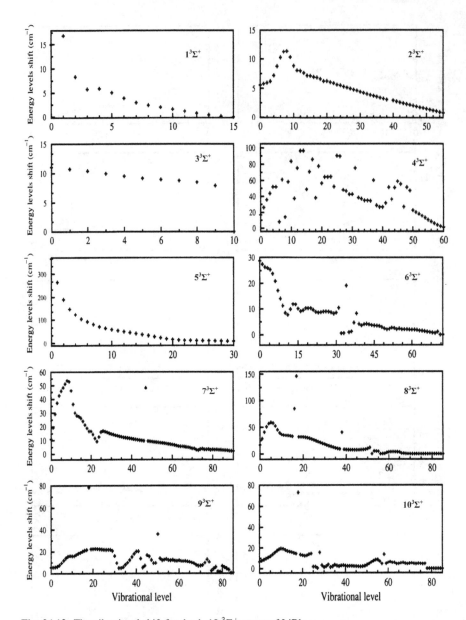

Fig. 24.12 The vibrational shift for the 1–10 $^3\Sigma^+$ states of LiRb

radial couplings and the adiabatic curves, like predissociation and collisions or radiationless transitions. The accurate adiabatic potential energy, radial coupling and adiabatic correction is available for interested researchers.

Acknowledgment We acknowledge support of this work by King Abdul Aziz City for Science and Technology (KACST) through the Long-Term Comprehensive National Plan for Science, Technology and Innovation program under Project No. 08-NAN148-7.

References

1. Thorsheim HR, Weiner J, Julienne PS (1987) Phys Rev Lett 58:2420
2. Fioretti A, Comparat D, Crubellier A, Dulieu O, Masnou-Seeuw F, Pillet P (1998) Phys Rev Lett 80:4402
3. Weiner J, Bagnato VC, Zilio S, Julienne PS (1999) Rev Mod Phys 71:1
4. Masnou-Seeuws F, Pillet P (2001) Adv Atom Mol Opt Phys 47:53
5. Bahns JT, Stwalley WC, Gould PL (2000) Adv Atom Mol Opt Phys 42:171
6. Rosker MJ, Rose TS, Zewail AH (1988) Chem Phys Lett 146:175
7. Engel V, Metiu H, Almeida R, Zewail AH (1988) Chem Phys Lett 152:1
8. Choi SE, Light JC (1989) J Chem Phys 90:2593
9. Grønager M, Henriksen NE (1998) J Chem Phys 109:4335
10. Balakrishnan N, Esry BD, Sadeghpour HR, Cornett ST, Cavagnero MJ (1999) Phys Rev A 60:1407
11. Igel-Mann G, Wedig U, Fuentealba P, Stoll H (1986) J Chem Phys 84:5007
12. Urban M, Sadlej AJ (1995) J Chem Phys 103:9692
13. Korek M, Allouche AR, Kobeissi M, Chaalan A, Dagher M, Fakherddin F, Aubert-Frecon M (2000) Chem Phys 256:1
14. Korek M, Younes G, AL-Shawa S. (2009) J Mol Struct (Theochem) 899:25
15. Jendoubi I, Berriche H, Ben Ouada H, Gadea FX (2011) J. Phys. Chem. A (submitted)
16. Jendoubi I, Berriche H, Ben Ouada H (2010) Unpublished work Conf Proc (in press)
17. Jensen JO, Yarkony DR (1988) J Chem Phys 89:975
18. Bishop DM, Cheung LM (1977) Phys Rev A 16:640
19. Kolos W, Wolniewicz L (1964) J Chem Phys 41:3663; (1965) 43:2429; 45 (1966) 509; 48 (1968) 3672; 49 (1968) 404; 50 (1969) 3228
20. Kolos W, Wolniewicz L (1963) Rev Mod Phys 35:473
21. Bishop DM, Cheung LM (1979) J Mol Spectrosc 75:462
22. Price RI (1978) Chem Phys 31:309
23. Koxos W, Wolniewicz L (1964) J Chem Phys 41:3663
24. Vidal CR, Stwalley WC (1982) J Chem Phys 77:883
25. Chan YC, Harding DR, Stwalley WC, Vidal CR (1986) J Chem Phys 85:2437
26. Berriche H (1995) Thèse Doctorat de l'Université Paul Sabatier, Toulouse
27. Gadea FX, Berriche H, Romero O, Villarreal P, Delgado Barrio G (1997) J Chem Phys 107:24
28. Gemperle F, Gadea FX (1999) J Chem Phys 110:11197
29. Gemperle F, Gadea FX (1999) EuroPhys Lett 48:513
30. Bishop DM, Cheung LM (1983) Chem Phys 78:1396
31. Bishop DM, Cheung LM (1983) Chem Phys 78:7265
32. Boutalib A, Gadéa FX (1992) J Chem Phys 97:1144
33. Gadéa FX, Boutalib A (1993) J Phys Atom Mol Opt Phys 26:61
34. Khelifi N, Oujia B, Gadéa FX (2001) J Chem Phys 117:879
35. Khelifi N, Zrafi W, Oujia B, Gadéa FX (2002) Phys Rev A 65:042513
36. Zrafi W, Oujia B, Berriche H, Gadéa FX (2006) J Mol Struct (Theochem) 777:87
37. Zrafi W, Oujia B, Gadéa FX (2006) J Phys B Atom Mol Opt Phys 39:1
38. Gadéa FX (1987) Tèse d'Etat, Université Paul Sabatier, Toulouse
39. Gadéa FX, Kuntz PJ (1988) Mol Phys 63:27
40. Gadéa FX (1987) Phys Rev A 36:2557
41. Gadéa FX (1991) Phys Rev A 43:1160

42. Gadéa FX, Pélissier M (1990) J Chem Phys 93:545
43. Mabrouk N, Berriche H, Gadea FX (2007) AIP Conf Proc 963:23
44. Mabrouk N, Berriche H (2009) AIP Conf Proc 1148:326
45. Johnson VA (1941) Phys Rev 60:373
46. www-lasim.univ-lyon1.fr/spip.php?article274

Part VII
Complex Systems, Solids, Biophysics

Chapter 25
Theoretical Studies on Metal-Containing Artificial DNA Bases

Toru Matsui, Hideaki Miyachi, and Yasuteru Shigeta

Abstract We have studied two topics about (i) the structural stabilities and electronic structures of metal-ion containing artificial DNA bases and (ii) conductivity of them. Before proceeding to the main topics, we have shown that a van der Waals corrected density functional method gives the stacking interaction, which agrees well with the reference value obtained by accurate methods in both cases for stacking two bases and two base pairs. We also investigated an origin of structural stability and electronic properties of several metal ion containing artificial DNA bases including chalcogen-substituted compounds. We estimated current-voltage characteristics of stacked natural and metal-containing artificial DNA bases by the scattering theory based on the non-equilibrium Green's function method. We found that the current-voltage characteristics dramatically change by capturing metal ion in the artificial DNA bases.

25.1 Introduction

Since the discovery of the double helix structure by Watson and Crick, DNA has been the most important substance in molecular biology. DNA consists of base, sugar, and phosphoric acid. There are four kinds of base in natural DNA, adenine (A), cytosine (C), guanine (G), and thymine (T) as shown in Scheme 25.1.

T. Matsui • Y. Shigeta (✉)
Institute of Picobiology, Graduate School of Lifescience, University of Hyogo,
Koto 3-2-1, Hyogo 678-1297, Japan

Core Research for Evolutional Science and Technology, Japan Science and Technology Agency
(CREST, JST), Saitama, Japan
e-mail: matsui@chem.sci.osaka-u.ac.jp; shigeta@cheng.es.osaka-u.ac.jp

H. Miyachi
Department of Applied Chemistry, School of Engineering, The University of Tokyo,
Hongo 7-3-1, Tokyo 678-1297, Japan
e-mail: z7m1101@students.chiba-u.jp

P.E. Hoggan et al. (eds.), *Advances in the Theory of Quantum Systems in Chemistry
and Physics*, Progress in Theoretical Chemistry and Physics 22,
DOI 10.1007/978-94-007-2076-3_25, © Springer Science+Business Media B.V. 2012

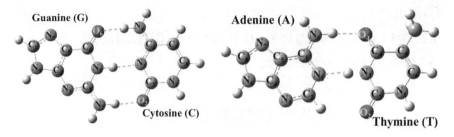

Scheme 25.1 The chemical structures of base pair (GC and AT pair), where the *broken lines* mean the hydrogen bonding

From the view of chemistry, DNA is one of the most remarkable examples for self-assembled materials existing in nature. There are two kinds of interaction, which stabilizes the whole structures of DNA base pairs. These interactions are the driving force of the self-assemble structure, which leads to beautiful double strand. One is the hydrogen bond, where an adenine-thymine (AT) and a guanine-cytosine (GC) base pair have two and three hydrogen bonds per pair, respectively. The hydrogen bonding energy is estimated as 13–17 kcal/mol for AT pair, 25–28 kcal/mol for GC pair, respectively [1]. The second is a base stacking interaction caused by $\pi - \pi$ interaction between the base pairs. This type of interaction mainly originates from the van der Waals (vdW) forces. According to accurate computation, this energy would be estimated as 10–16 kcal/mol between base pairs [2]. Although this energy is a little weaker than hydrogen bonding energy between bases, it is enough to form a stacked structure inside the strand.

DNA, which is the most fundamental material in biology, is also focused on the field in nano-science because of these electric properties. DNA would be a candidate for nano-size conductors, because electrons can move through $\pi - \pi$ stacking like in graphite. Moreover it is easy to imagine that DNA is one of the most available macromolecules taken from gene. The question whether DNA is conductive or not is hot topic all over the world. Many experiments and theoretical simulations have been performed for the purpose of understanding the conductivity of DNA. However, there are many papers with different conclusions: e.g. DNA cannot be conductive [3] or DNA is semi-conductive [4] or DNA is highly conductive [5] even superconductive as metals [6]. The reason why there exist experiments that DNA exhibit conductive is often explained by a hole transfer originated from guanine triplet [7]. In this case, guanine has been oxidized because guanine has h^+ so that GC pair has got to have positive charge. In addition, Giese and his coworker proposed that the oxidized guanine should lose its charge through the proton-transfer reaction between N_1 (G) and N_3 (C) [8]. Nevertheless recent experiments reveal that the conductivity of DNA strongly depends on the length, environments (solvent and solute ions), sequence, and so on. These facts indicate that the more sophisticated modification should be introduced to increase conductivity of DNA. One of the possible approaches is the metal binding to the DNA.

There were two patterns of interaction between metal complex and DNA duplex. One is the major groove binding. Metal cation attacks the major groove of DNA. The other type is called metal complexed DNA (M-DNA) [9]. For instance, a cisplatin, which is known as an antitumor drug, distorts the structure of the B-DNA by 90° [10, 11]. In this case, Pt(II) of the cislatin prefers to bind to N_7 of a guanine to that of an adenine. Similar to the platinum complexes, many metal complexes containing metal ions such as Ni(II), Zn(II), and Mg(II) bind to N_7 of the guanine [12–15]. In M-DNA, a zinc complex binds between a guanine and a cytosine (an adenine and a thymine), where the metal complex has a bridge structure.

New type of metal binding like artificial DNA has been taken much attention by many researchers. The artificial DNA consists of the analog of the DNA base (sometimes accompanied with metal cation), sugar and phosphoric acid. Artificial DNA can connect to the natural DNA so that it is possible to array metal cations in artificial DNA duplex by self-assemble. Moreover, in 2006, Miyake et al. succeeded in making Hg(II) binded to a thymine-thymine (**T–T**) mismatch, which resulted in a [**T–Hg(II)–T**] complex [16].

Quantum chemistry has an advantage of the computation for electronic structure so that it is appropriate to use the quantum chemical methods in order to understand the metal-DNA systems. Nevertheless, calculation methods of electronic structure of macromolecules also have many problems to be solved. For example, the computational cost is critical limitation. Density functional theory (DFT) enables us to compute rather large molecules, which consists of more than 100 atoms. While standard DFT has many defects, e.g., lack of the weak interaction such as dispersion forces, lack of consideration of the exchange in long-range region. To remedy them, Tsuneda and his coworkers have developed the long-range corrected (LC) scheme [17] combined with the correction of dispersion forces proposed by Anderson, Langreth and Lindqvist (hereafter we denote this correction as "ALL functional") [18, 19]. This method reproduces the reference value obtained by the high level *ab initio* calculation from the view of stacking energy and hydrogen bonding energy. However, no other test but benchmark set had been tried so that the application with the method toward understanding electronic structure of DNA and related systems is needed.

This review aims to understand the properties of artificial metal-DNA complexes from the view of theoretical chemistry. In Sect. 25.2, we introduce computational method to evaluate the stacking energy between base pairs by means of the DFT-based method and show the benchmark test for usefulness of the ALL functional with LC scheme [20]. In Sect. 25.3, we focused on the metal-containing artificial DNA and will discuss the structure of it by using of ALL functional with LC scheme and stability by a polarizable continuum model (PCM) [21, 22]. In Sect. 25.4, we investigated the electron conductivity of natural and artificial DNAs with simple model [23], which had been proposed by Luo et al. [24]. Finally, general conclusion is given in Sect. 25.5.

25.2 Theoretical Background

In general, quantum chemical investigation of the DNA bases in more than two base pairs requires highly accurate methods, because the system contains not only the hydrogen bonding interactions, but also the vdW interactions.

Florian et al. [25] computed the relative "solvation" free energies of mutation by using the molecular dynamics (MD) simulations to DNA decamer, which are in good agreement with experimental data. However, MD simulations use an empirical parameter to compute the vdW interaction, and the energy of electrons cannot be described in MD simulations. In order to give deeper understandings for stacked base pairs, it is important to compute the vdW interactions within the first principle methods.

In particular, electron correlation effects are crucial to describe the vdW interaction. To account for the electron correlation effects, density functional theory (DFT) is widely used. As the former interaction is mainly caused by electrostatic interaction between the base pairs, it is sufficient to describe it with the hybrid DFT methods. On the other hand, it is rather difficult to describe the latter interaction by means of the standard hybrid DFT because of the lack of the weak dispersion force. The latter interaction is so weak that the post Hartree–Fock (HF) theories such as the second order of Møller–Presset perturbation theory (MP2) and coupled-cluster method (CC) are at least required to describe it qualitatively. In 2004, Hobza et al. estimated accurate interaction energies between stacked bases by using MP2 based on the resolution of identity method (RI-MP2), with complete basis set (CBS) corrections [1,2].

However, the post-HF methods are limited to systems consisting of a few bases, because the costs of the MP2 and CCSD(T) scale as $O(N^5)$ and $O(N^7)$ with N (the number of basis functions), respectively. The less costly computation method is necessary to calculate electronic structures of molecules larger than two base pairs. Therefore one cannot investigate assembles of DNA bases by means of standard *ab initio* based method. It is desirable to utilize DFT for tackle with such large systems.

Recently, a new class of DFTs has attracted much attention for computing larger-scale molecules. To begin with the studies by Elstner's group [26] for the dispersion interaction in DFT, several groups proposed DFT with a dispersion correction to estimate the stacking energy [27–29]. Our coworkers have also developed a DFT-based method including the dispersion correction [30–32]. The method combines the long-range corrected (LC) DFT [17,33] with the Anderson–Langreth–Lundqvist (ALL) vdW functionals [18]. We have also showed that the Becke 88 (exchange functional) and one parameter progressive functional (OP) (correlation functional) combined with LC scheme (we call this method "LC-BOP")+ALL functional reproduced the results by post SCF method proposed by Hobza et al. The computational cost of the method scales as $O(N^4)$, like the conventional hybrid DFT methods. The ALL functional would be useful to investigate the effects of the stacking interactions on PT reactions as it gives dispersion correction accurately enough to reproduce the post SCF results with low cost.

25.2.1 Calculation of van der Waals Interaction Within LC-DFT

We here show the efficiency of the LC-DFT+ALL method to estimate interaction energies between two base pairs for the natural WC double strand structures by comparing several theoretical methods including the highly accurate *ab initio* method.

25.2.1.1 Computational Details

We performed the benchmark calculations of DFT+ALL methods in order to assess applicability. We used the CCst, GCst, and GGst proposed in JSCH-benchmark sets [34], where the most accurate computations were performed with approximated CCSD(T)/CBS-limit level of theory. We compared these references with the results of DFT+ALL methods in several basis sets. Pople-type basis sets $(6\text{-}31G, 6\text{-}31G(d,p), 6\text{-}31++G(d,p), 6\text{-}311+G(d,p), 6\text{-}311++G(df,pd))$ were used for this benchmark. We estimated the stacking energy between C (or G) and C (or G). By means of in the conventional DFT, HF, and MP2 methods, we estimated the stacking energy E^{int} defined as

$$E^{\text{int}} = E^{\text{tot}} - (E_{\text{B1}} + E_{\text{B2}}) + E^{\text{CP}}, \tag{25.1}$$

where E_{B1} and E_{B2} denote the energy of the base 1 and the base 2 (G or C). E^{CP} is a basis set superposition error (BSSE) obtained by the Boys–Bernardi counterpoise (CP) method [35]. In order to compare the results obtained by the conventional methods with those of the present DFT+ALL methods, we here denote the corrected energy $E^{\text{corrected}}$ obtained by adding the ALL energy functional as

$$E^{\text{corrected}} = E^{\text{int}} + E^{\text{ALL}}, \tag{25.2}$$

where E^{ALL} represents the energy contribution from the ALL functional. We examined HF, MP2, B3LYP, LC-BOP, HF+ALL, B3LYP+ALL, and LC-BOP+ALL. To compare with the result of the reference value, we divided the results into two groups, where one is obtained by standard DFT and HF and the other is by ALL-corrected DFT and HF. We used modified version of GAUSSIAN03 [36] for single point calculations.

25.2.1.2 Benchmark Test

JSCH Benchmark Set

Table 25.1 lists the results of the benchmark test. The reference values obtained by the CCSD (T)/CBS-limit are $-10.02\,\text{kcal/mol(CCst)}, -10.60\,\text{kcal/mol(GCst)}$ and $-12.67\,\text{kcal/mol(GGst)}$. HF and standard DFT cannot describe the stacking energy even qualitatively. On the other hand, our hybrid functional (LC-BOP+ALL)

Table 25.1 Benchmark of DFT+ALL in JSCH benchmark set (kcal/mol)

	DG[a]	DG(d,p)[a]	DG++(d,p)[a]	TG++(d,p)[b]	TG ++(df, pd)[b]
CCst: Reference value: CBS(RI-MP2): −11.00 kcal/mol CBS(T): −10.02 kcal/mol					
HF	−2.01	−0.95	−1.53	−1.42	−1.41
B3LYP	−0.96	−0.27	−1.35	−1.29	−1.28
LC-BOP	−3.74	−2.80	−4.12	−4.08	−4.11
MP2[c]	−4.18	−5.87	−8.19	−8.39	−8.76
HF+ALL	−7.00	−5.94	−7.14	−7.41	−7.46
B3LYP+ALL	−5.94	−5.26	−7.12	−7.37	−7.47
LC-BOP+ALL	−8.71	−7.79	−9.80	−10.13	−10.25
GCst: Reference value CBS(RI-MP2): −12.00 kcal/mol CBS(T): −10.60 kcal/mol					
HF	−0.50	−0.10	−0.41	−0.44	−0.44
B3LYP	−0.87	−0.46	−1.14	−1.18	−1.18
LC-BOP	−3.37	−2.85	−3.76	−3.81	−3.85
MP2[c]	−4.64	−6.79	−8.80	−9.22	−9.63
HF+ALL	−6.05	−5.65	−6.67	−7.14	−7.25
B3LYP+ALL	−6.40	−6.00	−7.65	−8.01	−8.16
LC-BOP+ALL	−8.89	−8.38	−10.15	−10.61	−10.65
GGst: Reference value CBS(RI-MP2): −14.80 kcal/mol CBS(T): −12.67 kcal/mol					
HF	0.17	0.65	0.02	−0.04	−0.11
B3LYP	0.05	0.41	−0.64	−0.67	−0.73
LC-BOP	−3.10	−2.50	−3.84	−3.87	−3.98
MP2[c]	−5.56	−8.19	−11.04	−11.48	−11.95
HF+ALL	−6.68	−6.17	−8.15	−8.15	−8.38
B3LYP+ALL	−6.76	−6.37	−9.09	−9.09	−9.13
LC-BOP+ALL	−9.90	−9.27	−12.10	−12.26	−12.33

[a]DG and DG++ mean Pople-type 6-31G and 6-31++G, respectively
[b]TG++ denotes 6-311++G
[c]We used frozen-core (FC) MP2 methods

well reproduces the results of reference values (see GCst in Table 25.1 especially). Note that the stacking energy obtained by LC-BOP+ALL is more negative than that of MP2 in the case of Pople-type basis sets. In general, MP2 tends to overestimate the vdW interaction. In this benchmark, MP2 considerably underestimate the stacking energy compared to the reference values (see the reference value of CBS (RI-MP2) in Table 25.1) because we used small basis sets.

We next discuss the dependence of basis function. The stacking energy without considering the effects of diffuse function is far from the reference value. Moreover, basis function 6-31G (d, p) gives the worst results in all basis sets. It may be important to consider the balance between polarization and diffuse in computing the stacking interaction with DFT+ALL method. Little change is found between 6-31++G (d, p) and 6-311++G (d, p) for all methods. This result indicates whether the basis function is double zeta or triple zeta has almost nothing to do with the stacking energy. The energy difference is not changed considerably between 6-311++G (d, p) and 6-311++G (df, pd) in all methods. Generally speaking, on

Table 25.2 Benchmark of 2
base pairs stacking interaction
with LC-BOP/6-31++G(d,p)
level calculation (kcal/mol)

	E^{int}	$E^{corrected}$	Ref[a]
GC 0/3.25	3.7	−13.2	−15.8
CG 0/3.19	−0.4	−17.0	−17.3
GG 0/3.36	5.1	−10.6	−11.2
GA 10/3.15	5.7	−10.9	−12.9
AG 8/3.19	3.3	−13.0	−12.5
TG 0/3.19	1.5	−14.8	−15.1
GT 10/3.15	5.8	−11.1	−13.4
AT 10/3.26	4.6	−11.6	−13.3
TA 8/3.16	3.7	−12.5	−12.8
AA 0/3.24	4.5	−11.9	13.1
AA 20/3.05	4.1	−13.6	−14.7

[a]The reference value is taken from ref [37]

the other hand, MP2 method depends strongly on the size of the basis functions in estimating the $\pi - \pi$ stacking. To compare $E^{corrected}$ with the E^{int}, the value of E^{ALL} does not strongly depend on the functional. These facts mean that the DFT+ALL method is advantageous over the other post HF methods such as MP2, since the present method is applicable to larger systems, which cannot be calculated by the post SCF methods.

Benchmark for 2 Base Pairs

In 2006, Šponer and his co-worker suggested the stacking energy between base pairs in 11 cases [37]. Their results are also based on the high level *ab inito* computation so that the reference value is reliable for benchmark.

We estimated the stacking energy between the CG pair and the GC pair (or two GC pairs), where the structure is taken from the higher layer of the optimized four base pairs model. The stacking energy E^{int} is estimated by

$$E^{int} = E^{tot} - (E_{B1B'1} + E_{B2B'2}) + E^{CP}, \qquad (25.3)$$

where E^{tot} is the total energy of the whole structure and $E_{B1B'1}$ and $E_{B2B'2}$ denote the energies of the base pair 1 and the base pair 2 (e.g., the CG or GC pair). Table 25.2 lists the results in the level of LC-BOP+ALL/6-31++G (d, p) and reference value. The notations of molecule are the same as the reference 9. We performed the single point calculation (which is available in the supporting information of their paper). The difference between computed value and the reference value is not more than 2.5 kcal/mol in every case. In the case of CG 0/3.19 and TA 8/3.16, the value of $E^{corrected}$ agrees well with the reference value. Judging from these results, we again found that LC-BOP+ALL has the validity of investigating the interaction energy between base pairs.

25.2.2 Summary of Sect. 25.2

LC-BOP+ALL method gives the stacking interaction, which agrees well with the reference value obtained by accurate methods in both cases for stacking two bases and staking two base pairs. We have been able to reproduce the reference value with a medium-sized basis set such as 6-31++G(d,p). Our method is expected to describe weak interactions such as vdW interaction and hydrogen bonding in molecules of a larger scale with reasonable accuracy, because the computational costs required for the DFT with the ALL functional correction are much lower than those for the post-Hartree Fock methods.

25.3 Structure and Properties of Metal-Containing Artificial DNA

25.3.1 Cu-Containing Artificial DNA

Recently, an artificial DNA, which consists of a DNA-like base sugar, and phosphoric acids, has taken much attention [38]. The artificial DNA has a selectivity of metal ions, which are captured by DNA-like bases, so that the metal ions could be arrayed hierarchically. In fact, many artificial DNAs with various metal ions have been synthesized and reported [39–41]. In 2004, Tanaka succeeded in arraying five [**H**–Cu(II)–**H**] (**H**: hydroxypyridone) into a DNA duplex [42,43]. Although details of the structure by NMR or X-ray experiments have not yet been available, they revealed that the distance between the copper ions are 3.7 ± 0.1Å with electron paramagnetic resonance (EPR) experiments at 1.5 K. Since the available structural information by the experiments is limited, the computational chemistry might contribute to get a deep understanding of the structure of the artificial DNA.

In this artificial DNA, it is obvious that metal-ligand interaction (in the case of [**H**–Cu(II)–**H**], the interaction between Cu(II) and O^-) takes over the hydrogen bonding interaction between two **H**s as an intra-base pair interaction. However, what kind of interaction works as the inter-base pair interaction to form a stable structure has not been clear. In spite of the repulsive Coulomb interaction among the copper ions and ligands, the copper containing artificial DNA is as stable as the natural B-DNA. Some experimental groups suggested that the spin-spin interactions among the metal ions should play an important role in the inter-base pair interaction. Since the order of spin-spin interaction is usually too small in comparison with a chemical bond, it is natural to assume that the inter-base pair interaction in the artificial DNA is the same as that in the natural B-DNA. Nevertheless, this possibility has not been explored for the artificial DNA yet.

Here we treat the two stacked [**H**–Cu(II)–**H**] in order to investigate a possible stable structure of them and an origin of the EPR signal.

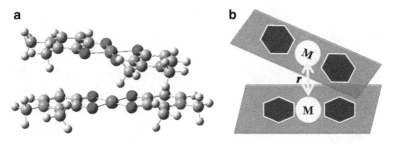

Fig. 25.1 (**a**) [H–Cu(II)–H] dimer used for calculation (**b**) Model of (**a**), where M represents metal cation and r means the distance between copper ions

25.3.1.1   Computational Details

Here we considered cis-type of Cu(II) coordinating hydroxypiridones referred as [H–Cu(II)–H] in the following. In the actual calculation, we replace the backbone molecules (deoxyribose-5′-phosphate (dP)) by hydrogen atoms or methyl groups for simplicity, and then we optimized the geometry of one artificial base pair [H–Cu(II)–H], which have planar structure like the base pairs in the natural B-DNA.

We next evaluated the stacking energy between the base pairs. It is difficult to optimize geometries of two base pairs, because no stable structure of two base pairs is found due to a lack of dispersion forces when one adopts the ordinary DFT and because it costs too much to compute gradients by means of the DFT with the vdW correction. Therefore, in order to get the potential energy surface, we performed the single point calculations by using the DFT with the vdW correction by fixing the distance between copper ions. In the calculation of two base pairs of [H–Cu(II)–H], the upper base is vertically located and twisted by 36° like the bases in the natural DNA base pair (see Fig. 25.1). In order to estimate the interaction energy, we adopted the same method explained in previous section. Basis sets used here are 6-311+G (d) for the copper atom and 6-31++G (d, p) for the other atoms. Throughout this section, we used a modified version of the GAUSSIAN03 program package [36].

25.3.1.2   Verification of Several Models

In order to confirm the appropriateness of the models for the artificial DNA, we first verify the model dependence of the inter-base pair interaction. We adopt three models, (a) a hydrogen atom model (dP=H) and (b) a methyl group model (dP=CH₃) (c) real model (dP=5′-deoxyribose) (Fig. 25.2). These models are often used in the analyses of the inter-base pair interaction between the bases in the natural B-DNA, where it is known that the backbone molecules do not contribute much to the stability of the duplex structure. We here also make the same assumption for the artificial DNA.

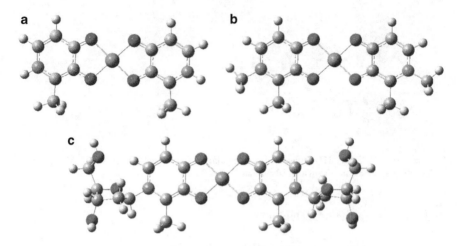

Fig. 25.2 The optimized structures of [**H**–Cu(II)–**H**], where (**a**) dP=H (hydrogen atom), (**b**) dP=CH$_3$ (methyl group) and (**c**) dP=backbone (5′-deoxyribose), respectively

Table 25.3 Important structural parameters taken from optimized geometries of [**H**-Cu(II)-**H**] and NBO charge and Mulliken spin on Cu atom	dP = H	dP=CH$_3$	dP=backbone
r(Cu–O)	1.94	1.94	1.94
r(CuO$^-$)	1.90	1.91	1.90
θ (O–Cu–O$^-$)	85.0	85.0	84.9
r(N–N)	9.86	9.90	9.91
α (NBO charge)	0.368	0.368	0.368
β (NBO charge)	1.066	1.066	1.066
Total	1.434	1.434	1.434
Mulliken Spin of Cu(II)	0.683	0.682	0.679

In order to understand the model dependency, we optimized one base pair with the backbone molecule replaced by some models. In this paragraph, we focused on the interaction between the base pairs so that the geometrical parameters of the Cu(II) and its surrounding oxygen atoms play an important role. Therefore, it is necessary to understand the geometry around them. Table 25.3 lists the optimized geometries in each model. There are not so remarkable changes in all models except the distance between nitrogen atoms. Table 25.3 also lists the natural bond orbital (NBO) charge for each spin and the total Mulliken spin of copper atom. Again no significant change is observed in all models. From these results, it is found that property and geometry around Cu atom do not depend on the model (the kind of dP) in 1 base pair. Therefore we used the hydrogen model and methyl model in further researches.

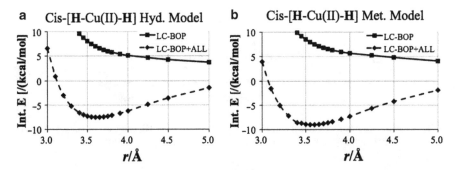

Fig. 25.3 The interaction energy between (**a**) hydrogen atom model (dP=H) and (**b**) methyl group model (dP=CH$_3$). In the figure, Int E of LC-BOP is obtained by E_{int} and LC-BOP+ALL is obtained by E_{int}^{total}

25.3.1.3   Interaction Energy Between Two Metal-Containing Base Pairs

Figure 25.3 shows the interaction energy between the hydrogen atom model and the methyl group model. Without the vdW correction, the interaction energy decreases as r becomes larger. The interaction seems to be repulsive over whole region mainly due to the repulsive Coulomb interaction between the bases, and no local minimum is found. With the vdW correction, on the other hand, the interaction energy becomes negative and [**H–Cu(II)–H**] planes attract each other. Moreover, the local minimum of the interaction energy was found around to be 3.55–3.65 Å. Thus, it is necessary to include the effect of the vdW interaction to form the DNA-like structure. Next, we discuss the difference of the stacking energy of (a) hydrogen atom model (dP=H) and (b) the methyl group model (dP=CH$_3$). The stacking energy of the methyl group model is 1.5–2 kcal/mol larger than that of the hydrogen atom model. The E^{vdW} of the methyl group model is larger than that of the hydrogen atom model due to the pseudo π electron delocalization in the methyl group model.

In 2002, Tanaka and his coworkers synthesized a base pair, [**H–H**], which does not have Cu(II) [44]. Here we compared [**H–H**] with the trans-[**H–Cu(II)–H**] in relation to the stacking energy. Hereafter we omit trans for simplicity. [**H–H**] is described as Scheme 25.2. We here adopt the methyl group model (dP=CH$_3$). [**H–H**] is stable because of the two hydrogen bonds similar to those in the adenine-thymine (AT) base pair. Therefore, we here consider [**H–H**] dimer as the analog of the AT base pair dimer. Šponer computed interaction energy between base pairs with high level *ab initio* calculation such as CCSD (T) [34, 37]. It is difficult to evaluate the stacking energy of the present system by the same quality, because the system of interest is too large. In order to check the reliability of the present method and understand how much is the error caused by constraint of whole structures, we evaluate the stacking energy of two AT pairs by using the LC-BOP+ALL method. We used the same structure of two AT pairs (which is available in the supporting information of their paper), where the r value fixed at 3.60 Å. Table 25.4 summarizes the results of the stacking energy. LC-BOP+ALL method agrees with the result of

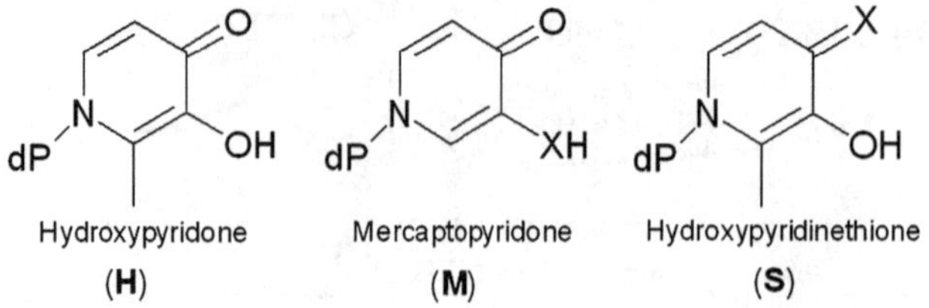

Hydroxypyridone   Mercaptopyridone   Hydroxypyridinethione
(**H**)   (**M**)   (**S**)

Scheme 25.2 Chemical structural formula of hydroxypyridone (**H**), mercaptopyridone (**M**) and hydroxypyridinethione (**S**). In reference [49], charlcogen atom X is set to sulfur

Table 25.4 Stacking energy at $r = 3.60$Å (in kcal/mol)

	E^{int}	E^{vdW}	E_{int}^{total}
[**H** –Cu(II)– **H**] ($r = 3.60$ Å), dP $= CH_3$	7.4	−16.6	−9.2
[H–H] ($r = 3.60$ Å), dP $= CH_3$	7.7	−16.9	−9.3
2 AT base pairs	4.1	−17.6	−13.6[a]

[a]The reference value is −14.7 kcal/mol

CCSD(T) for the two AT pairs. The stacking energy of [**H–H**] is smaller than that of the natural B-DNA by 4.3 kcal/mol. This energy difference originates from E^{int}, which can be explained by [**H–Cu(II)–H**], is not optimized. On the other hand, the dispersion correction E^{vdW} is almost same in all cases. Judging from these results, we have revealed that the chemical origin of the structural stability of [**H–Cu(II)–H**] is similar to that of [**H–H**] and of course that of the AT pair, which is mainly due to the vdW interaction in spite of the proposal by Tanaka et al.

25.3.1.4 Singlet-Triplet Energy Gap

According to the results of Tanaka et al., the spin state among Cu(II) ions is ferromagnetic. This fact is the ground of the existence of the spin-spin coupling. Due to the spin-spin coupling, the spin state of the [**H–Cu(II)–H**] dimer should be the triplet, which is observed in the EPR experiment. Although we have assumed that the spin state is the triplet in this study so far, the singlet and triplet states are very close to each other as shown below. Figure 25.4 shows the difference between singlet and triplet from the view of spin density and electron density. Judging from the spin density, we found that the spin concentrates on Cu(II) and its surroundings. In this case, singlet state also has spin density, which the spin of copper cation in one [**H–Cu(II)–H**] is opposite to that of the other [**H–Cu(II)–H**]. On the other hand, no change is found in electron densities as shown in Fig. 25.4.

We discuss here the energy gap between singlet and triplet with open shell DFT with the vdW correction and possibility of the spin state observed inthe actual

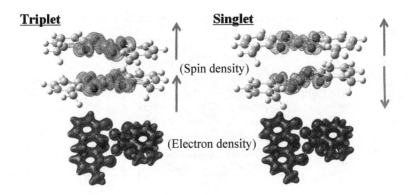

Fig. 25.4 The difference between triplet state and singlet state

experiments. According to the J-model proposed by Yamaguchi and his co-workers [45], the spin-spin interaction coupling term J_{ab} can be estimated by following formula,

$$J_{ab} = \frac{E_{LS} - E_{HS}}{<S^2>_{HS} - <S^2>_{LS}} \tag{25.4}$$

In this case, the whole system is anti-ferromagnetic when $J_{ab} < 0$ and is ferromagnetic when $J_{ab} > 0$. This model reproduces well the experimental data [46].

Table 25.5 lists the energies of the singlet and triplet states for [**H–Cu(II)–H**]. From the table, the energy differences between the triplet and singlet states are no more than $k_B T = 2\,(K)$, and spin-spin interaction are about $-1\,cm^{-1}$ in every case. The vdW correction does not change these tendencies so that it is difficult to find the remarkable singlet-triplet energy gap. The modeling of the backbone molecule (dP) is not a problem, because the backbone molecules are too far from Cu(II) to affect the spin state. These tendencies do not change with the other theoretical method (e.g. Hartee-Fock, MP2 or with other basis sets). Since the singlet-triplet energy gap is quite small, the singlet and triplet states should mix due to thermal excitations. If we assume the Boltzmann distribution of two spin states, there exists the triplet state for [**H–Cu(II)–H**] dimer even at 1.5 K, for example the distribution of the singlet and triplet species is about 7:3 for $k_B T = 1.35\,(K)$, whose triplet state signal may be observed in the actual experiment. Thus the present calculation does not contradict to the experimental evidence and we propose that singlet state exist as well as triplet state. Our researches have revealed that spin of copper atom in [**H–Cu(II)–H**] exists independently because the distance between copper atoms is too large to interact each other [47]. The same is true in the case of [(salen base)–Cu(II)–(salen base)] [48, 49]. More modification would be needed in order to let artificial DNA have magnetic properties.

Table 25.5 The energy difference between singlet and triplet and spin-spin coupling of [H–Cu(II)–H]

		Singlet (a.u.)	Triplet (a.u.)	Difference (K)	J_{ab} (cm^{-1})
Hyd/Gas	LC-BOP	−5025.600 450	−5025.600 445	1.65	−1.7
	E^{vdW}	−0.023 097	−0.023 097	0.00	0.0
Met/Gas	LC-BOP	−5182.316 655	−5182.316 650	1.70	−1.6
	E^{vdW}	−0.026 314	−0.026 314	0.00	0.0
Hyd/PCM	LC-BOP	−5025.709 254	−5025.709 250	1.40	−1.5
	E^{vdW}	−0.021 998	−0.021 998	0.00	0.0
Met/PCM	LC-BOP	−5182.398 660	−5182.398 656	1.35	−1.5
	E^{vdW}	−0.025 882	−0.025 882	0.00	0.0

25.3.2 Chalcogen Substitution for Metal-Containing Artificial DNA

As discussed before, the spin-spin interaction is not important for stacking of [**H**–Cu(II)–**H**]. This indicates low-spin metal such as Ni(II) and Pd(II) can be a candidate for array of metal ions in DNA duplex. The metal which has low ionization tendency and takes square-planar configuration is seemed to be applicable for arrayed metal-ion in DNA duplex. In 1999, Tanaka and his co-workers suggested artificial DNA, which contains palladium complex and platinum complex [38]. In 2008, Takezawa reported newly artificial DNA, which allows soft transition metal cation such as Ni(II), Pd(II), and Pt(II) to array in DNA duplex with a programmable manner [50]. In their paper, nickel and palladium cations were captured by mercaptopyridone (**M**) and platinum cation was captured by hydroxypyridinethione (**S**). The chemical structural formula of **H**, **M**, and **S** are shown in Scheme 25.2. These captions are explained by hard-soft acid base (HSAB) rule. It is interesting to investigate the chalcogen atom dependencies. If one uses selenium atom, pK_a becomes much smaller than one uses sulfur atom. The major topic of this section is metal-selectivity and chalcogen atom dependency from the view of quantum chemistry.

25.3.2.1 Computational Details and Model Compounds

Hereafter, we refer M(II) as metal cation and X as chalcogen atom. We simplified reactivity by considering only the energy difference between reactant and product. We adopted methyl model (dP=CH$_3$). X=S is synthesized in experiment in the case of [**M**–Pd(II)–**M**], [**M**–Ni(II)–**M**] and [**S**–Pt(II)–**S**]. We then examined X=Se in order to understand the dependency of chalcogen atoms. In this section, we used Becke 3 hybrid functional. As a basis function, we chose Def2-QZVP [51] for metal atoms because only this basis set has a polarization and diffuse function and is available for Pd(II) and Pt(II) atoms. We used 6-31++G (d, p) for the other atoms. We adopted polarizable continuum model (PCM) method for two reasons. One is

Table 25.6 Important structural parameters taken from optimized geometries in metal cation containing artificial DNA

		H (X=O)	**M** (X=S)	**M** (X=Se)	**S** (X=S)	**S** (X=Se)
Ni(II)	r(Ni–O)	1.88	1.88	1.88	1.88	1.88
	r(Ni–X)	1.88	2.25	2.36	2.24	2.35
	θ(X–Ni–O)	87.4	89.0	89.2	88.9	89.0
	r(N–N)	9.96	10.54	10.68	10.31	10.39
Cu(II)	r(Cu–O)	1.94	1.97	1.97	1.95	1.96
	r(Cu–X)	1.97	2.32	2.44	2.33	2.44
	θ(X–Cu–O)	85.2	87.2	87.5	86.9	87.2
	r(N–N)	10.10	10.67	10.82	10.45	10.54
Zn(II)	r(Zn–O)	2.01	2.12	2.13	2.06	2.07
	r(Zn–X)	2.07	2.36	2.46	2.40	2.50
	θ(X–Zn–O)	82.7	85.1	85.5	84.8	85.3
	r(N–N)	10.26	10.85	11.00	10.63	10.74
Pd(II)	r(Pd–O)	2.02	2.02	2.02	2.02	2.02
	r(Pd–X)	2.02	2.35	2.46	2.35	2.45
	θ(X–Pd–O)	83.4	86.0	86.5	85.8	86.3
	r(N–N)	10.23	10.76	10.91	10.56	10.64
Pt(II)	r(Pt–O)	2.03	2.03	2.03	2.03	2.03
	r(Pt–X)	2.03	2.35	2.45	2.34	2.44
	θ(X–Pt–O)	82.5	85.7	86.4	85.5	86.2
	r(N–N)	10.27	10.77	10.91	10.57	10.65

r denotes distance (in Å) and θ denotes the bond-angle (in degree)

that this reaction occurs in aqueous phase. The other is that a metal cation cannot be stable in gas phase. If we computed energy difference between reactant and product in gas phase, we could obtain even [**H**–Zn(II)–**H**] which cannot be stable in fact.

25.3.2.2 Metal Cations and Chalcogen Substitution

We chose Cu(II), Ni(II), Pd(II), and Pt(II) cations as a candidate metal for artificial DNA. As a reference we picked up Zn(II) cation, which cannot take square planar coordination actually. Table 25.6 lists the optimized geometries in each molecule. The distance M-O is not so changed by chalcogen in all metals. Especially in zinc cation, the distance is estimated larger than that in copper or nickel.

It indicates that M=Zn(II) is not stable in these structures The order of distance between nitrogen atoms is **M** (X=Se), **M** (X=S), **S** (X=Se), **S** (X=S) and **H**. Except for Zn(II), the largest value is 10.91 Å in [**M**–Pt(II)–**M**]. This distance is larger than that of natural DNA so that bond length between nitrogen atoms may affect the whole structure.

To estimate formation energy, we assumed the reaction 2**H** (or **M**,**S**) + M(II) → [**H**–M(II)–**H**]+2H$^+$. We compared the reaction energy as following formula.

$$G = G(\mathbf{H} - M(II) - \mathbf{H}) + 2G_p - \{2G(\mathbf{H}) + G(M(II))\} \qquad (25.5)$$

Table 25.7 Gibbs free energy (in 298.15 K) comparison of metal-cation containing artificial DNA (in kcal/mol)

	Ni(II)	Cu(II)	Zn(II)	Pd(II)	Pt(II)
H(X = O)	−19.6(○)	−61.1(∗)	66.9(×)	−2.3(▲)	23.5(×)
M(X = S)	−27.4(○)	−68.4(○)	62.0(×)	−22.0(∗)	0.5(▲)
M(X = Se)	−48.0(○)	−96.3(○)	40.2(×)	−43.7(○)	−22.0(○)
S(X = S)	−44.6(∗)	−92.0(○)	45.8(×)	−41.6(∗)	−18.2(∗)
S(X = Se)	−54.2(○)	−101.0(○)	34.2(×)	−50.4(○)	−23.8(○)

Marks in parentheses indicates the speculated possibility for the formation of metal ion containing artificial DNA (○ denotes "Supposed to be available", ▲ denotes "Depends on its condition" and × denotes "Supposed to be unavailable", respectively. On the other hand ∗ denotes experimentally synthetized one)

where G_p denotes the energy of proton (H^+). This reaction occurs in aqueous solution so that it is natural to consider two protons are dissociated as H_3O^+ ion. Therefore, we estimated G_p as the energy difference between H_2O and H_3O^+.

$$G_p = G(H_3O^+) - G(H_2O) \qquad (25.6)$$

Note here the reaction energy G is just an index parameter for stability of complex. In order to consider the stability in solution, we have to consider Gibbs energy. In principle, Gibbs energy can be obtained from a statistical partition function within a harmonic approximation, where one needs to compute vibrational frequencies. According to the Florian's work [25], dynamical effects adopting an explicit water model should be included to compute the solvation Gibbs energy in aqueous solution accurately, whereas our computation does not include the internal entropies in each compound. In this study, we assumed that the errors of Gibbs free energies would be cancelled out. Since the results agreed well with the experimental facts shown later, this assumption seems to be reasonable.

Table 25.7 list results of formation free energy. As expected, all the complexes that consist of Zn(II) are not stable. It is obvious that Zn(II) cannot make the artificial DNA in square planar configuration. The complexes, which contain Cu(II), are so stable that any complex would be synthesized. The difference between Ni(II) and Pd(II) is not so remarkable except for the case in X=O. Therefore, the tendency of metal mediate base pairing is not so changed whether the metal cation is Ni(II) or Pd(II). Table 25.7 also summarizes speculated possibilities of metal mediated base pairing for artificial DNA. Similarly, [**H**–Pt(II)–**H**] is not so stable that these structures cannot be synthesized, which can be understood by HSAB theory. The other structures are stable enough to be a candidate for metal-mediated base pairing. This selection rule of chalcogen atom is important in synthesizing complex especially in platinum atom. In particular, if selenium atom can be used, [**M**–Pt(II)–**M**] would be available.

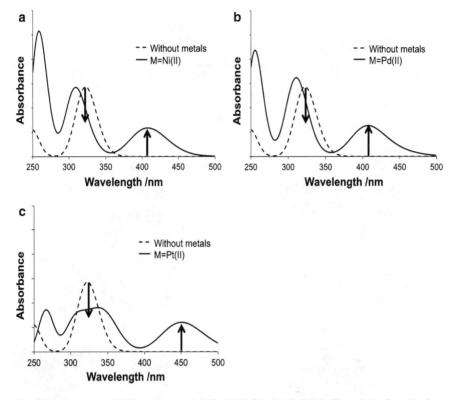

Fig. 25.5 Computed UV-Vis spectra of (**a**) [S–Ni(II)–S], (**b**) [S–Pd(II)–S], and (**c**) [S–Pt(II)–S]

25.3.2.3   UV-Vis Spectra of Newly Developed Artificial DNA

In order to confirm the existence of metal-mediated pairing, we have to compute the
UV-Vis spectra, whose experimental data are available. In this section, we computed
UV-Vis spectra by means of time-dependent (TD) DFT within implicit solvent by
using PCM. In describing the UV-Vis spectra, we let the peak have a Gaussian
type function with half-width 0.2 eV. Figure 25.5 shows both the experimental and
theoretical results. Like the experimental data, the decrease of peak around 350 nm.
From the Fig. 25.5b, the peak around 450 nm was observed, which corresponds
to the peak around 415 nm in Fig. 25.5a. Judging from Table 25.8, it is true
that we have specified the character of peak but the wavelength of the peak is
not discussed quantitatively in [S–Pt(II)–S]. There are two possibilities of error
in the wavelength of peak in [S–Pt(II)–S]; (1) defect of electron core potential
(ECP) (2) not considering relativistic effect. These effects will be considered in
further studies.

Figure 25.6 shows the molecular orbital, which mainly concerns with the excitation with large oscillator strength. Excitation of [S–2H$^+$–S] which does not have
metal cation is dominated by $\pi - \pi^*$ transition. In the case of [S–M(II)–S] (M=Ni,

Table 25.8 Absorbance wavelength ω of computed peak, which is concerned with the $\pi - \pi^*$ or d-π^* excitation in [S–M(II)–S] (in nm), and a difference in excitation energy between Se and S, i.e. $\Delta\omega = (\omega_{Se} - \omega_S)$

	Without metals	Ni(II)	Pd(II)	Pt(II)
X=S	352	400 (402)	407 (397)	456 (413)
X=Se	367	416	422	471
$\Delta\omega$	15	16	15	15

Numbers in parenthesis are experimental data

Without metals

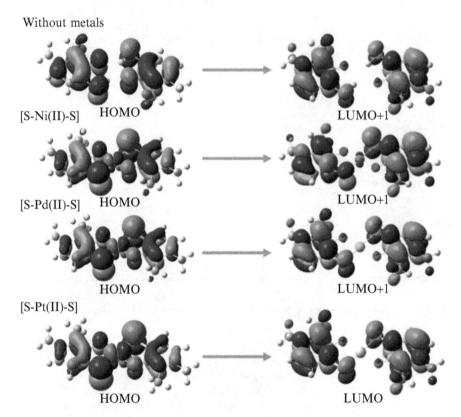

Fig. 25.6 Molecular orbitals concerned with the $\pi - \pi^*$ and d $- \pi^*$ excitation in each complex. [S–S] denotes hydroxypyridinethione dimer without metal cation

Pd, Pt), the same characters were found in the molecular orbital in **S**. On the other hand, the d function in the metal cation in HOMO almost disappears in the excited state (LUMO+1 or LUMO) so that d $- \pi^*$ transition identified as metal-to-ligand charge transfer (MLCT) was found in [S-M(II)-S] (M=Ni, Pd, Pt). Judging from the results of Table 25.8, the red shift from 350 nm in [S-2H$^+$-S] to 400–450 nm in [S–M(II)–S] can be explained by the effect of MLCT. The same tendency was found in the case of selenium. Especially, $4p$ functions of Se atom concerns with the $\pi - \pi^*$ transition so that a systematic red shift by about 15 nm was caused by replacement of $3p$ orbitals by $4p$ orbitals in substituting the charlcogen atoms. This change does not depend on nether the sort of the metal ion nor existence of the metalion.

25.3.3 Summary of Sect. 25.3

In order to investigate an origin of structural stability of a Cu(II)-containing artificial DNA, we evaluated the stacking energy between [**H–Cu(II)–H**] (**H**: hydroxypyridone) dimer by means of DFT+ALL method. The calculated distance between two Cu(II) is about 3.6 Å, which agrees well with the experimental data. Evaluated stacking energy is about 8–10 kcal/mol, which is slightly smaller than that of two base pairs in a natural B-DNA. This tendency does not change in [**H–H**], which does not contain Cu(II). These results indicate that the vdW interaction dominates the inter-base-pair interaction over spin-spin interaction, in contrast to a conjecture by an experimental group. According to the results by the open-shell DFT, antiferromagnetic (singlet) and ferromagnetic (triplet) states are almost degenerated when the two bases are vertically located and both bases have a planar structure as found in the B-DNA.

We also evaluated the stability and UV-vis spectra of [**S–M(II)–S**] (M=Ni, Pd, and Pt) (**S**: hydroxypyridinethione) using (TD) DFT. We calculated the formation energies of modified bases with possible combinations of chalcogen atoms and metal cations. The results confirmed that [**H–Ni(II)–H**] (**H**: hydroxypyridone), [**H–Cu(II)–H**], and [**S–Cu(II)–S**] would form stable metal-base pairing; on the other hand, [**H–Zn(II)–H**], [**S–Zn(II)–S**], and [**H–Pt(II)–H**] would not. We predicted UV-vis excitations at 400–410 nm, mainly dominated by a d-π^* transition accompanied by a $\pi - \pi^*$ transition in [**S–M(II)–S**], where a metal-to-ligand charge transfer shifts the peak of **S** (without metal cations).

25.4 Conductivity of Metal-Containing Artificial DNA

Chemical modification of DNA base pairing is one of the strategies to conductive DNA. It is well known that mutation of DNA such as base mismatch pair affects its conductivity [52]. Therefore, it is expected that proton-transfer reaction between base pairs also concerns with the conductivity. Rak's group investigated the influence of proton transfer to electron coupling in the two base pairs [53, 54]. On the other hand, the I-V characterization has not yet been investigated in these systems. The other strategy for conductive DNA is metal complex binding to DNA.

Originated from Mujica et al. [55], many theoretical studies have been reported for electronic transport properties in molecular device for benzenedithiol and DNA [56–58]. On the other hand, quantitative discussion for the conductance of molecular device needs much computational resources as well as quantum chemistry.

Here we investigate the conductivity of the artificial DNAs and compare the results with those of the natural DNAs by means of he scattering Green's function theory. In the following we briefly summarize the theory and apply it to the model DNA systems.

25.4.1 Theoretical Background

The molecular system supposed to be considered contains too many atoms to treat with quantitative manner. To understand the $I - V$ characteristics in theoretical chemistry is needed for conductivities of larger molecules. In 2002, Luo et al. succeeded in computing the current and conductivity of molecular wire such as benzendithiol with simple model based on theoretical chemistry [24]. This method uses the overlap matrix of model molecules and sulfur atoms and the probabilities that electron exists in the sulfur atoms in LUMO. This is also applicable for larger molecule such as DNA. Here we review the essence of the method and apply it to the natural and artificial DNAs.

25.4.1.1 Theoretical Background for I-V Characteristics Calculation

We summarized theoretical description for I-V curve of molecular junction proposed by Luo et al. Based on the scattering Green's function theory, matrix element T is expressed by a function of energy E_i (energy at which scattering process is observed) as following.

$$T(E_i) = \gamma_{1s}\gamma_{DN} \sum_{\eta} \frac{<1|\eta><\eta|N>}{Z - \varepsilon_n} \qquad (25.7)$$

In the above equation, a parameter z is complex variable $z = E_i + i\Gamma_i$. Therefore, absolute value of $T(E_i)$ is

$$|T(E_i)|^2 = \gamma_{1s}^2\gamma_{DN}^2 \sum_{\eta} \frac{|<1|\eta>|^2|<\eta|N>|^2}{(E_i - \varepsilon_n)^2 + \Gamma_{\eta}^2}, \qquad (25.8)$$

where $< 1|\eta >$, $< \eta|N >$ is the overlap between sulfur atom and extended molecule in each orbital. ε_{η} is the orbital energy. Here we restrict to use the oribitals from HOMO-9 to LUMO+9. γ_{1S}, γ_{DN} represents coupling element. Γ_{η} denotes escape rate given by Fermi's golden rule, i.e.,

$$\Gamma_{\eta} = \gamma_{1S}<1|\eta> + \gamma_{DN}<\eta|N> \qquad (25.9)$$

Coupling constant γ_{1S} and γ_{DN} is obtained by

$$\gamma_{1S} = V_{SL}d_{1L}$$

$$\gamma_{DN} = V_{SL}d_{NL} \qquad (25.10)$$

L refers to LUMO. The interaction energy V_{SL} is approximated as

$$V_{SL}^2 = (\Delta E_{HL} - \Delta E_{SL}^0)\Delta E_{SL}^0/2, \qquad (25.11)$$

where ΔE_{HL} is HOMO-LUMO gap of extended molecule, ΔE_{SL}^0 the difference between LUMO of bare molecule and HOMO of gold cluster. $d_{i,S}$ means the expansion coefficients of wave function of the S atoms, i.e.

$$d_{1L}^2 = \sum_i c_{1i}^2 \Big/ \sum_{a,i} c_{ai}^2. \qquad (25.12)$$

Using the linear response theory, i_{SD} is obtained by

$$i_{SD} = \frac{emk_BT}{2\pi^2\hbar^3} \sum_\eta \int_{eV_D}^{\infty} dE_q|T(E_q)|_\eta^2$$

$$\times \left[\ln\left\{ 1 + \exp\left(\frac{E_f + eV_D - E_q}{k_BT}\right) \right\} - \ln\left\{ 1 + \exp\left(\frac{E_f - E_q}{k_BT}\right) \right\} \right] \qquad (25.13)$$

E_f is Fermi energy of the system. We set the Fermi energy as the middle of HOMO-LUMO. T means temperature of system. To obtain the total current, we have to consider area of the contact. To summarize the results, total current I_{SD} is obtained by following factor.

$$I_{SD} = i_{SD} \times \pi \left(\frac{9\pi\hbar^3}{4}\right) \cdot \frac{1}{2mE_f} \qquad (25.14)$$

In this model calculation, the current depends strongly on the overlap between sulfur atoms around frontier orbitals and the coupling constant in LUMO.

25.4.2   I-V Characteristics for Natural and Artificial DNA Bases

We assumed that sulfur atom connects to hollow site of (111) orientation of gold clusters. Therefore, an extended molecule includes three gold atoms on each side. Geometries of bare molecule (in this case, base pair+sulfur atoms) were optimized by B3LYP. The basis set is chosen as LanL2DZ for metal atoms and 6-31 G (d) in geometry optimization and 6-31+G (d) in single point calculation for other atoms. Throughout this section, we used GAUSSIAN 03 program package [36].

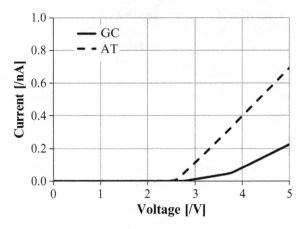

Fig. 25.7 Computed I-V characteristics of (**a**) AT and (**b**) GC pair monomer

Table 25.9 Distance between N_9 (A or G) and N_1 (T or C), coupling constant and site-orbital orverlap matrix elements of each molecule, where E_η is the orbital energy (in eV)

	AT			GC		
N_9-N_1 (Å)	8.96			9.04		
E_f (eV)	4.61			4.60		
Γ_{1S} (eV)	0.006			0.003		
γ_{DN} (eV)	0.075			0.134		
	E_η	$<1|\eta>$	$<\eta|N>$	E_η	$<1|\eta>$	$<\eta|N>$
HOMO−3	−7.15	0.017	0.895	−7.19	0.783	0.016
HOMO−2	−6.66	0.082	0.024	−6.05	0.049	0.005
HOMO−1	−5.43	0.133	−0.033	−5.51	0.000	0.443
HOMO	−5.41	−0.009	−0.449	−5.28	0.123	0.001
LUMO	−3.81	0.000	0.012	−3.91	0.000	0.011
LUMO+1	−3.76	0.082	−0.002	−3.70	0.002	0.105
LUMO+2	−3.60	0.197	−0.001	−3.62	0.091	0.000
LUMO+3	−3.57	0.003	−0.112	−3.47	0.211	0.003
LUMO+4	−2.13	0.023	0.967	−2.12	0.005	0.980
LUMO+5	−2.01	0.889	0.051	−1.99	0.000	0.125

25.4.2.1 In Plane I-V Characteristics of One Base Pair

Computed currents of a single base pair are shown in Fig. 25.7. In all cases, the current is not more than 1.0 nA. Table 25.9 lists the coupling constant and site-orbital matrix elements. Around Fermi energy, either $<1|\eta>$ or $<\eta|N>$ is almost 0 so that the transparent matrix element is also approximated to zero. Coupling constants are so small that these molecules cannot be conductive. In qualitatively, hydrogen bonding is not effective for electron carrier.

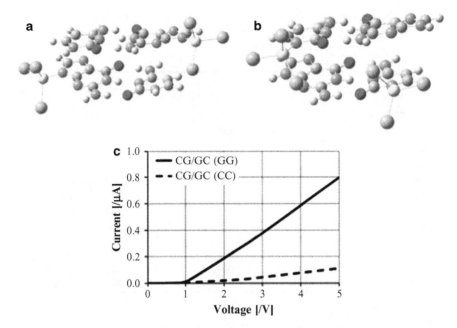

Fig. 25.8 The molecular modeling for I-V curve of CG/GC, where (**a**) electrodes connect to N_9 of guanine (**b**) electrodes connect to N_1 of cytosine. (**c**) I-V characteristics for system (**a**) and (**b**)

It is said that guanine has the lowest ionization potential among four bases. Therefore, guanine is the most unlikely to be negative (electron-carrier) so that the current of GC pair is less than that of AT pair. Through a proton transfer reaction, the current in 5 V changed slightly due to the change of hydrogen bonding length (see Table 25.9), which leads to the change of distance between N_9 (G or A) and N_1 (C or T).

25.4.2.2   Stacking Two Base Pairs

Next, we consider stacked GC pairs. We consider two patterns for gold electrode connection. In the former, the electrode is connected to N_9 of guanine. On the other hand, in the latter, the electrode is connected to N_1 of cytosine as shown in Fig. 25.8c also show the results for current-voltage characteristics of each molecule. High conductivity is observed in the case of 2 base pairs. If the electron pass contains $\pi - \pi$ stacking, the current increases greatly. According to the experimental results, 1,4-benzene-dithiol is estimated to be 0.3 µA when voltage is set to 5 V [59]. This implies that these molecules are more conductive than 1,4-benzene-dithiol. The current of CG/GC for the electrode that connects to N_9 of guanine is larger than that to N_1 of cytosine. For example Table 25.10 lists the coupling constant and site-orbital orverlap matrix elements of each molecule. When electrodes connect to

Table 25.10 Coupling constant and site-orbital orverlap matrix elements of each molecule, where E_η is the orbital energy (in eV)

	Connected to cytosine					
	CG/GC			C^+G^-/G^-C^+		
E_f(eV)	4.77			4.51		
γ_{1S} (eV)	0.078			0.026		
γ_{DN}(eV)	0.133			0.131		
	E_η	$\langle 1\|\eta\rangle$	$\langle\eta\|N\rangle$	E_η	$\langle 1\|\eta\rangle$	$\langle\eta\|N\rangle$
HOMO−3	−6.11	0.001	0.029	−5.96	0.008	0.003
HOMO−2	−6.09	0.009	0.016	−5.93	0.001	0.000
HOMO−1	−5.61	0.471	0.018	−5.34	0.040	0.156
HOMO	−5.61	0.086	0.100	−5.26	0.247	0.033
LUMO	−3.94	0.065	0.015	−3.75	0.130	0.416
LUMO+1	−3.94	0.005	0.170	−3.69	0.703	0.108
LUMO+2	−3.79	0.043	0.012	−3.53	0.090	0.228
LUMO+3	−3.79	0.052	0.009	−3.47	0.356	0.003
LUMO+4	−2.03	0.892	0.006	−2.10	0.068	0.909
LUMO+5	−2.02	0.917	0.009	−2.03	0.242	0.188

guanine side, the apparent difference between CG/GC and C^+G^-/G^-C^+ appears in coupling constant. Especially, the coupling constant γ_{1L} of CG/GC is ten times larger than that of C^+G^-/G^-C^+, which leads to 100 times difference in transition matrix element.

When electrodes connect to cytosine side, on the other hand, both coupling constant and the overlap around Fermi energy have large value so that the currents in 5 V are large in both CG/GC and C^+G^-/G^-C^+. In qualitatively, the G- of C^+G^-/G^-C^+ cannot take electron because guanine has already had a negative charge.

To summarize these discussions, DNA is conductive if the electrodes are directly connected to bases. Moreover the current greatly depends on which atom is connected to the electrode.

25.4.2.3 Metal-Containing Artificial DNA Bases

Next, we applied this model to the artificial DNA. We have already computed [**H–Cu(II)–H**]. In this section, we examined the I-V curve characteristic of [**H–M(II)–H**] (M=Ni, Pd) dimer in order to understand the difference between [**H–H**] (without metal cation) and [**H–M(II)–H**]. we set the distance between metal cations to 3.70 Å as the experimental data in [**H–Cu(II)–H**].

Figure 25.9 and Table 25.11 show the results for the artificial DNAs. The current in 5 V of [**H–H**] is so small that artificial DNA works as insurant without metal cations because γ_{1S} is almost 0. On the other hand, [**H–M(II)–H**] shows the conductive properties. The current, which begins to increase rapidly, appears

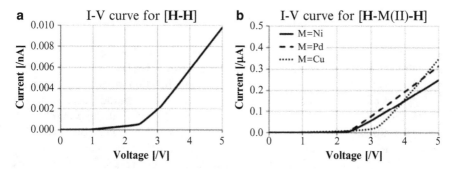

Fig. 25.9 I-V curve of (**a**) stacked [**H–H**] (in nA) and (**b**) stacked [**H–M(II)–H**] (M=Ni(II), Cu(II), and Pd(II)) (in μA)

Table 25.11 Coupling constant and site-orbital overlap matrix elements of each molecule, where E_η is the orbital energy (in eV)

	[H–H]			[H–Ni(II)–H]			[H–Pd(II)–H]		
E_f (eV)	4.47			4.43			4.43		
γ_{1S} (eV)	6.09×10^{-6}			0.020			0.024		
γ_{DN} (eV)	0.035			0.024			0.027		
	E_η	$<1|\eta>$	$<\eta|N>$	E_η	$<1|\eta>$	$<\eta|N>$	E_η	$<1|\eta>$	$<\eta|N>$
HOMO−3	−5.41	0.241	0.017	−5.34	0.165	0.018	−5.32	0.197	0.012
HOMO−2	−5.40	0.056	0.073	−5.31	0.168	0.022	−5.28	0.206	0.051
HOMO−1	−5.13	0.635	0.029	−5.20	0.268	0.108	−5.20	0.239	0.103
HOMO	−5.09	0.583	0.045	−5.14	0.265	0.141	−5.13	0.240	0.138
LUMO	−3.85	0.001	0.129	−3.72	0.069	0.036	−3.73	0.071	0.038
LUMO+1	−3.68	0.231	0.004	−3.72	0.073	0.036	−3.73	0.076	0.039
LUMO+2	−3.66	0.028	0.080	−3.56	0.131	0.021	−3.57	0.128	0.013
LUMO+3	−3.54	0.224	0.010	−3.54	0.126	0.017	−3.55	0.125	0.008
LUMO+4	−2.23	0.013	0.977	−2.06	0.132	0.817	−2.08	0.136	0.816
LUMO+5	−2.01	0.998	0.006	−2.05	0.128	0.821	−2.07	0.139	0.819

around 1.5 V for palladium, around 2.0 V for nickel, which can be explained by HOMO-LUMO gap. As a result, the current in 5 V is estimated as 0.25 μA and 0.31 μA for [**H–Ni(II)–H**] and [**H–Pd(II)–H**], respectively. Although this value is less than that of CG/GC (2 base pairs of natural DNA), [**H–M(II)–H**] does not depend on the connection to the electrodes. Our results clearly indicates that the metal coordination much enhances the intra base current in comparison with [**H–H**] (without metal cation) and natural bases as shown in previous paragraph. Therefore the metal-containing artificial DNA is one of good candidates for good conductor.

25.4.3 Summary of Sect. 25.4

We investigate the current-voltage $(I - V)$ characteristics between adjacent bases in both natural and artificial DNAs, i.e. hydroxypyridone (**H**) with Cu(II), Ni(II), and Pd(II) ions, using an elastic scattering Green's function method together with a density functional theory. We have found that the magnitude of the current of [**H**–M(II)–**H**] complex tends to become larger than that of the natural DNA and the artificial DNA without metal ions. Natural DNA cannot be conductive only in one base pair, but conductive in more than two base pairs due to the effect of $\pi - \pi$ stacking. The current of natural DNA changes whether electrodes connect to guanine or cytosine. In the case of artificial DNA, the current increases dramatically by the metal mediation. Especially, artificial DNA does not depend on the connection so that we will be able to obtain stable and high current. Therefore, the artificial DNA will be used for newly conductive nano-material.

25.5 Conclusions

We have studied two topics about (i) the structural stabilities and electronic structures of metal-ion containing artificial DNA bases and (ii) conductivity of them. In the Sect. 25.2, before proceeding the main topics, we have shown that the LC-BOP+ALL method gives the stacking interaction, which agrees well with the reference value obtained by accurate methods in both cases for stacking two bases and two base pairs. We have been able to reproduce the reference value with a medium-sized basis set such as 6-31++G(d,p). This fact is quite important for the investigation of newly developed artificial DNA, whose structures were not available.

In Sect. 25.3, we investigated an origin of structural stability of the Cu(II)-containing artificial DNA. We first confirmed appropriateness of our model system and evaluated the stacking energy between [**H**–Cu(II)–**H**] (**H:** hydroxypyridone) dimer by means of DFT+ALL. The calculated distance between two Cu(II) is about 3.6 Å, which agrees well with the experimental data. Evaluated stacking energy is about 8–10 kcal/mol, which is slight smaller than that of two base pairs in a natural B-DNA. This tendency does not change in [**H**–**H**]. These results indicate that the vdW interaction dominates the inter-base pair interaction over spin-spin interaction in contrast to a conjecture by an experimental group. According to the results by the open shell DFT, antiferromagnetic (singlet) and ferromagnetic (triplet) states are almost degenerated when the two bases are vertically located and both bases have a planer structure as found in the B-DNA. We also considered the metal ion and chalcogen substitution effects on the hydroxypridone artificial DNA bases. Our calculation results were in good agreement with the experimental facts and suggested a new variety of artificial DNA base, which contains Se atom.

In Sect. 25.4, we computed current-voltage characteristics of stacking natural and artificial DNA bases by the scattering theory based on the NEGF. The current is not observed if the electron path contains only hydrogen bonding such as 1 base pair. The current becomes larger if the electron path contains $\pi - \pi$ stacking molecule. On the other hand, the current changes dramatically whether the electrode connects to guanine or cytosine. In the case of artificial DNA, remarkable change for electron conductivity will be observed in the metal-mediated base pairing.

This review would help deep understanding of properties of several metal-containing artificial DNAs, i.e. structural properties, electronic and magnetic behavior, the UV-Vis spectra from the view of theoretical chemistry. According to the result of computed UV-Vis spectrum, artificial DNA can really exist because computed UV-Vis spectra agree well with the observed spectra. However, metal-containing artificial DNA has many challenges in present. For example, the magnetic properties are not available because high and low spin states are degenerating in room temperature. We expect salen base pair with manganese cation proposed by Carell et al. would be a good molecular magnetic. The electron conductivity would increase if we could find other artificial DNA, which has strong base pair interaction. The review also explored a possibility of the conductivity of metal-DNA complex. Especially, [**H**–**Cu(II)**–**H**] is observed in triplet states so that we will have to make it possible to apply NEGF theory to open shell systems. Both directions are needed for further understanding of nano-material science.

Acknowledgments This research was supported by the Core Research for Evolutional Science and Technology (CREST) Program High "Performance Computing for Multi-Scale and Multi-Physics Phenomena" of the Japan Science and Technology Agency (JST) and Grant-in-Aid for Young Scientists (A) (No. 22685003).

References

1. Jurečka P, Hobza P (2004) J Am Chem Soc 125:15608
2. Šponer J, Jurečka P, Hobza P (2004) J Am Chem Soc 126:10142
3. Braun E, Eichen Y, Sivan U, Ben-Yoseph G (1998) Nature 391:775
4. Porath D, Berzryadin A, de Vries S, Dekker C (2000) Nature 403:635
5. Fink H-W, Schönenberger C (1999) Nature 398:407
6. Yu Kasumov A, Kociak M, Gueron S, Reulet B, Volkov VT, Klinov DV, Bouchiat H (2001) Science 291:280
7. Giese B, Spichty M (2000) ChemPhysChem 1:195
8. Giese, B.; Wessely, S. Chem. Commun. 2001, 2108.
9. Zilberberg, I.L.; Avdeev G.M.; Zhidomirov, G.M. J. Mol. Struct. (Theochem) 1997, 418, 73.
10. Hush NS, Schamberger J, Bacskay GB (2005) Coord Chem Rev 249:299
11. Lee JS, Latimer LJP, Reid RS (1993) Biochem Cell Biol 71:162
12. Abrescia NGA, Malinina L, Fernandez LG, Huynh-Dinh T, Neidle S, Subirana JA (1999) Nucl Acids Res 27:1593
13. Baeyens KJ, DeBondt H, Pardi A, Holbrook SR (1996) Proc Natl Acad Sci 93:12851
14. Kankia BI (2000) Nucl Acids Res 28:911
15. Robinson H, Gao Y, Sabusgvill R, Joachimiak A, Wang AHJ (2000) Nucl Acids Res 28:1760
16. Miyake Y, Togashi H, Tashiro M, Yamaguchi H, Oda S, Kudo M, Tanaka Y, Kondo Y, Sawa R, Fujimoto T, Machinami T, Ono A (2006) J Am Chem Soc 128:2172

17. Iikura H, Tsuneda T, Yanai T, Hirao K (2001) J Chem Phys 115:3540
18. Andersson Y, Langreth DC, Lundqvist BI (1996) Phys Rev Lett 76:102
19. Kamiya M, Tsuneda T, Hirao K (2002) J Chem Phys 117:6010
20. Matsui T, Sato T, Shigeta Y, Hirao K (2009) Chem Phys Lett 478:238
21. Matsui T, Miyachi H, Sato T, Shigeta Y, Hirao K (2008) J Phys Chem B 112:16960
22. Matsui T, Miyachi H, Nakanishi Y, Shigeta Y, Sato T, Kitagawa Y, Okumura M, Hirao K (2009) J Phys Chem B 113:12790
23. Nakanishi Y, Matsui T, Shigeta Y, Kitagawa Y, Saito T, Kataoka Y, Kawakami T, Okumura M, Yamaguchi K (2010) Int J Quantum Chem 110:2221
24. Wang C-W, Fu Y, Luo Y (2001) Phys Chem Chem Phys 3:5021
25. Florian J, Goodman MF, Warshel A (2000) J Phys Chem B 104:10092
26. Elstner M, Hobza P, Frauenheim T, Suhai S, Kanxiras E (2001) J Chem Phys 114:5149
27. Zhao Y, Truhlar DG (2005) Phys Chem Chem Phys 7:2701
28. Antony J, Grimme S (2006) Phys Chem Chem Phys 8:5287
29. Becke AD, Johnson ER (2005) J Chem Phys 123:154101
30. Kamiya M, Tsuneda T, Hirao K (2002) J Phys Chem 117:6010
31. Sato T, Tsuneda T, Hirao K (2005) J Phys Chem 123:104307
32. Sato, T.; Tsuneda, T.; Hirao, K. J. Phys. Chem. 2007, 126, 234114.
33. Song J-W, Hirosawa T, Tsuneda T, Hirao K (2007) J Chem Phys 126:154105
34. Jurečka P, Šponer J, Černy J, Hobza P (2006) Phys Chem Chem Phys 8:1985
35. Boys SF, Bernardi F (1970) Mol Phys 19:553
36. Frisch, M. J. et al., Gaussian 03, Revision D.02, Gaussian, Inc., Wallingford CT, USA, 2004
37. Šponer J, Jurečka P, Marchan I, Luque EJ, Orozco M, Hobza P (2006) Chem Eur J 12:2854
38. Tanaka K, Shionoya M (1999) J Org Chem 64:5002
39. Meggers E, Holland PL, Tolman WB, Romesberg FE, Schultz PG (2000) J Am Chem Soc 122:10714
40. Switzer C, Sinha S, Kim PH, Heuberger BD (2005) Angew Chem Int Ed 44:1529
41. Tanaka K, Yamada Y, Shionoya M (2002) J Am Chem Soc 124:8802
42. Tanaka K, Tengeiji A, Kato T, Toyama N, Shionoya M (2003) Science 299:1212
43. Tanaka K, Shionoya M (2006) Chem Lett 35:694
44. Tanaka K, Tengeiji A, Kato T, Toyama N, Shiro M, Shionoya M (2002) J Am Chem Soc 124:12494
45. Yamaguchi K, Takahara Y, Fueno T (1986) In: Smith VH Jr., Schaefer HF III, Morokuma, K (eds) Applied quantum chemistry. Raidel, Boston, p. 155
46. Yamaguchi K, Okumura M, Maki J, Noro T, Namimoto H, Nakano M, Fueno T, Nakasuji K (1992) Chem Phys Lett 190:353
47. Nakanishi Y, Kitagawa Y, Shigeta Y, Saito T, Matsui T, Miyachi H, Kawakami T, Okumura M, Yamaguchi K (2009) Polyhedron 28:1714
48. Nakanishi Y, Kitagawa Y, Shigeta Y, Saito T, Matsui T, Miyachi H, Kawakami T, Okumura M, Yamaguchi K (2009) Polyhedron 28:1945
49. Clever GH, Reitmeier SJ, Carell T, Schiemann O (2010) Angew Chem Int Ed 49:4927
50. Takezawa Y, Tanaka K, Yori M, Tashiro S, Shiro M, Shionoya M (2008) J Org Chem 73:6092
51. Weigend F, Ahlrichs R (2005) Phys Chem Chem Phys 7:3297
52. Natsume T, Dedachi K, Tanaka S, Higuchi T, Kurita N (2005) Chem Phys Lett 408:381
53. Rak J, Makowska J, Voityuk AA (2006) Chem Phys 325:567
54. Sadowska-Sleksiejew A, Rak J, Voityuk AA (2006) Chem Phys Lett 429:546
55. Mujica V, Kemp M, Ratner MA (1994) J Chem Phys 101:6849
56. Asai Y (2003) J Phys Chem B 107:4647
57. Shimazaki T, Maruyama H, Asai Y, Yamashita K (2005) J Chem Phys 123:164111
58. Shimazaki T, Xue Y, Ratner MA, Yamashita K (2006) J Chem Phys 126:114708
59. Reed MA, Zhou C, Muller CJ, Burgin TP, Tour JM (1997) Science 298:252

Chapter 26
Systematic Derivation and Testing of AMBER Force Field Parameters for Fatty Ethers from Quantum Mechanical Calculations

M. Velinova, Y. Tsoneva, Ph. Shushkov, A. Ivanova, and A. Tadjer

Abstract Nontoxic drug delivery systems for efficient trans-membrane transport are central in the successful therapy of a number of diseases. Appropriate building blocks of reversible drug-carrying micelles are water-soluble surfactants, e.g. pentaethylene glycol monododecyl ether ($C_{12}E_5$). The present study aims to derive from first principles calculations and to test molecular mechanics parameters for such ethers to be used in subsequent all-atom simulations of micelle formation. Two monomers and one dimer with two different types of periphery, which are short-chain prototypes of the amphiphilic surfactant $C_{12}E_5$, are used as model systems. The geometry of low-energy conformers is obtained from conformational analysis with a modified OPLS force field and optimized at PBE and MP2 levels, with aug-cc-pVTZ basis sets in vacuum and in implicit solvent. The quantum-chemical calculations provide detailed information on the structural flexibility of the surfactant models and can be used as reference for MD simulations. Weak dependence of the parameters sought on the length of the oligomers and higher sensitivity to the type of periphery is found. Validation of the derived molecular mechanics parameters is carried out through comparison of the density, molecular volume, enthalpy of solvation and vaporization obtained from molecular dynamics simulations (Amber99/NPT/300 K) of diethyl ether to the existing experimental data. The two theoretical approaches yield similar results both at molecular level and as secondary thermodynamic output. Moreover, the derived set of molecular mechanics parameters is consistent with experiment and can be used for extensive molecular dynamics simulations of larger C_xE_y surfactant assemblies.

M. Velinova • Y. Tsoneva • Ph. Shushkov • A. Ivanova • A. Tadjer (✉)
Faculty of Chemistry, University of Sofia, 1 James Bourchier Ave., 1164 Sofia, Bulgaria
e-mail: maria.velinova@chem.uni-sofia.bg; yana_tsoneva@mail.bg; philip.shushkov@yale.edu; aivanova@chem.uni-sofia.bg; tadjer@chem.uni-sofia.bg

P.E. Hoggan et al. (eds.), *Advances in the Theory of Quantum Systems in Chemistry and Physics*, Progress in Theoretical Chemistry and Physics 22, DOI 10.1007/978-94-007-2076-3_26, © Springer Science+Business Media B.V. 2012

26.1 Introduction

The cell membrane is largely resistant to most current chemotherapeutics and to all alien proteins and DNA; therefore, use of the endocytic pathway for intracellular transfer seems unfeasible. The design of reliable, effective and risk-free artificial systems for transport across the cell membrane will pave the way to new pharmaceutical approaches for treatment of presently incurable diseases. Suitable candidates for such systems are micelles formed by ethers of oligo (ethyleneglycols) and fatty alcohols (C_xE_y) with controllable hydrophilic-hydrophobic balance. There is evidence that at very low concentrations the surfactant molecules preserve the self-assembly aptitude, especially in the subsurfacial layer [1]. This property, together with the lipid compatibility of oligo(ethyleneglycol) ethers, suggests that C_xE_y systems could be used as molecular reversible nanotransporters across the biomembranes.

The well-known nonionic surfactant $C_{12}E_5$ (Fig. 26.1) is used as a model drug delivery system in many studies. Balogh and Pedersen [2] have investigated with small-angle X-ray scattering the effect of adding a drug (lidocaine) to such a system. The surfactant system has been studied previously, amongst others, by Olsson and co-workers [3]. Therefore, SANS, NMR and light scattering data for drug-free $C_{12}E_5$ micelles are now available. In order to understand how the $C_{12}E_5$ micelles hold drugs, transport them in the bloodstream and release them in the target tissues, it is very important to clarify the structure of the micelles. Recently, molecular dynamics (MD) simulations have emerged as a powerful tool for investigating the static and dynamic structure of micelles. In order to apply them to C_xE_y systems, we need a molecular mechanics (MM) model of $C_{12}E_5$, which can be derived from *ab initio* estimates of the electronic and structural characteristics of smaller model systems allowing extraction of force field parameters for the target surfactant. The main problem is the parameterization of the head which contains ethylene glycol units. Also, poly (ethylene glycol), PEG, is one of the most widely used polymers in biotechnology and industry related to life-sciences [4]. The accurate knowledge of the structure and thermodynamic properties of PEG-based materials is of immense importance for most of the existing applications and, even more,

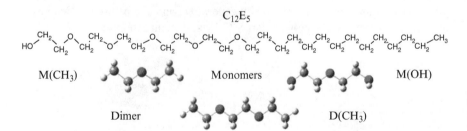

Fig. 26.1 Structural chemical formula of a C_xE_y fatty ether (*top*) and the models studied (*bottom*) with their notations used in the text

for the development of new ones. Ethylene glycol and PEG-containing molecules exhibit highly complex inter- and intra-molecular interactions, since they comprise both non-polar (i.e. $-CH_2-$) and polar (i.e. $-O-$) groups. Furthermore, the lowest-energy conformers have at least one torsion angle ($-O-C-C-O-$) corresponding to a *gauche* conformation, unlike many other polymer molecules where the *all-trans* conformation is that of lowest energy. This gauche-effect is also present in PEG oligomers [5].

Considerable effort has been devoted to the development of atomistic force fields that account accurately for such phenomena. Smith et al. developed a force field for PEG based on detailed *ab initio* calculations and used subsequently in molecular dynamics simulations of $HO-(CH_2CH_2O)_{12}-H$ [6]. Neyertz et al. [7] studied the crystalline region of poly (ethylene oxide) using MD simulations, while Lin et al. [8] employed MD to study the structural and dynamical properties of the amorphous regions of poly (ethylene oxide). Again Smith and co-workers developed a quantum-chemistry-based (MP2) force field for 1,2-dimethoxyethane (DME) and PEO in aqueous solution [9, 10] using several water models [11] They concluded that for dilute (water-rich) solutions, static and dynamic properties depend only weakly on the water model employed. Recently [12], they reported a MD study of the influence of hydrogen bonding and polar interactions on hydration and conformations of a PEO oligomer ($H-[CH_2-O-CH_2]_{12}-H$) and 1, 2-dimethoxyethane in dilute aqueous solutions.

Despite the extensive study of PEO and PEG, relatively little work has been done so far on its oligomers. This study focuses on PEG oligomers and derivation of force field parameters for them to be applied to systems such as $C_{12}E_5$.

26.2 Models and Computational Procedure

The present study focuses on the theoretical treatment of the monomers of poly (ethyleneoxide) (PEO) and poly (ethylene glycol) (PEG) with two types of periphery: CH_3-terminated $(M(CH_3))$ and OH-capped $(M(OH))$. A dimer with alkyl periphery is also modeled. These short-chained oligomers are regarded as minimalistic prototypes of the hydrophilic portion of the water-soluble amphiphilic surfactant $C_{12}E_5$. The selected molecular models are chosen in order to aid the elucidation of the dependence of the ether fragment MM parameters on the length and periphery of the oligomer chain. The effect of a polar medium (water) is investigated as another factor influencing the ether molecules behaviour in solution.

The molecular simulations were performed according to the computational protocol illustrated in Fig. 26.2. In order to find the possible stable conformers of the monomers and the dimer, exhaustive conformational analysis with simultaneous variation of all torsion angles between non-hydrogen atoms was carried out employing a version of the OPLS force field [13] with modified ether group parameters [14]. The conformational search was performed both in vacuum and in explicit aqueous medium using the TIP3P water model [15]. The accumulated

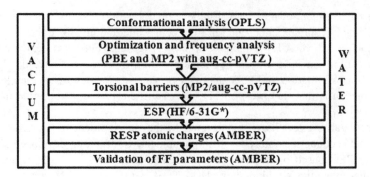

Fig. 26.2 Computational scheme followed in the derivation and validation of force field parameters

sets of conformations were examined and all conformers differing substantially (by more than 15°) in any of the torsion angles underwent geometry optimization. The latter was done with the gradient-corrected DFT functional PBE [16] and with the MP2 method [17] using the basis set aug-cc-pVTZ [18] in both approaches. The selected basis set is acknowledged as sufficiently large to ensure high precision of a variety of molecular characteristics, such as geometry, vibrational frequencies, and relative stability of the conformers. The implicit solvent model PCM [19] was applied whenever the geometry was optimized in aqueous medium. Energy minima of all optimized structures were checked by frequency analysis.

The torsional barriers for rotation about C–O bonds were quantified solely from MP2 calculations, since the accurate estimate of such energy differences requires explicit account of the electron correlation.

The electrostatic potential of the optimized molecules was generated with HF/6-31G* as recommended by the developers of AMBER95 [20] and used further on for obtaining the RESP charges. The latter were generated by multi-conformational fitting [20]. All calculations were done both in vacuum and water.

The force field parameters for simulation of amphiphilic fatty oligoethers obtained from the quantum mechanics calculations were tested by comparing a computed set of thermodynamic properties with available experimental data. In the current study the validity of the parameters was demonstrated with atomistic MD simulations of diethyl ether $(M(CH_3))$ in gas and liquid phase and in aqueous solution. The gas phase was represented by one molecule $M(CH_3)$ in vacuum. The liquid was modeled with 144 molecules of diethyl ether randomly distributed in a cubic periodic box. The dimensions of the box were chosen to ensure mass density of the liquid close to the experimental value at the same temperature – $0.714 \, \text{g/cm}^3$. The solution comprised one $M(CH_3)$ molecule immersed in a sufficient amount of water. Both the liquid and the solution were simulated by placing the repeating units in periodic boundary conditions. All simulations were performed with the force field parm 99 [21] appended with the derived MM parameters for the ether fragment in NPT ensemble at 300 K and 1 atm for the periodic calculations. The Berendsen method [22] was applied for the temperature and pressure couplings. MD trajectories were run utilizing the minimized structure as a starting input and the particle mesh Ewald (PME) [23] algorithm was used for evaluating the long-range electrostatic terms. The van der Waals interactions were truncated at a spherical cutoff of 1.0 nm (5.0 nm for the isolated molecule in vacuum) and compensated with long-range corrections. The time-step was 2 fs, and the simulation was 5 ns long. The equilibration and data production periods were generally greater than 1 ns and 4 ns, respectively. To prove that the stage of thermodynamic equilibrium was reached, the total energy, temperature, and pressure were calculated during the simulation (Fig. S1). The MD data revealed that these three characteristics oscillated moderately around a constant mean value, indicating that our samplings were statistically significant.

The water models used were TIP3P and TIP4P [15]. The molecular volume, mass density of the pure liquid, enthalpy of vaporization and enthalpy of solvation of diethyl ether were calculated to validate the force field parameter set. The trajectory was sampled by extracting frames every 200 fs for statistical analysis. The mean value variations were quantified by means of standard deviations.

The conformational search was performed with the module implemented in Hyperchem 7 [24]; the geometry optimization, frequency analysis, and torsional barriers scan were done with the program suite Gaussian 09 [25]; the RESP atomic charges were evaluated with the corresponding subroutine in the package Amber 8 [26]. The MD simulations were carried out with Amber 8.

26.3 Conformational Analysis

The number of structurally distinct low-energy conformations obtained from the conformational analysis and the number of the different structures after the subsequent geometry optimization with PBE/aug-cc-pVTZ in vacuum and in water are summarized in Table 26.1. The optimization with the MP2 method yields the same number of dissimilar conformers as the minimization with the DFT functional. Results from the conformational analysis reveal that both the periphery type and the presence of polar solvent influence the conformational flexibility of the molecules. The hydroxyl periphery allows more freedom of rotation about the skeletal C–O bonds. As expected, the molecules terminated by hydrophobic alkyl groups possess

Table 26.1 Number of structurally differing low-energy conformers found by the conformational search with OPLS and number of the dissimilar isomers after their geometry optimization with PBE/aug-cc-pVTZ in vacuum and water

Molecule	Vacuum		Water	
	OPLS	PBE	OPLS	PBE
$M(CH_3)$	4	3	3	1
$M(OH)$	35	22	53	38
$D(CH_3)$	55	26	36	10

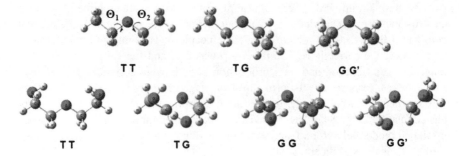

Fig. 26.3 Lowest-energy structures after PBE/aug-cc-pVTZ optimization of M(CH$_3$) and M(OH) in vacuum. *T* and *G* denote the *trans-* and *gauche*-conformations with regard to the two ether bonds, respectively. The symbols *G* and *G'* correspond to positions of the methyl/hydroxyl groups on the same or on opposite sides of the C–O–C plane

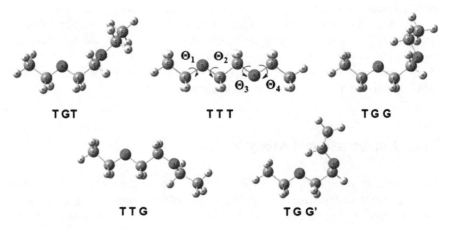

Fig. 26.4 Lowest-energy structures obtained from PBE/aug-cc-pVTZ geometry optimization of D(CH$_3$) in vacuum

more stable isomers in vacuum. In contrast, the set of stable structures with hydroxyl periphery becomes much larger in water.

Geometry optimization reduces the number of isomers found during conformational analysis but nevertheless a sizeable conformational diversity is retained, particularly for the dimer and the OH-terminated monomer. As in the conformational analysis, more optimized structures with alkyl periphery are stable in vacuum, while for the monomer with hydroxyl periphery the number of conformers stable in water is greater.

Comparing the structures obtained after energy minimization of the investigated systems with those found by the conformational search, one observes certain differences in the geometry (Figs. 26.3 and 26.4). The trans-conformation of the ether fragment is preferred both in the conformational analysis and after the geometry optimization for the monomers in vacuum. This finding is supported by several

Table 26.2 Values of the torsion angles C–O–C–C (Θ_1, Θ_2), their population and energy difference (ΔE [kcal/mol]) between the lowest-energy conformer and the respective structure in the same series found after PBE/aug-cc-pVTZ geometry optimization in vacuum and water of the E_x monomers; populations derived from analysis of gas-phase electron diffraction experiments [32]

		Vacuum						Water					
		$M(CH_3)$			$M(OH)$			$M(CH_3)$			$M(OH)$		
Conformer	Population	ΔE	Θ_1	Θ_2	ΔE	Θ_1	Θ_2	ΔE	Θ_1	Θ_2	ΔE	Θ_1	Θ_2
TT	0.69	0	180	180	0	180	180	–	–	–	0	180	180
TG	0.26	1.466	176	73	1.250	174	68	0	174	74	0.082	174	63
GG'	0.5	2.838	63	−63	3.359	−82	73	–	–	–	1.371	−62	70
GG	–	–	–	–	2.079	72	63	–	–	–	0.422	64	66

studies of diethyl ether, M(CH3), with IR [27–29] and Raman [30,31] spectroscopy. The normal coordinate analysis of the vibrational spectrum of M(CH3) shows that only the TT conformer exists in the solid state [27], whereas in gas and liquid phase at least two more conformers (TG and GG) are registered [27, 28, 31]. Figure 26.3 presents the lowest-energy M(CH3) conformers after PBE/aug-cc-pVTZ optimization in vacuum, which are in excellent agreement with the experimentally observed structures.

The results in Table 26.2 reveal that the TT conformer with dihedral angles C–O–C–C of 180° is the most stable, followed by the TG one. The least stable structure is GG', where the two methylene groups are in antiperiplanar alignment with respect to the ether group plane. Based on IR absorption intensity temperature dependence, Wieser et al. [28] estimated the enthalpy difference between TT and TG as 1.1 kcal/mol. This agrees well with the result obtained in this study.

The fourth experimentally established structure GG is not found in the simulations of M(CH3), whereas for M(OH) it turns out to be even stabler than GG'. It can be anticipated that the stability of this isomer is rooted in the possibility of intramolecular hydrogen bond formation between a hydrogen atom from one of the hydroxyl groups with an oxygen atom from the other.

In implicit aqueous solvent we find only one conformer for M(CH3), which features just one *gauche*-orientation. In contrast, M(OH) retains the conformational diversity, which most probably is due to various feasible interactions of the hydrophilic hydroxyl periphery with the polar water environment.

The energetically favorable isomers described above, however, do not exhaust the possible spatial orientations of studied molecules. In all cases the conformational assortment accumulated in the conformational search contracts upon geometry optimization but sizeable conformational variety is preserved. This is particularly prominent for M(OH) in water, where about 40 structures fall within an energy window of ~6 kcal/mol, and for D(CH3) in vacuum, for which about 30 isomers differ in energy by less than 3 kcal/mol. (Table S1).

Switching from DFT to MP2 has no effect on the conformational preference of the optimized molecules and their order of stability (Tables S2, S3, S4 and S5). Consequently, the two methods reproduce the conformational behavior for the molecules studied with negligible numerical differences in estimated parameters.

Table 26.3 Values of the torsion angles C–O–C–C (Θ_1 to Θ_4), their population and energy difference (ΔE [kcal/mol]) between the lowest-energy conformer and the respective structure in the same series found after PBE/aug-cc-pVTZ geometry optimization of the dimer of E_x in vacuum and water; populations derived from analysis of gas-phase electron diffraction experiments [32]

Conformer	Population	D(CH$_3$) vacuum					D(CH$_3$) water				
		ΔE	Θ_1	Θ_2	Θ_3	Θ_4	ΔE	Θ_1	Θ_2	Θ_3	Θ_4
TGT	0.23	0	179	84	176	172	0	177	79	173	165
TTT	0.13	0.178	180	180	180	180	0.671	179	177	179	176
TGG'	0.08	0.591	174	85	−75	−62	1.103	173	44	−74	−61
TGG	0.53	1.371	172	88	75	178	1.648	174	89	77	176
TTG	0.03	1.387	179	172	73	174	1.860	177	173	83	172

The geometry optimization modifies all the low-energy conformers of the dimer (Tables S4 and S5). Both inner and peripheral torsion angles change and even in vacuum there is already one *gauche*-orientation in the most stable conformer (Fig. 26.4). D(CH$_3$) has two important torsion angles C-C-O-C and O-C-C-O, which are present in the target oligomer too, and the correct parametrization of these torsions is vital for reliable simulation of C_xE_y. The conformers energies and populations of the dimer in gas and liquid phase and in solution were studied in detail experimentally [32–36] and theoretically [9, 11, 12, 37, 38] in order to gain deeper understanding of the behavior of such molecules. The experiment provides evidence that five principle conformers dominate in gas and liquid phase– TTT, TGT, TTG, TGG and TGG'. Our calculations match well this result. The nomenclature used by the experimentalists for the angles C–O–C–C, O–C–C–O and C–C–O–C is kept here as well but the torsions Θ_1, Θ_2, Θ_3 and Θ_4 discussed by us do not include O–C–C–O, since it has been parametrized already in the force field AMBER95. In Fig. 26.4 (Table S4) are shown the lowest-energy conformers obtained after PBE/aug-cc-pVTZ (MP2) optimization of D(CH$_3$) in vacuum. It is visible that the dimer exhibits marked *gauche*-effect, as the *gauche* conformation has unusually high population (Table 26.3), which has been registered experimentally [39], too. This class of molecules is a typical example of the *gauche*-effect – the polar substituents give preference to the *gauche*- rather than to the *trans*-conformation of the C–C bond. In ethers and polyethers this behavior is termed oxygen *gauche*-effect.

A comparison between the geometry of the dimers optimized in vacuum and in water makes clear that the number of *gauche*-kinks in water increases.

The analysis of solution experiments is complicated by the fact that the *gauche* conformer population is enhanced by interactions with strongly polar solvents such as water. Andersson and Karlstromg [40] estimate that water stabilizes the TGT versus the TTT conformer by 1.0–1.5 kcal/mol. Inomata and Abelo [41] used recently gas-phase NMR measurements as a source of structural information on D(CH$_3$). They reproduced the gas-phase NMR data using a RIS (Rotational Isomeric State) model where the TGT conformation is preferred over the TTT conformation by 0.4 kcal/mol. This minor energy difference requires a very careful

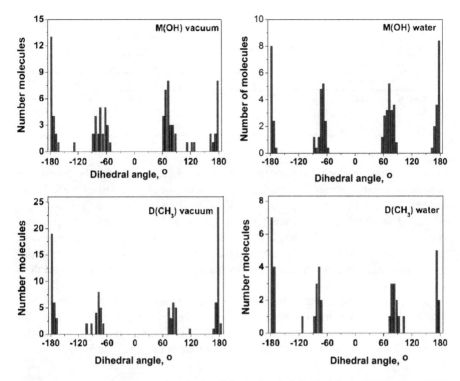

Fig. 26.5 Distribution of the torsion angles C–C–O–C and C–O–C–C in the PBE/aug-cc-pVTZ optimized D(CH$_3$) and M(OH) structures in vacuum (*left*) and water (*right*)

choice of computational protocol. Both methods employed in this study reproduce the experimental results with satisfactory accuracy (Tables 26.3, S4, S5). Juxtaposition of the energies of the most favorable conformers in vacuum and in water indicates that all molecules studied are more stable in polar medium. This is an expected result because all the compounds are water-soluble. For the species with alkyl periphery we observed a correlation between the energy raise and the number of *gauche*-bends of the torsions C–C–O–C in the respective molecule of a series: in all cases the energy grows with the increase of gauche-twists. However, in the dimer, where two angles of this type exist, no dependence between the energy and the mutual position (i.e., adjacent or disjoint) of the two *gauche*-orientations can be outlined in the set of stable isomers. No more than two *gauche*-turns are found for the dimer in vacuum, whereas in water a group of structures with three *gauche*-bends is witnessed. This indicates a tendency towards stabilization of the more compact geometry of these molecules in water. For M(OH) such dependence of the energy on the number of gauche-kinks can not be established. Concluding briefly, in water the ether molecules exist in a comparatively compact form and the molecular periphery is decisive for achieving such geometries.

Figure 26.5 illustrates the torsion angles distribution in vacuum and in water of the optimized dimers, D(CH$_3$), and monomers, M(OH). The data prove that

Table 26.4 Mean values and standard deviations of the bond lengths l_0 (C–O) and valence angles α_0 (C–O–C) obtained after PBE and MP2 geometry optimization with the basis set aug-cc-pVTZ in vacuum and in water, compared to experiment [43]

	MP2/aug-cc-pVTZ		PBE/aug-cc-pVTZ	
	l_0(C–O), Å	α_0(C–O–C), deg	l_0(C–O), Å	α_0(C–O–C), deg
Vacuum	1.420 ± 0.003	112.540 ± 0.815	1.427 ± 0.004	113.110 ± 0.788
Water	1.425 ± 0.001	112.710 ± 0.005	1.431 ± 0.002	113.090 ± 0.008
Exp.	1.411	112		

in all cases the antiperiplanar orientations are the most numerous. The abundance of *gauche*-conformations increases in water. Interestingly, they have fairly broad distributions spanning a range of about 30°. The mean value of the *gauche*-angle is shifted to absolute values greater than the typical for alkanes $\pm 0°$. Another peculiarity is that both M(OH) and D(CH$_3$) allow population of angles around $\pm 120°$, albeit occasionally. The summarized picture shows population of the entire conformational space, which implies comparatively low rotational barriers (see next section).

In addition to the overall shape of the monomers and the dimer, the particular structural parameters for implementation in the force field are analyzed, namely: the C–O bonds, C–O–C valence angles, and C–C–O–C dihedrals. Averaged values of the first two quantities for each conformer set are collected in Table 26.4.

26.4 Parameters of the Force Field

We aim at deriving parameters for ethers that are compatible with the AMBER force field. The potential energy function used in AMBER is:

$$
V(r^N) = \sum_{bonds} \frac{1}{2} k_b (l - l_0)^2 + \sum_{angels} \frac{1}{2} k_a (\alpha - \alpha_0)^2 + \sum_{torsions} \frac{1}{2} V_n [1 + \cos(nw - \gamma)]
$$
$$
+ \sum_{j=1}^{N-1} \sum_{i=j+1}^{N} \left\{ \varepsilon_{ij} \left[\left(\frac{\sigma_{ij}}{r_{ij}} \right)^{12} - 2 \left(\frac{\sigma_{ij}}{r_{ij}} \right)^{6} \right] + \frac{q_i q_j}{4 \pi \varepsilon_0 r_{ij}} \right\} \qquad (26.1)
$$

The five terms in Eq. 26.1 compute the energies of bond stretching, angle bending, torsion angles deformation, and non-bonded van der Waals (vdW) and electrostatic interactions, respectively. Detailed explanation of the parameters in the above equation can be found elsewhere [42]. In this study, we derive all parameters for the ether group (except for vdW, adopted directly from AMBER).

26.4.1 Bonded Parameters: Stretches and Bends

Table 26.4 contains the mean values and the standard deviations of the bond lengths l(C–O) and the valence angles α (C–O–C) obtained after averaging of the respective results yielded by PBE and MP2 geometry optimization of the studied model systems. Experimental data [43] are provided for reference as well. Apparently, the PBE/aug-cc-pVTZ optimization provides longer bonds than the empirical findings but the computed valence angles match very well the measured ones. The MP2/aug-cc-pVTZ minimization gives results closer to experiment than the DFT functional; yet the bond lengths are still overestimated.

Upon introduction of the aqueous environment minor extension of bond lengths occurs, while valence angles rest unaffected. Interestingly, the mean values of the bond lengths in the molecules with different periphery are essentially identical. The valence angles of the CH_3-terminated models remain insignificantly larger but in overall the parameters of the models with the two peripheries converge in water.

Detailed analysis of the computed geometries in both media show that the C–O bond lengths vary within 0.011 Å and the C–O–C valence angles – within 1.49°. These minor amplitudes demonstrate the very weak sensitivity of these bond lengths and angle magnitudes to the inter- and intramolecular environment. Least impact has the chain length. The role of the conformation in these two parameters is also immaterial, considering the small standard deviations. This signifies that the force field parameters can be obtained by means of averaging over the entire data set, irrespective of molecular size or periphery.

Other quantities needed for the parametrization of the bond-stretching and the angle-bending potentials are the respective force constants. These are obtained from the frequency analysis of the investigated molecules. The averaged characteristic frequencies of C–O stretching and C–O–C bending together with the corresponding force constants based on computations of the $M(CH_3)$ conformers in vacuum and water are presented in Table 26.5.

Table 26.5 Mean vibrational frequencies and force constants with standard deviations for the C–O bond stretching and the C–O–C angle bending, obtained from PBE and MP2 geometry optimization of $M(CH_3)$ in vacuum and water; experimental values [43] are shown for comparison

MP2/aug-cc-pVTZ				
	v(C–O) [cm^{-1}]	k_b(C–O) [kcal/mol.Å^2]	v (C–O–C) [cm^{-1}]	k_a(C–O–C) [kcal/mol.rad^2]
Vacuum	1081.00 ± 22.82	239.29 ± 10.21	467.00 ± 33.21	72.50 ± 12.26
Water	1054.00 ± 0.25	227.36 ± 0.22	504.00 ± 0.35	87.40 ± 0.35
PBE/aug-cc-pVTZ				
Vacuum	1085.00 ± 29.52	240.76 ± 15.86	451.00 ± 26.25	66.83 ± 10.50
Water	1083.00 ± 0.31	239.83 ± 0.39	487.00 ± 0.24	79.74 ± 0.54
Exp.	1120 1150	–	–	–

Both methods yield analogous results for the vibrational frequencies and force constants. The C–O bond stretching for all the molecules lies in the range 1000–$1150\,cm^{-1}$, which coincides with the experimentally identified range of ether bond vibration. The C–O–C bending is computed at lower frequency, in line with the quantitative relationship between stretching and bending deformations. It should be noted that in the dimer the bending can not be observed as an individual vibration but is always mixed with a torsional deformation, the letter being prevalent. An asymmetric stretching is registered in $M(CH_3)$, which is split into two bands in $D(CH_3)$ and even in three bands in $M(OH)$ because the vibrations of the two terminal C–O bonds occur at different frequencies (Table S6).

Replacing the vacuum with aqueous medium the stretching frequency decreases. Analysis of this frequency in the dimer (where the statistical data set is larger) reveals that its dependence on the molecular conformation is not significant. A frequency of $1120\,cm^{-1}$ is measured experimentally for the C–O stretching of diethyl ether in gas phase [43]. The results obtained here with the two computational approaches are lower than the experimentally established value, which is most probably related to the larger bond lengths estimated theoretically.

Regarding the force field parametrization, force constants based on averaging of the inner bonds should be taken, as the latter resemble to a greater extent their analogs in species with longer chains.

26.4.2 Bonded Parameters: Torsion Angles

In order to set the torsion potentials V_n in Eq. 26.1, we need the energy profile of rotation about the C–O bond. Thus, the next step is the simulation of these energy barriers. For the purpose, a scan of the potential energy surface upon stepwise variation of each torsion angle Θ_1 and Θ_2 (Figs. 26.3 and 26.4) with increment of $15°$ in the range $-180 \div 180°$ is performed. The C–C–O–C angles which could be varied in $M(CH_3)$ and $M(OH)$ are symmetrically equivalent but in $D(CH_3)$ there are two dihedral angles with dissimilar atomic surrounding – outer (Θ_1) and inner (Θ_2). Therefore, separate energy profiles for Θ_1 and Θ_2 are generated. Figure 26.6 shows the energy variation upon change of all considered dihedrals. The rotations are simulated with the MP2 method starting from of the PBE/aug-cc-pVTZ optimized geometries of the most stable conformers in each series. The absence of structural relaxation at fixed values of the dihedral angle allows the monitoring of energy changes invoked solely by the respective rotation.

The similarity of the profiles presented in Fig. 26.6 is the most striking impression one gets when looking at them. In all cases the curves are symmetric with respect to $0°$ with a deeper minimum at $180°$ and two shallower ones at $\pm 90°$. The *anti-gauche* transition passes through energy maxima at $\pm 120°$, and the eclipsed conformation has the highest energy. The barriers of the two transitions differ substantially in energy, the *anti-gauche* one being characterized by a smaller energy change, which is in keeping with the expectations from spatial standpoint.

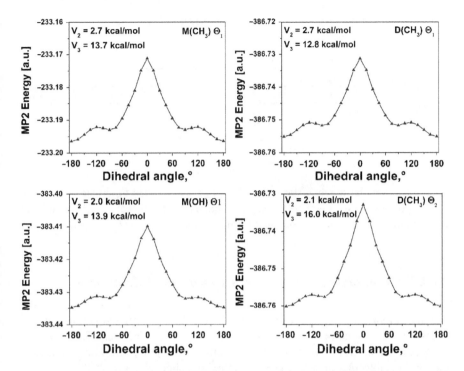

Fig. 26.6 MP2/aug-cc-pVTZ total energy profiles upon variation of selected C–C–O–C dihedral angles in the molecules with alkyl and hydroxyl periphery in vacuum; V_2 denotes the *anti- gauche* barrier, V_3 stands for transition from *gauche* to eclipsed conformation

The heights of the energy barriers respond very little to specific environments within or around the molecule. The effect of the chain length is again the weakest. In $M(CH_3)$ and $D(CH_3)$ the variation of Θ_1 requires 2.7 kcal/mol for transition to the *gauche* isomer and about 13 kcal/mol to adopt the eclipsed form. The *gauche*-orientation of the inner angle Θ_2 in $D(CH_3)$ proceeds through a slightly lower barrier at the expense of the *gauche*-eclipsed transition energy which increases by ca. 3 kcal/mol. Both barriers decrease insignificantly in water (Fig. S2).

The periphery type has no effect on the energy profiles of rotation about the C–C–O–C dihedrals (Fig. 26.6), since M(OH) features identical positions of the extrema. The *anti-gauche* barrier remains about 2 kcal/mol and the transition to the eclipsed stereoisomer costs ~2 kcal/mol less compared to its analog in $D(CH_3)$. The polar medium has practically no effect on the barrier heights which lower slightly in water (Fig. S2). The comparatively low *anti-gauche* barriers indicate that a minimal energy input can trigger interconversion between the two conformations.

The comparison of the averaged estimates of the rotational barriers (Table 26.6) with experimental data (2.6 kcal/mol) for the dimethyl ether obtained from rotational spectra [44] outlines the excellent consensus, which is a trustworthy validation of the chosen theoretical approach.

Table 26.6 Average torsion parameters for rotation around the ether bond; V_2 denotes the energy barrier for *anti-gauche* transition and V_3- the energy required for rotation from *gauche* to eclipsed form

MP2/aug-cc-pVTZ		
	V_2[kcal/mol]	V_3[kcal/mol]
Vacuum	2.464	13.467
Water	2.001	15.506
Exp. [44]	2.6	–

From molecular mechanistic point of view, the computed torsional profiles are sufficiently invariant with respect to the molecular environment and could be utilized for generation of torsional parameters for the rotation about the C-O bond in ethers.

26.4.3 Nonbonded Parameters: Electrostatic Interactions

Another important ingredient of the force field is the parametrized atomic charges necessary for estimation of the electrostatic interactions. The mean RESP atomic charges obtained from the multiconformational fit of the electrostatic potential of the studied molecules in vacuum and in water are collected in Fig. 26.7.

The computed values of the atomic charges correspond fully to the qualitative prognosis. The carbon atoms adjacent to ether and hydroxyl oxygen bear positive charge, while those from the peripheral methyl groups are negative. Much higher electron density is centered at oxygen atoms, the ether ones being less negative than the hydroxyl.

The quantitative differences stem from different sources. The effect of the water medium is expressed in the more pronounced polarization of M(CH$_3$) and depolarization of M(OH). For D(CH$_3$) such correlation can not be outlined: in water part of the molecule becomes more polarized whereas in the remaining fragment the atomic charges of opposite sign tend to convergence. One reason for such behavior might be the substantially different sizes of conformer sets used for atomic charges fit in the two media.

The RESP charges which were implemented in the force field are provided in Table S7.

With chain extension from monomer to dimer a surprising asymmetry in the charges of the methylene carbons occurs. While the carbons in the –CH$_2$– groups closer to the chain ends retain values greater than 0.2, those of the inner carbons decrease noticeably. Moreover, the carbon charges of two adjacent inner methylene groups (D(CH$_3$)) differ in magnitude, albeit not as much as the charges of an inner and an outer C-atom in M(OH) do. This is evidence that the conformational diversity has impact on the electrostatic potential in the core part of the dimers which is reflected in their atomic charges.

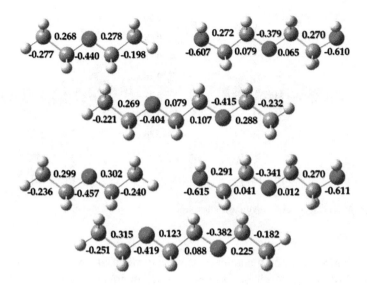

Fig. 26.7 RESP atomic charges of the non-hydrogen atoms obtained from fitting of the HF/6-31G* electrostatic potential generated in vacuum (*top*) and in water (*bottom*)

Table 26.7 Mean values and standard deviations of the mass density, enthalpy of vaporization (ΔH_{vap}), enthalpy of solvation (ΔH_{sol}), and molecular volume of diethyl ether obtained from MD simulations with the two sets of parameters derived from PBE and MP2 quantum mechanical calculations; experimental estimates [43,45] are provided where available

	Density [g/cm^3]	ΔH_{vap} [kJ/mol]	ΔH_{sol} [kJ/mol]		Molecular volume [Å^3]	
			TIP3P	TIP4P	TIP3P	TIP4P
MP2	0.717±0.035	20.07±1.25	−16.96±2.36	−19.78±1.89	162.98±5.01	151.38±2.07
PBE	0.716±0.034	20.18±0.46	−17.83±1.57	−21.97±2.51	167.71±0.62	154.81±1.84
Exp.	0.714	26	−26		148	

26.5 Validation of the Derived Force Field Parameters

Table 26.7 contains the calculated and the experimental values of the characteristics used for validation of the derived parameters.

26.5.1 Molecular Volume

The motivation in using NPT dynamics to calculate volume changes via this approach is to mimic photo-thermal experiments that also determine molecular volume changes on nanosecond timescales with similar precision [46,47].

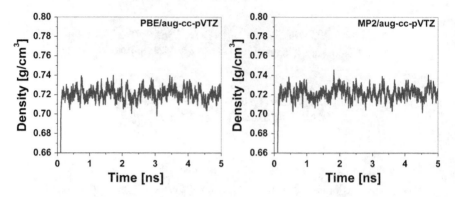

Fig. 26.8 Mass density of liquid diethyl ether as a function of the simulation time

The values in Table 26.7 show the obvious overestimation of the molecular volume obtained with both parameter sets compared to the experimental quantity [43], the MP2-based result being closer to the empirical one. A possible reason for the overvalued volume may be the larger bond lengths employed as structural parameters. Of the two water models utilized, TIP4P renders the better outcome.

26.5.2 Density

The time evolution of the density obtained from MD simulation is presented in Fig. 26.8. The mean values resulting from the two parameter sets are 0.717 and 0.716 g/cm^3 for the MP2 and PBE derived sets, respectively, and are in excellent agreement with the experimental measurement – 0.714 g/cm^3 [43].

Figure 26.8 and Table 26.7 show that the fluctuations of mass density during the MD simulations are negligible which demonstrates the stability of the system and the reliability of both parameter sets for the description of such models.

26.5.3 Torsion Angles – Population Analysis

Another exceedingly important validation step for emphatically flexible molecules is to test whether steric parameters derived by averaging the characteristics of a diverse conformational set are capable of reproducing the flexibility of species from the studied series at the MD level. Due to the unusual behavior of the torsion angles in this class of compounds (the pronounced *gauche*-effect), it is necessary to generate the probability distribution of the parametrized dihedral angle C–O–C–C and the symmetric C–C–O–C in the MD simulation of diethyl ether in liquid phase in order to check the performance of the modified force field.

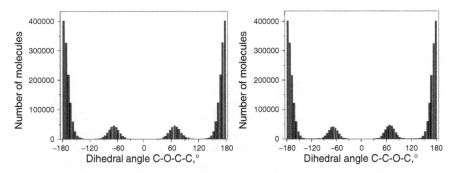

Fig. 26.9 Probability distributions of the dihedrals C–O–C–C (*left*) and C–C–O–C (*right*) of M(CH₃) obtained from a 5 ns MD simulation

Comparing Figs. 26.5 and 26.9 one easily sees the similarity between the MD statistics and the quantum mechanical calculations. In both approaches the *anti*-orientation is the most populated with sizeable presence of *gauche*-bends. The reproduction of the torsion angle distribution by the MD simulation supports the applicability of the derived parameters.

26.5.4 Enthalpy of Vaporization

The enthalpy of vaporization of one molecule from the liquid could be calculated according to the following expression [47]:

$$H_{VAP} = E_{G-L} + RT \tag{26.2}$$

This equation is based on the assumption that the gas phase may be treated as ideal gas and the volume of the liquid is negligible compared to that of the vapor. E_{G-L} represents the difference in the total energies of the gas (E_G) and the liquid (E_L) phase. The estimate of E_G is based on the MD non-periodic simulation of one M(CH₃) molecule in vacuum and the value of E_L is extracted from the MD simulation of the molecular liquid in periodic boundary conditions. The enthalpy of vaporization results obtained with the two parameter sets derived with different quantum mechanical methods (PBE and MP2) together with the experimentally assessed value are presented in Table 26.7. The two theoretical estimates are essentially identical. Comparison between theory and experiment is acceptable but needs improvement. More satisfying results would provide probably simulations of more realistic systems, e.g. the gas phase may be modeled more sophisticatedly.

26.5.5 Enthalpy of Solvation

Using the Hess law, the enthalpy of solvation can be represented as:

$$\Delta H_{SOL} = E_{SYS} - (E_G + E_{WAT}) \tag{26.3}$$

In this expression E_{SYS} is the total energy of the system built of one molecule diethyl ether and sufficient water molecules to mimic a solution. Once again, two water models are employed (TIP3P and TIP4P). The remaining terms in Eq. 26.3 are: E_G– average total energy (MD) of one molecule in vacuum; E_{WAT} – mean total energy (MD) from the simulation of the solvent without solute. The computational results and the experimental value can be found in Table 26.7. As for the other estimates, the PBE method and the TIP4P water model agree better with experiment.

26.6 Summary

This study comprises a theoretical investigation of all possible stable conformers of two monomers and a dimer containing alkyl-bound ether fragments simulated with two quantum chemical methods. Basic structural parameters, characteristic vibrational frequencies and atomic electron density distribution are evaluated. Rotation barriers about the ether bond are estimated. The mean values of these quantities averaged over the entire conformer ensemble are supplemented to the parameters of the molecular mechanics force field AMBER and tested by several molecular dynamics simulations of one of the model systems – diethyl ether.

The good agreement of the parameters computed with the quantum mechanical methods (MP2/aug-cc-pVTZ and PBE/aug-cc-pVTZ), compared both to each other and to the available experimental data, indicate that the molecular geometry of ether-group containing molecules, hence, their properties, have been described reliably using quantum chemistry. The *structural parameters* and the *vibrational frequencies* exhibit remarkable invariance to modification of the molecular medium allowing the set of assessed descriptors to be used successfully for molecular mechanics ether fragment parametrization in alkyl tail oligoethers.

Analysis of the obtained *RESP charges* shows that the charge set generated for the dimer should be used for simulation of longer oligomers. Moreover, the values generated in vacuum would be more adequate for implementation in AMBER, since the additional effects of polarization/depolarization should be accounted for by means of direct interactions with the explicit solvent molecules the charges of which are also based on gas phase calculations

The tests of the two sets of MP2- and PBE-derived parameters validate their applicability for simulation of systems from the class of C_xE_y as they reproduce with satisfactory accuracy the basic thermodynamic characteristics of the model molecules.

Acknowledgments The research was supported by Project DO-02-256/2008 of the National Science Fund of Bulgaria with partial funding from Projects DCVP-02-2/2009, DO-02-52/2008 and DO-02-136/2008; the Alexander von Humboldt Foundation is acknowledged for an Equipment Grant.

Supplementary Information Available online are: PBE and MP2 energy data (Table S1); MP2 torsion angles with conformer population in vacuum (Table S2) and in water (Table S3) for the monomers and analogous data for the dimer (Tables S4, S5); the PBE entries of Table 26.5 (Table S6); RESP charges (Table S7); MD trajectories of control parameters (Fig. S1); rotational barrier about Θ_2 of the dimer in water (Fig. S2).

References

1. (a) Tchoukov P, Mileva E, Exerowa D (2003) Langmuir 19:1215; (b) Mileva E, Tchoukov P, Exerowa D (2005) Adv Coll Interf Sci 47:114
2. Balogh J, Pedersen JS (2008) Prog Colloid Polym Sci 135:101
3. (a) Balogh J, Olsson U, Pedersen JS, Dispers J (2006) Sci Technol 27:497; (b) Balogh J, Olsson U, Dispers J (2007) Sci Technol 28:223; (c)Balogh J, Kaper H, Olsson U, Wennerstrm H (2006) Phys Rev E 73:041506
4. Glass JE, Ed. (1996) Hydrophilic Polymers: Performance with Environmental Acceptability, American Chemical Society, Washington, D.C., Vol. 248
5. (a) Smith GD, Yoon DY, Jaffe RL (1993) Macromolecules 26:5213; (b) Smith GD, Jaffe RL, Yoon DY (1996) J Phys Chem 100:13439
6. Smith GD, Yoon DY, Jaffe RL, Colby RH, Krishnamoorti R, Fetters LJ (1996) Macromolecules 29:3462
7. Neyertz S, Brown D, Thomas JO (1994) J Chem Phys 101:10064
8. Lin B, Boinske PT, Halley JW (1996) J Chem Phys 105:1668
9. Bedrov D, Pekny M, Smith GD (1998) J Phys Chem B 102:996
10. Smith GD, Borodin O, Bedrov D (2002) J Comput Chem 23:1480
11. Bedrov D, Smith GD (1999) J Phys Chem B 103:3791
12. Smith GD, Bedrov D (2002) Macromolecules 35:5712
13. (a) Jorgensen WL, Tirado-Rives J (1988) J Am Chem Soc 110:1657; (b) Pranate J, Wierschke S, Jorgensen WL (1991) J Am Chem Soc 113:2810
14. Jorgensen WL, Madura JD (1985) Mol Phys 56:1381
15. Jorgensen WL, Madura JD (1985) Mol Phys 56:1381
16. (a) Perdew JP, Burke K, Ernzerhof M (1996) Phys Rev Lett 77:3865; (b) Perdew P, Burke K, Ernzerhof M (1997) Phys Rev Lett 78:1396
17. Moller C, Plesset M (1934) Phys Rev 46:618
18. Kendall RA, Dunning TH Jr, Harrison RJ (1992) J Chem Phys 96:6796
19. Miertus S, Scrocco E, Tomasi J (1981) J Chem Phys 55:117
20. (a) Bayly CI, Cieplak P, Cornell WD, Kollman PA (1993) J Phys Chem 97:10269; (b) Cieplak P, Cornell WD, Bayly C, Kollman PA (1995) J Comput Chem 16:1357
21. Cornell WD, Cieplak P, Baily CI, Gould IR, Merz KM Jr, Ferguson DC, Fox T, Caldwell JW, Kollman PA (1995) J Am Chem Soc 117:5179
22. Berendsen JC, Postma JPM, van Gunsteren WF, DiNola A, Haak JR (1984) J Chem Phys 81:3684
23. Essmann U, Perera L, Berkowitz ML, Darden T, Lee H, Pedersen LG (1995) J Chem Phys 103:8577
24. HyperChem 7.0 (2002) Hypercube: Gainesville, FL
25. Frisch MJ et al (2009) Gaussian, Inc., Wallingford, CT
26. Case DA et al (2004) AMBER 8; University of California, San Francisco, CA

27. Snyder RG, Zerbi G (1967) Spectrochim Acta 23:391
28. Wieser H, Laidlaw WG, Krueger PJ, Fuhrer H (1968) Spectrochim Acta 24:1055
29. Perchard JP, Monier JC, Dixabo P (1971) Spectrochim Acta 27:447
30. Maissara M, Labenne JP, Devaure J (1985) J Chem Phys 82:451
31. Kanesaka I, Snyder RG, Strauss HL (1986) J Chem Phys 84:395
32. Kuze N, Kuroki N, Takeuchi H, Egawa T, Konaka S (1993) J Mol Struct 301:81
33. Astrup EE (1979) Acta Chim Scand A 33:655
34. Yoshida H, Tanaka T, Matsuura H (1996) Chem Lett 8:637
35. Goutev N, Ohno K, Matsuura H (2000) J Phys Chem A 104:9226
36. Yoshida H, Kaneko I, Matsuura H, Ogawa Y, Tasumi M (1992) Chem Phys Lett 196:601
37. Bedrov D, Borodin O, Smith GD (1998) J Phys Chem B 102:5683
38. Bedrov D, Smith GD (1998) J Chem Phys 109:8118
39. Ganguly B, Fuchs B (2000) J Org Chem 65:558
40. Andersson M, KarlstrBm G (1985) J Phys Chem 89:4967
41. Inomata K, Abe A (1992) J Phys Chem 96:7934
42. Cornell WD, Cieplak P, Bayly CI, Gould IR, Merz KM Jr, Ferguson DM, Spellmeyer DC, Fox T, Caldwell JW, Kollman PA (1995) J Am Chem Soc 117:5179
43. Kuchitsu K (1998) Structure of free polyatomic molecules basic data; Springer, Berlin. http://cccbdb.nist.gov/exp2.asp?casno=60297
44. (a) Favero LB, Caminati W, Velino B (2003) Phys Chem Chem Phys 5:4776; (b) Niidea Y, Hayashi M (2004) J Mol Spectrosc 223:152
45. Maikut OM, Makitra RG, Ya Palchikova E (2006) Russ J Gen Chem 76:170
46. (a) Hansen KC, Rock RS, Larsen RW, Chan SI (2000) J Am Chem Soc 122:11567; (b) Larsen RW, Langley T (1999) J Am Chem Soc 121:4495
47. Dai J, Li X, Zhao L, Sun H (2010) Fluid Phase Equilibria 289:156

Chapter 27
Anti-adiabatic State: Ground Electronic State of Superconductors

Pavol Baňacký

Abstract Based on the non-adiabatic *ab initio* theory of complex electronic ground states, treating electronic structure as explicitly dependent on nuclear dynamics, i.e. on instantaneous nuclear coordinates Q and momenta P, it has been shown that electron-phonon coupling in superconductors induces temperature-dependent electronic structure instability related to analytic critical point (ACP) fluctuation of bands across the Fermi level (FL). As ACP approaches FL, the adiabatic chemical potential μ_{ad} is substantially reduced to $\mu_{antad}(\mu_{ad} \gg \mu_{antad} < \hbar\omega)$ and the adiabatic Born-Oppenheimer approximation is violated. Due to the effect of nuclear dynamics, the system is stabilized as an antiadiabatic state of broken symmetry with a gap in its one-particle spectrum. Distorted nuclear structure, which is related to nuclei in the phonon mode r inducing transition to an anti-adiabatic state, has fluxional character. Geometric degeneracy of the antiadiabatic ground state enables formation of mobile bipolarons that can move over lattice in external electric potential as super-carriers without dissipation. Thermodynamic properties in the anti-adiabatic state correspond to thermodynamics of superconductors. It has been shown that Cooper-pair formation is not the primary reason for transition into superconducting state, but it is a consequence of anti-adiabatic state formation and represents correction to electron correlation energy. As illustrative examples, results of application of anti-adiabatic theory in study of superconductors MgB_2, YB_6, $YBa_2Cu_3O_7$, Nb_3Ge and corresponding (non superconducting) analogues are presented.

P. Baňacký (✉)
Faculty of Natural Sciences, Institute of Chemistry, Chemical Physics division, Comenius University, Mlynská dolina CH2, 84215 Bratislava, Slovakia
e-mail: pavol.banacky@fns.uniba.sk

P.E. Hoggan et al. (eds.), *Advances in the Theory of Quantum Systems in Chemistry and Physics*, Progress in Theoretical Chemistry and Physics 22, DOI 10.1007/978-94-007-2076-3_27, © Springer Science+Business Media B.V. 2012

27.1 Introductory Remarks to Theory of Superconductivity

Superconductivity, an amazing physical phenomenon was discovered nearly 100 years ago by Kamerlingh Onnes [1]. In spite of enormous attention which has been paid to this effect over a century, the microscopic mechanism of superconducting state transition remains unclear and represents an open challenge for theory.

Until the discovery of high-temperature superconductivity of cuprates by Bednorz and Muller in 1986 [2] and synthesis of first 90 K superconductor [3] in 1987, understanding of microscopic mechanism of superconducting (SC) state transition formulated within the BCS theory in 1957 [4] was generally accepted as a firm theoretical basis behind the physics of this phenomenon. The idea of Cooper-pair formation, i.e. formation of boson-like particles in momentum space, which are stable in a thin layer above the Fermi level and drive the system into more stable – superconducting state, is crucial in this case. Sufficient condition of pair formation is a weak, but attractive interaction between electrons. The possibility of effective attractive electron-electron (e-e) interactions was derived by Fröhlich [5, 6] as a consequence of electron-phonon (e-p) interactions.

The range of validity of the BCS theory for e-p interactions has been specified by Migdal [7] and Eliashberg [8]. It can be interpreted as Migdal's theorem and Eliashberg's restriction (ME approximation). The first is related to the validity of the condition $\omega\lambda/E_F \ll 1$ and the latter restricts the validity only for $\lambda \leq 1$, where λ is e-p coupling strength and ω and E_F are characteristic phonon and electron energy scales, respectively. Expressed explicitly, BCS-like theories are valid only for adiabatic systems that obey the adiabatic Born–Oppenheimer approximation (BOA), $\omega/E_F \ll 1$.

For conventional (low-temperature) superconductors, the BCS theory within the ME approximation (i.e. weak coupling regime) is an excellent extension of standard metal theory. In order to interpret high critical temperature and ensure pair condensation in case of high-temperature cuprates, other interaction mechanisms besides e-p interactions have been advocated (see e.g. [9–11]). Since copper, a transition metal with a partly-filled d-shell when chemically bound, is a central atom of high-T_c cuprates (formal charge $+2$ is usually considered), it is quite natural that attention has been focused on strong electron correlations (in standard Coulomb – repulsive e-e interactions), magnetic interactions and/or spin fluctuation effects. At present, the effect of Coulomb repulsion is usually incorporated via Coulomb pseudopotential μ^* and critical temperature is calculated according to the McMillan formula [12].

The e-p interactions, taken as responsible for electron pairing, driving transition into a superconducting state for classical low-T_c superconductors, have almost been abandoned and considered inappropriate to high-T_c cuprates [see e.g. 10]. Some aspects of d – wave superconductivity can be described by strongly correlated electron models e.g. Hubbard – like or t – J models (e.g. [9, 13–16]), even without explicit e-p interactions. The underlying leitmotiv in electron correlation treatments is understanding the phase diagram of high-T_c cuprates, i.e. the doping process. Introduction of charge carriers (holes or electrons) into the parent anti-feromagnetic insulator that causes transition to superconductor (or metal) has been generally accepted to be a universal feature of high-T_c cuprates and believed to be intimately related to superconductivity mechanisms. Bell-like shaped dependence of T_c on hole doping in the doping range $0.05 \leq x < 0.27$ for family of high-T_c cuprates is well known (see e.g. [17] and references therein). With the exception of YBCO ($x = 0.05$), the optimal hole doping with maximal T_c is $x \approx 0.16$. The electron doping is usually less favorable, but either hole or electron doping can induce superconductivity.

Doubtless, charge doping has perceptible impact on e-e interactions and influences, to some extent, subtler spin interactions. An open question is if these interactions represent the key factor behind the physics leading to a superconducting state transition upon doping? One must thus realize that like T_c, the lattice parameters and, very importantly, lattice dynamics is strongly influenced by doping. For instance, dependence of T_c on lattice parameter a in case of Hg-based cuprates follows similar dependence as T_c on hole doping [18]. Although the isotope effect coefficient α for optimally doped cuprates is tiny (i.e. $\alpha \approx 0.05$, an exception is optimally doped YBCO with $\alpha \approx 0.8$), doping in underdoped as well in overdoped regions for O-isotope effect throughout the cuprate family results in huge changes of isotope effect coefficient α, (e.g. $\alpha \approx 1$ for underdoped LSCO) [17–19]. It is important experimental evidence that doping induces major changes in lattice dynamics, in particular for CuO_2 layers. It implies both electron and nuclear degrees of freedom are involved which guides theory.

The results of high-resolution ARPES [20,21] of a large family of different high-T_c cuprates show that besides doping, an abrupt decrease in electron velocity near Fermi level, at about 50–80 meV in nodal direction, is the other feature common to high-T_c cuprates. The kink in nodal direction is temperature independent. More importantly for microscopic theory of superconductivity, seems to be the formation of a temperature-dependent kink on the momentum distribution curve (dispersion renormalization) close to Fermi level (~ 60 meV) in off-nodal direction at transition to superconducting state. It has been reported for Bi2212 [22–25]. Recently, presence of the kink in off-nodal direction has been observed at VUV ARPES study with sub-meV resolution of optimally doped untwined YBCO in superconducting state [26].

Formation of the off-nodal kink, has been ascribed by the authors [22] to coupling of electrons to bosonic excitations, preferably they consider a magnetic resonance mode such as observed in some inelastic neutron scattering experiments. The inconsistency in this interpretation has been pointed by Z-X. Shen and coworkers [23,25]. The main arguments are [25]: (a) magnetic resonance has not yet been observed by neutron scattering in such heavily doped cuprates, and (b) magnetic resonance has little spectral weight and may be too weak to cause the effect seen by ARPES. They agree, however, with the authors [22] that the renormalization effect seen by ARPES in cuprates may indeed be related directly to the microscopic mechanism of superconductivity. The authors [23, 25], instead of magnetic resonance mode, ascribe dispersion renormalization to coupling with phonon mode, in particular with B_{1g}-buckling mode of CuO_2 plane. The temperature dependence of dispersion renormalization they attribute to the DOS enhancement due to SC-gap opening and to the thermal broadening of the phonon self-energy in normal state.

These results along with those of neutron scattering [27, 28] indicate that for high-T_c cuprates the e-p coupling has to be considered as a crucial element of microscopic mechanism of SC state transition. For cuprates, the role of phonons at superconducting state transition must not be overlooked. As soon as a low-Fermi energy situation occurs ($\omega \leq E_F$) one can expect major contribution of non-adiabatic vertex corrections at the SC state transition. It is beyond the standard ME approximation and this problem has been studied within the non-adiabatic theory of superconductivity [29a, b, c]. On the other hand, as the ARPES results indicate, electron kinetic energy is decreased and proper treatment of e-e Coulomb interactions is essential. The competition between Coulomb vs. e-p interactions has been intensively studied within the Holstein – Hubbard models [30–34] as both interactions are short-range. The results are unsatisfactory, since heavy-mass polarons are formed that yield low values of T_c. It has been improved within the Frohlich – Coulomb model [35] that introduces long-range repulsion between charge-carriers and also long-range e-p interactions. The results show that there is a narrow window of parameters of Coulomb repulsion V_c and e-p interactions $E_P(V_c/E_P)$, resulting in the light-mass bipolaron formation. In this case, using bipolaron theory of superconductivity [36 a,b,c] their coherent motion represents supercarriers and high T_c may result.

McMillan's formula (a very good approximation for T_c of elementary metals and their alloys [37]), is often used for calculation of critical temperature of high-T_c superconductors within the BCS-generic framework. It has been shown [38] that in the strong-coupling regime $\lambda \gg 1$, T_c can be as large as $k_B T_c = \hbar \omega \lambda^{1/2}/2\pi$. However, there is problem with correct estimation of the Coulomb pseudopotential μ^* and with unrealistically large values of λ that would match high experimental T_c of novel superconductors. It has to be stressed, however, that strong coupling regime $\lambda > 1$ violates adiabatic condition $\omega/E_F \ll 1$ of the ME approximation, which is behind the derivation of the McMillan formula.

Moreover, new classes of superconductor, e.g. cuprates, fullerides and MgB_2 are systems that are rather pseudo-adiabatic with a high adiabatic ratio $\omega/E_F < 1$ [39], in contrast to elementary metals where adiabatic condition $\omega/E_F \ll 1$ is satisfied. This situation indicates the role of non-adiabatic contributions at calculation of e-p interactions within the BOA, an effect that is beyond the standard ME approach. As mentioned above, formulation of the nonadiabatic theory of superconductivity by Pietronero and coworkers [29a, b, c, 40a, b], which accounts for vertex corrections and cross phonon scattering (beyond ME approximation), has solved this nontrivial problem by generalization of Eliashberg equations. The theory, which is non-perturbative in λ and perturbative in $\lambda \omega_D/E_F$, has been applied to simulation and interpretation of different aspects of high-T_c superconductivity [41 a– e] . Basically, it accounts for non-adiabatic effects in a quasi-adiabatic state $\omega/E_F \leq 1$ and is able to simulate various properties of high-T_c superconductors, including high-values of

T_c, already for moderate e-p coupling, $\lambda \approx 1$. Moreover, it has also been shown that increased electron correlation is an important factor which makes corrections to vertex function positive, which is crucial for increasing T_c.

Nonetheless, sophisticated treatment of high-T_c superconductivity within the non-adiabatic theory faces serious problem related to possibility of polaron collapse of the band and bipolaron formation. According to bipolaron theory of Alexandrov [36a–c, 39, 42–44], polaron collapse occurs already at $\lambda \approx 0.5$ for uncorrelated polarons and even at a smaller value for a bare e-p coupling in strongly correlated systems. For $\omega/E_F \leq 1$, or $\lambda \geq 1$ and for $\omega/E_F \geq 1$ at any small value $\lambda \ll 1$, the nonadiabatic polaron theory has been shown to be basically exact [44]. Bipolarons can be simultaneously small and light in a suitable range of Coulomb repulsion and e-p interaction [45]. These results have important physical consequences. There are serious arguments that effect of polaron collapse cannot be covered through calculation of vertex corrections due to translation symmetry breaking and mainly, polaron collapse changes possible mechanism of pair formation, i.e. instead of BCS scenario with Cooper pair formation in momentum space, the BEC with mobile bipolarons (charged bosons) in real space becomes operative.

Discovery of superconductivity in a simple compound MgB_2 at $40\,K$ [46] has been very surprising and has started a new revitalization of superconductivity research in 2001. Besides the many interesting aspects, discovery of MgB_2 super-conductivity is crucial for general theoretical understanding of SC state transition on microscopic level. It is related to band structure (BS) fluctuation and dramatic changes of BS topology due to e-p coupling.

The σ bands split by coupling to E_{2g} mode in MgB_2 has been reported [47] already in 2001 but, with exception of possible impact of anharmonicity [48], no special attention has been paid to this effect. Superconductivity in MgB_2 has been straightforwardly interpreted [49] shortly after the discovery as a standard BCS-like, even of intermediate-strong coupling physics. For clumped nuclear equilibrium geometry, the BS is of adiabatic metal-like character. The E_F^σ of σ band electrons (chemical potential μ) is relatively small, $\approx 0.4\,eV$, but still large enough comparing to vibration energy of E_{2g} phonon mode ($\omega_{2g} \approx 0.07\,eV$). Though the adiabatic ratio $\omega/E_F \approx 0.15$ is sizable, it is small enough to interpret superconductivity within the adiabatic BCS-generic framework. It is supposed that nonadiabatic effects, anharmonic contributions and/or Coulomb interactions within generalized Eliashberg approach should be important in this case, however. On the other hand, the value of e-p coupling, $\lambda \approx 0.7$, indicates that polaron collapse can be expected and superconductivity should be of nonadiabatic bipolaron character rather than the BCS-like.

Nevertheless, things are even more complicated. It has been shown [50, 51] that the analytic critical point (ACP – maximum, minimum or saddle point of dispersion, in case of MgB_2 it is maximum) of the σ band at Γ point crosses Fermi level (FL) at vibration displacement $\approx 0.016\,A°/B$-atom, i.e. with amplitude $\approx 0.032\,A°$, which is smaller than root-mean square (rms) displacement ($\approx 0.036\,A°$) for zero-point vibration energy in E_{2g} mode. Thus, in vibrations when ACP approaches FL for distances less than $\pm\omega$, the adiabatic Born-Oppenheimer approximation (BOA) is not valid. Here, the Fermi energy (E_F^σ-chemical potential μ) of σ band electrons close to the Γ point is less than E_{2g} mode vibration energy $E_F^\sigma < \omega_{2g}$. When the ACP of the band touches Fermi level, the Fermi energy is basically reduced to zero, $E_F^\sigma \rightarrow 0$.

Moreover, shift of the ACP substantially increases the density of states (DOS) at FL, $n_\sigma(E_F) = (\partial\varepsilon_\sigma^0/\partial k)_{E_F}^{-1}$, and induces corresponding decrease of effective electron velocity $(\partial\varepsilon_\sigma^0/\partial k)_{E_F}$ of fluctuating band in this region of k-space. Physically, it represents the system transition from adiabatic $\omega \le E_F$ to a truly non-adiabatic $\omega > E_F$, or even to a strongly anti-adiabatic state with $\omega \gg E_F$. This effect has crucial theoretical impact. Not only is the ME approximation not valid (impossible to calculate non-adiabatic vertex corrections which represent off-diagonal corrections to the adiabatic ground state), but the adiabatic BOA itself does not hold.

The BOA is a crucial approximation of theoretical molecular as well as of solid state physics. It facilitates many-body problems.

In the BOA, the motion of the electrons is a function of the instantaneous nuclear coordinates Q, but is independent of the instantaneous nuclear momenta P. Usually, and in solid state physics basically always, only parametric dependence is considered – i.e. nuclear coordinates are just parameters to solve electronic problems within the clamped nuclei Hamiltonian. Nuclear coordinate-dependence, when explicitly treated, modifies nuclear potential energy by diagonal BO correction (DBOC) that reflects an influence of small nuclear displacements from equilibrium positions and corrects the electronic energy for clamped nuclei. The DBOC enters the potential energy term of nuclear motion (conserving nuclear kinetic energy term) and hence modifies vibration frequencies. The off-diagonal terms of the nuclear part of system Hamiltonian that mix electronic and nuclear motion through the nuclear kinetic energy operator term are neglected and it enables independent diagonalization of electronic and nuclear motion (adiabatic approximation). Neglecting the off-diagonal terms is justified only if these are very small, i.e. if the energy scales of electron and nuclear motion are very different and when adiabatic condition holds, i.e. $\omega/E \ll 1$. If necessary, small contribution of the off-diagonal terms can be calculated by perturbation methods as so called nonadiabatic correction to the adiabatic ground state.

Superconductors seem to be substantially different, at least in the case of MgB_2. There is considerable reduction of electron kinetic energy, which for antiadiabatic state results even for dominance of nuclear dynamics ($\omega \gg E_F$) in part of k-space. In such a case, it is necessary to study electronic motion as explicitly dependent on the operators of instantaneous nuclear coordinates Q as well as on the operators of instantaneous nuclear momenta P. It is a new aspect for many-body theory.

The electronic theory of solids has been developed assuming the adiabatic BOA. Thus naturally different theoretical – microscopic treatments of superconductivity based on model Hamiltonians, which focus on one or the other type of interaction mechanism, implicitly assume validity of the BOA and it is very seldom that possible BOA breakdown at transition to SC state is raised. The "nonadiabatic" effects in relation to electronic structure is commonly used for contributions of the off-diagonal matrix elements of interaction Hamiltonian (e.g. e-p coupling, e-e correlations,..) to the adiabatic ground state electronic energy calculated in second and higher orders of perturbation theory and excludes the true nonadiabatic – antiadiabatic situation, $\omega > E_F$.

In this connection, a lot of important questions arise, e.g.:

- How to treat the antiadiabatic state?
- Can a system be stable in an antiadiabatic state?
- Are the physical properties of the system in antiadiabatic state different from those in an adiabatic state?
- What is the driving force for adiabatic ↔ antiadiabatic state transition, i.e. which type of interaction mechanism triggers this type of transition and what are necessary conditions?
- How relevant is the adiabatic ↔ antiadiabatic state transition for SC state transition in MgB_2?
- Is the adiabatic ↔ antiadiabatic state transition an accidental effect at SC state transition which is present only in MgB_2, or an inherent physical mechanism in other superconductors?
- Is the adiabatic ↔ antiadiabatic state transition relevant for high, as well as for low-temperature superconductors?
- Phonons or strong electron correlations?
- What is the condensate nature – Cooper pairs or bipolarons?
- Is there any relation of the adiabatic ↔ antiadiabatic state transition to Cooper pairs formation?
- Cooper pairs or correction to electron correlation energy?

Theoretical aspects related to the above problems have been elaborated and discussed in detail in the "*Ab initio* theory of complex electronic ground state of superconductors", published recently [52a, b]. The main theoretical point is a generalization of the BOA by sequence of canonical – base function transformations. This formalism is equivalent to our original one, based on quasi-particle transformation treatment [53]. The final electronic wave function is explicitly dependent on nuclear coordinates Q and nuclear momenta P. Emerging new quasi-particles, i.e. nonadiabatic fermions, are explicitly dependent on nuclear dynamics. As a result, the effect of nuclear dynamics can be calculated as corrections to the clamped nuclei ground state electronic energy, the one-particle spectrum and the two-particle term, i.e. to the electron correlation energy[1].

---

[1]To avoid confusion, it should be stressed that telectron correlation energy as used in this paper stands for improvement of e-e interaction term contribution beyond the Hartree-Fock (HF) level, $E_{corr} = E_{exact} - E_{HF}(E_{exact} < E_{HF})$, as it can be calculated e.g. by (1/r)-perturbation theory in 2nd and higher orders, or by configurations interaction method. In condensed matter physics, electron correlation usually stands for an account for Coulomb e-e interaction at least on Hartree or HF level. On the HF level not only repulsive e-e term is present (like on Hartree level where spin is not considered at all), but also exchange term (fermion Coulomb-hole only for electrons with

It has been shown that due to e-p interactions, which drive the system from an adiabatic to antiadiabatic state, adiabatic symmetry is broken and system is stabilized in the antiadiabatic state at distorted geometry with respect to the adiabatic equilibrium high symmetry structure. Stabilization is due to participation of nuclear kinetic energy term, i.e. nuclear dynamics (P-dependence) which is absent in the adiabatic state within the BOA. The antiadiabatic ground state at distorted geometry is geometrically degenerate with fluxional nuclear configuration in the phonon modes that drive the system into this state. It has been shown that as long as it remains in an antiadiabatic state, nonadiabatic polaron – renormalized phonon interactions are zero in well defined k-region of reciprocal space. Along with geometric degeneracy of the antiadiabatic state it enables formation of mobile bipolarons (as polarized inter-site charge density distribution) that can move over lattice in external electric potential as super-carriers without dissipation. Moreover, it has been shown that due to e-p interactions at transition into antiadiabatic state, k-dependent gap in one-electron spectrum has been opened. Gap opening is related to the shift of the original adiabatic Hartree-Fock orbital energies and to the k-dependent change of density of states of particular band(s) at the Fermi level. The shift of orbital energies determines the one-particle spectrum uniquely (also thermodynamic properties). It has been shown that the resulting one-particle spectrum yields all thermodynamic properties that are characteristic of the superconducting state, i.e. temperature dependence of the gap, specific heat, entropy, free energy and critical magnetic field. The k-dependent change of the density of states at the Fermi level in transition from adiabatic (non-superconductive) into antiadiabatic state (superconductive) can be experimentally verified by ARPES or tunneling spectroscopy (spectral weight transfer at cooling superconductor from above T_c down to temperatures below T_c).

Results of the *ab initio* theory of the antiadiabatic state have shown that the Fröhlich's effective attractive e-e interaction term represents correction to electron correlation energy in transition from adiabatic into antiadiabatic state due to e-p interactions. Analysis of this term has shown that increased electron correlation is a consequence of stabilization of the system in superconducting electronic ground state, but not the reason of its formation.

In the present article, the key points of the antiadiabatic theory are demonstrated at study of real superconductive compounds (MgB_2, YB_6, $YBa_2Cu_3O_7$, Nb_3Ge) and the crucial antiadiabatic effects are shown to be absent in corresponding non-superconductive structural analogues (AlB_2, CaB_6, $YBa_2Cu_3O_6$, Nb_3Sb). An interested reader can find details of the theory in [52a, b, 53] or in recent review paper [54].

---

parallel spins). Correlation energy improves unbalanced treatment of e-e interaction for electrons with parallel and antiparalel spins on HF level.

27.2 Electronic Structure Instability – Transition to the Antiadiabatic State

27.2.1 Preliminaries

Crystal structures of the studied compounds are different. The MgB_2 crystallizes in hexagonal structure (hP3, *P6mmm*, #191; AlB_2 ω-type) – Fig. 27.1a. Cubic structure (cP7, *Pm3m*, #221- CaB_6 type) is characteristic for YB_6 – Fig. 27.1b. The family of high-temperature cuprate superconductors is represented by YBCO ($YBa_2Cu_3O_7$) with orthorhombic structure (oP14, *Pmmm*, #47, chain oxygen is in **b**-direction and vacancy in **a**-direction) – Fig. 27.1c. The family of classical low-temperature superconductors represents Nb_3Ge – superconductor with the highest T_c of this class of materials (A15 compounds – cP8, Pm3n, #223- Cr_3Si type) – Fig. 27.1d.

The band structures have been calculated by a computer code; Solid 2000. The code is based on the Hartree-Fock SCF method for infinite 3D-periodic cyclic cluster [55] with the quasi-relativistic INDO Hamiltonian [56]. Based on the results

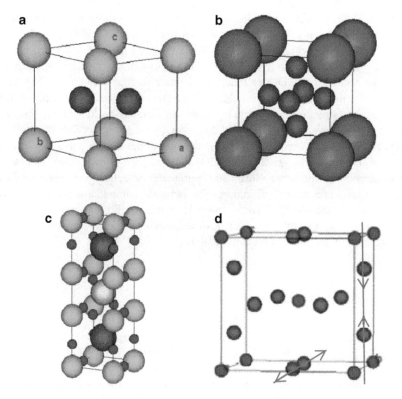

Fig. 27.1 Crystallographic cells of (**a**) MgB_2, (**b**) YB6, (**c**) $YBa_2Cu_3O_7$ and (**d**) Nb_3Ge with *arrows* indicating Nb atom vibration in $\Gamma_{1,2}$ phonon mode

of atomic Dirac-Fock calculations [57], the INDO version used in the SOLID package is parameterized for nearly all elements of the Periodic table of the elements. Incorporating the INDO Hamiltonian into the cyclic cluster method (with Born-von Karman boundary conditions) for electronic band structure calculations has many advantages and some drawbacks as well. The method is inconvenient for strong ionic crystals but it yields good results for intermediate ionic and covalent systems. The main disadvantage is overestimation of the total bandwidth. On the other hand, it yields satisfactory results for properties related to electrons at the Fermi level (frontier-orbital properties) and for equilibrium geometries [58–60].

In practical calculations, the basic cluster of the dimension ($N_a \times N_b \times N_c$), is generated by corresponding translations of the unit cell in the directions of crystallographic axes, a (N_a), b (N_b), c (N_c). On the basic cluster, the Born-von Karman boundary conditions are imposed and an "infinite" – 3D-periodic cyclic cluster structure is generated in calculation of matrix elements. In particular, the band structure calculations have been performed for the basic clusters $11 \times 11 \times 7$ for MgB_2, $9 \times 9 \times 9$ for YB_6, $5 \times 5 \times 5$ for $YBa_2Cu_3O_7$ and $11 \times 11 \times 11$ for Nb_3Ge. The scaling parameter 1.2 (1.0 for Nb_3Ge) has been used in calculations of the one-electron off-diagonal two-center matrix elements of the Hamiltonian (β-"hopping" integrals). The basic cluster of a given size generates a grid of ($N_a \times N_b \times N_c$) points in k-space. The HF-SCF procedure is performed for each k-point of the grid with the INDO Hamiltonian matrix elements that obey the boundary conditions of the cyclic cluster [55]. The Pyykko-Lohr quasi-relativistic basis set of the valence electron atomic orbitals (s,p-AO for Mg, B, Ba, O, Ge and s, p, d-AO for Cu, Y, Nb) has been used. The number of STO-type functions is unambiguously determined by that of valence AOs in atoms comprising the basic cluster. In general, the precision of the results of band structure calculation increases with increasing dimension of the basic cluster. It has been shown [55, 58–60], however, that there is an effect of saturation, a bulk limit beyond which the effect of increasing dimension on e.g. total electronic energy, orbital energies, HOMO-LUMO difference..., is negligible. In practice, the dimension of the basic cluster and parameter selection (e.g. for calculation of β integrals) is a matter of compromise between computational efficiency and convergence of calculated electronic properties and equilibrium geometry to some reference or experimental data. Note, however, that the basic efficiency and accuracy are restricted by the INDO parameterization.

27.2.2   Band Structures

In Fig. 27.2a, c, e, g(figures: left), are band structures (BS) of the studied compounds at equilibrium geometries. All the band structures are of adiabatic metal-like character with a relatively low density of states at the FL (indicated by a dashed line). Coupling to the respective phonon mode(s) in particular compounds seemingly does not change the metal-like character of BS. In all cases, however, e-p coupling induces BS fluctuation (see the pictures on the right), which is

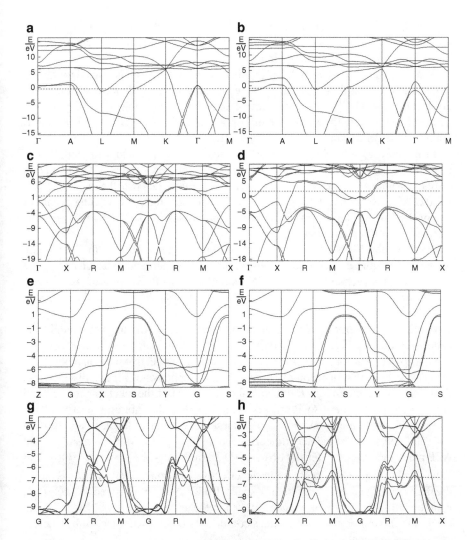

Fig. 27.2 Band structures of MgB_2 (**a**, **b**), YB_6 (**c**, **d**), $YBa_2Cu_3O_7$ (**e**, **f**), Nb_3Ge (**g**, **h**). Pictures on the *left* (**a**, **c**, **e**, **g**,) correspond to equilibrium high-symmetry structures. On the *right*: band structures (**b**, **d**, **f**, **h**) at distorted geometry with atom displacements in the respective phonon modes

characteristic by fluctuation of the analytic critical point (ACP) of some band across FL (cf. a-b, c-d, e-f, g-h).

In particular, for MgB_2 coupling to the E_{2g} phonon mode (in-plane stretching vibration of B-B) results in splitting of σ bands (p_x, p_y electrons of B atoms in **a-b** plane) in Γ point of the first Brillouin zone (BZ) – Fig. 27.2b. Related to band topology, the analytic critical point (ACP, maximum) of σ bands islocated at the

Γ point and, for a displacement $\approx 0.016 \text{A}°/\text{B}$-atom out of equilibrium position, the ACP crosses FL. This means periodic fluctuation of the BS between topologies $2a \leftrightarrow 2b$ in coupling to vibration in the E_{2g} mode.

The situation is similar for YB_6. In this case, BS fluctuation is related to the T_{2g} mode (valence vibration of B atoms in the basal **a-b** plane of B-octahedron). At the displacement $\approx 0.017 \text{ A}°/\text{B}$-atom out of equilibrium position, the ACP (saddle point) of the band with dominance of B-p and Y-d electrons crosses FL in the M point and the BS fluctuates between topologies $2c \leftrightarrow 2d$.

In the case of $YBa_2Cu_3O_7$, the BS fluctuation is associated with coupling to three modes, A_g, B_{2g}, B_{3g}, with the apical O(4) and CuO-plane O(2), O(3) atom displacements. At displacements $\approx 0.031 \text{ A}°$ of apical O(4) in the A_g mode and $\approx 0.022 \text{ A}°$ of O(2) and O(3) in the B_{2g}, B_{3g} modes, the ACP (saddle point) of one of the Cu-O plane (d-pσ) band in Y point crosses FL and undergoes periodic fluctuation between topologies $2e \leftrightarrow 2f$.

The situation for Nb_3Ge is presented in Fig. 27.2g, h. Coupling to Γ_{12} phonon mode (out-of phase vibration of Nb atoms in two perpendicular chains – see Fig. 27.1d, displacement $\approx 0.025 \text{ A}°/\text{Nb}$ atom) induces fluctuation of ACP (maximum) of $Nb(d_{x^2-y^2}, d_{z^2})$-bands at the R point across the FL, cf. the topology $2g \leftrightarrow 2h$.

In all cases presented in Fig. 27.2a–h, showing superconducting compounds, the band ACP crosses FL at a displacement which is less than the root-mean square (rms) displacement for zero-point vibration energy in respective phonon mode. This means, however, that in vibrations where the ACP approaches FL at a distance less than $\pm\omega$, the Fermi energy E_F (chemical potential μ) of the electrons in the band close to the point where ACP crosses FL is less than the vibration energy of the corresponding phonon mode, $E_F < \omega$. In these circumstances the adiabatic BOA is not valid and standard adiabatic theories cannot be applied. Moreover, shift of the ACP much increases the density of states (DOS) at FL, $n_\sigma(E_F) = (\partial\varepsilon_\sigma^0/\partial k)_{E_F}^{-1}$, and induces a corresponding decrease in the effective electron velocity $(\partial\varepsilon_\sigma^0/\partial k)_{E_F}$ of the fluctuating band in this region of k-space. Under these circumstances, the system is in the intrinsic nonadiabatic state, or even in the antiadiabatic state, $E_F \ll \omega$, and electronic motion depends on nuclear coordinates Q and is influenced by nuclear dynamics – momenta P.

Instability of the electronic structure at e-p coupling is absent in respective non-superconductive analogues, such as XB_2 ($X \equiv$ Al, Sc, Y, Ti, Zr, Hf, V, Nb, Ta, Cr, Mo, W, Mn,..), CaB_6, $YBa_2Cu_3O_6$ and Nb_3Sb. As an illustration, in Fig. 27.3 are band structures of these compounds at equilibrium high-symmetry structure (Fig. 27.3a, c, e, g) and at distorted geometry (Fig. 27.3b, d, f, h) with the same displacements in respective phonon modes as those in the case of corresponding superconductors at the transition in the antiadiabatic state.

In spite that for XB_2 compounds; coupling to the E_{2g} mode induces splitting of σ bands in the Γ point, the systems remain stable in the adiabatic state. For these systems, the ACP of the σ band does not fluctuate across FL. In Fig. 27.3 are band structures of AlB_2 at equilibrium high-symmetry structure (3a) and at distorted

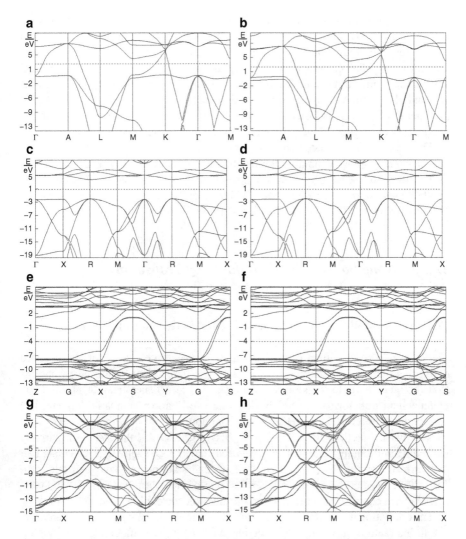

Fig. 27.3 Band structure of AlB_2, CaB_6, $YBa_2Cu_3O_6$, and Nb_3Sb at equilibrium geometry (**a, c, e, g**) and at distorted geometry (**b, d, f, h**)

geometry (3b) with the same B-atom displacements in the E_{2g} phonon mode as in the case of MgB_2 in transition to a superconducting state. In spite of σ bands splitting and nearly the same value of the e-p interaction strength (the calculated mean value is $\bar{u} \approx 1.01$ eV/u.cell) as that of MgB_2 ($\bar{u} \approx 0.98$ eV/u.cell), AlB_2 remains in e-p coupling in the adiabatic state as a non-superconductive compound. In this case, BS fluctuation (bands splitting in e-p coupling) does not decrease chemical potential. It remains in e-p coupling still larger than the vibration energy ($\mu_{ad} > \hbar\omega$) and, consequently, there is no driving force for transition into the antiadiabatic state.

In the case of deoxygenated YBCO that is without the chain oxygen [61] – YBa$_2$Cu$_3$O$_6$ (in contrast to the superconducting YBa$_2$Cu$_3$O$_7$) a combination of electron coupling to the A$_g$, B$_{2g}$ and B$_{3g}$ phonon modes leaves band structure without substantial change (Fig. 27.3e–f). In the case of YBa$_2$Cu$_3$O$_7$, the ACP (SP-saddle point) at Y point fluctuates across FL (see Fig. 27.2e–f), which yields substantial reduction of chemical potential $\rightarrow \mu_{antiad} < \hbar\omega$. For YBa$_2Cu_3O_6$ the SP does not fluctuate across FL and chemical potential remains larger than the phonon energy spectrum, $\mu_{ad} > \hbar\omega$, and the system remains in the adiabatic state.

The CaB$_6$ is an insulator and coupling to the T$_{2g}$ mode does not change this property – BS topology remains at e-p coupling without change (Fig. 27.3c–d).

The alloy Nb$_3$Sb of A15 class is non superconductive [62]. Metal-like character of Nb$_3$Sb and topology of BS remains without significant change at e-p coupling to Γ_{12} phonon mode of Nb atoms vibration –Fig. 27.3g, h.

27.2.3 Nonadiabatic Effects Induced by Transition into Antiadiabatic State

27.2.3.1 Formation of Antiadiabatic Ground State and Gap Opening

The main part of the effect of nuclear kinetic energy on electronic motion can be derived as diagonal correction by sequential Q,P-dependent base transformations [52] (or quasi-particle transformation [53]). This is a generalization of adiabatic Q-dependent transformation which yields the well-known adiabatic diagonal BO correction (DBOC) [63, 64]. Due to diagonal approximation with factorized form of total wave function, $\Psi_0(r,Q,P) = \Phi_0(r,Q,P).X_0(Q,P)$, the standard clamped nuclear Hamiltonian treatment can be used and the Q,P-effect is calculated in the form of corrections to the electronic ground-state energy (zero-particle term correction), corrections to orbital energies (one-particle term corrections) and two-particle term corrections (correction to the electron correlation energy).

The correction to the electronic ground-state energy in the k-space representation due to interaction of pair of states mediated by the phonon mode r can be written as [52–54],

$$\Delta E^0_{(na)} = 2 \sum_{\varphi_{Rk}} \sum_{\varphi_{Sk'}} \int_0^{\varepsilon_{k',max}} n_{\varepsilon_{k'}} (1 - f_{\varepsilon^0 k'}) d\varepsilon^0_{k'}$$

$$\int_{\varepsilon_{k,min}}^{\varepsilon_{kmax}} f_{\varepsilon^0 k} |u^r_{k-k'}|^2 n_{\varepsilon_k} \frac{\hbar\omega_r}{(\varepsilon^0_k - \varepsilon^0_{k'})^2 - (\hbar\omega_r)^2} d\varepsilon^0_k, \varphi_{Rk} \neq \varphi_{Sk'} \quad (27.1)$$

In general, all bands of 1st BZ of multiband system are covered. Coupling is of inter-band character, while $\varepsilon^0_k < \varepsilon_F$; $\varepsilon^0_{k'} > \varepsilon_F$.

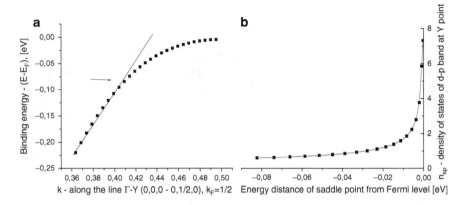

Fig. 27.4 Calculated dispersion of d-pσ band in ΓY direction with the kink formation indicated by the *arrow* (**a**) and increase of the d-pσ band DOS when ACP approaches FL (**b**) for $YBa_2Cu_3O_7$

Fermi-Dirac populations $f_{\varepsilon^0 k}$, $f_{\varepsilon^0 k'}$ make correction (27.1) temperature-dependent. Term $u^r_{k-k'}$ stands for matrix element of e-p coupling and $n_{\varepsilon k}, n_{\varepsilon k}$, are DOS of interacting bands at $\varepsilon^0_{k'}$ and ε^0_k. For adiabatic systems, such as metals, this correction is positive and negligibly small (DBOC). Only for systems in the antiadiabatic state the correction is negative and its absolute value depends on the magnitudes of $u^r_{k-k'}$ and $n_{\varepsilon k}, n_{\varepsilon k}$, at displacement for FL crossing. At the moment when ACP approach FL, the system not only undergoes transition to the antiadiabatic state but DOS of the fluctuating band is considerably increased at FL.

For all the studied systems at 0 K, the $\Delta E^0_{(na)}$ (which covers the effect of nuclear momenta) prevails in absolute value the electronic energy increase $\Delta E_{cr} = E_{d,cr} - E_{eq}$ at nuclear displacements $R_{d,cr}$ when ACP crosses FL as calculated for clumped nuclear adiabatic structures. For instance in case of $YBa_2Cu_3O_7$, calculated [65] increase at $R_{d,cr}$ is $\Delta E_{cr} = +170$ meV/unit cell, but correction to the total energy due to e-p coupling in antiadiabatic state is $\Delta E^0_{(na)} = -204$ meV/unit cell. The net effect of symmetry lowering (distortion) is the fermionic ground state energy stabilization. It means that due to effective nonadiabatic e-p coupling, the distorted structure for the specified displacements is by $(-204 + 170) = -34$ meV/unit cell more stable than undistorted – equilibrium structure on the BOA level. Under these circumstances, the system is stabilized in the antiadiabatic electronic ground state at broken symmetry with respect to the adiabatic equilibrium high-symmetry structure.

It can be identified by ARPES as a kink formation in the momentum distribution curve at FL, i.e. as band curvature at ACP when approaching FL – see the calculated results for $YBa_2Cu_3O_7$, (Fig. 27.4).

Due to translation symmetry of the lattice, the resulting antiadiabatic electronic ground state is degenerate when distorted, with a fluxional nuclear configuration in a given phonon mode(s) – see, e.g., Fig. 2 in Ref [50], or Fig. 27.6 below. The ground state energy is the same for different positions of the atoms involved (in phonon modes which drive the system into this state) in motion over circumferences of flux circles with radii equal to characteristic displacements $\Delta R_{cr} = |R_{eq} - R_{d,cr}|$ for FL crossing.

In transition to the antiadiabatic state, k-dependent gap $\Delta_k(T)$ in quasi-continuum of adiabatic one-electron spectrum is opened. The gap opening is related to shift $\Delta \varepsilon_{Pk}$ of the original adiabatic orbital energies ε^0_{Pk}, $\varepsilon_{Pk} = \varepsilon^0_{Pk} + \Delta \varepsilon_{Pk}$, and to the k-dependent change of DOS of particular band(s) at Fermi level. Shifts of orbital energies in band $\varphi_P(k)$ has the form [52–54],

$$\Delta \varepsilon(Pk') = \sum_{Rk'_1 > k_F} \left| u^{k'-k'_1} \right|^2 \left(1 - f_{\varepsilon^0 k'_1} \right) \frac{\hbar \omega_{k'-k'_1}}{\left(\varepsilon^0_{k'} - \varepsilon^0_{k'_1} \right)^2 - \left(\hbar \omega_{k'-k'_1} \right)^2}$$

$$- \sum_{Sk < k_F} \left| u^{k-k'} \right|^2 f_{\varepsilon^0 k} \frac{\hbar \omega_{k-k'}}{\left(\varepsilon^0_{k'} - \varepsilon^0_k \right)^2 - \left(\hbar \omega_{k-k'} \right)^2} \tag{27.2}$$

for $k' > k_F$, and

$$\Delta \varepsilon(Pk) = \sum_{Rk'_1 > k_F} \left| u^{k-k'_1} \right|^2 (1 - f_{\varepsilon^0 k'_1}) \frac{\hbar \omega_{k-k'_1}}{\left(\varepsilon^0_k - \varepsilon^0_{k'_1} \right)^2 - \left(\hbar \omega_{k-k'_1} \right)^2}$$

$$- \sum_{Sk_1 < k_F} \left| u^{k-k_1} \right|^2 f_{\varepsilon^0 k} \frac{\hbar \omega_{k-k_1}}{\left(\varepsilon^0_k - \varepsilon_{k_1 1^0} \right)^2 - \left(\hbar \omega_{k-k_1} \right)^2} \tag{27.3}$$

for $k \leq k_F$.

Replacement of discrete summation by integration, $\sum \ldots \to \int n(\varepsilon_k)$, introduces DOS $n(\varepsilon_k)$ into Eqs. 27.2 and 27.3, which is of crucial importance in relation to fluctuating band – see Fig. 27.4b. For corrected DOS $n(\varepsilon_k)$, which is the consequence of shift $\Delta \varepsilon_k$ of orbital energies, the following relation can be derived;

$$n(\varepsilon_k) = \left| 1 + (\partial(\Delta \varepsilon_k)/\partial \varepsilon^0_k) \right|^{-1} n^0(\varepsilon^0_k) \tag{27.4}$$

Term $n^0(\varepsilon^0_k)$ stands for uncorrected DOS of the original adiabatic states of particular band,

$$n^0(\varepsilon^0_k) = \left| (\partial \varepsilon^0_k / \partial k) \right|^{-1} \tag{27.5}$$

Close to the k-point where the original band, which interacts with fluctuating band, intersects FL, the occupied states near FL are shifted downward below FL and unoccupied states are shifted upward – above FL. The gap is identified as the energy between peaks in the corrected DOS above FL (half-gap) and below FL. The formation of peaks is related to the spectral weight transfer that is observed by ARPES or tunneling spectroscopy in cooling below T_c.

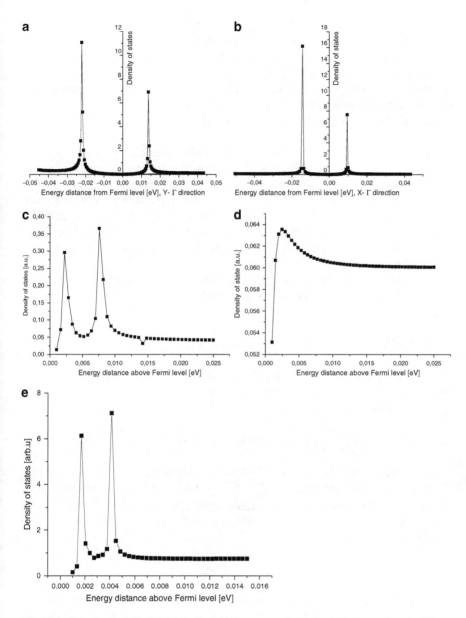

Fig. 27.5 Calculated (0 K) DOS in antiadiabatic state and gap formation near k point where particular bands intersect FL. The DOS are for YBCO in Y-Γ (**a**), X-Γ (**b**) direction, MgB$_2$ (**c**), YB$_6$ (**d**), and Nb$_3$Ge (**e**)

For the studied compounds, the calculated corrected DOS of particular band(s) with gap opening are in Fig. 27.5. In particular, YBa$_2$Cu$_3$O$_7$ exhibits an asymmetric gap in two directions: O1-pσ band gap is $\Delta_b(0) \approx 35.7$ meV in the $\Gamma - Y$ direction (5a) and $\Delta_a(0) \approx 24.2$ meV is in the $\Gamma - X$ direction (5b). The calculated asymmetry, i.e. the ratio $(\Delta_a(0)/\Delta_b(0))_{theor} \approx 0.68$ is very close to the experimental value (≈ 0.66) that has been recorded [66] for untwined $YBa_2Cu_3O_7$.

Two gaps, in σ and π band, are opened in $\Gamma - K(M)$ directions of MgB_2 – (5c): $\Delta_\sigma(0)/2 \approx 7.6\,meV$ and $\Delta_\pi(0)/2 \approx 2.2\,meV$. The result simulates tunneling spectra at positive bias voltage and calculated half-gaps are in a very good agreement with experimental high-precision measurements [67, 68].

A small gap opens on pd-band in the $\Gamma - X$ direction of YB_6 – (4d): $\Delta_{pd}(0)/2 \approx 2.2\,meV$.

In case of Nb_3Ge, we are not familiar with tunneling or ARPES spectra. However, recent results of point-contact spectroscopy obtained for superconductor of the same A15 group – $Nb_3Sn(T_c = 18.1\,K)$ [69], indicate that this system could be of two-gap character. Respective half-gaps are; $\Delta_1/2 \approx 3.92\,meV$ and much smaller gap is $\Delta_2/2 \approx 0.85\,meV$. Our calculation for $Nb_3Ge(T_c = 23.2\,K)$ is shown in Fig. 27.5e. Calculated DOS is also of two-gap character. The half gaps are; $\Delta_1/2 \approx 4.15\,meV$ and much smaller gap is $\Delta_2/2 \approx 1.7\,meV$. The gaps are opened in M-R direction of 1st BZ.

It should be stressed that this result is the first theoretical prediction of two-gap character for some superconductor of A15 family.

27.2.3.2 Critical Temperature T_c of Antiadiabatic State Transition

The corrections to orbital energies (2, 3) and to the ground state energy (1) are temperature dependent and decrease with increasing T. At a critical value T_c, the gap in one-particle spectrum [52–54],

$$\Delta(T) = \Delta(0)tgh[\Delta(T)/4k_B T] \tag{27.6}$$

as $\Delta(0)$ at 0 K, disappears – i.e. $\Delta(T_c) = 0$ (continuum of states is established at FL). At these circumstances holds $|\Delta E^0_{(na)}(T_c)| \leq \Delta E_{d,cr}$ and the system undergoes transition from the antiadiabatic into adiabatic state, which is stable for equilibrium high-symmetry structure above T_c. With respect to $\Delta(0)$, a simple approximate relation follows from Eq. 27.6,

$$T_c = \Delta(0)/4k_B \tag{27.7}$$

Calculated values of critical temperature for the set of studied compounds are presented in Table 27.1. As it can be seen, the values of T_c for transition into antiadiabatic state are in a good agreement with corresponding experimental values of T_c for superconducting state transition of particular compounds [3, 47, 70].

It should be noticed, however, that while there is a general consensus about importance of the e-p coupling to the A_g, B_{2g}, B_{3g} and E_{2g} phonon modes in case of $YBa_2Cu_3O_7$ and MgB_2 in transition to superconducting state, the situation with YB_6 is rather controversial. Recent studies [71, 72] advocate importance of e-p coupling to low-frequency (8–10 meV) phonon modes of Y-vibration for

Table 27.1 Calculated values of critical temperature for transition into antiadiabatic state (T_c-theor) and experimental values of critical temperature for transition into superconducting state (T_c-exp)

Compound	T_c-theor	T_c-exp
MgB_2	44 K	39.5 K
YB_6	12.2 K	8–10 K
$YBa_2Cu_3O_7$	103.5 K	92–94 K
Nb_3Ge	24 K	23.2 K

superconducting coupling. It is associated with overall value of dimensionless e-p coupling constant λ for calculation of T_c according to McMillan formula. For medium-strong coupling $\lambda \sim 1 - 1.4$ and Coulomb pseudopotential $\mu^* \sim 0.1 - 0.2$, the experimental $T_c \sim 6.2 - 9$ K is reproduced. In these circumstances, low-frequency Y- vibrations contribute by 84% to the overall value of λ. Our results show that coupling to Y-vibration does not induce the adiabatic-antiadiabatic state transition. This transition is connected to B-vibrations, in particular to T_{2g} mode. The value of dimensionless constant λ related to T_{2g} mode coupling, calculated in our study is, $\lambda_{T2g} \sim 0.1$. This value is in full agreement with decomposition of Eliashberg spectral function on contributions from the particular phonon modes in YB_6 calculated by Schell et al. [73]. The overall value of $\lambda \sim 0.48$ that accounts for nonlocal corrections on e-p coupling in the modes where B-octahedrons move as a whole is dominated by high-frequency (30–90 meV) B-vibrations. The authors [73] made conclusion that in transition to superconducting state in YB_6, the B-vibration phonon modes are essential. The conclusion is based on the fact that within McMillan formula, a small increase in overall $\lambda \sim 0.48$ can reproduce experimental T_c. In the present work, it is shown that in spite of the fact that coupling to T_{2g} mode is weak $(u^{k-k'} \sim 0.1\,\text{eV})$, transition in superconducting state and relatively high-value of T_c can be reached due to enormous increase of DOS of the fluctuating band in M point at FL, from the adiabatic value 0.06 states/eV to the value 1.09 states/eV at transition into antiadiabatic state.

27.2.3.3 Formation of Mobile Bipolarons in Real Space

From the theory of the antiadiabatic ground state [52a, b] follows that instead of Cooper pairs, formation of mobile bipolarons arise naturally as a result of translation symmetry breakdown at the antiadiabatic level. Bipolarons are formed as polarized inter-site charge density distribution, mobile on the lattice without dissipation due to degeneracy (fluxional structure) of the antiadiabatic ground state at distorted nuclearconfigurations. Formation of polarized inter-site charge density distribution at transition from adiabatic into antiadiabatic state is reflected by corresponding change of the wave function. For spinorbital (crystal orbital – band) $\varphi_{R(k)}$ holds [52]a,

$$|\varphi_S(x,Q,P)\rangle = a_S^+(x,Q,P)|0\rangle = \left(\bar{a}_S^+ - \sum_{rR} c_{SR}^r \bar{Q}_r \bar{a}_R^+ - \sum_{\check{r}R} \widehat{c}_{SR}^r P_{\check{r}} \bar{a}_R^+ \right.$$

$$\left. + O(\bar{Q}^2, \bar{Q}P, P^2)\right)|0\rangle = |\varphi_S(x,0,0)\rangle - \sum_{rR} c_{SR}^r \bar{Q}_r |\varphi_R(x,0,0)\rangle$$

$$- \sum_{\check{r}R} \widehat{c}_{SR}^r \bar{P}_{\check{r}} |\varphi_R(x,0,0)\rangle + \ldots\ldots \tag{27.8}$$

Expansion coefficients in (27.8) are coefficients of adiabatic Q-dependent canonical transformation c_{PQ}^r and of non-adiabatic P-dependent canonical transformation $\widehat{c}_{PQ}^r$,

$$c_{SR}^r = \frac{\partial c_{SR}(Q)}{\partial Q_r}; \quad \widehat{c}_{SR}^r = \frac{\partial \widehat{c}_{SR}(P)}{\partial P_r} \tag{27.8a}$$

Approximate solution [53] yields the following analytical forms,

$$c_{SR}^r = u_{SR}^r \frac{(\varepsilon_S^0 - \varepsilon_R^0)}{(\hbar\omega_r)^2 - (\varepsilon_S^0 - \varepsilon_R^0)^2}; \quad S \neq R \tag{27.8b}$$

$$\widehat{c}_{SR}^r = u_{SR}^r \frac{\hbar\omega_r}{(\hbar\omega_r)^2 - (\varepsilon_S^0 - \varepsilon_R^0)^2}; \quad S \neq R \tag{27.8c}$$

Bear in mind that for solids in reciprocal (quasi-momentum) space, the orbital energies are k-dependent, i.e. $\varepsilon_S^0 \equiv \varepsilon_S^0(k) \equiv \varepsilon_{S,k}^0$.

At transition into antiadiabatic state ($|\varepsilon_S^0(k_c) - \varepsilon_F^0|_{R_{eq}\pm Q} \ll \hbar\omega_r$), coefficients c_{RS}^r of Q-dependent transformation matrix (27.8b) become negligibly small and absolutely dominant for modulation of crude-adiabatic wave function are in this case coefficients $\widehat{c}_{RS}^r$ of P-dependent transformation matrix (27.8c). For simplicity, let us consider that transition into antiadiabatic state is driven by coupling to a phonon mode r with stretching vibration of two atoms (e.g. B-B in E_{2g} mode of MgB$_2$, valence T$_{2g}$ mode vibration of B-B atoms in basal **a-b** plane of B-octahedron in YB$_6$, vibration motion of O2, O3 in Cu-O planes – B$_{2g}$, B$_{3g}$ modes of YBCO, or Γ_{12} phonon mode vibration of Nb-Nb atoms in chains of Nb$_3$Ge in **a-b**, **a-c** or **b-c** plane). Let m_1 and m_2 are equilibrium site positions of involved nuclei on crude-adiabatic level and d_1 and d_2 are nuclear displacements at which crossing into antiadiabatic state occurs. At these circumstances the original crude-adiabatic wave function $\varphi_k^0(x,0,0)$, which corresponds to fluctuating crystal orbital (band) that crosses FL at e-p coupling, is changed in a following way,

$$\varphi_k(x,Q,P) \propto \left(1 + \sum_q u^{|q|} \frac{\hbar\omega_q}{(\hbar\omega_q)^2 - \left(\varepsilon_k^0 - \varepsilon_{k+q}^0\right)^2}\right.$$

$$\left. \times \left(P_1 e^{iq.[x-(m_1-d_1)]} + P_2 e^{iq.[x-(m_2+d_2)]}\right)\right) \varphi_k^0(x,0,0) \quad (27.9)$$

In (27.9), site approximation for momentum has been used, i.e. $P_q \propto (sign.q)\sum_m P_m e^{iq.m}$.

In the antiadiabatic state, for particular k and proper q values, nonadiabatic prefactors under summation symbol in (27.9) can be large. The prefactors, i.e. coefficients of P-dependent transformation matrix (27.8c), reflect influence of nuclear kinetic energy on electronic structure. At the dominance of these contributions (antiadiabatic state), strong increase in localization of charge density appears at distorted site-positions for x equal to $(m_1 - d_1)$ and $(m_2 + d_2)$. It induces (or increases) inter-site polarization of charge density distribution.

Schematic drawing illustrating these aspects in case of Nb$_3$Ge is presented in Fig. 27.6. The Γ_{12} phonon mode covers out-of phase stretching vibration of two perpendicular Nb chains in two planes – see Fig. 27.1d. For simplicity, drawing of only a single chain of Nb atoms in a plane (e.g. **b-c** plane) is sketched in Fig. 27.6. For equilibrium high-symmetry structure (R_{eq}) on the crude-adiabatic level, the highest electron density is localized at equilibrium position of Nb atoms in a chain – Fig. 27.6a. For distorted nuclear geometry ($R_{d,cr}$) in the Γ_{12} mode, electron density is polarized and the highest value is shifted into the inter-site positions– bipolarons are formed. The Fig. 27.6b corresponds to compression period in stretching vibration of Nb1-Nb2 which induces increase of Nb1-Nb2 inter-site electron density and decreases of Nb2-Nb3 electron density. For an expansion period, Fig. 27.6c, situation is opposite. Inter-site electron density is decreased for Nb1-Nb2 and increased for Nb2-Nb3. On the lattice scale, increase and decrease of electron density is periodic. On the adiabatic level, alternation of electron density is bound to vibrations at equilibrium nuclear positions (Fig. 27.6a–c).

As already mentioned, in the antiadiabatic state, ground state total electronic energy of system is degenerate. Distorted nuclear structure, related to a pair of nuclei in the phonon mode r (Γ_{12} phonon mode in case of Nb$_3$Ge), which induces transition into antiadiabatic state has fluxional character. There exist an infinite number of different – distorted configurations of this couple of nuclei in the phonon mode r and all these configurations, due to translation symmetry of the lattice, have the same ground state energy (Fig. 27.6b–e). Position of the involved displaced couple of nuclei is on the circumference of the flux circles with the centers at R_{eq} (equilibrium on crude-adiabatic level) with radii equal to $\Delta R_{cr} = |R_{eq} - R_{d,cr}|$. The $R_{d,cr}$ is distorted geometry at which ACP approaches FL and system undergoes transition from adiabatic into antiadiabatic state. Diameters of flux circles are $d_{cr} = 2\Delta R_{cr}$.

a-lattice parameter

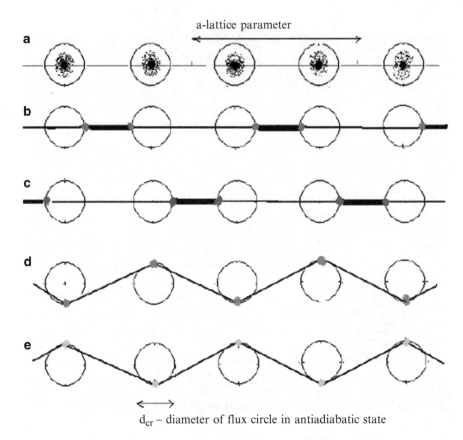

d_{cr} – diameter of flux circle in antiadiabatic state

Fig. 27.6 Schematic drawing of vibration periods and electron density in Nb-chain of A15 superconductors (Nb$_3$Ge, Nb$_3$Al,) Equilibrium geometry characterizes the line (**a**). *Lines* (**b**) and (**c**) represent compression and expansion period in vibration mode. The *circles* depict circumferences of flux-circles of degenerate antiadiabatic ground state. *Lines* (**d**), (**e**) represents cooperative transversal positions of Nb atoms at circumferential motion in antiadiabatic state – see text

Due to the geometric degeneracy of the ground state energy, the involved atoms can move over circumferences of the flux circles (Fig. 27.6b–e) without the energy dissipation.

The dissipation-less motion of the couple of nuclei implies, however, that e-p coupling of involved phonon mode and electrons of corresponding band has to be zero.

We have shown [74] that in the antiadiabatic state $(\hbar\omega_q/|\varepsilon_k^0 - \varepsilon_{k-q}^0| \to \infty)$ this aspect is fulfilled, i.e.

$$H'_{\langle e-p\rangle_0}(antad.state) \to 0 \qquad (27.10)$$

It means that for electrons which satisfy condition of extreme nonadiabaticity (antiadiabaticity with respect to interacting phonon mode r in particular direction of reciprocal lattice where the gap in one-electron spectrum has been opened), the electron (nonadiabatic polaron)-renormalized phonon interaction energy equals zero. Expressed explicitly, in the presence of external electric potential, dissipation-less motion of relevant valence band electrons (holes) on the lattice scale can be induced at the Fermi level (electric resistance $\rho = 0$). At the same time, the motion of nuclei remains bound to circumferential revolution over distorted, energetically equivalent, configurations. The electrons move in a form of itinerant-mobile bipolarons, i.e. as a polarized cloud of inter-site charge density distribution– sequence b, d, e, f, b, d, e, f. in Fig. 27.6. For temperature increase, thermal excitations of valence band electrons to conduction band induce sudden transition from the antiadiabatic state to adiabatic state at $T = T_c$, i.e. $|\Delta E^0_{(na)}(R_d)| \leq \Delta E_d(R_d)$ holds and the system is stabilized at equilibrium R_{eq} as it is characteristic for adiabatic structure.

In the adiabatic state, properties of the electrons are in sharp contrast with the properties of electrons in antiadiabatic state. The electrons are more or less tightly bound to respective nuclei and their motion is restricted to vibration at adiabatic equilibrium nuclear positions in a valence band and motion of electrons in conducting band is restricted by scattering with interacting phonon modes. It corresponds to situation at $T > T_c$.

For extreme adiabatic limit: $\hbar\omega_q/|\varepsilon^0_k - \varepsilon^0_{k-q}| \to 0$, it has been shown [74] that for e-p interaction energy holds,

$$H'_{\langle e-p\rangle_0}(ad.state) \to \sum_{qk} |u^q|^2 \frac{1}{\left(\varepsilon^0_k - \varepsilon^0_{k-q}\right)} \geq 0 \qquad (27.11)$$

Expression (27.11) represents basically energy of standard adiabatic polarons (small, self-trapped) that contributes to the total energy.

In Fig. 27.7, iso-density lines of highest electron density calculated for Nb_3Al on crude-adiabatic level by computer code SOLID2000 [55] are shown.

It represents cut of electron density in **b-c** plane of Nb_3Al crystal structure for 3×3 segments at different period of Nb atoms vibration in Γ_{12} phonon mode – Fig. 27.7a–f. The light spots correspond to regions of lowest electron density – positions of Al atoms. Situation for equilibrium nuclear configuration represents Fig. 27.7a with electron density localized at Nb atoms positions. Violet regions represent higher electron density which is dominated by contribution of $d_{x^2-y^2}$ and d_{z^2} AOs of Nb atoms. Iso-density lines encircle regions of highest electron density. Vibration of Nb atoms induces inter-site localization (compression and expansion period of vibration) – Fig. 27.7b, d and delocalization – Fig. 27.7c, e for transverse positions of Nb atoms in flux circles. It should be stressed that the figures correspond to electron density on crude-adiabatic level. Account for antiadiabatic situation increases substantially inter-site polarization due to contribution of the second – inter-site term with crucial antiadiabatic prefactor in Eq. 27.9.

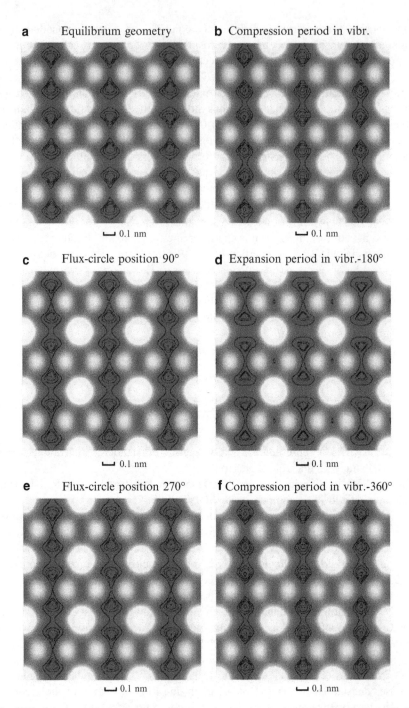

a Equilibrium geometry

b Compression period in vibr.

└─┘ 0.1 nm   └─┘ 0.1 nm

c Flux-circle position 90°

d Expansion period in vibr.-180°

└─┘ 0.1 nm   └─┘ 0.1 nm

e Flux-circle position 270°

f Compression period in vibr.-360°

└─┘ 0.1 nm   └─┘ 0.1 nm

Fig. 27.7 Calculated iso-density lines of highest electron density (cut in **b-c** plane) at equilibrium nuclear geometry R_{eq} of Nb$_3$Al (**a**) and at the distorted geometry $R_{d,cr}$ of Nb atoms in Γ_{12} phonon mode when nonadiabatic e-p interactions trigger transition into antiadiabatic state and inter-sites polarization is induced. Figures (**b–f**) represents different geometrical positions of Nb atoms on the circumferences of the flux circles

27.3   Discussion and Conclusion

Experimental results are always crucial for any theory which aims to formulate basic physics behind observed phenomenon or property. However, an experiment always cover much wider variety of different influences which have impact on results of experimental observation than any theory can account for, mainly if theory is formulated on microscopic level and some unnecessary approximations and assumptions are usually incorporated. On the other hand, interpretation of many experimental results is based on particular theoretical model. This is also the case of ARPES experiments at reconstruction of Fermi surface for electronic structure determination of high-Tc cuprates. Interpretation of experimental results is based on band structure calculated for particular compound. Methods of band structure calculations are always approximate, with different level of sophistication. Calculated band structure, mainly its topology at FL, is a kind of reference frame for assignment of particular dispersion of energy distribution curve (EDC) or momentum distribution curve (MDC) to particular band of studied compound at interpretation of ARPES. This is in direct relation with theoretical understanding of crucial aspects of SC-state transition in general.

In case of high-T_c cuprates, the crucial is considered to be the order parameter, i.e. symmetry of superconducting gap which should be of $(d_{x^2-y^2})$ symmetry. Within a simple orbital model consideration of CuO_2 plane in square lattice configuration, this symmetry is accepted to simulate **a-b** plane superconductivity of cuprates.

In this sense, unexpected results have been published recently [26]. The high-resolution (sub-meV) ARPES with VUV laser as a light source for optimally doped untwined YBCO reveals that in superconducting state, gap is opened in one-particle spectrum. The momentum dependence is very interesting, however. Like for other cuprates, strong band renormalization (kink formation at $\approx$60 meV) in off-nodal direction has been observed. The gap is opened in off-nodal (Γ-X and Γ-Y) directions with top of the dispersion at $\approx$20 meV below FL. In contrast to [66, 75], the results of [26] do not indicate presence of a/b asymmetry. What is surprising, however, is the finding that gap remains finite also in nodal Γ-S direction, with top of dispersion at $\approx$12 meV below FL. Presence of finite "nodeless" gap (i.e. in nodal direction) sharply contradicts accepted ideas of cuprates as superconductors of $(d_{x^2-y^2})$ symmetry. In normal state (100 K), the gap is closed in both, off-nodal and nodal directions.

The order parameter of $(d_{x^2-y^2})$ symmetry, assumes gap opening in a band with dominant contribution of Cu-based $(d_{x^2-y^2})$ orbital. In layered cuprates, it corresponds to $(d_{x^2-y^2} - p\sigma)$ bands of CuO_2 planes. Respected band structure calculation for YBCO is the DFT-based LDA published by Andersen et al. [76], to which the authors [26] refer. Inspection of BS in [76] reveals, however, that Cu2-O2O3- planes bands do not intersect FL in off-nodal Γ-X nor in Γ-Y direction, but the bands intersect FL in nodal Γ-S direction and in S –Y and S –X directions. It is an indication that experimentally detected gap [26], which is opened in off-nodal (Γ-X and Γ-Y) directions and also in nodal Γ-S direction, are not bands of CuO_2 planes.

In spite that there is a general agreement among different DFT-based band structure calculations in overall character of band structure for particular cuprate, there are small, but important differences in details concerning topology of some bands at FL. In particular for YBCO, DFT-based all-electron band structure calculation [77] using the FLAPW method yields important differences in topology of Cu-O chain-derived (d-pσ) band. In contrast to band structure [76] where the band intersects FL in S-X, Γ-Y and Γ-S directions, the BS [77] yields for Cu-O chain band intersections with FL in Γ-X, Γ-Y and Γ-S directions, respectively. Calculated Cu-O chain band topology [77] is in an excellent agreement with experimental electron-positron momentum density in optimally doped untwined YBCO detected [78] by 2D-angular correlation of electron-positron annihilation radiation (ACAR) technique. It should be stressed that ACAR is particularly sensitive for study of Cu-O chain Fermi sheets in YBCO. The character of ACAR, as experimentally detected, was predicted theoretically [79], and besides intersection of FL in Γ-X, Γ-Y directions, prediction has also been for FL intersection in Γ-T, Γ-U directions by Cu-O chain band.

With respect to topology of Cu-O chain band, as discussed above, this band should be the best candidate to be considered for gap opening in both, off-nodal (Γ-X, Γ-Y) and nodal (Γ-S) directions as seen in the ARPES results for YBCO [26]. One should stress, however, that also in case of gap opening in Cu-O chain band, expected $(d_{x^2-y^2})$ symmetry is lost. More over, the authors [26] attribute the effects seen in the ARPES to bands of Cu-O$_2$ planes and declare that in the spectra dispersion of Cu-O chain band is not present. There are a lot of experimental peculiarities of ARPES experiments with YBCO (twinning / untwining, surface states, CuO-chain problems, spectral dependence on light-source energy....), so it is difficult for me to make any comment about assignment of particular EDC/MDC to CuO$_2$ plane or Cu-O chain bands as done by the authors [26].

Nonetheless, an interpretation of the effects seen in the discussed ARPES from the stand-point of antiadiabatic theory should be of interest. The basic aspects concerning YBCO has been predicted [65] and shortly mentioned also in this paper. One should start with topology of the band structure[2] (Fig. 27.2e–f), in particular with dispersion of Cu-O chain band at FL. It can be seen that dispersion of this band at FL corresponds qualitatively to Cu-O chain band dispersion as it has been calculated by FLAPW method [77] and to its experimental character as seen at FL by ACAR method [78].

Coupling to A$_g$, B$_{2g}$, B$_{3g}$ phonon modes induces fluctuation (Fig. 27.2e $\leftrightarrow$ f, this paper) of the ACP (inflex point) of Cu-O$_2$ plane $(d_{x^2-y^2} - p\sigma)$ band at Y point across the FL. In the moment when the ACP approaches FL from the bonding side, strong renormalization of dispersion of this band (kink formation) could be seen by ARPES in off-nodal Y-Γ direction (Fig. 27.4a, this paper) at

---

[2]It should be stressed that HF-SCF method with semiempirical INDO Hamiltonian used at band structure calculation [55] overestimates bonding character and consequently band-width, which means that high-energy effects can hardly be studied, but for low-energy physics (like gap opening, kink formation,...) the method is reliable enough at least in a qualitative way.

about $k \approx 0.41/3.82 \approx 0.1\,A^{-1}$ (cf. with Fig. 2a(c) in [26]). At this situation, there is dramatic decrease of effective electron velocity and chemical potential ($\mu_{ad} \gg \mu_{antiad} < \hbar\omega$), while DOS is increased (Fig. 27.4b, this paper) at FL and system undergoes transition into antiadiabatic state. Influence of nuclear dynamics on electronic structure is now significant–according to antiadiabatic theory of e-p coupling as presented in this paper. System is stabilized in antiadiabatic state at distorted nuclear configuration $R_{d,cr}$ and gap is opened in one-particle spectrum. The gap (Sect. 27.2.3.1 – Fig. 27.5a, b, this paper) is opened only in the Cu-O chain (d-$p\sigma$) band, in off-nodal directions Γ-X and Γ-Y ($k_{\Gamma Y} \approx 0.053\,A^{-1}$, $k_{\Gamma X} \approx 0.016\,A^{-1}$) and also in nodal Γ-S direction. The half-gaps in the off-nodal directions are $\Delta_{\Gamma Y}/2, \approx 22\,meV$, $\Delta_{\Gamma X}/2, \approx 15\,meV$ and gap in the nodal direction is expected to be $\Delta_{\Gamma S}/2 \approx 15\,meV$ (cf. with Figs. 1, 7 and 8 in [26]). In contrast to [26], antiadiabatic theory yields a/b- gap asymmetry, which is in good agreement with other experimental results [66, 75]. At temperatures $T > T_c$, the gaps extinct. It should be manifested by disappearing of peaks on EDC (MDC) at FL in the ARPES spectra (see Fig. 8e in [26]). In this situation, the adiabatic DOS, with quasi-continuum of states at FL, is established and system is in adiabatic – non-superconductive state.

It should be mentioned that considerably smaller gap ($\Delta_{\Gamma T/U} \approx 5\,meV$) has been predicted [65] to be opened also in Γ-T/U directions. The prediction is related to the topology of the Cu-O chain (d-$p\sigma$) band in these directions (Figs. 10 and 11 in [65]). Theoretical calculation of the electron-positron momentum density [79] confirms this character of Cu-O chain topology. It is obvious that symmetry of the Cu-O chain band gap is not of ($d_{x^2-y^2}$) character. Symmetry of the gap is, in my opinion, the matter of band structure topology at FL, which is far more complicated than the one emerging from simple model of CuO_2 plane confined in a square lattice. It is a complex mater of crystal structure and chemical composition of particular cuprate. It should be reminded that YBCO is the only high-T_c cuprate with Cu-O chain in its structure. The gap opening in Cu-O chain band does not mean, however, that superconductivity in YBCO is realized in Cu-O chains. Superconductivity is realized by bipolaron mechanism in Cu-O_2 planes, no matter if gap symmetry is ($d_{x^2-y^2}$) or any other (antiadiabatic theory, see also [65]).

Based on the *ab initio* theory of complex electronic ground state of super-conductors, it can be concluded that e-p coupling in superconductors induces the temperature-dependent electronic structure instability related to fluctuation of analytic critical point (ACP – maximum, minimum or saddle point of dispersion) of some band across FL, which results in breakdown of the adiabatic BOA. When ACP approaches FL, chemical potential μ_{ad} is substantially reduced to $\mu_{antiad}(\mu_{ad} \gg \mu_{antiad} < \hbar\omega)$. Under these circumstances the system is stabilized, due to the effect of nuclear dynamics, in the antiadiabatic state at broken symmetry with a gap in one-particle spectrum. Distorted nuclear structure, which is related to couple of nuclei in the phonon mode r that induces transition into antiadiabatic state, has fluxional character. It has been shown that until system remains in antiadiabatic state, nonadiabatic polaron – renormalized phonon interactions are zero in well defined k-region of reciprocal lattice. Along with geometric degeneracy of the antiadiabatic ground state it enables formation of mobile bipolarons (in a form of polarized inter-site charge density distribution in real space) that can move over lattice in external electric potential as supercarriers without dissipation. With increasing T, the stabilization effect of nuclear kinetic energy to the electronic ground state energy decreases and at critical temperature T_c the gap(s) extinct and system is stabilized in the adiabatic metal-like state with a continuum of states at FL, which is characteristic by high-symmetry structure.

As it has been shown by analysis of e-p interaction Hamiltonian [52–54], an effective attractive e-e interaction, that is the basis of Cooper's pair formation, is in fact the correction to electron correlation energy at transition from adiabatic into antiadiabatic ground electronic state. In this respect, increased electron correlation is not the primary reason for transition into superconducting state, but it is a consequence of antiadiabatic state formation which is stabilized by nonadiabatic e-p interactions at broken translation symmetry. It has also been shown [52–54] that thermodynamic properties of system in the antiadiabatic state correspond to that of superconducting state.

Acknowledgments The author acknowledges support of the grants VEGA 1/0013/08; 1/0005/11 and Philip E. Hoggan for editing the English.

References

1. Onnes K (1911) Commun Phys Lab Univ Leiden 124c
2. Bednorz JG, Muller KA (1986) Z Phys B Condens Matter 64:189
3. Hor PH, Meng RL, Wang YQ, Gao L, Hung ZJ, Bechtold J, Forster K, Chu CW (1987) Phys Rev Lett 58:1191
4. Bardeen J, Cooper LN, Schrieffer JR (1957) Phys Rev 108:1175
5. Fröhlich H (1950) Phys Rev 79:845
6. Fröhlich H (1952) Proc R Soc A215:291
7. Migdal AB (1958) Zh Eksp Teor Fiz 34:1438; (1958) Sov Phys JETP 7:996
8. Eliashberg GM (1960) Zh Eksp Teor Fiz 38:996; (1960) 39:1437; (1960) Sov Phys JETP 11:696; (1960) 12:1000
9. Dagotto E (1994) Rev Mod Phys 66:763
10. Anderson PW (1997) The theory of superconductivity in high-Tc cuprates. Princeton University Press, Princeton
11. Kulic ML (2000) Phys Rep 338:1
12. McMillan WJ (1968) Phys Rev B 167:331
13. Maier Th, Jarrell M, Pruschke T, Keller J (2000) Phys Rev Lett 85:1524
14. Sorella S, Martins GB, Becca F, Gazza C, Capriotti L, Parola A, Dogotto E (2001) Phys Rev Lett 88:117002
15. White SR, Scalapino D (1998) Phys Rev Lett 80:1272; (1999) Phys Rev B 60:753
16. Paramekanti A, Randeria M, Trivedi N (2004) Phys Rev B 70:054504
17. Mourachkin A (2002) High-temperature superconductivity in cuprates. Kluwer Academic, Dordrecht
18. (a) Fukuoka A, Tokima-Yamamoto A, Itoh M, Usami R, Adachi S, Tamato S (1997) Phys Rev B 55:6612; (b) Frank J (1997) Physica C, 198:282–287
19. Pringle DJ, Williams GVM, Tallon JL (2000) Phys Rev B 62:12527

20. Lanzara A, Bogdanov PV, Zhou XJ, Kellar SA, Feng DL, Lu ED, Yoshida T, Eisaki H, Fujimori A, Kishio K, Shimoyama JI, Noda T, Uchida S, Hussain Z, Shen ZX (2001) Nature 412:510
21. Zhou XJ, Yoshida T, Lanzara A, Bogdanov PV, Kellar SA, Shen KM, Yang WL, Ronning F, Sasagawa T, Kakeshita T, Noda T, Eisaki H, Uchida S, Lin CT, Zhou F, Xiong JW, Ti WX, Zhao ZX, Fujimori A, Hassain Z, Shen ZX (2003) Nature 423:398
22. Gromko AD, Ferdorov AV, Chuang YD, Koralek JD, Aiura Y, Amaguchi YY, Oka KA, Youchi Ando, Dessau DS (2003) Phys Rev B 68:174520
23. Cuk T, Baumberger F, Lu DH, Ingle N, Zhou XJ, Eisaki H, Kanenko N, Hussain Z, Devereaux TP, Nagaosa N, Shen Z-X (2004) Phys Rev Lett 93:117003
24. Takahashi T, Sato T, Matsui H, Terashima K (2005) New J Phys 7:105
25. Zhou XJ, Cuk T, Evereaux TPD, Nagaosa N, Shen Z-X (2007) In: Schrieffer JR, Brooks JS (eds) Handbook of high-temperature superconductivity. Springer, New York, p 87
26. Okawa M, Ishizaka K, Uchiyama M, Tadamoto M, Masui T, Wang X-Y, Chen C-T, Watanabe S, Chainani A, Saitoh T, Schin S (2009) Phys Rev B 79:144528
27. Chung JH, Egami T, McQuinney RJ, Yethiraj M, Arai M, Yokoo T, Petrov Y, Mook HA, Endoch Y, Tajima S, Frost C, Dogan F (2003) Phys Rev B 67:014517
28. Pintschovius L, Reznik D, Reichardt W, Endoch Y, Hiraka H, Tranquada JM, Uchiama H, Masui T, Tajima S (2004) Phys Rev B 69:214506
29. (a) Pietronero L (1992) Europhys Lett 17:365; (b) Pietronero L, Strassler S (1992) Europhys Lett 18:627; (c) Pietronero L, Strassler S, Grimaldi C (1995) Phys Rev B 52:10516
30. Proville L, Aubry S (1999) Eur Phys J B 11:41
31. Bonca J, Trugman SA, Batistic I (1999) Phys Rev B 60:1633
32. Benedetti P, Zeyher R (1998) Phys Rev B 58:14320
33. Fehske H, Loos J, Wellein G (1997) Z Phys B 104:619
34. Bishop AR, Salkola M (1995) In: Salje EKH (ed) Polarons and bipolarons in High-tc superconductors and related materials. Cambridge University Press, Cambridge
35. Alexandrov AS, Kornilovitch PE (2002) J Supercond 15:403
36. (a) Alexandrov AS, Runninger J (1981) Phys Rev B 23:1796; (b) Alexandrov AS, Runninger J (1981) Phys Rev B 24:1164; (c) Alexandrov AS (1983) Zh Fiz Chim 57:273; (1983) Rus J Phys Chem 57:167
37. Scalapino DJ (1969) In: Parks RD (ed) Superconductivity, vol 1. Marcel Dekker, New York, p 449
38. Allen PB, Dynes RC (1975) Phys Rev B 12
39. Alexandrov AS (2001) Phys C 363:231
40. (a) Grimaldi G, Pietronero L, Strassler S (1995) Phys Rev B 52:10530; (b) Grimaldi G, Pietronero L, Strassler S (1995) Phys Rev Lett 75:1158
41. (a) Grimaldi G, Cappelluti E, Pietronero L (1998) Europhys Lett 42:667; (b) Grimaldi G, Pietronero L, Scottani M (1999) Europhys J B 10:247; (c) Cappelluti E, Grimaldi G, Pietronero L (2001) Phys Rev B 64:125104; (d) Botti M, Cappelluti E, Grimaldi G, Pietronero L (2002) Phys Rev B 66:054532; (e) Cappelluti E, Ciuchi S, Grimaldi G, Pietronero L (2003) Phys Rev B 68:174509
42. (a) Alexandrov AS (2003) Theory of superconductivity: from weak to strong coupling. IoP Publishing, Bristol/Philadelphia; (b) Alexandrov AS (2005) arXiv:cond-mat/0508769v4
43. Alexandrov AS, Edwards PP (2000) Phys C 331:97
44. Alexandrov AS (1992) Phys Rev B 23:2838
45. Hague JP, Kornilovich PE, Samson JH, Alexandrov AS (2007) J Phys Condens Matter 19:255214
46. Nagamatsu J, Nakagava N, Muranaka T, Zenitani Y, Akimitsu J (2001) Nature 410:63
47. An JM, Picket WE (2001) Phys Rev Lett 86:4366
48. Yilderim T, Gulseren O, Lynn JW, Brown CM, Udovic TJ, Huang Q, Rogado N, Regan KA, Hayward MA, Slusky JS, He T, Hass MK, Khalifah P, Inumaru K, Cava RJ (2001) Phys Rev Lett 87:037001
49. Kortus J, Mazin II, Belashchenko KD, Antropov VP, Boyer LL (2001) Phys Rev Lett 86:4656

50. Baňacký P (2005) Int J Quantum Chem 101:131
51. Boeri L, Cappelluti E, Pietronero L (2005) Phys Rev B 71:012501
52. (a) Baňacký P (2008) J Phys Chem Solids 69:2728; (b) Baňacký P (2008) J Phys Chem Solids 69:2696
53. Svrček M, Baňacký P, Zajac A (1992) Int J Quantum Chem 43:393; (1992) 43:415; (1992) 43:425; (1992) 43:551
54. Baňacký P (2010) Phonons and electron correlations in high-temperature and other novel superconductors, vol 2010. Hindawi Publ. Corp, New York. Advances in Condensed Matter Physics, special issue, Article ID 752943. http://www.hindawi.com/journals/acmp/2010/752943.html
55. Noga J, Baňacký P, Biskupič S, Boča R, Pelikan P, Zajac A (1999) J Comp Chem 20:253
56. Pople JA, Beveridge DL (1970) Approximate molecular orbital theory. McGraw-Hill Inc, New York
57. Boča R (1988) Int J Quant Chem 31:941; (1988) 34:385
58. Zajac A, Pelikán P, Noga J, Baňacký P, Biskupič S, Svrček M (2000) J Phys Chem B 104:1708
59. Zajac A, Pelikán P, Minar J, Noga J, Straka M, Baňacký P, Biskupič S (2000) J Solid State Chem 150:286
60. Pelikán P, Kosuth M, Biskupič S, Noga J, Straka M, Zajac A, Baňacký P (2001) Int J Quantum Chem 84:157
61. Baňacký P (2007) Phys C 460:1115
62. Lomnytska YF, Kuzma YB (2006) J Alloyes Comp 413:114
63. Kutzelnigg W (1997) Mol Phys 90:909
64. Svrček M, Baňacký P, Biskupič S, Noga J, Pelikan P, Zajac A (1999) Chem Phys Lett 299:151
65. Baňacký P (2009) In: Courtland KN (ed) Superconducting cuprates. Nova Science Pbl., New York, Chap 6, pp 187:212
66. Lu DH, Feng DL, Armitage NP, Shen KM, Damascelli A, Kim C, Ronning F, Shen ZX, Bonn DA, Liang R, Hardy WN, Rykov AI, Tajima S (2001) Phys Rev Lett 86:4370
67. Martinez-Samper P, Rodrigo JG, Rubio-Bollinger G, Suderow M, Vieira S, Lee S, Tajima S (2003) Phys C 385:233
68. Szabo P, Samuely P, Kacmarek J, Klein T, Marcus J, Fruchart D, Miraglia S, Mareinit C, Jensen AGM (2001) Phys Rev Lett 87:137005
69. Marz M, Goll G, Goldacker W, Lortz L (2010) Phys Rev B 82
70. Matthias BT, Gebale TM, Andres K (1968) Science 159:530
71. Lortz L, Wang Y, Tutsch U, Abe S, Meingart C, Popovich P, Knafo W, Schitsevalova N, Pademo YB, Junod A (2006) Phys Rev B 73:024512
72. Xu Y, Zhang L, Cui T, Li Y, Xie Y, Yu W, Ma Y, Zou G (2007) Phys Rev B 76:214103
73. Schell G, Winter H, Rietchel H, Gompf F (1982) Phys Rev B 25:1589
74. Baňacký P, Svrček M (1993) Int J Quantum Chem 46:475
75. Smilde HJM, Golubov AA, Ariando RG, Dekkers JM, Haskema S, Blank DHA, Rogalla H, Higenkamp H (2005) Phys Rev Lett 95:257001
76. Andersen OK, Lichtenstein AI, Jepsen O, Paulsen F (1995) J Phys Chem Solids 56:1573
77. Kouba R, Ambrosch-Draxl C, Zangger B (1999) Phys Rev B 60:9321
78. Haghighi H, Kaiser JH, Raynar S, West RN, Liu JZ, Shelton R, Howell RH, Solal F, Fluss ML (1991) Phys Rev Lett 67:382
79. Singh D, Pickett WE, van Stetten EC, Berko S (1990) Phys Rev B 42:2696

Chapter 28
Centre-of-Mass Separation in Quantum Mechanics: Implications for the Many-Body Treatment in Quantum Chemistry and Solid State Physics

Michal Svrček

Abstract We address the question to what extent the centre-of-mass (COM) separation can change our view of the many-body problem in quantum chemistry and solid-state physics. We show that the many-body treatment based on the electron-vibrational Hamiltonian is fundamentally inconsistent with the Born-Handy ansatz so that such a treatment can never fully account for the COM problem. The Born-Oppenheimer (B-O) approximation reveals a secret: it is the limiting case where the degrees of freedom can be treated classically. Beyond the B-O approximation they are in principle inseparable. The (unique) covariant description of all the equations, with respect to the individual degrees of freedom, leads to new types of interactions: in addition to the known vibronic (electron-phonon) ones the rotonic (electron-roton) and translonic (electron-translon) interactions arise. We have proved that as a result of the COM problem only the hypervibrations (hyperphonons, i.e. phonons + rotons + translons) have a general physical meaning in molecules and crystals; nevertheless, the use of pure vibrations (phonons) is a justified procedure only for so-called adiabatic systems. This state of affairs calls for a total revision of our contemporary view of general non-adiabatic effects, especially in connection with the Jahn-Teller effect and in formulating better approaches to superconductivity. Although the vibronic coupling is primarily responsible for the removal of the electron (quasi-) degeneracies the explanation of symmetry breaking and the formation of molecular and crystallic structures, rotonic and translonic couplings are necessary.

M. Svrček (✉)
Centre de Mechanique Ondulatoire Appliquée, CMOA Czech Branch,
Carlsbad, Czech Republic
e-mail: m.sv@o2active.cz

P.E. Hoggan et al. (eds.), *Advances in the Theory of Quantum Systems in Chemistry and Physics*, Progress in Theoretical Chemistry and Physics 22, DOI 10.1007/978-94-007-2076-3_28, © Springer Science+Business Media B.V. 2012

28.1   Introduction

As it is well known proper many body methods including Feynman diagrammatic techniques, developed in elementary particle physics, were transferred to solid-state physics many years ago. The introduction to quantum chemistry followed later, but only on the electronic level. So the question then appears: Is it possible to formulate the full quantum chemical electron-vibrational Hamiltonian in a second quantization formalism? The answer is negative. In fact the author did spend many years attempting to construct ideal representations by means of appropriate quasiparticle transformations (cf. equivalent Fröhlich type unitary transformations), but all variants, being either adiabatic- or nonadiabatic representations, did indeed fail. The reason lies actually on a deeper level than one would initially imagine.

The main scientific disciplines of quantum chemistry and solid-state physics were developed by way of a mathematical simplification or approximation of the Schrödinger equation, known as the Born-Oppenheimer (B-O) approximation [1]. It does not only give the basis of almost all quantum chemical calculations, but it also provides the very concept of molecular structure [2]. There are two main contemporary trends in quantum chemistry that put a question mark over the B-O approximation and its role in the definition of (molecular) structure: theories based on the incorporation of the centre-of-mass (COM) problem and applications in connection with the Jahn-Teller (J-T) effect.

There is no problem to include the COM problem in atomic calculations but its molecular implementation is very complicated. Monkhorst [3] did propose a simple model of molecular atoms for this purpose. The practical advantage of this approach, however was limited to the smallest molecules, is described in later works of Cafiero and Adamowitz [4], which were based on Monkhorst's ideas quoting: "We have the analogue of the nucleus with the heavy particle at the center of the internal coordinate system, and we have the analogues of electrons in the internal particles. The main difference between this model and an atom is that the internal particles in an atom are all electrons and in the "molecular atom" or "atomic molecule" the internal particles may be both electrons and nuclei (or, as we should more correctly say, pseudoparticles resembling the electron and the nuclei). Formally this difference manifests itself in the effective masses of the pseudoparticles and in the way the permutational symmetry is implemented in the wave function" [4].

This article contains an interesting note about the structure: "While molecular structure is a central concept in chemistry and physics, it should be remembered that for an isolated gas phase molecule in field-free space the most information that can be acquired is the average values of structural parameters (i.e., bond distances and angles). This point becomes apparent when molecular calculations are done without the B-O approximation – an almost universal approximation in quantum chemistry. While this approximation is extremely useful and has largely defined the terminology of modern spectroscopy, it also hides some simple quantum mechanical truths about the systems we study."

Moreover, the very limited applicability of the COM separation is not the only one problem; another difficulty arises from the introduction of the degrees of freedom for molecules, which are absent in atoms. Therefore Kutzelnigg [5] writes: "The adiabatic approximation with COM separation is good for atoms, because it takes care that the electrons participate in the COM motion. However, it is unbalanced for molecules, because it favours the COM motion with respect to other motions dominated by the nuclei, such as rotation and vibration, where the participation of the electrons is less trivial anyway than in the COM motion. The partial participation of the electrons in these motions is ignored in the adiabatic approximation both with and without COM separation."

The statement above is the reason why Kutzelnigg finally prefers the more pragmatic Born-Handy ansatz [6] proving its full equivalency with the COM separation: "Handy and co-workers have never claimed to have invented the ansatz referred to here as the "Born-Handy ansatz", but they certainly convinced a large audience that this ansatz is of enormous practical value, even if it has not been completely obvious why it leads to correct results. Handy and co-workers realized that the difficulties with the traditional approach come from the separation of the COM motion (and the need to define internal coordinates after this separation has been made). They therefore decided to renounce the separation."

This is without doubt a significant improvement since the application of the Born-Handy ansatz as a full replacement of the COM separation is not restricted by the size of the system under investigation. However, its applicability is unfortunately limited to adiabatic systems only. The Born-Handy formulation yields only the adiabatic corrections to the B-O results. Beyond this approach we enter the enigmatic region, which is usually denoted as the break-down of B-O approximation. Therefore our main goal is here to find an extension of the "Born-Handy formula", which is valid both in the adiabatic limit as well as beyond.

The most important consequence of the breakdown of the B-O approximation is indisputably the Jahn-Teller (J-T) effect [7], where the structure defined on the basis of this approximation does not in fact hold. The important role of the J-T effect is emphasized in Bersuker's book [8]: "Moreover, since the J-T effect has been shown to be the only source of spontaneous distortion of high-symmetry configurations, we come to the conclusion that the J-T effect is a unique mechanism of all the symmetry breakings in condensed matter." It is of course well known that problems related to the definition of crystallic structure in solid-state physics are mostly ignored assuming only BO structures. Nevertheless it is often criticised by scientists dedicated to studies of the J-T effect. For instance the proper understanding of superconductors should be evidently based on a solution of the non-adiabatic problem but the impact on the crystallic structure is neither reflected in the Fröhlich Hamiltonian [9, 10] nor in the BCS theory [11].

Bersuker writes in connection with the implications of the J-T effect on the superconductivity problem: "An illustration of the JT approach to electron–phonon coupling in solids may be found in the modern attempts to explain the origin of high-temperature superconductivity (HTSC). Experimental data show that the electron–phonon interaction is essential in this phenomenon." And continues "The existing BCS theory of superconductivity takes into account the electron–phonon interaction "in general" as an interaction of the electrons with the "bath of phonons" without detailed analysis of the local aspects of this interaction leading to the J-T effect. For broad-band metals with widely delocalized electrons the J-T electron–phonon coupling is weak and the J-T effect may be ignored. This is why the BCS theory explains the origin of superconductivity at low temperatures without taking into account the J-T effect. For narrower bands (which are characteristic of systems with HTSC) the J-T effect becomes significant, and the application of the achievements of the J-T effect theory to the HTSC problem seems to be most appropriate. This is indeed the subject of most current attempts to treat the HTSC yielding reasonable (reassuring) results."

At this stage the J-T approach to superconductivity is still problematic for basically two reasons: first, the J-T approach is not formulated in the many-body form as is the Fröhlich Hamiltonian, and second, this approach is ignorant with respect to the COM problem. Note, however, that the Fröhlich Hamiltonian and all theories based on it, including the BCS formulation, also suffer from the same omission.

Summarizing it is now desirable to find a unifying way incorporating both trends – the Born-Handy alternative of the COM separation and the J-T approach. However, this seems at first sight to be difficult requirement, because the Born-Handy formula accounts for the COM problem but it is only valid within the adiabatic limit. On the other hand, the J-T approach respects non-adiabaticity but does not comply with the COM problem.

As will be demonstrated here, this goal can be achieved through a revision of our previous understanding of the proper many-body second quantization formalism and by returning to the old question, i.e. "what will form molecular and crystalline structures when the B-O approximation breaks down?" Moreover, this reformulation will hopefully fulfil the views of many contemporary scientists: It will confirm the suggestion, due to Löwdin, that electron quasidegeneracies are not natural, and, the controversial idea, owing to Fröhlich, that superconductivity has to have a one-particle origin. Although the Fröhlich Hamiltonian was successfully incorporated in the BCS theory, Fröhlich did not accept this two-particle theory from the reasons mentioned above.

In the original development of quantum mechanics the belief in the Schrödinger equation, as applied to any number of nuclei and electrons, was simply that the calculation of molecular- and crystalline structure proceeded in a similar fashion as the calculation of the dynamics of complex objects by means of Newton mechanics. These "boring automatic applications of the Schrödinger equation" caused an outflow of many prominent scientists into other branches, e.g. nuclear physics, elementary particle physics, and later solid-state physics, with the field of quantum chemistry being "underestimated" for many years, rather than forming a platform (paradigmatic background) for true-many body treatments. Applications to systems slightly beyond the hydrogen atom, the simplest molecules, e.g. the hydrogen molecule played a decisive role in this development.

Solid-state physicists, on the other hand, got inspiration from the quantum field treatment of electron-photon interactions, and by "jumping over" nuclei, atoms, molecules, transferred these ideas into the field of solid-state condensed matter physics in the sophisticated form of a many-body treatment based on the electron-phonon interaction, in spite of the fact that its limits of validity were never investigated, particularly with respect to the COM problem.

The fundamentals of the approach to be presented here were already obtained in the authors PhD thesis in 1986 [12] and in an unpublished work from 1988 [13]. In full analogy with the solid-state electron-phonon interaction development, a similar apparatus for quantum chemistry was built utilizing the second quantization quasiparticle concept of the electron-vibrational Hamiltonian. This was a more complex operation than first anticipated. The formulation proceeded stepwise, i.e. first the crude representation, then the adiabatic, and finally the nonadiabatic one. As later recognized, quasiparticle transformations that leads to individual representations were in fact nothing but the full quantum chemistry equivalent of Fröhlich transformation used in solid-state physics.

Most of the equations and concepts, developed here, were subsequently published in a series of papers [14–21]. However, some very essential ingredients were missing in the original formulation, namely the inclusion of the COM problem. In particular, in the works [14–17] on the molecular adiabatic corrections were derived without correctly including the Born-Handy ansatz. Similarly, the concept of superconductivity, presented in [18–21], was restricted in the sense that the solution of Fröhlich's equation for the ground state assumed the structural instability, while the most important term – the trigger responsible for this structural instability – was missing. It goes without saying that these deficiencies must be remedied before a full picture of a truly COM compatible many-body treatment of nuclei and electrons on the same footing can emerge.

Therefore the author was confused when Biskupič, in 1998, during numerical tests performed on H2, HD and D2, using equations of the original aforementioned formulation, gave only a 20% contribution of the total groundstate adiabatic correction in comparison with the results of standard methods based on the Born-Handy formula [22]. Rather than being a bug in the program it was soon evident that this was a shortcoming of our theory. The first remedy was to convert the Born-Handy formula in a form compatible with the Coupled Perturbed Hartree-Fock (CPHF) [23,24] method just as the previous theory was expressed in the CPHF form, the step carried out in order to be able to make direct comparisons with the Born-Handy ansatz.

One learns directly that the converted Born-Handy formula leads to a curiosity, viz. the hydrogen molecule does not move and does not rotate. Nevertheless the final Born-Handy formula contains contributions from vibrational as well as from translational and rotational degrees of freedom in contrast to our previous theory, based on the quantization of the electron-vibrational Hamiltonian, which contained solely the contributions from the vibrational degrees. As it will be shown below, this understanding has a profound significance for all systems and phenomena beyond the Born-Oppenheimer approximation. Moreover, the interpretation of the

degrees of freedom in quantum mechanics will be discussed in detail and it will
be demonstrated that the presented improved theory gives these degrees quite a
different meaning compared to those of classical mechanics.

In this work we will emphasize those steps, which improves our previous
approach where the COM problem was ignored. By summarizing the conversion
of the Born-Handy ansatz to the CPHF compatible form [22], we will continue with
a presentation of the consequences of this result for correct quantizations of the
total Hamiltonian. This will be accompanied with a rederivation of the Born-Handy
formula from the newly developed theory, and it will be shown to be of significance
for the Jahn-Teller effect, for conductors, and for superconductors as well.

28.2   Conversion of the Born-Handy Formula in the CPHF Compatible Form

Let us start with the Born-Handy ansatz [6] for the groundstate electron wavefunction $\psi_0(\vec{R})$ where $\vec{R}$ represents nuclear coordinates. The adiabatic correction
ΔE_0 to the groundstate is expressed as a mean value of the nuclear kinetic
operator T_N [22],

$$
\begin{aligned}
\Delta E_0 &= \left\langle \psi_0(\vec{R}) \,\middle|\, T_N \,\middle|\, \psi_0(\vec{R}) \right\rangle_{R_0} \\
&= \sum_{i\alpha} \frac{\hbar^2}{2M_i} \left\langle \frac{\partial \psi_0(\vec{R})}{\partial R_{i\alpha}} \,\middle|\, \frac{\partial \psi_0(\vec{R})}{\partial R_{i\alpha}} \right\rangle_{R_0}
\end{aligned}
\tag{28.1}
$$

where after the integration per parts indexes i denote nuclei, α Cartesian coordinates
and M_i nuclear mass. In the adiabatic case the N-electron function $\psi_0(\vec{R})$ can be
expanded as a single Slater determinant though the one-electron functions $\varphi_I(\vec{R})$:

$$
\psi_0(\vec{R}) = \frac{1}{\sqrt{N!}} \left\| \prod_I^N \varphi_I(\vec{R}) \right\|
\tag{28.2}
$$

In this whole work we will use the following notation for spinorbitals: I, J, K, L –
occupied; A, B, C, D – virtual (unoccupied); P, Q, R, S – arbitrary ones. Substituting
(28.2) for $\psi_0(\vec{R})$ in (28.1) we get:

$$
\begin{aligned}
\Delta E_0 = \sum_{i\alpha} \frac{\hbar^2}{2M_i} \Bigg(&\sum_I \left\langle \frac{\partial \varphi_I}{\partial R_{i\alpha}} \,\middle|\, \frac{\partial \varphi_I}{\partial R_{i\alpha}} \right\rangle_{R_0} \\
&- \sum_{I \neq J} \left\langle \varphi_J \,\middle|\, \frac{\partial \varphi_J}{\partial R_{i\alpha}} \right\rangle \left\langle \frac{\partial \varphi_I}{\partial R_{i\alpha}} \,\varphi_I \right\rangle_{R_0} - \sum_{I \neq J} \left\langle \varphi_J \,\middle|\, \frac{\partial \varphi_I}{\partial R_{i\alpha}} \right\rangle \left\langle \frac{\partial \varphi_I}{\partial R_{i\alpha}} \,\varphi_J \right\rangle_{R_0} \Bigg)
\end{aligned}
\tag{28.3}
$$

The function $\varphi_P(\overrightarrow{R})$ can be expanded through the nuclear coordinates dependent coefficients $c_{PQ}(\overrightarrow{R})$ and the orthonormal set of one-electron wavefunctions defined in the equilibrium position R_0.

$$\varphi_P(\overrightarrow{R}) = \sum_Q c_{PQ}(\overrightarrow{R})\varphi_Q(R_0) \tag{28.4}$$

The Eq. 28.3 now reads:

$$\Delta E_0 = \sum_{i\alpha} \frac{\hbar^2}{2M_i} \left(\sum_{QI} |c_{QI}^{i\alpha}|^2 - \sum_{I\neq J} c_{JJ}^{i\alpha} c_{II}^{i\alpha*} - \sum_{I\neq J} |c_{IJ}^{i\alpha}|^2 \right) \tag{28.5}$$

where indices $i\alpha$ are related to first derivatives. Using the calibration for the diagonal coefficients in accordance with the CPHF formulation [22–24]

$$c_{II}^{i\alpha} = 0 \tag{28.6}$$

we arrive the very simple CPHF form of Born-Handy formula for the groundstate:

$$\Delta E_0 = \sum_{AIi\alpha} \frac{\hbar^2}{2M_i} |c_{AI}^{i\alpha}|^2 \tag{28.7}$$

This is the expression based on real coordinates $R_{i\alpha}$. But the analogical expression based on normal ones Q_r will be much more interesting. Therefore we introduce, again in accordance with the CPHF formulation, the expansion coefficients c_{PQ}^r.

$$c_{PQ}^r = \sum_{i\alpha} c_{PQ}^{i\alpha} \frac{\partial R_{i\alpha}}{\partial Q_r} = \sum_{i\alpha} c_{PQ}^{i\alpha} \alpha_{i\alpha}^r \tag{28.8}$$

In order to substitute for $c_{PQ}^{i\alpha}$ in (28.7), we need to know the inverse matrix $\beta_{i\alpha}^r$. Since it holds

$$\alpha^+ \beta = I \tag{28.9}$$

(28.7) can be rewritten as

$$\Delta E_0 = \sum_{AIrsi\alpha} \frac{\hbar^2}{2M_i} c_{AI}^r c_{AI}^{s\,*} \beta_{i\alpha}^{r\,*} \beta_{ia}^s \tag{28.10}$$

It is important to note, that the summation in (28.10) must be performed over all degrees of freedom, i.e. 3N, including 2 or 3 rotational and 3 translational ones! From the equiparticle theorem we know that the potential as well as the kinetic energy, standardly defined in accordance with the CPHF theory, contribute the same values – the halves of vibrational energy. But whereas diagonalizing the potential energy we get $3N - 5$ or $3N - 6$ nonzero values for vibrational modes and 5 or 6 zero values, the same procedure applied on the kinetic energy gives us 3N nonzero values, aside the vibrational modes we have 2 or 3 nonzero values for rotational modes and 3 nonzero values for translational ones.

$$\alpha^+ E_{pot} \alpha = \left\{ \frac{1}{2} \hbar \omega, 0, 0 \right\} \tag{28.11}$$

$$\beta^+ E_{kin} \beta = \left\{ \frac{1}{2} \hbar \omega, \rho, \tau \right\} = \sum_{i\alpha} \frac{\hbar^2}{4M_i} \beta_{i\alpha^*} \beta_{i\alpha} \tag{28.12}$$

With the help of (28.12) we can simplify the term inside of the Eq. 28.10,

$$\sum_{i\alpha} \frac{\hbar^2}{2M_i} \beta_{i\alpha}^{r*} \beta_{i\alpha}^{s} = \left\{ \hbar \omega_r |_{r\in V}, 2\rho_r |_{r\in R}, 2\tau_r |_{r\in T} \right\} \delta_{rs} \tag{28.13}$$

where ω_r represent the vibrational frequencies, whereas the ρ_r and τ_r represent the quanta of energy of the rotational and vibrational modes, extracted from the same secular equation for the kinetic energy (28.12) as the vibrational frequencies.

$$\Delta E_0 = 2 \sum_{AI} \left(\sum_{r\in V} \frac{1}{2} \hbar \omega_r + \sum_{r\in R} \rho_r + \sum_{r\in T} \tau_r \right) |c_{AI}^r|^2 \tag{28.14}$$

The final form of the Born-Handy formula consists of three terms: The first one represents the electron-vibrational interaction. I will not present the numerical details for H_2, HD and D_2 molecules here, it can be found in our previous work. The most important result here is that the electron-vibrational Hamiltonian is totally inadequate for the description of the adiabatic correction to the molecular groundstates; its contribution differs almost in one decimal place from the real values acquired from the Born-Handy formula. In the case of concrete examples – H_2, HD and D_2 molecules – the first term contributes only with ca 20% of the total value. The dominant rest – 80% of the total contribution – depends of the electron-translational and electron-rotational interaction [22]. This interesting effect occurs on the one-particle level, and it justifies the use of one-determinant expansion of the wave function (28.2). Of course, we can calculate the corrections beyond the Hartree-Fock approximation by means of many-body perturbation theory, as it was done in our work [22], but at this moment it is irrelevant to further considerations.

The Born-Handy ansatz [6] was verified on simple molecular systems many times in the last years [25–27], and especially interesting is the comparison of this simple pragmatic ansatz with the rigorous methods based on the separation of the centre-of-mass motion where one gets rather complicated expressions in terms of relative coordinates in a molecule-fixed frame. Kutzelnigg proved the validity of the Born-Handy ansatz by means of centre-of-mass analysis [5]. We can now ask: what happens in the case of the break-down of the adiabatic approximation? If the adiabatic case, beyond the B-O approximation, is a general centre-of-mass problem, then so is the break-down case much more the same problem. But where is there any reference to the centre-of-mass problem in the literature on e.g. the Jahn-Teller effect or superconductivity?

Centre-of-mass methods are generally based on the introduction of relative coordinates and relative effective masses and are hence very complicated. Their applicability in the case of the break-down of the adiabatic approximation is therefore questionable to say the least. Looking at the CPHF reformulation (28.14) we must agree that this equation is in contrast very simple. Equation 28.14 bounds together the 3N–6(5) vibration modes with 3(2) rotation and 3 translation ones. Therefore the new theory must be strictly built on the covariant notation regarding the individual degrees of freedom. They will be denoted 3N-dimensional hypervibrations in quantum chemistry and hyperphonons in solid-state physics.

Simultaneous use of the canonical transformations and introduction of degrees of freedom is unbalanced [5] and just this is a reason why rigorous methods eliminate their introduction at all [3, 4]. Nevertheless, if we insist on the introduction of degrees of freedom, we need to find an alternative to canonical transformations. To solve the COM problem on the many-body level therefore means to solve the compatibility problem of the many-body treatment with the Born-Handy ansatz, where the degrees of freedom are inseparable and have only a relative meaning.

28.3   Reconstruction of the Total Hamiltonian in the Second Quantization Formalism

In the beginning of this topic we pointed out some remarks about the cross-platform notation used in this paper. As one can see in the previous chapter we have used expansion coefficients c_{PQ} of the one-electron wavefunctions and eigenvectors $\alpha_{i\alpha}^r / \beta_{i\alpha}^r$ of the secular equations for the potential/kinetic oscillator energy defined on the set of complex numbers. In quantum chemistry this is often irrelevant, all the mentioned coefficients may be real, but in solid-state physics the complex number notation is necessary. In a similar way we will use the cross-platform notation for coordinate and momentum oscillator operators, namely $B_r = b_r + b_{\check{r}}^+$ and $\tilde{B}_r = b_r - b_{\check{r}}^+$. For systems that allow the real number solution of wavefunctions (quantum chemistry) it simply holds $r = \check{r}$. For system with translational symmetry where the conservation of quasimomentum holds (solid-state physics), the solution has complex number form, so that the meaning of spinorbitals P and vibrational modes r and $\check{r}$ after the transition in quasimomentum notation is: $P \to \mathbf{k}, \sigma, r \to \mathbf{q}, \check{r} \to -\mathbf{q}$. Since two modes with opposite sign of quasimomentum have the same energy, we assume that for any vibrational mode r there exists corresponding mode $\check{r}$ fulfilling the identity $\omega_r = \omega_{\check{r}}$.

The Born-Oppenheimer approximation leads in the final stage to the system of independent harmonic oscillators. The B-O vibrational Hamiltonian H_{BO} reads

$$H_{BO} = \sum_{r \in V} \hbar \omega_r \left(b_r^+ b_r + \frac{1}{2} \right) = \frac{1}{4} \sum_{r \in V} \hbar \omega_r (B_r^+ B_r + \tilde{B}_r^+ \tilde{B}_r) \qquad (28.15)$$

where r represents only the vibrational modes.

First we proceed to the introduction of hypervibrational modes. In accordance with secular equations (28.11) and (28.12) we get the following second quantization form for the potential and kinetic energies:

$$E_{pot} = \frac{1}{4} \sum_{r \in V} \hbar \omega_r B_r^+ B_r \qquad (28.16)$$

$$E_{kin} = \frac{1}{4} \left(\sum_{r \in V} \hbar \omega_r + 2 \sum_{r \in R} \rho_r + 2 \sum_{r \in T} \tau_r \right) \tilde{B}_r^+ \tilde{B}_r \qquad (28.17)$$

Let us denote the hypervibrational Hamiltonian as H_B.

$$H_B = E_{kin}(\tilde{B}) + E_{pot}(B) \qquad (28.18)$$

In order to get the covariant form of the Hamiltonian H_B, we will define the hypervibrational double-vector:

$$\omega = \begin{pmatrix} \omega_r \\ \tilde{\omega}_r \end{pmatrix} = \begin{pmatrix} \omega_r & 0 & 0 \\ \omega_r & \frac{2}{\hbar} \rho_r & \frac{2}{\hbar} \tau_r \end{pmatrix} \qquad (28.19)$$

This notation leads to the fully covariant expression for the Hamiltonian H_B with respect of all 3N hypervibrational modes.

$$H_B = \frac{1}{4} \sum_r (\hbar \omega_r B_r^+ B_r + \hbar \tilde{\omega}_r \tilde{B}_r^+ \tilde{B}_r) \qquad (28.20)$$

The essential problem of the Born-Oppenheimer approximation lies in the fact, that initially the electronic states are quantized whereas the motion of nuclei remains in classical form. Then the transition from the Cartesian to the normal coordinates is carried out on the basis of Newton mechanics, and finally the nuclear motion is quantized as the system of independent harmonic oscillators. This procedure represents the hierarchical type of quantization, which is a complete contradiction of the fundamental requirement of the second quantization procedure of the total Hamiltonian that must be simultaneous.

If we would try to find an ontological interpretation of the Born-Oppenheimer hierarchical type of quantization, surely the concept of quantization of atomic centers would be more adequate description than the concept of quantization of nuclei motion. The question arises how to retrace the simultaneous quantization of the unit system of electrons and nuclei. The author solved this problem in his

PhD thesis for the quantization of electron-vibrational Hamiltonian. Now a revised version will be presented for the complete electron-hypervibrational Hamiltonian which corrects the original version which totally failed during comparison tests with the results of the Born-Handy ansatz.

We cannot obviously start with the B-O approximation of moving electrons, following the motion of nuclei, but rather with the crude representation, i.e. the representation of fixed nuclear positions. The general form of the nonrelativistic Hamiltonian for any molecular and crystal system can be written in the form

$$H = T_N(\bar{\bar{B}}) + E_{NN}(\bar{B}) + \sum_{PQ} h_{PQ}(\bar{B})\,\bar{a}_P^+\bar{a}_Q + \frac{1}{2}\sum_{PQRS} v^0_{PQRS}\bar{a}_P^+\bar{a}_Q^+\bar{a}_S\bar{a}_R \qquad (28.21)$$

where T_N stands for kinetic energy of nuclei and E_{NN} for the potential energy of nuclei interactions. One-electron matrix elements h_{PQ} comprise electron kinetic energy and electron-nuclear interaction. The term v^0_{PQRS} represents two-electron interaction matrix elements. The operators marked with the bar are operators of "original" quasiparticles (electrons and hyperphonons) in the crude representation. The terms E_{NN} and h_{PQ} are defined through their Taylor expansion (see Eqs. 28.97–28.100 in Appendix 1 for details)

$$E_{NN}(\bar{B}) = \sum_{n=0}^{\infty} E_{NN}^{(n)}(\bar{B}) \qquad (28.22)$$

$$h_{PQ}(\bar{B}) = h^0_{PQ} + \sum_{n=1}^{\infty} u_{PQ}^{(n)}(\bar{B}) \qquad (28.23)$$

where h^0_{PQ} is one-electron term for fixed (equilibrium) nuclear coordinates and

$$u_{PQ}(\bar{B}) = \left\langle P \left| \sum_i \frac{-Z_i e^2}{|\mathbf{r} - \mathbf{R}_i|} \right| Q \right\rangle \qquad (28.24)$$

in terms of the second quantization represents the matrix elements of electron-hypervibrational interaction. We assume in (28.22 and 28.23) that the sums are convergent. It is important to emphasize that the operators $\bar{B}$ and $\bar{\bar{B}}$ in (28.21) refer to the whole set of hypervibrations.

The potential energy of the nuclear motion is defined through the quadratic part of internuclear potential plus some additive term representing the selfconsistent influence of electron-nuclear potential

$$E_{pot} = E_{NN}^{(2)}(\bar{B}) + V_N^{(2)}(\bar{B}) \qquad (28.25)$$

In the adiabatic limit the values of $V_N^{(2)}$ can be evaluated simply through the coupled perturbed Hartree-Fock method and the kinetic energy E_{kin} is identical with the kinetic energy of nuclei T_N. Now the crucial step is coming: At the case when the adiabatic approximation is not valid it is necessary to incorporate the new additive kinetic term originating from the kinetic energy of electrons. The resulting kinetic energy of the system has the form

$$E_{kin} = T_N(\bar{\bar{B}}) + W_N^{(2)}(\bar{\bar{B}}) \tag{28.26}$$

The total Hamiltonian (28.21) can be now divided into two parts

$$H = H_A + H_B \tag{28.27}$$

Where the first part H_A reads

$$H_A = E_{NN}(\bar{B}) - E_{NN}^{(2)}(\bar{B}) - V_N^{(2)}(\bar{B}) - W_N^{(2)}(\bar{\bar{B}}) + \sum_{PQ} h_{PQ}(\bar{B})\, \bar{a}_P^+ \bar{a}_Q$$

$$+\frac{1}{2} \sum_{PQRS} v_{PQRS}^0 \bar{a}_P^+ \bar{a}_Q^+ \bar{a}_S \bar{a}_R \tag{28.28}$$

and the second part H_B has the same form as (28.18)

$$H_B = \frac{1}{4} \sum_r (\hbar\omega_r \bar{B}_r^+ \bar{B}_r^+ \hbar\tilde{\omega}_r \bar{\bar{B}}_r^+ \bar{\bar{B}}_r) \tag{28.29}$$

The final electron-hypervibrational Hamiltonian in the second quantization formalism has now the form

$$H = E_{NN}(\bar{B}) - E_{NN}^{(2)}(\bar{B}) - V_N^{(2)}(\bar{B}) - W_N^{(2)}(\bar{\bar{B}}) + \sum_{PQ} h_{PQ}(\bar{B})\bar{a}_P^+ \bar{a}_Q$$

$$+\frac{1}{2} \sum_{PQRS} v_{PQRS}^0 \bar{a}_P^+ \bar{a}_Q^+ \bar{a}_S \bar{a}_R + \frac{1}{4} \sum_r \left(\hbar\omega_r \bar{B}_r^+ \bar{B}_r + \hbar\tilde{\omega}_r \bar{\bar{B}}_r^+ \bar{\bar{B}}_r \right) \tag{28.30}$$

It is necessary to notice that the crude representation (28.30) is the first and the last one where the quantization of nuclear motion can be accomplished by means of classical Newton mechanical separations of the degrees of freedom. All other representations will mix the vibrational, rotational and translational modes, and they will not be separable any more.

28.4 Unitary Transformations Applied to the Electron-Hypervibrational Hamiltonian

Our aim is now to find the most general group of quasiparticle transformations for the electron fermion and the hypervibration boson operators, binding individual representations of the total Hamiltonian. The author in his thesis on this topic [12] proposed two transformations – the first of the adiabatic type, dependent on nuclear coordinates Q, and the second of nonadiabatic type, dependent on the nuclear momenta P. When tried later to apply the results to solid-state physics, it was recognized that our transformations, applied to individual fermion and boson operators, were identical with the well-known Fröhlich transformation [10], applied to the whole Hamiltonian

$$H' = e^{-S(Q,P)} H e^{S(Q,P)} \tag{28.31}$$

namely with its decomposed part

$$H' = e^{-S_2(P)} e^{-S_1(Q)} H e^{S_1(Q)} e^{S_2(P)} \tag{28.32}$$

Although both expressions (28.31) and (28.32) are equivalent, they differ a somewhat in the individual orders of the Taylor expansion, since coordinate and momentum operators do not commute. This difference is not significant here but the latter one is more convenient due to the symmetric properties of the final expressions and it will be discussed hereafter.

The Fröhlich transformation is essential and it was applied to the superconductivity problem but it has not since been used in quantum chemistry problems.

Unfortunately Fröhlich could not know that the electron-phonon interaction was not the true interaction model and applied his transformation to the wrong Hamiltonian. Since the author did not initially involved the COM problem into the consideration, he made the same mistake as Fröhlich in all previous works referred to [12–20] with the aforesaid negative consequences for the later comparison tests [22] with the Born-Handy ansatz. Therefore we now present the solution which correctly incorporates all 3N degrees of freedom, unified under the conception of electron-hyperphonon interaction.

The advantage of the quasiparticle transformations lies in the fact that they are more transparent than the global transformation of the whole Hamiltonian. The first transformation in (28.32) with generator S_1 is equivalent to the adiabatic quasiparticle transformation from the crude into the adiabatic representation, defined through new quasiparticles in adiabatic representation with double bar

$$\bar{a}_P = \sum_Q c_{PQ}(\bar{\bar{B}}) \bar{\bar{a}}_Q \tag{28.33}$$

$$\bar{b}_r = \bar{\bar{b}}_r + \sum_{PQ} d_{rPQ}(\bar{\bar{B}}) \bar{\bar{a}}_P^+ \bar{\bar{a}}_Q \tag{28.34}$$

Analogous equations hold for the creation operators. The operators $c_{PQ}(\bar{\bar{B}})$ and $d_{rPQ}(\bar{\bar{B}})$ are defined trough their Taylor expansions and are limited through the unitarity conditions

$$\sum_R c_{PR} c_{QR}^+ = \delta_{PQ} \tag{28.35}$$

$$d_{rPQ} = \sum_R c_{RP}^+ [\bar{\bar{b}}_r, c_{RQ}] \tag{28.36}$$

The second transformation with generator S_2 is equivalent to the nonadiabatic transformation from the adiabatic representation into the final one, which we shall call "general", i.e. the representation that involves the adiabatic case as well as the nonadiabatic one. This representation is defined through new quasiparticles denoted simply without bar

$$\bar{\bar{a}}_P = \sum_Q \tilde{c}_{PQ}(\breve{B}) a_Q \tag{28.37}$$

$$\bar{\bar{b}}_r = b_r + \sum_{PQ} \tilde{d}_{rPQ}(\breve{B}) a_P^+ a_Q \tag{28.38}$$

where the operators $\tilde{c}_{PQ}(\breve{B})$ and $\tilde{d}_{rPQ}(\breve{B})$ are defined through their Taylor expansions and are limited through the unitarity conditions

$$\sum_R \tilde{c}_{PR} \tilde{c}_{QR}^+ = \delta_{PQ} \tag{28.39}$$

$$\tilde{d}_{rPQ} = \sum_R \tilde{c}_{RP}^+ [b_r, \tilde{c}_{RQ}] \tag{28.40}$$

For the Taylor expansion of both adiabatic and nonadiabatic unitarity conditions see Eqs. 28.101–28.104 in Appendix 1.

The form of the transformed Hamiltonian is very complex and the individual terms are put into Appendix 2. We demonstrate now only the main steps of treatment of the transformed Hamiltonian in the general representation.

At the first stage we will apply the Wick's theorem, as it is standardly defined in quantum chemistry, i.e. with respect to Fermi vacuum. For one-fermion terms the Wick's theorem results in

$$\sum_{PQ} \lambda_{PQ} a_P^+ a_Q = \sum_{PQ} \lambda_{PQ} N[a_P^+ a_Q] + \sum_I \lambda_{II} \tag{28.41}$$

and for two-fermion terms

$$\sum_{PQRS} \mu_{PQRS} a_P^+ a_Q^+ a_S a_R = \sum_{PQRS} \mu_{PQRS} N[a_P^+ a_Q^+ a_S a_R]$$

$$+ \sum_{PQI} (\mu_{PIQI} + \mu_{IPIQ} - \mu_{PIIQ} - \mu_{IPQI}) N[a_P^+ a_Q] + \sum_{IJ} (\mu_{IJIJ} - \mu_{IJJI}) \tag{28.42}$$

Analogous relations hold for three-fermion terms, which also occur in the transformed Hamiltonian. After complex application of the Wick's theorem on all fermion operators we get the normal form of the Hamiltonian in the general representation.

At the second stage we perform the very well known Moller-Plesset splitting [28] of the final Hamiltonian, i.e. the diagonalization of one-fermion terms in the normal form according to the formula

$$\sum_{PQ} \lambda_{PQ} N[a_P^+ a_Q] \rightarrow \sum_P \Lambda_P N[a_P^+ a_P] \tag{28.43}$$

First we obtain the well-known Hartree-Fock equation

$$f_{PQ}^0 = h_{PQ}^0 + \sum_I (v_{PIQI}^0 - v_{PIIQ}^0) = \varepsilon_P^0 \delta_{PQ} \tag{28.44}$$

Diagonalization of the terms which contain boson operators in the first order gives us equations for the first order coefficients of the unknown operators c and $\tilde{c}$ of quasiparticle transformations (28.33) and (28.37)

$$u_{PQ}^r + \left(\varepsilon_P^0 - \varepsilon_Q^0\right) c_{PQ}^r + \sum_{AI} \left[\left(v_{PIQA}^0 - v_{PIAQ}^0\right) c_{AI}^r - \left(v_{PAQI}^0 - v_{PAIQ}^0\right) c_{IA}^r \right] - \hbar \omega_r \tilde{c}_{PQ}^r$$

$$= \varepsilon_P^r \delta_{PQ} \quad c_{PP}^r = 0 \tag{28.45}$$

$$\left(\varepsilon_P^0 - \varepsilon_Q^0\right) \tilde{c}_{PQ}^r + \sum_{AI} \left[(v_{PIQA}^0 - v_{PIAQ}^0) \tilde{c}_{AI}^r - \left(v_{PAQI}^0 - v_{PAIQ}^0\right) \tilde{c}_{IA}^r \right] - \hbar \tilde{\omega}_r c_{PQ}^r$$

$$= \tilde{\varepsilon}_P^r \delta_{PQ} \quad \tilde{c}_{PP}^r = 0 \tag{28.46}$$

with the simplest chosen calibration for the diagonal terms.

Finally the set of equations in the second order of the Taylor expansion results in the ab-initio selfconsistent equations for hypervibrational frequencies ω and $\tilde{\omega}$, namely for unknown potential and kinetic matrix elements (28.25) and (28.26)

$$V_N^{rs} = \sum_I u_{II}^{rs} + \sum_{AI} [(u_{IA}^r + \hbar \omega_r \tilde{c}_{IA}^r) c_{AI}^s + (u_{IA}^s + \hbar \omega_s \tilde{c}_{IA}^s) c_{AI}^r] \tag{28.47}$$

$$W_N^{rs} = 2\hbar \tilde{\omega}_r \sum_{AI} c_{AI}^r \tilde{c}_{IA}^s \tag{28.48}$$

We can look at Eqs. 28.45–28.48 as the generalization of the CPHF [23, 24] equations for the case of general representation, i.e. including the cases of breakdown of the B-O approximation. We shall call them COM CPHF equations.

In the adiabatic limit where the coefficients $\tilde{c}$ equal zero we get

$$u_{PQ}^r + \left(\varepsilon_P^0 - \varepsilon_Q^0\right) c_{PQ}^r + \sum_{AI} \left[\left(v_{PIQA}^0 - v_{PIAQ}^0\right) c_{AI}^r - \left(v_{PAQI}^0 - v_{PAIQ}^0\right) c_{IA}^r \right]$$

$$= \varepsilon_P^r \delta_{PQ}; c_{PP}^r = 0 \tag{28.49}$$

$$V_N^{rs} = \sum_I u_{II}^{rs} + \sum_{AI} (u_{IA}^r c_{AI}^s + u_{IA}^s c_{AI}^r) \tag{28.50}$$

Adiabatic limit supposes that the vibrational frequencies ω are much smaller than the difference $\Delta\varepsilon^0$ between the first unoccupied and last occupied orbital. If we estimate the ratio $\tilde{c}/c$ from Eqs. 28.45 and 28.46) we can find out that the proportion holds

$$\tilde{c} \sim c\frac{\omega}{\Delta\varepsilon^0} \tag{28.51}$$

It means that the accuracy of adiabatic Eqs. 28.49 and 28.50 is limited up to the first order of the ratio $\omega/\Delta\varepsilon^0$. The adiabatic representation still links vibrations, rotations and translations into one set of nonseparable hypervibrations.

Now we can proceed from the adiabatic limit to the B-O limit. In both approximations the same Eqs. 28.49 and 28.50 hold. The only but remarkable difference is the classical concept of the separation of degrees of freedom in the latter one. It means that the coefficients r, s in these equations represent only the normal vibrational modes. And besides in this simplified form the Eqs. 28.49 and 28.50 are exactly identical with the standard Pople's CPHF equations [23, 24] after the formal rewrite from the fixed basis of atomic orbitals into the moving one, following the motion of nuclei. Since this is only a numerical problem, which does not affect the core of this topic, we only refer to preceding works [12, 17].

28.5 Derivation of the Extended Born-Handy Ansatz from the General Representation

Let us proceed to the fermion part of the general Hamiltonian, particularly the fermion part difference ΔH_F between the general Hamiltonian and the original crude one

$$H_F = H_{F(0)} + \Delta H_F \tag{28.52}$$

The Hamiltonian $H_{F(0)}$ consists of three well-known parts

$$H_{F(0)} = H_{F(0)}^0 + H_{F(0)}' + H_{F(0)}'' \tag{28.53}$$

where

$$H_{F(0)}^0 = E_0 = E_{NN}^0 + E_{SCF}^0 = E_{NN}^0 + \sum_I h_{II}^0 + \frac{1}{2}\sum_{IJ}(v_{IJIJ}^0 - v_{IJJI}^0) \tag{28.54}$$

contains the SCF energy of unperturbed electronic system and the nuclear potential energy,

$$H_{F(0)}' = \sum_P \varepsilon_P^0 N[a_P^+ a_P] \tag{28.55}$$

is one-electron spectrum as a result of diagonalization (28.44) and

$$H''_{F(0)} = \frac{1}{2} \sum_{PQRS} v^0_{PQRS} N[a^+_P a^+_Q a_S a_R]$$

(28.56)

represents two-electron Coulomb interaction in a normal product form.

The most interesting is the Hamiltonian ΔH_F consisting of four parts. As one can see in Appendix 3 the transformations produce also the three-fermion terms. Because they are irrelevant for further considerations we limit the study only to the three important parts:

$$\Delta H_F = \Delta H^0_F + \Delta H'_F + \Delta H''_F$$

(28.57)

For the correction to the ground state energy we get

$$\Delta H^0_F = \Delta E_0 = \sum_{AIr} \left(\hbar \tilde{\omega}_r |c^r_{AI}|^2 - \hbar \omega_r |\tilde{c}^r_{AI}|^2 \right)$$

(28.58)

The one-particle correction $\Delta H'_F$ is more complex and therefore we select only that terms which are decisive for excitation mechanism

$$\Delta H'_F = \sum_{PQr} \left[\hbar \tilde{\omega}_r \left(\sum_A c^r_{PA} c^{r*}_{QA} - \sum_I c^r_{PI} c^{r*}_{QI} \right) - \hbar \omega_r \left(\sum_A \tilde{c}^r_{PA} \tilde{c}^{r*}_{QA} - \sum_I \tilde{c}^r_{PI} \tilde{c}^{r*}_{QI} \right) \right]$$
$$N[a^+_P a_Q] + \sum_{PRr} \left[\left(\varepsilon^0_P - \varepsilon^0_R \right) \left(|c^r_{PR}|^2 + |\tilde{c}^r_{PR}|^2 \right) - 2\hbar \tilde{\omega}_r \mathrm{Re} \left(\tilde{c}^r_{PR} c^{r*}_{PR} \right) \right] N[a^+_P a_P]$$

(28.59)

The first part (28.59) is of a pure one-fermion origin and in the complete derivation (see Appendix 3) has a non-diagonal form. The second part is not of a pure one-fermion origin. It is a vacuum value of type $\langle 0|B_r B_s|0 \rangle$ and/or $\langle 0|\tilde{B}_r \tilde{B}_s|0 \rangle$ of the mixed fermion-boson terms, where the bosonic part is of the quadratic form of coordinate and/or momentum operators.

In a similar way we select from the correction $\Delta H''_F$ only the dominant term

$$\Delta H''_F = \sum_{PQRSr} (\hbar \tilde{\omega}_r c^r_{PR} c^{r*}_{SQ} - \hbar \omega_r \tilde{c}^r_{PR} \tilde{c}^{r*}_{SQ}) N[a^+_P a^+_Q a_S a_R]$$

(28.60)

If we proceed from the general to the adiabatic representation with zero $\tilde{c}$ coefficients we obtain exactly the Born-Handy ansatz (28.14) from the first principle derivation:

$$\Delta E_{0(ad)} = \sum_{AIr} \hbar \tilde{\omega}_r |c^r_{AI}|^2$$

(28.61)

Moreover, this expression is fully covariant with respect to all degrees of freedom, i.e. vibrations, rotations and translations.

Whereas the Born-Handy formula holds only in the framework of adiabatic approximation, the Eq. 28.58 is quite general and holds in the whole scale $\omega/\Delta\varepsilon^0$ including the non-adiabatic cases where the B-O approximation is broken. We denote it the generalized or extended Born-Handy formula. It is highly significant because, as we will see further, it defines the molecular and crystallic structure.

For a deeper insight into the properties of the extended Born-Handy formula we demonstrate its simple solution neglecting the two-electron terms. Then the Eqs. 28.45 and 28.46 have the analytical solution for the coefficients c and $\tilde{c}$:

$$c^r_{PQ} = u^r_{PQ} \frac{\varepsilon^0_P - \varepsilon^0_Q}{(\hbar\omega_r)^2 - (\varepsilon^0_P - \varepsilon^0_Q)^2} \tag{28.62}$$

$$\tilde{c}^r_{PQ} = u^r_{PQ} \frac{\hbar\tilde{\omega}_r}{(\hbar\omega_r)^2 - (\varepsilon^0_P - \varepsilon^0_Q)^2} \tag{28.63}$$

so that the extended Born-Handy formula can be expressed only by means of the matrix elements of electron-hypervibrational (electron-hyperphonon) interaction, one-electron energies and hypervibrational (hyperphonon) frequencies.

$$\Delta E_0 = \sum_{AIr} |u^r_{AI}|^2 \frac{\hbar\tilde{\omega}_r}{(\varepsilon^0_A - \varepsilon^0_I)^2 - (\hbar\omega_r)^2} \tag{28.64}$$

Rewriting this equation in the form of the sum of vibrational, rotational and translational parts, we obtain

$$\Delta E_0 = \sum_{AI, r \in V} |u^r_{AI}|^2 \frac{\hbar\omega_r}{(\varepsilon^0_A - \varepsilon^0_I)^2 - (\hbar\omega_r)^2}$$
$$+2 \sum_{AI, r \in R} |u^r_{AI}|^2 \frac{\rho_r}{(\varepsilon^0_A - \varepsilon^0_I)^2} +2 \sum_{AI, r \in T} |u^r_{AI}|^2 \frac{\tau_r}{(\varepsilon^0_A - \varepsilon^0_I)^2} \tag{28.65}$$

and so we have separate expressions for electron-vibrational (electron-phonon), electron-rotational (electron-roton) and electron-translational (electron-translon) contributions of the extended Born-Handy formula. In a full analogy with phonons – quasiparticles generated by the nuclear vibrations – we introduce similar quasiparticles for the nuclear rotations and translations, calling them simply rotons and translons.

In the case of adiabatic approximation where the inequality $\hbar\omega_r \ll \varepsilon^0_A - \varepsilon^0_I$ holds the Eq. 28.65 represents only a small correction to the energy of groundstate. But what about the non-adiabatic case? In the electron-phonon part of (28.65) there is a possible singularity when $\hbar\omega_r \approx \varepsilon^0_A - \varepsilon^0_I$. This case should affect the arising singularities in Eqs. 28.62 and 28.63 for the coefficients c and $\tilde{c}$. Fortunately this theory is fully selfconsistent and the extreme values of these coefficients should affect backward the frequencies ω_r defined through the Eqs. 28.47 and 28.48 so that the system has its own self-defense against such type of singularities, at least in molecules where the number of vibrational modes is finite. In crystals with infinite number of vibrational modes the integration valuer principal is used which eliminates this type of singularities.

The last two parts – electron-roton and electron-translon – of the Eq. 28.65 have singularities in the case of groundstate electron degeneracies. There are only two ways how to avoid them. Either all matrix elements u^r_{AI} for the degenerate states A and I and for rotational and translational modes have to equal zero, or the whole system has to change its structural arrangement in order to remove the degeneracy. The new equilibrium position of the nuclei naturally somehow increase the selfconsistent energy given by Eq. 28.54 but on the other hand the final energy of the ground state can be still smaller due to the first part of (28.65) which is negative in the case when both inequalities hold: $\varepsilon^0_A - \varepsilon^0_I > 0$ and $\varepsilon^0_A - \varepsilon^0_I < \hbar\omega_r$.

Now we can summarize previous considerations: In the case of the break down of the B-O approximation the electron-roton and electron-translon parts of the extended Born-Handy formula play the role of the trigger inducing a structural instability in the system. These parts are a direct consequence of the introduction of centre-of-mass problem into focus. They are responsible for the formation of molecular and crystallic structure. On the other hand, the electron-phonon part plays the role of a stabilizer of a new equilibrium position corresponding with new nuclear displacements. Therefore it is responsible for the formation of molecular and crystalline electronic structure.

28.6   Jahn-Teller Effect

The Jahn-Teller (J-T) effect is a direct consequence of the breakdown of the B-O approximation. At first this effect was studied only in a qualitative way on the basis of the group theory [7]. Nowadays there exist many extensive monographies dealing with the exact solutions of simple models where two degenerate or quasidegenerate levels are usually coupled with one or two vibrational modes [29, 30].

The J-T effect deals with molecular distortions due to electronically degenerate ground states. The J-T theorem was formulated as a statement: "For non-linear molecular entities in a geometry described by a point symmetry group possessing degenerate irreducible representations there always exists at least one non-totally symmetric vibration that makes electronically degenerate states unstable at this geometry. The nuclei are displaced to new equilibrium positions of lower symmetry causing a splitting of the originally degenerate states."

There were only a few articles devoted to the J-T effect in the 1930s before the World War II. Then the period of stagnation lasted almost two decades. Bersuker in his book [8] describes the reason:

Among other things Van Vleck [31] wrote that "it is a great merit of the J-T effect that it disappears when not needed." This declaration reflects the situation when there was very poor understanding of what observable effects should be expected as a consequence of the J-T theorem. The point is that the simplified formulation of the consequences of the J-T theorem as "spontaneous distortion" is incomplete and therefore inaccurate, and may lead to misunderstanding. In fact, there are several (or an infinite number of) equivalent directions of distortion, and the system may resonate between them (the dynamic J-T effect). It does not necessarily lead to observable nuclear configuration distortion, and this explains why such distortions often cannot be observed directly. Even in 1960 Low in his book [32] stated that "it is a property of the J-T effect that whenever one tries to find it, it eludes measurements."

Theoretical methods of the calculation of the J-T effect started to be developed in the 1950s, after the first experimental confirmations appeared. These methods are based on perturbation theory, in which the influence of the nuclear displacements via electron–vibrational (vibronic) interactions is considered as a perturbation to the degenerate states, and moreover, they are considered to be the proof of the J-T theorem.

Let us focus on the origin of the principal idea of the J-T effect. Before its final formulation by Jahn and Teller, first the Teller's student Renner [33] was inspired with the von Neumann–Wigner theorem about crossing electronic terms [34]: "Electronic states of a diatomic molecule do not cross, unless permitted by symmetry". Only if the states have different symmetry, they can cross.

Looking more carefully on the von Neumann-Wigner and the J-T theorems we can see significant differences: Whereas the first one is connected with the dissociation processes in molecules and the question of crossing or non-crossing potential curves, the second one concerns the rigid molecules and the question of their equilibrium nuclear positions. The J-T effect was formulated prematurely without the exact knowledge what forms the molecular structure and what the break down of B-O approximation really means. Two important factors were never incorporated in the J-T effect: the Fröhlich transformation and the centre-of-mass problem. These two factors are so profound that the J-T effect will never be satisfactorily explained without them.

After inclusion of the COM separation into our considerations, we can immediately recognize, from the Eq. 28.65, that vibronic coupling is really not responsible for the J-T effect; rather we find that the authentic J-T trigger is represented by electron-translational (translonic) and electron-rotational (rotonic) couplings.

Nowadays there are many attempts to implement the J-T effect into the problem of superconductivity. But first something specific related to superconductivity has to be implemented into the J-T effect, viz. the Fröhlich transformation. Fröhlich did propose his transformation [10] almost 20 years after the first formulation of the J-T effect [33]. Unfortunately, this transformation is mostly known in solid-state physics (and moreover used exclusively in the superconductivity problem) and after more than a half of century it has not been integrated in the domain of quantum chemistry. It is very important for several reasons first in the explanation of the hypervibronic coupling mechanism in the J-T effect. It further takes into account not only the dependence of electronic states on the nuclear coordinates, as it is usual in the adiabatic case, but also on the nuclear momenta, which is inherent in the non-adiabatic one. This type of transformation leads to new fermion quasiparticles that are able to describe the non-adiabatic case on the one-determinantal level, so that no secular problem with the crossing degenerate states arises as it is standardly used in the J-T calculations.

As we can see from the Eq. 28.65, after the translonic and rotonic coupling evokes the structural change, and therefore the small changes of the unperturbed energies ε^0, too, the vibronic coupling stabilizes the system in this new position. Let us look at the one-particle corrections $\Delta\varepsilon$ to these new unperturbed energies ε^0. For illustration we show only the diagonal corrections from the Eq. 28.59 in an analytical form after the neglecting of two-electron terms. Substituting from Eqs. 28.62 and 28.63 into (28.59) we get:

$$\Delta H'_F = \sum_P \Delta\varepsilon_P N[a_P^+ a_P]$$

$$= \sum_{Pr} \hbar\tilde{\omega}_r \left(\sum_{A\neq P} \frac{|u^r_{PA}|^2}{(\varepsilon_P^0 - \varepsilon_A^0)^2 - (\hbar\omega_r)^2} - \sum_{I\neq P} \frac{|u^r_{PI}|^2}{(\varepsilon_P^0 - \varepsilon_I^0)^2 - (\hbar\omega_r)^2} \right) N[a_P^+ a_P]$$

$$(28.66)$$

In order to have a better comparison of equations for one-electron energies (28.59), (28.66) and those for corrections to the ground state (28.58), (28.64) we introduce a symmetrical matrix Ω [13, 19].

$$\Omega_{PQ} = \sum_r \left(\hbar\tilde{\omega}_r |c^r_{PQ}|^2 - \hbar\omega_r |\tilde{c}^r_{PQ}|^2 \right) = \sum_r |u^r_{PQ}|^2 \frac{\hbar\tilde{\omega}_r}{(\varepsilon_P^0 - \varepsilon_Q^0)^2 - (\hbar\omega_r)^2}$$

$$\Omega_{PQ} = \Omega_{QP}; \quad \Omega_{PP} = 0 \qquad (28.67)$$

After the substitution of (28.67) into the aforementioned equations we get

$$\Delta E_0 = \sum_{AI} \Omega_{AI} \qquad (28.68)$$

$$\Delta\varepsilon_P = \sum_A \Omega_{PA} - \sum_I \Omega_{PI} \qquad (28.69)$$

Let us demonstrate the solution of the extended Born-Handy formula on an example of J-T effect with two degenerate electronic states. First rotonic and translonic coupling split them, so we obtain in a closed shell case one occupied orbital and one unoccupied (virtual) orbital with the unperturbed energies ε_o^0 and ε_u^0. Then the system finds its new equilibrium position via the vibronic coupling with the minimal value of the total energy. Therefore we have

$$\Delta E_0 = 2\Omega_{uo} < 0 \qquad (28.70)$$

and consequently the following relations hold

$$\Delta\varepsilon_o = \Omega_{uo} < 0; \qquad \Delta\varepsilon_u = -\Omega_{uo} > 0 \qquad (28.71)$$

We have proved, in this way, that the J-T one-electron energy splitting is a direct consequence of the solution of the extended Born-Handy formula.

The above-mentioned considerations have very important consequences. They imply that the extended Born-Handy formula, derived from the first principles, is the factual master equation for the explanation as well as for the calculation of the J-T effect. This effect was, however, discovered too soon without proper quantum mechanical knowledge. The inspiration given by the von Neumann-Wigner theorem was "out of the depth" in connection with this problem, and it could at most result in group theory formulations of the J-T effect. Since the Fröhlich transformation, accounted for the influence of nuclear momenta on the electronic states, and the center-of-mass problem, leading to the critical solution in the case of degenerate electronic states, were not taken into account, the J-T effect was falsely justified by the way of vibronic coupling. The authentic trigger of the J-T effect is not the vibronic, but rotonic and translonic coupling.

It is therefore necessary to reformulate the J-T effect, and not only in a version for molecules, but for both – molecules and crystals. The ontological statement emanating from the extended Born-Handy formula (28.65) is essential; all other considerations regarding the symmetrical properties of molecules and crystal, and of electronic states and vibration – rotation – translation modes follow as a consequence of the properties of this formula. Here is a new version of the reformulated J-T theorem:

Molecular and crystalline entities in the geometry of electronically degenerate ground states are unstable at this geometry except for the case when all matrix elements of electron-rotational and electron-translational interaction are equal zero.

28.7 Conductivity

Let us now focus on the case when all matrix elements of electron-rotational and electron-translational interaction equal zero. Then the system geometry of electronically degenerate ground state survives. This is exactly the case of conductors in solid-state physics. The electron-hypervibrational problem reduces to a simple classical electron-vibrational one. Equation 28.65 has then the form:

$$\Delta E_0 = \sum_{AI,r\in V} |u_{AI}^r|^2 \frac{\hbar\omega_r}{(\varepsilon_A^0 - \varepsilon_I^0)^2 - (\hbar\omega_r)^2} \tag{28.72}$$

The question arises, how to achieve the nullification of all electron-rotational and electron-translational terms u_{AI}^r in (28.65). If we try to find the solution in the form of the Bloch functions (which fully reflect the symmetry of the crystal), then since rotons and translons have zero quasimomentum values and virtual and occupied states A and I correspond to different quasimomentum values $\mathbf{k}$ and $\mathbf{k}$', naturally the above mentioned requirement for u_{AI}^r is fulfilled.

Now we can rewrite the Eq. 28.72 in solid-state notation ($r \rightarrow \mathbf{q}$; $I \rightarrow \mathbf{k}, \sigma$ with the occupation factor $f_{\mathbf{k}}$; $A \rightarrow \mathbf{k'}, \sigma'$ with the occupation factor $1 - f_{\mathbf{k'}}$; $\varepsilon_I^0 \rightarrow \varepsilon_{\mathbf{k}}^0$; $\varepsilon_A^0 \rightarrow \varepsilon_{\mathbf{k'}}^0$; $u_{AI}^r \rightarrow u_{\mathbf{k'k}}^{\mathbf{q}} = u^{\mathbf{k'-k}} = u^{\mathbf{q}}$).

$$\Delta E_0 = 2 \sum_{\mathbf{k},\mathbf{k'};\mathbf{k}\neq\mathbf{k'}} |u^{\mathbf{k'-k}}|^2 f_{\mathbf{k}}(1 - f_{\mathbf{k'}}) \frac{\hbar\omega_{\mathbf{k'-k}}}{(\varepsilon_{\mathbf{k'}}^0 - \varepsilon_{\mathbf{k}}^0)^2 - (\hbar\omega_{\mathbf{k'-k}})^2} \qquad (28.73)$$

This formula was derived by Fröhlich by means of the second order perturbation theory [9] and rederived by means of the unitary transformation [10]. Fröhlich in his derivations started with the generally accepted Hamiltonian in solid-state physics:

$$H = \sum_{\mathbf{k},\sigma} \varepsilon_{\mathbf{k}} a_{\mathbf{k},\sigma}^+ a_{\mathbf{k},\sigma} + \sum_{\mathbf{q}} \hbar\omega_{\mathbf{q}} \left(b_{\mathbf{q}}^+ b_{\mathbf{q}} + \frac{1}{2}\right) + \sum_{\mathbf{k},\mathbf{q},\sigma} u^{\mathbf{q}} \left(b_{\mathbf{q}} + b_{-\mathbf{q}}^+\right) a_{\mathbf{k}+\mathbf{q},\sigma}^+ a_{\mathbf{k},\sigma}$$

$$(28.74)$$

Here we stop to point out that just the Eq. 28.74 is the crucial problem. It involves only the electron-phonon terms and not the electron-hyperphonon ones which are necessary for the explanation of superconductors. This equation can be a good starting point for insulators, semiconductors and conductors, but never for superconductors. Fröhlich first believed that through the optimalization of occupation factors $f_{\mathbf{k}}$ in (28.73) he gets some decrease of the total energy and tried to interpret this new state as the state of superconductors, but later recognized that this solution leads to no experimentally detected gap. Although his transformations were unique and brilliant, they were unfortunately applied to the wrong Hamiltonian. Without knowing it, Fröhlich derived in his Eq. 28.73 exactly the correlation energy to the ground state of conductors.

After bypassing the trigger in (28.65), on one hand the crystal remains in the adiabatic state, but on the other the electrons from the last occupied (conducting) band are not part of the rigid system any more, they are quasi free and interact with the lattice only via the electron-phonon interaction without the backward influence on the lattice symmetry and nuclear displacements. The whole system is divided in two subsystems, the adiabatic "core" consisting of nuclei and electron valence bands, and the quasi free conducting electrons.

Therefore the Eq. 28.74 describes the crude representation with the energies of these two subsystems and the interaction terms between them. Justification of its use for conductors is not given a priori but as a consequence of the abnormal solution of the extended Born-Handy formula, which bypasses the trigger initiating the J-T effect, so that the general electron-hyperphonon problem can be reduced to the simple electron-phonon case. The explanation of conductivity is not at all so simple as it is universally believed. The COM problem plays here an important role and the conductivity represents only one possible solution of this problem. Fortunately, since this solution of the COM problem fully justifies the validity of the Hamiltonian (28.74) for conductors, all equations derived for them remain valid even though the COM problem was not included into the consideration.

Let us proceed to the one-electron corrections (28.59) when the electron-hypervibrational problem reduces to the electron-vibrational one. Since conductors have no gap, the one-particle derivation is more sensitive and we have to take into account also the second part of (28.59) which does not depend on the electron distribution defined by occupation factors as the first part but is exactly valid in the present form only for the vibrational vacuum.

The substitution for c and $\tilde{c}$ in (28.62) and (28.63) gives us

$$
\begin{aligned}
\Delta H'_F &= \sum_{P,r \in V} \left(\sum_{A \neq P} \frac{|u^r_{PA}|^2}{\varepsilon^0_P - \varepsilon^0_A - \hbar\omega_r} + \sum_{I \neq P} \frac{|u^r_{PI}|^2}{\varepsilon^0_P - \varepsilon^0_I + \hbar\omega_r} \right) N[a^+_P a_P] \\
&= \sum_{P,r \in V} \left(\sum_{R \neq P} |u^r_{PR}|^2 \frac{1}{\varepsilon^0_P - \varepsilon^0_R - \hbar\omega_r} - 2 \sum_{I \neq P} |u^r_{PI}|^2 \frac{\hbar\omega_r}{(\varepsilon^0_A - \varepsilon^0_I)^2 - (\hbar\omega_r)^2} \right) N[a^+_P a_P]
\end{aligned}
$$

(28.75)

and in the solid-state notation ($r \to \mathbf{q}; P \to \mathbf{k},\sigma, R \to \mathbf{k\text{-}q},\sigma, I \to \mathbf{k\text{-}q},\sigma$ with the occupation factor $f_{\mathbf{k\text{-}q}}$)

$$
\begin{aligned}
\Delta H'_F = \sum_{\mathbf{k},\mathbf{q},\sigma;\, \mathbf{q} \neq 0} |u^\mathbf{q}|^2 \frac{1}{\varepsilon^0_\mathbf{k} - \varepsilon^0_{\mathbf{k\text{-}q}} - \hbar\omega_\mathbf{q}} N[a^+_{\mathbf{k},\sigma} a_{\mathbf{k},\sigma}] \\
-2 \sum_{\mathbf{k},\mathbf{q},\sigma;\, \mathbf{q} \neq 0} |u^\mathbf{q}|^2 f_{\mathbf{k\text{-}q}} \frac{\hbar\omega_\mathbf{q}}{(\varepsilon^0_\mathbf{k} - \varepsilon^0_{\mathbf{k\text{-}q}})^2 - (\hbar\omega_\mathbf{q})^2} N[a^+_{\mathbf{k},\sigma} a_{\mathbf{k},\sigma}]
\end{aligned}
$$
(28.76)

The electron energies $\varepsilon^0_\mathbf{k}$ with the corrections (28.76) represent the well-known quasiparticles – polarons that were originally derived on the basis of Lee-Low-Pines transformation [35]. Now it is clear how the polarons can be directly derived from the general representation as a special solution of the COM problem where the trigger inducing the structural instability is bypassed. Whereas the first part of (28.76) concerns only individual polarons, the general representation yields also the second part of the corrections (28.76), which must be added to the polaron energies. Put differently, every polaron "feels" an effective field of other polarons, ergo, dressed polarons are created.

28.8 Fröhlich Hamiltonian and the BCS Theory

We will discuss now the two-particle term (28.60) for conductors. Again, the electron-hyperphonon problem reduces to the electron-phonon one, so after substitution for c and $\tilde{c}$ in (28.62) and (28.63) we get:

$$
\Delta H''_F = \sum_{PQRSr} \sum_{P \neq R, Q \neq S} u^r_{PR} u^{r*}_{SQ} \frac{\hbar\omega_r \left[\left(\varepsilon^0_P - \varepsilon^0_R\right)\left(\varepsilon^0_S - \varepsilon^0_Q\right) - (\hbar\omega_r)^2 \right]}{\left[\left(\varepsilon^0_P - \varepsilon^0_R\right)^2 - (\hbar\omega_r)^2 \right] \left[\left(\varepsilon^0_S - \varepsilon^0_Q\right)^2 - (\hbar\omega_r)^2 \right]} N\left[a^+_P a^+_Q a_S a_R \right]
$$

(28.77)

In solid-state notation this term reads ($r \to \mathbf{q}; P \to \mathbf{k}+\mathbf{q}, \sigma, Q \to \mathbf{k'}, \sigma'; R \to \mathbf{k}, \sigma,$ $S \to \mathbf{k'}+\mathbf{q}, \sigma'$):

$$\Delta H'' = \sum_{\mathbf{k},\mathbf{k'},\mathbf{q},\sigma,\sigma'} {}_{\mathbf{q}\neq 0} \frac{|u^{\mathbf{q}}|^2 \, \hbar\omega_{\mathbf{q}} \left[\left(\varepsilon_{\mathbf{k}+\mathbf{q}}^0 - \varepsilon_{\mathbf{k}}^0\right)\left(\varepsilon_{\mathbf{k'}+\mathbf{q}}^0 - \varepsilon_{\mathbf{k'}}^0\right) - (\hbar\omega_{\mathbf{q}})^2\right]}{\left[\left(\varepsilon_{\mathbf{k}+\mathbf{q}}^0 - \varepsilon_{\mathbf{k}}^0\right)^2 - (\hbar\omega_{\mathbf{q}})^2\right]\left[\left(\varepsilon_{\mathbf{k'}+\mathbf{q}}^0 - \varepsilon_{\mathbf{k'}}^0\right)^2 - (\hbar\omega_{\mathbf{q}})^2\right]}$$

$$N\left[a_{\mathbf{k}+\mathbf{q},\sigma}^+ a_{\mathbf{k'},\sigma'}^+ a_{\mathbf{k'}+\mathbf{q},\sigma'} a_{\mathbf{k},\sigma}\right] \tag{28.78}$$

When Fröhlich was unsuccessful with his derivation of the ground state energy correction (28.73), regarding the desired gap measured in superconductors, he declared in the last sentence of his second famous paper [10] that the theoretical treatment of superconductivity effects has to wait for the development of new methods for dealing with two-particle effective interaction, based on his transformation. He published it as a challenge that somehow by means of the true many-body treatment, going beyond the Hartree-Fock approximation, the expected gap could be achieved. He derived the following two-particle expression, known as the Fröhlich Hamiltonian:

$$\Delta H''_{F(Fr)} = \sum_{\mathbf{k},\mathbf{k'},\mathbf{q},\sigma,\sigma'} {}_{\mathbf{q}\neq 0} |u^{\mathbf{q}}|^2 \frac{\hbar\omega_{\mathbf{q}}}{\left(\varepsilon_{\mathbf{k}+\mathbf{q}}^0 - \varepsilon_{\mathbf{k}}^0\right)^2 - (\hbar\omega_{\mathbf{q}})^2} a_{\mathbf{k}+\mathbf{q},\sigma}^+ a_{\mathbf{k'},\sigma'}^+ a_{\mathbf{k'}+\mathbf{q},\sigma'} a_{\mathbf{k},\sigma}$$

$$\tag{28.79}$$

Comparing the Eqs. 28.78 and 28.79, we can see that they are different in two details. Our derivation contains the normal product of the creation and annihilation operators; therefore it is the two-particle correction to the one-particle solution represented by selfconsistent polarons (28.76). Fröhlich Hamiltonian does not contain the normal product; it refers directly to electron corrections. But this detail is not important.

The more interesting fact is the difference in the terms containing the electron and vibrational energies caused by application of various transformations (28.31) and (28.32). The first remarkable consequence of this fact is the symmetrical relation between indices $\mathbf{k}$ and $\mathbf{k'}$ in (28.78) that is not fulfilled in the expression (28.79). Wagner was the first who pointed out this problem in the Fröhlich's expression and therefore proposed the effective two-electron interaction gained on the basis of pure adiabatic transformation with the generator $S_1(Q)$ [36]. Later Lenz and Wegner [37] analysed in details the ambiguity of the form of the Fröhlich Hamiltonian by means of the continuous unitary transformations.

This ambiguity problem is also reflected in the reduced form of both Hamiltonian (28.78) and (28.79), used in the BCS theory [11]. Whereas our form of the reduced Hamiltonian is fully attractive,

$$\Delta H_{red} = -2 \sum_{\mathbf{k},\mathbf{k'};\mathbf{k}\neq\mathbf{k'}} |u^{\mathbf{k'}-\mathbf{k}}|^2 \frac{\hbar\omega_{\mathbf{k'}-\mathbf{k}}[(\varepsilon_{\mathbf{k'}}^0 - \varepsilon_{\mathbf{k}}^0)^2 + (\hbar\omega_{\mathbf{k'}-\mathbf{k}})^2]}{[(\varepsilon_{\mathbf{k'}}^0 - \varepsilon_{\mathbf{k}}^0)^2 - (\hbar\omega_{\mathbf{k'}-\mathbf{k}})^2]^2} a_{\mathbf{k'}\uparrow}^+ a_{-\mathbf{k'}\downarrow}^+ a_{-\mathbf{k}\downarrow} a_{\mathbf{k}\uparrow}$$

$$\tag{28.80}$$

Fröhlich's reduced Hamiltonian

$$\Delta H_{red(Fr)} = 2 \sum_{\mathbf{k},\mathbf{k}';\mathbf{k}\neq\mathbf{k}'} |u^{\mathbf{k}'-\mathbf{k}}|^2 \frac{\hbar\omega_{\mathbf{k}'-\mathbf{k}}}{(\varepsilon_{\mathbf{k}'}^0 - \varepsilon_{\mathbf{k}}^0)^2 - (\hbar\omega_{\mathbf{k}'-\mathbf{k}})^2} a_{\mathbf{k}'\uparrow}^+ a_{-\mathbf{k}'\downarrow}^+ a_{-\mathbf{k}\downarrow} a_{\mathbf{k}\uparrow} \quad (28.81)$$

has both attractive and repulsive parts.

Although the problem of the correct derivation of the Fröhlich Hamiltonian has been thoroughly discussed in the past, the much more important problem of the possibility of the creation of an energy gap by means of an effective attractive two-electron interaction was never re-examined, in spite of the fact that Fröhlich, who first derived this effective two-electron Hamiltonian finally never accepted the two-particle Cooper-pair based theory and claimed that the superconductivity has to be of one-particle origin.

We have studied the influence of two-particle interaction on the removing the degeneracy in continuous spectrum [21] and our results are surprising: This degeneracy can never be removed by a two-particle mechanism. The two-particle mechanism can only decrease the total energy but does not open any gap. It represents only the correlation energy. The detailed analysis was performed in our previous paper [21].

The most important argument against the explanation of supeconductivity on the basis of the effective two-electron Hamiltonian follows from the article [22] where Biskupič with his numerical test on H_2, HD and D_2 molecules confirmed that the Fröhlich based transformations contribute only with ca 20% of the total value of the adiabatic correlation energy. The error due to the neglecting of the COM problem is 400% whereas the error of the Hatree-Fock approach is only ca 7%. This fact implies our most important objection: The role of the COM problem in non-adiabatic cases, as e.g. superconductivity, is much more emergent than the two-particle treatment beyond the Hartree-Fock approach. The true many-body has to be primarily build on the electron-hyperphonon mechanism, and this consequently disqualifies the Fröhlich Hamiltonian and all theories build on it, including the BCS one. They cannot lead to any gap since they describe only the correlation energy of conductors.

28.9 State of Superconductivity

Whereas the extended Born-Handy formula (28.65) has a unique solution for small systems (molecules), for the large systems (solids) its solution is ambiguous. We have shown that the solution via bypassing the trigger leads to conductors. Now we will deal with another solution with an active trigger causing the change of the system geometry and removing the electron degeneracy.

Let us consider the conductor with the half-filled conducting band. Rotonic and translonic coupling first splits the initial lattice into two sublattices, so that the new arising system indicates only the half symmetry in respect to the initial one. This implies the splitting of the initial band into two new bands, overlapping on the unperturbed level. We denote the unperturbed energies of the lower valence band as $\varepsilon^0_{v,\mathbf{k}}$, and those of the higher conducting band as $\varepsilon^0_{c,\mathbf{k}}$. In a similar way we get twice as many hyperphonon branches – innerband with acoustical branches, and interband containing only the optical branches, but moreover rotons and translons. We denote the frequencies of the former set as $\omega_{a,\mathbf{q}}$ and the frequencies of the latter set as $\omega_{o,\mathbf{q}}$. Finally the vibronic coupling via the optical phonon modes stabilizes the whole system in this new configuration. After the rewriting of the Eq. 28.65 in solid-state notation $(r \to o, \mathbf{q}; I \to v, \mathbf{k}, \sigma; A \to c, \mathbf{k}', \sigma'; \varepsilon^0_I \to \varepsilon^0_{v,\mathbf{k}}; \varepsilon^0_A \to \varepsilon^0_{c,\mathbf{k}'}; u^r_{AI} \to u^{\mathbf{q}}_{\mathbf{k}'\mathbf{k}} = u^{\mathbf{k}'-\mathbf{k}} = u^{\mathbf{q}})$ we get

$$\Delta E_0 = 2 \sum_{\mathbf{k},\mathbf{k}'} |u^{\mathbf{k}'-\mathbf{k}}|^2 \frac{\hbar \omega_{o,\mathbf{k}'-\mathbf{k}}}{(\varepsilon^0_{c,\mathbf{k}'} - \varepsilon^0_{v,\mathbf{k}})^2 - (\hbar \omega_{o,\mathbf{k}'-\mathbf{k}})^2}$$

$$+ 4 \sum_{\mathbf{k},r \in R} |u^r|^2 \frac{\rho_r}{(\varepsilon^0_{c,\mathbf{k}} - \varepsilon^0_{v,\mathbf{k}})^2} + 4 \sum_{\mathbf{k},r \in T} |u^r|^2 \frac{\tau_r}{(\varepsilon^0_{c,\mathbf{k}} - \varepsilon^0_{v,\mathbf{k}})^2} \qquad (28.82)$$

Note that Eq. 28.82 totally differs from the (28.73) for conductors, which was derived by Fröhlich. His equation could never describe superconductors since it supposes only the B-O level of structure typical of conductors. On the other hand, the Eq. 28.82 fully respects the J-T splitting of bands. All unperturbed energies $\varepsilon^0_{v,\mathbf{k}}$ and $\varepsilon^0_{c,\mathbf{k}}$ with the same quasimomentum $\mathbf{k}$ have to differ in some small nonzero values. Instead of Cooper pairing of two electrons with opposite quasimomenta and spins, as it is stated in the BCS theory, we obtain the pairing between occupied valence and unoccupied conducting band electronic states with the same quasimomenta and spins, i.e. the coherent process over the whole crystal. This leads to a configuration with the single-valued occupancy of states: they are either occupied and belong to the valence band, or are unoccupied and belong to the conducting band. It seams that it is a similar solution, which is typical of insulators or semiconductors. On the contrary, the Fröhlich Eq. 28.73 leads to partial occupancy of states and is optimized with respect to the occupation factors, which is typical of conductors, whereas in the Eq. 28.82 the only optimized parameter is the J-T displacement of the former sublattice with respect to the latter one.

Now we have to answer the question whether the optimalization process of the ground state energy (28.82) is able to open an energy gap. The diagonal form of the J-T one-particle excitation expression (28.66) is fully justified in solid-state physics where the translational symmetry is supposed. Since we have two bands, in solid-state notation the one-particle Hamiltonian (28.66) reads

$$\Delta H'_F = \sum_{\mathbf{k},\sigma} (\Delta \varepsilon_{v,\mathbf{k}} + \Delta \varepsilon_{c,\mathbf{k}}) N[a^+_{\mathbf{k},\sigma} a_{\mathbf{k},\sigma}] \qquad (28.83)$$

so that we have two sets of one-particle corrections, one set for valence band electronic corrections and the latter set for conducting ones.

$$\Delta\varepsilon_{v,\mathbf{k}} = \sum_{\mathbf{q}\neq 0} |u^{\mathbf{q}}|^2 \left(\frac{\hbar\omega_{o,\mathbf{q}}}{(\varepsilon_{v,\mathbf{k}}^0 - \varepsilon_{c,\mathbf{k}-\mathbf{q}}^0)^2 - (\hbar\omega_{o,\mathbf{q}})^2} - \frac{\hbar\omega_{a,\mathbf{q}}}{(\varepsilon_{v,\mathbf{k}}^0 - \varepsilon_{v,\mathbf{k}-\mathbf{q}}^0)^2 - (\hbar\omega_{a,\mathbf{q}})^2} \right)$$

$$+2\sum_{r\in R} |u^r|^2 \frac{\rho_r}{(\varepsilon_{v,\mathbf{k}}^0 - \varepsilon_{c,\mathbf{k}}^0)^2} + 2\sum_{r\in T} |u^r|^2 \frac{\tau_r}{(\varepsilon_{v,\mathbf{k}}^0 - \varepsilon_{c,\mathbf{k}}^0)^2} \qquad (28.84)$$

$$\Delta\varepsilon_{c,\mathbf{k}} = -\sum_{\mathbf{q}\neq 0} |u^{\mathbf{q}}|^2 \left(\frac{\hbar\omega_{o,\mathbf{q}}}{(\varepsilon_{c,\mathbf{k}}^0 - \varepsilon_{v,\mathbf{k}-\mathbf{q}}^0)^2 - (\hbar\omega_{o,\mathbf{q}})^2} - \frac{\hbar\omega_{a,\mathbf{q}}}{(\varepsilon_{c,\mathbf{k}}^0 - \varepsilon_{c,\mathbf{k}-\mathbf{q}}^0)^2 - (\hbar\omega_{a,\mathbf{q}})^2} \right)$$

$$-2\sum_{r\in R} |u^r|^2 \frac{\rho_r}{(\varepsilon_{c,\mathbf{k}}^0 - \varepsilon_{v,\mathbf{k}}^0)^2} - 2\sum_{r\in T} |u^r|^2 \frac{\tau_r}{(\varepsilon_{c,\mathbf{k}}^0 - \varepsilon_{v,\mathbf{k}}^0)^2} \qquad (28.85)$$

We can take notice of innerband frequencies $\omega_{a,\mathbf{q}}$ that are not involved in the ground state energy equation but are present in one-particle correction terms. These terms are the same as those in the reduced Fröhlich's Hamiltonian (28.81), i.e. the denominators of them can achieve both positive and negative values. On the other hand the terms with interband optical frequencies $\omega_{o,\mathbf{q}}$ are optimized by means of the Eq. 28.82, therefore the negative denominators will be prevailing. This will result in negative values of $\Delta\varepsilon_{v,\mathbf{k}}$ and positive values of $\Delta\varepsilon_{c,\mathbf{k}}$. Of course, from the general form of the Eqs. 28.84 and 28.85 we cannot uniquely predicate the existence of a gap. Not all conductors become necessary superconductors at absolute zero. It depends on many factors but the most important factor is the bandwidth. It is apparent from (28.84) and (28.85) that the narrow bands (high T_C superconductors) result in greater gaps than broad bands (low T_C superconductors).

The most important fact is that the Eqs. 28.84 and 28.85 for the superconducting gap and entirely unlike polaron equations (28.76) for conductors are two different solutions of one common Eq. 28.59, as well as the ground state Eqs. 28.82 for superconductors and (28.73) for conductors are two solutions of one extended Born-Handy formula (28.58). This strongly contradicts the BCS theory, which seems to be "a better ground state" for conductors.

The privileged position of the extended Born-Handy formula can be seen also in the derivation of the main thermodynamical properties of superconductors. We need not know anything specific about superconductors; the pure assumption of the J-T like solution of this formula is sufficient.

Let us start with the temperature dependent form of the Eq. 28.69 [13, 19].

$$\Delta\varepsilon_P(T) = \sum_{A(T)} \Omega_{PA} - \sum_{I(T)} \Omega_{PI} = \sum_Q \Omega_{PQ}(1 - 2f_Q(T)) \qquad (28.86)$$

Fermions in the general representation naturally obey the Fermi-Dirac statistics and therefore the occupation probability for the state Q is given by the well-known expression

$$f_Q(T) = \frac{1}{e^{\frac{\varepsilon_Q(T)-\mu}{kT}} + 1} \qquad (28.87)$$

where ε_Q is the energy of the fermion state Q (i.e. $\varepsilon_Q^0 + \Delta\varepsilon_Q$). Then the Eq. 28.86 can be rewritten after substitution from the expression (28.87) as:

$$\Delta\varepsilon_P(T) = \sum_Q \Omega_{PQ} \text{tgh} \frac{\varepsilon_Q(T) - \mu}{2kT} \qquad (28.88)$$

In order to get a reasonable analytical result let us adopt a simplified model where for any virtual state we suppose (in solids this corresponds to an ideal narrow band case):

$$\varepsilon_A(T) - \mu = \Delta\varepsilon(T) \qquad (28.89)$$

and for any occupied state:

$$\varepsilon_I(T) - \mu = -\Delta\varepsilon(T) \qquad (28.90)$$

Then (28.88) has the form

$$\Delta\varepsilon_P(T) = \Delta\varepsilon_P(0)\text{tgh} \frac{\Delta\varepsilon(T)}{2kT} \qquad (28.91)$$

Further we omit the index P according to the simplifying conditions (28.89) and (28.90) and will search for the critical temperature T_c at which the energy gap vanishes. Because the energy gap Δ_0 at the zero temperature is given as:

$$\Delta_0 = 2\Delta\varepsilon(0) \qquad (28.92)$$

we finally get the ratio between the energy gap and the critical temperature

$$\frac{\Delta_0}{kT_c} = 4 \qquad (28.93)$$

For comparison, in the BCS theory this ratio is 3,52. In relative values both the BCS and our dependence of the energy gap on the temperature are exactly the same (i.e. the dependences of Δ/Δ_0 on T/T_c). The study of other physical properties, such as specific heat, is published in our previous paper [19]. Let us note that the Eq. 28.93 was derived without any specific requirements for the detailed mechanism of superconductivity in comparison with the BCS theory. It reflects the thermodynamical properties of non-adiabatic systems in a more general form, solely as a consequence of the solution of the extended Born-Handy formula.

As it was mentioned above, the Eq. 28.82 leads to the ground state, which is distinctive of insulators and semiconductors. How superconductors differ from them? There is one important difference: classical insulators are based on the structure defined by means of the B-O approximation, i.e. the structure with only one real ground state corresponding to the uniquely defined geometry for the minimum total energy of the system. On the other hand, the Eq. 28.82 is based on the J-T splitting of the original lattice of the conducting state into two sublattices. This splitting is never single-valued but there always exist several (or an infinite number) of equivalent directions of distortion. Therefore we can define superconductors as multigroundstate insulators with several equivalent ground states that correspond to different nuclear positions – Jahn-Teller equivalent configurations.

28.10 Effect of Superconductivity

We shall distinguish two fundamental attributes of superconductivity – the state of superconductivity and the effect of superconductivity – that lead to two complementary descriptions of superconductors. On one side the state of superconductivity is characterized by the state of a conducting material, which, after the Jahn-Teller condensation, becomes an insulator with several equivalent ground states. The state of superconductivity determines all statical properties of superconductors: energy gap, its temperature dependence, specific heat, density of states near the Fermi surface etc. On the other side the effect of superconductivity determines all dynamical properties of superconductors: supercurrent, Meissner effect, quantization of magnetic flux, etc. We shall devote in this section just to the problem of effect of superconductivity.

The fact that the superconductor cannot be defined unambiguously on the microscopical level, i.e. that it is characterized by the occurrence of several equivalent groundstates, implies the possibility of spontaneous transition from one ground state into another one. This process, known as the dynamic J-T effect, represents a new degree of freedom of the whole system, which is orthogonal to other degrees of freedom and is also independent on them. It means that this new degree of freedom is quite nondissipative. The transition process has a cooperative long range order property, i.e. the sublattices cannot be deformed (otherwise the conception of two bands would be disturbed) and can only move one with respect to the other. Because the transition from one state into another is conditioned by the overcoming of the potential barrier between two neighbouring ground states we shall speak about the tunnelling process. In this respect we can find a quantum chemical analogy – molecules with two ground states (right torque and left torque). There is also a spontaneous tunnelling transition from one configuration to the other one.

The effect of superconductivity is therefore caused by nuclear microflows through equivalent ground states. There is a question if this nuclear motion and the lattice symmetry lowering can be detectable. Because all the equivalent ground states are symmetrically localized around the symmetrical central point (i.e. the point corresponding to the ground state of material above T_c) there are the same probabilities of the occurrence of the system in each of these states. The resulting effect is therefore symmetrical. The experimentally measured nuclear form factors indicate the rotational ellipsoids originating from the vibrational degrees of freedom. There is a possibility that this new nondissipative "rotational" degree of freedom is hidden in the abovementioned rotational ellipsoids. According to our theory the rotational ellipsoids would be enhanced at the phase transition below T_c. And indeed, the recent investigation of structure and superconducting properties of Nb_3Sn ($T_c = 18.5\,K$) by X-ray diffraction [38] fully confirms the theory presented here. On the studied low-T_c compound Nb_3Sn, where the Jahn-Teller effect at the transition from the normal to superconducting state has not been assumed before, a discontinuous increase of the isotopic Debye temperature factors of niobium and tin has been observed in the temperature dependence at cooling near to T_c. Maybe the finer experiments show in future some changes in formfactor values of further low- and high-T_c superconductors near the critical temperature.

We have mentioned the state of superconductivity formed by means of the pairing between occupied valence and unoccupied conducting band electronic states with the same quasimomenta and spins, as a consequence of the electron-translon and electron-roton interaction. This is the first pairing process relating to superconductivity. Then we have mentioned the effect of superconductivity caused by the dynamic J-T effect, which on the crystal level with translational symmetry induces the temporary pairing of the neighbouring nuclei. This is the second pairing process. Now the question arises, what is the origin of the superconducting flow of electrons.

It is clear that the dynamic J-T effect affects not only nuclear positions but also the electron distributions. Due to the translational symmetry in crystals this dynamic J-T effect has two levels: on the former level the tunnelling process occurs between the equivalent ground states, causing the movement of two sublattices, and on the latter level the tunnelling process arises between the electron distributions. Therefore we shall speak about the double-level dynamic J-T effect. Whereas the tunnelling of nuclei is limited within the meaning of "there and back", the tunnelling of electron distributions has more abilities – "there and back", "only backwards", and "only forwards". Since the electron distribution of superconductors – multi-ground-state insulators – is always of the closed shell form, the minimum tunnelling electron distribution consists of two electrons with the same quasimomenta and the opposite spins. And this is the third pairing process, which explains the supercurrent with the minimum charge $2e$. Since the double-level dynamic J-T effect was never investigated before, there is no experience how to treat it exactly. We only know that both levels of this double-level effect induce two new nondissipative degrees of freedom: the former degree for the tunnelling of nuclei (two sublattices) and the latter one for the tunnelling of two-electron pairs. From the preliminary considerations we can only estimate the maximum supercurrent velocity of each electron (i.e. the velocity in the "only forwards" mode). If we denote the frequency of the nuclei relating to the former new degree of freedom as ω_N and the original full symmetry (i.e. before the J-T splitting) lattice constant as $\mathbf{a}$, the maximum velocity $\mathbf{v}_{max}$ will be defined as

$$\mathbf{v}_{max} = \frac{\omega_N}{2\pi}\mathbf{a} \qquad (28.94)$$

Let us note that only both electrons from each electron pair have to tunnel with the same velocity but the velocities of various pairs are not correlated.

The existence of the latter new degree of freedom has the most important consequence in a fact that the quasimomenta of the tunnelling electron pairs belong to some orthogonal space relating to that one, in which the quasimomenta of electrons in the valence and conducting bands of superconductors – multi-ground-state insulators are defined. Therefore the quasimomenta of the tunnelling pairs cannot be expressed via the $\mathbf{k}$-space any more, but we have to introduce the orthogonal $\mathbf{l}$-space. Each electron from the pair defined via the "statical" quasimomentum $\mathbf{k}$ and spin $\pm\sigma$ moves then as a de Broglie wave with the quasimomentum $\mathbf{l_k}$. The values of $\mathbf{l_k}$ are only limited by the maximal value

$$\mathbf{l_k} \in \langle -\mathbf{l}_{max}, \mathbf{l}_{max} \rangle \tag{28.95}$$

where the maximal value of $\mathbf{l}_{max}$ can be expressed by means of the Eq. 28.94:

$$\mathbf{l}_{max} = \frac{\mathbf{p}_{max}}{\hbar} = \frac{m}{\hbar}\mathbf{v}_{max} = \frac{m\omega_N}{h}\mathbf{a} \tag{28.96}$$

Thus, we have shown that the simple Cooper pair based mechanism cannot explain the origin of superconductivity and that three different pairing mechanisms are necessary to its full understanding. The first and initiating pairing mechanism is related only to electronic states and not to real particles. This type of pairing is responsible for the state of superconductors alias multi-ground-state insulators. Since the excitation mechanism is one-particle, the whole theory describing the state of superconductivity has to be indispensably one-particle. On the other hand, the double-level dynamic J-T effect induces two new nondissipative degrees of freedom accompanying with the pairing of real particles during the tunnelling process – temporary pairing of neighbouring nuclei and the pairing of two electrons with the same quasimomenta and opposite spins. This is the final effect o superconductivity described on two-particle basis.

The above-mentioned conclusions influence the concept of correspondence between macrostates and microstates. It is commonly believed that any macrostate of superconductor with a certain value of supercurrent corresponds to one appropriate microstate described by a certain value of charge carrier quasimomentum. According to our theory the macrostate with zero supercurrent corresponds to several microstates, i.e. microscopical configurations representing equivalent ground states, and any other macrostate with nonzero supercurrent corresponds to a certain transition process between these microscopical configurations.

Further we mention the conception of two phases: superconducting and conducting. This conception originates from the phenomenological idea of parallel coexistence of two phase components – superconducting (x) and conducting $(1 - x)$. It is motivated by the classical thermodynamics where in a similar way e.g. the coexistence of liquid and gaseous phases of the same matter is described. This macroscopical phenomenological conception was later incorporated in microscopical theories. So, in compliance with the BCS theory, the Cooper-paired electrons representing the superconducting phase coexist with free non-paired electrons representing the conducting phase in a parallel way. On the contrary to this our theory considers these two phases to be not parallel but orthogonal in the ontological sense. What does this important difference mean?

In the two-particle theories based on the Cooper pair idea two different entities are identified: the entity responsible for the condensation and excitation mechanism leading to the gap formation and the entity responsible for the transfer of super-current. Cooper pairs are the Bose condensation, which decay into free conducting electrons through the excitation mechanism, and simultaneously they are carriers of superconducting current.

In our theory we sharply distinguish these two entities. The former one corresponds to the one-electron J-T excitations. The condensation process represents the creation of the multi-ground-state insulator with fully occupied valence band and empty conducting band. The excitation mechanism is one-particle in principle. The conducting phase of the superconductor in this sense resembles the conductance of thermally excited insulator (semiconductor). The condensation and excitation mechanism is a subject of investigation of the state of superconductivity.

The latter entity corresponds to the tunnelling of two-electron distributions (in the delocalised terminology) or double occupied binding orbitals (in the localised terminology), which are the carriers of the supercurrent. By this process one set of paired nuclei decays and another one arises. The tunnelling process is two-particle in principle, is connected with two new nondissipative degrees of freedom, one for sublattices and one for paired electrons, and is orthogonal with respect to the electron-hyperphonon interaction mechanism, which is responsible for the one-particle gap formation. The phenomenological nature of the carriers of the supercurrent is the further subject of the investigation of the effect of superconductivity.

28.11 Conclusion

The main goal of this work was the implementation of the COM problem into the many-body treatment. The many years experience with the inconvenience of the direct COM separation on the molecular level and its consequent replacement with the Born-Handy ansatz as a full equivalent was taken into account. It was shown that the many-body treatment based on the electron-vibrational Hamiltonian is fundamentally inconsistent with the Born-Handy ansatz so that such a treatment can never respect the COM problem.

The only way-out insists in the requirement, to take into account the whole electron-vibration-rotation-translational Hamiltonian. It means, that the total Hamiltonian in the crude representation, expressed in the second quantization formalism, has explicitly to contain not only the vibrational energy quanta, but also the rotational and translational ones, which originate from the kinetic secular matrix.

We shall call these new quasiparticles – rotational and translational quanta – as rotons and translons, in a full analogy with phonons in solid-state physics. This is the background of the true many-body treatment in quantum chemistry and solid-state physics, which we shall call COM many-body theory. It leads to a revised concept of degrees of freedom, which are inseparable and have only a relative meaning.

The quasiparticle transformations, binding individual representations of the total Hamiltonian, are then the generalization of the original Fröhlich transformations in such a way that they contain, besides the electron-vibrational (vibronic or electron-phonon) interaction, additionally the electron-rotational (rotonic or electron-roton) and the electron-translational (translonic or electron-translon) interactions. In order to achieve a unique covariant description of all equations with respect to individual degrees of freedom, we have introduced the concept of hypervibrations (hyper-phonons), i.e. vibrations + rotations + translations together, and the consequent concept of electron-hypervibrational (hypervibronic or electron-hyperphonon) interaction. We have proved that due to the COM problem only the hypervibrations (hyperphonons) have true physical meaning in molecules and crystals; nevertheless, the use of pure vibrations (phonons) is justified only in the adiabatic systems, i.e. the case when electron energies are much greater than the vibrational ones. This fact calls for a total revision and reformulation of our contemporary knowledge of all non-adiabatic systems.

The most important equation, derived in this work, is the extended Born-Handy formula, valid in the adiabatic limit as well as in the case of break down of the B-O approximation. Since due to the many-body formulation the extended Born-Handy formula can be expressed in the CPHF compatible form, the extended CPHF equations, describing the non-adiabatic systems, will immediately follow from the presented theory. We shall call them COM CPHF equations. Whereas in the adiabatic limit the extended Born-Handy formula represents only small corrections to the system total energy, in non-adiabatic systems it plays three important roles: (1) removes the electron degeneracies, (2) is responsible for the symmetry breaking, and (3) forms the molecular and crystalline structure.

The first role – removal of electron degeneracies – is fulfilled via the vibronic coupling. The second role – the symmetry breaking – is caused by the rotonic and translonic coupling. Finally the third role – forming of structure – is a result of optimalization where all three types of coupling participate. Only in the adiabatic limit the forming of molecular and crystallic structure reduces to the standard one, defined by the B-O approximation. Moreover, at finite temperatures the extended Born-Handy formula plays yet another role: it defines all thermodynamic properties of the non-adiabatic systems, as was demonstrated on the derivation of the critical temperature of superconductors.

Since the J-T effect was always studied without the inclusion of the COM problem, only vibronic coupling was taken into account, and therefore the symmetry breaking and forming of structure were misunderstood. The trigger causing the system instability has the origin in rotonic and translonic coupling. It is necessary to reformulate the J-T theorem in a new way. One possible formulation is proposed here: "Molecular and crystallic entities in a geometry of electronically degenerate ground state are unstable at this geometry except the case when all matrix elements of electron-rotational and electron-translational interaction equal to zero."

As was mentioned in the introduction, modern attempts to explain HTSC by means of the J-T approach operate with the term "strong vibronic coupling" in order to advocate the presence of the J-T effect. On the other hand they allow for the BCS theory applied to LTSC where only "week vibronic coupling" occurs, which means that the J-T effect may be ignored. However, after the inclusion of the COM problem we have come to the conclusion that the question of strong or week vibronic coupling is absolutely irrelevant for the applicability of the J-T effect. Only symmetry breaking, stimulated by rotonic and translonic couplings, acting as a trigger, plays a decisive role. Either the trigger is bypassed, and then the crystal remains in conducting state; or it is switched on and the J-T effect is active, and this leads to the superconducting state.

It is a fundamental problem in solid-state physics, that due to a misleading many-body treatment the true nature of the crystalline structure beyond the B-O approximation was never revealed. This is the reason why the BCS theory, in spite of the fact that Fröhlich was critical to it and disregarded it, survives more than a half of century up till now. The BCS theory is based on the naive belief that the structure of superconductors is the same as the structure of conductors, i.e. that it is defined through the B-O approximation. As we have shown in our previous work [21], there is no mechanism, which could split the degenerate electronic spectrum of conductors and open an energy gap at the adiabatic level.

Fröhlich applied the unitary transformation on the Hamiltonian describing conductors, but his attempt to remove the degeneracy failed. Then he proposed the "true" many body treatment. Bardeen with Cooper and Schrieffer continued to fulfil Fröhlich's idea, and with the full multiconfiguration method used on the Fröhlich-transformed Hamiltonian, they attempted to remove the degeneracy. After 2-years of intensive work they had no positive solution. At the last moment Bardeen accepted the trial function proposed by Schrieffer, inconsistent with the particle conservation law, and leading to the concept of the Cooper pair based theory, known as BCS. Nevertheless, the solid-state Hamiltonian does not contain information of superconductivity, so that Fröhlich as well as Bardeen in fact only calculated the correlation energy of conductors.

Since quantum field many body techniques are not directly transferable into quantum chemistry dealing with small molecular systems, they are not fully transferable into the solid-state physics dealing with great systems (crystals) either. As it was explained in detail in this work, only the COM many-body formulation is applicable in non-adiabatic cases. For non-adiabatic crystals the state of conductivity and superconductivity are two possible solutions of the extended Born-Handy formula. This is a quite different view from that using only the classical many body (without COM). The non-adiabatic treatment of crystals leads always to the splitting into two subsystems. In the case of conductors the first subsystem is the "adiabatic core" consisting of nuclei and all valence bands, and the second subsystem is the "fluid" of quasi-free conducting electrons. The explanation of conductors on the basis of a COM true many-body treatment is not so simple as in the case of the classical many body theory. Whereas for the classical many body the conducting state is a real ground state (and superconductors are something like a "better ground state" after the Bose condensation of Cooper pairs), the COM many body presents the conducting state as an abnormal solution of the extended Born-Handy formula, as some excited state, representing the crystal analogy of ionized molecular systems where the ionized electrons are not the part of the molecular "ion core" as well. In other words, the conductors "fling away" one set of electrons so that the rest – the "core", consisting of nuclei and valence bands, is not degenerate any more and is therefore adiabatic with the well-defined structure on the basis of the B-O approximation. The "off-cast" electrons are not the integral part of the adiabatic "core" anymore, but still belong to the system and interact with the "core" via the standard electron-phonon interaction.

In spite of the fact that the effective Hamiltonian of the type adiabatic core + electronic fluid + electron-phonon interaction between them describes conductors correctly, the origin of conductivity was misunderstood, and consequently, the theory superconductivity was misinterpreted as well. The state of superconductivity is not a "better ground state" as the BCS theory explains it, but the real ground state, and moreover, the multi ground state due to the J-T effect. This is the only difference between superconductors and insulators: whereas the latter are occupying a simple ground state, the former are of multi ground state character. The electron-phonon mechanism can never describe superconductivity; we need the complete COM many body, i.e. the electron-hyperphonon mechanism. The rotonic a translonic coupling splits the system into two subsystems – two sublattices, causes the symmetry lowering, defines new non-adiabatic structure, and creates the pairs from all occupied valence and unoccupied conducting band electronic states with the same quasimomenta and spins, with the cooperative behaviour over the whole crystal. This is only a pairing of states and not of real particles; therefore the theory of superconductivity is one-particle in principle, as Fröhlich demanded. Finally the vibronic coupling via the optical phonon modes stabilizes the whole system in this new configuration and opens an energy gap. No set of electrons is "flung away" as in the case of conductors, all electrons are located due the symmetry lowering in the fully occupied valence bands, exactly as in the case of insulators.

We shall distinguish two fundamental attributes of superconductivity – the state and the effect of superconductivity – that lead to two complementary descriptions of superconductors. On one side the state of superconductivity is characterized by the state of a conducting material, which, after the Jahn-Teller condensation, becomes an insulator with several equivalent ground states. The state of superconductivity determines all statical properties of superconductors: energy gap, its temperature dependence, specific heat, density of states near the Fermi surface etc. On the other side the effect of superconductivity determines all dynamical properties of superconductors: supercurrent, Meissner effect, quantization of magnetic flux, etc. Whereas the state of superconductivity is of one-particle nature, the effect is two-particle in principle. This is a rather subtle relationship since behind this result lies the precise conditions for the onset of Yang's Off-Diagonal Long-Range Order,

ODLRO [39]. The multi ground state character of superconductors leads to the double-level dynamic J-T effect, which induces two new nondissipative degrees of freedom accompanying with the pairing of real particles during the tunnelling process – temporary pairing of neighbouring nuclei and the pairing of two electrons with the same quasimomenta and opposite spins. The new electronic degree of freedom implies a new quasimomentum space orthogonal to the one where the electrons are described in the state of superconductor – multi ground state insulator. This is the final effect of superconductivity described on two-particle basis, which explains the supercurrent with the minimum charge 2e.

Appendices

Appendix 1

We present here some useful relations for the expansions of Eqs. 28.21–28.24 only up to the second order of the Taylor expansion since we will not take the anharmonic terms into account in this paper. Maybe the following relations are trivial but it is worth to mention them due to the cross-platform notation.

$$h_{PQ}^0 = h_{QP}^{0*} \qquad v_{PQRS}^0 = v_{QPSR}^0 = v_{SRQP}^{0\,*} = v_{RSPQ}^{0\,*} \qquad (28.97)$$

$$E_{NN}^r = E_{NN}^{\breve{r}\,*} \qquad E_{NN}^{rs} = E_{NN}^{sr} = E_{NN}^{\breve{r}\breve{s}\,*} = E_{NN}^{\breve{s}\breve{r}\,*} \qquad (28.98)$$

$$u_{PQ}^r = u_{QP}^{\breve{r}\,*} \qquad u_{PQ}^{rs} = u_{PQ}^{sr} = u_{QP}^{\breve{r}\breve{s}\,*} = u_{QP}^{\breve{s}\breve{r}\,*} \qquad (28.99)$$

Let us suppose that for every spinorbital X there exists a unique spinorbital $\hat{X}$ so that the following symmetrical relations hold:

$$h_{PQ}^0 = h_{\hat{Q}\hat{P}}^0 \qquad v_{PQRS}^0 = v_{\hat{R}\hat{S}\hat{P}\hat{Q}}^0 \qquad u_{PQ}^r = u_{\hat{Q}\hat{P}}^r \qquad (28.100)$$

In quantum chemistry where the real wave functions are used and the identity $X = \hat{X}$ is supposed, the Eqs. 28.100 hold trivially. In solid-state theory where the assignment $X \to \mathbf{k}, \sigma$ and $\hat{X} \to -\mathbf{k}, \pm\sigma$ ($\mathbf{k}$ is the electron quasimomentum and σ is the spin) is done, these equations lead to the well-known symmetrical relations.

In a similar way, we can expand up to the second order the unitary conditions (28.35 and 28.36) of the adiabatic transformation

$$c_{PQ}^0 = \delta_{PQ} \quad c_{PQ}^r + c_{QP}^{\breve{r}\,*} = 0 \qquad c_{PQ}^{rs} + c_{QP}^{\breve{r}\breve{s}\,*} = -\sum_R \left(c_{PR}^r c_{QR}^{\breve{s}\,*} + c_{PR}^s c_{QR}^{\breve{r}\,*} \right) \quad (28.101)$$

$$d_{rPQ}^0 = c_{PQ}^{\breve{r}} \qquad d_{rPQ}^s = c_{PQ}^{\breve{r}s} + \sum_R c_{RP}^{\breve{s}\,*} c_{RQ}^{\breve{r}} \qquad (28.102)$$

and the unitary conditions (28.39) and (28.40) for the nonadiabatic transformation

$$\tilde{c}^0_{PQ} = \delta_{PQ} \quad \tilde{c}^r_{PQ} - \tilde{c}^{\breve{r}*}_{QP} = 0 \quad \tilde{c}^{rs}_{PQ} + \tilde{c}^{\breve{r}\,\breve{s}*}_{QP} = \sum_R \left(\tilde{c}^r_{PR} \tilde{c}^{\breve{s}*}_{QR} + \tilde{c}^s_{PR} \tilde{c}^{\breve{r}*}_{QR} \right) \tag{28.103}$$

$$\tilde{d}^0_{rPQ} = -\tilde{c}^{\breve{r}}_{PQ} \quad \tilde{d}^s_{rPQ} = -\tilde{c}^{\breve{r}s}_{PQ} + \sum_R \tilde{c}^{\breve{s}}_{RP} \tilde{c}^{\breve{r}}_{RQ} \tag{28.104}$$

Appendix 2

The terms of the electron-hypervibrational Hamiltonian up to the second order of Taylor expansion in the general representation are presented here in details. The notation $H^{n(k,l)}_X$ is used where X denotes the terms originating from transformation of the part A or B of the crude Hamiltonian, n represents the power of the Taylor expansion, k the power of the coordinate operator B_r and l the power of the momentum operator $\tilde{B}_r$.

$$H^0_A = E^0_{NN} + \sum_{PQ} h^0_{PQ} a^+_P a_Q + \tfrac{1}{2} \sum_{PQRS} v^0_{PQRS} a^+_P a^+_Q a_S a_R \tag{28.105}$$

$$H^{1(1,0)}_A = \sum_r E^r_{NN} B_r + \sum_{PQr} \left[u^r_{PQ} + \sum_r \left(h^0_{PR} c^r_{RQ} + h^0_{RQ} c^{\breve{r}*}_{RP} \right) \right] B_r a^+_P a_Q$$

$$+ \sum_{PQRSTr} \left(v^0_{PQTS} c^r_{TR} + v^0_{TQRS} c^{\breve{r}*}_{TP} \right) B_r a^+_P a^+_Q a_S a_R \tag{28.106}$$

$$H^{1(0,1)}_A = \sum_{PQRr} \left(h^0_{PR} \tilde{c}^r_{RQ} - h^0_{RQ} \tilde{c}^{\breve{r}*}_{RP} \right) \tilde{B}_r a^+_P a_Q$$

$$+ \sum_{PQRSTr} \left(v^0_{PQTS} \tilde{c}^r_{TR} - v^0_{TQRS} \tilde{c}^{\breve{r}*}_{TP} \right) \tilde{B}_r a^+_P a^+_Q a_S a_R \tag{28.107}$$

$$H^{2(2,0)}_A = -\frac{1}{2} \sum_{rs} V^{rs}_N B_r B_s + \sum_{PQrs} \left[\frac{1}{2} u^{rs}_{PQ} + \sum_R \left(\frac{1}{2} h^0_{PR} c^{rs}_{RQ} + \frac{1}{2} h^0_{RQ} c^{\breve{r}\,\breve{s}*}_{RP} + u^r_{PR} c^s_{RQ} + u^r_{RQ} c^{\breve{s}*}_{RP} \right) \right.$$

$$\left. + \sum_{RS} h^0_{RS} c^{\breve{r}*}_{RP} c^s_{SQ} \right] B_r B_s a^+_P a_Q$$

$$+ \frac{1}{2} \sum_{PQRSTrs} \left\{ v^0_{PQTS} c^{rs}_{TR} + v^0_{TQRS} c^{\breve{r}\,\breve{s}*}_{TP} + \sum_U \left[v^0_{PQTU} c^r_{TR} c^s_{US} + v^0_{TURS} c^{\breve{r}*}_{TP} c^{\breve{s}}_{UQ} \right. \right.$$

$$\left. \left. + 2 \left(v^0_{TQUS} - v^0_{TQSU} \right) c^{\breve{r}*}_{TP} c^s_{UR} \right] \right\} B_r B_s a^+_P a^+_Q a_S a_R \tag{28.108}$$

$$H_A^{2(0,2)} = -\frac{1}{2}\sum_{rs} W_N^{rs}\tilde{B}_r\tilde{B}_s + \sum_{PQRrs}\left(\frac{1}{2}h_{PR}^0\tilde{c}_{RQ}^{rs} + \frac{1}{2}h_{RQ}^0\tilde{c}_{RP}^{r\,s\,*} - \sum_S h_{RS}^0\tilde{c}_{RP}^{r\,*}\tilde{c}_{SQ}^s\right)\tilde{B}_r\tilde{B}_s a_P^+ a_Q$$

$$+\frac{1}{2}\sum_{PQRSTrs}\left\{v_{PQTS}^0\tilde{c}_{TR}^{rs} + v_{TQRS}^0\tilde{c}_{TP}^{r\,s\,*} + \sum_U\left[v_{PQTU}^0\tilde{c}_{TR}^r\tilde{c}_{US}^s + v_{TURS}^0\tilde{c}_{TP}^{r\,*}\tilde{c}_{UQ}^{s\,*}\right.\right.$$

$$\left.\left.+2\left(v_{TQSU}^0 - v_{TQUS}^0\right)\tilde{c}_{TP}^{r\,*}\tilde{c}_{UR}^s\right]\right\}\tilde{B}_r\tilde{B}_s a_P^+ a_Q^+ a_S a_R \qquad (28.109)$$

$$H_A^{2(1,1)} = \frac{1}{2}\sum_{PQRrs}\left\{u_{PR}^r\tilde{c}_{RQ}^s - u_{RQ}^r\tilde{c}_{RP}^{s\,*} + \sum_S\left[\left(h_{PR}^0 c_{RS}^r + h_{RS}^0 c_{RP}^{r\,*}\right)\tilde{c}_{SQ}^s\right.\right.$$

$$\left.\left. - \left(h_{SR}^0 c_{RQ}^r + h_{RQ}^0 c_{RS}^{r\,*}\right)\tilde{c}_{SP}^{s\,*}\right]\right\}\left(B_r\tilde{B}_s + \tilde{B}_s B_r\right)a_P^+ a_Q$$

$$+\frac{1}{2}\sum_{PQRSTUrs}\left[\left(v_{PQTS}^0 c_{TU}^r - v_{PQTU}^0 c_{TS}^r\right)\tilde{c}_{UQ}^s + \left(v_{TURS}^0 c_{TQ}^{r\,*} - v_{TQRS}^0 c_{TU}^{r\,*}\right)\tilde{c}_{UP}^{s\,*}\right.$$

$$\left.+\left(v_{TQUS}^0 - v_{TQSU}^0\right)\left(c_{TP}^{r\,*}\tilde{c}_{UR}^s - c_{UR}^r\tilde{c}_{TP}^{s\,*}\right)\right]\left(B_r\tilde{B}_s + \tilde{B}_s B_r\right)a_P^+ a_Q^+ a_S a_R \quad (28.110)$$

$$H_A^{2(0,0)} = 2\sum_{PQr} E_{NN}^r \tilde{d}_{rPQ}^0 a_P^+ a_Q + \sum_{PQRr}\left\{\left[u_{PR}^r + \sum_S\left(h_{PS}^0 c_{SR}^r + h_{SR}^0 c_{SP}^{r\,*}\right)\right]\right.$$

$$\tilde{d}_{rRQ}^0 + \left[u_{RQ}^r + \sum_S\left(h_{RS}^0 c_{SQ}^r + h_{SQ}^0 c_{SR}^{r\,*}\right)\right]\tilde{d}_{rPR}^0\right\}$$

$$a_P^+ a_Q + 2\sum_{PQRSr}\left[u_{PR}^r + \sum_T\left(h_{PT}^0 c_{TR}^r + h_{TR}^0 c_{TP}^{r\,*}\right)\right]\tilde{d}_{rQS}^0 a_P^+ a_Q^+ a_S a_R$$

$$+\sum_{PQRSTUr}\left[\left(v_{PQTS}^0 c_{TU}^r - v_{PQTU}^0 c_{TS}^r\right)\tilde{d}_{rUR}^0 + \left(v_{TQRS}^0 c_{TU}^{r\,*} - v_{TURS}^0 c_{TQ}^{r\,*}\right)\right.$$

$$\left.\tilde{d}_{rPU}^0 + \left(v_{TQUS}^0 - v_{TQSU}^0\right)\left(c_{TP}^{r\,*}\tilde{d}_{rUR}^0 + c_{UR}^r\tilde{d}_{rPT}^0\right)\right]a_P^+ a_Q^+ a_S a_R$$

$$+2\sum_{PQRSTUVr}\left(v_{PQVT}^0 c_{VS}^r + v_{VQST}^0 c_{VP}^{r\,*}\right)\tilde{d}_{rRU}^0 a_P^+ a_Q^+ a_R^+ a_U a_T a_S \qquad (28.111)$$

$$H_B^0 = \frac{1}{4}\sum_r\left(\hbar\omega_r B_r^+ B_r + \hbar\tilde{\omega}_r \tilde{B}_r^+ \tilde{B}_r\right) = \frac{1}{4}\sum_r\left(\hbar\omega_r B_r B_r - \hbar\tilde{\omega}_r \tilde{B}_r \tilde{B}_r\right) \qquad (28.112)$$

$$H_B^{1(1,0)} = \sum_{PQr}\hbar\omega_r \tilde{d}_{rPQ}^0 B_r a_P^+ a_Q \qquad (28.113)$$

$$H_B^{1(0,1)} = -\sum_{PQr} \hbar\tilde{\omega}_r d^0_{\underset{r}{\sim}PQ} \tilde{B}_r a_P^+ a_Q \tag{28.114}$$

$$H_B^{2(2,0)} = 0 \tag{28.115}$$

$$H_B^{2(0,2)} = -\sum_{PQRrs} \hbar\tilde{\omega}_r \left(d^0_{\underset{r}{\sim}PR} \tilde{c}^s_{RQ} - d^0_{\underset{r}{\sim}RQ} \tilde{c}^{s*}_{RP} \right) \tilde{B}_r \tilde{B}_s a_P^+ a_Q \tag{28.116}$$

$$H_B^{2(1,1)} = \frac{1}{2} \sum_{PQrs} \left(\hbar\omega_r \tilde{d}^s_{\underset{r}{\sim}PQ} - \hbar\tilde{\omega}_s d^r_{\underset{s}{\sim}PQ} \right) \left(B_r \tilde{B}_s + \tilde{B}_s B_r \right) a_P^+ a_Q \tag{28.117}$$

$$H_B^{2(0,0)} = \sum_{PQRr} \left(\hbar\omega_r \tilde{d}^0_{\underset{r}{\sim}PR} \tilde{d}^0_{\underset{r}{\sim}RQ} - \hbar\tilde{\omega}_r d^0_{\underset{r}{\sim}PR} d^0_{\underset{r}{\sim}RQ} \right) a_P^+ a_Q$$

$$+ \sum_{PQRSr} \left(\hbar\omega_r \tilde{d}^0_{\underset{r}{\sim}PR} \tilde{d}^0_{\underset{r}{\sim}QS} - \hbar\tilde{\omega}_r d^0_{\underset{r}{\sim}PR} d^0_{\underset{r}{\sim}QS} \right) a_P^+ a_Q^+ a_S a_R$$

$$\tag{28.118}$$

Appendix 3

Here is the detailed derivation of the fermion part of the Hamiltonian in the general representation. ΔH_F from the Eq. 28.52 can be expressed as a sum of two contributions

$$\Delta H_F = \Delta H_{[F]} + \Delta H_{\langle F \rangle} \tag{28.119}$$

where $\Delta H_{[F]}$ is of a pure fermion origin, i.e. the part that is invariant against boson (hypervibrational) excitations, and $\Delta H_{\langle F \rangle}$ represents the effective part dependent on boson excitations. We take into account only the boson vacuum mean values:

$$\langle 0|B_r|0 \rangle = \langle 0|\tilde{B}_r|0 \rangle = \langle 0|B_r\tilde{B}_s + \tilde{B}_s B_r|0 \rangle = 0 \tag{28.120}$$

$$\langle 0|B_r B_s|0 \rangle = -\langle 0|\tilde{B}_r \tilde{B}_s|0 \rangle = \delta_{\underset{r}{\sim}s} \tag{28.121}$$

If we introduce the discrete particle-hole occupation factors

$$h(A) = 0, h(I) = 1, p(A) = 1, p(I) = 0 \tag{28.122}$$

and a simplifying notation for the adiabatic derivatives v^r_{PQRS} of the coulomb interaction v^0_{PQRS}

$$v^r_{PQRS} = \sum_T \left(v^0_{PQTS} c^r_{TR} + v^0_{PQRT} c^r_{TS} - v^0_{TQRS} c^r_{PT} - v^0_{PTRS} c^r_{QT} \right) \tag{28.123}$$

we finally get the individual contributions for one- and two-fermion terms. The surprising fact is the occurrence of three-fermion term, in spite of the fact the crude representation contains only one- and two-electron terms.

$$
\Delta H'_F = \sum_{PQr} \left[\hbar\tilde{\omega}_r \left(\sum_A c^r_{PA} c^{r*}_{QA} - \sum_I c^r_{PI} c^{r*}_{QI} \right) - \hbar\omega_r \left(\sum_A \tilde{c}^r_{PA} \tilde{c}^{r*}_{QA} - \sum_I \tilde{c}^r_{PI} \tilde{c}^{r*}_{QI} \right) \right] N[a^+_P a_Q]
$$

$$
-2 \sum_{PQr} E^{r*} \tilde{c}^r_{PQ} N[a^+_P a_Q] + \sum_{PQr} \left[(h(P) - p(P))\varepsilon^r_P + (h(Q) - p(Q))\varepsilon^{r*}_Q \right] \tilde{c}^r_{PQ} N[a^+_P a_Q]
$$

$$
- \sum_{PQAIr} \left[(v^r_{PIQA} - v^r_{PIAQ}) \tilde{c}^{r*}_{IA} + (v^r_{PAQI} - v^r_{PAIQ}) \tilde{c}^{r*}_{AI} \right] N[a^+_P a_Q] \tag{28.124}
$$

$$
\Delta H''_{[F]} = \frac{1}{2} \sum_{PQRS} v^0_{PQRS} N[a^+_P a^+_Q a_S a_R] + \sum_{PQRSr} \left(\hbar\tilde{\omega}_r c^r_{PR} c^{r*}_{SQ} - \hbar\omega_r \tilde{c}^r_{PR} \tilde{c}^{r*}_{SQ} \right) N[a^+_P a^+_Q a_S a_R]
$$

$$
-2 \sum_{PQSr} \varepsilon^r_P \tilde{c}^{r*}_{SQ} N[a^+_P a^+_Q a_S a_P] + \sum_{PQRSTr} \left\{ \sum_I \left[v^0_{PQTS} c^r_{TI} - v^0_{PQTI} c^r_{TS} \right. \right.
$$

$$
\left. + (v^0_{TQSI} - v^0_{TQIS}) c^r_{PT} \right] \tilde{c}^{r*}_{RI} + \sum_I \left[v^0_{TIRS} c^r_{QT} - v^0_{TQRS} c^r_{IT} + (v^0_{IQTS} - v^0_{IQST}) c^r_{TR} \right] \tilde{c}^{r*}_{IP}
$$

$$
- \sum_A \left[v^0_{PQTS} c^r_{TA} - v^0_{PQTA} c^r_{TS} + (v^0_{TQSA} - v^0_{TQAS}) c^r_{PT} \right] \tilde{c}^{r*}_{RA}
$$

$$
- \sum_A \left[v^0_{TARS} c^r_{QT} - v^0_{TQRS} c^r_{AT} + (v^0_{AQTS} - v^0_{AQST}) c^r_{TR} \right] \tilde{c}^{r*}_{AP} \right\} N[a^+_P a^+_Q a_S a_R]
$$

$$
\tag{28.125}
$$

$$
\Delta H'''_{[F]} = -2 \sum_{PQRSTUVr} (v^0_{PQVT} c^r_{VS} - v^0_{VQST} c^r_{PV}) \tilde{c}^{r*}_{UR} N[a^+_P a^+_Q a^+_R a_U a_T a_S] \tag{28.126}
$$

$$
H'_{\langle F \rangle} = \frac{1}{2} \sum_{Pr} u^{r\breve{r}}_{PP} N[a^+_P a_P] + \sum_{PRr} \left[(\varepsilon^0_P - \varepsilon^0_R)(|c^r_{PR}|^2 + |\tilde{c}^r_{PR}|^2) - 2\hbar\tilde{\omega}_r Re(\tilde{c}^r_{PR} c^{r*}_{PR}) \right] N[a^+_P a_P]
$$

$$
+ \sum_{PAIr} \left\{ \sum_R \left[(v^0_{PRPA} - v^0_{PRAP})(c^r_{IR} c^{r*}_{IA} + \tilde{c}^r_{IR} \tilde{c}^{r*}_{IA}) - (v^0_{PRPI} - v^0_{PRIP})(c^r_{AR} c^{r*}_{AI} + \tilde{c}^r_{AR} \tilde{c}^{r*}_{AI}) \right] \right.
$$

$$
\left. + \frac{1}{2} \left[(v^0_{PIPA} - v^0_{PIAP})(c^{r\breve{r}}_{AI} - \tilde{c}^{r\breve{r}}_{AI}) - (v^0_{PAPI} - v^0_{PAIP})(c^{r\breve{r}}_{IA} - \tilde{c}^{r\breve{r}}_{IA}) \right] \right\} N[a^+_P a_P]
$$

$$
\tag{28.127}
$$

$$
H''_{\langle F \rangle} = \frac{1}{2} \sum_{PQRSTr} \left\{ v^0_{PQTS} \left(c^{r\breve{r}}_{TR} - \tilde{c}^{r\breve{r}}_{TR} \right) + v^0_{TQRS} \left(c^{r\breve{r}}_{TP} - \tilde{c}^{r\breve{r}}_{TP} \right) \right.
$$

$$
- \sum_U \left[v^0_{PQTU} \left(c^r_{TR} c^{r*}_{SU} + \tilde{c}^r_{TR} \tilde{c}^{r*}_{SU} \right) + v^0_{TURS} \left(c^r_{PT} c^{r*}_{UQ} + \tilde{c}^r_{PT} \tilde{c}^{r*}_{UQ} \right) \right.
$$

$$
\left. \left. +2 \left(v^0_{TQSU} - v^0_{TQUS} \right) \left(c^r_{PT} c^{r*}_{RU} + \tilde{c}^r_{PT} \tilde{c}^{r*}_{RU} \right) \right] \right\} N[a^+_P a^+_Q a_S a_R] \tag{28.128}
$$

Acknowledgements The author wishes to express his gratitude to E. Brändas for his valuable advice during compilation of this paper, to O. Šipr for critical reading of the manuscript and useful suggestions and to V. Žárský for constant help and encouragement.

References

1. Born M, Oppenheimer R (1927) Ann Phys (Leipzig) 84:457
2. Primas H, Müller-Herold U (1984) Elementare Quantenchemie. Teubner, Stuttgart, p 147 ff
3. Monkhorst HJ (1999) Int J Quant Chem 72:281
4. Cafiero M, Adamowicz L (2004) Chem Phys Letters 387:136–141
5. Kutzelnigg W (1997) Mol Phys 90:909
6. Handy NC, Lee AM (1996) Chem Phys Lett 252:425
7. Jahn HA, Teller E (1937) Proc R Soc Lond A 161:220
8. Bersuker IB (2006) The Jahn-Teller effect. Cambridge University Press, Cambridge, England
9. Fröhlich H (1950) Phys Rev 79:845
10. Fröhlich H (1952) Proc R Soc Lond A215:291
11. Bardeen J, Cooper LN, Schrieffer JR (1957) Phys Rev 108:1175
12. Svrček M (1986) Faculty of mathematics and physics. PhD thesis, Comenius University, Bratislava
13. Svrček M (1988) The break down of Born-Oppenheimer approximation, the unifying formalism for quantum chemistry and solid-state theory, unpublished
14. Hubač I, Svrček M (1988) Int J Quant Chem 23:403
15. Hubač I, Svrček M, Salter EA, Sosa C, Bartlett RJ (1988) Lecture notes in chemistry, vol 52. Springer, Berlin, pp 95–124
16. Svrček M, Hubač I (1991) Czech J Phys 41:556
17. Svrček M (1992) Methods in computational chemistry. In: Molecular vibrations, vol 4. Plenum Press, New York, pp 145–230
18. Svrček M, Baňacký P, Zajac A (1992) Int J Quant Chem 43:393
19. Svrček M, Banacký P, Zajac A (1992) Int J Quant Chem 43:415
20. Svrček M, Baňacký P, Zajac A (1992) Int J Quant Chem 43:425
21. Svrček M, Baňacký P, Zajac A (1992) Int J Quant Chem 43:551
22. Svrček M, Baňacký P, Biskupič S, Noga J, Pelikán P, Zajac A (1999) Chem Phys Lett 299:151
23. Gerratt J, Mills JM (1968) J Chem Phys 49:1719–1730
24. Pople JA, Raghavachari K, Schlegel HB, Binkley JS (1979) Int J Quant Chem Symp 13:225
25. Kołos W, Wolniewicz W (1964) J Chem Phys 41:3663
26. Wolniewicz W (1993) J Chem Phys 99:1851
27. Kleinman LI, Wolfsberg M (1974) J Chem Phys 60:4749
28. Moller C, Plesset MS (1934) Phys Rev 46:618, Sosa
29. Köppel H, Domcke W, Cederbaum LS (1984) Adv Chem Phys 57:59
30. Bersuker IB, Polinger BZ (1983) Vibronic interactions in molecules and crystals (in Russian). Nauka, Moscow
31. Van Vleck JH (1939) J Chem Phys 7:61
32. Low W (1960) Paramagnetic resonance in solids. Academic Press, New York
33. Renner R (1934) Z Phys 92:172
34. von Neumann J, Wigner E (1929) Phys Z 30:467
35. Lee TD, Low FE, Pines D (1953) Phys Rev 90
36. Wagner M (1981) Phys Stat Sol B107:617
37. Lenz P, Wegner F (1996) Nucl Phys B 482:693–712
38. Hanic F, Baňacký P, Svrček M, Jergel M, Smrčok L, Koppelhuber B, unpublished
39. Yang CN (1962) Rev Mod Phys 34:694

Chapter 29
Delocalization Effects in Pristine and Oxidized Graphene Substrates

Dmitry Yu. Zubarev, Xiaoqing You, Michael Frenklach, and William A. Lester, Jr.

Abstract It is natural to consider graphene as a polyaromatic hydrocarbon (PAH). This name suggests that delocalized bonding should be a useful concept if one aims to gain insights into structure-property relationships in graphene. Aromatic/antiaromatic nature of small PAH can be established in a straightforward manner according to a multitude of techniques such as Clar's rules and various measures of aromaticity. Large PAHs that are considered as realistic models of graphene can raise challenges to the aforementioned approaches due to the cost of associated calculations and conceptual difficulties. There is an apparent need for systematic studies of local and global delocalization phenomena in graphene. The present account summarizes some of the recent findings that consider certain properties of pristine and oxidized graphene substrates in the context of formation of Mobius or Huckel aromatic systems. Emergence of anti-ferromagnetic diradical states, relative stability of PAH oxyradicals, and onset of patterns of local aromaticity are discussed. Robustness of several popular approaches to characterization of delocalization effects is assessed. The harmonic oscillator model

D.Yu. Zubarev
Kenneth S. Pitzer Center for Theoretical Chemistry, University of California, Berkeley, CA, USA
e-mail: dmitry.zubarev@berkeley.edu

X. You
Department of Mechanical Engineering, University of California, Berkeley, CA, USA
e-mail: xiaoqing.you@berkeley.edu

M. Frenklach
Department of Mechanical Engineering, University of California, Berkeley, CA, USA

Environmental Energy Technologies Division, Lawrence Berkeley National Laboratory, Berkeley, CA, USA
e-mail: myf@me.berkeley.edu

W.A. Lester, Jr. (✉)
Kenneth S. Pitzer Center for Theoretical Chemistry, University of California, Berkeley, CA, USA

Chemical Sciences Division, Lawrence Berkeley National Laboratory, Berkeley, CA, USA
e-mail: walester@lbl.gov

P.E. Hoggan et al. (eds.), *Advances in the Theory of Quantum Systems in Chemistry and Physics*, Progress in Theoretical Chemistry and Physics 22,
DOI 10.1007/978-94-007-2076-3_29, © Springer Science+Business Media B.V. 2012

of aromaticity (HOMA) is shown to be extremely suitable for investigation of large substrates. The described results suggest that further studies of peculiarities of delocalization effects in PAHs can lead to substantial progress in development of models appealing to chemical intuition and capturing the most relevant aspects of the electronic structure of graphene.

29.1 Introduction

Graphene is a quasi two-dimensional material with honeycomb structure that can be seen as a result of conjugation of benzene-like fragments. Unique properties of graphene are widely recognized. The potential of this material for emergent technologies stimulates both experimental and theoretical research efforts world-wide [1–3]. Much interest is attracted to peculiarities of the electronic structure of graphene. This topic is pertinent to many issues of fundamental and applied relevance [4–18]. One of important tasks of electronic structure studies is to discern the role of edges of finite-size graphene substrates in the formation of unusual electronic states and in rendering their chemical activity. Theoretical studies and computational modeling are of great help especially in cases when it is hard to collect experimental data or to conduct experiments in a reproducible and controllable manner. Polyaromatic hydrocarbons (PAH) usually serve as models of graphene because they have the same structure as the interior. The chemical nature of pure graphene edges [19–22] is a subject of ongoing research but in systematic model studies it is reasonable to consider saturated, i.e., hydrogenated edges of PAHs. Computational studies of large substrates can be demanding up to the point of being infeasible if very accurate theoretical approaches are invoked. Consideration of small PAHs is more attractive in this regard but might be inadequate because the number of atoms forming the edge is comparable to the number of atoms forming the interior. As a result, the role of edges could be exaggerated.

Numerous studies have been performed to identify the nature of the electronic ground states of molecules related to graphene. Predictions were made that linear polyacenes with more than eight rings have triplet ground states [23] and later that oligoacenes longer than hexacene are diradical singlet in their ground state [24]. It was demonstrated [25] that the ground states of linear acenes and polyacenes with more than seven rings are antiferromagnetic but are not necessarily diradical. The major contribution of the zigzag edges to the localization of the spin-polarized states was shown along with the growth of magnetization with the size of the acene. The electronic properties of graphene nanoribbons with varying widths have been extensively studied computationally [26–28] and experimentally [29,30]. Both wide graphene ribbons and large rectangular PAHs were reported to have antiferromagnetic ground states [25, 31]. Computational studies showed [32, 33] that no zero-energy states exist in finite-length zigzag nanoribbons and that trigonal zigzag nanodisks have degenerate zero-energy states and show ferromagnetism. Energy gaps of graphene nanoribbons with zigzag or armchair edges were found to decrease with the width of the systems increasing [28]. Rectangular PAHs exhibiting both armchair and zigzag edges were extensively studied to identify the smallest system yielding spin-polarization [34]. It was found that at least three consecutive units long at zigzag edges yields spin polarization if the width is 1 nm or wider. High level ab initio studies showed the possibility of stable high spin states of coronene and corannulene [35] and singlet polyradical states of long acenes [36].

Among many reactions of PAHs their oxidation becomes important in the view of the connection to oxidation of fossil fuels and formation of soot [37]. The latter is a hazardous substance and there is urgency in studies of its oxidative destruction. Experimental studies of oxidative resistance of soot are complicated because it is difficult to control soot composition which is not well determined and is dependent on many experimental factors [38–42]. Past theoretical studies of soot oxidation have focused on the model cases of oxidation reactions of one-ring aromatics [43–46] and oxygen chemisorption at selective sites of two- and three-ring aromatics [47–50]. PAHs constitute fundamental structural moieties of many other carbonaceous materials such as graphite, char, carbon black, fullerenes, carbon nanotubes, and, most recently, graphene sheets. Therefore, studies of oxidation of PAHs can contribute to development of technology of chemical modification of graphene especially in the context of improvement of efficiency of battery anodes and design of novel semi-conducting materials.

The present review summarizes some of the recent results obtained in theoretical studies of pristine and oxidized graphene substrates. It shows that global and local delocalization effects can be seen as a common grounds for explanation of certain peculiar properties of these systems. In the former case it leads to a rationale for the origin of stabilization of diradical singlet states and in the latter case it provides a very simple and intuitive perspective on relative stability of oxyradicals formed. Before discussing these findings in details, it is useful to overview pertinent approaches to the assessment of aromaticity.

29.2 Aromaticity Measures

Aromaticity is associated with delocalization of bonds and like the concept of a chemical bond itself, it is ill-defined. Various properties of molecules can be used to construct measures of aromaticity [51–53]. Rigorous approaches estimating energy of the resonance stabilization are computationally expensive. A magnetic-property-based criterion of aromaticity called nuclear-independent chemical shift index (NICS) is a widely accepted measure of local aromatic character of a molecule [54]. NICS characterizes magnetic shielding within a ring structure and is usually calculated at the geometric center of the ring and at some displacement above the ring. Its performance along with alternative techniques has been reviewed recently [55]. For anti-aromatic, non-aromatic, and aromatic systems NICS takes values from positive, to zero, to negative, respectively. Both closed- and open-shell systems can be treated [56]. In case of planar rings the out-of-plane component of the NICS tensor ($NICS_{zz}$) has advantage over the total NICS [57]. Recent studies of small aromatic, anti-aromatic, and non-aromatic organic molecules demonstrated that $NICS_{zz}$ is equal to the z-component of the induced magnetic field [58].

The harmonic oscillator model of aromaticity [59] assesses the amount of stabilization or destabilization in a benzene-like ring from the deviation of inter-atomic distances from the perfect carbon-carbon bonds of benzene. It is defined by,

$$\text{HOMA} = 1 - \frac{a}{n} \sum_{i=1}^{N} (r_{CC} - r_{CC-\text{benzene}})^2 \qquad (29.1)$$

where a was chosen to be 257.7 so that $\text{HOMA} = 0$ for a Kekule form of benzene and $\text{HOMA} = 1$ for the aromatic form of benzene, n is the number of bond lengths in the ring, r_{CC} is the carbon-carbon bond length in the system under consideration, and $r_{CC-\text{benzene}}$ is set to the experimental carbon-carbon bond length of benzene. Deviation of HOMA from 1 is a signature of deviation of aromatic character of the ring from that of benzene. Global aromaticity of a molecule can be assessed as a sum of HOMA for individual rings [60–62]. It has been shown for polyacenes that HOMA and NICS give similar results regarding global and local aromaticity, but NICS tends to overestimate the latter [63, 64]. Also, additional computational efforts are required in order to assess NICS. They can be substantial in the case of large systems, e.g. pertinent to graphene chemistry, whereas HOMA can be readily obtained from the equilibrium geometry. It has been noted that energetic and magnetic criteria of aromaticity need not be consistent with each other.

Both NICS and HOMA are suitable for quantitative studies of delocalization effects. There are also qualitative approaches, such as Huckel electron counting rules and Clar's sextet model [65] that lead to the assignment of aromaticity or antiaromaticity. Local aromatic fragments can be identified from bonding analysis if patterns of bonds are found that are similar to the prototypical organic aromatic systems. For example, adaptive natural density partitioning (AdNDP) [66, 67] can be used to separate bonding objects with a high degree of localization of electron pairs, such as lone pairs and two center-two electron bonds, from delocalized bonding objects, e.g., π-orbitals of benzene. It was shown previously for PAHs [67] that AdNDP and Clar's sextet assignment are in agreement only if the electronic structure of the molecule can be described using a single Clar structure. In case of resonance of Clar structures, there is no direct correspondence between these two approaches.

There are over 50 various criteria proposed and used to make a judgment about aromatic character of molecular systems (see, for example, Ref. [55] and references therein). This large number is due to a conceptual difficulty of defining aromaticity. Aromaticity manifests itself in many properties, including energetic, magnetic, structural, and chemical, so each can give rise to a separate aromaticity index. Typically, one seeks a consistency between several measures and a balance between their conceptual rigor and computational feasibility. These considerations led to choosing HOMA and NICS_{zz} in the study of linear PAHs and HOMA only in the study of larger two-dimensional PAHs. Only qualitative approaches, i.e. electron counting and orbital symmetry analysis, were used in the study of PAH diradicals.

29.3 Delocalization and Diradical States of Graphene Substrates

Insights into formation of diradical states in graphene substrates were gained in a study combining a perfect-pairing approximation to the coupled-cluster theory (PP-CC) and density functional theory (DFT) calculations using hybrid functional and generalized gradient approximation (GGA) [68]. Three rectangular PAHs were considered (Fig. 29.1). The substrates display both zigzag and arm-chair edges and can be seen as four, five, and six rows of linear acenes containing four, five, and six rings, respectively. The stoichiometric formula for these substrates is $C_2(n_a n_z + n_a + n_z)H_2(n_a + n_z + 1)$, where $n_a, n_z = 4, 5, 6$, so that the substrates $C_{48}H_{18}$, $C_{70}H_{22}$, and $C_{96}H_{26}$ are designated "4a4z", "5a5z", and "6a6z". The first and the third systems belong to C_{2h} point-group, the second is of C_{2v} symmetry.

DFT calculations provided relative energies of the spin-states of interest which included closed-shell singlets, diradical open-shell singlets, and triplets. PP-CC calculations were used to obtain occupation numbers of alpha and beta orbitals, which were used to assess diradical character of each substrate. Results of the calculations are summarized in Table 29.1. DFT cannot really resolve which state is lower in energy. One sees that 4a4z closed- and open-shell singlets are effectively degenerate, the 5z5a and 6z6a open-shell singlet and triplet are effectively degenerate. At the PP-CC level 4a4z is not a diradical; 6a6z is; 5a5z – is in between.

Additional information about the nature of diradical open-shell singlet states can be extracted from PP-CC calculations. As an active space approach, PP-CC accounts for the contribution of additional electron configurations that becomes significant when strong electron correlation effects are encountered in systems with small energy gaps between occupied and virtual orbital spaces. The Boys localization procedure that precedes PP-CC computations shows clear difference between the structure of molecular orbitals of the substrates with noticeable and negligible diradical character. Only well-localized orbitals are produced in the case of 4a4z substrate. Both 5a5z and 6a6z substrates have one delocalized orbital remaining after localization. This delocalized orbital is in fact the highest occupied molecular orbital (HOMO) of the PP-CC calculations and it is the one involved in

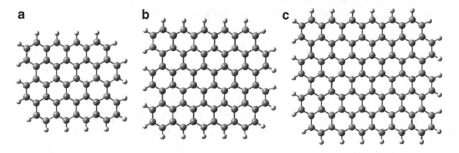

Fig. 29.1 Geometries of (**a**) 4a4z PAHs, (**b**) 5a5z PAHs, and (**c**) 6a6z PAHs

Table 29.1 Relative energies and diradical character of 4a4z, 5a5z, and 6a6z graphene patches

Spin State	E_{rel}, kcal/mol B3LYP	E_{rel}, kcal/mol PW91	γ_0^a PP–CC
"4a4z" $C_{48}H_{18}$			
$S=0, <S^2>=0.0$	0.26	0.00	
$S=1, <S^2>=2.1$	6.61	9.10	
$S=0, <S^2>=0.3$	0.00		0.01
"5a5z" $C_{70}H_{22}$			
$S=0, <S^2>=0.0$	8.10	1.60	
$S=1, <S^2>=2.1$	0.20	0.80	
$S=0, <S^2>=1.1$	0.00	0.00	0.53
"6a6z" $C_{96}H_{26}$			
$S=0, <S^2>=0.0$	15.27	4.50	
$S=1, <S^2>=2.1$	0.54	0.20	
$S=0, <S^2>=1.3$	0.00	0.00	0.91

For detailed discussion of the computational methodology and references see Ref. [67]
[a] diradical character calculated from PP–CC orbital occupation numbers

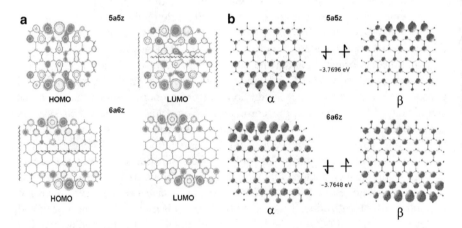

Fig. 29.2 (**a**) Isosurfaces of orbitals involved in formation of singlet diradical states of 5a5z and 6a6z in PP-CC calculations; *wavy lines* show sites of sign inversion; (**b**) isosurfaces of the spin density and diagrams of Kohn-Sham singly occupied levels of the singlet diradical states of a 5a5z and 6a6z substrates from plane wave DFT calculations

the formation of the diradical due to the strong electron correlation with the lowest unoccupied molecular orbital (LUMO) (Fig. 29.2a).

It is established that in addition to Huckel aromaticity which is an extra stabilization of a molecule due to delocalized bonding involving $4N+2$ electrons, there is also Mobius aromaticity [69, 70]. The latter is associated with 4N electrons involved in delocalized bonding in molecules that exhibit similarity between the nodal structure of their molecular orbitals and orbitals of Heilbronner's polyenes with Mobius twist [71]. Mobius strip topology is not a prerequisite of Mobius aromaticity. The 4N electron counting rule is due to a peculiar structure of an orbital energy diagram where doubly degenerate orbitals are not preceded by a non-degenerate completely bonding orbital as in the case of Huckel aromaticity. Alternatively, the number of out-of-phase overlaps in orbitals contributing to delocalized bonding can be considered [69, 70]. It is important to keep in mind that the counting rules are applied to electrons and orbitals contributing to delocalized bonds which have to be separated from the localized bonding objects. Among the three substrates considered, 4a4z does not have a globally delocalized bonding component because all of its orbitals can be localized. Globally delocalized bonding exists in 5a5z and 6a6z and is rendered by their HOMOs that remain delocalized after the localization procedure. Formation of diradical states in 5a5z and 6a6z is, therefore, due to the orbitals contributing to delocalized bonding, i.e., HOMO and a strongly correlated LUMO. In diradical states, these orbitals become singly occupied and degenerate with the orbital energy diagram of Mobius type (Fig. 29.2b). Analysis of out-of-phase overlap shows an odd number with sites of sign inversion (Fig. 29.2a), consistent with the structure of a Mobius orbital array [69, 70]. Therefore, provided the open-shell singlet states of 5a5z and 6a6z are multiconfigurational, these systems should be considered as multi-configurational Mobius aromatic.

29.4   Local Aromaticity and Stability of Graphene Oxyradicals

Studies of stability of oxyradicals of various PAHs have been performed in order to characterize the role of edges in graphene oxidation and the influence of the oxidized edge on the electronic structure of the interior [72, 73]. Theoretical approaches ranging from second order Moeller-Plesset perturbation theory (MP2) to GGA and hybrid DFT to semi-empirical PM6 method were used. Treatment of large substrates is challenging for any ab initio method except for GGA. It is also risky to rely on a single theoretical framework, so semi-empirical studies should be considered as viable alternatives to DFT. Detailed discussion of the computational methodology and references can be found in Refs. [72] and [73]. Analysis of the delocalization effects pertinent to the topic of the present contribution has been performed using AdNDP, NICS, and HOMA in the case of linear substrates and HOMA in the case of rectangular substrates.

29.4.1   Linear Substrates

Schematic representation of bonding in the pentacene molecule and its oxyradicals is based on the results of AdNDP analysis (Fig. 29.3) [72]. Oxyradicals II-V are open-shell systems so the analysis was performed separately for α- and β-components of the electron density and results superimposed.

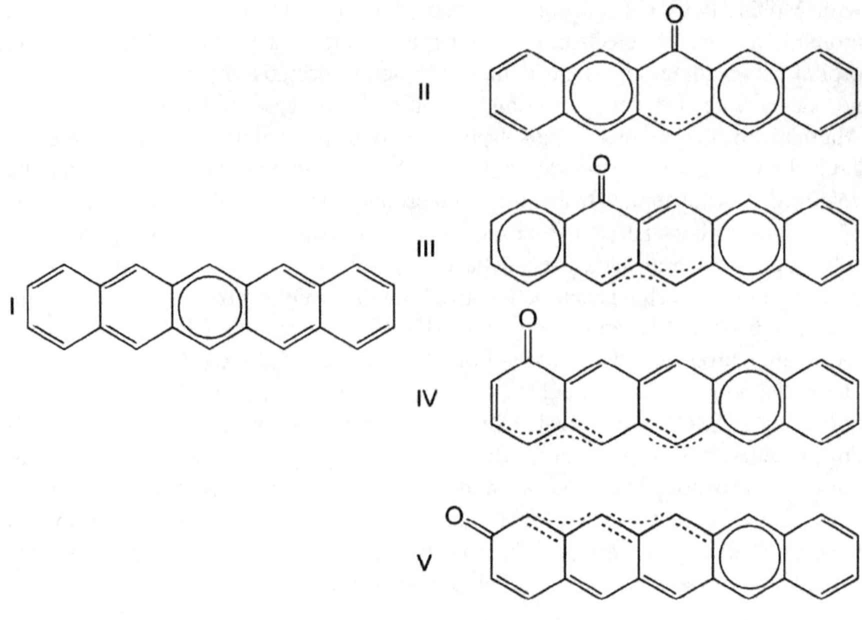

Fig. 29.3 Schematic representation of chemical bonding in systems I – V. For the open shell systems II – V the bonds found simultaneously in α- and β-spaces are depicted as regular nc-2e bonds. Non-coinciding nc-1e bonds from α- and β-spaces are superimposed (α are *dashed lines* on the interior, β – on the exterior). Oxygen lone-pairs and 2c-2e CH σ-bonds are omitted

Both hybrid DFT (B3LYP) and MP2 calculations showed higher stability of oxyradicals with O bound by "interior" rings. Standard Gibbs free energies of pentacene oxyradicals II-V were calculated to assess their thermodynamic stability as a function of temperature (Fig. 29.4). The temperature dependencies indicate that below 1000 K the relative thermodynamic stability of II-V is the same as the ordering of their relative energies at both B3LYP and MP2 levels of theory, i.e., is in the order: II > III > IV > V. Above 1000 K oxyradical II becomes less stable than III due to the larger entropy of III and increasing contribution of the entropy contribution to the Gibbs free energy with temperature.

The relative stability of four pentacene oxyradicals can be related to the strength of delocalization effects in their π-electron systems. The quantitative (NICS, HOMA) and qualitative (AdNDP) approaches agree in their assessment of local aromaticity of the rings of pentacene and its oxyradicals (Fig. 29.3 and Table 29.2). When O binds to a particular ring forming a double bond it destroys local π-aromaticity and renders the ring non-aromatic. Furthermore, it excludes the ring from the globally delocalized π-bonding system and leads to a fragmentation of the latter. Two separated locally π-aromatic fragments are formed in II and III and only one in IV and V, hence the higher stability of the former two. The patterns of conjugation in the remaining rings is another factor. Orthobenzoquinone-like

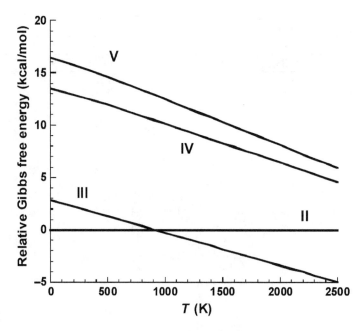

Fig. 29.4 Standard Gibbs free energy of pentacene oxyradicals relative to that of pentacene oxyradical II as a function of temperature at B3LYP/6-311G(d,p) level of theory

conjugation revealed in IV should have higher stability than the parabenzoquinone-like conjugation in V (Fig. 29.3). The role of π-delocalization in oxyradicals for their stability can be clearly seen from the nearly linear dependence of the relative energies of II-V and their cumulative HOMA (Fig. 29.5). Cumulative HOMA of a PAH is calculated as sums of HOMA of each individual ring under the assumption of the additive nature of global aromaticity of a PAH with respect to local aromaticity of its benzene-like rings. In the view of fragmentation of the globally delocalized π-bonding in oxyradicals, their relative energies are additive with respect to the stabilization energy of independent locally aromatic fragments. This is the reason for the simple correlation between relative energies and cumulative HOMA.

29.4.2 Rectangular Substrates

From the investigation of the linear PAH as a model of a zigzag graphene edge in reaction with O^{70} it was concluded that the relative stability of linear oxyradicals is controlled by fragmentation of the delocalized π-electron system and formation of locally aromatic fragments. Fragmentation is the only viable option for the rearrangement of π-bonds in linear PAHs. Two-dimensional substrates should be more flexible in accommodating changes in their electronic structure

Table 29.2 Assessment of local aromaticity: values of HOMA and NICS$_{zz}$ indices for structures I – V and prototypical aromatic benzene and naphthalene molecules

Structure	Ring i HOMA	Ring i NICS$_{zz}^a$	Ring ii HOMA	Ring ii NICS$_{zz}^a$	Ring iii HOMA	Ring iii NICS$_{zz}^a$	Ring iv HOMA	Ring iv NICS$_{zz}^a$	Ring v HOMA	Ring v NICS$_{zz}^a$
I	0.49	−4	0.57	−17	0.59	−22	0.57	−17	0.49	−4
		−13		−27		−31		−27		−13
		−21		−33		−37		−33		−21
II	0.77	−7	0.71	−1	−0.17	25	0.71	−1	0.77	−7
		−16		−11		17		−11		−16
		−24		−20		4		−20		−24
III	0.89	−3	−0.12	24	0.57	2	0.69	−10	0.67	−7
		−12		16		−7		−19		−16
		−20		3		−17		−27		−24
IV	0.10	19	0.51	2	0.60	−9	0.65	−15	0.59	−6
		12		−7		−19		−24		−15
		0		−17		−27		−31		−23
V	0.03	16	0.43	10	0.58	−3	0.67	−11	0.63	−7
		8		1		−12		−21		−15
		−3		−10		−21		−28		−23
Benzene	0.99	−14								
		−23								
		−29								
Naphthalene	0.79	−13	0.79	−13						
		−22		−22						
		−29		−29						

^{a} Three NICS$_{zz}$ values are calculated at 0.0 Å, 0.5 Å, and 1.0 Å above the ring, yielding NICS(0.0), NICS(0.5) and NICS(1.0), respectively

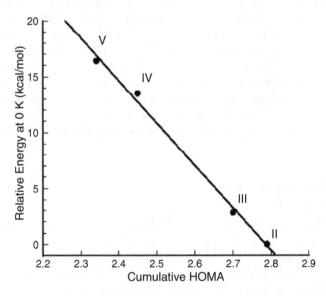

Fig. 29.5 ZPE-corrected relative energies of oxyradicals II-V (at the UB3LYP/6-311G(d,p) level of theory) plotted against cumulative HOMA. The *straight line* reflects nearly linear dependence

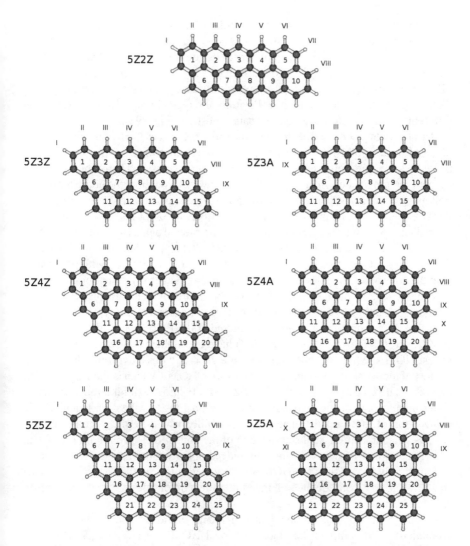

Fig. 29.6 Structures of two-dimensional PAH substrates used to form oxyradicals. Roman numerals label position of oxygen in corresponding oxyradicals. Arabic numerals label six-atomic rings

due to oxidation. Measures of local aromaticity should be informative and robust tools in the assessment of such changes. Considering the size of the substrates of interest, HOMA is perceived as the optimal choice as it does not require any quantum chemical computations beyond geometry optimization and readily provides information about local and global localization.

Analysis of stability of two families of oxyradicals formed from two-dimensional substrates (Fig. 29.6) has been performed at GGA DFT and semi-empirical PM6 levels of theory [73] to ascertain whether the previously observed relationship between local aromaticity and relative stability of linear PAH oxyradicals holds for two-dimensional substrates. The substrates are designated according to the size and the shape of the edges. Two types of substrates are investigated, 5ZmZ and 5ZnA, where Z stands for "zigzag", A for "armchair", $m = 2$–5 and $n = 3$–5. Therefore, 5Z3A is a substrate of rectangular shape with five rings along the zigzag edge and three rings along the armchair edge. Position of oxygen in the edge is used to label different oxyradicals formed from the same substrate (Fig. 29.6).

As a double C-O bond is formed instead of a single C-H bond in the edge of a substrate molecule the respective C cannot contribute to the π-system in the corresponding ring rendering the latter antiaromatic. Therefore, oxidation triggers rearrangements in the π-bonding framework throughout the entire molecule. Following the analysis performed in the study of pentacene oxyradicals, cumulative HOMA was used to quantify delocalization effects in 5ZmZ and 5ZnA families of oxyradicals. Relative energies of oxyradicals are plotted against cumulative HOMA in Fig. 29.7. The linear trend appears to persist. Certain peculiarities are noticeable, though. First, cumulative HOMA of oxyradicals varies noticeably with position of O in the edge at the PM6 level. These pronounced changes suggest strong influence of edge oxidation on the aromaticity of the entire oxyradical. DFT trends are steeper, showing that cumulative HOMA does not depend strongly on the position of oxidation. Second, outliers are encountered at the PM6 level of theory. Isomers I-5Z2Z and I-5Z3A (Figs. 29.6 and 29.7) have the highest relative energy and the lowest HOMA, which is consistent with the generally observed trend. Isomer I of the rest of the substrates has an anomalously high HOMA which is inconsistent with its high relative energy. Isomers VI-5Z5Z and VII-5Z5A exhibit similar anomaly. Weak involvement of the respective rings in global π-delocalization is a plausible explanation of these results. The relative stability would be due to other factors such as structural strains and stabilization of the unpaired electron. For example, in I-5Z(3,4,5)A and I-5Z(4,5)Z, ring 1 is indeed antiaromatic, but its HOMA value is consistently higher than the HOMA of oxidized rings in other oxyradicals. This can be seen as a consequence of incomplete relaxation of C-C bonds leading to structural strain.

Another way to look at the rearrangements of bonds and change of local aromaticity is to compute the change of local HOMA of oxyradicals relative to the substrate and to each other (see detailed discussion in Ref. [73]). At both levels of theory, the ring interacting with O becomes anti-aromatic. DFT shows uniformly high values of local HOMA (above 0.5) and, respectively, appreciable degree of delocalization, in the remaining rings. Therefore, the loss of aromaticity by an oxidized ring appears to be a local effect which does not significantly affect other rings. PM6 shows drastic changes of local aromaticity from ring to ring in each substrate. It also shows significant redistribution of aromatic fragments as a response to oxidation of different rings. Upon oxidation HOMA variations of individual rings at the DFT level are typically by an order of magnitude smaller than those at the PM6 level. For this reason, PM6 results reveal more clearly patterns of local aromaticity in the oxyradicals studied and, therefore, will be discussed further in greater details.

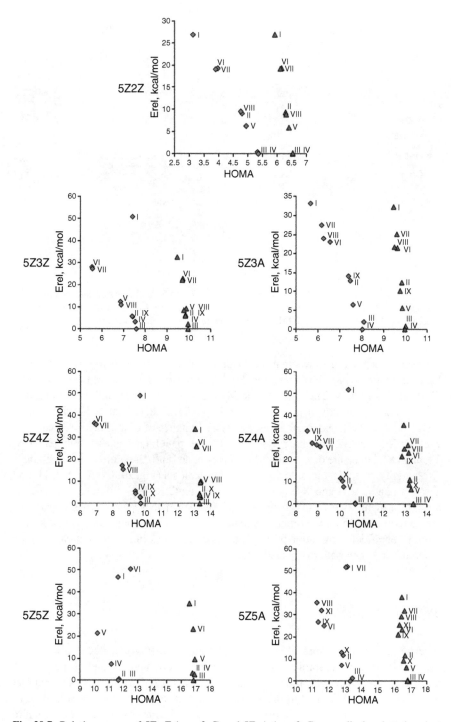

Fig. 29.7 Relative energy of 5Z*m*Z (*m* = 2–5) and 5Z*n*A (*n* = 3–5) oxyradicals, plotted against cumulative HOMA: *red squares* – PM6, *blue triangles* – DFT; roman numbers label the isomers according to the position of oxygen along the edge (see Fig. 29.6)

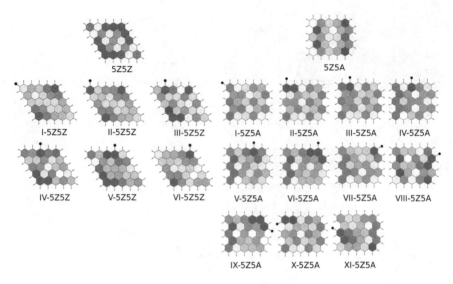

Fig. 29.8 An example of HOMA pattern classes in 5Z5Z and 5Z5A substrates and corresponding oxyradicals (PM6 level). HOMA is represented by color changing from *blue* for HOMA = 0.0, to *white* for HOMA = 0.5, to *red* HOMA = 1.0. *Black solid dot* marks oxygen position

Further inspection shows that the oxidation at the edge leads to three classes of local aromaticity patterns. Oxyradicals of 5Z5Z and 5Z5A substrates at Fig. 29.8 illustrate this finding. The one, "checker-board-like", is related to the Clar structure of coronene with highly aromatic disjoint rings (Fig. 29.8, left panel: III; right panel: I, II, III, IV, and X), and another resembles a resonance of Clar structures leading to cycles of six weakly aromatic rings (Fig. 29.8, left panel: I, II, and VI; right panel: V, VI, and IX), hence the designations "Clar coronene" and "superaromatic coronene". Aromaticity of a considerable number of oxyradicals (Fig. 29.8, left panel: IV and V; right panel: VIII and XI) is in the "intermittent" regime which can be seen as a mixture of the two forms. It remains to be seen if these three types of patterns are general and emerge in other graphene-based systems. Their formation does not seem to be affected by the nature of the substrate edges in the present case, so the assumption of generality is plausible.

29.5 Conclusions

Investigation of the nature of the diradical open-shell singlet states in a family of rectangular PAHs exemplifies relevance of delocalization effects and the concept of aromaticity to theoretical studies of electronic structure of graphene. It showed that diradical states are formed in the substrates that have a globally delocalized component in their bonding framework. This delocalized component bears the signature of Mobius aromaticity. Given the multiconfigurational nature of open-shell singlet states, PAH diradicals can be classified as multiconfigurational Mobius aromatic.

The interplay between local aromaticity of benzene-like rings of graphene substrates and the global aromaticity of the latter has led to a simple explanation of the trends in relative stability of two families of linear and two-dimensional PAH oxyradicals. For any substrate, oxidation leads to the loss of aromaticity by the corresponding ring and eliminates its contribution to inter-ring conjugation. As a result, the relative stability of linear oxyradicals is controlled by fragmentation of the globally delocalized π-electron system. It mainly depends on the nature and the amount of locally π-aromatic fragments formed. Relative energies of linear oxyradicals show linear dependency of the cumulative HOMA aromaticity measure. This linear trend is essentially preserved in oxyradicals of two-dimensional substrates. No fragmentation but rather rearrangement of the pattern of local aromaticity occurs upon oxidation in the edge. It appears to lead to three classes of local aromaticity patterns. The first class is checker-board-like. It is related to the Clar structure of coronene with highly aromatic disjoint rings. The second class resembles a resonance of Clar structures leading to cycles of six weakly aromatic rings, hence the designations "Clar coronene" and "superaromatic coronene". Also, there is an "intermittent" class which can be seen as a mixture of the previous two. It remains to be seen if these three types of patterns are of general relevance and emerge in other graphene-based systems. Their formation does not seem to be affected by the nature of the substrate edges in the present research, so the assumption of generality is plausible. This shows, that a very local event, such as oxidation of the edge, has a strongly non-local effect on the entire framework of C–C bonds.

Computational studies of graphene and related systems are naturally focused on obtaining accurate quantitative data. In order to rationalize huge amount of quantitative information and place it in the context of chemical theory it is important to utilize and further develop chemically relevant models and concepts that are covered by the term "chemical bonding". Aromaticity is an extremely powerful example of a qualitative model with explanatory and predictive power. The results discussed in the present contribution show that indeed it can be fruitfully applied in the study of graphene. To date, results of a mostly descriptive nature are obtained. There is a confidence that they can be further extended to yield simple predictive models appealing to chemical intuition.

Acknowledgements WAL, and MF were supported by the Director, Office of Energy Research, Office of Basic Energy Sciences, Chemical Sciences, Geosciences and Biosciences Division of the US Department of Energy, under Contract No. DE-AC03-76F00098. XY, and MF were supported by the US Army Corps of Engineers, Humphreys Engineering Center Support Activity, under Contract No. W912HQ-07-C-0044. DYZ was supported by the National Science Foundation under grant NSF CHE-0809969. This research used resources of the National Energy Research Scientific Computing Center (NERSC), which is supported by the Office of Science of the U.S. Department of Energy under Contract No. DE-AC02-05CH11231.

References

1. Novoselov KS, Geim AK, Morozov SV, Jiang D, Zhang Y, Dubonos SV, Grigorieva IV, Firsov AA (2004) Science 306:666–669.
2. Geim AK, Novoselov KS (2007) Nature Mater 6:183–191.
3. Lee EJH, Balasubramanian K, Weitz RT, Burghard M, Kern K (2008) Nature Nanotech. 3:486–490.
4. Fujita M, Wakabayashi K, Nakada K, Kusakabe K J Phys Soc Jpn (1996) 65:1920–1923;
5. Wakabayashi K, Sigrist M, Fujita M J Phys Soc Jpn (1998) 67:2089–2093.
6. Wakabayashi K, Fujita M, Ajiki H, Sigrist M Phys Rev B (1999) 59:8271–8282.
7. Kusakabe K, Maruyama M Phys Rev B (2003) 67:092406.
8. Yamashiro A, Shimoi Y, Harigaya K, Wakabayashi K Phys Rev B (2003) 68:193410.
9. Lee H, Son YW, Park N, Han S, Yu J Phys Rev B (2005) 72:174431.
10. Zhang Y, Tan YW, Stormer HL, Kim P Nature (2005) 438:201–204.
11. Berger C, Song ZM, Li X, Wu X, Brown N, Naud C, Mayou D, Li T, Hass J, Marchenkov AN, Conrad EH, First PN, de Heer WA Science (2006) 312:1191–1196.
12. Son YW, Cohen ML, Louie SG Nature (2006) 444:347–349.
13. Peres NMR, Castro Neto AH, Guinea F Phys Rev B (2006) 73:195411.
14. Peres NMR, Castro Neto AH, Guinea F Phys Rev B (2006) 73:241403(R).
15. Ezawa M Phys Rev B (2006) 73:045432.
16. Barone V, Hod O, Scuseria GE Nano Lett (2006) 6:2748–2754.
17. Han MY, Özyilmaz B, Zhang Y, Kim P Phys Rev Lett (2007) 98:206805.
18. Srinivas G, Zhu Y, Piner R, Skipper N, Ellerby M, Ruoff R Carbon (2010) 48:630–635.
19. Stein SE, Brown RL Carbon (1985) 23:105–109.
20. Stein SE, Brown RL J Am Chem Soc (1987) 109:3721–3729.
21. Radovic LR, Bockrath B J Am Chem Soc (2005) 127:5917–5927.
22. Philpott MR, Kawazoe Y Chem Phys (2009) 358:85–95.
23. Houk KN, Lee PS, Nendel N J Org Chem (2001) 66:5517–5521.
24. Bendikov M, Duong HM, Starkey K, Houk KN, Carter EA, Wudl F J Am Chem Soc (2004) 126:7416–7417.
25. Jiang DE, Dai S J Phys Chem A (2008) 112:332–335.
26. Fujita M, Wakabayashi K, Nakada K, Kusakabe K J Phys Soc Jap (1996) 65:1920–1923.
27. Nakada K, Fujita M, Dresselhaus M, Dresselhaus MS Phys Rev B (1996) 54:17954.
28. Son YW, Cohen ML, Louie SG Phys Rev Lett (2006) 97:216803.
29. Kobayashi Y, Fukui K, Enoki T, Kusakabe K, Kaburagi Y Phys Rev B (2005) 71:193406.
30. Niimi Y, Matsui T, Kambara H, Tagami K, Tsukada M, Fukuyama H Phys Rev B (2006) 73:085421.
31. Jiang DE, Sumpter BG, Dai S J Chem Phys (2007) 127:124703.
32. Ezawa M Phys Rev B (2007) 76:245415.
33. Ezawa M Physica E (2008) 40:1421–1423.
34. Hod O, Barone V, Scuseria GE Phys Rev B (2008) 77:035411.
35. Vergés JA, Chiappe G, Louis E, Pastor-Abia L, SanFabián E Phys Rev B (2009) 79:094403.
36. Hachmann J, Dorando JJ, Aviles M, Chan GKL J Chem Phys (2007) 127:134309.
37. Frenklach M Phys Chem Chem Phys (2002) 4:2028–2037.
38. Nagle J, Strickland-Constable RF Proc Fifth Carbon Conf (1962) 1:154.
39. Kennedy IM Prog Energy Combust Sci (1997) 23:95–132.
40. Stanmore BR, Brilhac JF, Gilot P Carbon (2001) 39:2247–2268.
41. Higgins KJ, Jung HJ, Kittelson DB, Roberts JT, Zachriah MR J Phys Chem A (2002) 106:96–103.
42. Kim CH, Xu F, Faeth GM Combust Flame (2008) 152:301–316.
43. Brezinsky K Prog Energy Combust Sci (1986) 12:1–24.
44. Olivella S, Sole A, Garcia-Raso A J Phys Chem (1995) 99:10549–10556.
45. Fadden MJ, Barckholtz C, Hadad CM J Phys Chem A (2000) 104:3004–3011.

46. Xu ZF, Lin MC J Phys Chem A (2006) 110:1672–1677.
47. Montoya A, Mondragon F, Truong TN Fuel Proc Technol (2002) 77:125–130.
48. Sendt K, Haynes BS Proc Combust Inst (2005) 30:2141–2149.
49. Sendt K, Haynes BS Combust Flame (2005) 143:629–643.
50. Sendt K, Haynes BS J Phys Chem A (2005) 109:3438–3447.
51. Special edition on aromaticity. Schleyer PvR Ed.; Chem Rev (2001) 101, No. 5
52. Special edition on heterocycles. Katrizky A Ed.; Chem Rev (2004) 104, No. 5
53. Special edition on delocalization pi and sigma. Schleyer PvR Ed.; Chem Rev (2005) 105, No. 10
54. Schleyer PV, Maerker C, Dransfeld A, Jiao HJ, Hommes N J Am Chem Soc (1996) 118:6317–6318.
55. Chen ZF, Wannere CS, Corminboeuf C, Puchta R, Schleyer PV Chem Rev (2005) 105:3842–3888.
56. Hemelsoet K, Van Speybroeck V, Marin GB, De Proft F, Geerlings P, Waroquier M J Phys Chem A (2004) 108:7281–7290.
57. Fallah-Bagher-Shaidaei H, Wannere CS, Corminboeuf C, Puchta R, Schleyer PV Org Lett (2006) 8:863–866.
58. Merino G, Heine T, Seifert G Chem Eur J (2004) 10:4367–4371.
59. Kruszewski J, Krygowski TM Tetrahedron Lett (1972) 36:3839–3842.
60. Schleyer PV, Jiao HJ Pure & Appl Chem (1996) 68:209–218.
61. Aihara J, Kanno H J Phys Chem A (2005) 109:3717–3721.
62. Cyranski MK, Krygowski TM, Katritzky AR, Schleyer PV J Org Chem (2002) 67:1333–1338.
63. Cyranski MK, Stepien BT, Krygowski TM Tetrahedron (2000) 56:9663–9667.
64. Portella G, Poater J, Bofill JM, Alemany P, Sola M J Org Chem (2005) 70:2509–2521.
65. Clar E The Aromatic Sextet. J. Wiley & Sons: London, (1972)
66. Zubarev DY, Boldyrev AI Phys Chem Chem Phys (2008) 10:5207–5217.
67. Zubarev DY, Boldyrev AI J Org Chem (2008) 73:9251–9258.
68. Wang J, Zubarev DY, Philpott MR, Vukovic S, Lester WA, Cui T, Kawazoe Y Phys Chem Chem Phys (2010) 12:9839–9844.
69. Zimmerman HE J Am Chem Soc (1966) 88:1564–1565.
70. Zimmerman HE Acc Chem Res (1971) 4:272–280.
71. Heilbronner E, Tetrahedron Lett (1964) 5:1923–1928.
72. Zubarev DY, Robertson N, Domin D, McClean J, Wang J, Lester WA, Whitesides R, You X, Frenklach M J Phys Chem C (2010) 114:5429.
73. Zubarev DY, You X, McClean J, Lester WA, Frenklach M (2011) J Mater Chem 21:3404–3409.

Chapter 30
20-Nanogold Au$_{20}$(T_d) and Low-Energy Hollow Cages: Void Reactivity

E.S. Kryachko and F. Remacle

Abstract Five 20-nanogold low-energy hollow cages are identified at the density functional level by performing a computational search on the corresponding potential energy surfaces in the different charge states. Their structures and stabilities are investigated and compared with the tetrahedral ground-state and space-filled cluster Au$_{20}$(T_d). Special attention is devoted to the bifunctional reactivity of the studied Au$_{20}$ hollow cages: the outer, exo-reactivity and the inner, void reactivity. The void reactivity results in endohedrality, i.e. in the existence of @-fullerenes of gold. We analyze the general features of the voids of the reported 20-nanogold fullerenes. The values of ionization potentials and electronaffinities, the molecular electrostatic potential and HOMO and LUMO patterns are invoked for this purpose and compared with those of C$_{60}$ that has a similar void size. This is on the one hand. On the other, as already known in the literature, the space-filled Au$_{20}$(T_d) reveals a perfect confinement for some guest atoms. The mechanism of the formation of void of Au$_{20}$(T_d) that enables to trap a guest is illustrated by using a guest gold atom which is repelled by the so called 'interior' atoms of Au$_{20}$(T_d). The computed repulsion energy provides a rough estimate of the energy needed to form a void inside this cluster.

E.S. Kryachko (✉)
Bogolyubov Institute for Theoretical Physics, Kiev-143, 03680 Ukraine and Department of Chemistry, Bat. B6c, University of Liège, B-4000 Liège, Belgium
e-mail: eugene.kryachko@ulg.ac.be

F. Remacle
Department of Chemistry, Bat. B6c, University of Liège, B-4000 Liège, Belgium
e-mail: fremacle@ulg.ac.be

P.E. Hoggan et al. (eds.), *Advances in the Theory of Quantum Systems in Chemistry and Physics*, Progress in Theoretical Chemistry and Physics 22,
DOI 10.1007/978-94-007-2076-3_30, © Springer Science+Business Media B.V. 2012

30.1   Introduction

The discovery, in the 1980s, of the buckyball C_{60} and larger fullerenes [1] which possess a spatially closed inner and void space or, in other words, have the shape of a hollow cage led in the following decades to the identification of 'endohedral' or @-fullerenes whose voids trap guest heteroatoms, ions, and molecules [2–4] and whose number continues to grow every year. One of the first among them was La@C_{60} [2]. These @-fullerenes exhibit rather unusual features, primarily in their stability and reactivity. Nowadays, many endohedral C_{60}-fullerenes have been synthesized and characterized, both experimentally and theoretically. These mainly host or confine atoms of metals, alkali-metals in particular [4, 5], and of noble gases due the strong electron affinity $EA(C_{60}) = 2.65 - 2.69\,eV$ [6] – trapping of molecules, though, have been reported to a lesser extent. Among the molecules which can be trapped in fullerenes' voids are polar diatomic molecules, such as LiF and LiH – which often increase their stability [4 a]. Moreover, the stability can be con controlled, as was, for example, with LiF@C_{60}, via manipulation with hydrogens exo-attached to the external side of C_{60} [4b].

The concept of endohedrality was naturally extended from carbon to other chemical elements, giving thus rise to similar fullerene-like structures or hollow cages, or shortly, 'fullercages' [7]. There was also gold, the noblest atom, that is well known for its bulky chemical inertness [8, 9], gross catalytic activity [10] and color [11] in nano-dimensions, and strong relativistic effects [12]. The latter are strikingly manifested in a quite unusual shape of gold clusters Au_N that turns to be preferentially three-dimensional (3D) when the cluster size N is larger than nine [13].

Generally speaking, any 3D molecular shape can either be space-filled, compact, or can admit the existence of some void, emptiness, partial 'no-pair direct bonding', in some sense [14], that results in a sort of fullerene-like or hollow cage shape [15–18]. The definition of "fullerene likeness" relies on that of fullerene two versions of which exist in the literature. According to the IUPAC definition [19 a]: "Fullerenes are defined as polyhedral closed cages made up entirely of n three-coordinate carbon atoms and having 12 pentagonal and $(n/2 - 10)$ hexagonal faces, where $n \geq 20$. Other polyhedral closed cages made up entirely of n three-coordinate carbon atoms shall be known as quasi-fullerenes." The CAS defines [19b] fullerenes as "the even-numbered, closed spheroidal structures of 20 or more carbon atoms, in which every atom is bonded to three other atoms." Mathematically (see e.g.[19c]), fullerene is a convex simple 3D-polytope with the maximum δ_3 where δ_3 is the total number of inner diagonals, that is, segments that join two vertices of polytope and that exist, except for their ends, within the polytope relative interior. By another definition, a given cluster is a hollow cage if it has, in addition to its outer space, a spatially closed inner void space which diameter – also called void diameter (see Ref. [20] for the definition) – exceeds double van der Waals radii. For gold, it is equal to 3.32 Å. It is obviously that a void of any hollow cage is separated from a cage's outer space by at least a single-atom layer.

A void may enable to accommodate some heteroatom or molecule – a dopant, speaking generally, – which size plays a decisive role in a stable incarceration. In general, gold cage clusters can be partitioned into two classes [21]: (i) shell-like or oblate flat cages, such as for instance the anionic clusters Au_{15a}^-, Au_{15b}^-, Au_{15d}^-, Au_{16c}^-, and Au_{17c}^- considered in [22], and (ii) spherical-like hollow cages – depending on whether the Au-Au lines (inner diagonals), which connect the gold atoms through the cage's void, range within $3-\sim 5$Å or exceed 5Å. The shell-like flat cages therefore enable to accommodate a small heteroatom, such as the hydrogen, without major structural distortions. This is a typical example of confinement. This is e.g. the lowest-energy neutral cage Au_{15} reported in [23]. Does it mean that a type of cage is a function of charge state or, speaking in general, whether a shape, space-filled or hollow, can be charge-state dependent? Other examples of shell-like flat cages are the cages $Au_{N=16-19}$, derived from the ground-state tetrahedral gold cluster $Au_{20}(T_d)$ by removing 4, 3, 2, and 1 vertex atoms [23b]. In contrast, the spherical-like hollow cages can be doped by a larger atom, such as gold, for instance.

Historically, golden fullerenes Au_N begun with $Au_{12}(I_h)$ [24] which was predicted in 2002 and have since been spread out to $Au_{14,16,18}$ [22, 25, 26], Au_{20} [25, 27, 28], Au_{24-28} [29], $Au_{32}(I_h)$ [30], Au_{38-56} [31–34], and nowadays reached $Au_{72}(I)$ [35]. Au_{20} is a magic gold cluster whose ground state on the potential energy surface (PES) is a tetrahedral space-filled structure $Au_{20}(T_d)$ in the charge states Z = 0, ±1. Though it by many features resembles the hollow cage C_{60} [27] (for recent works see [36]): for instance, its external diameter reaches approximately 0.7 nm – that is, $Au_{20}(T_d)$ is of the same nanometer size as C_{60}, $Au_{20}(T_d)$ has no void. In contrast, the ground-state structure $Au_{32}(I_h)$ is a golden fullerene. Pursuing this motif, the hollow cages Au_{38}, Au_{42}, Au_{44}, and Au_{56} are less stable than their space-filled counter partners.

However, it seems quite inessential, immaterial, even irrelevant, from the viewpoint of endohedrality, to differentiate gold clusters into space-filled or hollow cages. Why? As demonstrated for the first time by Molina and Hammer [37], the space-filled cluster $Au_{20}(T_d)$ enables, contrary to the common point of view, to trap the dopant atoms X = H, Li, K, and Na^+, thus resulting in the endohedral golden fullerenes $X@Au_{20}$ with endo-binding energies equal to −0.85, −0.41, −1.39, and −0.59 eV, respectively. Note that more accurate endo-binding energies of $H@Au_{20}(T_d)$ and $Li@Au_{20}(T_d)$ are correspondingly equal to −0.53 and −2.49 eV [38]. What is the mechanism behind this phenomenon, the phenomenon of transforming a space-filled structure to a hollow cage under doping? Why doping enables to create a void in a space-filled structure gold cluster? Is there any sort of selectivity implying that some dopants make a void, whereas the others do not? In some sense, this makes $Au_{20}(T_d)$ and C_{60} even more alike. Speaking generally, in what are the inner voids of golden fullerenes similar to those of fullerenes? This question turns out to be reformulated as: Are there any fundamental properties of the chemical reactivity of molecular cages that lead to atom encapsulation? To be rigorous, herein, the chemical reactivity is broadly defined as a capability of a given molecular species to form chemical bonding patterns while interacting or

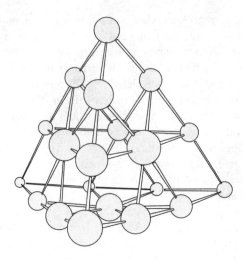

Fig. 30.1 The tetrahedral space-filled structure $\mathbf{I} \equiv \mathrm{Au}_{20}(T_d)$ in the charge state $Z = 0$

contacting with other atoms, ions, or molecules. Therefore, a molecular hollow cage endows two kinds of chemical reactivity: the inner or void, or endo-reactivity that characterizes a capability of a hollow cage to bind within its void – indicated by @; the outer or exo-reactivity that characterizes a capability of a hollow cage to bind in its outer space – indicated hereafter by &. Do the hollow cages exhibit such features which are not achievable by their space-filled counterparts and which endow them with the ability to trap atoms? These questions are addressed in the present work by invoking the example of the 20-nanogold Au_{20} in different charge states since $N = 20$ is definitely a benchmark, magic number for gold.

The present work is composed of five Sections. The computational methodology is outlined in the next Sect. 30.2. In Sect. 30.4, we thoroughly examine the general properties of low-energy hollow cages in the different charge states Z and compare them with the tetrahedral space-filled structure $\mathbf{I} \equiv \mathrm{Au}_{20}(T_d)$ which is shown, for the reason of completeness and closeness of the present work, in Fig. 30.1 and which, in the charge states $Z = 0, \pm 1$, occupies the ground electronic state. Section 30.4.1 focuses on the concept of void reactivity of the studied cages in terms of the molecular electrostatic potential. Section 30.4.2 ends with general discussion.

30.2 Computational Methodology

All computations of gold clusters were carried out with the DF potentials B3LYP, BP86, and PW91PW91 (PW91 for short) in conjunction with the energy-consistent $19\text{-}(5s^2 5p^6 5d^{10} 6s)$ valence electron relativistic effective core potential (RECP) of the Los Alamos double-zeta type LANL2DZ [39]. The GAUSSIAN 03 package of quantum chemical programs [40] was used throughout the present work. All geometrical optimizations were carried out with the keywords "Tight" and "Int=UltraFine". The harmonic vibrational frequencies and concomitant zero-point energies (ZPE) were calculated in order to locate the energy-minimum structures and to distinguish them from the saddles. Enthalpies and entropies, which are also reported in the present work, were obtained from the partition functions calculated at room temperature (298 K) by means of using the Boltzmann's thermostatistics and the rigid-rotor-harmonic-oscillator approximation.

30.3 20-Nanogold Hollow Cages in Various Charge States: Basic Features

Applying the aforementioned computational methodology, we have studied the low-energy section of the PES of 20-nanogold where five hollow cages **II-VI**, located closer to the ground-state Au$_{20}$(T_d) in the neutral charge state, have been identified. The neutral and Z-charged cages **II-VI** with Z = 0, ±1, -2 are definitively minima on the 20-gold PESs with real harmonic vibrational frequencies only. For Z = 0, they all fall within 9–175 cm^{-1}. Their basic properties are summarized in Tables 30.1–30.5. In the charge state Z = 0, these cages of Au$_{20}$ are characterized by external diameters of ca. 0.8–1.2 nm, which are slightly larger that, 0.7 nm, of C$_{60}$ (see Figs. 30.2–30.6). These cages are essentially hollow in the neutral state with void diameters of $\sim$4.5–7.0Å.

Structurally, cage **II** resembles a helix. **III** exhibits a bilayer motif via forming by two planar Au$_{10}$(D_{2h}) clusters twisted w. r. t. one another by a dihedral angle of 33.1° and bonded to each other by means of 20 additional metallic bonds. Due to this quite specific bilayer shape, synthesis of cage **III** might be accessible. The specific binding energy, E$_{binding*}$, that defines the energy of the formation of cage **III** from two Au$_{10}$(D_{2h}) monomers in the 0-charge state is, according to Table 30.2, strongly DF-dependent: e.g., the B3LYP estimate of -2.34 eV is the lowest, by the absolute value, among the reported, whereas the PW91 yields the highest one equal to -4.34 eV. Comparing to the Au$_{10}$(D_{2h}) monomer, the cage formation downshifts the HOMO by 0.7–0.8 eV, narrows the HOMO-LUMO energy gap, and increases the electron affinity by 0.85 eV. In terms of bonding patterns, the cage formation is performed due to appearance of new inter-monomeric, "glue" bonds which arise only between 3-coordinated atoms, so that in the dimeric cluster, they become 5-coordinated. Cage **III** therefore relies on 52 bonds in the 0-charge state.

Comparing to cage **II**, **III** is less stable in the neutral charge state and correspondingly placed above the former by approximately 0.05(-PW91;0.11–BP86, 0.37–B3LYP) eV (see Tables 30.1 and 30.2). The last: due to its double-layer structural motif, cage **III** is anticipated to be aromatic. According to Table 30.2, its NICS(0) [43] at the cage center is equal to -14.7 ppm that is rather close to that of Au$_{20}$(T_d). Being neutral, the hollow Au$_{20}$ isomers are closed-shell structures which, as reported in Tables 30.1–30.5, are highly stable w. r. t. the atomization channel

Table 30.1 The basic properties of the structure **II**, the low-energy hollow cage Au$_{20}^{II}$ or, shortly, cage **II** (Fig. 30.2) in the different charge states Z = 0, ±1, −2

Cage II	B3LYP				BP86			PW91		
	Z = 0	−1	+1	−2	Z = 0	−1	+1	Z = 0	−1	+1
Rotational constants	0.01892 0.01494 0.01225	0.01859 0.01459 0.01242	0.01854 0.01512 0.01250	0.01769 0.01398 0.01286	0.01952 0.01556 0.01260	0.01931 0.01535 0.01273	0.01922 0.01575 0.01278	0.01961 0.01564 0.01262	0.01941 0.01544 0.01276	0.01931 0.01584 0.01281
ΔE^Z	1.48	0.59	0.82	−0.05	1.56			1.60	0.77	0.99
ε_{HOMO}	−5.71	−2.58 −2.86	−8.50 −9.07	−0.16	−5.48	−2.41 −2.62	−8.33 −8.83	−5.36	−2.30 −2.50	−8.21 −8.70
ε_{LUMO}	−4.52	−1.12 −1.83	−7.41 −7.68	1.77	−5.08	−1.78 −2.31	−7.99 −8.23	−4.97	−1.68 −2.20	−7.87 −8.11
Δ	1.19	1.46 1.03	1.09 1.39	1.93	0.40	0.63 0.31	0.34 0.60	0.39	0.62 0.30	0.34 0.59
ΔE_{ST}^a	0.28				0.22			0.22		
Lowestv[b]	11 (26)	10	10	17	13 (28)	20	13	12 (29)	20	12
Highestv[b]	154 (161)	153	152	163	162 (172)	159	161	163 (172)	160	161

(continued)

Table 30.1 (continued)

Cage II	B3LYP				BP86			PW91		
	$Z=0$	-1	$+1$	-2	$Z=0$	-1	$+1$	$Z=0$	-1	$+1$
IR$_{max}$	154{3}^{c}	153{2}	152 (2)	110{9}	162{2}	159{1}	161{1}	163{2}	160{1}	161{1}
$-$E$_{binding}$	33.94				41.73			43.20		
IE$_1$/VHDE	6.692 (7.398)^b/5.413				6.852			6.730(7.336)^b/6.657		
EA$_1$/VDE	3.543 (2.606)^b/3.6454				3.748/3.812			3.630(2.803)^b/3.691		
EA$_2$	1.163(0.525)^b							(0.433)^b		
μ	0.68				0.50			0.51		
Void diameter	4.97–5.94	5.07–6.00	5.04–6.00		4.84–5.81	4.92–5.86	4.87–5.86	4.82–5.80	4.90–5.84	4.85–5.85

E$_{binding}$ (in eV) is defined as the ZPE-corrected binding energy of the neutral cage, [cage]$^{Z=0}$, taken with respect to (w. r. t.) 20 non-interacting gold atoms. ΔE^Z (in eV) is the relative ZPE-corrected energy of [cage]^Z w. r. t. the ground-state cluster Au$_{20}^{Z=0}$(T_d). Since [cage]$^{Z=0,-2}$ is the closed-shell structure in the zero-charge state and the open-shell one for $Z=\pm 1$, its MO eigenvalues ε_{HOMO} of the HOMO, ε_{LUMO} of the LUMO, and their energy gap $\Delta = \varepsilon$ LUMO $-$ ε HOMO (all in eV) are either spin-independent or spin-dependent. In the case of $Z=\pm 1$, they are given for the spin-up and down electrons. Note that in our early work [36b], we have not succeeded to reach convergence using the BP86 density functional for the charged clusters Au$_{20}^{\pm 1}$(T_d), either starting from the optimized BP86 neutral geometry or from the B3LYP geometries of Au$_{20}^{\pm 1}$(T_d). The singlet-triplet splitting ΔE$_{ST}$ is given in eV. The vertical detachment energies, VEDE and VHDE, first and second ionization energies, IE$_1$ and IE$_2$, and adiabatic electron affinities, EA$_1$ and EA$_2$, are given in eV. Also note that a comparison of the energies of the reported cages should take into account also called density functional theory (DFT) error which is about 0.2 eV [14d, o, p.] The lowest and highest unscaled vibrational frequencies (in cm^{-1}) of the studied cages, together with the vibrational modes corresponding to the maximal IR intensities (in curved brackets, in km · mol^{-1}), are also included. Rotational constants are given in GHz.
^{a}Notice that, as resulted from the experimental autodetachment studies of the Au$_{20}^-$(T_d), the gap between the ground singlet state and the first excited triplet state is equal to 1.77 eV [41,42]
^{b}The corresponding property of Au$_{20}^Z$(T_d) cluster is indicated in parentheses
^{c}The mode of Au$_{20}^{Z=0}$(T_d) with IR$_{max}$ is 3-degenerate and describes the collective stretch of the vertex atoms centered at 119cm^{-1}\{5\} (B3LYP), 130cm^{-1}\{4\} (BP86), and 131 cm^{-1}\{3\} (PW91). The experimental maximum of the IR intensity is reached at 148cm^{-1} [36e]

Table 30.2 The basic properties of the low-energy hollow cage **III** (Fig. 30.3)

Cage III	B3LYP Z = 0	−1	+1	−2	BP86 Z = 0	−1	+1	PW91 Z = 0	−1	+1
Rotational constants	0.01958	0.02074	0.01955	0.02113	0.02028	0.02031	0.02020	0.02038	0.02044	0.02029
	0.01667	0.01486	0.01685	0.01370	0.01727	0.01706	0.01738	0.01734	0.01717	0.01745
	0.01208	0.01252	0.01218	0.01200	0.01229	0.01224	0.01243	0.01231	0.01224	0.01247
$-E^*_{binding}Z$	2.34	3.19	3.38		3.94	4.36	4.99	4.34	4.77	5.40
ΔE^Z	1.85	0.68	0.99	0.34	1.67			1.65	0.95	0.92
ε_{HOMO}	−5.50	−2.52	−8.34	+0.01	−5.32	−2.31	−8.21	−5.20	−2.19	−8.09
	$(−6.33)^a$	−2.83	−9.01		$(−6.11)^a$	−2.48	−8.84	$(−5.99)^a$	−2.37	−8.73
ε_{LUMO}	−4.37	−1.10	−7.14	+1.52	−4.93	−1.95	−7.75	−4.79	−1.83	−7.61
	$(−4.07)^a$	−1.74	−7.58		$(−4.75)^a$	−2.20	−8.12	$(−4.62)^a$	−2.09	−8.00
Δ	1.13	1.42	1.20	1.51	0.40	0.36	0.46	0.42	0.36	0.48
	$(2.27)^a$	1.09	1.43		$(1.36)^a$	0.28	0.72	$(1.37)^a$	0.28	0.73
ΔE_{ST}	0.30									
Lowestν	12	20	16	11	16	13	18	17	15	19
Highestν	144	137	140	141	153	144	151	155	144	153
IR_{max}	$100\{1\}^b$	100,134{1}	16,57 {1}	99{3}	117{1}114 {1},	115 {2}	62 {1}	118 {1}115 {1},	117 {2}	63 {1}
$-E_{binding}$	33.62				41.62			43.15		
$IE_1/VHDE/IE_2$	$6.539/6.537/9.374\ (7.398/8.226)^c$				6.722/6.666			$6.602/6.546(7.336/9.963)^c$		
EA_1/VDE	$3.776(2.927)^a/3.609$				3.628/3.730			3.502/3.607		
EA_2	0.863									
μ	0.00				0.00			0.00		
NICS(0) (ppm)	−14.7				−17.6			−14.3		
(at the center)	$−17.7^d$				$−36^e$					
Void diameter	4.2–4.5	4.8–5.9								

Due to a specific double-layer shape of cage **III**, $E^*_{binding}Z$ (in eV) is defined as the ZPE-corrected binding energy of [cage **III**]^Z w. r. t. two non-interacting planar clusters $Au_{10}(D_{2h})$ (see Fig. 30.3) and $Au_{10}^Z(D_{2h})$. For other notations see Table 30.1

[a] The corresponding property of the planar $Au_{10}(D_{2h})$ cluster

[b] The collective stretch of the glue bonds

[c] The corresponding property of $Au_{20}^Z(T_d)$ cluster is indicated in parentheses [d] Ref. [36b]

[e] BP86/TZVPP(2f)

Table 30.3 The basic properties of the low-energy hollow cage **IV** (Fig. 30.4)

Cage **IV**	B3LYP				BP86			PW91		
	$Z=0$	-1 (III^{-1})	$+1$	-2 (III^{-2})	$Z=0$	-1 (III^{-1})	$+1$	$Z=0$	-1 (III^{-1})	$+1$
Rotational constants	0.02048		0.01953		0.02122		0.01576	0.02135		0.02032
	0.01434		0.01528		0.02021		0.01361	0.01483		0.01583
	0.01327		0.01342		0.01477		0.01370	0.01364		0.01372
ΔE^Z	1.42		0.66		1.40			1.43		0.84
ε_{HOMO}	-5.81		-8.55		-5.64		-8.43	-5.52		-8.31
			-9.04				-8.83			-8.71
ε_{LUMO}	-4.21		-6.94		-4.82		-7.59	-4.71		-7.48
			-7.76				-8.32			-8.21
Δ	1.59		1.61		0.82		0.83	0.81		0.83
			1.28				0.52			0.50
ΔE_{ST}	0.48									
Lowestν	15		19		22		20	22		19
Highestν	140		144		152		156	154		158
IR$_{max}$	140{4}^{a}		144{1}		152{2}		156{1}	154{2}		158{1}
$-E_{binding}$	34.05				41.89			43.37		
IE$_1$/VHDE	6.647/6.648				6.862			6.744/6.750		
EA$_1$	3.340				3.359			3.280		
μ	1.01				0.95			0.94		
Void diameter	5.2–5.5									

For other notations see Table 30.1
^{a}Ref.[36b]

Table 30.4 The basic properties of the low-energy hollow cage **V** (Fig. 30.5)

Cage **V**	B3LYP				BP86			PW91		
	$Z=0$	-1	$+1$	-2	$Z=0$	-1	$+1$	$Z=0$	-1	$+1$
Rotational constants	0.01630	0.01615	0.01645	0.01682	0.01679	0.01696	0.01694	0.01687	0.01703	0.01702
	0.01514	0.01484	0.01524	0.01445	0.01562	0.01564	0.01569	0.01568	0.01572	0.01575
	0.01513	0.01397	0.01524	0.01294	0.01562	0.01525	0.01569	0.01568	0.01530	0.01575
ΔE^Z	0.81	0.48	0.37	0.26						
$HOMO$	−5.94	−2.16	−8.75	−0.28	−5.72	−1.80	−8.59	−5.60	−1.68	−8.48
		−9.22				−2.88	−8.99		−2.77	−8.88
ε_{LUMO}	−3.70	−0.91	−6.56	+1.66	−4.34	−1.61	−7.22	−4.22	−1.50	−7.10
		−8.00				−1.66	−8.59		−1.55	−8.39
Δ	2.24	1.25	2.19	1.94	1.38	0.19	1.37	1.38	0.18	1.38
			1.22			1.22	0.40		1.22	0.49
Lowest ν	24	12	24	10	24	15	23	23	14	22
Highest ν	154	165	151	156	164	161	163	165	163	165
IR_{max}	125 {3}	123 {5}	51,124 {1}	122 {7}	134 {2}	57,134 {1}	131 {2}	136 {2}	136 {1}	130 {1}
$-E_{binding}$	34.66									
IE_1	6.963				7.107			6.993		
EA_1	2.930				3.071			2.953		
EA_2	0.753									
μ	0.57				0.58			0.62		
Void diameter	5.7–6.9				5.7–6.9			5.7–6.9		

For other notations see Table 30.1

Table 30.5 The basic properties of the low-energy hollow cage **VI** (Fig. 30.6)

	B3LYP				BP86			PW91		
Cage **VI**	Z = 0	−1	+1	−2	Z = 0	−1	+1	Z = 0	−1	+1
Rotational	0.01590	0.01612	0.01614		0.01639	0.01621	0.01664	0.01645	0.01631	0.01678
constants	0.01529	0.01504	0.01548		0.01583	0.01566	0.01596	0.01589	0.01574	0.01610
	0.01494	0.01446	0.01489		0.01549	0.01554	0.01539	0.01558	0.01560	0.01556
ΔE^Z	0.99	0.63	0.40	II^{-2}		0.94	0.71	1.17	0.94	0.71
HOMO	−5.84	−2.01	−8.60		−5.63	−1.86	−8.46	−5.52	−1.73	−8.30
		−3.04	−9.11			−2.85	−8.84		−2.74	−8.68
ε_{LUMO}	−3.84	−0.99	−6.66		−4.45	−1.52	−7.30	−4.33	−1.41	−7.19
		−1.25	−7.79			−1.74	−8.35		−1.62	−8.19
Δ	2.00	1.02	1.94		1.18	0.34	1.16	1.19	0.32	1.11
		1.79	1.32			1.11	0.49		1.12	0.49
Lowest ν	9	16	11		17	15	11	16	15	11
Highest ν	166	162	172		174	169	172	175	170	172
IR$_{max}$	127 {3}	126 {4}	127 {1}		138 {2}	128,136 {3}	139 {1}	134,140 {2}	131 {3}	142 {1}
$-E_{binding}$	34.47							43.63		
IE$_1$	6.809				6.992			6.875		
EA$_1$	2.925				3.152			3.028		
EA$_2$	1.207				0.566			0.436		
μ	1.45				1.37			1.42		
Void diameter	4.8–5.6									

For other notations see Table 30.1

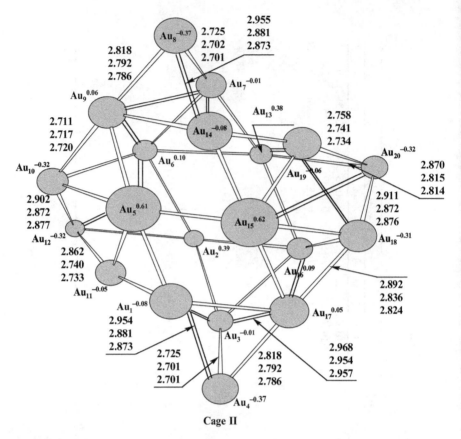

Cage II

Fig. 30.2 The golden hollow cage **II** in the neutral charge state Z = 0. Cage **II** resembles a helix with the void diameter of ∼5.9Å. Its external diameter is ∼1.2nm. Selected bond lengths are given in Å (reading from *top to bottom*: B3LYP, BP86, and PW91 DF potentials). The B3LYP Mulliken charges of some gold atoms are indicated by *superscripts*. Gold atoms of cage **II** are partitioned into the following three groups: two vertex 3-coordinated atoms, $Au_4^{-0.37}$ and $Au_8^{-0.37}$ (cf.: four in $Au_{20}(T_d)$), whose distance define the external diameter of cage **II**; two vertex 4-coordinated atoms, $Au_{12}^{-0.32}$ and $Au_{20}^{-0.32}$; two edge 5-coordinated atoms, $Au_{10}^{-0.32}$ and $Au_{18}^{-0.32}$; and 14 face-centered 6-coordinated atoms (cf.: twelve in $Au_{20}(T_d)$ together with four 9-coordinated), that in total yields 54 Au-Au bonds of the bond lengths ∈ (2.7Å, 3.0Å), that is by 6 less than of $Au_{20}(T_d)$. As follows from Table 30.1, the energetic 'cost' of such, say, 'less bonding' amounts to 1.5–1.6eV. Let also notice a low polarity of **II** comprising of 0.5–0.7D

$Au_{20}^{\text{II-VI}} \Rightarrow 20\,Au$. This stability is corroborated by high absolute values, ∼33–34eV, of the corresponding energies, $-\Delta E_f$, and enthalpies, $-\Delta H_f$, of formation, despite a large entropy effect. The latter increases the Gibbs free energies, ΔF_f^{298}, of formation at T = 298 K and correspondingly lowers the cages' stabilities by 6–7eV, but they still remain strongly bound.

Mappings of $Au_{20}(T_d)$ and the cages **II – VI** onto different charge states results in the following B3LYP stability patterns (in eV):

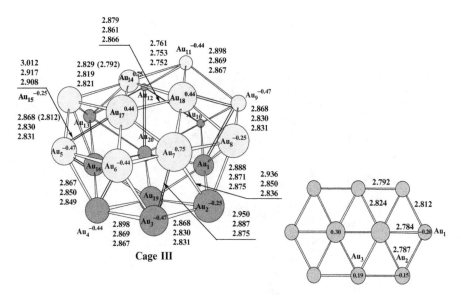

Fig. 30.3 *Left panel*: The nonpolar golden hollow cage **III** in the neutral charge state Z = 0. The void diameter of the cage **III** is approximately equal to 0.84 nm, i. e. it slightly exceeds that of Au$_{20}$(T_d). As readily seen in this Figure, **III** is composed of the two planar Au$_{10}$(D_{2h}) clusters colored correspondingly in *yellow* (*light* in b/w) and *brown* (correspondingly *dark*). The angle between the longest diagonals Au$_5$ – Au$_9$ and Au$_3$ – Au$_{12}$, each of which connects the two vertex gold atoms mostly separated from each other, amounts to 33.1°. Selected bond lengths are given in Å (reading from *top to bottom*: B3LYP, BP86, and PW91 DF potentials). The B3LYP Mulliken charges of some gold atoms are indicated by *superscripts*. *Right panel*: The nonpolar Au$_{10}$(D_{2h}) cluster. Its ε_{HOMO} = −6.33, −6.11, −5.99 eV; ε_{LUMO} = −4.07, −4.75, −4.62 eV; and Δ = 2.27, 1.36, 1.37 eV are respectively obtained at the B3LYP, BP86, and PW91 computational levels. The selected bond lengths (in Å) correspond to the B3LYP one. The B3LYP Mulliken charges are shown for some gold atoms

$$\mathbf{Z = 0}: \mathrm{Au_{20}}^0(T_d) \overset{0.81}{<} \mathbf{V}^0 \overset{0.18}{<} \mathbf{VI}^0 \overset{0.43}{<} \mathbf{IV}^0 \overset{0.06}{\leq} \mathbf{II}^0 \overset{0.37}{<} \mathbf{III}^0$$

$$\mathbf{Z = -1}: \mathrm{Au_{20}}^{-1}(T_d) \overset{0.48}{<} \mathbf{V}^{-1} \overset{0.11}{<} \mathbf{II}^{-1} \overset{0.04}{\leq} \mathbf{VI}^{-1} \overset{0.05}{\leq} \mathbf{III}^{-1} = \mathbf{IV}^{-1}$$

$$\mathbf{Z = +1}: \mathrm{Au_{20}^{+1}}(T_d) \overset{0.37}{<} \mathbf{V}^{+1} \overset{0.03}{\leq} \mathbf{VI}^{+1} \overset{0.26}{<} \mathbf{IV}^{+1} \overset{0.16}{<} \mathbf{II}^{+1} \overset{0.17}{<} \mathbf{III}^{+1}$$

$$\mathbf{Z = -2}: \mathbf{II}^{-2} = \mathbf{VI}^{-2} \overset{0.05}{\leq} \mathbf{I}^{-2} \overset{0.26}{<} \mathbf{V}^{-2} \overset{0.08}{<} \mathbf{III}^{-1} = \mathbf{IV}^{-2}$$

where the quantity above the sign indicates the difference in stability of the left-side structure over the right-side one. The change of the cage properties under the charge alternation is presented in Tables 30.1–30.5.

The neutral cages **II-VI** are energetically placed within 0.8–1.85 eV relative to the ground-state cluster Au$_{20}^{Z=0}$(T_d). On the other hand, on the cationic sheet of the 20-gold PES, **V**$^{+1}$ becomes closer to Au$_{20}^{+1}$(T_d) by 0.37 eV. The gap of 0.48 eV

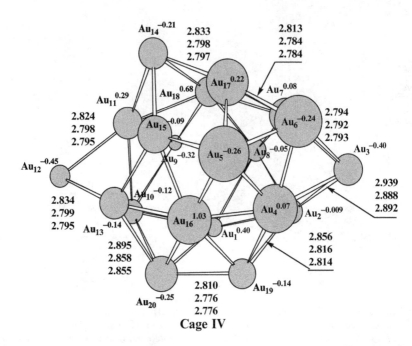

Cage IV

Fig. 30.4 The golden hollow cage **IV** in the neutral charge state $Z = 0$. Cage **IV** is characterized by a void diameter, $R(Au_9 - Au_{16})$, of 5.5Å. Its external diameter associated with $R(Au_3 - Au_{12})$ amounts to 0.98 nm. Selected bond lengths are given in Å (reading from *top to bottom*: B3LYP, BP86, and PW91 DF potentials). The B3LYP Mulliken charges of some gold atoms are indicated by *superscripts*

separates $Au_{20}^{+1}(T_d)$ from these anionic cages. Both energy gaps can likely be achievable in experiments.

The dianionic $Z = -2$ charge state is different – as already mentioned in [36b,38], due to Coulomb repulsion, $Au_{20}^{-2}(T_d)$ evolves, as seen in Fig. 30.7, to a void shape structure. $Au_{20}^{-2}(T_d)$ shares the very bottom of the dianionic 20-gold PES with $\mathbf{II}^{-2}$, also displayed in Fig. 30.7, though actually (within the DFT computational error) the latter lies slightly below $Au_{20}^{-2}(T_d)$ by ~0.05 eV. Therefore, cage **II** could be easily detected experimentally, at least in the dianionic charge state. Relative to the asymptote comprising of $18Au + 2Au^-$, $\mathbf{II}^{-2}$ is essentially stable that is manifested by $\Delta E_f = -34.32$ eV and a Gibbs free energy $\Delta F_f^{298} = -27.56$ eV at the B3LYP computational level.

The HOMO and LUMO eigenvalues and the corresponding HOMO-LUMO gap, Δ, the first adiabatic electron affinities (EA_1) and ionization potentials (IE_1) are the key features that determine chemical reactivity. For cages **II-VI**, they are presented in Tables 30.1–30.5. The HOMO's energies are ca. 0.4–0.9 eV higher the HOMO of $Au_{20}(T_d)$. Cages **III** and **IV** exhibit rather narrow gaps, $\Delta \approx 0.4$ eV (BP86 and PW91), compared to $Au_{20}(T_d)$, which are however comparable with Δ of $Au_{42}(I_h)$. On the contrary, the HOMO-LUMO gap of the cage **V** is twice

Cage V in Charge States Z = 0, + 1, −1, and −2

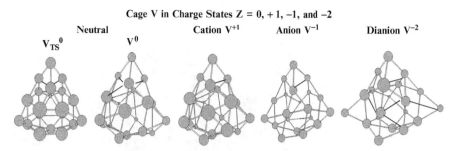

Neutral   Cation V^{+1} Anion V^{-1} Dianion V^{-2}

V$_{TS}^0$ V^0

Fig. 30.5 Cage **V** in different charge states Z $=$ 0, $+1$, -1 and -2 where it is placed above Au$_{20}^Z$(T_d) by 0.81, 0.48, 0.37, and 0.26 eV, correspondingly. For Z $=$ 0, ±1, **V**^{Z} is the closest hollow cage, among the known ones in the literature, to the ground-state structure. It is readily seen in this Figure that the void enlarges from the neutral charge state to the anionic and dianionic ones (see Table 30.4). In the anionic state, **V**$^{-1}$ is quite symmetric. Unlike the Z $=$ -1 charge state, the neutral cage **V**^{0} is far from being symmetric and is directly linked to a quasi symmetric structure that is actually the first-order saddle **V**$_{TS}^0$. The latter lies energetically close – by only 0.052 eV – to **V**^{0}. **V**$_{TS}^0$ is practically nonpolar: its total dipole moment is only 0.05 D. As the transition structure, **V**$_{TS}^0$ has a single imaginary frequency of 8 i cm^{-1}. Interestingly, **V**$_{TS}^0$ arises either under encaging LiF into cage **II** or cage **V** that both result in the endohedral complex LiF@**V**$_{pre-TS}^0$ where the cage **V**$_{pre-TS}^0$ is energetically close to **V**$_{TS}^0$. The standard Pople's basis set $6-311++$G(d,p) is used for the non-gold atoms

Cage VI in Charge States Z = 0, +1, −1, and −2

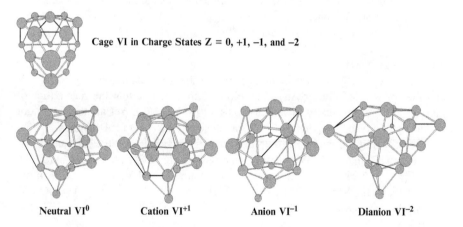

Neutral VI^0 Cation VI^{+1} Anion VI^{-1} Dianion VI^{-2}

Fig. 30.6 Cage **VI** in different charge states. The cage **VI**^{0} was obtained by choosing the hollow cage Au$_{16}$ [26], which is shown in the *left-top* insert, adding perpendicularly one atom of gold to each of its four hexagon faces with a central gold atom, and relaxing the resultant geometry. For Z $=$ -2, **VI**$^{-2}$ coincides with **II**$^{-2}$

wider, that is, reaches approximately the same width as in the larger cages, such as e.g. Au$_{24}$, Au$_{27}$, and Au$_{28}$. The EA$_1$ of the studied cages are significantly large, $\in$ (3.776 eV–**II**; 2.925 eV–**VI** $\approx$ 2.930 eV–**V**), thus exceeding the EA$_1^{B3LYP}$(Au$_{20}$(T_d)) $=$ 2.606 eV (notice that EA$_1$(C$_{60}$) $=$ 2.57 eV) by ca. 0.5–1.2 eV.

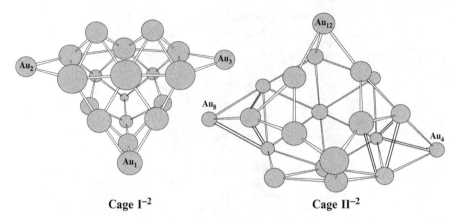

Cage I^{-2} Cage II^{-2}

Fig. 30.7 The golden hollow cage clusters **I**$^{-2}$ and **II**$^{-2}$ in the dianionic charge state $Z = -2$. **I**$^{-2}$ is characterized by the B3LYP distances $R(Au_1–Au_2) = R(Au_2–Au_3) = 7.863$ and $R(Au_1–Au_3) = 10.517$Å. Its void diameter is equal to ~ 5.3Å. The structure of the **II**$^{-2}$ is characterized by the B3LYP distances $R(Au_4–Au_8) = 10.157$, $R(Au_4–Au_{12}) = R(Au_8–Au_{20}) = 8.116$, $R(Au_4–Au_{20}) = R(Au_8–Au_{12}) = 7.198$, and $R(Au_1–Au_{16}) = 4.932$Å

Actually, these EA_1 get closer to the EA_1 of Au_{16} that is about 4 eV, and hence, by analogy with Au_{16}^-, these cages are not likely to easily bind molecular oxygen.

The IE_1 of **III**, **IV**, and **V** are lower – $\in (6.963\,\text{eV–V}; 6.539\,\text{eV–III})$– compared to IE_1^{PW91} of $Au_{20}(T_d) = 7.34$ eV and $IE_1^{B3LYP}(Au_{20}(T_d)) = 7.398$ eV. Summarizing:
EA: $Au_{20}^0(T_d) < \textbf{V} = \textbf{VI} < \textbf{IV} < \textbf{II} < \textbf{III}$ and, on contrary, $\textbf{IE}_1$: $Au_{20}^0(T_d) > \textbf{V} > \textbf{VI} > \textbf{II} > \textbf{IV} > \textbf{III}$.

Altogether, the above distinctions between the studied hollow cages **II** − **IV** and the space-filled cluster $Au_{20}(T_d)$ allow us to conclude that the reactivities of these cages are drastically different from that of $Au_{20}(T_d)$. Yet another indicator of the reactivity of a given molecule or cluster is the chemical hardness η that is approximately defined as a half of the difference between the IE_1 and EA_1, viz., $\eta \approx (IE_1–EA_1)/2$. Within this reactivity scale, **IV** with lowest hardness, $\eta^{B3LYP}(\textbf{IV}) = 1.38$ eV, is more reactive than both **III** and **V**, with II being more reactive than cage **V**: $\eta^{B3LYP}(\textbf{IV}) < \eta^{B3LYP}(\textbf{III}) = 1.57$ eV $< \eta^{B3LYP}(\textbf{V}) = 1.65$ eV. All cages are more reactive than $Au_{20}(T_d)$, characterized by $\eta^{B3LYP}(Au_{20}(T_d)) = 2.40$ eV.

By analogy with the concept of the vertical electron detachment energy, let us define the vertical hole detachment energy, VHDE, of complex X as the total energy difference between the cationic and neutral systems, both taken, without the ZPE, in the equilibrium geometry of cation. The VHDEs of cages **II**-**IV** are given in Tables 30.1–30.3.

30.4   Void Reactivity of 20-Nanogold Cages: Few Approaches for Measuring

According to a rather broad definition, the chemical reactivity of molecule or molecular cluster is its capability to form, while interacting or contacting with other atoms, ions, or molecules, some chemical bonding patterns. These bonding patterns are established between the atoms of interacting molecules which are typically placed in the outer space of each interacting partner. In this sense cages are different – by definition, they possess atoms which are assumed to be capable to form chemical bonds with molecules from the outer space as well as from the inner one (inside a cage, in voids), i.e. with those which are encapsulated or confined within a cage. This implies a bifunctionality of the chemical reactivity – the exo- and endo-reactivity – of some atoms which constitute cages and thus suggests potential routes to control both chemical reactivities.

The traditional approach to measure the reactivity of a given molecule or molecular cluster consists in estimating its ionization energy and the electron affinity that govern the electron transfer which, according to the Sanderson's principle [44], is the integrable part of the formation of chemical bonds when this molecule or cluster is brought into contact with the other one. The ionization energies and electron affinities of cages **II-VI** are discussed in Sect. 30.4. They indicate how a given cage reorganizes either upon removal or upon addition of an electron. They are, however, global characteristics of cage that cannot be partitioned into those, which may be solely ascribed either to the outer or to the void reactivity, to differentiate the reactivity of hollow cages as cages **II-VI** and to predict what either each of them or both are. Obviously, the chemical hardness is global too.

The void regions of the cages **II-VI** are in fact spatially confined areas which can accommodate some dopant(s). A remarkable feature of these cages is that they all have only one-atom-layer that separates the void from the outer surface and may thus facilitate the direct control of the outer reactivity from the inner one. A typical point of view that often prevails is that the void reactivity of a cage is a direct consequence of the spatial confinement, i.e., the dopant feels the cage's boundary, and therefore, the size of dopant plays a decisive role for a stable encapsulation. Put in other words, if the size of the void is of the same order of magnitude or comparable with that of the dopant, one may anticipate that doping is stable. Obviously, doping influences the outer reactivity.

On the one hand, it is well known that the patterns of the molecular orbitals, particularly of the HOMO and LUMO, and the concept of the molecular electrostatic potential (MEP) [45, 46] are crucial for an understanding of chemical reactivity. For example, the buckyball fullerene C_{60} exhibits two different behaviors of the MEP [47]. It is positive in the entire void region, where C_{60} is capable to encage atoms, ions, and some molecules, and the positive part of the MEP reaches the outer central regions of the pentagon rings. On the contrary, the MEP is negative in the outer region and most negative at the midpoints of the bonds linking two hexagonal rings.

30.4.1 HOMO and LUMO Patterns

The patterns of the HOMOs and LUMOs of cages **II**, **III**, and **IV** are shown in Figs. 30.8 and 30.9. The HOMO of **II** is mainly composed of: (i) the $6s$ AOs of the 'top' and 'bottom' gold atoms $Au_4^{-0.37}$ and $Au_8^{-0.37}$ where the superscript indicates the Mulliken charge; (ii) the $6p$ AOs of $Au_1^{-0.06}$, $Au_3^{-0.01}$, $Au_7^{-0.01}$, $Au_{13}^{0.38}$, and $Au_{14}^{-0.08}$, and largely of $Au_5^{0.62}$ and $Au_{15}^{0.62}$; and (iii) the $5d$ AOs localized on Au_3, Au_5, Au_7, Au_{14}, and Au_{15}. The latter two AOs partly protrude into the void of cage **II** where a hole can therefore partially appear under the ionization. In this sense, the void of cage **II** can be treated as a polarizable 'sphere', by analogy with the C_{60} cage [46]. In contrast, the LUMO of cage **II** lies substantially 'outdoors', in the outer space, thus emphasizing that the electron attachment may primarily contributes to the outer reactivity of cage **II**. This LUMO is mainly composed of the $6s$ AOs localized on almost all gold atoms, except Au_3, Au_{5-7}, Au_1, Au_{16-18}, and Au_{20}, and $6p$ AOs on $Au_{2,3}$, $Au_{5,7}$, $Au_{11,13}$, and $Au_{15,20}$. A small part of the LUMO, determined by the $5d$ AOs of Au_3, Au_7, Au_8, Au_{12}, and Au_{17}, is however placed within the void.

The shapes of the HOMO and LUMO of cage **III** are rather spectacular. They are mostly composed of AOs $Au_{1,7,14,16}^{0.75}(6p_x^{0.19})$, $Au_{2,8,13,15}^{-0.25}(6s^{0.21})$, $Au_{3,5,9,12}^{-0.45}(6s^{0.48})$, and the 6-folded $Au_{17,18,19,20}^{0.41}(6p_y^{0.11}5d_{-2}^{0.14})$ for the HOMO and $Au_{1,7,14,16}(6s^{0.43}6p_y^{0.14})$, $Au_{4,6,10,11}(6s^{0.22})$ and $Au_{17,18,19,20}(6p_y^{0.16})$ for the LUMO. The HOMO and LUMO are mostly localized in the outer space around the $Au_{3,5,9,12}$ and $Au_{1,4,6,7,10,11,14,16}$, respectively. Juxtaposing the HOMO and LUMO of cage **III** in Fig. 30.8, one readily concludes that the void portion of its HOMO is larger than the LUMO one.

In addition, the HOMO $-$ 2–HOMO $-$ 4 of cage **III** which, due to their $5p$ and $5d$ AOs, that are largely localized in its void region are displayed in Fig. 30.9. Since these HOMO-2 – HOMO-4 lie within 0.8–1.6 eV from the HOMO, they participate in the first – at least, HOMO-2 – and second ionization processes of cage **III**. These are therefore the processes where the void reactivity of cage **III** can be detected. Distinguishably different are the HOMO and LUMO of cage **IV**. The outer part of its HOMO is sharply localized on the opposite vertex gold atoms $Au_3^{-0.40}(6s^{0.62})$ and $Au_{12}^{-0.45}(6s^{0.68})$. The rest of the HOMO lies in the void. Its LUMO is considerably localized on the side vertex atoms outward.

30.4.2 Molecular Electrostatic Potential Patterns

The MEPs of the cages **II**, **III**, and **IV** are shown in Fig. 30.10 for the different charge states $Z = 0$, ± 1, and -2, where, for the latter, the MEP of $\mathbf{I}^{-2}$ is added for comparison, bearing in mind Fig. 30.7. It is seen in this figure that the MEPs of the cages' outer space are nonnegative – that is indicated by blue regions converging to the green ones. The void MEP of cage **II** is of both signs: negative is shown in

HOMO   LUMO

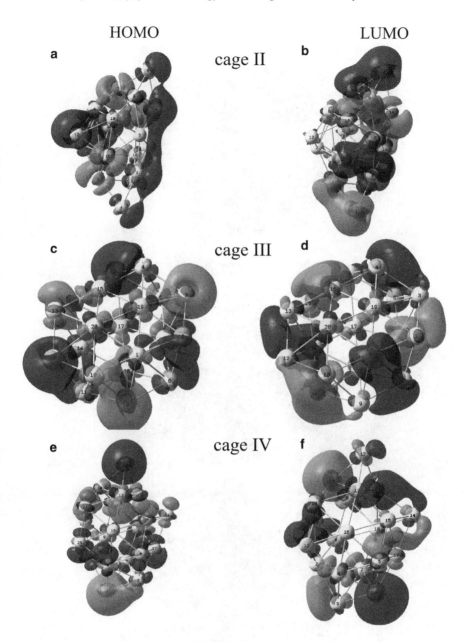

Fig. 30.8 The HOMOs and LUMOs of cages **II-IV**: (**a**) and (**b**) plot the HOMO and LUMO of cage **II**, respectively; (**c**) and (**d**) the HOMO and LUMO of cage **III**; and (**e**) and (**f**) plot the HOMO and LUMO of cage **IV**

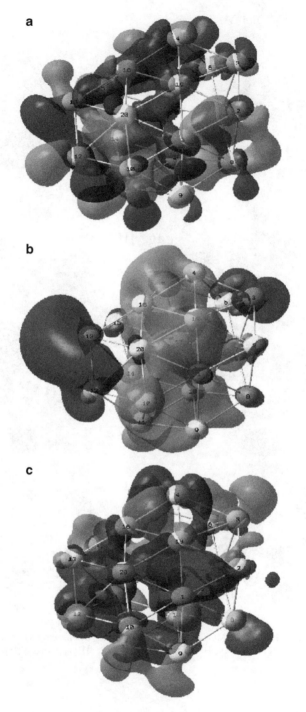

Fig. 30.9 The HOMO-2 (**a**), HOMO-3 (**b**), and HOMO-4 (**c**) of cage **III**. Their orbital eigenvalues are correspondingly equal to -6.29, -6.42, and $-7.06\,\mathrm{eV}$

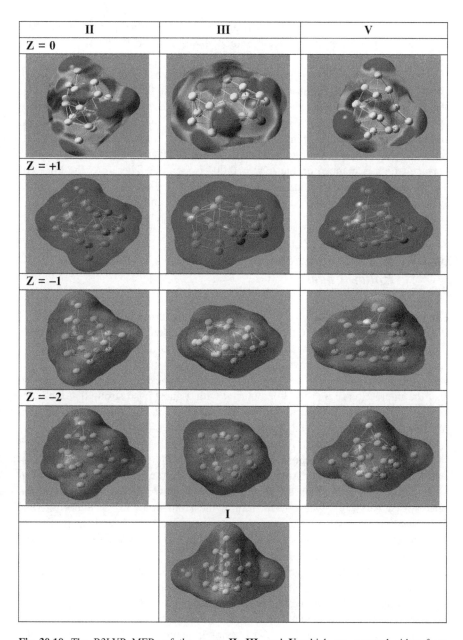

Fig. 30.10 The B3LYP MEPs of the cages **II**, **III**, and **V** which are mapped either from $-0.01(red)$ to $+0.01(blue)|e|/(4\pi\varepsilon_0 a_0)$ for the $Z = 0$ charge state or from $-0.1(red)$ to $+0.1(blue)|e|/(4\pi\varepsilon_0 a_0)$ for the $Z = \pm 1$ and -2 (including, as the reference, the MEP of the hollow cage $\mathbf{I}^{-2}$) charge states onto $0.001|e| \cdot \text{Å}^{-3}$ isosurface of the one-electron density $\rho(\mathbf{r})$

red, as e.g. in the neighborhood of Au_1 and Au_5, and positive, in blue, as e.g. in a small and deep 'pocket' in the neighborhood of Au_{17}. Hence, the latter may confine some atom.

The MEP of **III** is positive outside and takes both signs in the void. In Fig. 30.10, there exists a pair of symmetric negative MEP regions in the neighborhood of Au_{12}, Au_{14}, Au_{20}, and Au_3, Au_7, and Au_{19}. And there are the other two regions, close to Au_{13}, Au_{17}, and Au_{10}, Au_{18}, which are positive. This implies the existence of two different 'pockets' for trapping. For cage **IV**, in addition, there appear a 'pocket' with a slightly negative charge in the center of the MEP and the essentially positive surface which is mapped on the $\rho(\mathbf{r}) = 0.004$ isocontour. Therefore, the MEPs of **II**, **III** and **IV** definitely demonstrate a capability of their voids to trap neutral, as well as both, positively and negatively, charged atomic and molecular guests.

30.4.3 Endohedrality: Space-Filled $Au_{20}(T_d)$ vs. Au_{20} Hollow Cages

Due to a space-filled shape of the ground-state structure $Au_{20}(T_d)$ on the neutral 20-gold PES, the latter is seemingly not able to confine any guest atom or molecule. This is not consistent with the existence of some endohedral fullerenes $X@Au_{20}(T_d)$ that was computationally proven in [37,38]. Actually, there is no contradiction. True, any hollow cage enables to trap, by the definition, a guest. It may however happen that the interaction of a space-filled cluster with a guest is so strong that the latter pushes aside the cluster interior, creates there a void and becomes trapped therein. We illustrate this statement ad absurdum, in some sense.

Let consider in Fig. 30.11 two stable structures, both composed of 21 gold atoms and both initially chosen as endohedral: the left-hand one as $Au@Au_{20}^{II}$ and the right-hand as $Au@Au_{20}(T_d)$. That is, in the other words, one gold atom was initially trapped either in the void of the hollow cage Au_{20}^{II} or inside the space-filled $Au_{20}(T_d)$, where it substitutes the atom of Li in the endo-fullerene $Li@Au_{20}(T_d)$ discussed in Introduction. As a result of optimization, the former remains endohedral, i.e. as $Au@Au_{20}^{II}$, where the trapped gold atom $Au^{+5.29}$, ca. $+5$ positively charged according to its Mulliken charge, forms nine void bonds with Au_{20}^{II} via transferring its nearly five electrons to the latter cluster. This encaging gains the energy of 1.645 eV. In contrast, the latter structure converts to the exo-bonded $Au\&Au_{20}(T_d)$, implying that the interior of $Au_{20}(T_d)$ repels the guest gold atom to the outer space where it becomes bonded by three bonds. This effect of repulsion of Au by the $Au_{20}(T_d)$ interior is naturally anticipated because the energy that is needed to distort the interior of $Au_{20}(T_d)$, roughly estimated from the frequencies of the Au-Au bonds forming it and being equal to $\sim$90–100 cm^{-1}, is approximately the same as the energy of interaction between the initially trapped gold atom and those atoms of $Au_{20}(T_d)$. That is why this atom was repelled and

$Au@Au_{20}^{II}$ $Au\&Au_{20}(T_d)$

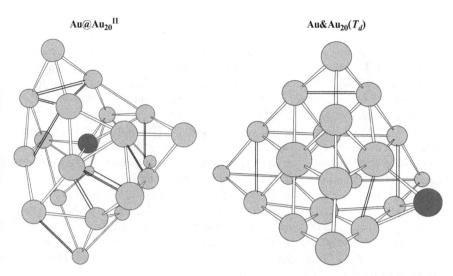

Fig. 30.11 *Left*: The endohedral golden fullerene $Au@Au_{20}^{II}$, formed by trapping of the guest gold atom (*blue circle*) in the void of cage **II**. The trapped atom $Au^{+5.29}$ bears a Mulliken charge of $+5.29$ that implies that the nine void bonds it forms with Au_{20}^{II} by means of electron transfer to cage **II** which becomes negatively charged. The bond lengths fall within $\{2.90, 3.05\}$ Å. $Au@Au_{20}^{II}$ is slightly polar, of 0.74 D. *Right*: The exo-bonded $Au\&Au_{20}(T_d)$. Its dipole moment is equal to 0.25 D. The exo-bonding includes three Au-Au bonds with bond lengths of 2.85Å. The energy gap $\Delta = 1.714$ and $2.017\,eV$ for spin-up and spin-down electrons

the resultant structure is $Au\&Au_{20}(T_d)$ with three exo-bonds, characterized by the binding energy of $-1.149\,eV$, that is less than $Au@Au_{20}^{II}$, though it is placed below the latter by $1.030\,eV$.

30.5   Summary and Conclusions

Due to their exceptional reactivity, gold nano-particles in all their diverse size, shape, and charge state are currently at the forefront of theoretical and experimental nanoscience. The gold hollow cages – golden fullerenes – turn out to be interesting systems with bifunctional reactivity–the void and the outer ones which can be manipulated by doping. Obviously, manipulation with reactivity does actually demand a way or ways to measure it. Unfortunately, this problem has not been rigorously and consistently formulated so far and that is why remains not well-defined.

In the present work, we have identified five 20-nanogold low-energy hollow cages at the BP86, B3LYP, and PW91PW91 density functional levels by a detailed examination of the 20-nanogold PESs in the different charge states and their thorough comparison. As primarily thought, the PES search was performed in two directions. One is to build hollow cages in the neutral charge state directly, from blocks, such as 10-nanogold Au_{10}, and to investigate their stability and closeness, in energy, to the ground-state tetrahedron $Au_{20}(T_d)$. This way the hollow cage III has been created. The other is to proceed to the dianionic PES, which is experimentally accessible (see [38] and references therein), and to search it for the ground state – this direction was explored in [38] and it was resulted in the hollow cage **II**. It then appeared that under the vertical electron detachment (VED), the reorganization energy of cage **III** becomes negative: $\Delta E_{reorg}^{B3LYP}(\mathbf{III}|VED) = -0.17\,eV$, either implying a collapse of void or an existence of a new stable isomer that must lie on the neutral PES below cage **III**. The latter was actually the case that gives rise to the cage **IV** which, as neutral, is placed below cage **III** by $2.5–3.6\,kcal \cdot mol^{-1}$. On the anionic PES, cages **III** and **IV** coincide with each other. The origin of cage **V** is outlined in the legend to Fig. 30.5 as implying the stable endohedral cage under encaging LiF into cage **II**. The history of appearance of the last cage **VI** is described in the legend to Fig. 30.6 and in [26]. All these cages are stable in the charge states $Z = 0, \pm 1$, and -2 since, according to Tables 30.1–30.5, their owest modes are strictly positive.

In the present work, we have suggested few different approaches to measure the void reactivity of 20-nanogold cages: on the one hand, these are ionization energy and electron affinity, which are definitely the global characteristics, and on the other, the local, such as the HOMO and LUMO, and the MEP. All of them have been compared with C_{60}. It has been demonstrated that the MEP, which patterns of the studied hollow cages look quite different from that of C_{60}, is useful to assess the electrostatic nature of possible dopants. We thus anticipate that this concept might be rather useful in designing golden fullerene-type nanomaterials with the tailored void and doping-controlled properties. It is definitely useful for metal $M@Au_N^{-1}$ golden fullerenes where metal atom is inside the anionic cage and where the bonding scenario is largely governed by the MEP since the metal – cage interaction is dominantly ionic (see [48] and references therein) and determines the metal position.

It has also been fully answered the question why the space-filled cluster $Au_{20}(T_d)$ enables to trap guest atoms [48], answered in a manner that makes the partition of a 3D golden shape either in a space-filled or hollow one rather smeared, quite inapplicable or even ill-defined. The reason of that is rather simple – it lies in a relative softness of its 'interior' bonds, which stretches fall within the interval of $\{90\,cm^{-1}, 100\,cm^{-1}\}$. This softness has been probed by the guest atom of gold. Since its interaction with these bonds of $Au_{20}(T_d)$ is of approximately the same magnitude as the bond dissociation energies, the guest gold atom is repelled by the 'interior' and becomes exo-bonded. The situation between the hollow cage **II** of Au_{20} and the guest gold atom is different: the latter is naturally trapped in a void and forms therein nine rather strong void bonds.

Acknowledgments This work was partially supported by the AIP 'Clusters and Nanowires' Project of the Belgian Federal Government and the EC FET proactive NanoICT Project 'MOLOC'. One of the authors, E. S. K., gratefully thanks FNRS (Belgium) and the FRFC project 2.4.594.10.F for supporting his stay at the University of Liège and the Organizing Committee of the QSCP-XV, in particular the Chair Philip E. Hoggan, for the kind invitation, the generous hospitality, and the excellent organization. E. S. K. also thanks Benjamin Soulé de Bas, Mike J. Ford, Alessandro Fortunelli, Uzi Landman, Pekka Pyykkö, Gernot Frenking, and the reviewer for the valuable suggestions and comments.

References

1. (a) Kroto HW, Heath JR, OBrien SC, Curl RF, Smalley RE (1985) Nature (London) 318:162; (b) Heath JR, Zhang Q, O'Brien SC, Curl RF, Kroto HW, Smalley RE (1987) J Am Chem Soc 109:359; (c) Kroto HW, Heath JR, OBrien SC, Curl RF, Smalley RE (1987) Astrophys J 314:352
2. Heath JR, O'Brien SC, Zhang Q, Lui Y, Curl RF, Kroto HW, Tittel FK, Smalley RE (1985) J Am Chem Soc 107:7779
3. See e. g.: (a) Elkind FD, O'Brien SC, Carl RF, Smalley RE (1988) J Am Chem Soc 110:4464; (b) Cox DX, Trevor DJ, Reckmann KC, Kaldor A (1986) J Am Chem Soc 108:2457; (c) Ross MM, Callaham JH (1991) J Phys Chem 95:5720; (d) Alvare MM, Gillan EG, Holczer K, Kaner RB, Min KS, Whetten RL (1991) J Phys Chem 95:10561; (e) Weisker T, Bohme DK, Hrusak J, Kratschmer W, Schwarz H (1991) Angew Chem Int Ed Engl 30:884; (f) Saunders M, Jimenez-Vazquez HA, Cross RJ, Poreda RJ (1993) Science 259:1428; (g) Shinohara H (2000) Rep Prog Phys 63:843; (h) Nishibori E, Takata M, Sakata M, Tanaka H, Hasegawa M, Shinohara H (2000) Chem Phys Lett 330:497; (i) Peres T, Cao BP, Cui WD, Khing A, Cross RJ, Saunders M, Lifshitz C (2001) Int J Mass Spectrom 210:241; (j) Murata Y, Murata M, Komatsu K (2003) J Am Chem Soc 125:7152; (k) Shiotani H, Ito T, Iwasa Y, Taninaka A, Shinohara H, Nishibori E, Takata M, Sakata M (2004) J Am Chem Soc 126:364; (l) Hu YH, Ruckenstein E (2005) J Am Chem Soc 127:11277; (m) Klingeler R, Kann G, Wirth I (2001) J Chem Phys 115:7215
4. (a) Cioslowski JJ (1995) Am Chem Soc 117:2553; (b) Hu YH, Ruckenstein E (2005) Am Chem Soc 127:11277; (c) Belash IT, Bronnikov AD, Zharikov OV, Palnichenko AV (1990) Synth Met 36:283; (d) Dunlap BI, Ballester JL, Schmidt PP (1992) J Phys Chem 96:9781; (e) Guha S, Nakamoto K (2005) Coord Chem Rev 249:1111; (f) Turker L, Gumus S (2006) Poly Aromat Comp 26:145; (g) Jantoljak H, Krawez N, Loa I, Tellgmann R, Campbell EEB, Litvinchuk AP, Thomsen CZ (1997) Phys Chem (Munchen) 200:157; (h) Koltover VK (2006) J Mol Liq (2006) 127:139; (i) Gromov A, Krawez N, Lassesson A, Ostrovskii DI, Campbell EEB (2002) Curr Appl Phys 2:51; (j) Popok VN, Azarko II, Gromov AV, Jönsson M, Lassesson A, Campbell EEB (2005) Solid State Commun 133:499; (k) Johnson RD, de Vries MS, Salem J, Bethune DS, Yannoni CS (1992) Nature 355:239; (l) Zhao YL, Pan XM, Zhou DF, Su ZM, Wang RS (2003) Synth Met 135:227; (m) Cioslowski J, Fleischmann ED (1991) J Chem Phys 94:3730
5. Odom TW, Nehl CL (2008) The so called 3M's principle: make, measure, model. ACS Nano 2:612
6. (a) Wang L-S, Conceicao J, Jin CM, Smalley RE (1991) Chem Phys Lett 182:5; (b) Boltalina OV, Siderov LN, Sukhanova EV, Sorokin ID (1993) Rapid Commun Mass Spectrom 7:1009; (c) Brink C, Andersen LH, Hvelplund P, Mather D, Volstad JD (1995) Chem Phys Lett 233:52; (d) Chen G, Cooks RG, Corpuz E, Scott LT (1996) J Am Soc Mass Spectrom 7:619; (e) Wang XB, Ding LF, Wang LS (1999) J Chem Phys 110:8217; (f) For current reference see also Betowski LD, Enlow M, Riddick L, Aue DH (2006) J Phys Chem 110:12927
7. Generally speaking, the existence of endohedral fullerenes alone does not mean the independent existence of the corresponding hollow cages. See e. g.: Kareev IE, Kuvychko IV, Shustova NB, Lebedkin SF, Bubnov VP, Anderson OP, Popov AA, Boltalina OV, Strauss SH (2008) Angew Chem Int Ed 47:6204
8. Heiz U, Landman U (2006) Nanocatalysis. Springer, New York

9. (a) Hammer B, Nørskov JK (1995) Nature (London) 376:238; (b) Valden M, Lai X, Goodman DW (1998) Science 281:1647; (c) Sanchez A, Abbet S, Heiz U, Schneider W-D, Häkkinen H, Barnett RN, Landman U (1999) J Phys Chem A 103:9573; (d) Schmid G, Corain B (2003) Eur J Inorg Chem 3081

10. (a) Haruta M, Kobayashi T, Sano H, Yamada N (1987) Chem Lett 405;(b) Haruta M, Yamada N, Kobayashi T, Iijima S (1989) J Catal 115:301; (c) Haruta M, Tsubota S, Kobayashi T, KageyamaH, Genet MJ, Delmon B (1993) J Catal 144:175; (d) Haruta M (1997) Catal J Today 36:153; (e) Iizuka Y, Tode T, Takao T, Yatsu KI, Takeuchi T, Tsubota S, Haruta M (1999) Catal J Today 187:50; (f) Shiga A, Haruta M (2005) Appl Catal A Gen 291:6; (g) Date M, Okumura M, Tsubota S, Haruta M (2004) Angew Chem Int Ed 43:2129

11. Gardea-Torresday JL, Parson JG, Gomez E, Peralta-Videa J, Troiani HE, Santiago P, Yacaman MJ (2002) Nano Lett 2:397

12. (a) Pyykkö P (2004) Angew Chem Int Ed 43:4412; (b) Pyykkö P (2005) Inorg Chim Acta 358:4113; (c) Pyykkö P (2008) Chem Soc Rev 37:1967

13. (a) Häkkinen H, Landman U (2000) Phys Rev B 62:R2287; (b) Häkkinen H, Moseler M, Landman U (2002) Phys Rev Lett 89:033401; (c) Häkkinen H, Yoon B, Landman U, Li X, Zhai HJ, Wang LC (2003) J Phys Chem A 107:6168; (d) Bonačić-Koutecký V, Burda J, Mitric R, Ge MF, Zampella G, Fantucci P (2002) J Chem Phys 117:3120; (e) Furche F, Ahlrichs R, Weis P, Jacob C, Gilb S, Bierweiler T, Kappes MM (2002) J Chem Phys 117L:6982; (f) Gilb S, Weis P, Furche F, Ahlrichs R, Kappes MM (2002) J Chem Phys 116:4094; (g) Lee HM, Ge M, Sahu BR, Tarakeshwar P, Kim KS (2003) J Phys Chem B 107:9994; (h) Wang JL, Wang GH, Zhao JJ (2002) Phys Rev B 66:035418; (i) Xiao L, Wang L (2004) Chem Phys Lett 392:452; (j) Olson RM, Varganov S, Gordon MS, Metiu H, Chretien S, Piecuch P, Kowalski K, Kucharski S, Musial M (2005) J Am Chem Soc 127:1049; (k) Koskinen P, Häkkinen H, Huber B, von Issendorff B, Moseler M (2007) Phys Rev Lett 98:015701; (l) Han VK (2006) J Chem Phys 124:024316; (m) Fernández EM, Soler JM, Garzón IL, Balbás LC (2004) Phys Rev B 70:165403; (n) Fernández EM, Soler JM, Balbás LC (2004) Phys Rev B 73:235433; (o) Remacle F, Kryachko ES (2004) Adv Quantum Chem 47:423; (p) Remacle F, Kryachko ES (2005) J Chem Phys 122:044304; (q) Johansson MP, Lechtken A, Schooss D, Kappes MM, Furche F (2008) Phys Rev A 77:053202; (r) Häkkinen H (2008) Chem Soc Rev 37:1847; (s) Huang W, Wang L-S (2009) Phys Rev Lett 102:153401

14. This term was borrowed from: (a) McAdon MH, Goddard WA III (1988) J Phys Chem 92:1352; (b) Glukhovtsev MN, Schleyer PVR (1993) Isr J Chem 33:455; (c) Danovich D, Wu W, Shaik S (1999) J Am Chem Soc 121:3165; (d) de Visser SP, Kumar D, Danovich M, Nevo N, Danovich D, Sharma PK, Wu W, Shaik S (2006) J Phys Chem A 110:8510; (e) See also Ritter SK (2007) C&EN January 29:37

15. Historically, the concept of a hollow cage is rooted to the fifth century B.C. when Democritus postulated the existence of immutable atoms characterized by size, shape, and motion. A motion of atoms requires the existence of a free, unoccupied, or empty space, or a void (nothingness) as a real entity [16]. The concept of the free space (free volume) was exploited 135 years ago by J. D. van der Waals in his PhD thesis [17] (see also Ref. [18] as current reference), as the volume which complements the volume excluded by a molecule

16. (a) See e. g. the online Edition of the Encyclopaedia Britannica; (b) Also: Prigogine I, Stengers I (1984) Order out of Chaos. Mans new dialogue with nature. Bantam Books, Toronto, p 3

17. van der Waals JD (1873) Continu"ıteit van den Gas en Vloeistoftoestand (The English translation: "On the Continuity of the Gas and Liquid State"), PhD thesis. University of Leiden, Leiden

18. Kryachko ES (2008) Int J Quantum Chem 108:198

19. (a) A Fullerene Work Party (1997) Pure Appl Chem 69:1411; (b) Chemical Abstracts, Index Guide 19921996, Appendix IV 162163; (c) Miyazaki T, Hiura H, Kanayama T (2002) Theoretical study of metal-encapsulating Si cage clusters: revealing the nature of their peculiar geometries. ArXiv: cond-mat/0208217v1. Accessed 12 Aug 2002; (d) See also: Ward J (1984) The artifacts of R. Buckminster Fuller: a comprehensive collection of his designs and drawings, vol 3. Garland, New York

20. Hoyer W, Kleinhempel R, Lörinczi A, Pohlers A, Popescu M, Sava F (2005) J Phys Condens Matter 17:S31. "The void size is defined as the diameter of the sphere of maximum size that can be introduced in an interstice without intersecting any surrounding atom defined by its radius. The position of the centre of a void is obtained by moving the starting position inside an interstice in small aleatory steps and retaining only those movements that increase the radius of the sphere that can be introduced in the interstice"
21. Bulusu S, Zeng XC (2006) J Chem Phys 125:154303
22. Bulusu S, Li X, Wang L-S, Zeng XC (2006) Proc Natl Acad Sci USA 103:8326
23. (a) Wang JL, Wang GH, Zhao JJ (2002) Phys Rev B 66:035418; (b) Fa W, Luo C, Dong JM (2005) Phys Rev B 72:205428
24. Pyykkö P, Runeberg N (2002) Angew Chem Int Ed 41:2174
25. Gao Y, Bulusu S, Zeng XC (2005) J Am Chem Soc 127:156801
26. (a) Xing X, Yoon B, Landman U, Parks JH (2006) Phys Rev B 74:165423; (b) It is interesting to note that the hollow cage Au$_{16}$ was obtained in [23b] by removing the four vertex atoms of the ground-state gold cluster Au$_{20}$(T_d and by a further relaxation of the resultant one. Absolutely the reverse procedure to that proposed in the present work to obtain the hollow cage VI
27. Li J, Li X, Zhai H-J, Wang L-S (2003) Science 299:864
28. (a) Apra E, Ferrando R, Fortunelli A (2006) Phys Rev B 73:205414; (b) Ref. [13m]; (c) de Bas BS, Ford MJ, Cortie MB (2004) J Mol Struct (Theochem) 686:193; (d) Fernndez EM, Soler JM, Balbs LC (2006) Phys Rev B 73:235433; (e) Wang J, Bai J, Jellinek J, Zeng XC (2007) J Am Chem Soc 129:4110; (f) Krishnamurty S, Shafai GS, Kanhere DG, de Bas BS, Ford MJ (2007) J Phys Chem A 111:10769
29. Xing X, Yoon B, Landman U, Parks JH (2006) Phys Rev B 74:165423
30. Johansson MP, Sundholm D, Vaara J (2004) Angew Chem Int Ed 43:2678
31. Schmid G (2008) Chem Soc Rev 37:1909, and references therein
32. Schweikhard L, Herlert A, Vogel M (1999) Philos Mag B 79:1343
33. (a) The NIST, http://webbook.nist.gov/, reports only two of them dealing with the first electron affinity; (b) Taylor KJ, Pettiette-Hall CL, Cheshnovsky O, Smalley RE (1992) J Chem Phys 96:3319; (c) See also: von Issendorff B, Cheshnovsky O (2005) Annu Rev Phys Chem 56:549; (d) Bulusu S, Zeng XC (2006) J Chem Phys 125:154303
34. See e.g. Last I, Levy Y, Jortner J (2002) Proc Natl Acad Sci USA 99:9107 and references therein
35. Karttunen AJ, Linnolahti M, Pakkanen TA, Pyykkö P (2008) Chem Commun 465
36. (a) King RB, Chen Z, Schleyer PVR (2004) Inorg Chem 43:4564; (b) Kryachko ES, Remacle F (2007) Int J Quantum Chem 107:2922; (c) See also: (a) Kryachko ES, Remacle F (2007) J Chem Phys 127:194305; (d) Kryachko ES, Remacle F (2008) Mol Phys 106:521; (e) Gruene P, Rayner DM, Redlich B, van der Meer AFG, Lyon JT, Meijer G, Fielicke A (2008) Science 321:674
37. Molina LM, Hammer B (2005) J Catal 233:399
38. Kryachko ES, Remacle F (2010) J Phys Conf Ser 248:012026
39. (a) Hay PJ, Wadt WR (1985) J Chem Phys 182:270, 299; (b) Wadt WR, Hay PJ (1985) J Chem Phys 82:284
40. Frisch MJ, Trucks GW, Schlegel HB, Scuseria GE, Robb MA, Cheeseman JR, Montgomery JA Jr, Vreven T, Kudin KN, Burant JC, Millam JM, Iyengar SS, Tomasi J, Barone V, Mennucci B, Cossi M, Scalmani G, Rega N, Petersson GA, Nakatsuji H, Hada M, Ehara M, Toyota K, Fukuda R, Hasegawa J, Ishida M, Nakajima T, Honda Y, Kitao O, Nakai H, Klene M, Li X, Knox JE, Hratchian HP, Cross JB, Adamo C, Jaramillo J, Gomperts R, Stratmann RE, Yazyev O, Austin AJ, Cammi R, Pomelli C, Ochterski JW, Ayala PY, Morokuma K, Voth GA, Salvador P, Dannenberg JJ, Zakrzewski VG, Dapprich S, Daniels AD, Strain MC, Farkas O, Malick DK, Rabuck AD, Raghavachari K, Foresman JB, Ortiz JV, Cui Q, Baboul AG, Clifford S, Cioslowski J, Stefanov BB, Liu G, Liashenko A, Piskorz P, Komaromi I, Martin RL, Fox DJ, Keith T, Al-Laham MA, Peng CY, Nanayakkara A, Challacombe M, Gill PMW, Johnson B, Chen W, Wong MW, Gonzalez C, Pople JA (2004) GAUSSIAN 03 (Revision C.02). Gaussian Inc., Wallington

41. (a) Curl RF, Smalley RE (1988) Science 242:1017; (b) Cai Y, Guo T, Jin C, Haufler RE, Chibante LPF, Fure J, Wang L, Alford JM, Smalley RE (1991) J Phys Chem 95:7564; (c) Wang L-M, Bulusu S, Zhai HJ, Zeng XC, Wang LS (2007) Angew Chem Int Ed 46:2915

42. (a) Gu X, Ji M, Wei SH, Gong XG (2004) Phys Rev B 70:205401; (b) Fa W, Zhou J, Luo C, Dong J (2006) Phys Rev B 73:085405; (c) Jalbout AF, Contreras-Torres FF, Prez LA, Garzn IL (2008) J Phys Chem A 112:353

43. (a) Schlyer PVR, Maerker C, Dransfeld A, Jiao H, Hommes NJRvE (1996) J Am Chem Soc 118:6317; (b) Chen Z, Wannere CS, Corminboeuf C, Puchta R, Schleyer PVR (2005) Chem Rev 105:3842

44. (a) Sanderson RT (1951) Science 114:670; (b) Idem (1955) Science 121:207; (c) Idem (1952) J Am Chem Soc 74:272

45. (a) Scrocco E, Tomasi J (1973) Topics Curr Chem 42:95; (b) Politzer P, Truhlar DG (eds) (1981) Chemical applications of atomic and molecular electrostatic potentials. Plenum, New York; (c) Pullman A, Pullman B (1981) Chemical applications of atomic and molecular electrostatic potentials. Plenum, New York, p 381; (d) Murray JS, Sen K (eds) (1996) Molecular electrostatic potentials, concepts and applications, theoretical and computational chemistry, vol 3. Elsevier, Amsterdam

46. (a) Tielens F, Andrés J (2007) J Phys Chem C 111:10342; (b) Wang D-L, Sun X-P, Shen H-T, Hou D-Y, Zhai Y-C (2008) Chem Phys Lett 457:366

47. (a) Claxton TA, Shirsat RN, Gadre SR (1994) J Chem Soc Chem Commun 6:731; (b) Mauser H, Hirsch A, van Eikema Hommes NJR, Clark T (1997) J Mol Model 3:415

48. (a) In fact, our statement is in concord with the following statement: "It has been generally accepted that extractable EMFs [EMF=endohedral metallofullerenes] take on endohedral structures. However, definitive proof of the structure must be performed for each EMF". Ref. [48b]; (b) Yamada M, Akasaka T, Nagase S (2010) Acc Chem Res 43:92

Chapter 31
A Theoretical Study of Complexes of Crown Ethers with Substituted Ammonium Cations

Demeter Tzeli, Ioannis D. Petsalakis, and Giannoula Theodorakopoulos

Abstract The electronic and geometric structures of the complexes of dibenzo-18-crown-6 ether and of dibenzo-18-crown-6 ether of fulleroN-methylpyrrolidine with diphenylammonium cation, $Ph_2NH_2^+$, and its derivative with π-extended tetrathiafulvalene, π-exTTF, were investigated by employing density functional theory. We calculate geometries, complexation energies and some absorption spectra of the lowest energetic minima of the above complexes in the gas phase as well as in $CHCl_3$ solvent. The complexation energies, corrected for basis set superposition error reach up to 2.2 eV in the gas phase and up to 1.3 eV in the $CHCl_3$ solvent, at the M06-2X/6-31G(d,p) level of theory. In the complexes, the cations and the crown ethers are deformed to maximize the number of the hydrogen bonds. The presence of fulleroN-methylpyrrolidine, attached to the crown ethers, increases the complexation energies by up to 0.2 eV due to additional interactions. The complex of fullerene crown ethers with a π-exTTF derivative of $Ph_2NH_2^+$ presents charge transfer transitions in the absorption spectrum and may serve as candidate for organic photovoltaics.

31.1 Introduction

There is great interest in the study of crown ethers since their discovery [1, 2] because they are highly adaptable hosts for a large number of guests [3–5]. The complexation of crown ethers with many guests, both neutral [4] and cationic [3,6] and their high degree of selectivity have been investigated both experimentally and theoretically [3, 4, 6, 7]. The conformations of the crown ethers, the size of the

D. Tzeli (✉) • I. D. Petsalakis • G. Theodorakopoulos
Theoretical and Physical Chemistry Institute, National Hellenic Research Foundation,
48 Vassileos Constantinou Ave., Athens 116 35, Greece
e-mail: dtzeli@eie.gr; idpet@eie.gr; ithe@eie.gr

P.E. Hoggan et al. (eds.), *Advances in the Theory of Quantum Systems in Chemistry and Physics*, Progress in Theoretical Chemistry and Physics 22,
DOI 10.1007/978-94-007-2076-3_31, © Springer Science+Business Media B.V. 2012

guest, the nature of the intermolecular interactions and in solution the solvating ability of the solvent with respect to the cation and the complex determine the strength of their binding with various species and its specificity. As a result they are used in many diverse fields such as in catalysis, in enantiomer resolution, in membrane separation of cationic species etc. [8].

Moreover, supra-molecular systems involving crown ethers, fullerene and π-extended systems have been achieved that can mimic the photosynthetic process [9–14]. The fullerene C_{60} has been used successfully as an electron acceptor in the construction of model photosynthetic systems [9], the π-extended systems, such as porphyrins [12], phthalocyanines [13], π-extended tetrathiafulvalene (π-exTTF) derivatives [9, 10], which are utilized as electron donors, while the crown ethers act as a bridge between the electron donor and acceptor. In the absorption spectrum of the complexes, the absorption maxima are associated experimentally and theoretically with the formation of charge-transfer states [14–16]. Consequently, these supramolecular systems have potential for applications in photonic, photocatalytic, and molecular optoelectronic gates and devices [9–14]. As a result, the study of the conformations and the complexation behavior of crown ethers and their derivatives are motivated both by scientific curiosity regarding the specificity of their binding and by potential technological applications.

The present work is a continuation of our previous studies on the electronic and geometric structures of four crown ethers and their complexes with $(CH_3)_x NH_{4-x}^+, x = 0 - 4$ in the gas phase and in $CHCl_3$ solvent [7] and on the complexes of the dibenzo-18-crown-6 ether of fullero-N-methylpyrrolidine with a π-exTTF derivative [16]. In the present work a theoretical study on the complexes of dibenzo-18-crown-6 ether and of dibenzo-18-crown-6 ether of fullero-N-methylpyrrolidine with the diphenylammonium cation, $Ph_2NH_2^+$, were investigated by employing density functional theory. In what follows, we describe the computational approach in Sect. 31.2, we discuss our results in Sect. 31.3, and we summarize our findings in Sect. 31.4.

31.2   Computational Approach

We used the density functional theory, at the M06-2X/6-31G(d,p) level of theory, to study the electronic and geometric structures of the complexes of dibenzo-18-crown-6 ether and of dibenzo-18-crown-6 ether of fullero-N-methylpyrrolidine with diphenylammonium cation, $Ph_2NH_2^+$. M06-2X [17] is a hybrid meta exchange correlation functional, a highly-nonlocal functional with double the amount of nonlocal exchange and is recommended for the study of non-covalent interactions [18], such as the present interactions. The decision to employ this functional was based on the conclusions of our previous studies about the applicability of the B3LYP, M05-2X, M06-2X, MPWB1K and B2PLYP-D functionals in conjunction with three

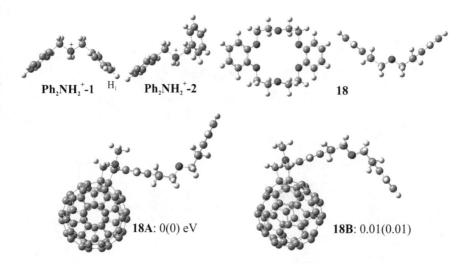

Fig. 31.1 $Ph_2NH_2^+$: the lowest isomers of the diphenylammonium cation; **18**: dibenzo-18-crown-6 ether (*top and side view*); **18A** and **18B**: the lowest isomers of the dibenzo-18-crown-6 ether of fullero-N-methylpyrrolidine; BSSE corrected energy differences from the most stable structure are shown in the gas phase (in $CHCl_3$ solvent) at the M06-2X/6-31G(d,p) level of theory

basis sets on complexes of crown ethers with $(CH_3)_xNH_{4-x}^+$, $x = 0 - 4$ [7] and the applicability of the B3LYP, CAM-B3LYP, M06-HF and M06-2X on the benzene dimer and the fullerene-benzene system [16] and on complexes of fullerene crown ethers with a π-exTTF derivative [16]. We concluded that the M06-2X functional in conjunction with the 6-31G(d,p) basis set [19] is a good choice for complexes having both hydrogen bonds and very weak vdW interactions (dispersion forces between nonpolar species) such as interactions between phenyl groups or between phenyl group and C_{60} [16].

In the present study, we calculated two energetically degenerate minima of the diphenylammonium cation, i.e., $Ph_2NH_2^+-1$ and $Ph_2NH_2^+-2$, see Fig. 31.1. The lowest minimum of the dibenzo-18-crown-6 ether, **18**, is given in Fig. 31.1. while the second lowest minimum lies 0.16 eV above the global minimum [7]. The two lowest isomers, **18A** and **18B**, of the dibenzo-18-crown-6 ether of fullero-N-methylpyrrolidine are practically degenerate and differ in the direction of the crown ether, namely up (**A**) or down (**B**) with respect to fullero-N-methylpyrrolidine, [7] see Fig. 31.1.

Different isomers of the complexes of $Ph_2NH_2^+-1$ and $Ph_2NH_2^+-2$ with **18** and **18A** or **18B** were determined and optimized. The full optimization of these structures led to two and eight low-lying energy minima of the complexes of cations with the crown ether **18** and the fullerene crown ether **18A** or **18B** in the gas phase, respectively, see Figs. 31.2 and 31.3. A derivative of the $Ph_2NH_2^+-1$ isomer was

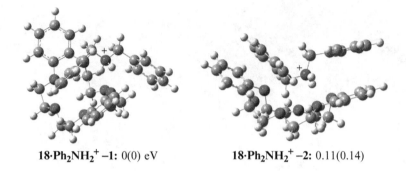

18·Ph$_2$NH$_2^+$ –1: 0(0) eV   18·Ph$_2$NH$_2^+$ –2: 0.11(0.14)

Fig. 31.2 The energetically lowest conformers of the **18 · Ph$_2$NH$_2^+$** complex, i.e., complex of dibenzo-18-crown-6 ether with diphenylammonium cation; BSSE corrected energy differences from the most stable structure are shown in the gas phase (in CHCl$_3$ solvent) at the M06-2X/6-31G(d,p) level of theory

obtained when the H$_1$ atom was substituted with a π-exTTF through a C$\equiv$C group, see Fig. 31.4. The energetically lowest complexes of **T** with **18A** and **18B**, i.e., **18A · T** and **18B · T** are depicted in Fig. 31.4. For the calculation of their absorption spectra of the two calculated complexes, 50 singlet-spin excited electronic states have been calculated by Time Dependent DFT (TDDFT) calculations [20]. It was necessary to include a large number of excited states, about 50 excited states, in order to reach the **T** absorbing states because there are many fullerene excited states at lower excitation energies [16].

Single point calculations at the gas phase optimum geometry of all structures have been carried out in CHCl$_3$ solvent. As we showed in our previous study on the complexation of the present crown ethers and others with $(CH_3)_x NH_{4-x}^+$, $x = 0 - 4$ cations, full optimization in CHCl$_3$ solvent of the optimum gas phase geometry of the complexes results only in a slight change in geometry and an increase of the complexation energy by less than 0.02 eV [7]. Thus, we did not carry out full optimization in CHCl$_3$ solvent. The calculations in the solvent were carried out employing the polarizable continuum model [21]. This model is divided into a solute part lying inside a cavity, surrounded by the solvent part represented as a structureless material characterized by its macroscopic properties, i.e., dielectric constant and solvent radius. This method reproduces well solvent effects [22, 23].

For all minima determined, the complexation energy (CE$_u$) and the corrected values with respect to the basis set superposition error (CE) in the gas phase and in CHCl$_3$ solvent (CE$_{sol}$) were calculated. The basis set superposition error (BSSE) corrections were made using the counterpoise procedure [24] since such corrections are especially important for van der Waals (vdW) systems [25, 26] which is the case of the complexes calculated here.

All calculations were performed using the Gaussian 09 program package [27].

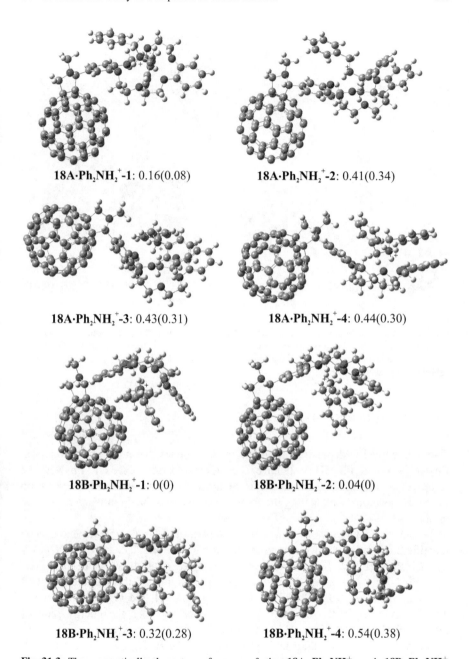

18A·Ph$_2$NH$_2^+$-1: 0.16(0.08)   **18A·Ph$_2$NH$_2^+$-2**: 0.41(0.34)

18A·Ph$_2$NH$_2^+$-3: 0.43(0.31)   **18A·Ph$_2$NH$_2^+$-4**: 0.44(0.30)

18B·Ph$_2$NH$_2^+$-1: 0(0)   **18B·Ph$_2$NH$_2^+$-2**: 0.04(0)

18B·Ph$_2$NH$_2^+$-3: 0.32(0.28)   **18B·Ph$_2$NH$_2^+$-4**: 0.54(0.38)

Fig. 31.3 The energetically lowest conformers of the **18A** · Ph$_2$NH$_2^+$ and **18B** · Ph$_2$NH$_2^+$ complexes. Gas phase (CHCl$_3$ solvent) BSSE corrected energy differences from the most stable structure are shown at the M06-2X/6-31G(d,p) level of theory

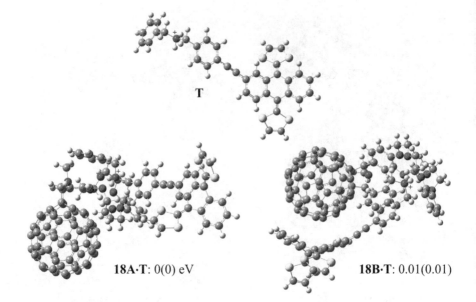

T

18A·T: 0(0) eV   **18B·T**: 0.01(0.01)

Fig. 31.4 Minimum structures of the **T** cation (π-exTTF derivative of diphenylammonium cation) and **18A · T** and **18B · T** complexes of **18A** and **18B** with **T**. BSSE corrected energy differences from the most stable structure are shown in the gas phase (in CHCl$_3$ solvent) at the M06-2X/6-31G(d,p) level of theory

31.3 Results and Discussion

Two energetically degenerate minima of the diphenylammonium cation, i.e., **Ph$_2$NH$_2^+$–1** and **Ph$_2$NH$_2^+$–2**, have been identified, see Fig. 31.1. **Ph$_2$NH$_2^+$–2** results from **Ph$_2$NH$_2^+$–1** when the H of the CH$_2$ group and the phenyl group exchange positions. Calculating the frequencies, we find that both isomers are true minima.

In the present study 10 low energy supramolecular complexes have been identified; two minima of the complex of **18** with Ph$_2$NH$_2^+$, i.e., **18 · Ph$_2$NH$_2^+$–1** and **18 · Ph$_2$NH$_2^+$–2**, four minima of the complex of **18A** with Ph$_2$NH$_2^+$, i.e., **18A · Ph$_2$NH$_2^+$–1**, **–2, –3**, and **–4**, and four minima of the complex of **18B** with Ph$_2$NH$_2^+$, i.e., **18B · Ph$_2$NH$_2^+$–1**, **–2, –3**, and **–4**, see Figs. 31.2 and 31.3. The last number in the name of the complexes specifies the energy rank of the **18 · Ph$_2$NH$_2^+$, 18A · Ph$_2$NH$_2^+$**, and **18B · Ph$_2$NH$_2^+$** species at the M06-2X/6-31G(d,p) level of theory after taking into account the BSSE correction. The vdW bond distances of the complexes are given in Table 31.1 and the complexation energies, CE, in Table 31.2.

Table 31.1 Van der Waals bond distances R(Å) of the complexes of the **18, 18A,** and **18B** crown ethers with ammonium cations at the M06-2X/6-31G(d,p) level of theory

Complex	$R_{H_N...O}$	$R_{H_C...O}{}^a$	$R_{H_C...O}{}^b$	$R_{ph...ph}{}^c$	$R_{ph...F}{}^d$	$R_{H_C...ph}{}^e$
$18 \cdot Ph_2NH_2^+ -1$	1.969, 1.917	2.473, 2.559	2.419, 2.403	3.710		2.468
$18 \cdot Ph_2NH_2^+ -2$	1.838, 2.242	2.153, 2.292/ 2.402^f	2.615	3.747		2.863, 2.899
$18A \cdot Ph_2NH_2^+ -1$	1.821, 1.845	2.316	2.330	3.569		
$18A \cdot Ph_2NH_2^+ -2$	1.754, 2.545	2.089, 2.175	2.592, 2.658			2.920
$18A \cdot Ph_2NH_2^+ -3$	1.944, 2.055/2.056^f	2.588	2.332, 2.673	3.695		2.415, 3.063
$18A \cdot Ph_2NH_2^+ -4$	1.759, 2.301	2.290, 2.109	2.496	3.712, 3.911		3.284
$18B \cdot Ph_2NH_2^+ -1$	1.853, 1.930/2.523^f	2.126	2.678	3.574	3.345	2.504, 2.941
$18B \cdot Ph_2NH_2^+ -2$	1.952, 2.079/2.366^f	2.188, 2.615	2.396			2.367, 2.396
$18B \cdot Ph_2NH_2^+ -3$	2.140/2.326^f 1.916/2.258^f	2.291, 2.403/ 2.418^f	2.444, 2.627			
$18B \cdot Ph_2NH_2^+ -4$	1.671, 1.713		2.040, 2.241, 2.543	3.728	3.123	2.250

^{a} H_C atoms of the CH_2 groups
^{b} H_C atoms of phenyl groups
^{c} Distance between the two centers of the phenyl groups
^{d} Distance between the center of a phenyl group and the nearest C atoms of fullerene
^{e} Distance between the H_C atoms of the CH_2 or phenyl groups and the center of phenyl group
^{f} One H atom interacts with two O atoms

31.3.1 $18 \cdot Ph_2NH_2^+$ Complexes

The two lowest minima of the complex of **18** with $Ph_2NH_2^+$ were obtained from the complexation of **18** with each of the energetically degenerate minima of the diphenylammonium cation, i.e., $\mathbf{Ph_2NH_2^+ -1}$ and $\mathbf{Ph_2NH_2^+ -2}$. However, the $\mathbf{Ph_2NH_2^+ -1}$ cation interacts more strongly with the **18** crown ether than with the $\mathbf{Ph_2NH_2^+ -2}$ species, by 0.1 eV both in the gas phase and in CHCl$_3$ solvent, showing some selectivity of the crown ether. The CE(CE$_{sol}$) of the $18 \cdot Ph_2NH_2^+ -1$ is 1.97 (1.17) eV, see Table 31.2.

Different types of van der Waals interactions are observed in the complexes. Hydrogen bonds formed between the ammonium H_N atoms and the crown ether O atoms. The strongest $H_N \cdots O$ bonds in complexes of dibenzo-18-crown-6 ether with ammonium cation have a CE of 0.9 eV per bond [7] and bond distances of 1.8 Å. The H atoms of the methylene and/or the phenyl group interact with the O atoms of ethers. The weakest interactions observed are between H atoms and phenyl groups as well as between two phenyl groups. In addition, the interactions are not always 1

Table 31.2 Complexation Energies CE(eV) in the gas phase and in CHCl$_3$ solvent, CE$_{sol}$, of the complexes of the **18**, **18A**, and **18B** crown ethers with ammonium cations at the M06-2X/6-31G(d,p) level of theory

Complex	CE$_u^a$	CE^b	CE$_{sol}^b$
18 · Ph$_2$NH$_2^+$ −1	2.36	1.97	1.17
18 · Ph$_2$NH$_2^+$ −2	2.22	1.86	1.04
18A · Ph$_2$NH$_2^+$ −1	2.43	2.05	1.23
18A · Ph$_2$NH$_2^+$ −2	2.15	1.80	0.97
18A · Ph$_2$NH$_2^+$ −3	2.16	1.78	1.00
18A · Ph$_2$NH$_2^+$ −4	2.15	1.77	1.01
18B · Ph$_2$NH$_2^+$ −1	2.73	2.21	1.31
18B · Ph$_2$NH$_2^+$ −2	2.57	2.17	1.31
18B · Ph$_2$NH$_2^+$ −3	2.28	1.89	1.03
18B · Ph$_2$NH$_2^+$ −4	2.12	1.67	0.93
18A · T^c	2.40	2.02	1.27
18B · T^c	2.50	1.99	1.18

^{a}BSSE uncorrected values
^{b}BSSE corrected values
^{c}Reference [16]

to 1 and in many cases one H atom interacts with two O atoms. All vdW distances are given in Table 31.1.

31.3.2   18A · Ph$_2$NH$_2^+$ and 18B·Ph$_2$NH$_2^+$ Complexes

Eight low lying minima of the complex of dibenzo-18-crown-6 ether of fullero-N-methylpyrrolidine with the Ph$_2$NH$_2^+$ cation, four with the **18A** isomer of crown ether and four with the **18B** isomer have been determined, see Fig. 31.3. The **18B** isomer forms the most stable structures, because the cation can be captured between fullerene and crown ether and additional vdW bonds formed can further stabilize the complex. The complexation energies of minima range from 2.21 to 1.67 eV in the gas phase and from 1.31 to 0.93 eV in CHCl$_3$ solvent at the M06-2X/6-31G(d,p) level of theory, see Table 31.2. It is worth noting that the BSSE corrections are up to 0.5 eV, however, the relative stability of the structures does not change.

In all eight minima, hydrogen bonds are formed between the H$_N$ atoms, which are attached to the N atoms, and the O atoms of the crown ethers with bond distances ranging from 1.7 to 2.5 Å. Additional vdW bonds between the H$_C$ atoms of the CH$_2$ or phenyl groups and the O atoms of the crown ether are formed, and the distances range from 2.0 to 2.7 Å. The above three types of interactions are not always 1 to 1 and one H atom can interact with two O atoms, see Table 31.2. Moreover, π-stacking interactions between phenyl groups are formed in all cases with the exception of **18B · Ph$_2$NH$_2^+$ −2** and **18B · Ph$_2$NH$_2^+$ −3**, with distances of 3.6–3.9 Å, while interactions between C$_{60}$ and phenyl group appear only in **18B · Ph$_2$NH$_2^+$ −1** and **18B · Ph$_2$NH$_2^+$ −4**, with distances of 3.3 and 3.1 Å, respectively. Finally, hydrogen vdW interactions between H atoms and phenyl groups are formed in all cases with the exception of **18A · Ph$_2$NH$_2^+$ −1** and **18B · Ph$_2$NH$_2^+$ −3**.

The two lowest minima are the $\mathbf{18B \cdot Ph_2NH_2^+ -1}$ and $\mathbf{18B \cdot Ph_2NH_2^+ -2}$ isomers and they contain the $\mathbf{Ph_2NH_2^+ -2}$ isomer of the cation, but it is deformed and twisted, respectively. These minima are further stabilized by the existence of the fullerene, see above. Without the fullerene, the structures are not stable minima. Further optimization of these structures lead to significant change of the complexation in both isomers and finally lead to the $\mathbf{18 \cdot Ph_2NH_2^+ -2}$ complex. The CE values of $\mathbf{18B \cdot Ph_2NH_2^+ -1}$ and $\mathbf{18B \cdot Ph_2NH_2^+ -2}$ are $\mathrm{CE(CE_{sol})} = 2.21(1.31)$ and $2.17(1.31)$ eV, respectively. In $CHCl_3$ solvent, the minima are energetically indistinguishable.

The third lowest minimum is formed by the $\mathbf{18A}$ isomer, $\mathbf{18A \cdot Ph_2NH_2^+ -1}$. It consists of the $\mathbf{Ph_2NH_2^+ -1}$ minimum and the crown ether is deformed with respect to the free $\mathbf{18}$ crown ether. However, even though the fullerene does not interact with the cation as in the above cases, without the fullerene this structure is not stable. The CE values are $\mathrm{CE(CE_{sol})} = 2.05(1.23)$ eV.

In the remaining five isomers, the $\mathbf{Ph_2NH_2^+ -1}$ minimum is included in both $\mathbf{18A \cdot Ph_2NH_2^+ -3}$ and $\mathbf{18B \cdot Ph_2NH_2^+ -4}$, while in the last isomer the $\mathbf{Ph_2NH_2^+ -1}$ cation is deformed. The $\mathbf{Ph_2NH_2^+ -2}$ minimum is part of $\mathbf{18A \cdot Ph_2NH_2^+ -2}$ and $\mathbf{18A \cdot Ph_2NH_2^+ -4}$. Finally, the $\mathbf{18B \cdot Ph_2NH_2^+ -3}$ isomer does not include any stable minimum of the $Ph_2NH_2^+$ cation.

The two lowest minima $\mathbf{18 \cdot Ph_2NH_2^+ -1}$ and $\mathbf{18 \cdot Ph_2NH_2^+ -2}$ of the complex of the $\mathbf{18}$ crown ether are part of the $\mathbf{18A \cdot Ph_2NH_2^+ -3}$ or $\mathbf{18B \cdot Ph_2NH_2^+ -4}$ and $\mathbf{18A \cdot Ph_2NH_2^+ -4}$, respectively and not of the two lowest minima of the complex of the fullerene crown ether, i.e., $\mathbf{18B \cdot Ph_2NH_2^+ -1}$ and $\mathbf{18B \cdot Ph_2NH_2^+ -2}$. Comparing the CE values of the $\mathbf{18 \cdot Ph_2NH_2^+ -1}$ and $\mathbf{18A \cdot Ph_2NH_2^+ -3}$, which differ only in the existence of the fullerene, the first one has a CE value larger by 0.2 eV than the second one. However, the existence of the fullerene causes another minimum to be the global one, as mentioned above, which is more stable by 0.2 eV than the global $\mathbf{18 \cdot Ph_2NH_2^+ -1}$ minimum of the complex of the $\mathbf{18}$ crown ether.

31.3.3 $\mathbf{18A \cdot T}$ and $\mathbf{18B \cdot T}$ Complexes

Substituting the H_1 atom of the $\mathbf{Ph_2NH_2^+ -1}$ isomer with a π-exTTF through a $C \equiv C$ group, the $\mathbf{T}$ cation is obtained, see Fig. 31.4. The energetically lowest complexes of $\mathbf{T}$ with $\mathbf{18A}$ and $\mathbf{18B}$, i.e., $\mathbf{18A \cdot T}$ and $\mathbf{18B \cdot T}$, [16] are depicted also in Fig. 31.4 These two minima correspond to $\mathbf{18A \cdot Ph_2NH_2^+ -1}$ and to a slightly deformed $\mathbf{18B \cdot Ph_2NH_2^+ -3}$, respectively. Again, the lowest in energy $\mathbf{18B \cdot T}$ does not correspond to the lowest minimum $\mathbf{18B \cdot Ph_2NH_2^+ -1}$. Moreover, in $\mathbf{18B \cdot T}$ additional interactions are formed between the fullerene and the π-exTTF of $\mathbf{T}$. The $\mathrm{CE(CE_{sol})}$ values of $\mathbf{18A \cdot T}$ and $\mathbf{18B \cdot T}$ are $2.02(1.27)$ and $1.99(1.18)$ eV, respectively.

The $\mathbf{18A \cdot T}$ and $\mathbf{18B \cdot T}$ isomers form a dyad each consisting of an electron donor (fullerene), an electron acceptor (π-exTTF of $\mathbf{T}$) and a crown ether as a bridge between them. Their absorption spectra are given in Fig. 31.5. The spectrum

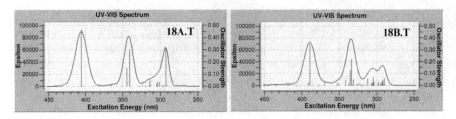

Fig. 31.5 Absorption spectrum of **18A · T** and **18B · T** at the M06-2X/6-31G(d,p) level of theory

of **18A · T** presents three major features at about 400, 340, and 290 nm which correspond to transitions from the **T** cation also to **T**. However, charge transfer transitions from **T** to the fullerene are observed very close to the **T** → **T** transitions. On the other hand, the absorption spectrum of **18B · T** presents also three major absorption peaks at about 390, 330, and a double peak around 300 nm with smaller oscillator strengths. The first and the second major peaks correspond to transition from **T** to **T**/fullerene, i.e., the excited orbital has electron density in both **T** and fullerene species. That happens because the fullerene interacts with the π-exTTF of **T**, see Fig. 31.5. We can label these transitions as charge transfer transitions. The last double peak corresponds to transitions from **T** to fullerene, namely charge transfer transitions. Moreover, the absorption spectra of five other isomers of **18A · T** and **18B · T** complexes are similar to spectrum of **18A · T** and the use of the CAM-B3LYP functional leads to similar major absorption peaks, transitions and conclusions [16]. Thus, the studied complex may serve as a candidate for molecular optoelectronics applications.

31.4 Remarks and Conclusions

The complexes of dibenzo-18-crown-6 ether and of dibenzo-18-crown-6 ether of fullero-N-methylpyrrolidine with the diphenylammonium cation, $Ph_2NH_2^+$, and its derivative with π-extended tetrathiafulvalene, π-exTTF, were investigated by employing density functional theory. We calculated geometries, complexation energies and some absorption spectra of the lowest energetic minima of the above complexes in the gas phase as well as in $CHCl_3$ solvent. A summary of our main results follows:

1. The complexation energies, corrected for basis set superposition error, reach up to 2.2 eV in the gas phase and up to 1.3 eV in the $CHCl_3$ solvent, at the M06-2X/6-31G(d,p) level of theory.
2. The minima of the cations and of the crown ethers are deformed to maximize the number of the hydrogen bonds formed and present the largest complexation energies. Bonds are formed between the H atom of ammonium, of the methylene

and/or the phenyl group and the O atoms of ethers. Moreover, π-stacking interactions arise between two phenyl groups or phenyl group with fullerene.

3. The presence of fullero-N-methylpyrrolidine, attached to the crown ether, results in minima, where the cation is captured between fullerene and crown ether. Additional interactions that can further stabilize the complex are formed.

4. The attachment of the fullero-N-methylpyrrolidine to the dibenzo-18-crown-6 ether or the attachment of π-exTTF to $Ph_2NH_2^+$ cation changes the complexation of the global minimum. As a result, it is not safe to suppose that the attachment of a group even if it is away from the complexation area will not change the type of the global minimum complex.

5. The complex of fullerene crown ethers with a π-exTTF derivative of $Ph_2NH_2^+$ presents charge transfer transitions in its absorption spectrum and may have potential for applications in organic photovoltaics and molecular electronic devices.

Acknowledgments Financial support from the EU FP7, Capacities Program, NANOHOST project (GA 201729) and the NATO grant, CBP.MD.CLG.983711 are acknowledged.

References

1. Pedersen CJ (1967) J Am Chem Soc 89:7017
2. Pedersen CJ (1970) J Am Chem Soc 92:391
3. Izatt RM, Pawlak K, Bradshaw JS (1991) Chem Rev 91:1721
4. Izatt RM, Bradshaw JS, Pawlak K, Bruening RL, Tarbe BJ (1992) Chem Rev 92:1261
5. Lehn J-M (1995) Supramolecular chemistry. VCH, Weinheim
6. Izatt RM, Bradshaw JS, Nielsen SA, Lamb JD, Christensen JJ (1985) Chem Rev 85:271
7. Tzeli D, Petsalakis ID, Theodorakopoulos G (2011) Phys Chem Chem Phys 13:954 and references therein
8. Ha YL, Chakraborty AK (1993) J Phys Chem 97:11291 and references therein
9. Illescas BM, Santos J, Díaz MC, Martín N, Atienza CM, Guldi DM (2007) Eur J Org Chem 5027
10. Santos J, Grimm B, Illescas BM, Guldi DM, Martín N (2008) Chem Commun 5993
11. Maligaspe E, Tkachenko NV, Subbaiyan NK, Chitta R, Zandler ME, Lemmetyinen H, D'Souza F (2009) J Phys Chem A 113:8478
12. Nierengarten J-F, Hahn U, FigueiraDuarte TM, Cardinali F, Solladié N, Walther ME, Van Dorsselaer A, Herschbach H, Leize E, Albrecht-Gary A-M, Trabolsi A, Elhabiri M (2006) C R Chimie 9:1022
13. D'Souza F, Chitta R, Sandanayaka ASD, Subbaiyan NK, D'Souza L, Araki Y, Ito O (2007) J Am Chem Soc 129:15865
14. D'Souza F, Maligaspe E, Sandanayaka ASD, Subbaiyan NK, Karr PA, Hasobe T, Ito O (2009) J Phys Chem A 113:8478
15. Gayarthi SS, Wielopolski M, Pérez EM, Fernádez G, Sénchez L, Viruela R, Orti E, Guldi DM, Martin N (2009) Angew Chem Int Ed 48:815
16. Tzeli D, Petsalakis ID, Theodorakopoulos G (2011) Phys Chem Chem Phys 13:11965
17. Zhao Y, Truhlar D (2008) Theor Chem Acc 120:215
18. Zhao Y, Truhlar D (2008) Acc Chem Res 41:157
19. Curtiss LA, McGrath MP, Blandeau J-P, Davis NE, Binning RC Jr, Radom L (1995) J Chem Phys 103:6104

20. Marques MAL, Gross EKU (2004) Annu Rev Phys Chem 55:427
21. Cozi M, Scalmani G, Rega N, Barone V (2002) J Chem Phys 117:43
22. Tomasi J, Mennucci B, Cammi R (2005) Chem Rev 105:2999
23. Pedone A, Bloino J, Monti S, Prampolini G, Barone V (2010) Phys Chem Chem Phys 12:1000
24. Boys SF, Bernardi F (1970) Mol Phys 19:553
25. Tzeli D, Mavridis A, Xantheas SS (2002) J Phys Chem A 106:11327
26. Jeziorski B, Moszynski R, Szalewicz K (1994) Chem Rev 94:1887
27. Frisch MJ, Trucks GW, Schlegel HB, Scuseria GE, Robb MA, Cheeseman JR, Scalmani
    G, Barone V, Mennucci B, Petersson GA, Nakatsuji H, Caricato M, Li X, Hratchian HP,
    Izmaylov AF, Bloino J, Zheng G, Sonnenberg JL, Hada M, Ehara M, Toyota K, Fukuda R,
    Hasegawa J, Ishida M, Nakajima T, Honda Y, Kitao O, Nakai H, Vreven T, Montgomery
    JA Jr, Peralta JE, Ogliaro F, Bearpark M, Heyd JJ, Brothers E, Kudin KN, Staroverov VN,
    Kobayashi R, Normand J, Raghavachari K, Rendell A, Burant JC, Iyengar SS, Tomasi J, Cossi
    M, Rega N, Millam JM, Klene M, Knox JE, Cross JB, Bakken V, Adamo C, Jaramillo J,
    Gomperts R, Stratmann RE, Yazyev O, Austin AJ, Cammi R, Pomelli C, Ochterski JW, Martin
    RL, Morokuma K, Zakrzewski VG, Voth GA, Salvador P, Dannenberg JJ, Dapprich S, Daniels
    AD, Farkas O, Foresman JB, Ortiz JV, Cioslowski J, Fox DJ (2009) Gaussian 09, Revision A.1.
    Gaussian, Inc., Wallingford

Chapter 32
A Review of Bonding in Dendrimers and Nano-Tubes

M.A. Whitehead, Ashok Kakkar, Theo van de Ven, Rami Hourani, Elizabeth Ladd, Ye Tian, and Tom Lazzara

Abstract Geometric characterization of two 1,3,5-triethynylbenzene (TEB) based dendrimers containing tin and platinum linking agents, and a 3,5- dihydroxy-benzylalcohol (DHBA) based dendrimer containing a siloxane linking used the Semi-Empirical (PM3) Molecular Orbital Method and the Density Functional Theory Method (DFT). The theoretical results showed the increased rigidity of the backbone going from DHBA Silicon, Si, structure to TEB Tin, Sn, structure to TEB Platinum, Pt, dendrimer makes the dendrimer structure less globular, more planar and elongated.

The self-assembly of poly(styrene-*alt*-dimethyl-*N*,*N*-propylamide) (SMI) polymers into nanotubes was studied by PM3. Ordered polymer self-assembly resulted from π-stackingof styrenes and van der Waals interactions between the

M.A. Whitehead (✉) • A. Kakkar • E. Ladd
Chemistry Department, McGill University, 801, Sherbrooke Street West Montreal QC, Canada H3A 2K6
e-mail: tony.whitehead@mcgill.ca; ashok.kakkar@mcgill.ca

T.V. de Ven
Pulp and Paper Research Center, McGill University, 3420, University Street Montreal QC, Canada H3A 2K7
e-mail: theo.vandeven@mcgill.ca

R. Hourani
Berkley University, 160 Hearst Memorial Mining Building Berkley CA, USA
e-mail: rfj.hourani@gmail.com

Y. Tian
Chemistry Department, McGill University, 801, Sherbrooke Street West, Montreal QC, Canada H3A 2K6
e-mail: ye.yian@mail.mcgill.ca

T. Lazzara
Institut für Organische und Biomolekulare Chemie Gottingen University, Tammannstr. 237077 Göttingen, Germany
e-mail: thomas.lazzara@gmail.com

P.E. Hoggan et al. (eds.), *Advances in the Theory of Quantum Systems in Chemistry and Physics*, Progress in Theoretical Chemistry and Physics 22, DOI 10.1007/978-94-007-2076-3_32, © Springer Science+Business Media B.V. 2012

maleimide chains. Every styrene, and half of the maleimide chains, in a racemo-di-isotactic SMI polymer form π-stacks and chain-chain pairs with neighboring racemo-di-isotactic polymer. Racemo-di-isotactic polymers in bent associations form a minimum-energy nanotube structure. Nanorods were observed experimentally with diameters of 5 nm.

32.1 Introduction

Dendrimers are an important class of macromolecules whose unique properties led to applications in a wide-range of areas [1–5]. These unique properties include a hyper branched mono-disperse structure where higher generations often contain well defined internal cavities. A two dimensional representation of a dendrimer is in Fig. 32.1 [6]. Key structural features include the core unit, and linking agents, the branching groups, the monomer repeat unit building block, and surface terminal groups. Each part of the dendrimer can be independently varied to tailor the properties of the final product. Dendrimers are synthesized using highly controlled reaction sequences through either convergent or divergent synthesis. If prepared convergently individual dendrons are first synthesized and then attached to the core unit: in divergent techniques the entire structure is grown outwards from the core unit [6].

As each layer of dendrons is added the generation number of the dendrimer increases, as well as the globularity of its structure [6].

The structure and properties of three specific first generation dendrimers, are in Fig. 32.2. Structures (b) and (c) are both based on 1,3,5-triethynylbenzene (TEB), used as both core unit and building block. The structures differ only in the metal unit as a linking agent, dendrimer (c) employs a platinum, Pt, unit, while dendrimer

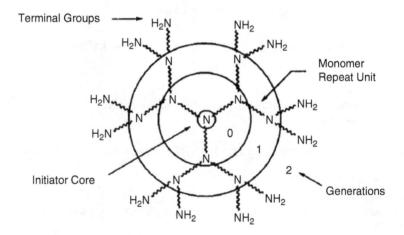

Fig. 32.1 Example of a dendrimers structure in 2Dimensions [1]

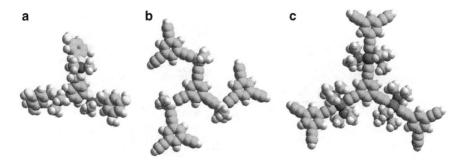

Fig. 32.2 Optimized structures of the dendrimers of interest having (**a**) Si (**b**) Sn and (**c**) Pt linking units

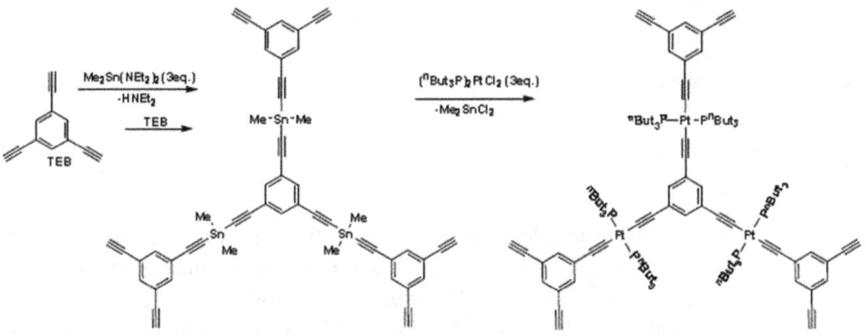

Scheme 32.1 Synthesis of TEB based tin and platinum dendrimers [7]

(b) is tin, Sn, based. The synthesis of dendrimers (b) and (c), are both prepared through the divergent methodology, Scheme 32.1. [7].

The dendrimer in Fig. 32.2a uses a 3,5-dihydroxybenzylalcohol (DHBA) unit as a core and building block and a siloxane linking agent. The synthesis performed in a divergent manner is Scheme 32.2. [8].

Determining the three-dimensional structures of dendrimers is important. Altering the metal linking agent and the building blocks of a dendrimer affect the size and shape of the dendrimer. A detailed investigation of the properties of three different dendrimers containing differing metal linking agents and building blocks (Fig. 32.2), is reported, using Semi-Empirical PM3method and DFT, which makes the ground state electronic properties of a system dependent on the Electron Density of the system. [9].

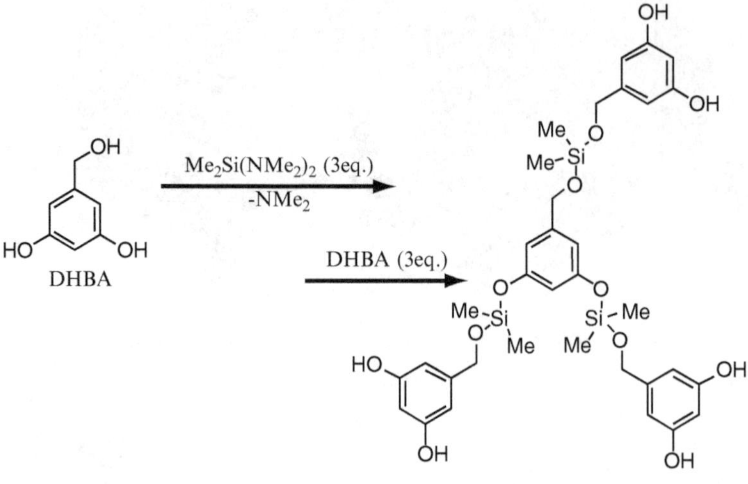

Scheme 32.2 Synthesis of DHBA based siloxane dendrimer [8]

32.2 Theoretical Methods

The large increase in size of each dendrimer generation needs a theoretical model capable of geometric characterization of such large molecules. Therefore the Molecular Mechanics Method with the MM+ Force Field was used. Geometry optimization used the Polak-Ribiere Conjugate Gradient, set to terminate at an RMS gradient of $0.01 \, \text{kcal} \, \text{Å}^{-1} \, \text{mol}^{-1}$. The Semi-Empirical (PM3) [10], was then used with the Gaussian98 program [11]. Semi-Empirical optimizations were carried out under standard convergence criteria (max force $= 4.5 \times 0^{-4}$ Hartrees bohr^{-1}; RMS force $= 3.0 \times 10^{-4}$ Hartrees Bohr^{-1}; max displacement $= 1.8 \times 10^{-3}$ Å; RMS displacement $= 1.2 \times 10^{-3}$ Å) [12, 13]. The structure of the Platinum dendrimer could not be optimized using PM3 because the Platinum atom is not parametrized. Consequently the Density Functional Theory (DFT) (B3LYP) [14] was used with a LANL2DZ [15] basis set to optimize this structure. All the structures were also optimized using DFT to ensure that structures optimized using PM3 and DFT models can be directly compared. This was proven by comparing bond lengths and angles as well as overall diameter of the structures which showed the DFT results comparable to the PM3 in the overall structure of the Tin or siloxane dendrimers. Therefore it can be assumed that the DFT optimized structure of the platinum dendrimer is similar to what would have been obtained if a PM3 calculation could have been performed. This was later checked with a PM6 calculation having Pt in the basis set. The various conformations were explored by varying the torsional axes to discover the Minimum Global Energy Conformation [16], which was conformed both by the Gaussian98 program as well as the use of the Tree-Branch Method [17]. The Tin and siloxane dendrimers were then re-optimized using the DFT model, again with the help of the Gaussian98 program.

The association between SMI polymers was modelled using SMI hexamers. Linear associations between SMI polymers were previously studied. New additional calculations on bent associations were found and compared to previous results. Semi-empirical PM3 calculations were used as above. The stabilization energy and stacking geometry of the styrenes agree with results for similar systems. All calculations were performed using the Gaussian 03 program.

32.3   Results and Discussion

The Bond Lengths and Angles of one of the dendrons on each of the structures was measured, Tables 32.1–32.3. Because the arms are degenerate only the bond lengths and angles of one dendron on each structure need be measured. In the siloxane dendrimer even though there is an extra CH_2 group on one of the dendrons because a different core molecule is used, the arms still remain effectively degenerate [12]. The numbers assigned to the atoms of each dendron are shown below in Fig. 32.2.

From the Bond Angles and Lengths in Tables 32.2 and 32.3 it is clear that the DFT calculations do not significantly alter the overall structure of the Tin or siloxane dendrimers. This data supports our assumption that the DFT optimized structure of the Platinum dendrimer correlates with what would have been obtained if it had been possible to perform PM3 calculations. The subsequent PM6 calculations proved this assumption correct.

Table 32.1 Bond lengths and angles for platinum dendrimer

	Bond lengths		Bond angles	
Atoms	Bond	Length (A)	Bond	Angle (°)
0-1	C–C	1.439	0-1-2	179.9
1-2	C≡C	1.240	1-2-3	179.9
2-3	C–Pt	2.019	2-3-4	88.1
3-4	Pt–P	2.367	2-3-5	89.8
3-5	Pt–P	2.367	2-3-6	179.8
3-6	Pt–C	2.014	3-6-7	180.0
6-7	C≡C	1.240	6-7-8	180.0
7-8	C–C(aromatic)	1.436	7-8-9	120.7
8-9	C–C(aromatic)	1.417	8-9-10	121.0
9-10	C–C(aromatic)	1.413	9-10-11	119.7
10-11	C–C(aromatic)	1.413	10-11-12	120.1
11-12	C–C(aromatic)	1.413	11-12-13	119.7
12-13	C–C(aromatic)	1.413	12-13-8	121.0
13-8	C–C(aromatic)	1.417	10-14-15	180.0
10-14	C–C	1.438	9-10-14	120.2
14-15	C≡C	1.224	11-12-16	120.1
12-16	C–C	1.438	12-16-17	180.0
16-17	C≡C	1.224		

Table 32.2 Bond lengths and angles for tin dendrimer

		Bond lengths			Bond angles	
		PM3	DFT		PM3	DFT
Atoms	Bond	Length (A)		Bond	Angle (°)	
0-1	C–C	1.418	1.437	0-1-2	180.0	179.8
1-2	C ≡ C	1.199	1.234	1-2-3	179.8	179.4
2-3	C–Sn	2.007	2.081	2-3-4	109.3	109.1
3-4	C–Sn	2.092	2.129	2-3-5	109.3	109.3
3-5	C–Sn	2.093	2.129	2-3-6	107.7	106.6
3-6	C–Sn	2.007	2.084	3-6-7	179.7	179.5
6-7	C ≡	1.199	1.234	6-7-8	180.0	179.7
7-8	C–C (aromatic)	1.418	1.437	7-8-9	119.8	120.4
8-9	C–C (aromatic)	1.398	1.414	7-8-13	119.7	120.2
9-10	C–C (aromatic)	1.398	1.413	8-9-10	119.5	120.6
10-11	C–C (aromatic)	1.398	1.413	9-10-11	120.5	119.5
11-12	C–C (aromatic)	1.398	1.413	10-11-12	119.5	120.4
12-13	C–C (aromatic)	1.398	1.413	11-12-13	120.5	119.6
13-8	C–C (aromatic)	1.398	1.414	9-10-14	119.7	120.2
10-14	C–C	1.418	1.437	10-14-15	180.0	180.0
14-15	C ≡ C	1.192	1.224	11-12-16	119.7	120.2
12-16	C–C	1.418	1.437	12-16-17	180.0	180.0
16-17	C ≡ C	1.192	1.223			

Table 32.3 Bond lengths and angles for siloxane dendrimer

		Bond lengths			Bond angles	
		Length (A)			Angle (°)	
Atoms	Bond type	PM3	DFT	Bond	PM3	DFT
0-1	C–O	1.353	1.383	0-1-2	123.9	137.4
1-2	O–Si	1.709	1.720	1-2-3	115.0	111.5
2-3	Si–C	1.891	1.873	1-2-4	102.3	109.0
2-4	Si–C	1.895	1.881	1-2-5	109.1	103.2
2-5	O–Si	1.706	1.700	2-5-6	119.6	128.9
5-6	C–O	1.392	1.451	5-6-7	112.7	113.0
6-7	C–C	1.511	1.523	6-7-8	120.0	119.3
7-8	C–C (aromatic)	1.392	1.411	6-7-12	119.5	120.3
8-9	C–C (aromatic)	1.402	1.408	7-8-9	119.3	119.4
9-10	C–C (aromatic)	1.399	1.404	8-9-10	121.5	121.2
10-11	C–C (aromatic)	1.400	1.406	9-10-11	117.9	118.4
11-12	C–C (aromatic)	1.400	1.406	10-11-12	121.4	121.5
12-7	C–C (aromatic)	1.396	1.405	8-9-13	116.0	122.5
9-13	C–O	1.368	1.402	10-9-13	122.5	116.3
11-14	C–O	1.368	1.400	12-11-14	122.8	116.7
				10-11-14	115.8	121.8

Table 32.4 Diameters

Dendrimer	Optimization model	Diameter range (Å)
Tin	PM3	21.9–22.0
	DFT	22.0–22.4
Silicon	PM3	16.1–19.2
	DFT	13.7–17.4
Platinum	DFT	26.7–30.0

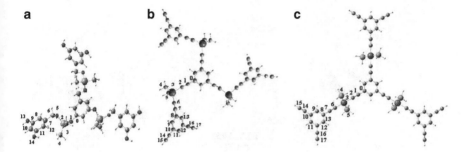

Fig. 32.3 Numbers assigned to the atoms of each non-degenerate dendron to facilitate the report of bond lengths and angles, shown in Table 32.1 for dendrimers with (**a**) Si (**b**) Sn and (**c**) Pt linking units

The diameters of each of the dendrimers was also measured, this data is in Table 32.4. Because these structures are not perfectly spherical, multiple diameters were measured for each structure and a range of values is presented.

The values presented for the PM3 optimized siloxane [12] Si and Tin [7] Sn dendrimers have previously been reported.

The relatively narrow range of diameters of the Tin dendrimer is caused by its more regular, globular shape compared to the other two structures. This is caused by both the tetrahedral geometry enforced by the Tin centre, as well as the rigidity of the TEB backbone. The tetrahedral shape enforced by the Tin, moiety causes it to be more globular than the Platinum dendrimer as the arms fold back on themselves in a Turbine Shape, where the Square Planar arrangement caused by the Platinum centre leads to a more spread out planar conformation, shown by the side views of the dendrimers, Fig. 32.3. The rigidity of the TEB backbone gives the Tin, and Platinum dendrimers a more rigid and regular shape unlike Silicon dendrimer. Also, the third arm in the Silicon dendrimer which contains an extra CH_2 group further decreases the regularity. Although all three structures should be about the same size the Platinum dendrimer is clearly the most elongated, as evidenced by its larger diameter, Table 32.4. This is caused by both the rigid square planar geometry caused by the Platinum centre, as well as the rigidity of the TEB backbone, both of which prevent the arms of this dendrimer from folding back.

The PM3 method gave the Delocalized Molecular Orbitals (DLMO) of the Tin and siloxane dendrimers. Those for the siloxane dendrimer have previously been reported [12] and will not be discussed here. The three Degenerate Highest

Fig. 32.4 *Side views* of dendrimers with (**a**) Si (**b**) Sn and (**c**) Pt linking units

DLMO 251
−0.36Hr

DLMO 252
−0.36Hr

DLMO 253
−0.36Hr

Fig. 32.5 The Highest Occupied Molecular Orbitals (HOMO) of the tin dendrimer along with their numbers and energies

Occupied Molecular Orbitals (HOMO) of the Sn structure are shown in Fig. 32.4. The Degeneracy of these three Valence Orbitals shows the equal reactivity of each arm of the dendrimer.

32.4   Interactions Between SMI Polymers

Ithas been shown [18–24], that the methods used predict π bonding correctly for many different molecular structures, in the gas phase and when hydrated. The theoretical predictions were proved by experiment. π-Stacking interactions between styrenes and the van der Waals forces between maleimide chains cause SMI polymer aggregation. Racemo-di-isotactic SMI polymers have an ordered distribution of styrenes along a main axis, with maleimide chains at 70° to the styrenes. In contrast, atactic polymers structures are not periodic, preventing ordered association [25] Two possible association conformations occur: head-to-tail in which the polymers are in identical orientation and head-to-head where they are in opposite orientation [25]. SMI polymers can join with different association angles, and three limiting geometries exist: when there is no rotation between polymers (linear association) and two limiting cases if the rotation is ±60° (bent associations). Association distances and stabilizaion energies have been previously calculated for the linear associations 2 and 5 using a series of constrained optimizations, followed by relaxing the system [25]. Here, the associations 1, 3, and 4 = 6 are calculated using the methods for 2 (Fig. 32.6).

Head-to-Head

Head-to-Tail

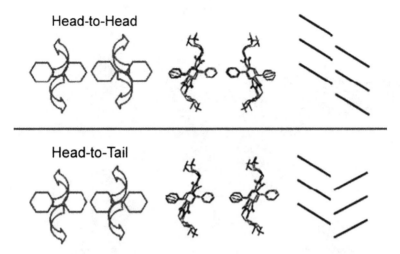

Fig. 32.6 (**a**) Lateral and cross-section view of a racemo-di-isotactic SMI polymer [25] (hydrogens removed for clarity). (**b**) The six associations studied: 2, 5 are linear and 1, 3, 4, 5 are bent associations. Associations 4 and 6 are equivalent, and associations 2 and 5 have been studied previously [25]

Evaluating the energy as a function of inter-phenyl distance (r') allows comparison of the six associations (Fig. 32.7). The average r' is between 4.2 Å and 5.5 Å. The bent associations 1, 3, 4, and 6 are the most stable, with 20 kJ/mol of stabilization energy for each π-stacking monomer pair (for two hexamers, there are three π-stacking pairs). Figure 32.6 summarizes the PM3 results obtained when the constraints are released. The stabilization energy from π-stacking was previously determined by the same type of calculation on the linear association complexes 2 and 5 [25]. Chain-chain interactions are negligible for linear associations, and the stabilization energy represents only π-stacking: 13 kJ/mol per π-stack. The increased stabilization energy for bent associations compare to linear ones comes from chain-chain stabilizing van der Waals interactions between the maleimide chains, and this contributes about 7 kJ/mol per monomer pair. Both π-stacking and van der Waals interactions are present in bent associations and produce more stable complexes than linear ones. If maleimide chains are shorter, the stabilization energy for bent associations does not change much compared to linear ones. Complexes with no chain-chain interactions would become insensitive to changes in association angles.

32.5   Nanotubes from Self-assembled SMI Polymers

Associating more than two hexamers (500 atoms) was built up from smaller units. The optimal inter-phenyl distances and association angles for the most stable bent associations were used to build larger complexes. The inter-phenyl distance

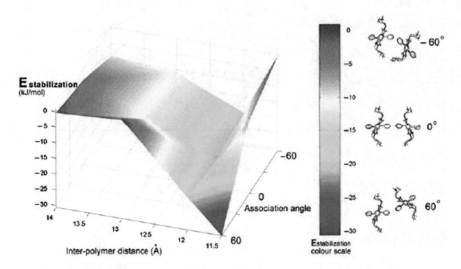

Fig. 32.7 Stabilization energy as a function of r' for the different associations. The stabilization energies are given for two hexamers and therefore for three π-stacking SMI styrene pairs

is about 5 Å with association angles of 60°. The bent association (1, 3, 4 = 6) have equivalent energies, are equally probable, and give polymer sheets, from combining bent associations and grown by addition of racemo-di-isotactic polymers. When the sheets grow, non-associated peripheral styrenes remain. The system is more stable when additional π-stacks form between peripheral styrenes. The curvature forms a closed tube because the bent associations already form a 60° angle. In a closed loop, the hydrophobic styrenes and half of the maleimide chains are no longer in contact with the hydrophilic solvent, but interact with a more hydrophobic environment inside the nanotube walls. A closed octagonal structure consisting of eight SMI polymers forms a short polymer nano-tube segment. Figure 32.8 shows an SMI nano-tube from eight polymers in the head-to-head conformation. The inter-phenyl distances and association angles of the optimized values were used to build the 3-dimensional nano-tube. Using different periphery atoms as reference points gave an outer diameter of 4.8 ± 0.2 nm and an inner diameter of 1.7 ± 0.2 nm. An equivalent structure is also possible for SMI polymers associating in a head-to-tail conformation and a mixture of the two conformations. Stabilization energies, inter-phenyl distances, and association angles are comparable between head-to-head and head-to-tail conformations. The nano-tube forms with racemo-di-isotactic polymers.

Polymers are polydisperse, and π-stacking is rarely perfect between associated polymers, the short closed segment can have protruding polymers. These ends offer nucleation points where addition of further racemo-di-isotactic polymers can occur. The nanotube linear growth is illustrated in Fig. 32.9. SMI nanotubes are not expected to associate in bundles, as was the case for SMA nanotubes, but to give individual rods. The styrenes π-stack inside the walls of the nanotubes and are unavailable for further interaction between nanotubes. Unfavorable solvent-styrene interactions are decreased, and favorable hydrophobic styrene-styrene and chain-chain interactions are increased in polar solvents such as water unlike SMA [26].

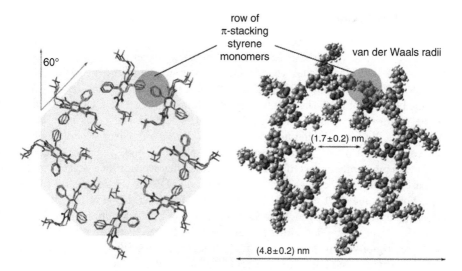

Fig. 32.8 SMI nanotube (cross section, perpendicular to association plane): (*top*) the nanotube has an octagonal shape, made from eight racemo-di-isotactic SMI polymers in the head-to-head conformation; (*bottom*) SMI nanotube shown with van der Waals radii

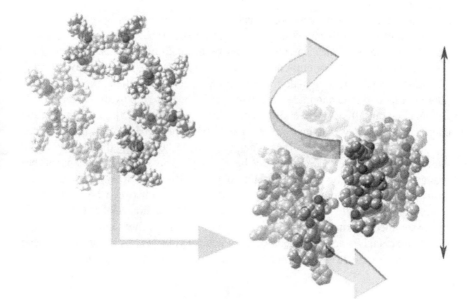

Fig. 32.9 Proposed linear growth mechanism for SMI nanotubes. SMI polymers self-assemble at the edges of an initially closed structure, and the addition makes the nanotube grow in length (*arrows*) (*lighter shade* atoms are further behind the plane of view)

32.6 Conclusions

Theoretical calculations using the Semi-Empirical Parameterization Model 3 (PM3 and PM6) Molecular Orbital Theory and the Density Functional Theory (DFT) proved useful to compare dendrimers with different metal linking centres and organic building blocks. The structure optimizations showed how the overall shape of the dendrimers change when the backbone is varied. The dendrimer which employed a Platinum linking unit was shown to be a rigid planar structure, while that containing Tin was more turbine shaped. However both of these structures have much more inflexible conformations than the Silicon dendrimer, because of the increased rigidity of the TEB backbone compared to DHBA. This is caused by the triple bonds in the TEB structure. The optimized structures also gave insight about the size of the dendrimers which ranged from 21.9 Å to 22.4 Å for the Tin dendrimer, 13.7–19.2 Å for silicon dendrimer and 26.7–30.0 Å for the Platinum structure. The increasing rigidity of the backbone in going from the DHBA Silicon structure to the TEB Tin structure and finally to the TEB Platinum dendrimer decreases the number of permutations in the angle at which the arms attach to the core and restricts the possible conformations to more spread out structures. It is clear that as the rigidity of the backbone is increased, the dendrimer structure becomes less globular and more planar and elongated.

Association between racemo-di-isotactic SMI polymers was investigated theoretically and gave rod-shaped aggregates. The bent associations are more stable because of van der Waals interactions between maleimide chains. Multiple bent associations form a minimum-energy nanotube structure. Although only the self-assembly of relatively long maleimide chains has been studied here, the conclusions apply to other styrene and maleimide copolymers because removing the maleimide chains does not affect the overall geometry of the polymer [25]. Additionally, the chemical structure of maleimide chains can modify the van der Waals interaction.

This study shows that by functionalizing SMA the size of the styrene-based alternating copolymer nanotubes can be changed. The shape of SMI nanotubes remains octagonal, the outer diameter increases from 4.4 to 4.8 nm, the inner diameter decreases from 2.0 to 1.7 nm, and aggregation between SMI nanotubes is not possible. In the sample studied, a very small fraction of the SMI actually self-assembled in nanotubes because polymer chirality occurs randomly, and only a low percentage is actually racemo-di-isotactic. Synthesis of racemo-di- isotactic SMI and their derivatives would be very interesting, if achievable [27].

References

1. Vassilieff, T.; Sutton, A; Kakkar, A. *J. Mater. Chem.* **2008**, 18, 4031.
2. R. van Heerbeek, P. C. J. Kamer, P. W. N. M. van Leeuwen and J. N. H. Reek, *Chem. Rev.*, **2002**, 102, 3717.
3. A. Andronov and J. M. J. Frechet, *Chem. Commun.*, **2000**, 1701.

4. D. Astruc and F. Chardac, *Chem. Rev.*, **2001**, 101, 2991.
5. J. M. Lupton, I. D. W. Samuel, P. L. Burn and S. Mukamel, *J. Chem. Phys.*, **2002**, 116, 2, 455.
6. Zeng, F. and Zimmerman, S. C.; *Chem. Rev.* **1997**, 97, 1681
7. Hourani, R.; Whitehead, M. A.; Kakkar, A.; *Macromolecules*, **2008**, 41, 508.
8. Bourrier, O.; Kakkar, A.; *J Mater Chem*, **2003**, 13, 1306
9. Koch, W.; Holthausen, M.C.; *A Chemists Guide to Density Functional Theory*, **2000**, Wiley-VCH
10. (a) J.J.P. Stewart, J. Comput. Chem., **1989**, 10, 2, 209. (b) J.J.P. Stewart, J. Comput. Chem., **1989**,10, 2, 221.
11. M.J. Frisch, G.W. Trucks, H.B. Schlegel, G.E. Scuseria, M.A. Robb, J.R. Cheeseman, V.G. Zakrzewski, J.A. Montgomery, R.E. Stratmann, J.C. Burant, S. Dapprich, J.M. Millam, A.D. Daniels, K.N. Kudin, M.C. Strain, O. Farkas, J. Tomasi, V. Barone, M. Cossi, R. Cammi, B. Mennucci, C. Pomelli, C. Adamo, S. Clifford, J. Ochterski, G.A. Petersson, P.Y. Ayala, Q. Cui, K. Morokuma, D.K. Malick, A.D. Rabuck, K. Raghavachari, J.B. Foresman, J. Cioslowski, J.V. Ortiz, B.B. Stefanov, G. Liu, A. Liashenko, P. Piskorz, I. Komaromi, R. Gomperts, R.L. Martin, D.J. Fox, T. Keith, M.A. Al-Laham, C.Y. Peng, A. Nanayakkara, C. Gonzalez, M. Challacombe, P.M.W. Gill, B.G. Johnson, W. Chen, M.W. Wong, J.L. Andres, M. Head-Gordon, E.S. Replogle and J.A. Pople, Gaussian 98W (revision A.5), Gaussian Inc., Pittsburgh, PA, **2003**.
12. Hourani, R.; Kakkar, A.; Whitehead, M. A.; *J Mater Chem*, **2005**, 15, 2106.
13. Hourani, R.; Kakkar, A.; Whitehead, M. A.; *Theochem*, **2007**, 807, 101
14. a) Becke, A.D. *J. Chem. Phys.* **1993**, 98, 5648. b) Lee, C.; Yang, W.; Parr, R. G.; *Phys. Rev. B*, **1988**, 37, 785.
15. Hay, P. J.; Wadt, W. R. *J. Chem. Phys.* **1985**, 82, 270.
16. HyperChem release 5.11. For Windows molecular modelling system, Hypercube Inc., 419 Philip Street, Waterloo, Ont., Canada, N2L 3X2, **1999**.
17. Villamagna, F.; Whitehead, M.A.; *J. Chem. Soc., Faraday Trans.*, **1994**, 90, 1, 47.
18. M.A. Whitehead, Theo van de Ven and Cecile Malardier-Jugroot, Journal of Molecular Structure: THEOCHEM. (2004), **679**, 171–177.
19. Cécile Malardier-Jugroot, T.G.M. van de Ven and M.A. Whitehead, Proceedings of the First Applied Pulp & Paper Molecular Modelling Symposium, Cascades Ltd. (2006), 257–270.
20. C. Malardier-Jugroot, M.A. Whitehead and Theo van de Ven, Proceedings of the First Applied Pulp & Paper Molecular Modelling Symposium, Cascades Ltd. (2006), Poster
21. Thomas D. Lazzara, Theo G.M. van de Ven, M.A. (Tony) Whitehead, IPCG Newsletter-February **2008**
22. Theo. G.M. van de Ven, Thomas Dominic Lazzara and M.A (Tony) Whitehead, The Proceedings ot the Fundamental and Applied Pulp & Paper Modelling Symposium, 2008. Cascades Inc. **2009**, 63–75.
23. M.A (Tony) Whitehead, Ye Tien, Rami Hourani, Ashok Kakkar, Thomas D. Lazzara, Theo. G.M. van de Ven, Joseph Kinghorn Taenzer and Intakhab Alam Zeeshan, The Proceedings ot the Fundamental and Applied Pulp & Paper Modelling Symposium, 2008. Cascades Inc. **2009**, 23–48.
24. Thomas D. Lazzara, Michael A. Whitehead and Theo G. M. van de Ven, European Polymer Journal, 45, 1883–1890, **2009**.
25. Lazzara, T. D.; Whitehead, M. A.; van de Ven, T. G. M. J. *Phys. Chem. B* **2008**, 112, 16, 4892
26. M.A. Whitehead, Cecile Malardier-Jugroot and Theo. T.G.M. van de Ven, J. Phys. Chem. B. (2005), **109**(15), 7022–7032.
27. Thomas D. Lazzara , Theo G.M. van de Ven, M. A. (Tony) Whitehead Macromolecules, **2008**, 41, 674

Index

CPSIA information can be obtained
at www.ICGtesting.com
Printed in the USA
LVHW081931210620
658633LV00003B/16